13th High-Tech Plasma Processes Conference

(HTPP 2014)

Journal of Physics: Conference Series Volume 550

Toulouse, France
22 – 27 June 2014

Editors:

A. Gleizes
J. Schein
P. Teulet

ISBN: 978-1-63439-664-6
ISSN: 1742-6588

TABLE OF CONTENTS

ARC-MATERIALS INTERACTION AND METALLURGY

LIGHTNING IN AERONAUTICS 1
F Lago

MODELLING LIGHTNING INITIATION AND ATTACHMENT TO AIRCRAFT 9
E Tam, J J Lowke, A B Murphy

STUD ARC WELDING IN A MAGNETIC FIELD – INVESTIGATION OF THE INFLUENCES ON THE ARC MOTION 16
K Hartz-Behrend, J L Marqués, G Forster, A Jenicek, M Müller, H Cramer, A Jilg, H Soyer, J Schein

INFLUENCE OF THE ARC PLASMA PARAMETERS ON THE WELD POOL PROFILE IN TIG WELDING 23
A Toropchin, V Frolov, A V Pipa, R Kozakov, D Uhrlandt

EXPERIMENTAL CHARACTERIZATION OF THE WELD POOL FLOW IN A TIG CONFIGURATION 33
M Stadler, M Masquère, P Freton, X Franceries, J J Gonzalez

TIME RESOLVED THOMSON SCATTERING DIAGNOSTIC OF PULSED GAS METAL ARC WELDING (GMAW) PROCESS 41
M Kühn-Kauffeldt, J L Marquès, J Schein

INVESTIGATION OF THE ANODE ATTACHMENT PROCESS IN PLASMA ARC CUTTING 48
S Eichler, N Hussary, E Siewert, J Schein

PROPERTIES OF THERMAL AIR PLASMA WITH ADMIXING OF COPPER AND CARBON 58
S Fesenko, A Veklich, V Boretskij, Y Cressault, A Gleizes, Ph Teulet

LONG ARC STABILITIES WITH VARIOUS ARC GAS FLOW RATES 66
K Maruyama, K Takeda, M Sugimoto, Y Noguchi

INTERACTION OF A FREE BURNING ARC WITH REGENERATIVE PROTECTIVE LAYERS 72
D Uhrlandt, S Gorchakov, V Brueser, St Franke, A Khakpour, M Lisnyak, R Methling, Th Schoenemann

TIME RESPONSE OF ARC DRIVEN BY ALTERNATING MAGNETIC FIELD 82
K Takeda, H Okubo, M Sugimoto

SPECTROSCOPIC STUDY AND HIGH SPEED IMAGING OF A TRANSIENT ARC 92
P Ratovoson, F Valensi, M Razafinimanana, T Tmenova

STRANGE COLLECTIVE BEHAVIORS OF CATHODE SPOTS IN LOW VACUUM ARC 101
M Sugimoto, S Takasugi

INVESTIGATION OF PLASMA FLOW IN VACUUM ARC WITH HOT CATHODE 107
R Amirov, N Vorona, A Gavrikov, G Lizyakin, V Polistchook, I Samoylov, V Smirnov, R Usmanov, I Yartsev

PLASMA TORCHES AND SPRAYING

COMPARISON OF CO_2 AND OXYGEN DC SUBMERGED THERMAL PLASMAS FOR DECOMPOSITION OF CARBOXYLIC ACID IN AQUEOUS SOLUTION 115
S Safa, A Hekmat-Ardakan, G Soucy

LARGE-EDDY SIMULATION OF SUBSONIC-SUPERSONIC FLOW AND HEAT TRANSFER IN A HYBRID GAS-WATER STABILIZED ARC 121
J Jeništa, H Takana, H Nishiyama, M Bartlová, V Aubrecht, P Krenek

ADVANCED OXIDE POWDERS PROCESSING BASED ON CASCADE PLASMA 131
O P Solonenko, A V Smirnov

DEVELOPMENT OF A 100 KW PLASMA TORCH FOR PLASMA ASSISTED COMBUSTION OF LOW HEATING VALUE FUELS 141
S Takali, F Fabry, V Rohani, F Cauneau, L Fulcheri

EFFECT OF SUSPENSION CHARACTERISTICS ON IN-FLIGHT PARTICLE PROPERTIES AND COATING MICROSTRUCTURES ACHIEVED BY SUSPENSION PLASMA SPRAY 151
E Aubignat, M P Planche, A Allimant, D Billières, L Girardot, Y Bailly, G Montavon

PULSED LAMINAR ARC JET WITH SYNCHRONIZED SUSPENSION INJECTION-SPECTROSCOPIC STUDIES 161
J Krowka, V Rat, C Chazelas, J-F Coudert

ANALYSIS OF PROCESSES IN DC ARC PLASMA TORCHES FOR SPRAYING THAT USE AIR AS PLASMA FORMING GAS ... 168
V Frolov, D Ivanov, A Toropchin
MECHANISMS OF HEAT TRANSFER BETWEEN A DROPLET AND A PLASMA JET IN SUSPENSION PLASMA SPRAYING ... 174
F Girard, E Meillot, S Vincent, L Bianchi, J P Caltagirone

SYNTHESIS OF POWDERS AND NANOMATERIALS

SYNTHESIS OF CARBON NANOTUBES BY HIGH CURRENT DIVERGENT ANODE-CHANNEL PLASMA TORCH ... 184
R H Amirov, E Kh Isakaev, M B Shavelkina, T B Shatalova
UV FE I ABSORPTION IN THE CARBON-IRON ARC PERIPHERY UNDER FORMATION OF CARBON-ENCAPSULATED IRON NANOPARTICLES ... 191
H Lange, O Labedz, M Bystrzejewski
EXPERIMENTAL TESTS FOR CARBON NANOMATERIAL SYNTHESIS USING DC PLASMA JET ... 199
H Lange, O Labedz, I Tylska, A Huczko, M Bystrzejewski
TWO-DIMENSIONAL SPECTROSCOPIC OBSERVATION OF A PULSE-MODULATED INDUCTION THERMAL PLASMA TORCH FOR NANOPOWDER SYNTHESIS ... 207
N Kodama, K Kita, Y Tanaka, Y Uesugi, T Ishijima, S Watanabe, K Nakamura
THERMAL PLASMA SOURCE FOR PROCESSING OF MOO_3 CRYSTALS ... 217
A Lebid, A Veklich, V Boretskij, S Savenok, O Andreev
SIMILARITY CRITERIA OF HOSP FORMATION AT PLASMA PROCESSING OF AGGLOMERATED PARTICLES ... 222
O P Solonenko

DEPOSITION AND SURFACE TREATMENT

DEVELOPMENT OF SUPER-HYDROPHOBIC PTFE AND PET SURFACES BY MEANS OF PLASMA PROCESSES ... 232
S Zanini, R Bami, R Della Pergola, C Riccardi
NUMERICAL SIMULATION OF BOSCH PROCESSING FOR DEEP SILICON PLASMA ETCHING ... 242
P Moroz, D J Moroz
ECWR PLASMA ENHANCED CHEMICAL VAPOUR DEPOSITION OF MICROCRYSTALLINE SILICON THIN FILMS ... 252
E Farsari, A G Kalampounias, E Amanatides, D Mataras
MORPHOLOGICAL AND CHEMICAL ANALYSIS OF PP FILM TREATED BY DIELECTRIC BARRIER DISCHARGE ... 260
E C Dell'Orto, A Vaccaro, C Riccardi
COMPARISON OF ICP AND DBD PLASMA PROCESS FOR THE DEGREASING OF METAL REINFORCEMENTS ... 269
F Judee, E Duffour, F Perisse, S Menecier, D Vacher
PHOTOINDUCED SUPERHYDROPHILICITY OF AMORPHOUS TIO_X-LIKE THIN FILMS BY A SIMPLE ROOM TEMPERATURE SOL-GEL DEPOSITION AND ATMOSPHERIC PLASMA JET TREATMENT ... 279
V E Vrakatseli, E Pagonis, E Amanatides, D Mataras
MANUFACTURING OF COMPOSITE TITANIUM-TITANIUM NITRIDE COATINGS BY REACTIVE VERY LOW PRESSURE PLASMA SPRAYING (R-VLPPS) ... 287
B Vautherin, M-P Planche, A Quet, L Bianchi, G Montavon
SPATIAL AND TEMPORAL EVOLUTION OF MICRODISCHARGES IN SURFACE DIELECTRIC BARRIER DISCHARGES FOR AERONAUTICAL APPLICATIONS PLASMAS ... 297
R Barni, I Biganzoli, C Riccardi

NON-EQUILIBRIUM PLASMAS

OPTICAL DIAGNOSTICS OF STREAMERS: FROM LABORATORY MICRO-SCALE TO UPPER-ATMOSPHERIC LARGE-SCALE DISCHARGES 305
M Šimek, T Hoder, V Prukner, P F Ambrico

EXPERIMENTAL INVESTIGATION OF FILAMENTARY AND NON-FILAMENTARY REGIMES IN A SURFACE DIELECTRIC BARRIER PLASMA ACTUATOR 315
I Biganzoli, R Barni, A Gurioli, R Pertile, C Riccardi

EXPERIMENTAL INVESTIGATION OF LISSAJOUS FIGURE SHAPES IN PLANAR AND SURFACE DIELECTRIC BARRIER DISCHARGES 325
I Biganzoli, R Barni, A Gurioli, R Pertile, C Riccardi

ON THE USE OF PULSED DIELECTRIC BARRIER DISCHARGES TO CONTROL THE GAS-PHASE COMPOSITION OF ATMOSPHERIC PRESSURE AIR PLASMAS 335
R Barni, I Biganzoli, E Dell'Orto, C Riccardi

TOWARD THE DESIGN OF MULTI ASYMMETRIC SURFACE DIELECTRIC BARRIER DISCHARGE (ASDBD) ACTUATORS 345
M Zadeh, V Rohani, F Cauneau, F Fabry, L Fulcheri

CHARACTERIZATION OF A LOW PRESSURE SUPERSONIC PLASMA JET 352
S Caldirola, R Barni, C Riccardi

ION ENERGY DISTRIBUTION FUNCTIONS IN A SUPERSONIC PLASMA JET 362
S Caldirola, H E Roman, C Riccardi

MAXIMIZATION OF THE WORKING DOMAIN OF AN ATMOSPHERIC PRESSURE TOWNSEND DISCHARGE (APTD) USING A CURRENT-SOURCE STATIC CONVERTER 372
M C Bouzidi, X Bonnin, N Naudé, H Piquet, A Belinger, N Gherardi

DETERMINATION OF N AND O-ATOMS, OF $N_2(A)$ AND $N_2(X, V>13)$ METASTABLE MOLECULES AND N_2^+ ION DENSITIES IN THE AFTERGLOWS OF N_2-H_2, AR-N_2-H_2 AND AR-N_2-O_2 MICROWAVE DISCHARGES 381
H Zerrouki, A Ricard, J P Sarrette

ANALYSIS OF THE MEASURED 1S-2S STARK SPECTRA OF HYDROGEN ISOTOPES USED FOR E-FIELD STRENGTH DETERMINATION IN A HOLLOW CATHODE DISCHARGE 390
M I De La Rosa, C Pérez, K Grützmacher, L M Fuentes

EXPERIMENTAL CHARACTERIZATION OF THE EXCITATION STATE OF PICOSECOND LASER-INDUCED TUNGSTEN PLASMAS 395
B Pérès, V Morel, A Bultel, A Benyagoub, I Monnet, E Gardès, G Godard, C Gobin, C Jouen, A Hideur, I Schneider, Z Mezei

ACHIEVEMENT OF LOCAL THERMODYNAMIC EQUILIBRIUM FOR NS LASER-INDUCED PLASMAS ON ALUMINIUM SAMPLE AT DIFFERENT WAVELENGTHS 404
V Morel, A Bultel

ELABORATION OF COLLISIONAL-RADIATIVE MODELS APPLIED TO EARTH AND MARS ENTRY PROBLEMS 414
J Annaloro, A Bultel, P Omaly

MODELLING RADIATIVE PROPERTIES OF PARTICIPATING SPECIES IN A MICROWAVE PLASMA REACTOR FOR DIAMOND DEPOSITION 421
S Prasanna, Ph Rivière, A Soufiani

COMPUTATIONAL SIMULATION OF NANOSECOND PULSED DISCHARGE FOR PLASMA ASSISTED IGNITION 431
H Takana, I V Adamovich, H Nishiyama

ORDERING OF THE FLAME TRACK IN THE RING MODE OF THE TRICHEL PULSE NEGATIVE CORONA DISCHARGE 440
R H Amirov, S A Barengolts, E V Korostylev, N V Pestovskii, A A Petrov, I S Samoylov, S Yu Savinov

Author Index

Journal of Physics: Conference Series **550** (2014) 011001 doi:10.1088/1742-6596/550/1/011001

PREFACE

The High-Tech Plasma Processes – 13th European Plasma Conference (HTPP-2014) was held in Toulouse (France) on 22–27 June 2014.

The conference series started in 1990 as a thermal plasma conference and has gradually expanded to include other related topics. Now the High-Tech Plasma Processes - European Plasma Conference (HTPP) is an international conference organised in Europe every two years with topics encompassing the whole field of plasma processing science. The aim of the conference is to bring different scientific communities together, to facilitate contacts between science, technology and industry and to provide a platform for the exploration of both the fundamental topics and new applications of plasmas.

For this edition of HTPP, as was the case for the last, we have acheived a well balanced participation from the communities of both thermal and non-thermal plasma researchers. 142 people from 17 countries attended the conference with the total number of contributions being 155, consisting of 8 plenary and 8 invited talks plus 51 oral and 88 poster contributions. We have received numerous papers corresponding to the contributions of HTPP-2014 that have been submitted for publication in this volume of Journal of Physics: Conference Series. Each submitted contribution has been peer reviewed (60 referees with at least two reviewing each paper) and the Editors are very grateful to the referees for their careful support in improving the original manuscripts.

In total, 52 manuscripts have been accepted for publication covering a range of topics of plasma processing science from plasma fundamentals to process applications through to experiments, diagnostics and modelling. We have grouped the papers into the following 5 topics:

- Arc-Materials Interaction and Metallurgy
- Plasma Torches and Spraying
- Synthesis of Powders and Nanomaterials
- Deposition and Surface Treatment
- Non-Equilibrium Plasmas

We deeply thank the authors for their enthusiastic and high-grade contributions and we are convinced that this volume of Journal of Physics: Conference Series will be interesting for our community.

Finally, we would like to thank the conference chairmen, the members of the steering committee, the international scientific committee, the local organizing committee, the organizing secretariat and the financial support from the conference sponsors that allowed the success of HTPP-2014.

The Editors of the HTPP-2014 Proceedings

Dr Alain Gleizes, chairman of HTPP-2014
Prof. Jochen Schein, head of the ISC
Prof. Philippe Teulet

Toulouse, 14th October 2014

Journal of Physics: Conference Series **550** (2014) 011001 doi:10.1088/1742-6596/550/1/011001

HTPP 2014

MAIRIE DE TOULOUSE

RÉGION MIDI-PYRÉNÉES

UNIVERSITÉ TOULOUSE III PAUL SABATIER

Laplace

INP ENSEEIHT

cnrs

AAE

High-Tech Plasma Processes

13th European Plasma Conference

22-27 June 2014 – Toulouse, France

Scope

The conference series started in 1990 as a thermal plasma conference and gradually expanded to include other topic fields as well. Now the High-Tech Plasma Processes Conference (HTPP) is an international conference based in Europe with topics encompassing the whole area of plasma processing science. This conference is open to all the international community involved in plasma science and plasma technology, and not limited to the European participants. The aim of the conference is to bring different scientific communities together, facilitate the contacts between science, technology and industry and provide a platform for the exploration of both fundamental topics and new applications of plasmas.

The symposium is organized to include plenary lectures, invited and contributed oral presentations, and poster sessions.

International Scientific Committee

Jochen Schein (Germany) *Head of ISC*
Mikhail Benilov (Portugal)
Annemie Bogaerts (Belgium)
Giovanni Bolelli (Italy)
Peter Bruggeman (USA)
David Cameron (Finland)
Isabelle Choquet (Sweden)
Gilles Flamant (France)
Francesco Fracassi (Italy)
Laurent Fulcheri (France)
Vasco Guerra (Portugal)
Milan Hrabowsky (Czech Republic)
Mounir Laroussi (USA)
Marc Leparoux (Switzerland)
Jean-Luc Meunier (Canada)
Jerzy Mizeraczyk (Poland)
Christian Oehr (Germany)
Yasunoti Tanaka (Japan)
Dirk Uhrlandt (Germany)
Jaime de Urquijo (Mexico)
Joost van der Mullen (the Netherlands)
Anatoly Veklich (Ukraine)

Local Organizing Committee

Alain Gleizes *(Chair)*
Jean-Jacques Gonzalez *(co-chair)*
Yann Cressault *(co-chair)*

Marie-Ange Albouy
Marie-Claude Bordage
David Buso
Myriam Dziadowiec
Pierre Fort
Pierre Freton
Laurent Garrigues
Agnès Gaunie-Picart
Nicolas Gherardi
Mathieu Masquère
Viviane Neff
Manitra Razafinimanana
Philippe Teulet
Flavien Valensi

Steering Committee

Vittorio Colombo (Italy)
Pietro Favia (Italy)
Alain Gleizes (France)
Dimitrios Mataras (Greece)
Daniel Morvan (France)
Jochen Schein (Germany)
Armelle Vardelle (France)

Plenary speakers

Jan Benedikt *(Ruhr-Universität Bochum, Germany)*
Cold atmospheric plasmas for synthesis of good quality materials? Understanding plasma-surface interaction and thin film growth

Sylvain Coulombe *(McGill University, Canada)*
Plasma-Assisted Synthesis of Nanostructures: A Key Enabler for Green Technologies

Frédéric Lago *(DGA-TA, France)*
Lightning in Aeronautics

Javad Mostaghimi *(Toronto University, Canada)*
Towards a Complete Model of DC Plasma Spray Process

Tony Murphy *(CSIRO, Australia)*
The critical role of plasma–surface interactions in thermal plasma processes

Jean-Michel Pouvesle and Eric Robert *(Université d'Orléans, France)*
Antitumor effect of low temperature plasmas: recent results and perspectives

Milan Simek *(Academy of Sciences, Praha, Czech Republic)*
Optical diagnostics of streamers: from laboratory micro-scale to upper-atmospheric large-scale discharges

Takayuki Watanabe *(Kyushu University, Japan)*
Multi-phase AC Arc for Innovative Glass Melting

http://htpp2014.sciencesconf.org

Peer review statement

All papers published in this volume of *Journal of Physics: Conference Series* have been peer reviewed through processes administered by the proceedings Editors. Reviews were conducted by expert referees to the professional and scientific standards expected of a proceedings journal published by IOP Publishing.

Journal of Physics: Conference Series **550** (2014) 012001 doi:10.1088/1742-6596/550/1/012001

Lightning in aeronautics

F Lago
DGA Techniques aéronautiques, 47 rue Saint-Jean, BP n°53123, 31131 BALMA Cedex, France

E-mail: frederic.lago@intradef.gouv.fr

Abstract. It is generally accepted that a civilian aircraft is struck, on average, once or twice per year. This number tends to indicate that a lightning strike risk is far from being marginal and so requires that aircraft manufacturers have to demonstrate that their aircraft is protected against lightning. The first generation of aircrafts, which were manufactured mainly in aluminium alloy and had electromechanical and pneumatic controls, had a natural immunity to the effects of lightning. Nowadays, aircraft structures are made primarily with composite materials and flight controls are mostly electronic. This aspect of the "more composite and more electric" aircraft demands to aircraft manufacturers to pay a particular attention to the lightning protection and to its certification by testing and/or analysis. It is therefore essential to take this risk into account when designing the aircraft. Nevertheless, it is currently impossible to reproduce the entire lightning phenomenon in testing laboratories and the best way to analyse the lightning protection is to reproduce its effects. In this context, a number of standards and guides are produced by standards committees to help laboratories and aircraft manufacturers to perform realistic tests. Although the environment of a laboratory is quite different from those of a storm cloud, the rules of aircraft design, the know-how of aircraft manufacturers, the existence of international work leading to a better understanding of the lightning phenomenon and standards more precise, permit, today, to consider the risk as properly controlled.

1. The lightning phenomenon

Lightning finds its origin in the clouds of vertical development such as cumulonimbus. Within this kind of cloud, complex phenomena cause the water droplets and ice crystals to charge electrically by friction. These particles are distributed, within the cloud, mainly into two groups of opposite electrical charges. When the electric field reaches a threshold value (of the order of a hundred kilovolts per meter) an arc is formed between the cloud and the ground (20% of cases) or between two clouds (80% of cases).

In the case of lightning strike on an aircraft, the aircraft is flying in an area where the ambient electric field reaches a value of several tens of kilovolts per meter. This field can vary very slowly (about one second) when it is produced by the storm cloud or quickly (tens of milliseconds) when initiated by a natural lightning develops. Two different processes can lead to a lightning discharge as presented in the figure 1:

- In about 90% of the cases, the aircraft, by its presence, triggers the lightning. Indeed, around the aircraft, the electric field is amplified by the geometry of the object. On its surface, the electric field can exceed the ionization value of the air. It then develops from the aircraft, two electrical discharges: one called "positive" because it moves in the direction of the field, the other called "negative" as it moves in the opposite direction.

- In about 10% of the cases, the aircraft intercepts an arc of lightning triggered naturally.

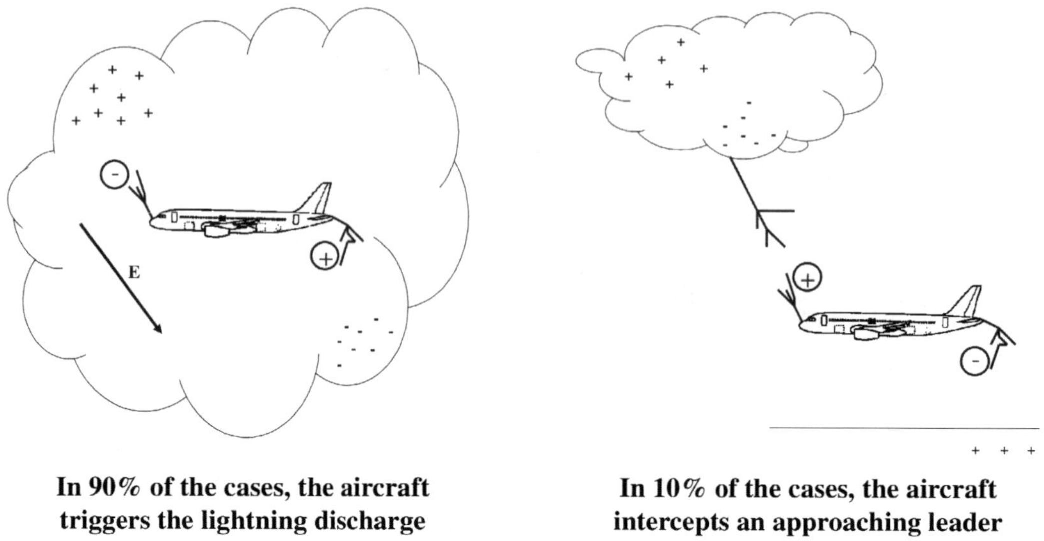

Figure 1. Scenarios of lightning strike in flight.

This initiation phase is concluded by the junction of these bidirectional discharges with those coming from the cloud and/or the ground. The circuit is then closed and a large amount of electrical charges can then flow into the channel thus formed: the established current phase. It consists of a continuous hundred amperes with a duration of hundreds of milliseconds that on which are superimposed impulse discharges current of several tens of thousands of amperes with a duration of hundreds of microseconds.

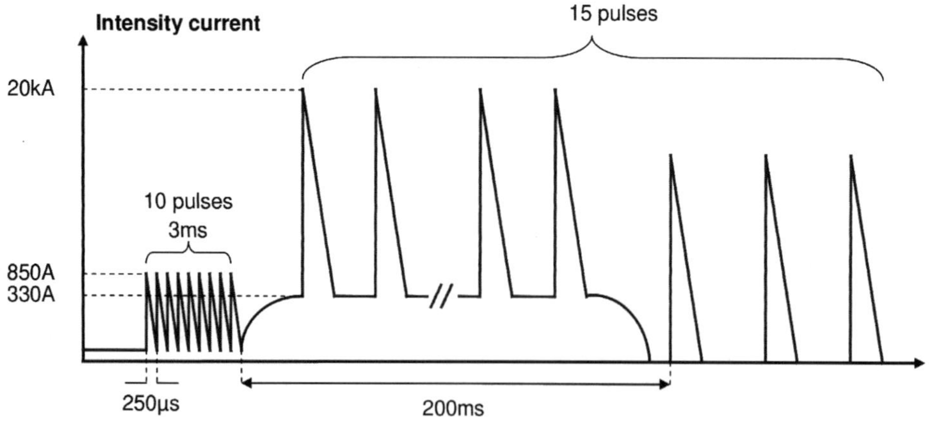

Figure 2. Typical current waveform recorded during a lightning measurement campaign in flight [1].

Furthermore, during the phase of a lightning strike, the aircraft continues to move at a certain speed while the lightning channels remain motionless in the air. This phenomenon, called sweeping is the result of the displacement of the lightning arc on the fuselage of the aircraft in a discontinuous manner.

2. The standardized environment

Journal of Physics: Conference Series **550** (2014) 012001 doi:10.1088/1742-6596/550/1/012001

The complexity of the lightning phenomenon and its interaction with a plane is impossible to simulate, in its entirety, in testing laboratories. The best way to demonstrate the resistance against lightning of an aircraft is not to reproduce the phenomenon, but is to reproduce these effects. The effects of lightning on an aircraft are divided into two types:

- Direct effects: they represent all the thermomechanical damages to the aircraft structure and external equipments caused by the direct attachment of the lightning channel and/or conduction of the current lightning.
- Indirect effects represent all electrical transients induced by lightning inside conductive components in an aircraft such as electrical circuits.

Civil certification authorities, such as EASA (European Aviation Safety Agency) and FAA (Federal Aviation Administration), require, through text indicating the basic recommendations, to protect aircraft against catastrophic effects of lightning. Standard committees, such as EUROCAE (European Organization for Civil Aviation Equipment) and SAE (Society of Automotive Engineers), establish guides and normative documents that explain how to achieve these recommendations. Among these works, there may be cited:

- ED-84/ARP 5412A [2] which defines the waveforms (current, voltage, external and internal environment) used to reproduce the effects of lightning.

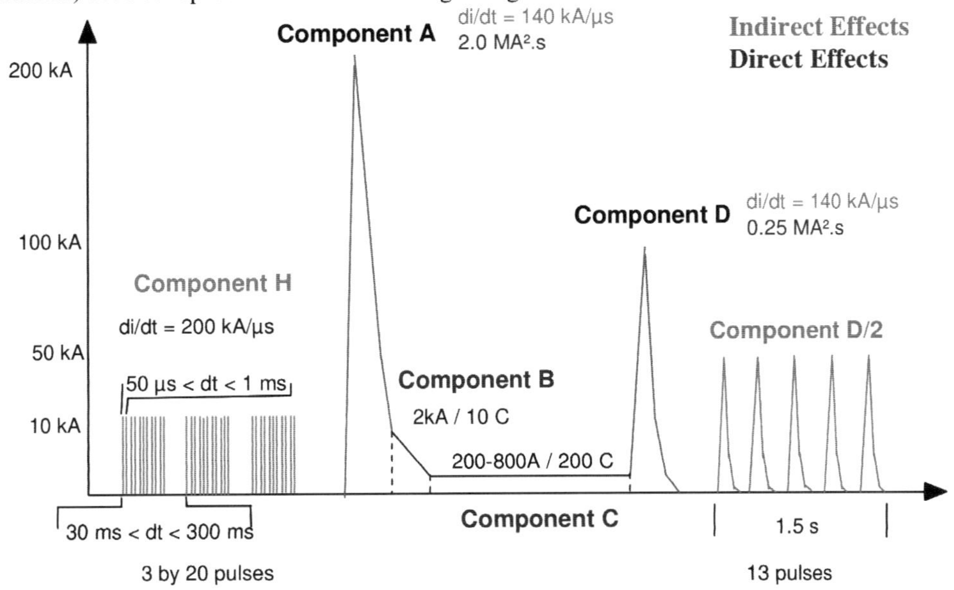

Figure 3. External standardized current waveform representative of an aircraft lightning strike.

- ED-91/ARP 5414A [3] which defines the zoning concept: the plane is divided into zones (see figure 4) where a lightning threat will be applied according to the fact that the location will be the site of a direct attachment of the lightning arc, of the sweeping process or also of the conduction of the lightning current. The concept of zoning permits to adapt the lightning protection to the zone of interest and so to optimize the cost and weight of the solution.
- ED-105 / ARP 5416A [4] which defines the test methods to reproduce the effects of lightning.
- But also other guides as AC20-53B, the AC25-981C dedicated to fuel areas or the ED-113 which defines the all the steps that are required for certification to direct effects purpose.

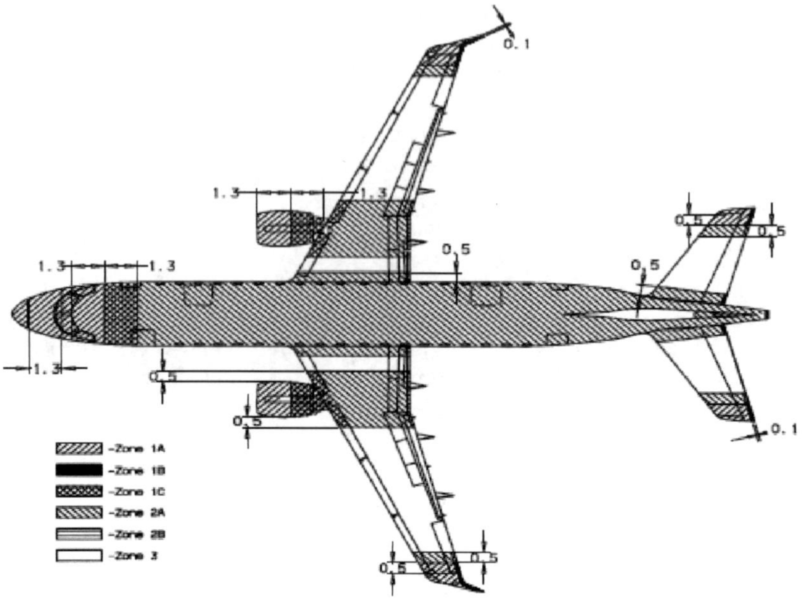

Figure 4. Typical zoning of an aircraft [3].

3. Lightning tests

The tests are one way to demonstrate the resistance of the plane to the effects of lightning. Because of the scale difference between the thundercloud and the facility, tests have to be conducted in a rigorous manner (in accordance with the rules defined in various standard documents) and an inappropriate experimental setup can generate unrealistic results. For each kind of the lightning arc effects, one or more tests will be associated with. To demonstrate the resistance to direct effects, high voltage and high current tests will be performed. Similarly, to demonstrate resistance to indirect effects, equipment and system tests will be conducted.

3.1. Direct effects tests

Due to the electrical power limitation of generators in facilities, it is actually impossible to reproduce simultaneously the voltage and the current damages on aircraft. So, it has been decided to divide these tests into two large categories in order to reproduce the effects of lightning: high voltage testing and high current tests.

3.1.1. High voltage tests. These tests are performed with standardized voltage waveforms [3] representing the variations of the electric field during a lightning strike. Their purpose is to:

- Determine the primary attachment points on a structure and permit to calculate the zoning of the aircraft. Zoning can also be considered as a first barrier because it permits aircraft manufacturer to associate a lightning threat to a zone and so, to adapt the protection that will be associate.
- To test the dielectric parts of an aircraft (such as radome, antenna ...) to possible puncture.

Two voltages waveforms are mainly used for these tests: a faster waveform A dedicated to reproduce the reattachment of the arc channel in zone 2 and the slowest waveform D more appropriated for initial attachment regions in zone 1 (see figure 5).

These tests need to generate arcs of some meters which are often produced by Marx generators. The current flowing in these arcs such generated is often limited to few thousand of amperes and it is not enough to assess the resistance to high current damages.

Journal of Physics: Conference Series **550** (2014) 012001 doi:10.1088/1742-6596/550/1/012001

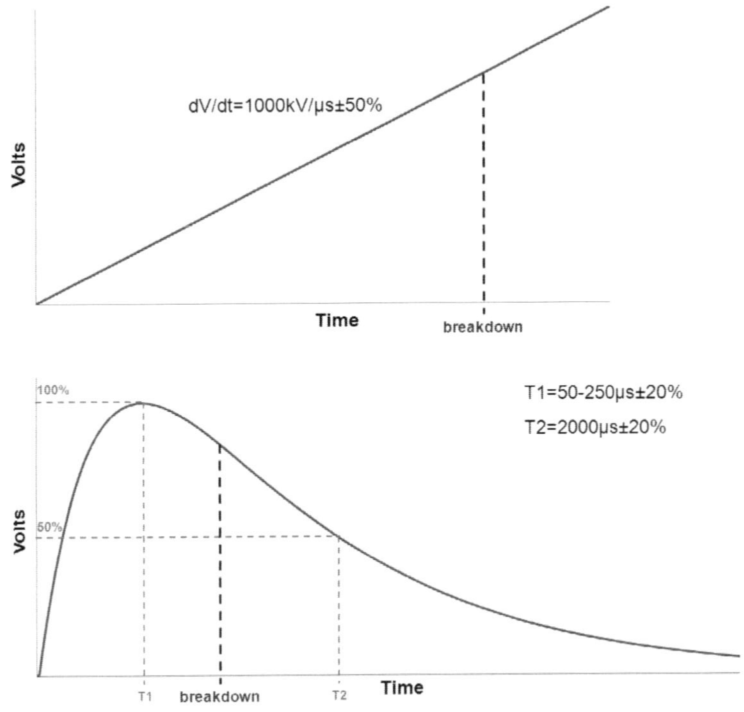

Figure 5. Voltage waveform A (on the top) and voltage waveform D (on the bottom).

The length of the arcs that can be generated and the volume of the testing facilities are limited that is why it is necessary to perform the tests on mock-ups representative of the aircraft in terms of geometry. Nowadays, these tests tend to be abandoned and replaced by an electric field modelling more reliable.

Tests on antennas and radomes are performed using two waveforms representative of a lightning strike phase:

- A waveform representing the slow variations of the electric field during an arc attachment to the structure.
- A waveform representing the fast variations of the electric field during an arc reattachment of the structure during the sweeping process.

These tests permit to evaluate the lightning protections associated to each part of the aircraft, e.g., lightning strips diverters for radomes.

3.1.2. High current tests. These tests are performed according to the normative current waveforms A, B, C and D defined in Figure 3 which permits to reproduce the power and energy associated to each zone in order to evaluate the thermomechanical damages. Their purpose is to:

- Generate arcs in order to study the behaviour of the structure during an arc attachment. These tests are called arc attachment tests.
- Generate a current flow within a sample. These tests are called current conducting tests.

Each waveform is associated with a specific damage. Continuous waveforms (B and C) can cause surface damages due to the energy deposition by the root arc and this damage is mainly effective on metallic materials. The pulse waveforms (A and D) can cause damages by Joule's effect, acoustic shock wave and strong magnetic forces but can also generate sparkings which could ignite fuel vapours in the tanks. The damage after test must be correlated with specifications and criteria defined by the aircraft manufacturer: mechanical strength, acceptable level of puncture...

The measurements associated with this kind of test are very important because they have to measure currents, voltages, temperatures, mechanical deformations, energy from spark...

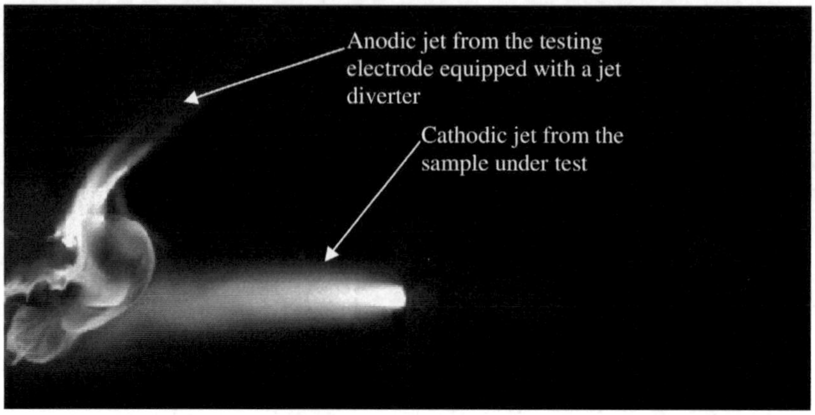

Figure 6. Neutral filtered picture of a 400A continuing arc.

During these tests, the currents which are typically generated are about tens to hundreds of thousands of amperes and the voltages involved are of the order of tens to hundreds of thousands of Volts. These voltages cannot generate electrical arcs with important lengths but are sufficient to breakdown small dielectric thicknesses (such as paints). The arcs generated have lengths of 5 to 10cm and the test electrode has a jet diverter: an insulating ball made with araldite which deflects the anodic jet and thus avoids the contamination of the specimen by the shooting electrode.

3.2. Indirect effects tests

These tests are applied to the equipments for which a protection is required. Their philosophy is based on the diagram below:

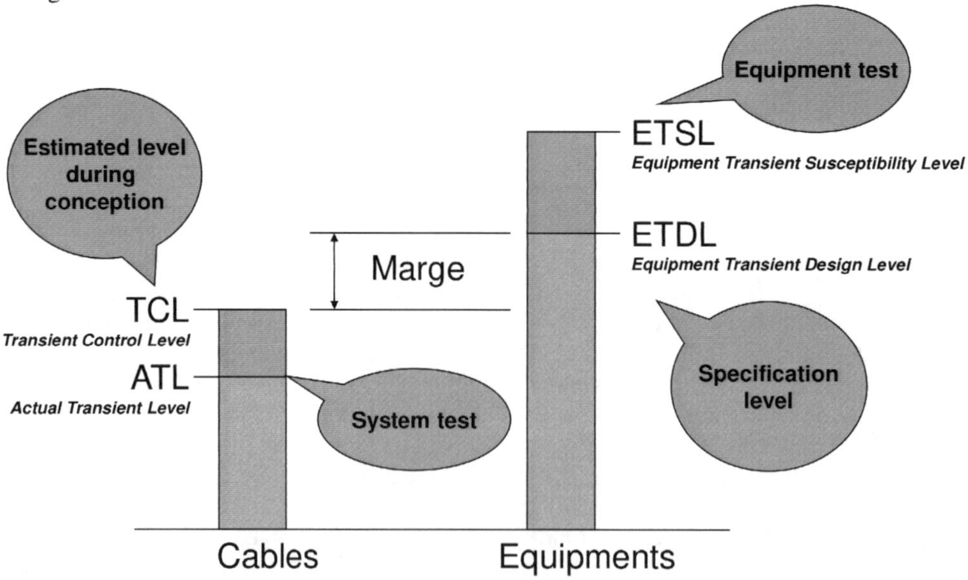

Figure 7. Principle of the lightning indirect effects tests.

At the beginning of the aircraft design, the manufacturer defines a TCL (Transient Control Level): This is a stress lightning level that the equipment could be submitted during a worst-case lightning strike of the aircraft. This level is estimated by calculation, analysis and/or experience from another aircrafts. Then, the aircraft manufacturer applies a margin in order to determine the ETDL (Equipment Transient Design Level) which is the level of certification required to the equipment manufacturer. Tests permit to complete the certification principle. It exists:

- Equipment tests in order to ensure that the equipment under test is able to sustain the ETDL but also to define the susceptibility or faulty level: it is the ETSL.
- System tests in order to measure levels on wiring and equipment and to verify that the measured level (ATL) is lower than the specified (TCL) level.

This principle permits to certify equipments in parallel with the development of the aircraft. The final verification on systems is used to validate the hypothesis.

3.2.1. Tests on systems. These tests occur at the end of the development of the aircraft. Lightning current injections are made directly on the equipped aircraft and permit to define the ATL through measurement of transfer functions. For each equipment, the "worst case" lightning strike is identified: the entry point and the exit point, the type of waveform to be applied depending on the spectral content. At each injection, measurements of current and/or voltage are conducted on the bundles and cables inside the aircraft.

Figure 8. Lightning indirect effects tests on a system equipped with a coaxial current return.

The injections are performed with intensity current generally lower than those imposed by the standardized documents. This is due to:

- The difficulty by the current generators to generate a waveform simultaneously with a fast rising front coupled with a large amount of current.
- The reluctance of aircraft manufacturers to inject a severe current waveform on a plane.

The calculation according to the normalized waveforms is made by extrapolation assuming that the answer of the system is linear. Even more, it is also currently impossible to generate repetitive pulse component such as H or D/2.

In addition, a coaxial current return is used in order to be as close as possible to a lightning strike in flight.

3.2.2. Tests on equipments. For equipments, an internal lightning environment is also defined in the same way that the waveforms shown in Figure 3. These voltage and current waveforms come from analysis and experience but also take into account the location of the equipment in question, its criticality in respect of the safety of flight.

In this context, two types of tests are performed:

- Damage tolerance tests in order to validate the protections.
- Functional tests following predefined acceptance criteria.

4. Conclusion

Lightning is a phenomenon which has to be considered by aircraft manufacturers when developing the aircraft. Nowadays, the demonstration of its capacity to withstand the effects of lightning is largely performed by ground testing. Normative documents are used to guide manufacturers and testing facilities to ensure representative tests and thus reproduce the effects of lightning.

However, many questions still arise. Due to the technological limitation of generators laboratory, it is not possible to simultaneously reproduce all the electrical parameters of a strike: for example, what is the influence of di/dt on the damage or on the appearance of a spark? What could be the influence of repetitive lightning waveforms on the answer of the system in terms of transfer function? Many studies are still needed to better understand the phenomena and perhaps refine the margins of protection taken by manufacturers.

Nowadays, more predictive modelling [5], [6] can help manufacturers and laboratories to better define the lightning tests.

5. Bibliography

[1] Larsson A 2002 The interaction between a lightning flash and an aircraft in flight *C. R. Physique* ***3*** 1423–1444
[2] ED-84 1997 Aircraft lightning environment and related test waveforms (EUROCAE)
[3] ED-91 1998 Aircraft lightning zoning (EUROCAE)
[4] ED-105 2005 Aircraft lightning test methods (EUROCAE)
[5] Lago F, Gonzalez J J, Freton P, Uhlig F, Lucius N and Piau G P 2006 A numerical modelling of an electric arc and its interaction with the anode: part III. Application to the interaction of a lightning strike and an aircraft in flight *J. Phys. D: Appl. Phys.* ***39*** 2294–2310
[6] Chemartin L, Lalande P, Delalondre C, Lago F and Cheron B 2009 Simulated Lightning Attachment on Unpainted Aluminium Panels during continuing current stage: effects of the jet diverting electrode *International Conference On Lightning and Static Electricity*

Journal of Physics: Conference Series **550** (2014) 012002 doi:10.1088/1742-6596/550/1/012002

Modelling Lightning Initiation and Attachment to Aircraft

E. Tam[1], J. J. Lowke[1] and A. B. Murphy[1]

[1] CSIRO Materials Science and Engineering, P.O. Box 218, Lindfield NSW 2070, Australia

E-mail: eugene.tam@csiro.au

Abstract. We present initial calculations of the formation of streamers on an aircraft. A two-dimensional model has been used to determine electric field strengths and charge densities around solids of various geometries and electrical conductivities. The calculations take into account the distortion of the background electric fields by the solid material and the production and motion of charged species. Detailed time-dependent visualizations of the streamer initiation and propagation are presented. The effects of size and aspect ratio of the gross features of the aircraft on the development of the streamers is discussed.

1. Introduction

In recent years, aircraft manufacturers have been motivated to produce aeroplanes largely made out of carbon composite materials rather than aluminium, due to the significant reduction in weight for comparable or superior mechanical properties. These materials however, have significantly reduced electrical conductivities, which makes lightning protection more difficult. Planes are struck by lightning about once every year on average, so a thorough understanding of the initiation of lightning by aircraft, and in particular which regions of the aeroplane are likely to be struck by lightning, is important.

There have been a number of studies in the past in which the development of streamers has been modelled [1–5]. Earlier attempts used simplified one dimensional models with an assumed diameter of the lightning leader [1–3]. In more recent years, there have been some studies in more complex chemical models (i.e. with nitrogen and oxygen species considered separately) have been calculated in two dimensions.[4, 5]. However the ionization rates, which have an exponential dependence on the ratio of the electric field strength to the number density of the gas, were kept either constant or artificially low, possibly due to numerical instabilities and the method used to solve the equations.

A two-dimensional axisymmetric model has been developed to determine electric field strengths and charge densities within and around solids of various geometries and conductivities. The calculations include the distortion of background electric fields by the metal and dielectric materials in the aircraft and the initiation of corona discharges on the aircraft. A finite-difference approach incorporating a direct method is used to solve the system of equations in a coupled fashion. By solving all of the equations simultaneously, as opposed to using a segregated approach, we are able to overcome numerical instabilities associated with an ionization rate which grows exponentially with the electric field strength.

2. Model

In the atmosphere, there is a fairly continuous production of free electrons, produced for example by cosmic rays. We represent the effect by including a constant production term S_e. Assuming air has an extremely low conductivity in the absence of an electric field (e.g., 10^{-17} S/m) and knowing the mobilities of the charged species, one can estimate the equilibrium charge densities without the presence of an electric field; this is our initial condition. The computed solution is not sensitive to the initial conditions for the charged species densities for the parameters used in these calculations, as the peak electric field far exceed the breakdown value of air and the charged species density grows rapidly in the initial stages. Electrons in the atmosphere slowly attach to form negative ions, in particular O_2^- ions. The rate of this attachment is represented by η/N, where N is the gas number density and the attachment coefficient η is the fraction of electrons attached in moving a unit distance forward in the electric field. In regions of high electric field, electrons are of sufficient energy to ionize air molecules to form a new electron and a positive ion. This rate is represented by α/N, where the ionization coefficient α is the fraction of the electrons that produce a new electron by ionization in moving a unit distance forward in the electric field. Thus the net increase in electrons in unit time at a given position is $\partial n_e/\partial t = n_e(\alpha - \eta)W_e$, where W_e is the electron drift velocity. Additional terms are required in the electron continuity equation to account for the recombination of electrons and positive ions, and to represent the effects of convective flow of the charged particles. Lastly, a diffusion term is included, where the diffusion coefficient, D_e, is estimated by using the Einstein–Smoluchowski relation, i.e., $D_e/\mu = k_B T/e$, where $\mu = W_e/E$ is the electron mobility, k_B is the Boltzmann constant, e is the charge of an electron and T is the electron temperature.

Predictions of the onset of corona (streamer) discharges at typical electric potentials are obtained by using a control volume approach to solve the equations

$$\nabla^2\phi = \frac{e}{\epsilon_0}(n_+ - n_- - n_e) \tag{1}$$

$$\frac{\partial n_e}{\partial t} = \alpha W_e n_e - \eta W_e n_e - \gamma n_e n_+ - \nabla\cdot(\boldsymbol{W_e} n_e) + \nabla\cdot(D_e\nabla n_e) + S_e \tag{2}$$

$$\frac{\partial n_+}{\partial t} = \alpha W_e n_e - \gamma n_+(n_e + n_-) - \nabla\cdot(\boldsymbol{W}_+ n_+) + \nabla\cdot(D_+\nabla n_e) + S_+ \tag{3}$$

$$\frac{\partial n_-}{\partial t} = \eta W_e n_e - \gamma n_- n_+ - \nabla\cdot(\boldsymbol{W}_- n_-) + \nabla\cdot(D_-\nabla n_-) \tag{4}$$

where t is the time, n_x is the number density of electrons, positive and negative ions for $x = e$, $+$ and $-$ respectively and γ is the recombination coefficient.

The ionization and attachment coefficients are each a function of the electric field; that is α/N, η/N and W depend on E/N. The first two of these variables are shown in Fig. 1. The values are taken from the theoretical predictions of [6], and are consistent with other theoretical investigations and also with experimental values of both ionization and attachment coefficients [7]. Recombination coefficients and ion mobilities were taken from [6].

3. Results

The particular configuration of our calculations is illustrated in Fig. 2. The thundercloud is represented as imposing a potential of -100 MV at the upper plane of the computational domain. Frequently, thunderclouds have been calculated to be at higher potentials, for example at -500 MV. The height of this potential surface is taken to be at 2 km. Aircraft frequently fly at higher altitudes, but most lightning strikes to aircraft seem to occur as the aircraft approaches the earth. We choose the model aircraft to be a vertical cylinder of radius 25 cm and length 50 m. We choose the cylinder to be vertical as this will be a position for which there is maximum enhancement of the electric field from the cloud at the tips of the cylinder. The length of 50 m

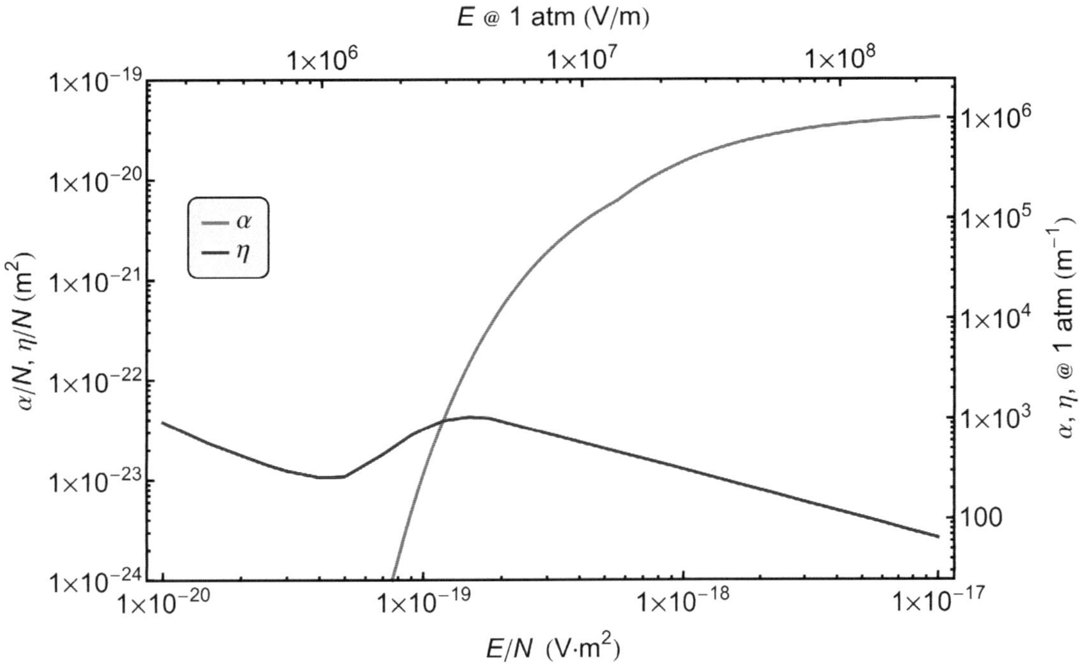

Figure 1. ionization and attachment coefficients, normalised to the number density, as a function of the ratio of electric field strength and number density in dry air. Top and right scales indicate values at 1 atmosphere.

is chosen to be representative of lengths of typical large aircraft. The radius of 25 cm is chosen to approximate the radii of curvature occurring at wing tips and the tail of aircraft. Then, the ratio of the cylinder's length (measured from its centre to its tip) of 25 m, divided by its radius of 25 cm, gives an aspect ratio of 100.

There are three stages in the development of the electric field. First, there is the background field from the thundercloud of 50 kV/m produced from the 100 MV postulated for the cloud potential at a height of 2 km. Second, there is distortion by the cylinder or aircraft fuselage, as illustrated by Fig. 3, where as yet there are no electrical charges in the air at the high field tips of the cylinder. Third, there are distortions due to the space charge produced by the discharge.

Fig. 3 shows the electric field strength in the immediate surroundings of the conductive cylinder. The electric field strength inside the conductor is calculated to be less than 10 V/m, significantly lower than the average field strength of 50 kV/m. The electric fields in the gas in the immediate surroundings of the cylinder tips can be seen to be significantly enhanced, with field strengths of the order of 5 MV/m, in agreement with the approximate analytic prediction.

The left-hand side of Fig. 4 shows, in more detail, the electric field in the vicinity of the top tip of the cylinder shown in Fig. 3. The peak electric field in this figure exceeds 2.5 MV/m, thus streamer formation is expected as the electric field exceeds the breakdown field of air. The right-hand side of Fig. 4 shows typical calculated field distributions once the space charge effects have become important. This field distribution changes with time as the discharge develops and becomes larger, there being a significant maximum of the electric field at the leading edge of the discharge.

The distributions of electron, positive-ion, negative-ion and net-charge densities for the

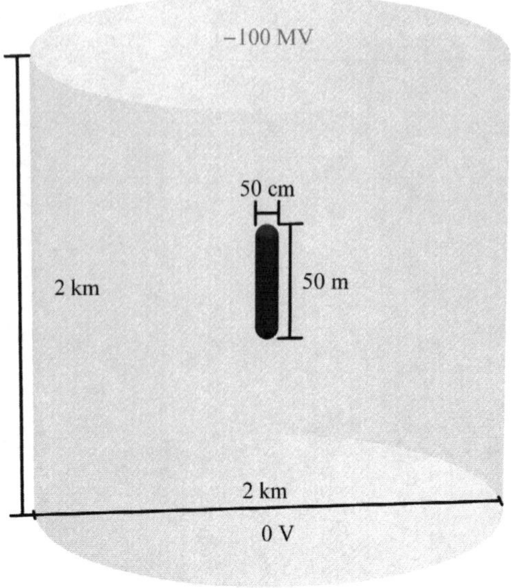

Figure 2. Configuration of the calculations, with the aircraft represented by a cylinder of length 50 m and radius 25 cm halfway between the earth and a thundercloud at 100 MV and height 2 km. (Figure not to scale).

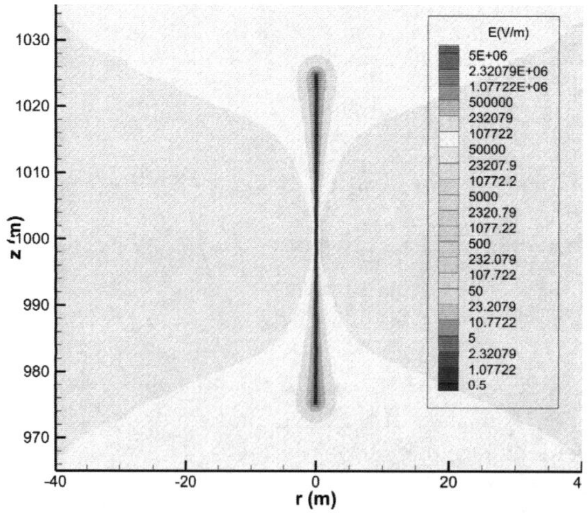

Figure 3. Electric field at an early stage (t $\approx$ 100 ns), showing distortion of the background field by the cylinder. The difference between cylinders of different conductivities cannot be distinguished at this magnification. Note the scale is logarithmic in this figure and the background field is 50 kV/m.

conditions of Fig. 4 are shown in Fig. 5. In these figures, the values of the densities in the metal are set to zero for clarity, as the electron and positive ion densities are several orders of

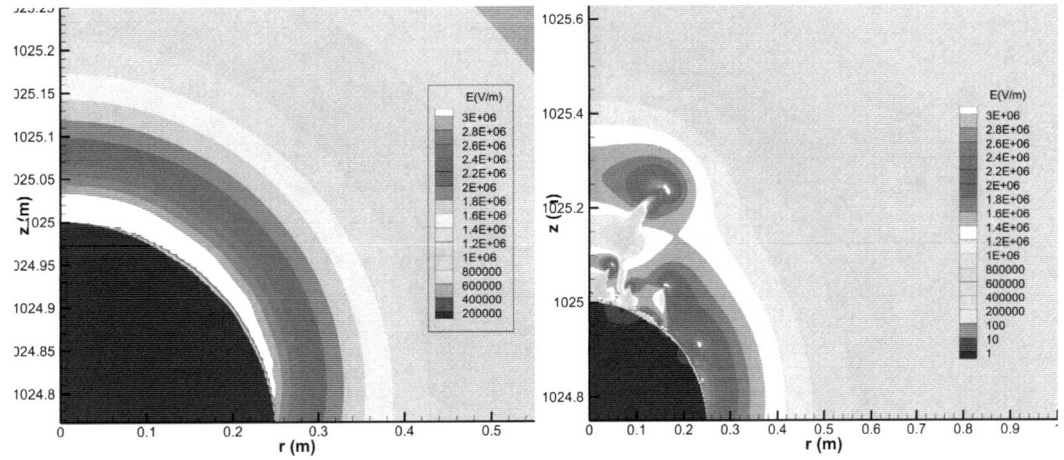

Figure 4. Left: Electric field at an early stage (t ≈ 100 ns) near the top cap of a cylinder, showing distortion of the background field by the cylinder. Right: As with left at t = 1.84 μs.

magnitude larger than in the gas.

The maximum electron densities on the left of Fig. 5 occur in the regions of maximum electric field strength in Fig. 4. When the field drops below 2.5 MV/m, there is a rapid transition from net electron production to net electron loss due to attachment of electrons to the neutral gas molecules. Electron attachment to neutrals will reduce the brightness of a stepped leader.

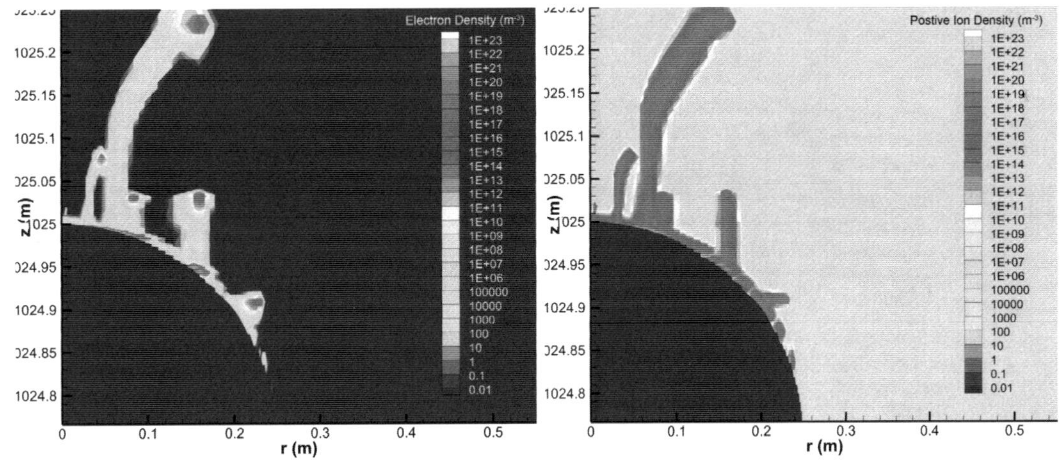

Figure 5. Electron density (left) and positive ion density near the top cap of the cylinder at t = 1.84 μs.

The right-hand side of Fig. 5 shows the positive-ion density distributions. The negative-ion density, which is not shown, has approximately the same distribution. These densities are generally larger that the electron density. Despite this fact, the net charge density is still large, as shown in Fig. 6. This indicates that there is a significant degree of distortion of the electric field by space charge.

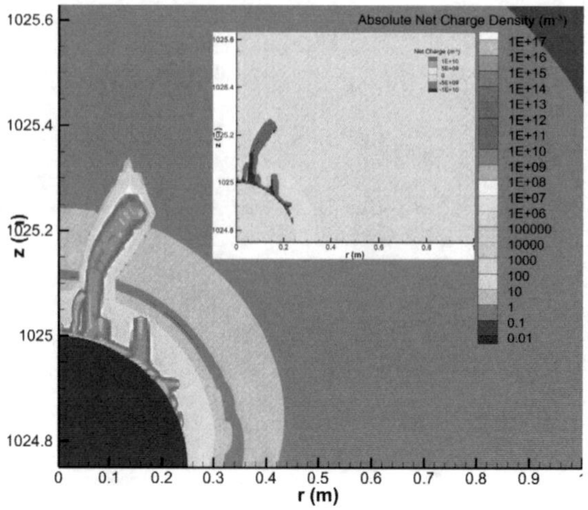

Figure 6. Absolute value of net charge density near the top cap of the cylinder at t = 1.84 μs. The inset shows the net charge, indicating the regions which are negatively or positively charged.

The drift velocity of the electrons with the calculated electric fields is of the order of 10^4 m/s. Within the time frame, ~ 2 μs of Fig. 5 the streamers have grown to approximately 25 cm, which means the streamers grow at a rate of 10^5 m/s, which is an order of magnitude larger than the drift velocity of the electrons (which in turn is two orders of magnitude larger than the drift velocity of the heavy ionized species). This indicates that the growth of the streamers is predominately due to the production of charges rather than the charged species rearranging themselves at a rapid pace. In areas where the electric field is large, the charged species production is also large which then quickly leads to a large density of charge species. Once the number of charged species is sufficiently large, the distance each charged species needs to move to make a significant differences to the net charge density (and therefore the electric field) is reduced and therefore the electric field within the streamers are significantly smaller than at the edges of the streamers, where there are fewer charged species to alter the field.

The results presented here were only for the top half (positive end) of the cylinder. The results on the other end are similar; however, the streamer grows at a steady rate on the negative end in contrast to the positive end, which has a stop–start behaviour.

4. Conclusion

We have simulated the initiation of streamer discharges from a capped cylinder, of dimensions similar to that of an aeroplane fuselage, in an electric field representing that between the thundercloud and ground. The calculations were made in two dimensions because of the available computational resources. This means that the streamers are effectively conical in shape. Nevertheless, we illustrate the interaction between electric fields of charged species that lead to corona initiation and streamer progress.

We have achieved significantly higher charge densities than were possible in a previous study [8], in which values of the ionization coefficient were capped to avoid numerical instabilities. Because such capping is no longer necessary with our new code, we are able to achieve higher charge densities, and accurately predict the faster formation of streamers that occurs in nature.

This model is and does not take into account the role of ice or water. Ice is widely believed

Journal of Physics: Conference Series **550** (2014) 012002 doi:10.1088/1742-6596/550/1/012002

to play a role in the formation of lightning. There are two mechanisms through which ice is believed to assist the formation of lightning. The first is that collisions between ice particles are believed to cause non-inductive charge transfer which then separate forming a large electric field [9]. The second is the fact that ice can have sharp edges which, combined with its relatively high electrical conductivity compared to air, can enhance the electric field locally around the ice particles. Neither of these mechanisms has been considered, as explaining the high potential of the cloud was outside the scope of the paper, whose main focus is investigating how lightning is triggered from the aircraft. Water in the form of vapour modifies both the ionization and attachment coefficients. The effect on the formation of streamers is complex and will be a subject of future work.

References

[1] Morrow R and Lowke J J 1997 *J. Phys. D: Appl. Phys.* **30** 614-627
[2] Aleksandrov N L and Bazelyan E M 1996 *J. Phys. D: Appl. Phys.* **29** 740-752
[3] Babaeva N Y and Naidis G V 1996 *J. Phys. D: Appl. Phys.* **29** 2423-2431
[4] Luque A, Ratushnaya V and Ebert U 2008 *J. Phys. D: Appl. Phys.* **41** 234005
[5] Liu N., Kosar B, Sadighi S, Dwyer J R and Rassoul H K 2012 *Phys. Rev. Lett.* **109** 025002
[6] Lowke J J 1992 *J. Phys. D: Appl. Phys.* **25** 202-210
[7] Phelps A V 1987 *Excitation and ionization coefficients, Gaseous Electronics V* (New York: Pergamon)
[8] Lowke J J 2004 *IEEE Trans. Plasma Sci.* **32** 4-17
[9] Mansell E R, MacGorman D R, Ziegler C L and Straka J M 2005 *J. Geophys. Res.* **110** D12101

Stud arc welding in a magnetic field – Investigation of the influences on the arc motion

K Hartz-Behrend[1], J L Marqués[2], G Forster[1], A Jenicek[3], M Müller[3], H Cramer[3], A Jilg[4], H Soyer[4] and J Schein[1]
[1] Universität der Bundeswehr München, Institut für Plasmatechnik und Mathematik, Neubiberg, Germany
[2] Universität der Bundeswehr München, Institut für Automatisierungs- und Regelungstechnik, Neubiberg, Germany
[3] GSI mbH, Niederlassung SLV München, München, Germany
[4] Heinz Soyer Bolzenschweisstechnik GmbH, Wörthsee, Germany

E-mail: karsten.hartz-behrend@unibw.de

Abstract. Stud arc welding is widely used in the construction industry. For welding of studs with a diameter larger than 14 mm a ceramic ferrule is usually necessary in order to protect the weld pool. Disadvantages of using such a ferrule are that more metal is molten than necessary for a high quality welded joint and that the ferrule is a consumable generally thrown away after the welding operation. Investigations show that the ferrule can be omitted when the welding is carried out in a radially symmetric magnetic field within a shielding gas atmosphere. Due to the Lorentz force the arc is laterally shifted so that a very uniform and controlled melting of the stud contact surface as well as of the work piece can be achieved. In this paper a simplified physical model is presented describing how the parameters welding current, flux density of the magnetic field, radius of the arc and mass density of the shielding gas influence the velocity of the arc motion. The resulting equation is subsequently verified by comparing it to optical measurements of the arc motion. The proposed model can be used to optimize the required field distribution for the magnetic field stud welding process.

1. Introduction

Stud arc welding, also called stud welding, is an arc welding process in which a metal fastener (stud) is welded onto a metal work piece. It is a welding process widely used in the construction industry [1]. The type of arc ignition, the welding duration, the weld pool shielding and the type of power source define different process types [2]. In this paper the drawn-arc stud welding process with a ceramic ferrule will be considered. In general, the use of a ceramic ferrule is necessary when studs with a diameter larger than 14 mm are to be welded. The ceramic ferrule is used for protecting the weld pool but the application of a ferrule has the disadvantage that more metal is molten than necessary for a high quality welded joint, i.e. the energy load during the welding process is higher than required. Additionally the ferrule is a consumable usually thrown away after the welding operation and thus it would be an advantage if the ferrule could be omitted.

About 5 years ago a new process form of stud welding was presented: radially symmetric magnetic field stud welding. In this process a radially symmetric magnetic field is used to laterally shift the arc by means of the Lorentz force such a way that a very uniform and controlled melting of the stud and the work piece can be achieved [3]. Main advantages of this new process variant are that the ceramic ferrule can be omitted [3] and that the energy load is reduced by about 70 percent, resulting in lower thermal stress inside the work piece and therewith a reduced work piece distortion. By means of radially symmetric magnetic field stud welding it is possible to weld studs with a diameter larger than 12 mm. High quality welded joints for studs with a diameter of 16 mm have been achieved.

The aim of this paper is to present a simplified theoretical model describing the effect of important process parameters in the magnetic field stud welding process on the velocity of the arc motion. Such a model can be very useful for both modeling and optimizing the magnetic field stud welding process. The equation resulting from this model agrees quite well with optical measurements of the arc motion.

The paper is organized as follows: In the next two sections the conventional stud welding process and the magnetic field stud welding process without ferrule are described. In the section that follows a simplified physical model for the arc motion under a magnetic field is derived. In order to experimentally verify the resulting equations an optical measurement setup has been developed, as described in the fifth section. In the last section the experimental results as well as their good agreement with the theoretical model is discussed. The paper concludes with a summary and an outlook.

2. Drawn-arc stud welding with a ceramic ferrule

The drawn-arc stud welding process with a ceramic ferrule consists of four steps (figure 1):

- In the first step, a stud is loaded into the chuck of a welding gun, a ceramic ferrule is placed around the stud, and the stud is positioned against the work piece to close an electrical circuit with a current of about 100 A.
- Afterwards a trigger is pressed and the welding gun lifts up the stud in the ceramic ferrule. An initial pilot arc is ignited, creating an electric path for the welding current. After a short fixed time of about 100 ms the actual welding electric current is switched on and the resulting welding arc melts the stud surface as well as the work piece.
- During step three the stud is plunged into the molten weld pool.
- Finally the welding is completed and the welding gun is lifted off. The ferrule is mechanically removed.

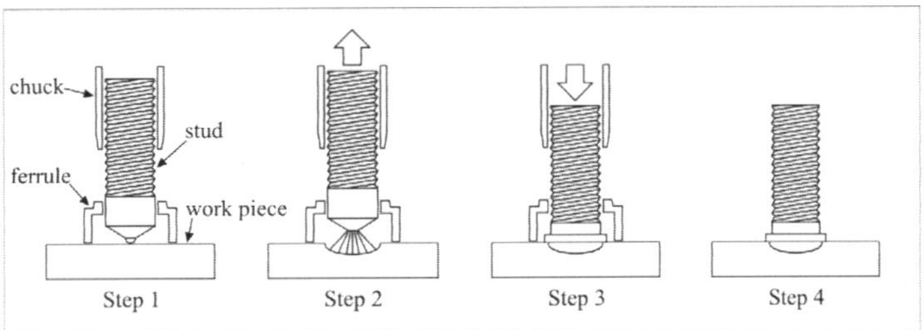

Figure 1. The four steps of a drawn-arc stud welding process

In the next section the magnetic field stud welding process is described, as well as its differences to the welding process just described in this section.

3. Magnetic field stud welding

For the magnetic field stud welding process no ceramic ferrule is used. In order to protect the weld pool a shielding cap is surrounding the welding area (figure 2). This cap is automatically lowered once a trigger has been pressed and subsequently a shielding gas is introduced into the cap. After a fixed period of time the stud is lifted up and a pilot arc is ignited. The process steps following the lifting of the stud are similar to those for the welding process with a ferrule, the only difference being the applied magnetic field. This field is generated by an electric coil located in the shielding cap. Due to the magnetic field a Lorentz force is acting perpendicularly on the arc resulting in a lateral arc motion. As already mentioned a very uniform and controlled melting of the stud surface and the work piece can be achieved due to this arc motion.

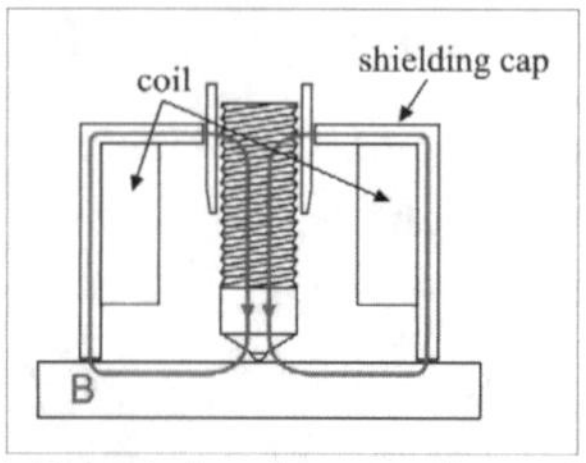

Figure 2. Principle of magnetic stud welding

4. A simplified theoretical model for the magnetic field stud welding

In order to model and/or optimize the magnetic field stud welding process it is very helpful to assess the arc motion as a function of the most relevant process parameters. In this section a simplified model is presented which allows to correlate the velocity of the arc motion to such parameters as welding current, applied magnetic field, arc radial extension and mass density of the cold shielding gas. Let us consider a vertical plasma cylinder of constant radius r_{core}, carrying a stationary electric current I_{el} directed upwards (figure 3). We assume in a first approximation that the whole cylinder displays a homogenously constant temperature and that the electric current is also homogeneously distributed across the cylinder cross section such that the resulting electric current density within the plasma core is equal to

$$j_{el} = \frac{I_{el}}{\pi r_{core}^2} . \quad (1)$$

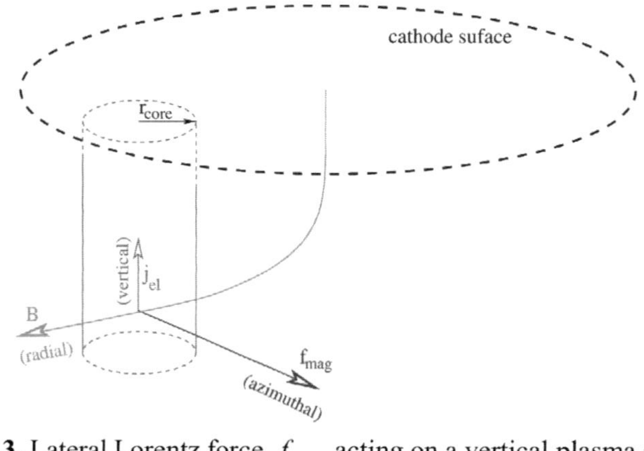

Figure 3. Lateral Lorentz force f_{mag} acting on a vertical plasma cylinder

The cathode spot of this plasma column is located at some off-center position at the cathode surface and a stationary magnetic field B produced by the coil displays a mainly radial component at the lower part of the plasma column, pointing outwards (see figure 3). As a result a magnetic force per volume of magnitude

$$f_{mag} = j_{el} B \quad (2)$$

is produced in azimuthal direction, dragging tangentially the plasma core and generating thus an approximately circular motion of the plasma column. The mechanical work W spent during a core displacement of r_{core} to a position where previously only surrounding cold shielding gas existed leads to the following relation for the work per volume:

$$\frac{W}{Volume} = f_{mag} r_{core} = j_{el} B r_{core} \,. \tag{3}$$

The main contribution to W arises from the displacement of the more massive cold shielding gas volume surrounding the plasma column. Thus the mechanical work (per volume) mainly goes into an increase in kinetic energy (per volume) to the final arc velocity v_{arc}. This can be expressed by

$$j_{el} B r_{core} \approx \frac{1}{2} \rho_{gas}(T_{cold}) v_{arc}^2 \Rightarrow v_{arc} \approx \sqrt{\frac{2 I_{el} B}{\pi\, r_{core}\, \rho_{gas}(T_{cold})}} \tag{4}$$

or equivalently

$$v_{arc}^2 \propto \frac{I_{el} B}{r_{core}\, \rho_{gas}(T_{cold})}, \tag{5}$$

whereas $\rho_{gas}(T_{cold})$ represents the mass density of the cold shielding gas. A more detailed derivation of a similar relation can be found in [4]. It is worth noting that the simple model leading to equation (4) does not include the partial heating of some peripheral regions of the plasma column, produced by the electric current across the plasma core and required for its existence. Hence the kinetic energy variation of the plasma core actually contains an additional contribution depending on the temperature of the plasma column but independent of the applied magnetic field. Furthermore the magnetic tangential force particularly acts on the lower part of the plasma core, slightly bending the plasma column and producing thus an additional radial component in the arc motion.

Figure 4. High-speed stereo optical system (left: measuring principle; right: experimental setup)

Since the derivation of the relations (4) and (5) is based on several simplifying assumptions about the plasma arc, it should be verified if these relations also hold for a real magnetic field stud welding process. The experimental verification of equation (5) is carried out by means of a high-speed stereo optical investigation of the arc motion, as presented in the next section.

5. Description of the high-speed stereo optical setup

In the developed experimental setup, the moving arc is observed at an angle of 90° by using 2 mirrors and a mirror prism (figure 4). By means of the mirror prism, the 2 images of the arc are projected simultaneously onto the CCD-chip of a high-speed camera (model "dimax HD" from the company PCO). Due to the different viewing angle in the 2 images the arc position can be determined. Two typical examples of the recorded measurements are given in figure 5: initial ignition of the pilot arc (above) and full developed welding arc (below).

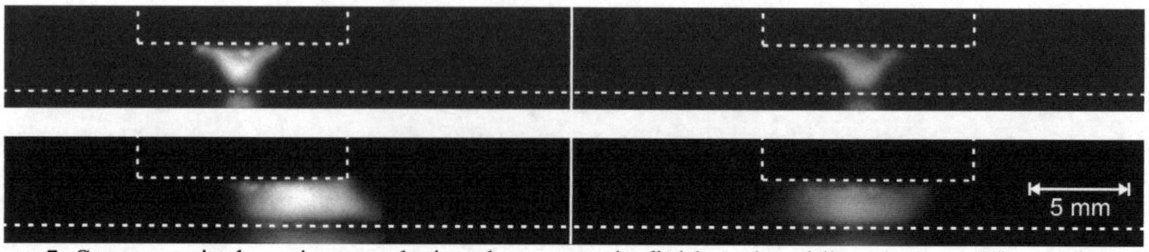

Figure 5. Stereo optical arc images during the magnetic field stud welding process (above: ignition of pilot arc; below: actual welding arc; left/right: images obtained with the left/right mirror of the stereo optical system). The dotted lines indicate the position of the stud and the work piece, respectively.

6. Investigation of the arc motion by means of high-speed stereo optical setup

In this section the experimental results for the arc motion and their agreement with the theoretical model are discussed. During the experiments the arc was recorded at a frame rate of 10,000 fps and a resolution of 960x100 pixels. The exposure time was set at 2 μs. For each investigated parameter the measurement was repeated 3 times. The mean arc motion velocity $\bar{v}_{arc}$ is determined by averaging the position change of the arc between two consecutive images. Studs of diameter 12 mm and 16 mm were used, with a two-component shielding gas mixture comprising 82 % argon and 18 % carbon dioxide (Corgon 18). First, the welding current I_{el} was set to 600 A, 800 A, 1000 A and 1200 A, with a constant coil current of 705 mA and studs with a diameter of 16 mm. In this case the recorded images of the high-speed camera were used to determine both the arc position as well as the mean radial extension $\bar{r}_{bright}$ of the bright area of the welding arc. This latter radius is directly proportional to actual radius $\bar{r}_{core}$ of the plasma arc.

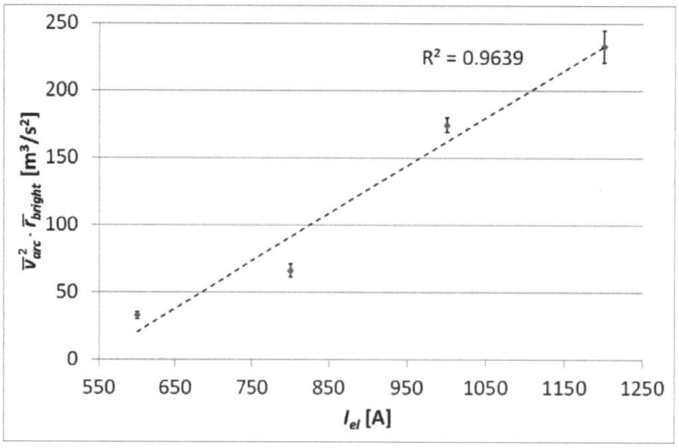

Figure 6. Nearly linear increase with the welding current I_{el} of the squared mean velocity of the arc motion $\bar{v}^2_{arc}$, multiplied with the radial extension $\bar{r}_{bright}$ of the bright area of the welding arc (proportional to the mean radius $\bar{r}_{core}$ of the arc). Each measurement point corresponds to the mean value of three measurements and the error bars indicate one standard deviation from that mean value.

As displayed in figure 6 there exists a very good linear correlation between the squared arc velocity multiplied by the $\bar{r}_{bright}$ and the electric current flowing through the arc:

$$\bar{v}_{arc}^2 \cdot \bar{r}_{bright} \propto I_{el} \,. \tag{6}$$

Hence

$$\bar{v}_{arc}^2 \propto \frac{I_{el}}{\bar{r}_{core}} \tag{7}$$

just as expected from the simplified theoretical model. The vertical offset of the straight line shown in figure 6 is related to the additional contribution to the kinetic energy density discussed at the end of the previous section. In figure 6 the standard deviation at each measurement point is between 3 and 7 percent of $\bar{v}_{arc}^2$, multiplied with $\bar{r}_{bright}$.

In a second experimental investigation the applied magnetic field was modified by selecting the coil current I_{coil} to 210 mA, 330 mA, 450 mA, 1100 mA and 1500 mA. The welding current I_{el} was kept constant at a value of 700 A and studs with a diameter of 12 mm were used. As shown in figure 7 the following linear relation

$$\bar{v}_{arc}^2 \propto I_{coil} \tag{8}$$

holds in a good approximation. Since the strength of the horizontally applied magnetic field B in the gap between the stud and the work piece is proportional to the coil current I_{coil} we conclude

$$\bar{v}_{arc}^2 \propto B \,, \tag{9}$$

which corresponds to the equation derived in the simplified model. In figure 7 the standard deviation at each measurement point is between 7 and 13 percent of $\bar{v}_{arc}^2$.

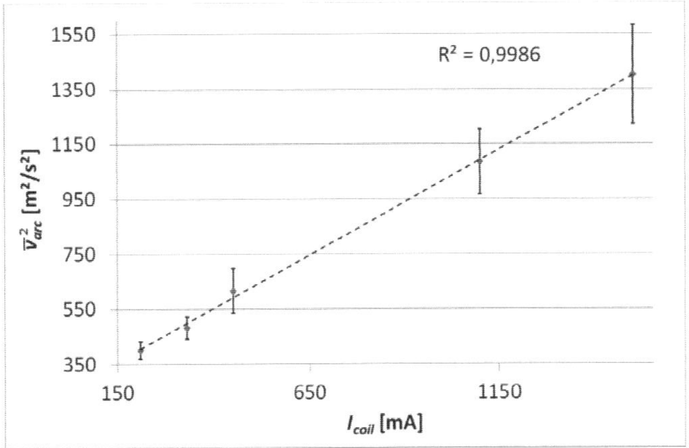

Figure 7. Nearly linear increase with the coil current I_{coil} of the squared mean velocity of the arc motion $\bar{v}_{arc}^2$. Each measurement point corresponds to the mean value of three measurements and the error bars indicate one standard deviation from that mean value.

In order to investigate the influence of the shielding gas density on the velocity of the arc motion stud welding experiments with helium as shielding gas were performed. The ratio of density between

Corgon 18 and helium is approximately a factor of 10. Due to relation (5) for the mean velocity $\bar{v}_{arc}$ of the arc motion a difference of approximately a factor 3 can be expected. The experiments display an average arc velocity of $\bar{v}_{arc}$ = 22 m/s for Corgon 18 and $\bar{v}_{arc}$ = 65 m/s for helium, with constant arc welding current of I_{el} = 700 A and coil current of I_{coil} = 330 mA. This is again in good agreement with the theoretical model.

7. Summary and outlook

In this paper the magnetic field stud welding process has been investigated. The main advantage of this welding process, compared to a process using a ceramic ferrule, is that this consumable ferrule can be omitted. As a consequence the energy load is reduced, less metal is molten and the thermal stress induced in the work piece is reduced. Based on a physical theoretical model a simple equation has been derived relating the motion velocity of the welding arc to parameters such as welding current, flux density of the applied magnetic field, radius of the arc and mass density of the shielding gas. This equation has been experimentally verified by recording the arc motion with a high-speed stereo camera. Such equation can be useful for both modeling and optimizing the magnetic field stud welding process. In future work a detailed simulation of the local magnetic field distribution will be carried out in order to determine how the geometry of the welding gun and coil influence the motion of the welding arc.

Acknowledgements

This work is funded by the Bayerische Forschungsstiftung BFS (Bavarian Research Foundation), funding number AZ 959-11. This support is gratefully acknowledged.

References

[1] Chambers H A 2001 *PCI journal* **46** 46
[2] Jenicek A and Cramer H 2003 *Welding and Cutting* **55** 105
[3] Jenicek A, Forster G, Hartz-Behrend K, Müller M, Reiter T, Schein J and Soyer H 2013 Reproduzierbare Schweissqualitaet durch Bolzenschweissen mit Hubzuendung und magnetisch bewegtem Lichtbogen *DVS Congress 2013, DVS-Berichte Band* **296** 289
[4] Maecker H H and Stablein H G 1986 *IEEE Transactions on Plasma Science* **PS-14** 291

Influence of the arc plasma parameters on the weld pool profile in TIG welding

A Toropchin[1], V Frolov[1], A V Pipa[2], R Kozakov[2], and D Uhrlandt[2]

[1]Saint Petersburg State Polytechnical University, 29 Polytechnicheskaya, 195251 Saint-Petersburg, Russia,

[2] Leibniz Institute for Plasma Science and Technology (INP Greifswald), Felix-Hausdorff-Str. 2, 17498 Greifswald, Germany

E-mail: `frolov.eed@gmail.com`

Abstract. Magneto-hydrodynamic simulations of the arc and fluid simulations of the weld pool can be beneficial in the analysis and further development of arc welding processes and welding machines. However, the appropriate coupling of arc and weld pool simulations needs further improvement. The tungsten inert gas (TIG) welding process is investigated by simulations including the weld pool. Experiments with optical diagnostics are used for the validation. A coupled computational model of the arc and the weld pool is developed using the software ANSYS CFX. The weld pool model considers the forces acting on the motion of the melt inside and on the surface of the pool, such as Marangoni, drag, electromagnetic forces and buoyancy. The experimental work includes analysis of cross-sections of the workpieces, high-speed video images and spectroscopic measurements. Experiments and calculations have been performed for various currents, distances between electrode and workpiece and nozzle diameters. The studies show the significant impact of material properties like surface tension dependence on temperature as well as of the arc structure on the weld pool behaviour and finally the weld seam depth. The experimental weld pool profiles and plasma temperatures are in good agreement with computational results.

1. Introduction

Tungsten Inert Gas (TIG) welding was developed in 1940s, and since the beginning it is widely applied in industry and a continuously improved technology. It is based on the process of the heat transfer by the electric arc from the non-consumable tungsten cathode to the workpiece-anode. Straight polarity is suitable for steels, investigated in the present article, to maximize the welding penetration and minimize the cathode heating. For aluminium, for example, the alternating current is more suitable for surface oxide layer damaging in the reverse polarity period and tungsten cathode cooling with maximum workpiece heating in the straight polarity period. The reverse polarity is never used to avoid melting of the tungsten electrode and the consequent contamination of the weld pool.

The process of TIG welding is operated with arc and weld pool shielding by an inert gas. As inert gas welders use argon or helium. Choice of the inert gas is based on the industrial requirements and purposes. The helium arc produces deeper penetration, but helium is more expensive than argon. That is the reason for applications of argon and helium mixtures. One of the advantages of TIG welding is that addition of filler in the weld pool is optional. Use of filler rods is mandatory in particular for thick workpieces (above 34 mm) to make a seam with required quality and mechanical characteristics.

The TIG process is typically ignited by a high-voltage and high-frequency oscillator to avoid contamination of the weld pool by molten tungsten. In contrast, other types of welding processes, gas metal arc welding (GMAW) for example, can be ignited simply with the touch of the two electrodes.

TIG welding processes use arc plasmas with maximum temperatures in the range from 12000 to 20000 K depending on current and voltage. These values are high enough to melt the metal and weld two metal pieces.

The present article is focused on the classic TIG welding, but also investigates and compares the results considering a setup similar to plasma welding. Therefore, the tungsten cathode is placed inside a narrow ceramic nozzle. However, the second nozzle typically used in plasma welding was not applied in the present study. Plasma welding processes are characterized by a compressed arc with elevated temperatures. The compression also increases the depth of weld penetration and decreases the heat-affected zone. Mechanical characteristics of the seams welded by plasma welding can be improved in comparison with classic TIG welding.

Welding as other fields of industry is on the way of continuous progress. Adding new materials in the cathode material, like lanthanum or thorium, to increase it heat resistance, new nozzles with improved inert gas flow etc are one side of the progress. The goal of welding process modeling is to minimize the experimental part for improving the construction of plasma torches or other process improvements in industry or in laboratories. The models must be reliable and include a large number of physical processes to consider the forces acting on the arc, on the motion of the melt inside and on the surface of the pool. The present investigation is focussed on the examination of one of these forces, the necessity to consider the Marangoni force acting on the weld pool profile. Marangoni force, its mathematical formulation and its influence on the shape of the weld pool, depending on the value of the surface tension of the metal have been considered in several papers [1, 2, 3]. Present article describes a method of surface tension calculation of steel depending on the temperature of the metal. Also the investigations of various welding modes, such as arc current and arc length, and various nozzles tip have been done to estimate the efficiency of process and accuracy of simulation model.

2. Simulation model

The time dependent two-dimensional coupled computational model of the arc and the weld pool was developed using software ANSYS CFX. The main governing equations include mass continuity (1), momentum conservation (2),(3), energy conservation (4):

$$\frac{\partial}{\partial z}(\rho V_z) + \frac{1}{r}\frac{\partial}{\partial r}(r\rho V_r) = 0 \tag{1}$$

$$\rho\left(V_z\frac{\partial V_r}{\partial z} + V_r\frac{\partial V_r}{\partial r}\right) = -\frac{\partial p}{\partial r} + \left[\vec{J}\times\vec{B}\right]_r + \frac{\partial}{\partial z}\left(\mu\frac{\partial V_r}{\partial z}\right) + \frac{1}{r}\frac{\partial}{\partial r}\left(r\mu\frac{\partial V_r}{\partial r}\right) - \mu\frac{V_r}{r^2} \tag{2}$$

$$\rho\left(V_z\frac{\partial V_r}{\partial z} + V_r\frac{\partial V_r}{\partial r}\right) = -\frac{\partial p}{\partial z} + \left[\vec{J}\times\vec{B}\right]_z + \frac{\partial}{\partial z}\left(\mu\frac{\partial V_z}{\partial z}\right) + \frac{1}{r}\frac{\partial}{\partial r}\left(r\mu\frac{\partial V_z}{\partial r}\right) + \rho g_z \tag{3}$$

$$\rho V_z\frac{\partial H}{\partial z} + \frac{1}{r}\rho V_r\frac{\partial H}{\partial r} = \sigma E^2 - U_{\mathrm{rad}} + \frac{\partial}{\partial z}\left(\frac{\lambda}{C_{\mathrm{p}}}\frac{\partial H}{\partial z}\right) + \frac{1}{r}\frac{\partial}{\partial r}\left(r\frac{\lambda}{C_{\mathrm{p}}}\frac{\partial H}{\partial r}\right), \tag{4}$$

where ρ is mass density, V_z and V_r are axial and radial velocity components, p is pressure, $\vec{J}$ is electric current density, $\vec{B}$ is magnetic field, μ is viscosity, g_z is gravitational acceleration, H is enthalpy, σ is electrical conductivity, E is electric field strength, U_{rad} is energy losses due to radiation, λ is thermal conductivity and C_{p} is heat capacity at constant pressure. Along with these the current continuity coupled with Ohms law and Maxwells equations similar to those in [1] are used. The boundary conditions were set according to experimental values to meet gas flow rate, electric current and ambient temeprature.

Also, the forces acting on the weld pool, such as drag force by the arc, buoyancy, electromagnetic force and Marangoni force are included in the model [1, 2, 3]. Such effects as metal vapour, turbulence of plasma flow and the movement of the weld pool surface are neglected due to its lower impact on the main behavior of the TIG arc under the conditions considered in this work. The radiative, thermodynamic and transport properties are taken from [4, 5]. The inhomogeneous mesh is used to run the simulations faster. The mesh is different for various geometric configurations and has 30000 cells in average. The smallest cell is 0.03 mm x 0.1 mm in the domain of arc burning and workpiece melting, the biggest one 0.2mm x 2mm in the domain of gas outflow. The most important force affecting the profile of the weld pool is the Marangoni force, which is the phenomenon of mass transfer along the interface between two fluids that occurs due to the presence of the gradient of the surface tension. The higher the surface tension of the liquid, the higher force tighten the surface. Therefore, the liquid moves to the domain with the greater coefficient of surface tension [10]. The Marangoni force is given by

$$M_A = \frac{\partial}{\partial z}\left(\frac{\partial \gamma}{\partial T}\frac{\partial T}{\partial r}\right) \tag{5}$$

where γ is the surface tension of the molten pool and T the workpiece temperature [1]. The influence of the Marangoni force is best illustrated by the following example: comparing a material with falling surface tension characteristic and a material like mild steel with rising characteristic. The dependence of weld pool profile and direction of molten metal movement on the surface tension is shown in figure 1 [2].

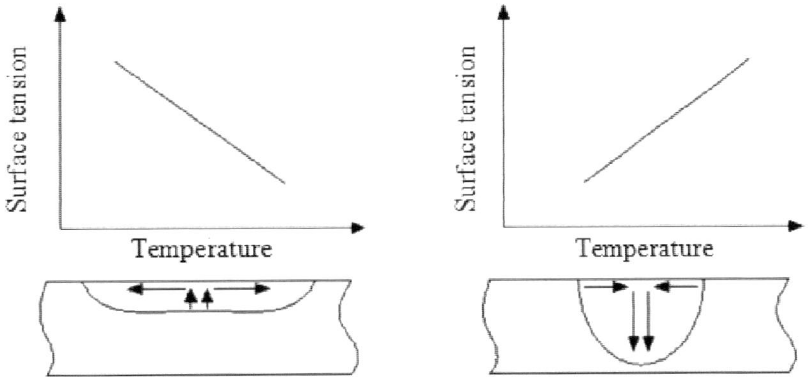

Figure 1. Schematic illustration of convection in the weld pool driven by Marangoni force resulting from the temperature dependence of the surface tension.

As the workpiece material in the present study steel S235 with following composition is considered: C=0.17%, Mn=1.4%, P=0.035%, S=0.035%, N=0.012%, Fe=97.8%. The surface tension of steel S235 in liquid condition is required. Using the reference data [11, 12], the parameters of capillary activity F_{1873} were obtained for the temperature 1873 K.

From the values of capillary activity, atomic weight and percentage content of the elements presented in table 1 the surface tension of the alloy was calculated using the following equation [11]

Table 1. Element data of steel S235

Element	C	Mn	P	S	N	Fe
F_{1873}	2.0	5.0	1.5	500.0	150.0	1
Atomic weight	12	55	31	32	24	56
Number of moles, n_i	0.0140	0.0260	0.0010	0.0010	0.0009	1.746
$\sum n_i$			1.807			
Atomic part in steel, $X_i = \frac{n_i}{\sum n_i}$	0.0080	0.0140	0.0006	0.0006	0.0005	0.966
$X_i \cdot F_{1873}$	0.016	0.070	0.001	0.275	0.072	0.966
$\sum X_i \cdot F_{1873}$			1.4			

$$\gamma = \gamma_{\mathrm{Fe}} - 2000 \log \left(\sum X_i F_T \right) \tag{6}$$

where γ_{Fe} is the surface tension of pure iron for the respective temperature. Thus, the surface tension value for steel S235 at 1873 K is $\gamma_{1873} = 1.551$ N/m. The capillary activity and the surface tension were calculated for other temperatures using the following equation [11].

$$F_T = F_{1873}^{\frac{1873}{T}} \tag{7}$$

The surface tension of steel S235 as a function of temperature is shown in figure 2. It decreases down to 1.431 N/m at 2300 K, for example.

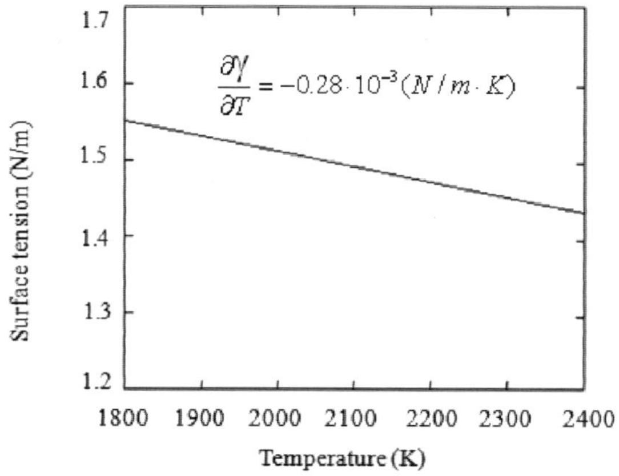

Figure 2. Surface tension of steel S235 as a function of temperature.

For experiments and simulations of the TIG welding process various geometric configurations are considered, but for all of them the following parameters were the same:

- tungsten cathode of 3.2 mm diameter,
- steel workpiece S235 of 10 mm thickness with falling surface tension characteristic,
- shielding gas is argon of the highest quality with the flow rate 12 l/min.
- time interval of simulated arc burning is 10 seconds

Figure 3 gives schematic illustration of the first pair of investigated domains with a ceramic nozzle of 11 mm diameter and two distances between electrodes: 6 mm (a) and 10 mm (b). Figure

3 (c) shows the geometry using a nozzle of 8.1 mm diameter. A welding process with tungsten cathode inside the nozzle, similar to plasma welding, was investigated with this geometry. The distance between cathode and workpiece is 16 mm.

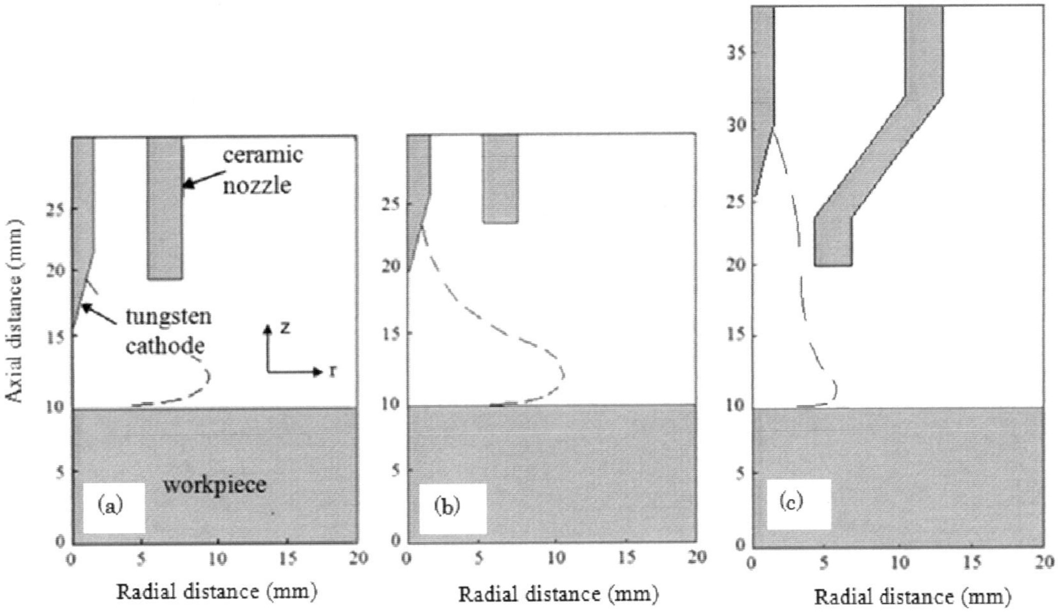

Figure 3. Investigated geometric configurations.

3. Experimental set up

For confirmation of the reliability of calculation results the following scheme of the experimental investigations has been considered: a commercial power supply is connected to the tungsten cathode surrounded by the ceramic nozzle and to the steel workpiece anode. Both, cathode and anode are fixed. The arc is ignited by the high-voltage oscillator. Argon was used as shielding inert gas with the flow rate 12 l/min. The time of anode heating is 10 seconds.

Spectroscopic measurements of plasma temperature are based on the investigations of absolute intensities of the argon spectrum line 696.54 nm [6, 7, 8] over the arc cross section at three axial positions: in the centre of the arc column, 1 mm in front of the cathode and 1 mm in front of the workpiece-anode. The arc was imaged by means of a spherical mirror on the entrance slit of 0.75 m monochromator with intensified CCD camera (PIMAX-2, Roper Scientific) providing spectral resolution 0.05 nm and spatial resolution 0.012 mm/pixel. Abel inversion algorithm has been used to reconstruct the radial distribution of the argon line emission coefficient. The radial plasma temperature profile has been calculated using equilibrium composition calculations [9] for pure argon. The experimental study also includes the analysis of cross-sections of the workpieces and high-speed video images of the arc with spatial resolution of 0.035 mm/pixel.

4. Calculation results

Calculation results for the geometric configurations presented in figure 3 (a) and (b) (ceramic nozzle of 11 mm diameter, cathode tip location outside the nozzle) are shown exemplary for the

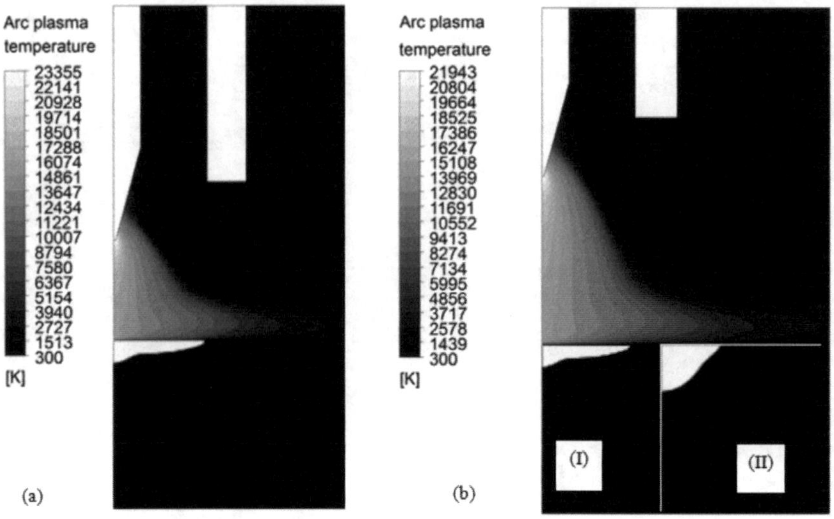

Figure 4. Weld pool profile of steel S235 and two-dimensional calculation of arc plasma temperature for an arc current of 200 A for distances between electrodes 6 mm (a) and 10 mm (b). The white area in the workpiece region indicated the molten part. In figure (b) two weld pools are calculated – with (I) and without (II) inclusion of Marangoni force.

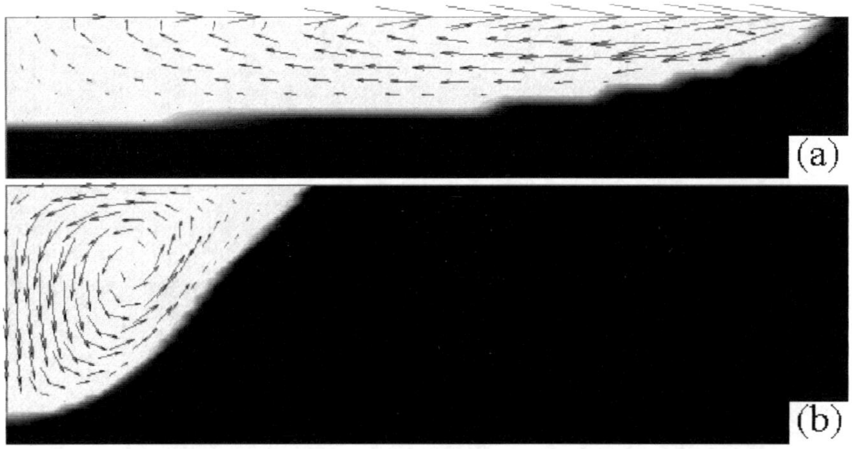

Figure 5. Flow vectors in the weld pool with (a) and without (b) inclusion of Marangoni force in the calculation model. Results are shown for mild steel S235 and current 150A.

current of 200 A in figures 4 (a) and (b), for 150 A in figure 6 (a) and for 100 A in figure 6 (b). For an arc current of 200 A and distance between electrodes 6 mm the width of weld pool is 11.5 mm, penetration depth is 1.4 mm. For the same current, but for distance 10 mm the width is 10 mm and depth is 1.4 mm. Figure 4 (b) also shows two variations of weld pool profile: I - is the weld pool calculation with Marangoni force addition and II-without it. The flow vectors diagrams for both cases are shown in figures 5(a) and (b) respectively. The observed directions of the flow correspond to those in figure 1 and confirm the data obtained in [1, 2, 3].

Decreasing the arc current to 150 A (figure 6 (a)) and to 100 A (figure 6 (b)) the width of

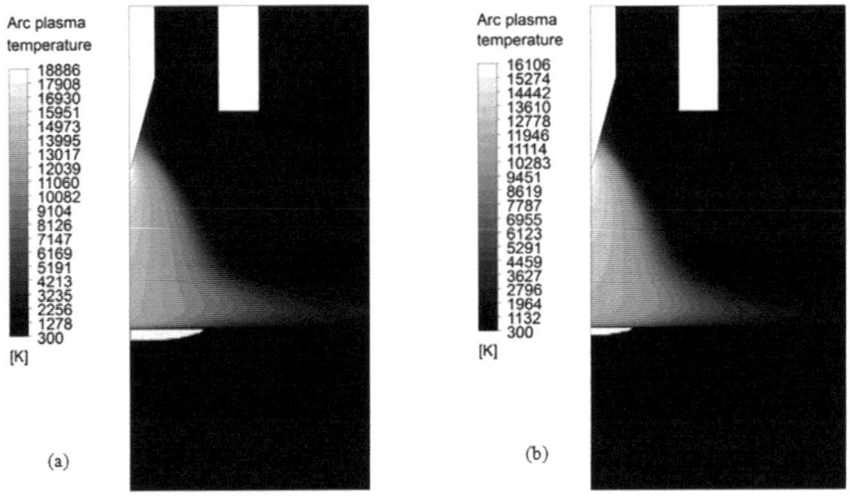

Figure 6. Weld pool profile and arc plasma temperature as in figure 4(b) for an arc current of 150 A (a) and 100 A (b) and 10 mm distance between electrodes.

the weld pool is decreased down to 8.5 mm and 5 mm, the depth down to 0.9 mm and 0.5 mm, respectively.

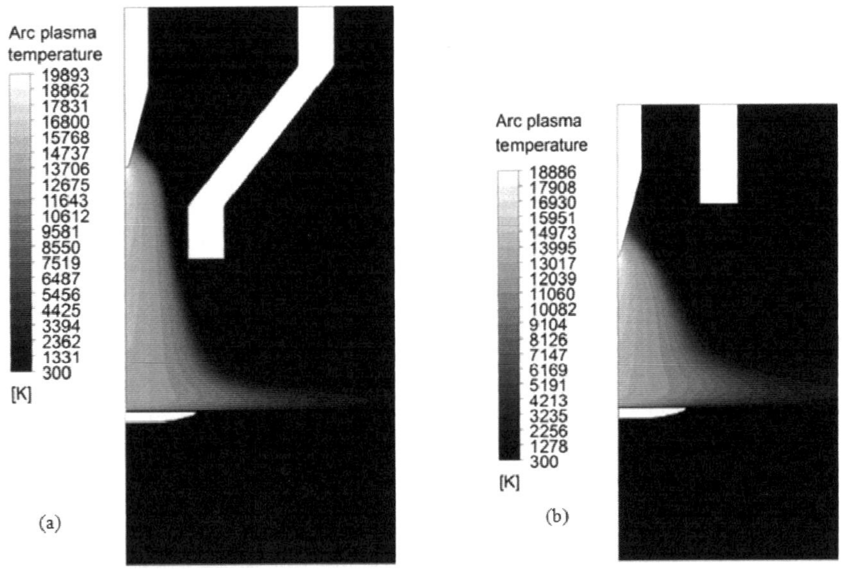

Figure 7. Weld pool profile and arc plasma temperature as in figure 4 for an arc current of 150 A for plasma arc (a) (configuration as in figure 3 (c)) and for TIG arc (b) (configuration as in figure 3 (b)).

The calculated arc temperature and weld pool profile for the geometry presented in figure 3 (c) (ceramic nozzle of 11 mm diameter, cathode tip location inside the nozzle) are shown for an arc current of 150 A in figure 7 (a). Comparing these results with figure 6 (a) (represented also in figure 7 (b) for better visibility) shows that the plasma arc is considerably straighter and more focussed than the TIG arc. Consequently, the arc temperature is increased.

The process described in figure 7 (a) has similarities with plasma welding. However, it didnt give the expected weld pool parameters: the width of weld pool of plasma and TIG welding is the same and penetration is just a little bit deeper for plasma welding. This is due to the fact that the process was similar but not real plasma welding and the second nozzle, producing the cooling of weld pool border, wasnt used.

(a) (b)

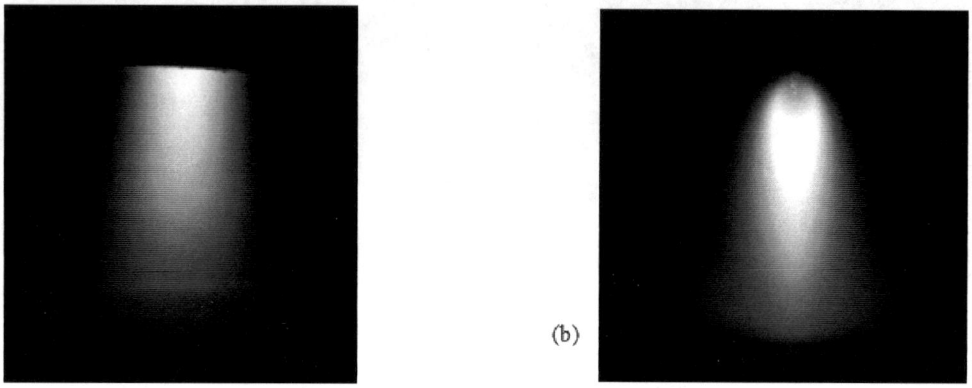

Figure 8. High speed images of the plasma arc (a) and the TIG arc (b).

The images of plasma and TIG welding arc from high-speed cameras are shown in figure 8. The general shape of the arc corresponding to the argon radiation is in good agreement with the shape of the hot regions in the calculated temperature distributions shown in figure 7.

A more detailed validation of the calculations and the model is obtained from the comparison with the results of optical emission spectroscopy for the configuration in figure 3(a) (200 A, 6 mm electrode distance). The radial distributions of plasma temperature reconstructed from the argon line radiation are shown in figure 9 as wide strip. Width of the strip corresponds to estimated experimental error. There are two main sources of experimental uncertainty. First is related to slight asymmetry of the arc which leads to certain ambiguity of the Abel transformation. Second is related to instability of the Abel inversion when emission coefficient passes normal maximum and is only relevant for temperatures above 15000 K. The lines in figure 9 correspond to the calculated two-dimensional distribution in figure 4 (a). A very good agreement has been observed within the accuracy range of the experimental results indicated in the figure.

The cross-sections of the workpieces for a TIG welding arc current of 200 A (a), 150 A (b), and 100 A (c) and distance between electrodes 10 mm are shown and compared with calculated results in figure 10. The form of the weld pool changes with rising electric current. Apart from increasing size one additional feature can be observed. The deepening in the center of the weld pool appears at 200 A. The explanation of this behaviour lies in the mutual balance of the forces which are acting on the melt. As we said before, four forces are acting on the weld pool and Marangoni force is the most important. For higher currents the efficiency of the electromagnetic force affecting the profile of the weld pool is increasing. Thus for currents higher than 200 A the deepening in the centre of weld pool appears.

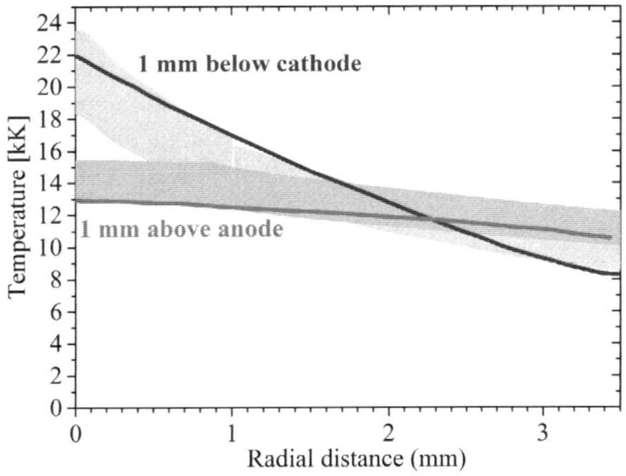

Figure 9. Comparison of TIG welding arc temperature profiles form optical emission spectroscopy (coloured strips) with calculated result (solid curves) for an arc current of 200 A and electrode distance 6 mm.

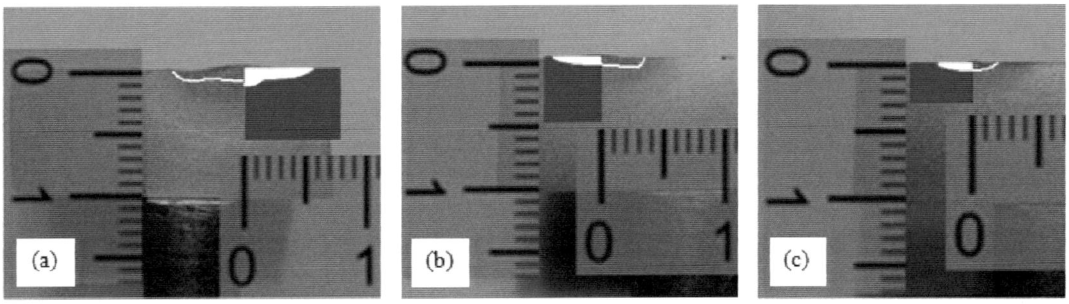

Figure 10. Cross-sections of the workpieces for a TIG welding arc current of 200 A (a), 150 A (b), 100 A (c) and electrode distance 10 mm. Comparison with calculated results.

5. Summary

Comparison of the experimental arc temperature distribution, cross-sections of the workpieces, and images from high-speed cameras shows good agreement with computational results. Herewith the computational model of a coupled magneto-hydrodynamic simulation of the welding arc and the fluid description of the weldpool has been verified succesfully. The surface tension, calculated by simple equations popular in moulding industry gave the reliable value for mild steel in a TIG welding process. The necessity of consideration of the Marangoni force in the model is shown.

Reasons of remaining insignificant differences between model and experimental results are asymmetry of the real plasma arc and neglecting of metal vapour, turbulence of plasma flow and the movement of weld pool surface in the model. For better calculation quality these effects should be considered in future work.

References

[1] Tanaka M and Lowke J J 2007 Predictions of weld pool profiles using plasma physics *J. Phys. D: Appl. Phys.* **40** R1 R23
[2] Tanaka M, Yamamoto K, Tashiro S, Nakate K, Yamamaoto E, Yamazaki K, Suzuki K, Murphy A B and

Lowke J J 2010 Time-dependent calculations of molten pool formation and thermal plasma with metal vapour in gas tungsten arc welding *J. Phys. D:Appl. Phys.* **43** 434009 (11 pp)

[3] Mougenot J, Gonzalez J-J, Freton P and Masquère M 2010 Plasmaweld pool interaction in tungsten inert-gas con-figuration *J. Phys. D:Appl. Phys.***46** 135206 (14 pp)

[4] Murphy A B and Arundell C J 1994 Transport coefficients of argon, nitrogen, oxygen, argon-nitrogen and argon-oxygen plasmas *Plasma Chem. Plasma process* **14** 451

[5] Menart J, Heberlein J V R and Pfender E 1999 Theoretical radiative transport results for a free-burning arc using a line-by-line technique *J. Phys. D:Appl. Phys.* **32** 55

[6] Olsen H N 1963 The Electric arc as a light source for quantitative spectroscopy *J. Quant. Spectrosc. Radiat. Transfer.* **3** 305-333

[7] G. N. Haddad G N and Farmer A J D 1984 Temperature determinations in a free-burning arc: 1. Experimental techniques and results in argon *J . Phys. D: Appl. Phys.* **17** 1189-1196

[8] G. Wilhelm G, Kozakov R, Goett G, Schoepp H and Uhrlandt D 2012 Behaviour of the iron vapour core in the arc of a controlled short-arc GMAW process with different shielding gases *J. Phys. D: Appl. Phys.* **45** 085202 (11pp)

[9] Wendt M 2011 Net emission coefficients of argon iron plasmas with electron Stark widths scaled to experiments *J. Phys. D: Appl. Phys.* **44** 125201

[10] Toropchin A, Frolov V and Uhrlandt D 2013 Calculation of the weld pool profile coupled with the arc simulation in a TIG process using software ANSYS CFX *Proc. XXth Symposium on Physics of Switching Arc* (Brno 2013) Editors: Aubrecht V, Bartlova M 301-304

[11] Popel S I, Sotnikov A I 1986 Theory of metallurgical processes *Metallurgy* (Moscow) 139

[12] Popel S I 1994 Surface effects in liquids *Metallurgy* (Moscow) 432

Experimental characterization of the weld pool flow in a TIG configuration

M Stadler[1], M Masquère[1], P Freton[1], X Franceries[2,3] and JJ Gonzalez[1]
[1] Université de Toulouse ; UPS, INPT, CNRS ; LAPLACE (Laboratoire Plasma et Conversion d'Energie) ;
118 route de Narbonne, F-31062 Toulouse cedex 9, France.
[2]INSERM, UMR1037 CRCT, F-31000 Toulouse, France
[3]Université Toulouse III-Paul Sabatier, UMR1037 CRCT, F-31000 Toulouse, France

E-mail: marine.stadler@laplace.univ-tlse.fr

Abstract. Tungsten Inert Gas (TIG) welding process relies on heat transfer between plasma and work piece leading to a metallic weld pool. Combination of different forces produces movements on the molten pool surface. One of our aims is to determine the velocity on the weld pool surface. This provides a set of data that leads to a deeper comprehension of the flow behavior and allows us to validate numerical models used to study TIG parameters. In this paper, two diagnostic methods developed with high speed imaging for the determination of velocity of an AISI 304L stainless steel molten pool are presented. Application of the two methods to a metallic weld pool under helium with a current intensity of 100 A provides velocity values around 0.70 m/s which are in good agreement with literature works.

1. Introduction

Tungsten inert gas (TIG) welding is widely used in industry. This technique performs an electric arc between a tungsten electrode and a piece of metal. This process involved a shielding gas as argon, helium or mixtures at atmospheric pressure. The heat transfer between the arc and the work piece leads to a metallic weld pool. The molten pool movements may be explained by the combination of four forces [1]: Marangoni [2], Laplace, drag and gravity forces. Weld pool shape is related to the molten zone motion during the process. This flow motion and the surface velocity are influenced by parameters such as arc length [3], shielding gas [4, 5], current or sulfur content of the metal work piece [6, 7]. The depth and shape of the pool are of greatest importance as they determine the thickness and the strength of the weld. It is thus essential to understand the flow behavior to control the weld quality. This understanding involves the characterization of the weld pool surface movement and the associated velocity.

Literature provides several works using modeling to describe either the material [8, 9] or the whole process with the interaction [10, 11]. Anyway, based on our knowledge, there is a lack of experimental works regarding the velocity estimation in a TIG configuration. This issue is treated with particle tracking velocimetry [12] or particle imaging velocimetry [13]. In these works, application of these techniques on a 316L stainless steel anode in interaction with an argon arc provides velocity values of 0.7 and 0.13 m/s. The development of a diagnostic method appears as a necessity as much to the weld pool phenomena understanding as model validation.

In this work, two diagnostic methods based on high speed imaging and on pixel intensity evolution in a time series of images are presented. The development of these methods and their application to the estimation of velocity on molten pool surface are described. The first method relies on Hlina works [14] and uses phase angle difference coming from Fast Fourier Transform of pixel intensity to estimate velocity whereas the second one consists in tracking bright pixels in time on a selected area of the image. These techniques are introduced in the section 2. Section 3 describes the applications of the two methods on a validation case. In the last section, both methods are applied to an arc transferred configuration and velocity of a stainless steel weld pool under helium is obtained, compared and discussed.

2. Diagnostic methods developed

Until now, weld pool velocity was measured in our work by analyzing the wave front displacement of weld pool oscillations on successive frames extracted from high speed camera acquisitions. This method gave access to a good estimation of velocity values since there were in good agreement with those provided with numerical modeling or with the few experimental works founded in literature. Anyway, this method had the drawback of being really time consuming since it was done manually: video acquisitions own at least 3000 frames. Automation of the method needed to be developed. Nevertheless, it remains a reliable method to us and it is used as a reference value to validate the two diagnostic methods developed and is described below (section 3).

2.1. Method 1: Phase angle difference for velocity estimation

Direct observation during the weld process pointed out wavelet formations on the surface of the molten pool which can be assumed as oscillations. As Hlina reported in its work, the Fast Fourier Transform (FFT) was used to evaluate the gas velocity on a plasma torch thanks to the arc jet oscillations. The optical emission of the plasma jet was detected and recorded with photodiodes along the jet axis leading to the detection of the time shift between two diodes [14]. As sketch in figure 1, the method is based on the measurement of the phase angle difference from two points separated with a distance d along the plasma flow [14]. By the mean of the FFT, a linear dependence between the phase angle difference extracted from two signals ($s_1(t)$ and $s_2(t)$) and the frequency is assumed. The slope $\frac{\Delta\varphi}{\Delta f}$ of the linear arrangement of the phase angle difference allows then to estimate the related velocity through $v = 2\pi d \frac{\Delta f}{\Delta\varphi}$ (1), where d is the distance between the two signals extracted from two points of the acquisition, Δf the frequency and $\Delta\varphi$ the phase increment.

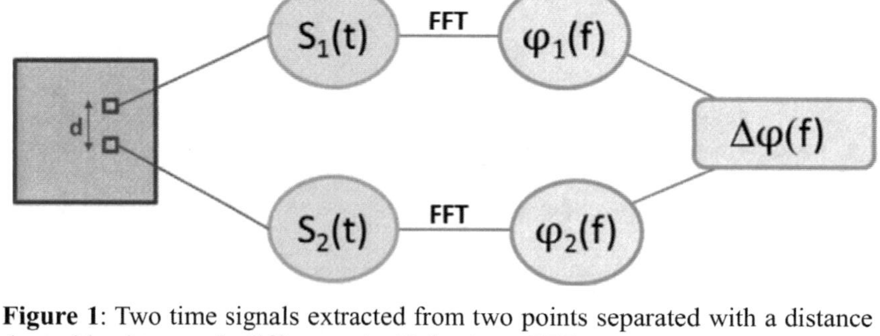

Figure 1: Two time signals extracted from two points separated with a distance d provide phase angle difference $\Delta\varphi(f)$ thanks to FFT

While it was used for gas oscillations in [14], this method has been transposed to molten pool oscillations.

2.2. Method 2: Bright pixel tracking for velocity estimation

Direct observation during the process pointed out displacement of bright pixel on the molten pool surface characteristic of the wavelet movement. The second method uses this pixel intensity evolution with time. As presented in figure 2, a segment [AB] of the image is selected on each frame and a map representing pixel intensity with time is obtained. This map displays maxima of intensity which are typical of the wave front displacement. Line joining several maxima gives access to the wavelet displacement velocity with the ratio $\frac{\Delta pix}{\Delta t}$.

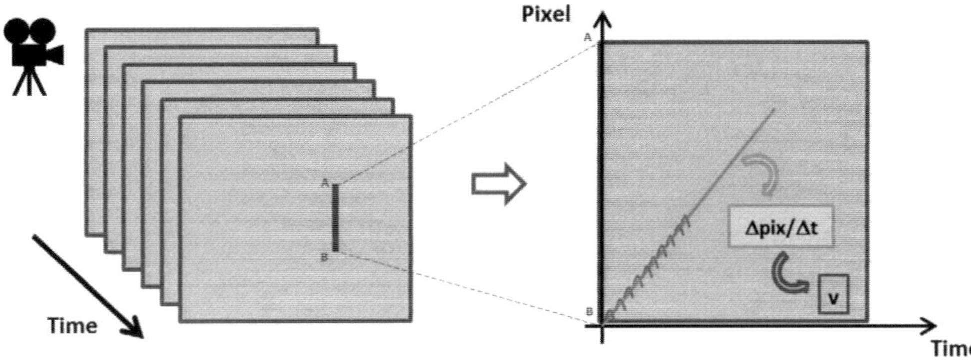

Figure 2: A selected area [AB] extracted on each frames gives access to an intensity map as a function of time

3. Validation case : wavelet velocity induced by a water drop

The two methods have been first performed on a simplified case in order to test them. The metallic weld pool movement has been assimilated to oscillations created by successive drops on water surface (figure 3). Velocity value has been deduced with the two methods and compared to the reference value obtained with direct observation of our videos. It allows us to follow the wave front displacement on successive frames. Suitable calibration provides the velocity value with the simple relation $v = \frac{d}{t}$ (2). In this case, acquisitions have been done with a frame rate of 10 000 fps and a spatial resolution of 100 μm. Direct observation gives access to a velocity value of 0.20 ± 0.02 m/s.

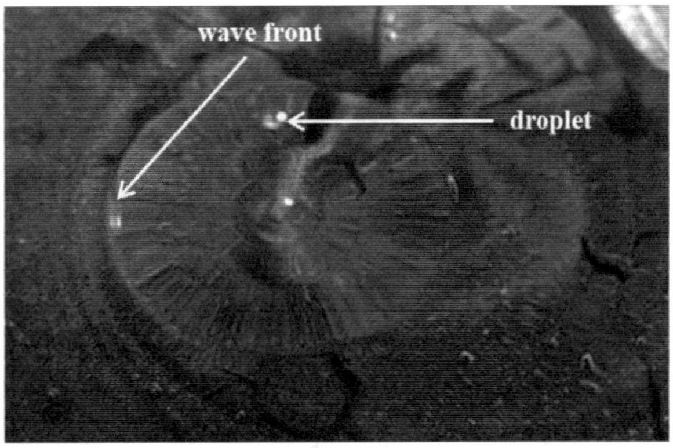

Figure 3: Frame extracted from video acquisition to visualize the water wave front induced by a water drop

3.1 Method 1: Phase angle difference for velocity estimation

Intensity of two signals extracted from high speed camera acquisitions and separated with a distance of 2 pixels has been extracted on the whole recorded frames. FFT provides phase angles of the two signals. Phase angle difference has then been drawn in function of the frequency

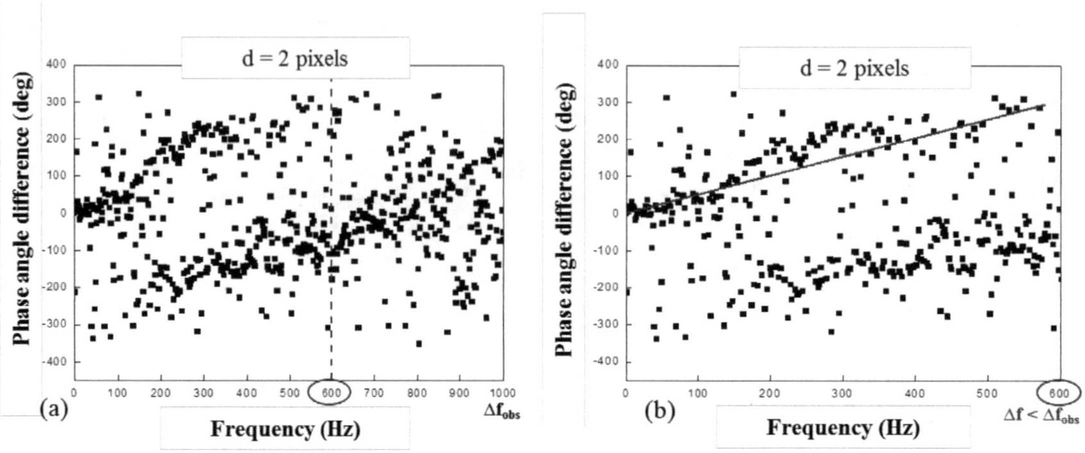

Figure 4 : Validation case: phase angle difference as a function of the frequency for two signals separated with a distance of 2 pixels (a) on the theoretical frequency range Δf_{obs} calculated and (b) on a reduced range of frequency

The ratio $\frac{\Delta\varphi}{\Delta f}$ is given by the slope of the linear arrangement of the phase angle difference (drawn on solid line in figure 4(b)). Points displaying between -100 and -360 deg correspond to phase jump when the phase is unwrapped.

Distance d is measured with the image calibration (1 pixel = 100 μm) done before the acquisition. With this ratio, the velocity value is estimated at 0.24 ± 0.02 m/s which is in good agreement with velocity estimated with our reference method.

In figure 4 (a), the phase angle difference is observed on a specific range of frequency (Δf_{obs}) on which the phase angle is unwrapped. It is strongly linked to the chosen distance between the two signals extracted and the assumed velocity. As the distance rises between the signals, for a given velocity, the observation frequency range decreases. However, phase angle difference often needs to be observed on a frequency range smaller than Δf_{obs} because of the reduction of information in the highest frequencies noticed in the analysis of the frequency spectrum. This is the case in the figure 4(a) where the theoretical calculus gives a observation frequency range from 0 to 1 kHz but where the information is clearer if the range is limited to 600 Hz (figure 4(b)). Furthermore, another parameter seems essential. It has been noticed that for a too high distance between considered pixels, reduction of the observation frequency range seems no longer sufficient. It seems that beyond few pixels, signals are less correlated and it gets more difficult to provide reliable information. Figure 5(a) displays the phase angle difference for two signals extracted with a distance of 6 pixels. Theoretical calculus give a maximal frequency of 333 Hz but it is clearly shown that information is harder to get than the previous case presented in figure 4: here (figure 5(a)), a linear arrangement is supposed on the first hundreds of hertz. If the frequency range is reduced to 100 Hz (figure 5(b)) as explained before, the ratio might be obtained with the slope (solid line displays in the figure) and the velocity deduced is 0.26 m/s ± 0.02 m/s. While the obtained value is still close to the attempted one, it seems more careful to choose a smaller distance to get a number of points higher and a linear arrangement more obvious, as presented in figure 4. Moreover, if we look closely to the extracted signal in time, an absence of correlation between them when the distance gets too important appears: figure 5(c) (resp. figure 5(d)) displays intensity for two signals (solid line and dash line) separated with a distance of 2 pixels (resp. 6 pixels). It is clearly shown that signals 1 and 2 in figure 5(c) presents the same pattern shifted in time whereas figure 5(d) shows a lack of correlation as mentioned above.

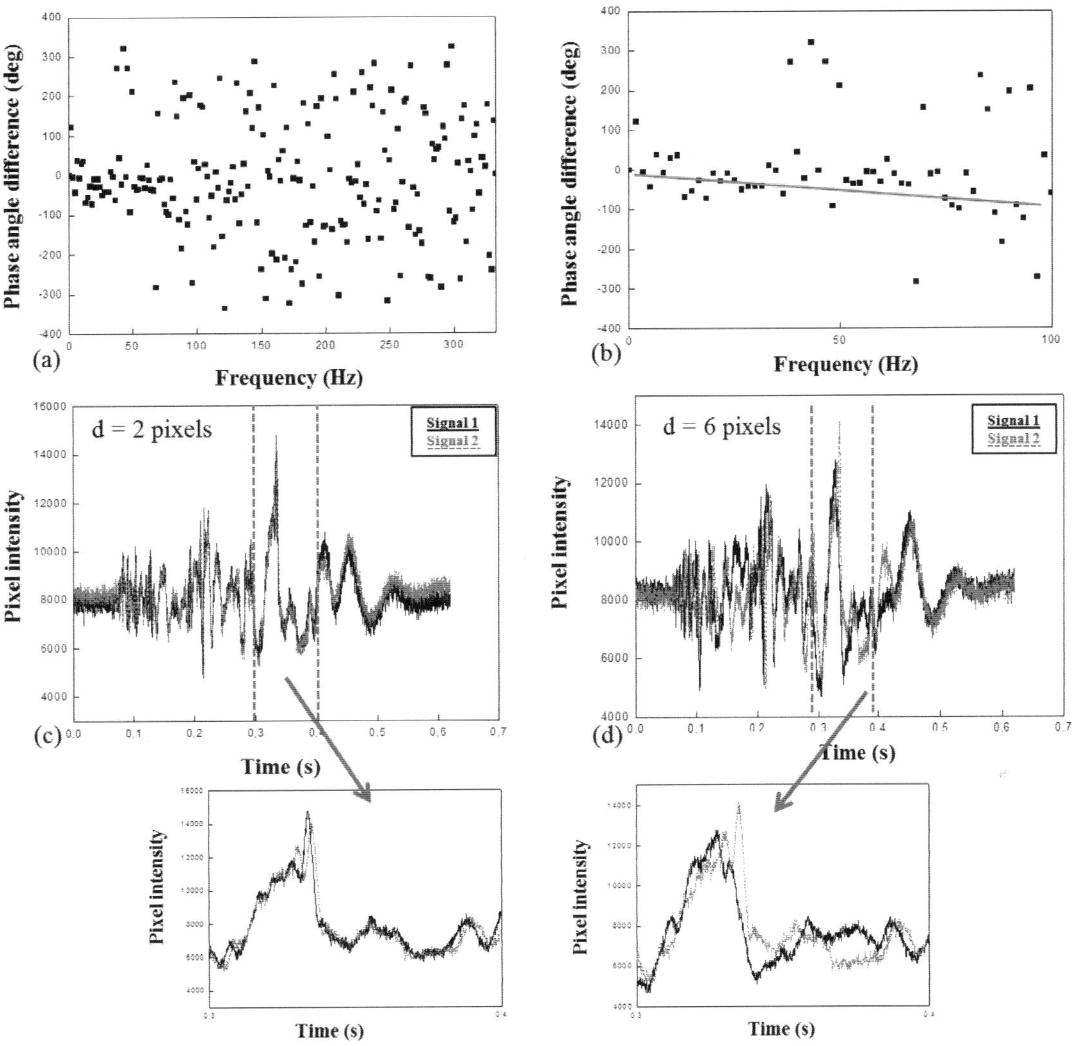

Figure 5 : (a),(b) phase angle difference as a function of the frequency for two signals separated with a distance of 6 pixels ; pixel intensity as a function of time for two signals separated of (c) 2 pixels and (d) 6 pixels

From this first test we can conclude that the distance and the observation frequency range must be chosen carefully for a good use of the method.

3.2 Method 2: Bright pixel tracking for velocity estimation

Intensity of a section of the image (100 pixels in length) has been extracted on the whole recorded frames and juxtaposed to obtain an intensity map with time. The map related to the water drop case is displayed in figure 6. The ratio $\frac{\Delta pix}{\Delta t}$ is deduced from the lines (drawn in solid line in the figure) joining maxima. Distance d is estimated with our image calibration (1 pixel = 100 µm) and number of frames provides time with the frame rate used during acquisition (10 000 fps). The velocity value is estimated at 0.21 ± 0.02 m/s which is also in good agreement with velocity estimated with our reference method.

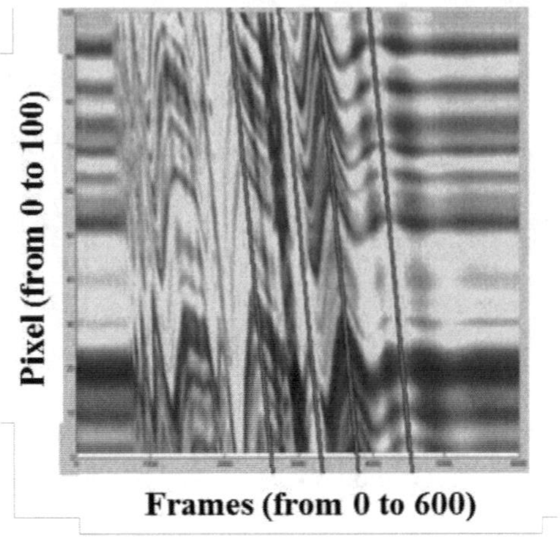

Figure 6: Intensity map as a function of frames related to the water drop case and line (solid line) joining maxima used to velocity estimation

4. Results and discussion

The two methods have then been performed on metallic weld pool obtained with the process described below.

An arc transferred configuration which is close to a TIG configuration is available in our team (figure 7). The electric arc is created between a thoriated tungsten cathode with a tip angle of 60° and a cylindrical anode (5 mm in height and 50 mm in diameter) of 304L stainless steel considered as the work piece. The length of the arc is adjustable but set at 5 mm for this experiment. Time of interaction is limited to 2 minutes. The device is placed in a hermetic and water cooled reactor within the gas is injected. The inert atmosphere is assured by helium with a flow rate of 10 Nl/min. Current intensity is set at 100 A. The measurements are performed with a high-speed camera set at 15 000 frames per second with a spatial resolution of 46 µm. This rate is chosen to be high enough to see the weld pool movements and to get a suitable resolution to observe the entire weld pool.

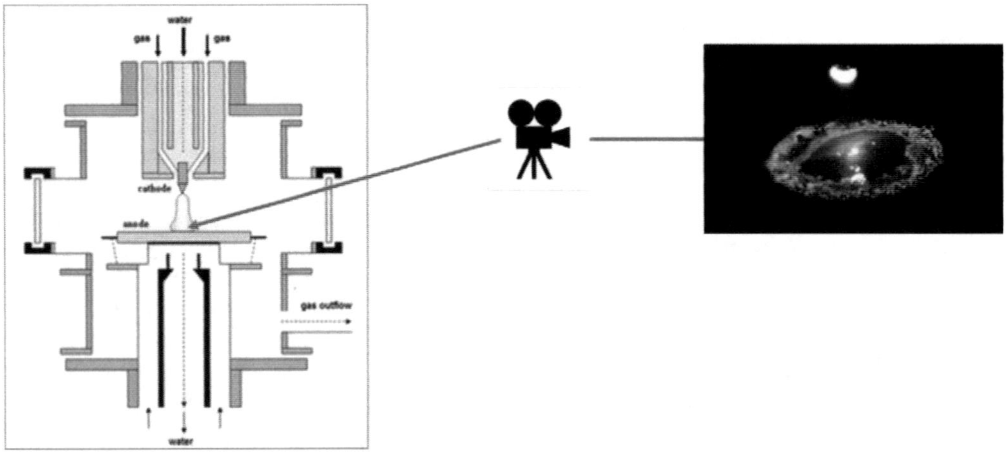

Figure 7: Experimental set up used (left) and frame extracted from video acquisition of a stainless steel weld pool under helium (right)

Figure 8 presents results obtained with the two methods. Application of method 1 (figure 8(a),(b)) displays the linear arrangement of the phase angle difference from two signals extracted with a distance of 2 pixels. As presented in figure 8(a), the observation frequency range is reduced to 2500 Hz (figure 8(b)). The velocity values obtained is 0.74 ± 0.02 m/s. Application of method 2 provides an intensity map as a function of frames (Figure 8(c)). The ratio $\frac{\Delta pix}{\Delta t}$ is measured from the solid lines drawn in figure 8(d). Only frames from 1300 to 1600 are presented here to ease the observation but the ratio has been calculated on the entire map. Velocity estimated with the application of method 2 is 0.71 ± 0.02 m/s. Reference velocity given by direct observation is 0.70 ± 0.02 m/s.

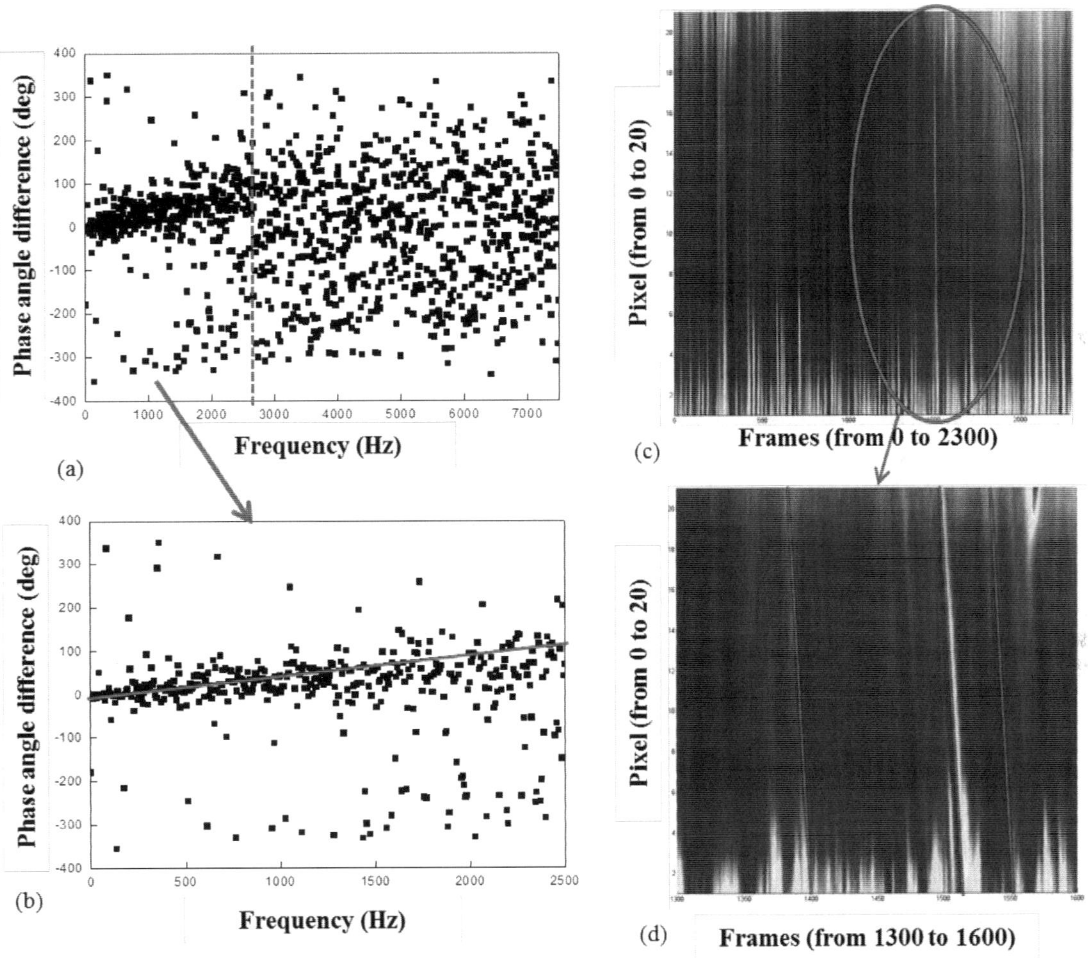

Figure 8 : Applications of (a)(b) method 1 and (c)(d) method 2 to a metallic weld pool under helium with a current intensity of 100 A

This first test shows a good agreement between the two methods and our reference velocity value. Modelling of the interaction of a 5 mm arc and a 304L stainless steel anode under helium with an intensity current set as 150 A allow to estimate a velocity on the surface of the weld pool of 0.59 m/s after 20 second of interaction [15]. Tanaka et al.[16] pointed out in their work that velocity values changed in time and with the content in sulfur of the material, so we remain careful since to our knowledge, any literature work provides velocity values with the exact same conditions as ours (intensity, time after interaction).

5. Conclusion

This work is a first step in velocity estimation in weld pool. Two diagnostic methods based on high speed imaging have been presented here and validated on a simple case. First results on a weld pool configuration have then been displayed. Results seem encouraging. Nevertheless, these methods need to be improved and results obtained need to be compared: further experiments will be carried out with operational conditions closer to those found in literature. Once these results compared, this work will lead to further investigations: influence on current intensity, nature of the shielding gas and nature of anode material on velocity in weld pool.

6. References

[1] Tanaka M *et al* 2002 *Metall. Mater. Trans. A.* ***33A*** 2043
[2] Mills KC et al 1998 *Philos. Trans. R. Soc. London*, Ser. A **356** 911
[3] Ushio M et al 2004 *IEEE Trans. Pl. Sc* **32** 108
[4] Mougenot J et al 2013 *J.Phys. D : Appl. Phys.* **46** 495203
[5] Tanaka M et al 2007 *Sci. Technol. Weld. Joi.* **12** 2
[6] Mills KC and Keene BJ 1990 *Int. Mater. Rev.* **35** 185
[7] Shirali AA and Mills KC 1993 *Weld. J.* ***71*** 347s
[8] Kim WH et al 1997 *Numer. Heat Transfer Part A* **32** 633
[9] Dong W et al 2011 *Int. J. Heat Mass Transfer* **54** 1420
[10] Lu F et al 2006 *Comp. Mater. Sci.* **35** 458
[11] Mougenot J et al 2013 *J.Phys. D: Appl. Phys.* **46** 135206
[12] Henrikson P 2005 *Math. Mod. Weld Ph.* **7** 125
[13] Zhao CX et al 2009 *The 8th International Conference Trends in Welding Research* (Pine Mountain, ASM international)
[14] Hlina J et al. 1996 *Acta Tech. CSAV* **41** 373
[15] Murphy AB et al. 2009 *J. Phys. D : Appl. Phys* **42** 194006
[16] Tanaka M and Lowke JJ 2007 *J. Phys. D : Appl. Phys.* **40** R1

Time resolved Thomson scattering diagnostic of pulsed gas metal arc welding (GMAW) process

M Kühn-Kauffeldt[1], J L Marquès[2] and J Schein[1]

[1]Universitaet der Bundeswehr München, Institut für Plasmatechnik und Mathematik, Neubiberg, Germany

[2]Universitaet der Bundeswehr München, Institut für Mess- und Automatisierungstechnik, Neubiberg, Germany

E-mail: marina.kuehn-kauffeldt@unibw.de

Abstract. In this work a Thomson scattering diagnostic technique was applied to obtain time resolved electron temperature and density values during a gas metal arc welding (GMAW) process. The investigated GMAW process was run with aluminum wire (AlMg 4,5 Mn) with 1.2 mm diameter as a wire electrode, argon as a shielding gas and peak currents in the range of 400 A. Time resolved measurements could be achieved by triggering the laser pulse at shifted time positions with respect to the current pulse driving the process. Time evaluation of resulting electron temperatures and densities is used to investigate the state of the plasma in different phases of the current pulse and to determine the influence of the metal vapor and droplets on the plasma properties.

1. Introduction

Gas metal arc welding (GMAW) is one of the most common industrial welding processes. An electric arc ignited between the workpiece and the consumable wire electrode is used to heat and melt the wire. Metal droplets fall on the workpiece and form the join. In order to achieve a controlled metal transfer to the workpiece a pulsed current can be used, which allows to obtain a so called one-droplet-per-pulse mode. Due to their reproducibility pulsed GMAW processes are interesting for industrial application. Although GMAW processes have been widely studied over past years there is still a need for experimental investigations in order to obtain a comprehensive model of the process [1].

Thomson scattering is a well-established diagnostic for simultaneous measurement of electron temperature and density without the assumption of LTE in pure argon plasmas [2]. This technique does not require the knowledge of the gas composition, so it can be applied for diagnostics of gas mixtures. Although widely applied to gas tungsten arc welding (GTAW) [3, 4] only few works investigated atmospheric discharges in presence of metal vapor by means of Thomson scattering [5, 6]. However it has not been used to investigate GMAW processes so far.

The plasmas containing metal vapor are often optically thick, which makes the application of optical diagnostic particularly difficult. The advantage of Thomson scattering is, that this technique requires an optically thin plasma only in a small spectral range. GMAW processes which use aluminum as the wire electrode are in general suitable for the Thomson scattering diagnostic in the wavelength range of $532nm$. Since neither aluminum nor argon plasma emit strong lines in the spectral range of $532 \pm 5nm$, the plasma can be considered as optically thin.

This work presents a Thomson scattering setup for time resolved measurements of pulsed GMAW processes. The goal of the investigation is to determine electron temperature and density in different current phases of the process.

2. Theory of Thomson Scattering

The radiation scattered by free non relativistic electrons in a plasma is proportional to the spectral density function $S(k,\omega)$, which indicates the fluctuation of the electron density along the scattering wave vector $k = |\mathbf{k_s} - \mathbf{k_i}| = \frac{4\pi}{\lambda_i}\sin(\theta/2)$. Here $\mathbf{k_i}$ is the wave vector of the incident laser radiation and $\mathbf{k_s}$ the wave vector of the scattered light, λ_i is the laser wavelength and θ the scattering angle. $\omega = \omega_i - \omega_s$ indicates the frequency shift of the scattered light from the incident wavelength [7]. $S(k,\omega)$ is mainly influenced by the free electrons fluctuations and the electrostatic influence of the ions on the free electrons. These influences are described in the so called electron and ion feature of the scattered radiation spectrum.

In the case of plasmas present in welding arcs electron and ion velocity can be described by the Maxwellian distribution function $f(v) = \frac{1}{(\pi v_0)^{3/2}}\exp\left(-\frac{v^2}{v_0^2}\right)$, with $v_0 = \sqrt{2k_BT/m}$ the mean thermal speed, T the particle temperature, m the particle mass and k_B the Boltzmann constant. Moreover the plasma is assumed to be collisionless. If in addition the ratio between the electron and ion temperature is $T_e/T_i \approx 1$, which is the case here, then the Salpeter approximation [8] can be used to calculate the spectral density function. In this case the electron contribution of $S(k,\omega)$ can be described as

$$S_e(k,\omega) \cong \frac{2\pi^{1/2}}{v_{0,e}k}\frac{\exp(-x_e^2)}{|\epsilon|^2} \tag{1}$$

where $\epsilon = (1 + \alpha^2 Rw(x_e))^2 + (\alpha^2 Iw(x_e))^2$ is the dielectric plasma function. For the real dimensionless frequency variable defined as $x_e = \omega/k_B v_{0,e}$ $Rw(x_e)$ and $Iw(x_e)$ (the Landau damping term) are the given by

$$Rw(x_e) = 1 - 2x_e\exp(-x_e^2)\int_0^{x_e}\exp(p^2)dp, \qquad Iw(x_e) = i\pi^{1/2}x_e\exp(-x_e^2) \tag{2}$$

Depending on the density and temperature of the investigated plasma one can distinguish between coherent and non coherent Thomson scattering, which is described by the parameter α

$$\alpha = \frac{1}{k\lambda_D} = \frac{1}{4\pi * \sin(\theta/2)}\frac{\lambda_i}{\lambda_D} \tag{3}$$

with λ_D the Debye length given as $\lambda_D \approx \sqrt{\frac{\epsilon_0 k_B T_e}{e^2 n_e}}$, electron density n_e, temperature T_e, permittivity of free space ϵ_0 and the electron charge e. In plasmas investigated in this paper coherent Thomson scattering ($\alpha \geq 1$) is expected.

Electron behavior in a plasma can be influenced by collisions with heavy particles. In this case the composition of the plasma becomes relevant. However as shown by [9], in a TIG arc plasma, the electron heavy particles collision frequency is several orders of magnitude lower than the plasma frequency. In the GMAW process similar conditions are expected. Consequently in this case the electron feature of the scattering spectrum does not depend on plasma composition.

In general laser radiation may influence initial conditions of the plasma and heat up the free electrons [10]. However as was previously shown [11] this effect can be neglected under the experimental conditions which were chosen for this work.

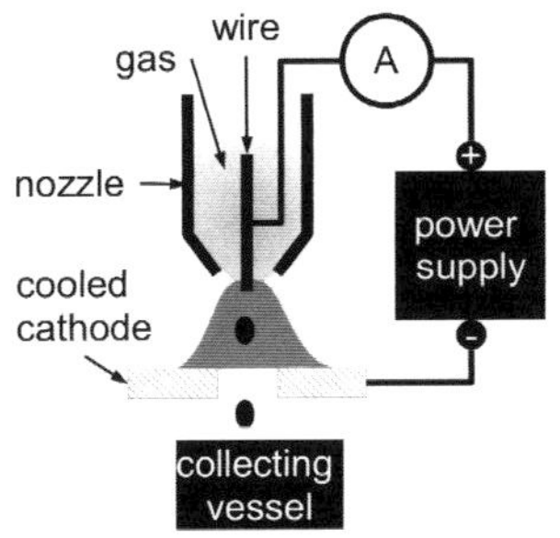

Figure 1. Schematic of the GMAW process investigated by Thomson scattering.

3. Experimental setup

The gas metal plasma of the GMAW process was generated at atmospheric pressure by a welding gun as shown in figure 1. The arc was ignited between an aluminum wire (AlMg 4,5 Mn) serving as an anode and a water cooled donut shaped copper cathode with a $100mm$ outer diameter. On the central axis the copper cathode was provided with a hole of $7mm$ in diameter. The wire had a diameter of $1.2mm$ and was fed at a velocity of $3m/min$. The metal droplets fall through the hole into a collecting vessel and so prevent metal accumulation on the cathode surface. Hence the cathode did not have to be moved during the process. As a shielding gas argon 4.6 (99,9 % purity) at a flow rate of $25slm$ was used. The anode stickout was set to $17.5mm$ and was kept constant during the whole process. OTC DW 300 was used as power supply in the pulsed DC mode. The process current was monitored by a hall sensor. The region investigated by the Thomson scattering method was located $5.5mm$ above the cathode surface.

The principle of the Thomson scattering measurement setup is shown in figure 2. As the radiation source a pulsed Nd:YAG laser with a central wavelength of $532nm$ was used (ML II, Continuum). The laser energy was set to $25mJ$ per pulse with the pulse duration of $3-5ns$ and the repetition rate of maximum $15Hz$. The radiation beam was focused on the arc column. The beam waist within the plasma was approximately $500\mu m$. The scattering light was detected at

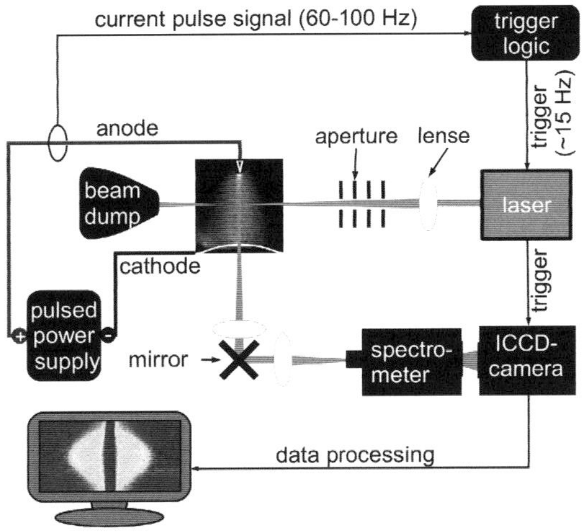

Figure 2. Schematic of the Thomson scattering setup for measurements in a GMAW process.

an angle of $\theta = 90°$. The image of the measurement volume was projected on to the slit of a Jobin Yvon Spectrometer (Fastier-Ebert design) with a focal length of $250mm$ and a diffraction grating with $1500g/mm$ using two mirrors for image rotation and an imaging optics. The resulting spectrum was recorded by the ICCD camera Pi-MAX (Princeton Instruments, GEN III image intensifier). The camera was synchronized with the laser using the Q-switch output of the laser. The spatial resolution of the setup in the radial direction of the arc was $0.4mm$.

The OTC power supply ensures a constant welding quality by regulating the voltage, current level and frequency of the arc in order to obtain a constant arc length. Since the current pulse frequency varies during the welding process, an additional trigger logic is needed in order to implement time resolution. Therefore a combination of a hall sensor (SS94A1F) for current pulse tracking and a micro-controller (Arduino Due board with Atmel SAM3X8E ARM Cortex-M3 MCU processor) are used. The analog digital converter of the micro-controller board is used to detect a predefined level of the current pulse. With a sampling rate of 1 MHz and a 12 bit resolution it is fast and precise enough to deliver reproducible trigger points for the GMAW process. The counter functionality of the micro controller allows to implement a temporal delay with a precision of $25ns$. The delay function is used to temporally shift the outgoing laser trigger impulse, which initiates the Thomson scattering measurement.

In general it is possible to obtain measurements from a single laser pulse using the described setup. However in order to improve the signal to noise ratio the scattering signal generated by several laser shots is accumulated on the CCD chip of the camera. On the short time range the GMAW process is sufficiently reproducible. However on the long time range the behavior of the arc may change. Hence in order to ensure comparable results the signal was integrated over 5 laser pulses during the high current phase and 20 laser pulse during the low current phase data acquisition. The image intensifier was activated for $5ns$ per pulse in order to minimize the plasma background radiation. In order to minimize the CCD readout noise the on-chip accumulation function of the ICCD camera was used. Additionally the background radiation is recorded by keeping the experimental conditions and the trigger time of the GMAW pulse but by blocking the path of the laser radiation.

A supplementary high speed camera was installed for imaging of the arc. The images were use for the arc shape visualization. Thus it was possible to optically verify that the arc shape did not considerably change during the signal integration procedure.

4. Results and discussion

Figure 3 shows an image of the spatially resolved Thomson scattering spectrum of the GMAW processes recorded by the ICCD camera. Here the aluminum plasma background radiation, the electron feature of the Thomson scattering spectrum (white dotted line) and the central laser wavelength are visible. The intensity of the latter depends on the ion feature of the Thomson scattering spectrum, the Rayleigh scattering on the neutral atoms and reflection of the laser beam on the cathode and anode surface.

A theoretical Thomson scattering spectrum, which was fitted to the measured spectrum at a specific spatial position is shown in figure 4. For the fitting procedure the background radiation was subtracted from the original signal. As previously described in [11] lookup tables which correlate the shape parameters of theoretical Thomson scattering spectra under consideration of the instrumental function of the spectrometer with the corresponding electron temperatures and densities were used to assign temperatures and densities to the measured spectral profiles.

An example of a typical temperature and density profile evaluated from the Thomson scattering measurement in the high current phase is depicted in figure 5. In figure 6 the trigger time and typical current and voltage characteristic of the arc are plotted. At the a current value of $400A$ the maximum temperatures in the range of $14000K$ and maximum density in the range of $1.6 \times 10^{23} m^{-3}$ are reached in the center of the arc. Both electron temperature and density

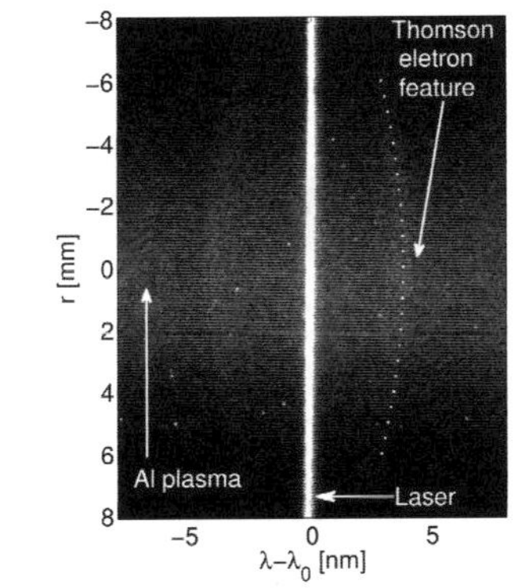

Figure 3. Image of a spatially resolved Thomson scattering spectrum recorded by the ICCD camera. The white dotted line indicates the electron feature

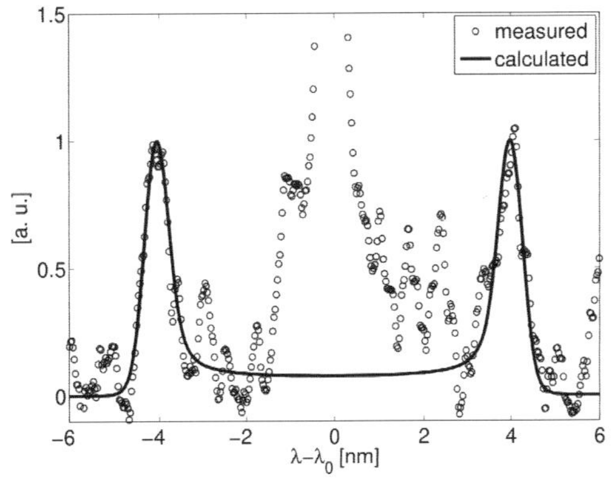

Figure 4. Measured and fitted Thomson scattering spectra resolved at a single spatial position.

diminish with an increasing distance to the arc center. Temperatures under $5000K$ and densities under $1 \times 10^{22}m^{-3}$ could not be resolved by the used setup. For temperatures and densities under the given limit the expected spectral shift of the scattered radiation would lie below the order of magnitude of the instrumental width of the spectrometer.

In order to clearly arrange the evaluation of the time-resolved results only the temperatures and densities in the central region of the arc are plotted. A typical temperature and density evolution along the current pulse can be seen in figure 7(c). Figure 7(b) shows the measurement time relative to the current pulse whereas in figure 7(a) the images of the arc shape at the time of the measurement are depicted. The relative intensities in the images are scaled in order to visualize the shape of the arc in an optimal way. Hence the image intensities at different t_{meas} are not comparable. The white horizontal bar in the images indicates the position of the incident laser beam.

In first approximation it can be concluded that the electron temperature and density are

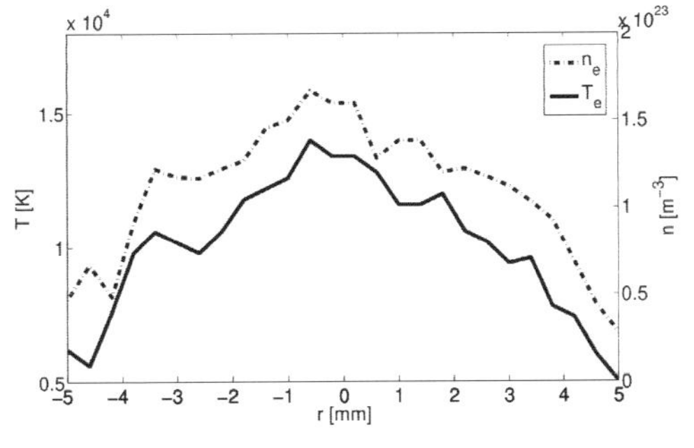

Figure 5. Radial electron temperature and density measured in the GMAW process at $5.5mm$ above the cathode surface.

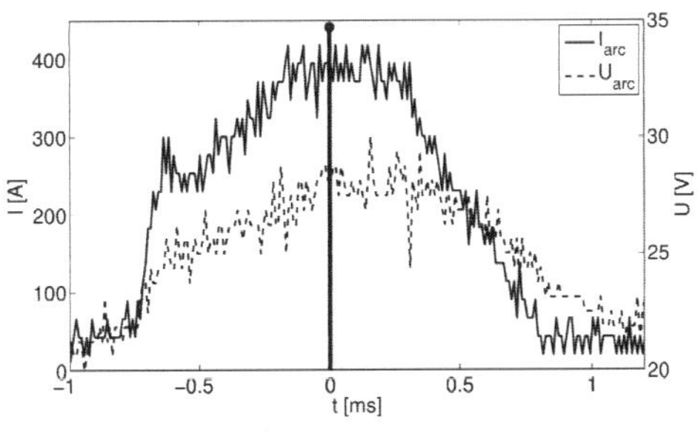

Figure 6. Typical current pulse during GMAW process using Al wire. $t = 0ms$ marks the trigger position for the Thomson scattering measurement.

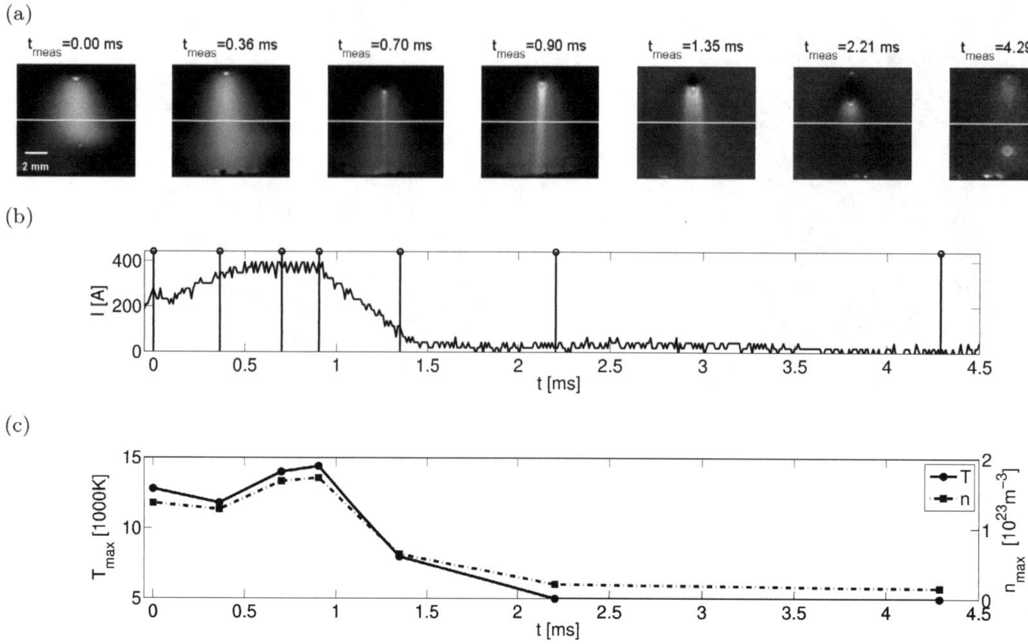

Figure 7. (a) Images of the GMAW arc recorded at the time t_{meas} simultaneously to the Thomson scattering measurements; The white horizontal bar indicates the laser beam position;(b) process current pulse with the measurement time t_{meas} marked by the vertical bars; (c) time resolved electron temperature and density in the center of the GMAW arc.

clearly related to the arc current. The highest values are reached by the end of the high current phase, whereas temperature as well as density fall with the decreasing current.

The last two measurements at $t_{meas} = 2.21ms$ and $t_{meas} = 4.29ms$ are made in the low current phase, where arc current was in the range of $10A$ to $30A$. As already seen on the images of the arc, almost no plasma is visible in this phase. The lighter spots positioned above the laser beam at $t_{meas} = 2.21ms$ and under the beam at $t_{meas} = 4.29ms$ are aluminum metal droplets detached from the wire anode. The scattering signal in these two measurements was considerably lower than in the preceding ones. Thus only signals close to the center of the arc could be evaluated. Nevertheless it can be concluded that the electron temperature in this phase does not exceed $5000K$. The maximal electron density lies in the range of $1.5 - 2 \times 10^{22}m^{-3}$.

In these two measurements almost no difference between the measured plasma parameters can be seen before and after the metal droplet has passed the measurement volume. However it has to be considered that those values lay in the lower resolution limit of the applied setup. Moreover due to the weak signal the relative error increases in comparison to the measurements in the high current phase.

When assuming local thermal equilibrium (LTE), Saha equation and the ideal gas law [11] can be applied to estimate the ionization rate. It is further assumed that plasma is composed from argon and aluminum atoms and ions. For the conditions measured in the low current phase the estimation delivers an ionization rate, which does not exceed 1.5%. In contrast ionization rate reaches from 5% at the edge of the plasma column up to 50% on the central arc axis in the high current phase. This explains a relatively week scattering signal in the low current phase. In order to be able to investigate the influence of the droplets more in detail either the Thomson scattering has to be adapted to resolve lower temperatures and densities or different methods such as emission spectroscopy should be used.

5. Conclusions and Outlook

In this work time resolved Thomson scattering was successfully performed to determine electron temperature and density in the GMAW process run with aluminum wire and pure argon as a shielding gas. It was shown that the setup is suitable to investigate electron temperatures and densities in the high current phase of the process. As a next step validation of the measurement using spectroscopic methods such as Stark broadening will be performed for the GMAW process.

Acknowledgments

The authors greatly appreciate the support of the Institute for Thermodynamics at the University of Federal Armed Forces in Munich for providing the Pi-MAX ICCD camera. The technical realization of the measurement equipment was widely supported by the technical staff of the Lab for Plasma Technology from the University of Federal Armed Forces, Munich. This work was funded in part by the DFG grant SCHE 428\6-3

References

[1] Murphy A B 2010 The effects of metal vapour in arc welding *Journal of Physics D: Applied Physics* **43** 434001
[2] Murphy A B 2001 Thermal plasmas in gas mixtures *Journal of Physics D: Applied Physics* **34** R151
[3] Gregori G, Kortshagen U, Heberlein J, Pfender E 2002 Analysis of Thomson scattered light from an arc plasma jet *Phys. Rev. E* **65** 046411
[4] Snyder S C, Lassahn G D, Reynolds L D 1993 Direct evidence of departure from local thermodynamic equilibrium in a free-burning arc-discharge plasma *Phys. Rev. E* **48** 4124 - 4127
[5] Tomita K, Yoshitake S, Uchino K, Takenaka D, Toda H, Hikita M, Suzuki K 2013 Measurements of Electron Density and Electron Temperature of Arc Discharge Plasmas Containing Metallic Vapors using Laser Thomson Scattering *IEEJ Transactions on Fundamentals and Materials* **133** 458 - 464
[6] Terasaki H, Tanaka M, Ushio M 2002 Effects of metal vapor on electron temperature in helium gas tungsten arcs *Metallurgical and Materials Transactions A* **33** 1183 - 1188
[7] Froula D H, Glenzer S H, Luhmann Jr N C, Sheffield J 2011 *Plasma Scattering of Electromagnetic Radiation* Academic Press, Elsevier
[8] Salpeter E E 1960 Electron Density Fluctuations in a Plasma *Phys. Rev.* **120** 1528 - 1535
[9] Kopainsky J, Vilsmeier F 1975 Effect of electron-neutral collisions on the satellites of the Thomson scattering spectrum in arc plasmas *Applied physics* **8** 223 - 228
[10] Snyder S C, Reynolds L D, Fincke J R, Lassahn G D, Grandy J D, Repetti T E 1994 Electron-temperature and electron-density profiles in an atmospheric-pressure argon plasma jet *Phys. Rev. E* **50** 519 - 525
[11] Kühn-Kauffeldt M, Marques J -L, Forster G, Schein J 2013 Electron temperature and density measurement of tungsten inert gas arcs with Ar-He shielding gas mixture *Journal of Instrumentation* **8** C10017

Investigation of the Anode Attachment Process in Plasma Arc Cutting

Stefan Eichler[1], Nakhleh Hussary[2], Erwan Siewert[3] and Jochen Schein[1]

[1] Universität der Bundeswehr München, Institut für Plasmatechnik und Mathematik, Neubiberg, Germany
[2] Victor Technologies, West Lebanon, New Hampshire, USA
[3] Linde AG, Unterschleißheim, Germany

E-mail: stefan.eichler@unibw.de
jochen.schein@unibw.de
erwan.siewert@linde-gas.com
nakhleh_hussary@victortechnologies.com

Abstract. The anode attachment process in plasma arc cutting is still not well understood in spite of decades of industrial use. Previously, several approaches were made to analyze the attachment mechanisms including imaging, discharge current and voltage measurements as well as the use of thermocouples. In this paper a different approach is described to evaluate the attachment position. Six electrically separated water-cooled copper plates arranged in layers are used as an anode emulating a workpiece. The current through each layer is measured individually using current Hall sensors. The thus obtained information about the current distribution across each plate is used to deduce the anode attachment position inside the workpiece. This diagnostics allows a quick determination of the influence of process parameter variations like the cutting current, gas flow rate, cutting speed or the torch distance on the current distribution inside the workpiece. Using this setup, it is observed that no single attachment appears; the current is divided to flow through all anode segments. The torch distance and cutting speed proved to have the biggest influence on the anode current distribution. Comparison between measurements conducted with the new setup and an experiment using steel plates instead of copper plates is provided.

1. Introduction

Plasma Arc Cutting (PAC) is a commercial process used to cut nearly every conducting material. The advantages of this process are the high cutting speed, the high quality of the cut and the low equipment cost. The process is competing with laser cutting and oxygen fuel cutting.

In plasma cutting a transferred arc is used where the material to be cut becomes the anode, while the cathode is located inside the so-called cutting torch. The anode energy is mainly transferred by electrons and convection as well as by chemical reactions when using oxygen containing gases. Their impact changes with the process parameters [1-6]. Moreover the position of the anode attachment is an important factor influencing the cutting quality as well as cutting speed. Some scientific publications [1, 4-11] describe investigations of the anode attachment of a cutting arc by various methods. Mostly video techniques (ultraviolet [8] and high speed imaging [1]) are used to identify the position of the attachment but also thermocouples [4] are employed to locate the hottest spot inside the workpiece and also the cutting discharge voltage as a measure of the effective length of the arc is recorded [7]. All these publications are based on experiments done cutting mild or stainless steel.

In [4] *Nemchinsky* and *Severance* used thermocouples which are placed at various locations inside the workpiece (10 mm mild steel), thereby identifying the location which is the hottest. By measuring the temperature they found the hottest spot in the upper third of the workpiece and concluded that the arc must have its attachment there. The disadvantages of this technique are the low spatial resolution and slow response time as well as the fact that the temperature is influenced by the kerf geometry, which is irregular.

Osterhouse used in [7] the cutting voltage to determine the attachment position. By cutting mild steel sheets of different thickness he found out that the voltage did not change up to a certain thickness and that the arc attachment is essentially stationary. He came to the conclusion that the anode attachment is in the upper third of the workpiece. For a faster cutting speed he measured a lower cutting voltage and concluded that the arc is shorter and most likely has an attachment point closer to the workpiece surface. Moreover he found, that the nozzle-to-workpiece-distance has no strong influence on the arc anode attachment position. Furthermore he assumed that an increase in plasma and shielding gas pressure raise the downward force on the arc attachment. The fact that the arc voltage is also influenced by a change in vaporization is not considered.

Teulet et al. used in [1] a normal CCD camera observing the arc from behind. They used neutral density filters to be able to identify a bright spot and attributed that the arc attaches here. He determined, that the arc root position is not strongly affected by operating conditions and that the anode spot is always close to the middle of the plate thickness. Just for high cutting speeds the attachment point moves toward the top and a second spot or an oscillating spot close to the workpiece surface appears.

Bemis used in [8] a CCD camera in combination with an ultraviolet bandpass filter to determine the attachment point. The camera he used is placed below and on the side of the workpiece looking towards the cutting gap. He concludes that the brightest zone corresponds to the anode arc attachment and determines single/multiple spot attachments as well as a diffuse attachment of the arc.

Colombo et al. [9] as well as *Kavka et al.* [5] determined in contrast to [7] and [8] that the arc attachment is characterised by either a single or a multiple attachment moving from the top to the bottom. *Davydenko* [10] found that two anode attachments exist; the upper one moves with a high frequency between 15 – 50 kHz whereas the lower with 0.5 – 1 Hz. *Kavka et al.* [5] found an anode attachment mainly located in the upper part of the plate close to the top of the material.

In summary the attachment mode (spot, single / multiple or diffuse) of the arc as well as the movement frequency differ in the current studies. However detailed information about the exact position of the anode attachment are needed for modelling the process [11,12], where up to now only not-validated assumptions are made.

The use of the location of a bright spot as it is performed by [1,5, 8-10] to identify the attachment position might not show the actual current path. Investigations of the GMAW process by *Schick* [13] and *Murphy* [14] show that metal vapour which is produced may radiate brighter than the actual current conducting plasma. Therefore a method to determine the actual current path is needed.

2. New Experimental Setup - Segmented Anode

In this paper a new approach is made to determine the anode attachment position. Here it is assumed that the position of the arc attachment is synonymous with the position of the strongest current flow. This setup allows a quick and highly reproducible determination of the influence of various parameters.

2.1. Experimental Design

A state of the art, dual gas cutting torch (Victor Technologies XT 300) as shown in figure 1 is used. As displayed in the figure, the plasma gas is flowing around the electrode and the shielding gas around the plasma gas nozzle. Both plasma and shielding gas have a radial velocity component (vortex flow).

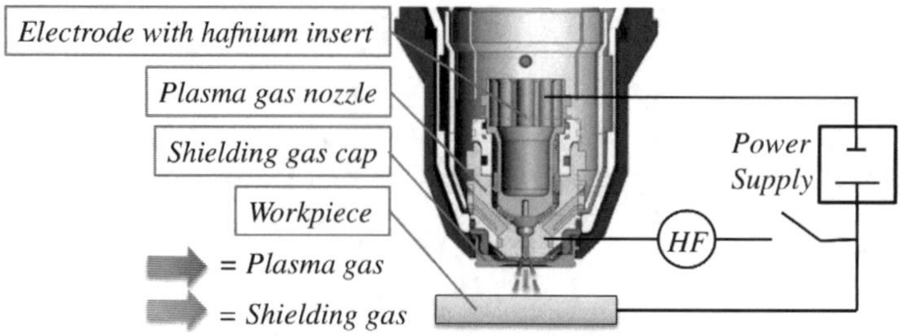

Figure 1. Schematic drawing of the PAC dual gas process.

The ignition of the arc is achieved by a high-voltage-high-frequency source to establish the pilot arc which is established between the electrode and the plasma gas nozzle. The pilot arc ionizes the plasma gas allowing the ignition of the main arc between electrode and workpiece without using high voltage. The gases used are oxygen as plasma gas and air as shielding gas.

2.2. The Anode Attachment

In order to be able to find an alternative way to determine the location of the anode attachment a new approach with a segmented copper anode has been developed to reproducibly investigate different process parameters in a well-defined setting.

The new experimental anode consists of six 1.5 mm thick water-cooled copper plates, as shown in figure 2. The plates are held together by three non-conducting screws. Between the copper plates sheets of aluminium oxide are located in order to insulate the plates from each other. Each plate is connected individually by three 1.5 mm^2 copper cables to allow a current measurement through each plate via hall sensors (LEM LF 205 S/SP3 with a bandwidth of 100 kHz).

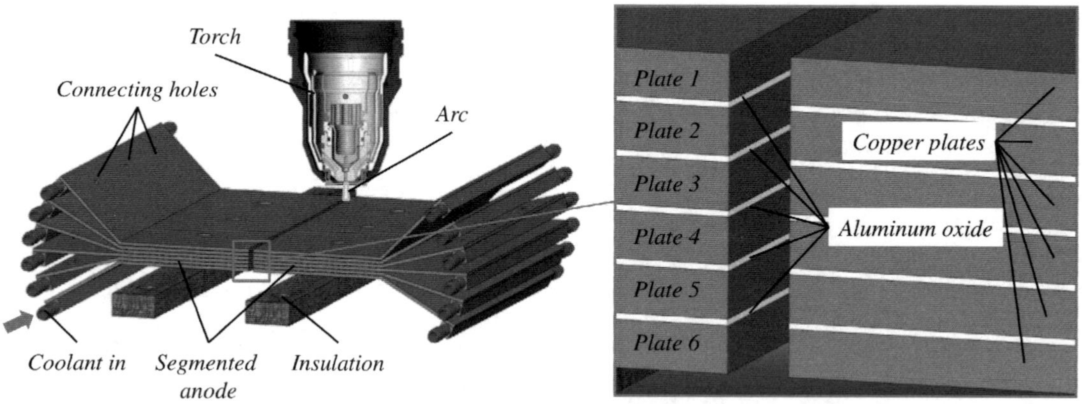

Figure 2. Experimental setup for the anode attachment trials.

All individual plate cables are connected to the power supply (Victor Technologies Ultra Cut 300). Additionally the voltage drop across the arc is measured by a voltage transducer (LEM DLV 250 with a bandwidth of 16 kHz). The voltage is measured between torch and the connection point of all cables. All signals are recorded with a sampling frequency of 50 kHz by the National Instrument data acquisition board USB 6363 (16 bit) connected to a computer, as shown in figure 3. In the experiments the torch is fixed while the segmented anode is moved by a linear unit.

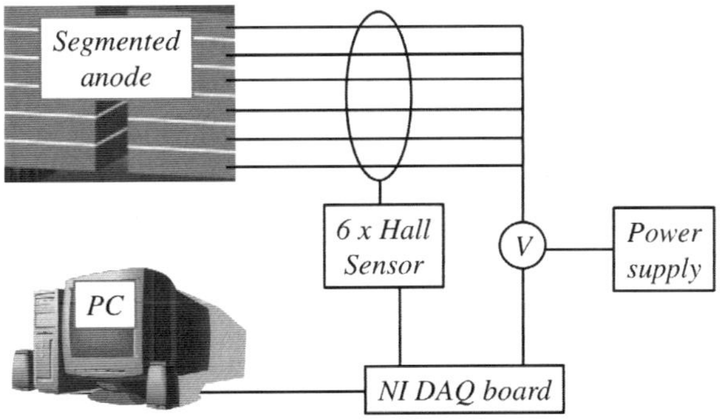

Figure 3. Schematic setup of the measurement equipment for the segmented anode trials.

2.3. Verification Measurements

The conditions achieved with the water cooled copper plates described in section 2.2 differ from a real cutting process due to a reduction in evaporating as well as a different thermal conductivity of the anode material. Therefore verification measurements are performed to validate the results obtained with the segmented copper anode to draw conclusions about the influence of thermal conductivity and metal vapour. The general experimental setup and procedure for the verification measurement is the same as the one used with the segmented copper anode but the copper plates were replaced by 1.5 mm thick mild steel plates. Also the steel plates are electrically separated by 0.25 mm thick layers of aluminium oxide. The experimental setup is displayed in figure 4.

Figure 4. Experimental setup for the verification trials.

Each plate is connected by three 1.5 mm² copper cables that are welded to the sides of the mild steel plates.

3. Trials And Results

To characterize the anode attachment and to verify the measurements done, a total of three campaigns have been performed.

- Reproducibility trials (copper plates).
- Influence of different process parameters (copper plates).
- Verification trials (steel plates).

The first two are done with the water cooled copper anode described in section 2.2 and the verification trials are performed with mild steel plates as described in section 2.3.

3.1. Reproducibility

The reproducibility trials have been conducted with the standard parameter settings used in practice to cut 10 mm mild steel plates, listed in table 1.

Table 1. Standard parameter for 10 mm mild steel.

Parameter	Value
cutting current (A)	100
torch distance (mm)	2.8
cutting speed (mm s^{-1})	36.33
plasma gas	oxygen
plasma gas pressure (Pa)	760000
shielding gas	air
shielding gas pressure (Pa)	400000

To show the reproducibility of the diagnostic technique three tests are carried out for the same settings. The current flow through the six plates is displayed in figure 5. The colours for the current traces through the individual plates from top to bottom are red, green, dark blue, violet, light blue and grey. The basic shape does not change between the trials. Most of the current is running over the top three plates (blue, red and green). Also the amount of total current stays almost the same.

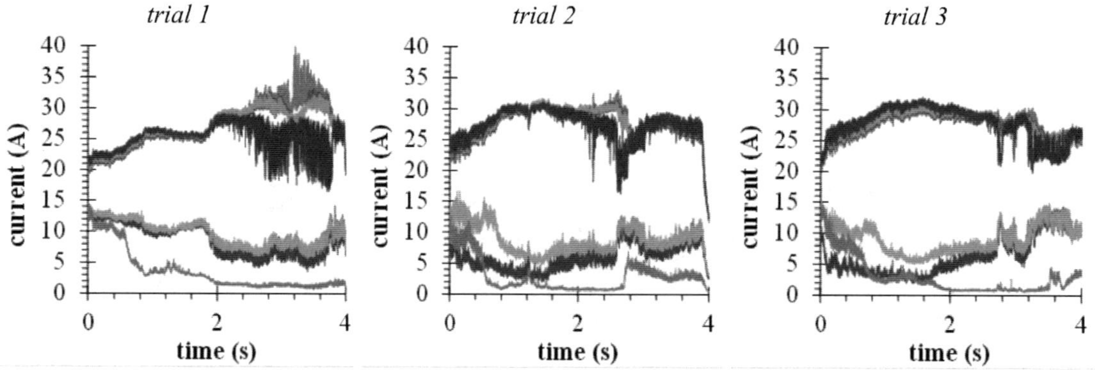

Figure 5. Current flow over the six plates at the same set of parameters (plates from top to bottom: red, green, dark blue, violet, light blue and grey).

In order to compare the three trials, the arithmetic mean value of the whole run is taken into consideration. The result is displayed in figure 6.

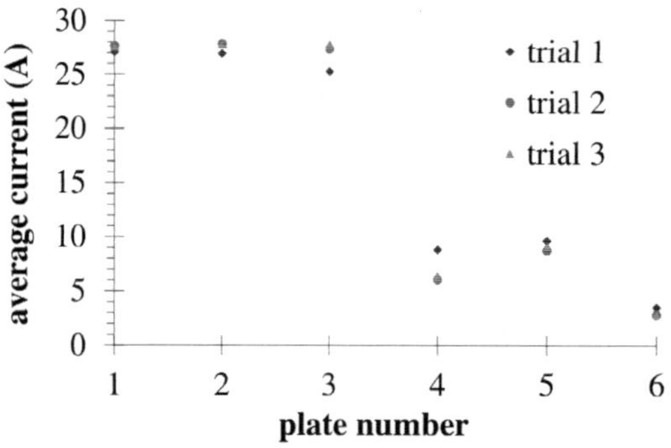

Figure 6. Average cutting current over the six anode plates

Trial two and three show almost the same behaviour regarding the current flow over the individual plates. Trial one is slightly different (value of the average current of plate 3 and 4 changes) but still shows the same tendency. The highest standard deviation in current is measured at plate number four with 1.51 A and the lowest at plate number one with 0.33 A. These fluctuations are still quite small and thus these results should allow for a reasonable allocation of the current distribution.

3.2. Different Process Parameters

In this section, the influence of process parameters on the anode attachment is observed. Only one parameter at a time is changed while the other ones are the standard parameters listed in table 1. The parameters changed are:

- Cutting speed.
- Torch distance.
- Plasma gas pressure.
- Shielding gas pressure.
- Total cutting current.

3.2.1. Influence of the cutting speed. Four cutting speeds were set in the range between 16.67 mm s^{-1} and 41.67 mm s^{-1}. The average current running over each plate, depending on the cutting speed, is displayed in figure 7a.

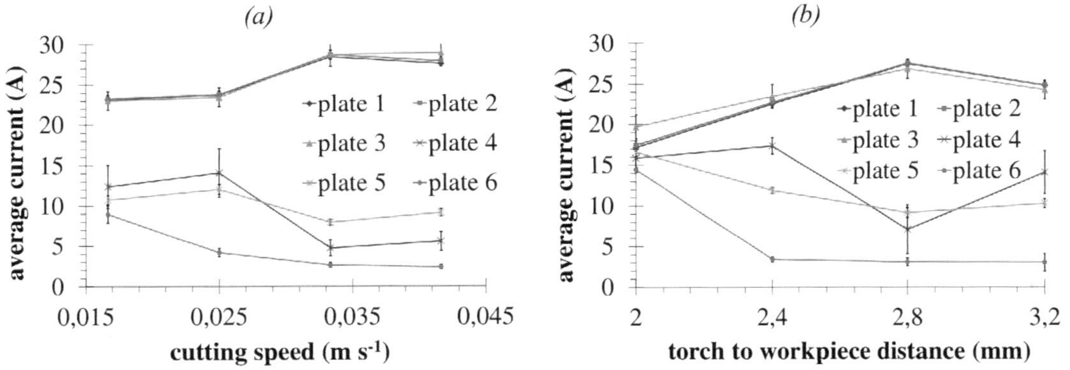

Figure 7. Average cutting current over the six anode plates as a function of the cutting speed (a) and the torch distance (b).

The current flow across the top three plates shows a clear tendency; it increases with higher cutting speeds thus confirming the observations of [7]. The current flow ranges from 23.03 A at 100 cm min^{-1} to 28.96 A at 250 cm min^{-1}. The tendency of current flow through plate number six (bottom plate) is different: it decreases with an increasing cutting current from 8.97 A to 2.38 A. The current flowing across plates four and five vary as shown in figure 7a; but the tendency is a decreasing current flow at an increasing cutting speed. Summing up a faster cutting speed pushes the current distribution closer to the top of the workpiece.

3.2.2. Influence of the torch distance. In this section, the influence of the torch distance (torch to plate standoff) on the anode attachment is observed. Four trials with distances of 2.0 mm, 2.4 mm, 2.8 mm and 3.2 mm were performed. The current flow across the top three plates rises until a distance of 2.8 mm is reached and drops slightly for a longer arc. The current across plate four displays an oscillating behaviour while plate five and six are showing a decrease in current with increasing torch distance, see figure 7b. For a small distance of 2.0 mm the amount of current flowing over each plate is almost the same. With increasing cutting distance the current distribution is moved towards to the top of the workpiece.

3.2.3. Influence of the plasma gas and shielding gas pressure. The influence of the plasma gas inlet pressure on the current distribution is displayed in figure 8a. The current flowing over the top three plates is again similar and makes up over two thirds of the total current. It has its maximum at a pressure of 760 kPa. Plate four and five show an opposite behaviour while the current running over plate six stays nearly constant.

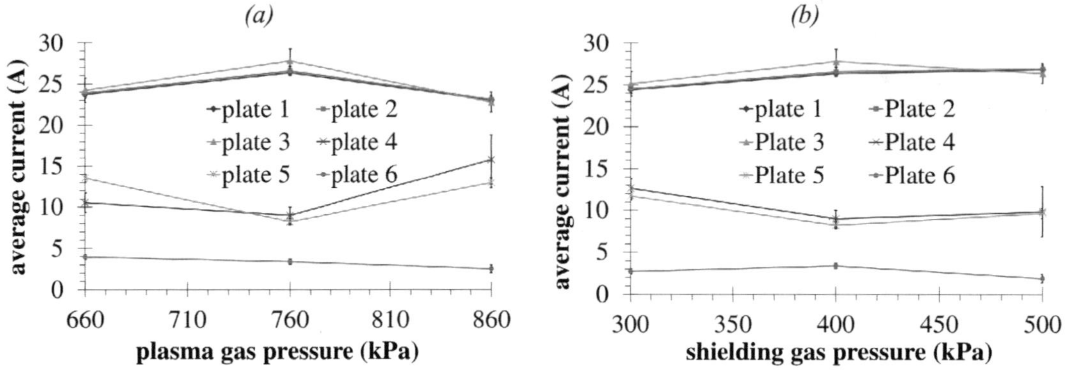

Figure 8. Average cutting current over the six anode plates as a function of the plasma gas pressure (a) and the shielding gas pressure (b).

Overall the changes of the current flow are much smaller compared to the parameters observed before.

The shielding gas pressure on the current distribution is observed at three different inlet pressures. The current flows over all six plates are displayed in figure 8b. The current flow over the first three plates does not change valuable. It slightly increases with a rising pressure. Plate four and five behave the opposite way. Plate six has a maximum at about 400 kPa. The change in current distribution is even smaller than the one depending on the plasma gas pressure.

3.2.4. Influence of the arc current. The influence of the total cutting current on the current distribution over the six plates of the segmented anode is described in this section. For a total current of 50 A and 75 A the percentage of current flowing over the top three plates varies however, at 100 A they show a similar distribution, see figure 9. The current distribution stays constant for 50 A and 75 A and changes slightly for the 100 A.

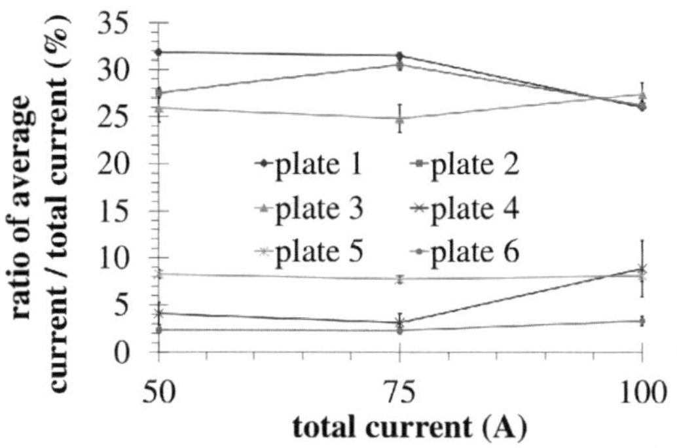

Figure 9. Average cutting current over the six anode plates as a function of the total current

To sum it up, the change in the current distribution as a function of total current is not really significant.

3.3. Verification Trials

The verification trials have been done with standard settings used also for the reproducibility trials, see table 1.

The verification trials show a current flow over all six 1.5 mm mild steel plates just as it is the case for the copper anode setup (see figure 10a). Due to the fact, that the steel plates are connected by the molten metal after a short period of time only the first 0.2 s of the experimental run are taken into consideration. A second trial with this setup is due to the shortcut not possible. In contrast to the copper plate setup most of the current is flowing just through the top two plates (red and green in figure 10a). Comparing the mean values of the current flow over each plate for the reproducibility trials (copper) and the verification trial (mild steel) it is clear that there is a common tendency. In both cases the amount of current is decreasing with an increasing plate number as seen in figure 10b. Still there are major differences looking at the actual values, which are closer to the trials at a lower cutting speed or a lower torch distance.

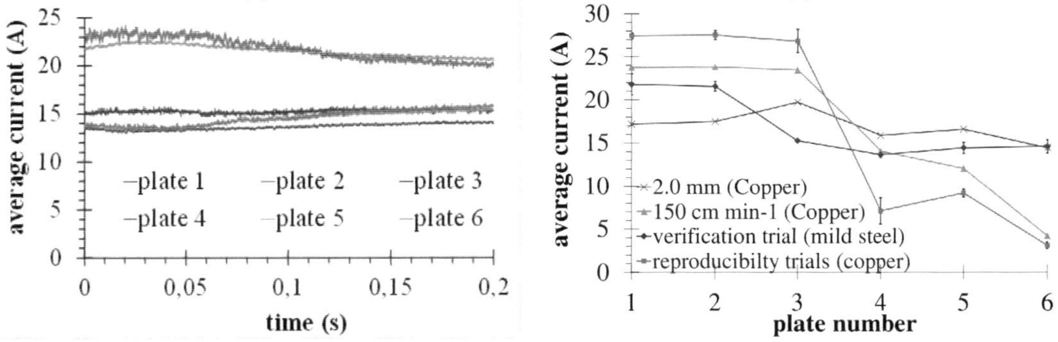

Figure 10. Current flow over the six plates of mild steel (a) and mean value of the current flow for different parameter settings (b).

Comparing the arc discharge voltage of the verification trial (mild steel) and one of the reproducibility trials (copper) shows that the voltage of reproducibility trial (copper) is slightly higher than the one of the verification trial (mild steel). The average voltage over a time period of 1.5 s shows an average voltage of 163.2 V for the verification trial (red plot of figure 11) and an average of 172.1 V for the reproducibility trial (green plot).

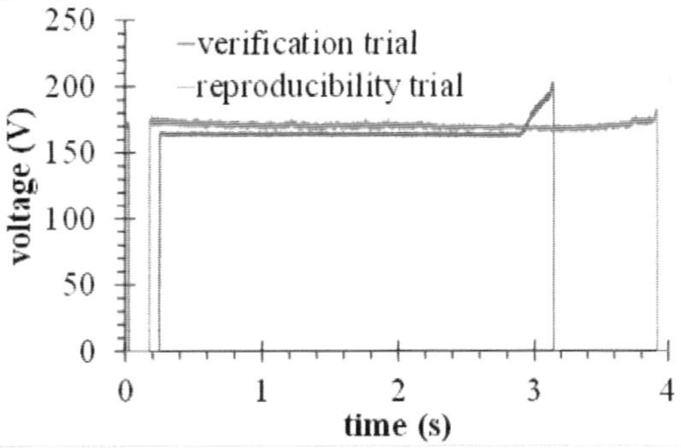

Figure 11. Discharge voltage of the arc of a verification and a reproducibility trial

4. Discussion

In all trials described in this paper a current flow over all six plates was detectable. Therefore it appears that a single spot attachment of the arc does not exist. It seems as if the current is distributed across the thickness of the complete workpiece more or less concentrated in certain areas.

The measurements for the reproducibility trials differ depending on the time. Due to the high temperature gradient between the solid copper and the plasma jet of the cutting torch the presence of metal vapour cannot be ruled out. This metal vapour could cause the differences in the attachment behaviour and also the occasional fluctuations. Another effect could be the temperature of the copper anode which is not in a stationary condition due the short measurement time of a few seconds. An increasing cutting speed leads to an increase of the amount of current running over the top three plates. Therefore it can be concluded that the current conduction concentrates more in the upper region and confirms *Osterhouse* results [7]. The effect of the cutting speed was also discovered by *Teulet* in [1]. He discovered an attachment point moving towards the workpiece surface with an increasing cutting speed.

In the trials done also the torch distance was varied. The results show, that with a smaller distance between torch and workpiece, the arc divides more evenly across all six plates. The current flowing over each plate is almost the same for a distance of 2 mm. A possible explanation for this behaviour could be that due to the small distance no cold gas from the atmosphere is entrained into the cutting gap. The lack of cooling might allow the arc to attach over the whole workpiece thickness. Due to the sonic conditions of the arc, as described by *Ramakrishnan et al.* [15] or *Freton et al.* [11][12] it might also be possible that a shock layer of the under expanded flow forces the arc to attach over the hole thickness.

The deviations and tendencies of the change in the current flow distribution by varying the gas pressures and the total cutting current are not as significant as the two other parameters mentioned above. The low influences of the gas pressure also agree with the results of *Teulet* in [1].

The verification trials show the same tendencies as the trials with the copper anode. In both cases a current flow over all six plates can be detected; ruling out a single spot attachment of the arc. The high effort needed to perform a 0.2 s experiment with mild steel plates show that the copper setup is much more efficient.

In [5] *Kavka* presents the frequency spectrum of the cutting voltage of his cutting trials showing peaks at 50 kHz and at 200 kHz, those fluctuation cannot be detected by the presented experimental setup. The plasma cutting system *Kavka* is using operates with a vapour mixture of water and alcohol as plasma gas instead of oxygen as used in the trials presented here. It might be possible that this fact leads to the different attachment behaviour. It cannot be ruled out, that there might be fluctuations of the anode attachment that are beyond the capability of this setup. But the current recorded does not show significant fluctuations which may show a tendency towards diffuse attachment of the arc instead of a moving spot attachment.

5. Summary

In this paper, the anode attachment behaviour of the PAC arc was observed. This was done with a segmented copper anode consisting of six electrically separated water cooled copper plates emulating a workpiece. The parameters varied were:

- Cutting Speed.
- Torch distance.
- Plasma gas pressure.
- Shielding gas pressure.
- Total cutting current.

This method has been validated by observing the (1) reproducibility of the results, (2) actual cutting trials and (3) by comparing the discharge voltage of the arc. The following conclusions can be drawn based on this new method of anode attachment characterization:

- The trials are reproducible.
- The verification shows, that the adopted method is able to present tendencies of different cutting parameters.
- Most of the current flows across the top three plates.
- The biggest influence on the anode attachment is caused by a change of the torch distance and the cutting speed.
- An increasing cutting speed, leads to a concentration of the anode attachment toward the top surface of the workpiece.
- The smaller the torch distance the more equally the anode attachment distributes over the six plates.
- The gas pressures of the plasma and shielding gas as well as the total cutting current have only a little influence on the anode attachment.

Similarities to the results given in the literature can be drawn, because most of the current flows through the top three plates of the segmented anode which can be compared well with the findings of [1,3]. Additionally, the influence of the cutting speed agrees well with the measurements done by [2,4].

References

[1] Teulet P H, Girard L, Razafinimanana M, Gleizes A, Bertrand P H, Camy-Peyret F, Baillot E and Richard F 2006 *J. Phys. D: Appl. Phys.* **39** 1557–73
[2] Quigley M B C, Richards P H, Swift-Hook D T and Gick A E F 1973 *J. Phys. D: Appl. Phys.* **6** 2250–59
[3] Nemchinsky V A 1997 *J. Phys. D: Appl. Phys.* **30** 2566–72
[4] Nemchinsky V A and Severance W S 2006 *J. Phys. D: Appl. Phys.* **39** 423–38
[5] Kavka T, Chumak O, Šonský J, Heinrich M, Stehrer T and Pauser H 2013 *J. Phys. D: Appl. Phys.* **46** 065202
[6] Hackett C M and Garg S 2010 *Eur. Phys. J. Special Topics* **182** 145–159
[7] Osterhouse D J, Lindsay J W and Heberlein J V R 2013 *J. Phys. D: Appl. Phys.* **46** 224013
[8] Bemis B L and Settles G S 1999 *IEEE Trans. Plasma Sci.* **27** 44–5
[9] Colombo V, Concetti A, Ghedini E, Rotundo F, Sanibondi P, Boselli M, Dallavalle S, Gherardi M, Nemchinsky V and Vancini M 2011 *Plasma Chem Plasma Process* DOI 10.1007/s11090-011-9338-8
[10] Davydenko V G, Kharitonov E P, Shapiro I S and Shimanovich V D 1974 *Inzhenerno-Fizicheskii Zhurnal* **26** 451–6
[11] Freton P, Gonzalez J J, Gleizes A, Peyret F C, Caillibotte G and Delzenne M 2002 *J. Phys. D: Appl. Phys.* **35** 115–131
[12] Freton P, Gonzalez J J, Camy Peyret F and Gleizes A 2003 *J. Phys. D: Appl. Phys.* **36** 1269–1283
[13] Schnick M, Füssel U, Hertel M, Spille-Kohoff A and Murphy A B 2010 *J. Phys. D: Appl. Phys.* **43** doi:10.1088/0022-3727/43/2/022001
[14] Murphy A B 2010 *J. Phys. D: Appl. Phys.* **43** doi:10.1088/0022-3727/43/43/434001
[15] Ramakrishnan S, Gershenzon M, Polivka F, Kearney T N and Rogozinski M W 1997 *IEEE Transactions on Plasma Science* **25** 937–46
[16] Hartmann R and Heberlein J V R 2001 *J. Phys. D Appl. Phys.* **34** 2972–78

Properties of thermal air plasma with admixing of copper and carbon

S Fesenko[1], A Veklich[1], V Boretskij[1], Y Cressault[2], A Gleizes[2] and Ph Teulet[2]

[1] Taras Shevchenko National University of Kyiv, Faculty of Radio Physics, Electronics and Computer Systems, 4g Glushkova Ave., Kyiv, Ukraine
[2] Université de Toulouse; UPS, INPT; LAPLACE; 118 route de Narbonne, F-31062 Toulouse cedex 9, France

E-mail: van@univ.kiev.ua

Abstract. This paper deals with investigations of air plasma with admixing of copper and carbon. Model plasma source unit with real breaking arc was used for the simulation of real discharges, which can be occurred during sliding of Cu-C composite electrodes on copper wire at electromotive vehicles. The complex technique of plasma property studies is developed. From one hand, the radial profiles of temperature and electron density in plasma of electric arc discharge in air between Cu-C composite and copper electrodes in air flow were measured by optical spectroscopy techniques. From another hand, the radial profiles of electric conductivity of plasma mixture were calculated by solution of energy balance equation. It was assumed that the thermal conductivity of air plasma is not depending on copper or carbon vapor admixtures. The electron density is obtained from electric conductivity profiles by calculation in assumption of local thermodynamic equilibrium in plasma. Computed in such way radial profiles of electron density in plasma of electric arc discharge in air between copper electrodes were compared with experimentally measured profiles. It is concluded that developed techniques of plasma diagnostics can be reasonably used in investigations of thermal plasma with copper and carbon vapors.

1. Introduction

Usually copper wire and various types of inserts, which are fixed on the contact surface of the pantograph, are widely used in power supply circuits of electric transport. Graphite, coke, copper-graphite or copper alloys can be used as materials in producing of such inserts. Each of these types of inserts has some advantages and disadvantages [1].

In particular, in spite of good lubricating properties of graphite and coke inserts, they have a comparatively high electrical resistivity. So, deterioration of tribological parameters of copper trolley wire takes place due to annealing in a result of significant Joule heating [2]. Moreover, such inserts have a sufficiently low hardness, so their service life is reduced.

The metal contact inserts on a base of copper alloys are characterized by a significantly less resistivity and a longer life. However, as a result of their affinity with the copper wire, the last one is significantly abraded by such inserts. Additionally, it must be noted that cost of copper wire is significantly higher in comparison with the cost of inserts.

Therefore, design engineers of such inserts usually recommend a copper-graphite composite as an optimal material for their produce [1]. Such composite has good lubricating properties due to its graphite component. The additions of copper in these materials provide the much higher electrical conductivity in comparison with graphite inserts. To avoid an affinity of these composite inserts with a conducting wire, the copper content in such material is insignificant.

On the other hand, an electric arc between a conducting wire and a pantograph contact frequently occurs, which leads to a significant erosion of the material. This phenomenon is greatly enhanced when icing of contact wire takes place. Understanding of a mass transfer processes in discharge gap and knowledge of plasma parameters will help to optimize the techniques with the aim to reduce the erosion of contact and wire. So, it will be contribution to improvement of economic performance of contact material productions.

The aim of this work is a study of plasma properties of model electric arc ignited in air atmosphere between Cu-C composite or copper electrodes. This paper deals with experimental investigations of parameters of thermal plasma mixture and calculations of electrical conductivity as well as electron density in discharge column.

2. Experimental investigations

2.1. Arc discharge arrangement

The free burning electric arc was ignited in air between the end surfaces of Cu-C composite non-cooled vertically arranged electrodes. At the fabrication technology the copper content in this composition was controlled around 20% [3]. This copper content was additionally verified by mass spectroscopy studies. The peculiarities of surface condition and erosion rate of composite electrodes under influence of arc discharge and plasma composition as well will be study in detail by additional investigation using metallography and optical spectroscopic techniques. Nevertheless, it was strongly controlled that both electrodes were not significantly consumed during experimental investigations. Additionally arc discharge between non-cooled copper electrodes in air flow 6.4 slpm was studied.

The diameter of the rod electrodes was 6 mm, the discharge gap was 8 mm, arc current was 3.5 or 30 A. To avoid the metal droplet appearing, a pulsing high current mode was used: namely, the rectangular current pulse of 30A was put on the "duty" low-current (3.5 A) discharge. The high-current pulse duration was of 30ms. The registration of arc plasma emission was performed at 7 ms after current pulse rise when a steady-state mode of electric arc discharge was realized. We found in our previous investigation that in such steady-state mode of arc discharge in air between copper containing electrodes the local thermodynamic equilibrium (LTE) is realized [4].

A more detail description of experimental setup and measurement procedure one can find in [5,6].

2.2. Experimental techniques

2.2.1. Temperature measurement. Boltzmann plot method was applied for plasma temperature determination. The one-pass tomographic recording of the spatial distribution of spectral line intensities was used in the first techniques [7]. Monochromator MDR-12 with 3000-pixels CCD linear image sensor (B/W) Sony ILX526A accomplished fast scanning of spatial distribution of radial intensity. Due to the instability of the discharge, statistical averaging of the recorded spatial distributions of the radiation characteristics was carried out.

In alternative techniques the diagnostics with simultaneous registration of spectral and spatial distribution of emission intensity was used. Spectra registration was performed by digital camera on CCD matrix in a grating spectrometer in such approach [6].

Selection of Cu I spectral lines and analysis of their spectroscopic data was previously carried out in [8]. So, copper spectral lines 510.5, 515.3, 521.8, 570.0 and 578.2 were used for plasma diagnostics in present work.

2.2.2. Electron density measurement. Radial profiles of electron density were obtained from half-width of spectral line Cu I 515.3 nm in assumption of dominating quadratic Stark effect. The spectral device combined with Fabry-Perot interferometer in etalon mode was used for registration of spectral line profiles at 30 A [9]. A special program graphical user interface in such experimental techniques was used.

Unfortunately, such Fabry-Perot interferometer can't be used for half-width determination in a case of 3.5 A current. So, electron density in this case was calculated by algorithm based on previously obtained plasma parameters, namely: temperature distributions for 3.5 and 30 A, electron density for 30 A, intensities ratio of Cu I spectral lines. Comprehensive description of the algorithm is described in paper [8].

3. Results and discussions

3.1. Results of temperature measurements

Radial temperature profiles T(r) of electric arc discharges are shown in figure 1. Experimentally obtained data (Exp) and error bars are approximated by Gaussian curves T, T_{Sup} and T_{Inf} accordingly. Measurements were carried out in middle section of arc column.

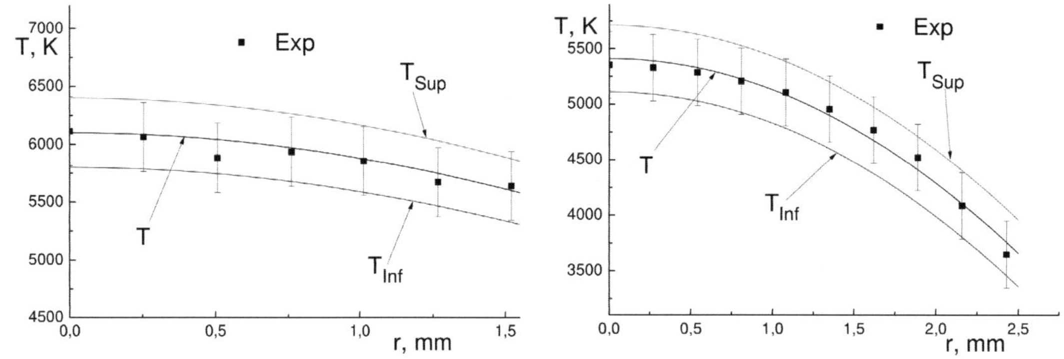

a

b

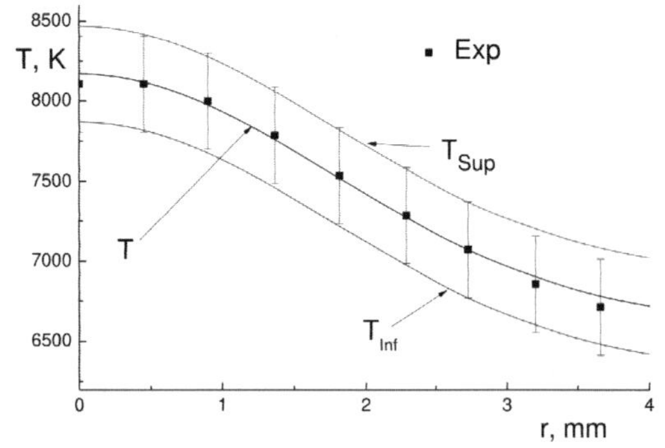

c

Figure 1. Experimentally obtained data (Exp) and approximation curves of temperature radial profiles T, T_{Sup} and T_{Inf} in middle section of discharge gap of electric arcs between Cu-C electrodes at current 3.5 A in air (a) and copper electrodes in air flow at currents 3.5 A (b) and 30 A (c).

3.2. Calculation of plasma parameters

3.2.1. Transport properties of arc discharge plasma. The real thermal and electrical conductivity radial profiles of air-copper plasma mixtures were calculated in detail in our previous investigations [10]. Electric arc between copper electrodes in air flow at current 3.5 A was studied. It was found that the copper admixture almost has no influence on the thermal conductivity of such plasma mixture in experimental temperature range 3000 K<T<6000 K (see figure 2). It is well known, in contrast to this parameter, the electric conductivity is very sensitive to the metal admixture in plasma.

So, the electrical conductivity σ of air plasma with admixing of copper can be calculated by solution of energy balance equation (1). In this approach the thermal conductivity λ of air without any admixtures was used. The electric field E of arc plasma column was measured additionally by discharge length L varying (see figure 3).

$$\sigma E^2 + \frac{1}{r}\left[\frac{d}{dr}\left(r\lambda\frac{dT}{dr}\right)\right] = 0\,. \tag{1}$$

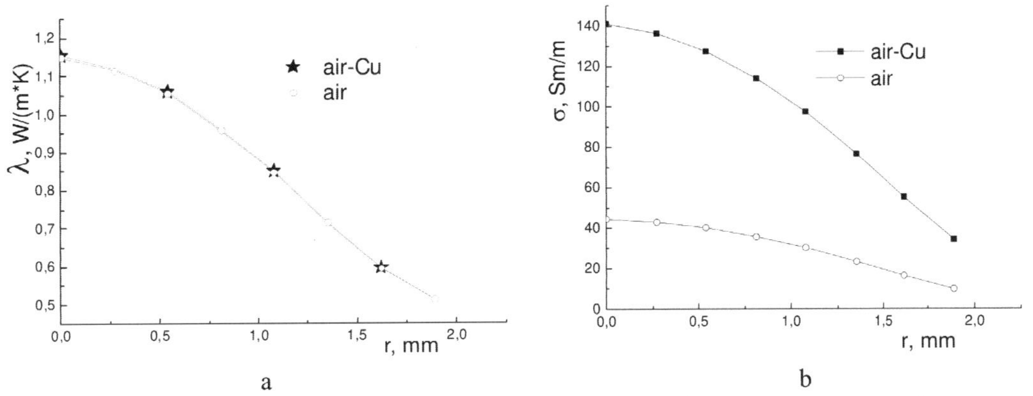

Figure 2. Radial profiles of thermal (a) and electrical (b) conductivity of plasma in arc discharge in air flow with copper admixture (filled) and without it (open).

It must be noted that in equation (1) radiation losses term is neglected. Our preliminary estimation of radiation losses, caused by emission of copper spectral lines, showed that influence of this mechanism is insignificant for the temperatures up to 8500 K. Additionally we can note that just similar conclusion was made in paper [11], where author showed that radiation losses is negligible in the current range up to 30 A.

So, radial distribution of electrical conductivity can be calculated in the following manner:

$$\sigma(r) = -\frac{1}{rE^2}\left[\frac{d}{dr}\left(r\lambda\frac{dT}{dr}\right)\right]. \tag{2}$$

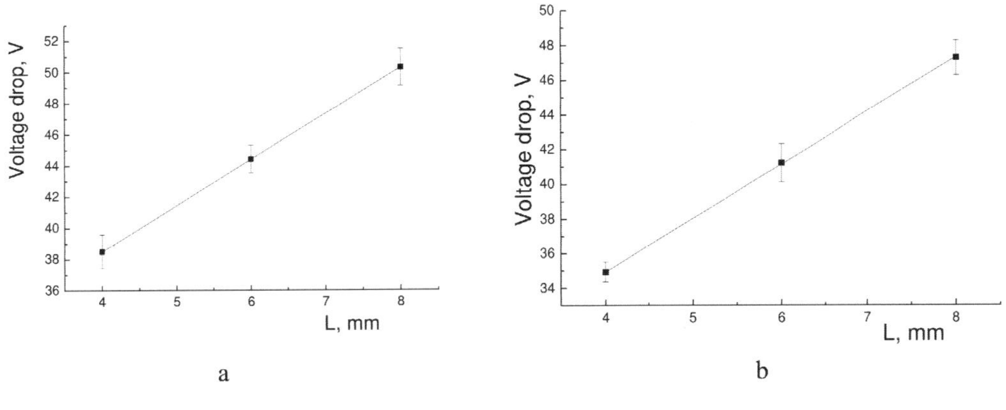

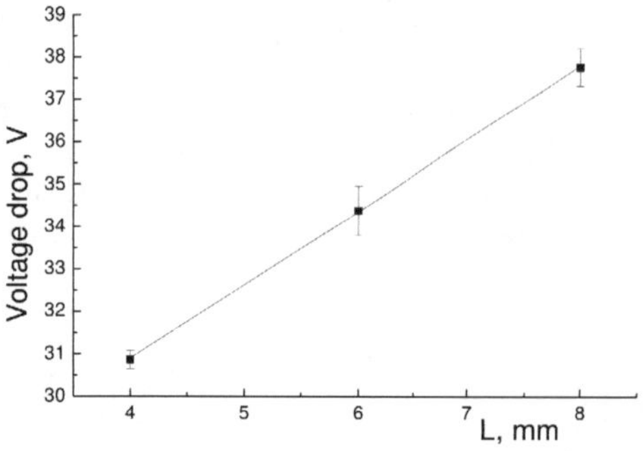

c

Figure 3. Voltage drop on discharge gap of electric arcs between Cu-C electrodes in air at current 3.5 A (a) and between copper electrodes in air flow at currents 3.5 A (b) and 30 A (c).

In figure 4 electrical conductivity radial profiles of electric arc plasma are shown. Calculated data (from equation (2)) was approximated by Gaussian curves according to three different temperature profiles T, T_{Sup} and T_{Inf}. The procedure of approximation was carried out under strong requirements of true value of experimental discharge current.

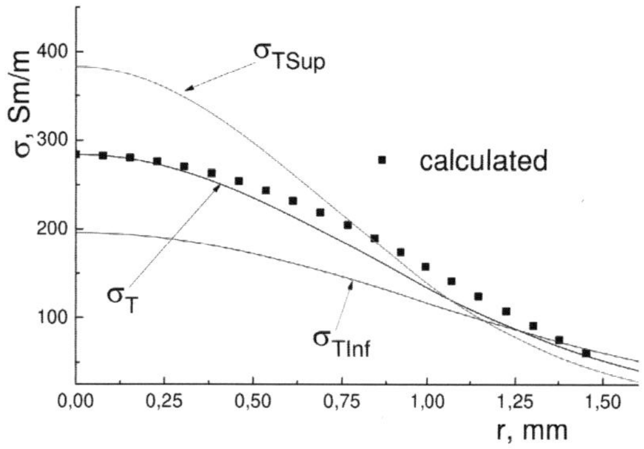

a

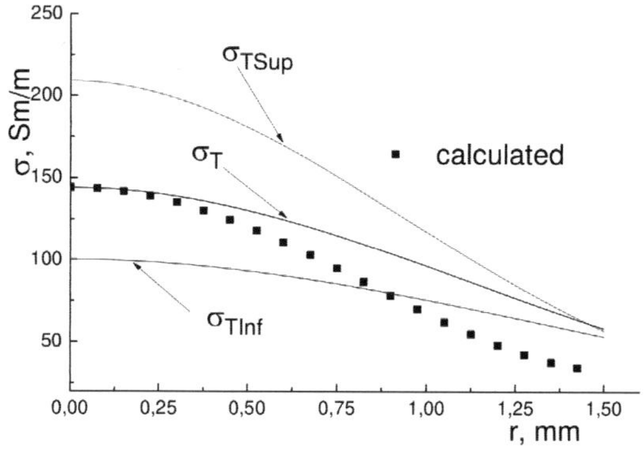

b

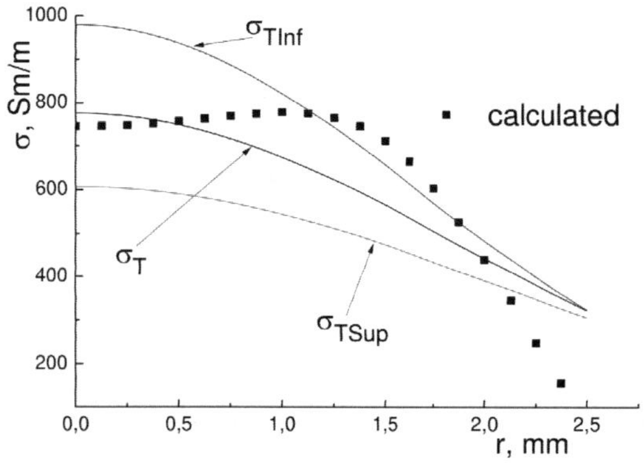

c

Figure 4. Radial profiles of electrical conductivity calculated from equation (2) and approximation curves σ_T, σ_{TSup} and σ_{TInf} in discharge gap of electric arcs between Cu-C electrodes at current 3.5 A in air (a) and between copper electrodes in air flow at currents 3.5 A (b) and 30 A (c).

At the next step of investigation the radial profiles of electron density in arc discharge were obtained from calculated profiles of electrical conductivity.

3.2.2. Electron density calculation. Since the mobility of electron is much higher than the ion mobility, the ion current in plasma can be neglected. Then plasma electrical conductivity can be written as:

$$\sigma(r) = eN_e(r)\mu_e(r), \tag{3}$$

where e is the elementary electrical charge, $N_e(r)$ – electron density distribution, μ_e – electron mobility, which can be expressed in terms of the frequency of electron collisions with other particles:

$$\mu_e(r) = \frac{e}{m_e \nu_t(r)}, \tag{4}$$

where m_e is the mass of electron, ν_t – electron frequency collisions with particles of all sorts, which can be written as:

$$\nu_t(r) = \sum_p \nu_{e-p}(r), \tag{5}$$

$$\nu_{e-p}(r) = N_p(r) \cdot u_e \cdot Q_{e-p}, \tag{6}$$

where ν_{e-p} is the frequency of electron collisions with particles of sort "p", $N_p(r)$ – the radial distribution of particle density of sort "p" and $u_e Q_{e-p}$ – the product of the average electron velocity and collision cross section for particles of sort "p". It was assumed that plasma is in LTE, so Maxwell distribution was used in calculation of the average value u_e in equation (6).

Thus, electron densities can be obtained from equations (3) – (6). Peculiarities of necessary calculation of plasma composition and the cross sections determination are discussed in detail in paper [10].

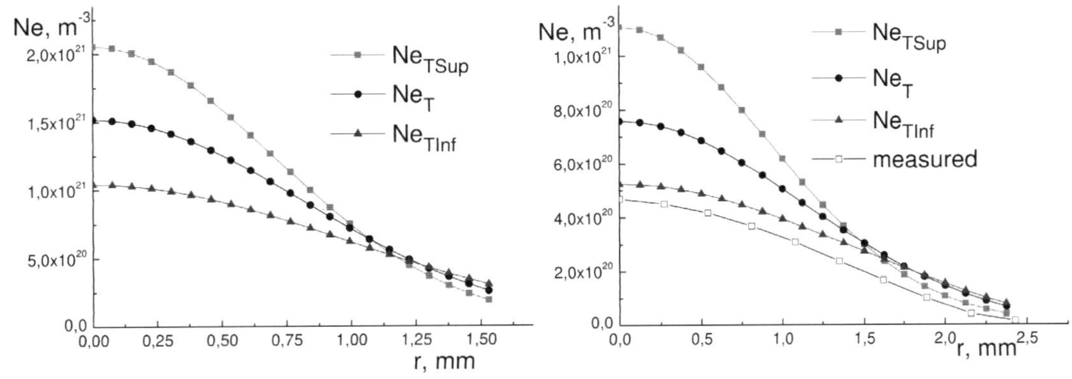

a b

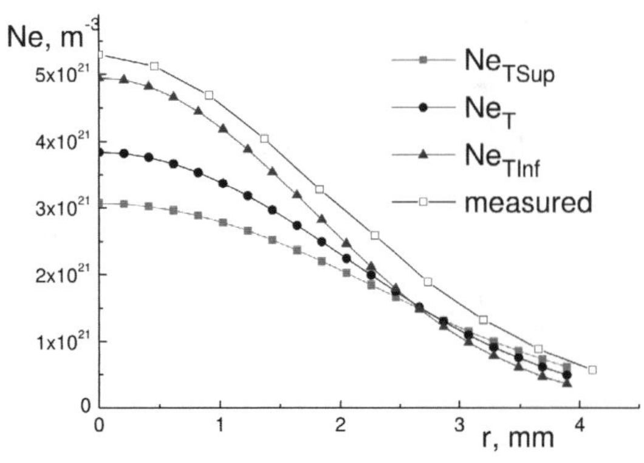

c

Figure 5. Radial profiles of electron densities obtained from σ_T, σ_{TSup} and σ_{TInf} in discharge gap of electric arcs between Cu-C electrodes in air at current 3.5 A (a) and copper electrodes in air flow at currents 3.5 A (b) and 30 A (c). Experimentally measured profiles are shown as curves with open symbols (b, c).

In figure 5 radial profiles of electron densities in electric arc plasma column are shown. Simulations were carried out on the base of calculated electrical conductivities according to three different temperature profiles T, T_{Sup} and T_{Inf}. In case of electric arcs between copper electrodes in air flow at currents 3.5 A and 30 A experimentally measured radial profiles of electron density in plasma between copper electrodes in air flow are additionally shown in figure 5 (b,c) [12].

One can conclude that agreement between experimental and calculated data is more or less acceptable. Unfortunately, experimental validation of electron densities in electric arc between composite Cu-C electrodes was complicated due to the strong continuum radiation. Nevertheless, calculated radial profiles of electron density in electric arc between these electrodes seem to be quite reasonable. At the next step of investigation another kind of experimental techniques must be used to obtain such profiles.

Conclusions

The complex technique of plasma property studies is suggested. From one hand, the radial profiles of temperature and electron density in plasma of electric arc discharge in air between Cu-C composite and copper electrodes in air flow were measured by optical spectroscopy techniques. From another hand, the radial profiles of plasma electric conductivity were calculated by solution of energy balance equation. The electron density was obtained from the electric conductivity profiles by calculation in LTE assumption and taking into account that thermal conductivity is not affected by electrode material admixtures.

The good agreement between experimental and calculated in such way plasma electron densities in electric arc discharge between copper electrodes in air flow is found. So such approach can be recommended for low temperature thermal plasma diagnostics.

Acknowledgement

The authors would like to thank to Prof. Valentin Ya. Berent for the helpful assistance in the provision of composite electrodes. This research has been partially supported by project UKE2-31103-KV-12 of Civilian Research and Development Foundation (CRDF).

References

[1] Berent V.Ya., Gnezdilov S.A. 2009 Improvement of performance of current collectors on the carbon base *Friction & Lubrication in machines and mechanisms* **2** 18-23

[2] Miedzinski B., Dzierzanowski W., Wandzio J., Grechanyuk N., Shoffa V.N. 2012 Dynamics of sliding contacts in mine slow-speed railway transportation *Electrical contacts and electrodes. – Kyiv: "Frantsevich Institute for Problems of Materials Science"* 63-69

[3] Berent V.Ya., Gnezdilov S.A. 2008 Enhancement of current collectors on the carbon base of electric trains *Friction and lubrication of machinery* **8** 9–15

[4] Babich I.L., Boretskij V.F., Veklich A.N., Semenyshyn R.V. 2014 Spectroscopic data and Stark broadening of Cu I and Ag I spectral lines: Selection and analysis *Advances in Space Research* **54** 1254-1263

[5] Semenyshyn R.V., Veklich A.N., Babich I.L., Boretskij V.F. 2014 Spectroscopy peculiarities of thermal plasma of electric arc discharge between electrodes with Zn admixtures *Advances in Space Research* **54** 1235-1241

[6] Veklich A.N., Lebid A.V., Soroka P.V., Boretskij V.F., Babich I.L. 2013 Investigations of thermal plasma with metal impurities. Part II: Peculiarities of spectroscopy by WI, MoI, CuI spectral lines *Problems of Atomic Science and Technology. Series: Plasma Physics* **19** 213-215

[7] Veklich A.N., Osidach V.Ye. 2003 Technique of spectroscopic diagnostic of electric arc plasma *Bulletin of Taras Shevchenko National University of Kyiv. Radiophysics & Electronics* **5** 12-14

[8] Babich I.L., Boretskij V.F., Veklich A.N., Ivanisik A.I., Semenyshyn R.V., Kryachko L.A., Minakova R.V. 2010 Spectroscopy of electric arc plasma between composite electrodes Ag-CuO *Electrical contacts and electrodes. – Kyiv: "Frantsevich Institute for Problems of Materials Science"* 82-115

[9] Veklich A.N., Osidach V.Ye. 2004 The determination of electron density in electric arc discharge plasma *Bulletin of Taras Shevchenko National University of Kyiv. Series: Physics & Mathematics.* **2** 428-435

[10] Boretskij V.F., Cressault Y, Teulet Ph., Veklich A.N. 2011 Plasma of electric arc discharge in carbon dioxide with copper vapours *XIX th Symposium on Physics of Switching Arc (FSO 2011) 5-9 September 2011, Brno, Czech Republic* 121 –124

[11] Lowke J.J. 1979 Simple theory of free-burning arcs *J. Phys. D: Appl. Phys.* **12** 1873–1886

[12] Boretskij V., Veklich A., Cressault Y., Gleizes A., Teulet Ph. 2012 Non-equilibrium plasma properties of electric discharge in air between copper electrodes *Problem of Atomic Science and Technology. Series: Plasma Physics* **18**, 181-183

Long arc stabilities with various arc gas flow rates

K Maruyama[1], K Takeda[1], M Sugimoto[1] and Y Noguchi[2]

[1] Faculty of Systems Science and Technology, Akita Prefectural University, 84-4 Tsuchiya-Ebinokuchi, Yurihonjo, Akita 015-0055, Japan
[2] Sanei-Kikai Co. Ltd., 146-3 Kawaguchi-Yazuma, Yurihonjo, Akita 015-0051, Japan

E-mail: M15A010@akita-pu.ac.jp

Abstract. A new arc torch for use in magnetically driven arc device was developed with a commercially available TIG welding arc torch. The torch has a water-cooling system to the torch nozzle and has a nozzle nut to supply a swirling-free plasma gas flow. Its endurance against arc thermal load is examined. Features of its generated arc are investigated.

1. Introduction

Electric arcs, which are widely used as high-intensity heat sources in various industry activities, have power output that is easily controllable by adjustment of the arc current. Among different modes of arc operation, a transferred arc is often used in many material processes such as welding, cutting, melting, and refining metals because of its high-efficiency heat transfer. Thermal and electromagnetic pinch effects act on the arc, which concentrates the heat flow into a small area. This property is desirable to obtain high temperatures in a narrow zone of the material, but it is not convenient for heating wide areas of materials.

Several trials were conducted to expand the arc heating area [1–3]. The authors obtained a wide heat source by imposing an alternating magnetic field perpendicularly to a transferred arc [4–6]. The arc is driven by an electromagnetic force induced from the interaction between the magnetic field and arc current, which is flowing into the arc column. Therefore, the alternating magnetic field enables the arc to oscillate. A high-speed oscillating arc can be regarded as an expanded heat source with width of the oscillation amplitude. In addition to conventional arc characteristics, a magnetically driven arc presents the following benefits:

(1) The heat-source width, the width is namely the width of the oscillation amplitude of the arc, can be varied easily by adjusting the imposed magnetic field.
(2) The heat flux density distribution can be controlled by varying the waveform of the imposed magnetic field.

For example, if the waveform of the magnetic field is rectangular, then the heat flow is concentrated at both ends of the oscillation.

In a recent investigation, a magnetically driven arc technology was applied to steel plate heat treatment [7]. The authors have developed a novel heat treatment system for steel product using a magnetically driven arc. This equipment is able to raise the steel plate heat treated area hardness up to 2–3 times that of a non-heat-treated surface.

With this equipment, however, swirling gas flow was used to stabilize the arc plasma, which makes the arc oscillation surface inclined because the movement of a magnetically driven arc is twisted by Magnus force. Therefore, the heat treatment surface edge is also inclined, which causes an error in the heat treatment and which is inconvenient for the application [8]. In addition to this swirling gas problem, the optimum gas flow rate for the using torch nozzle diameter to treat the surface has never been reported in the literature. This misunderstanding often engenders a cratered surface on heat- treated steel plate.

In the present study, a novel plasma torch is developed for heat treatment, using a low-cost (commercial) arc torch with a swirling-free nozzle. Furthermore, characteristics of the arc generated with this torch are investigated. The optimum arc gas flow rates for the new nozzle for application for heat treatment are determined.

2. Development of a swirling-free gas flow torch

A swirling-free arc torch for a magnetically driven arc device was developed using a commercially available torch for TIG welding. Figure 1 shows the torch (WP-18SP; by La Mer Co. Ltd.).

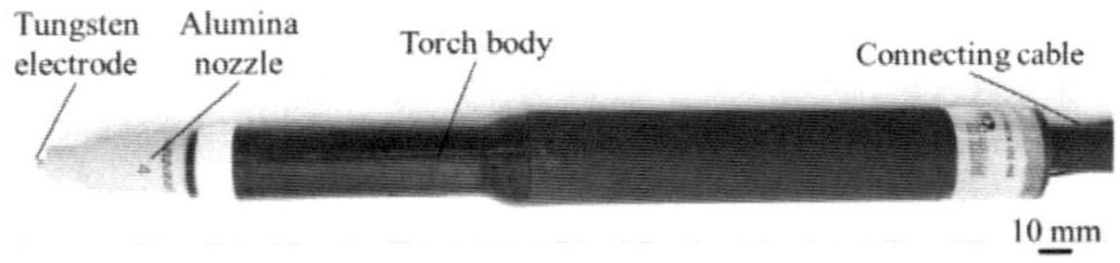

Figure 1. Photograph of a commercially available TIG welding torch used in this study.

This torch has an alumina nozzle that gives no-swirling flow to plasma gas. However, when we used it for a preliminary study of a magnetically driven arc, the alumina nozzle was melted because a strong thermal energy flows from the arc plasma, as shown in figure 2.

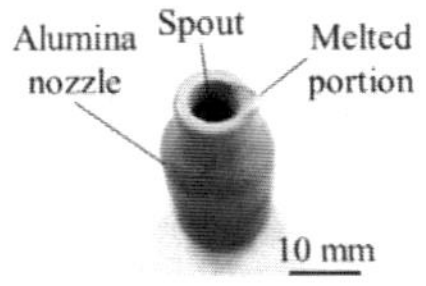

Figure 2. Photograph of the melted alumina nozzle.

To avoid melting down of the nozzle, a cooling system made of copper was equipped on the torch as shown in figure 3(a).

(a) Overall view. (b) Nozzle nut.

Figure 3. Photograph of a torch nozzle cooling system: (a) is the overall picture of nozzle cooling system, and (b) is the exchangeable nozzle nut.

Figure 3(b) shows a copper nozzle nut included in the new cooling system. The nozzle nut can be exchanged. Therefore the arc diameter is variable. In this development, four nozzles are prepared with diameters of 3.2 mm, 4.0 mm, 5.0 mm, and 6.0 mm.

3. Experimental method

Figure 4 presents an experimental arrangement for long straight-through arc generation.

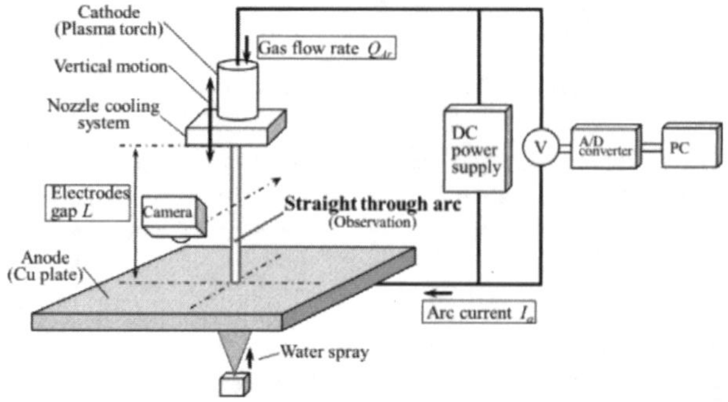

Figure 4. Schematic apparatus of experimental equipment.

A DC power supply, which outputs constant current, is connected to plasma torch and anode. The plasma forming gas is argon, which is fed to the plasma torch. A transferred arc, which is straight-through form, is generated between the plasma torch and the anode. The anode, a copper plate of 5 mm thickness, is cooled by water sprayed onto the lower side of the anode. The arc voltage is measured using a voltage meter connected in parallel to the power supply. Therefore, the obtained value contains some errors, although it is presumably negligible compared to the actual arc voltage. The obtained data flow through an A/D converter and are continuously recorded on the linked PC. The A/D converter outputs the analysis data length of 0.8 s because the memory can store only 2048 samples though the sampling rate is 2560 Hz. Photographs of the arc profile are taken using a camera as shown in figure 4.

Table 1 presents the experimental conditions.

Table 1. Experimental conditions.

Parameters	Symbols	Values
Arc current	I_a	60 A
Gas flow rate	Q_{Ar}	3–12 NL/min
Electrodes gap	L	50 mm
Nozzle diameter	d	5.0 mm

4. Experimental results and discussion

4.1. Arc profile transitions with various arc gas flow rates

Figure 5 presents photographs of arc profile variations as the arc gas flow rate is increased from 3 NL/min to 12 NL/min.

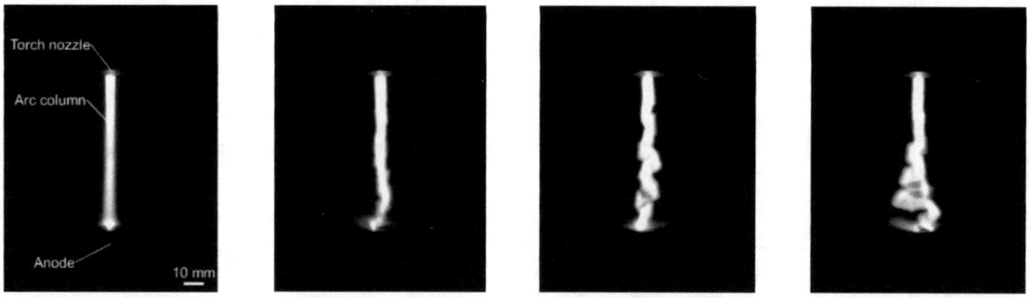

(a) 3 NL/min (b) 6 NL/min (c) 9 NL/min (d) 12 NL/min

Figure 5. Photographs of the arc profiles in various arc gas flow rates. The camera exposure time is 1/4000 s.

Journal of Physics: Conference Series **550** (2014) 012009 doi:10.1088/1742-6596/550/1/012009

As might be apparent from figures 5(a) and 5(b), the arc column is almost straight through when the arc gas flow rate is small. However, as the gas flow rate increases, it turns to the right and the left. The turn length also increases as shown in (c) and (d). We refer to the arc profiles of (a) and (b) as the "Laminar flow arc", and to those of (c) and (d) as "Turbulent flow arc". Figure 5 also shows that the total arc column length increases as the gas flow rate increases. Therefore, it can be expected that the arc voltage increases corresponding to the arc column extension. The average of the measured voltage values over 0.8 s is calculated, which is considered DC component of the arc voltage here. Figure 6 presents variations of the DC component in the arc voltage for various arc gas flow rates. The values are almost constant between 3 NL/min and 5 NL/min, which agrees with the arc profile in straight form depicted in figure 5(a). However, the arc voltage increases linearly with the gas flow rate of more than 5 NL/min, which is explainable by the reasons presented above.

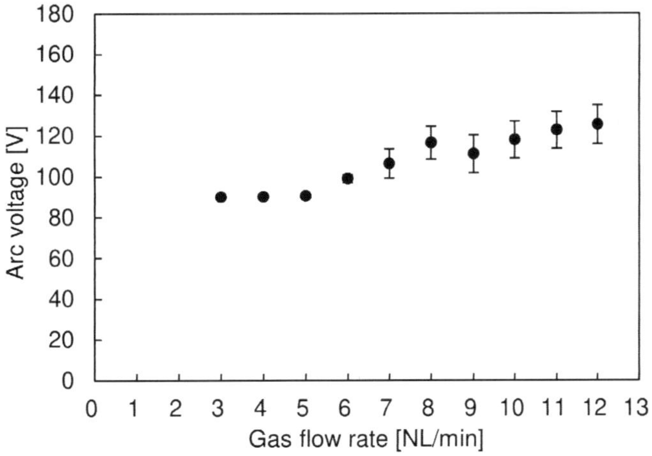

Figure 6. Variations of DC component in the arc voltage for various arc gas flow rates. The error bars indicate arc voltage fluctuations.

4.2. Measurement of fluctuation in arc voltage with various arc gas flow rates

Figure 7 shows temporal variation of arc voltage fluctuation for various arc gas flow rates corresponding to those of figure 5.

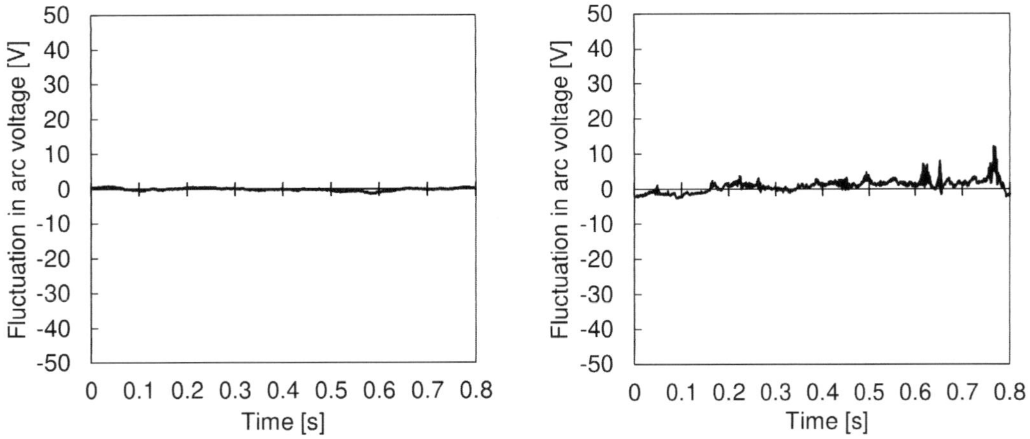

(a) 3 NL/min

(b) 6 NL/min

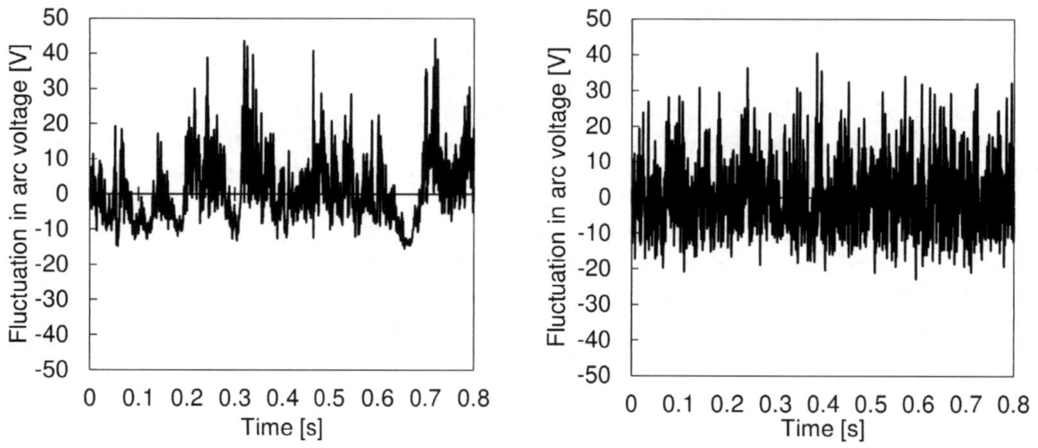

(c) 9 NL/min (d) 12 NL/min

Figure 7. Variations of the arc voltage fluctuation.

In figures 7(a) and 7(b), which are the "Laminar flow arc", the fluctuation in arc voltage is within ± 10 V. In contrast, the fluctuation in arc voltage is over ±10 V in (c) and (d), which are "Turbulent flow arc". This result implies that the fluctuation in arc voltage results from the rapid profile change of the arc column. In the case of the "Turbulent flow arc", extensions and contractions of the arc column possibly occur in an extremely short time. The degree of such changes increases with the arc gas flow rate.

Then, we calculate the root-mean-square (RMS) value of the fluctuation in arc voltage for each of the arc gas flow rates by the following equation:

$$V_{RMS} = \frac{\sum \sqrt{\left(V_{fluctuation}\right)^2}}{N}.$$

Here, $V_{fluctuation}$ is the fluctuation in arc voltage. Figure 8 presents the variations of the RMS values of arc voltage fluctuation for various arc gas flow rates.

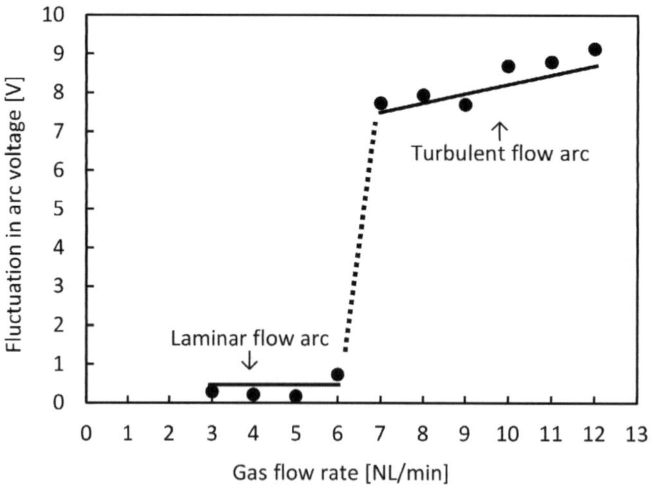

Figure 8. Variations of RMS values of arc voltage fluctuation for various gas flow rates.

In figure 8, the closed circles denote the calculated values of fluctuation in arc voltage from experimentally obtained results. The two solid lines and a broken line indicate inclination of their variation. When the gas flow rate is that for "Laminar flow arc", the RMS fluctuation value is about 0.5 V and is not so changed. However, when the gas flow rate is that for "Turbulent flow arc", the RMS fluctuation value slightly increases from 7.8 V to 9.3 V. The most interesting feature in this result is that the RMS fluctuation value increases drastically between 6 NL/min and 7 NL/min. More detailed knowledge about this steep increment is necessary. We intend to study that subject in our future work.

5. Conclusions

A swirling-free plasma torch with a water-cooling system for the nozzle part for a magnetically driven arc device was developed. Then the arc characteristics were investigated. A long stable arc was generated for various arc gas flow rates. The following are the remarkable results obtained from the gas flow rate changing experiment.

(1) The torch nozzle was confirmed as undamaged, although the arc column is not oscillated yet.
(2) The arc profile changes drastically from a laminar type to a turbulent one as the arc gas flow rate increases.
(3) The DC components of arc voltage change linearly with the arc gas flow rates.
(4) Fluctuation in arc voltage also steeply changes with the arc gas flow rates, which corresponds to the arc profile change.

These results suggest that the optimum gas flow rate for heat treatment application is less than 6 NL/min.

References

[1] Arata Y and Maruo H 1972 Magnetic control of arc plasma and its application for welding *Technology Reports of the Osaka University* **22** pp 135–45
[2] Harry E J and Goodwin D 1978 Surface heat treatment using a plasma torch with a magnetically traversed arc *Proc. of 4th Int. Conf. on Advances in Welding Processes* (Cambridge: Welding Institute) **181**
[3] Takeda K 1990 Oscillating plasma arc by external AC magnetic field *Proc. Int. Workshop on Plasma Jets in the Development of New Materials Technology (Frunze, USSR, 3–9 September 1990)* pp 485–92
[4] Toh T, Tanaka J, Muraki J, Yamamoto Y and Takeda K 2005 Magneto-hydrodynamic simulation of DC arc plasma under AC magnetic field *ISIJ Int.* **49** pp 1845–49
[5] Yamamoto T, Takeda K, Toh K and Tanaka J 2007 Distribution of heat flux transported by a magnetically driven arc *Thin Solid Films* **515** pp 4228–33
[6] Yamamoto T 2010 Fundamental study of the arc control for heating over large surface *Doctoral Thesis of Akita Prefectural University* (Japan)
[7] Akiho R, Sugimoto M, Takeda K, Noguchi Y and Miura T 2013 The development of a heat treatment system using a magnetically driven arc *Trans. JSME (in Japanese) Series* C **79** 806 pp 3979–92
[8] Akiho R, Takeda K, Sugimoto M and Noguchi Y 2014 Magnus effect on arc driven by alternating magnetic field in swirling plasma gas *ISIJ Int.* **54** pp 774–80

Journal of Physics: Conference Series **550** (2014) 012010 doi:10.1088/1742-6596/550/1/012010

Interaction of a free burning arc with regenerative protective layers

D Uhrlandt, S Gorchakov, V Brueser, St Franke, A Khakpour, M Lisnyak, R Methling and Th Schoenemann

Leibniz Institute for Plasma Science and Technology (INP Greifswald), Germany

E-mail: uhrlandt@inp-greifswald.de

Abstract. The possible use of protective layers made of ceramic powders for walls in thermal plasma applications is studied. A stable free burning arc of currents up to 5 kA between copper-tungsten electrodes is used to analyse the arc interaction with samples coated by mixtures of $CaCO_3$, $MgCO_3$, and $Mg(OH)_2$ with plaster. By means of optical emission spectroscopy the maximum arc temperature and the radiation impact on the surfaces are estimated to be around 15000 K and 20 MWm^{-2}, respectively. Thermographic measurements confirm the efficient protection of substrates by all layer materials. Layers containing $CaCO_3$ lead to the lowest heating of ceramic samples which may be caused by a strong evaporation of the layer material.

1. Introduction

The interaction of the arc with electrodes or other walls is the key issue of operation and lifetime of the devices in many applications of thermal plasmas. The arc can be close to chamber walls where the impact by heat and radiation leads to erosion. On the one hand, this erosion can finally damage the chamber and limit the lifetime. On the other hand, the processes of wall erosion can be also beneficial and are used as a main mechanism for pressure build-up and arc extinction in self-blast circuit breakers, for example. In any case, the erosion of electrodes and other walls leads to a modification of the gas properties and therefore has often strong influence on arc ignition and the properties of the established arc. There is a large number of studies of the contact erosion and wall ablation in switching devices due to a dominant role of these processes for the functionality. Most studies are focussed on the impact on dielectric recovery (e.g. in high-current circuit breakers with ablation dominated arcs - see e.g. [1, 2]) or on arc motion and extinction (e.g. in low-voltage circuit breakers with magnetically blown arcs - see e.g. [3, 4, 5]). Progress has been made in particular in the understanding and modelling of the arc interaction and ablation of polytetrafluoroethylene (PTFE) nozzle walls [6]. Here, simple relations of arc energy and arc radiation to the ablation rate are known. Gas-producing polymeric materials as replacements for PTFE are also intensively studied [1, 4, 5].

The unintentional and often crucial erosion of isolating walls has been studied mainly for low-voltage circuit breakers. Here, arc interaction with polymeric materials as well as ceramics and their impact on breaking capacity have been investigated. Less attention has been paid up to know to possibilities of a wall protection with coatings together with a fundamental study of their erosion processes.

This paper describes first steps for an experimental study of such protective layers and their capability to reduce heat impact on walls and erosion. Ideal coatings should have a high robustness and adhesion at the surface and should be cheap and easy to coat. In this study ceramic powder layers of 2 mm thickness are considered which have the additional advantage of cooling during heat impact and self-regeneration after the arc impact. The three substances $CaCO_3$, $MgCO_3$, and $Mg(OH)_2$ are used which build up oxides of CO_2 or H_2O due to thermal separation. Under appropriate conditions, for example in a closed chamber, the opposite reactions of the oxides back to carbonates or hydroxides are possible. Hence, over longer times after arc impact the carbonates or hydroxides can be deposited back on the surfaces, and a partly regeneration of the layers can be reached. $Mg(OH)_2$ has the additional advantage to be a fire suppressant medium because of the dissociation to MgO and H_2O. A comparison of the behaviour of the three substances is also motivated by the differences of their melting points and thermal conductivity.

First aim of the work is the evaluation of an appropriate setup to study the interaction of the coatings with high-current arcs under reproducible conditions. In contrast to other works, the focus is on the arc impact, the vaporisation and the final thermal impact on the coated wall and not on the changes of the gas medium and its resulting dielectric strength. A free burning arc experiment is considered which allows for optical investigations. High-speed images, spectroscopy of the arc and surface temperature measurements should reveal the different behaviour of the three coatings and indicate the potential protective functionality.

The electrical setup, the substrate properties and the observation methods are given in section 2. The analysis of the arc by optical emission spectroscopy is presented in section 3. Results of thermographic studies of the behaviour of substrates with different coatings after arc interaction are given and discussed in section 4. Section 5 gives final conclusions and an outlook on the next steps.

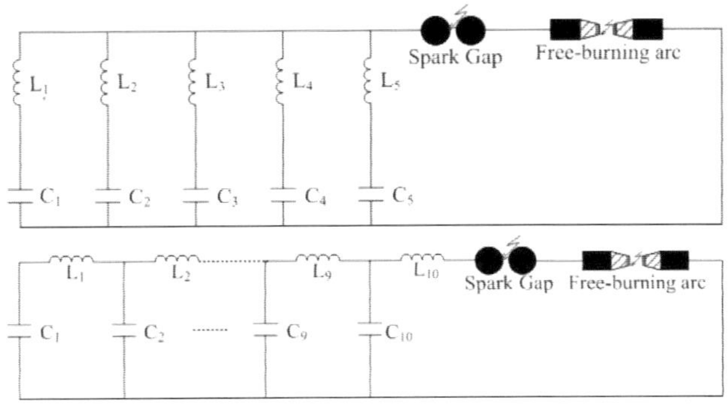

Figure 1. Electrical circuit arrangements for a sinusoidal current at 50 Hz (upper part) and for rectangular current pulses (lower part).

2. Experimental setup and materials

Electrical setup Two options of arc operation have been used for the test of arc interaction with coated substrates. Beside a sinusoidal current at 50 Hz a specific pulsed current generator is used to provide rectangular pulses with peak currents up to 5 kA over 5 ms. The electrical circuits for both arrangements are given in Figure 1. For the sinusoidal current LC elements with C_1 to C_5=525 μF and L_1 to L_5=4700 μH are coupled in parallel. LC elements with constant capacity C_1 to $C_1$0= 262.5 μF and stepwise decreasing inductances from L_1=610 μH, L_2=380 μH, down to L_9=260 μH and L_{10}=180 μH are coupled in series to produce the rectangular pulses. A spark

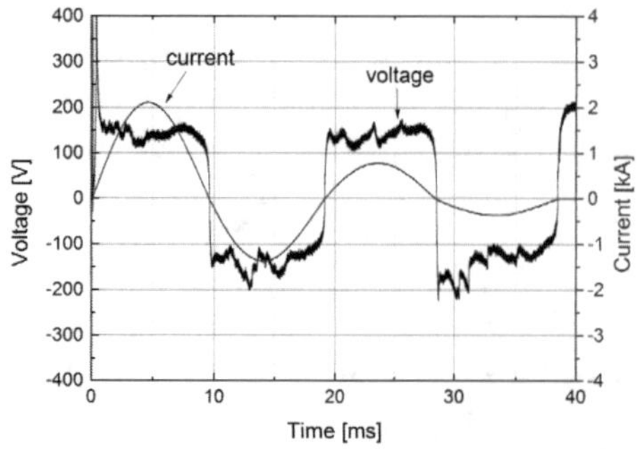

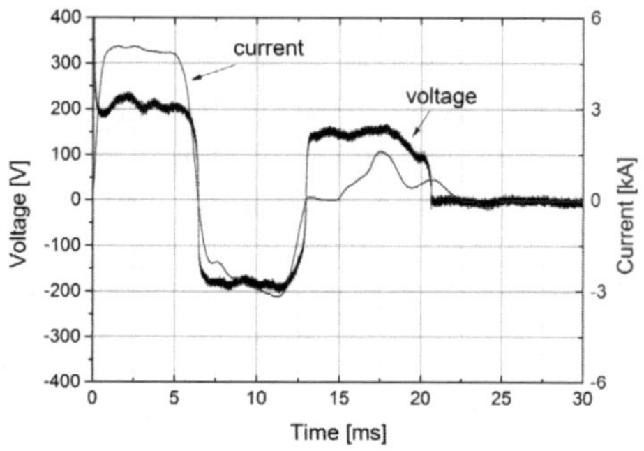

Figure 2. Typical current and voltage courses of the AC operation at 50 Hz (left side) and for the rectangular pulses of 5 ms duration (right side).

gap is used as the switch. The current is measured near to the arc gap by a current probe (Pearson monitor). Rectangular pulses are considered to ensure constant arc impact over time as far as possible. Typical waveforms are given in Figure 2. The current amplitude is damped, so that four half waves or pulses of decreasing peak current are measured.

The arc is operated between copper-tungsten rod electrodes of 10 mm diameter in a distance of about 50 mm in ambient air. Ceramic nozzles are used to insulate the sides of the rod electrodes and to ensure a flat and limited arc attachment area. The electrode erosion at current densities larger than 10^8 Am^{-2} promotes an almost stable arc operation at least near the electrodes by the establishment of copper vapour jets (see [7] for comparison). However, the interaction of these jets from cathode and anode can cause instabilities. The electrode setup is shown in Figure 3. The arc is ignited by a thin copper wire between the electrodes. High-speed videos confirm a vaporisation of the wire in a time period of up to 1 ms. Therefore, the pulse is considered to be nearly unaffected by the wire ignition.

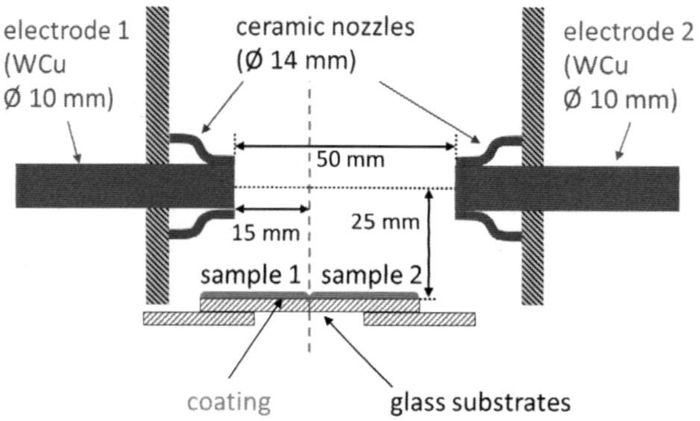

Figure 3. Setup of the arc experiment (top view).

Layer materials and samples Glass substrates of 3 mm thickness and ceramic substrates of 1 mm thickness are coated with powder layers of approximately 2 mm thickness. Plaster ($Ca[SO_4]{\cdot}2H_2O$) has been used as a simple binding agent. Hence, the layers consist of mixtures of the three substances $CaCO_3$, $MgCO_3$, and $Mg(OH)_2$ with plaster (approximately 30%). The

Figure 4. Glass substrate with a $CaCO_3$ layer (side and front view).

coating with plaster is rather fragile and not very robust. Experiments have been done also with organic binders. It turned out, that all coatings show strong evaporation during arc interaction which may be dominated by the binder material. For a better comparability of the behaviour of the powder material with respect to evaporation and cooling of the substrates, the binding with plaster has been favoured. The arc interaction with a pure plaster layer has been considered for comparison. Figure 4 shows an example of a glass substrate with a $CaCO_3$ layer.

The substrates are positioned sideward to the arc in distances of 25 mm from the arc axis (electrode 1) as illustrated in Figure 3. Two samples are tested in one pulse in parallel. Because of the axial variation of the arc properties the contact point of the two samples is considered as a reference position in axial direction. This position is marked by a dashed vertical line in both Figures 3 and 5. Only small variations of the arc radiation in the vicinity of this line are expected. Hence, areas on both samples near to the line can be considered as comparable impact zones. Nevertheless, experiments are always repeated with changed positions of the samples.

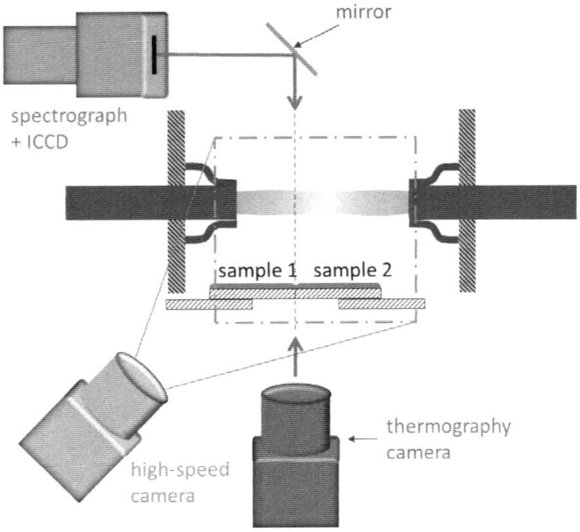

Figure 5. Setup of the optical diagnostics.

Optical setup Arc and surface erosion are observed by high speed cameras with typically 10000 fps and an acquisition time of 2 μs. (see the observation frame in Figure 5 - dash-dotted line). One camera (Motion pro Y4, black&white) has been positioned to have a front view on the surfaces (line of sight perpendicular to arc axis and sample surface). A second camera (Motion pro Y6, color) has been used to record from the side (perpendicular to arc axis and parallel to sample surface, top view).

Optical emission spectroscopy of copper lines is used to estimate the arc temperature [7] and the radiation impact. Therefore, the arc cross section at the axial reference position is imaged to the slit of a 0.5 m spectrograph coupled with an ICCD camera. The camera is triggered at 5 ms after arc ignition with 50 μs acquisition time.

The upper surface of the coated layers as well as the backside of both samples are observed by a thermography camera with a spectral sensitivity range from about 7.5 to 12 μm and 50 fps (acquisition time of about 20 ms, 640x480 pixel array, spatial resolution 3 pixel/mm).

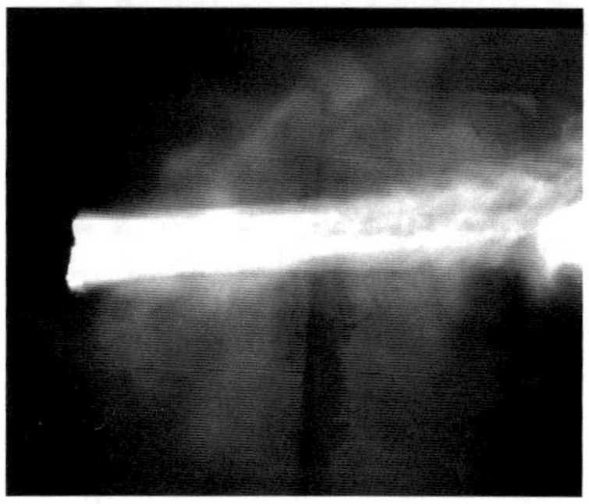

Figure 6. High-speed image of the arc above two samples coated with $CaCO_3$, front view.

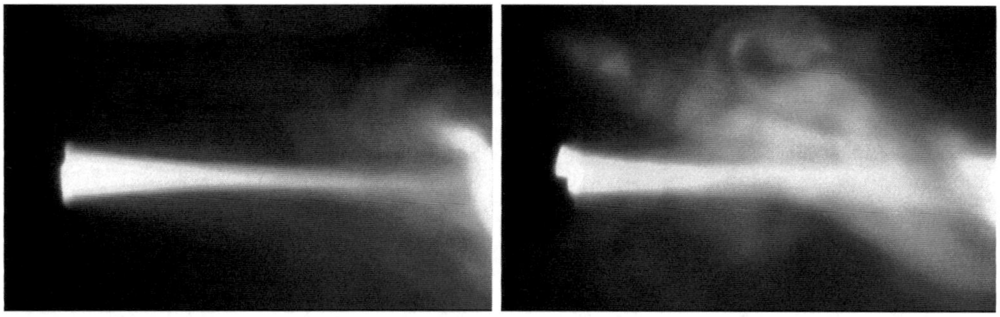

Figure 7. High-speed images (top view) of the arc recorded in parallel to the sample surfaces with a layer of $Mg(OH)_2$ (left side) and of $CaCO_3$ (right side).

3. Optical and spectroscopic analysis of the arc

The general behaviour of the arc and the arc interaction with the coated substrates has been studied by the high-speed cameras. Examples are given in the Figs. 6 and 7. Both examples are recorded at the end of the first pulse during rectangular current pulse. The cathode can be seen in each picture on the left side. The establishment of a copper vapour jet becomes clearly visible in the figures as a result of considerable copper radiation. The jet is almost stable along a distance of about 2 cm from the cathode and above sample 1. The jet becomes more irregular when interacting with a corresponding jet from the anode side over the second half of the arc length. Notice, that the jet on the cathode side is typically longer and more stable than that from the anode side in the considered arrangement. However, because of the current course with damped amplitude, the first pulse period with the cathode jet in front of the samples causes most of the heat load on the sample surfaces.

The different interaction with layers of $Mg(OH)_2$ (left side) and of $CaCO_3$ (right side) is illustrated in Fig. 7. Whereas on the left side picture radiation from the copper vapour jet can be seen only, additional radiation occurs from vapour outside of the jet in the right picture. The observation from the movies confirms a strong evaporation of the $CaCO_3$ layer which results in the additional radiation. Spectroscopic analysis confirms that this additional radiation is not scattered light from the copper jet and includes line radiation of Ca.

A spectral range between 505 and 518 nm is considered in the emission spectroscopy of the arc in the copper vapour jet. As it has been shown in [7] this range provides a number of strong copper atomic line (Cu I) and copper ionic lines (Cu II). Currents of several kA generate typically hot arcs which are dominated by vapour from electrode erosion. Under this conditions strong line broadening can considerably hinder the application of conventional methods of plasma temperature determination from the intensity of optical thin lines (e.g. single line analysis or Boltzmann plot analysis). As an alternative method, rough estimations of the arc temperature can be obtained by comparison of the measured spectra with simple spectrum simulations. The spectral radiance resulting from Cu I and Cu II lines in the considered spectral range under the assumption of a homogeneous arc in copper vapour has been simulated for varying temperatures and pressures in a preceding work (see details in [7]).

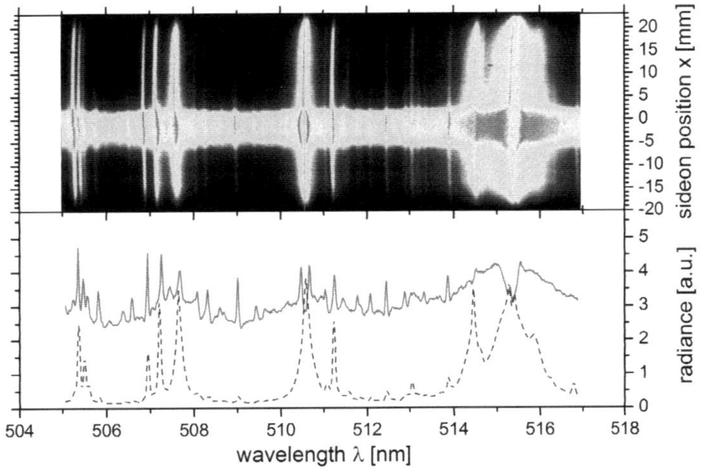

Figure 8. Example of the two-dimensional spectral radiance as recorded by the ICCD camera (upper part) in the middle of the first rectangular current pulse for the case of samples with $CaCO_3$ and pure plaster layers. Mean values of the spectral radiance over strips in the arc core (red solid line) and in the outer arc range (blue dashed line in the lower part).

In the present study the spectrum of the arc has been recorded at an axial position 15 mm in front of the cathode in almost all experiments. The obtained spectra are very similar and reflect the establishment of the copper vapour jet as already illustrated in the high-speed images. Exceptions occur if the region of the stagnation point between the electrode jets overlaps in the spectrum observation area. An example of the 2D image of the ICCD is shown in Fig. 8 (upper part). The signal has been averaged over a stripe covering few mm in the range of the arc core (copper vapour jet) as well as in a selected range outside the core. The resulting spectral radiances (in arbitrary units) are shown in the lower part of Fig. 8. The spectrum is dominated

Journal of Physics: Conference Series **550** (2014) 012010 doi:10.1088/1742-6596/550/1/012010

by copper atomic lines as expected, in particular line groups near 515.5 nm, 510.5 nm and 507.5 nm. Copper ionic lines near to 506.5 nm, 509 nm and 512.5 nm with lower intensity can be identified. The atomic lines are strongly broadened. In the arc core the lines near 510.5 and 515.5 nm show self-absorption.

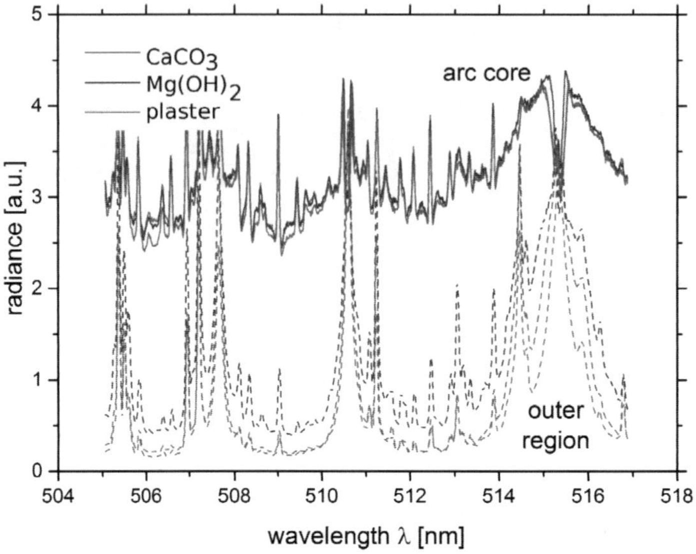

Figure 9. Examples of mean values of the spectral radiance over strips in the arc core (solid lines) and in the outer arc range (dashed lines) for three different cases with layers of $CaCO_3$, $Mg(OH)_2$ and pure plaster.

Mean values of the spectral radiance of the arc at peak currents in three different cases (different coatings) are compared in Fig. 9. As illustrated in the figure, the spectrum in the arc core is very similar in all experiments. There are larger differences in the spectrum outside the arc core. However, almost the same structure of lines is observed here.

The measured spectral radiances in arbitrary units have been compared with simple spectrum simulations as detailed in [7]. An example is shown in Fig. 10. The radiance measured outside the arc core shows best correspondence with simulations for a homogeneous arc of 14000 K at a copper pressure of 2 bar. Here, the intensity relation between atomic and ionic lines as well as the line width of the atomic lines have been considered in particular. The radiance measured in the arc core cannot be compared successfully with the simulations because of two reasons. Continuum radiation was not included in the simple model and may play a strong role in the arc core. Strong absorption up to self-absorption of lines cannot be simulated with the model of a homogeneous arc. However, the spectrum in the arc core seems to be dominated by these effects. Nevertheless, the intensity relation of atomic and ionic lines and its strong temperature dependence allows for conclusions about the temperature range of the arc. Because of the very low intensity of the considered CuII lines a maximum temperature between 14000 and 16000 K is estimated. The large broadening of lines indicates a large overpressure in the jet compared with atmospheric pressure.

Within the limited accuracy of the presented analysis there are no distinct differences of the arc in the first rectangular pulse in pulsed operation and at peak current in the sinusoidal operation.

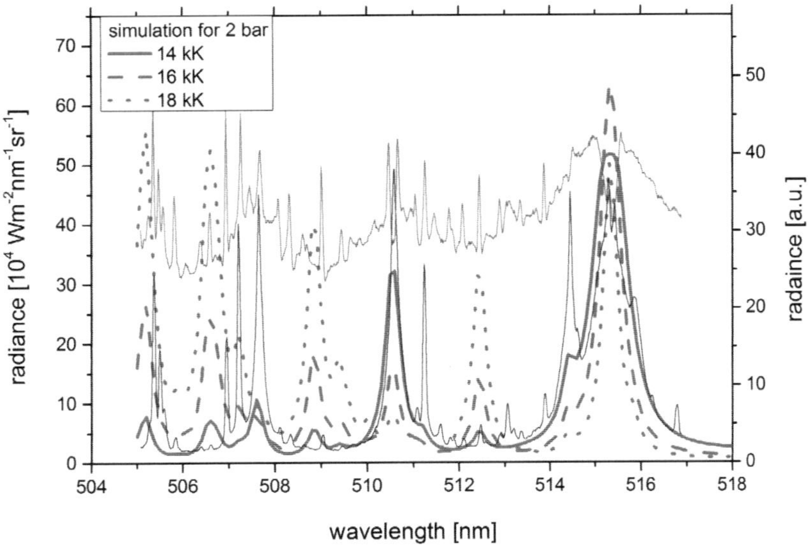

Figure 10. Spectral radiance from Fig, 8 (lower part) in comparison with simulated radiances from a homogeneous copper arc at 2 bar pressure and temperatures 14 to 18000 kK.

The estimation of the arc temperature and arc radius allows for estimation of the total arc radiation and of the expected heat load at the layer surfaces from the arc radiation absorption. The first rectangular current pulse over the first 5 ms is considered again as a example. The arc radius can be estimated from the typical radius of the copper jet radiation which is quite similar to the electrode radius (5 mm). Measurements of 200 V arc voltage and 4.7 kA current result in a total electric power of about 870 kW in the arc column if a cathode fall voltage of about 15 V is excluded. The net emission coefficient for pure copper for a homogeneous arc of 5 mm in radius at 15000 K is 1.2×10^{10} $\text{Wm}^{-3}\text{sr}^{-1}$ [8]. Assuming an arc of 50 mm length the total emitted radiation is about 300 kW and approximately 30% of the electric power. The part of absorbed electrical power on the surfaces of surrounding walls have been investigated for polytetrafluoroethylene nozzles in ablation dominated arcs at currents of the same order of magnitude then considered here. About 20% are reported in [6]. Considering this fraction of electric power and the distance of 25 mm from the arc axis to the sample surface in the experiments, an energy density of about 20 MWm^{-2} can be roughly estimated for the thermal load of the layer surfaces by plasma radiation.

4. Study of surface temperatures by thermography

The records of the thermography camera are illustrated by the examples in Fig. 11. The backside of coated glass substrates has been investigated in first experiments. However, because of the low heat conduction and the thickness of 3 mm of the substrates, no significant heating of the backsides could be observed.

Next measurements have been focussed on the upper surface of the layers as illustrated in the left side in Fig. 11. Records during the arc-surface interaction period could not be used because of the disturbances of remaining hot vapours. In records at about 20 to 40 ms after arcing pulse the interaction regions on the surfaces localized near to stable arc parts and with a more regular temperature development have been identified. Such regions are indicated by the red circles in Fig. 11. Mean values of the temperature in the indicated area have been recorded for each sample in steps of 20 ms.

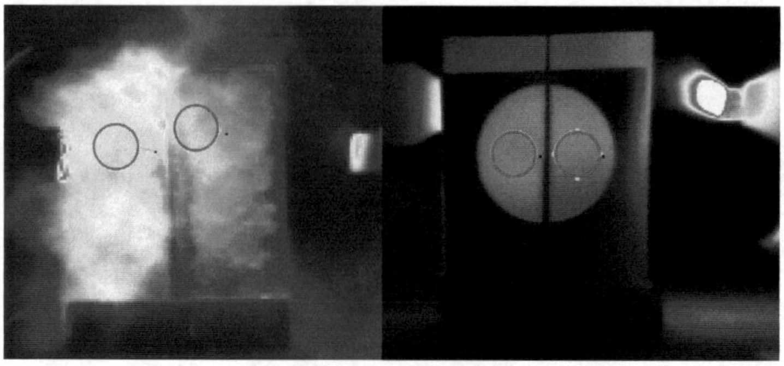

Figure 11. Examples of thermography camera images of the surfaces of the layers about 40 ms after arc interaction (left side) and of the backside of ceramic substrates with different layers about 4 s after arc interaction (right side). The red circles indicate the area for temperature averaging in both cases.

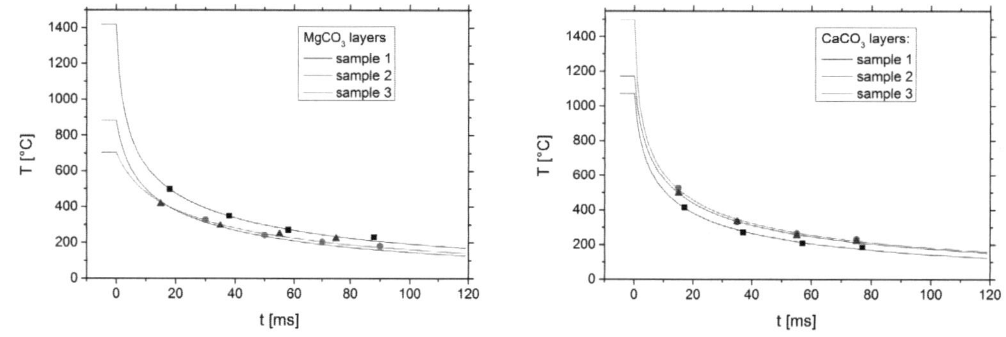

Figure 12. Temporal courses of the surface temperature of the layers after arc interaction for two different materials and three experiments each: symbols are measurements and lines are fit curves according to the Stefan-Boltzmann law.

The cooling of the layer surface after arc interaction should be dominated by radiation at least in the first second, and thermal conduction can be neglected. The time dependence of the surface temperature can be described by the Stefan-Boltzmann law:

$$\frac{\mathrm{d}T}{\mathrm{d}t} = A\left(T^4(t) - T_0^4\right) \tag{1}$$

where T_0 is the room temperature and the proportionality factor A includes Stefan-Boltzmann constant, surface emissivity ε as well as quantities defined by specific heat of the layer which are not known for the layer materials. The emissivities of the investigated materials have been estimated utilizing a constant temperature heat plate comparing the surface temperatures with that of a reference probe at different temperatures. The reference probe was covered with a coating of known emissivity. It revealed that the emissivity of all the investigated materials is close to one. This is reasonable for the IR spectral region of the thermography camera due to the porous and rough structure of the sample surfaces. No temperature dependence of the emissivity has been observed.

To get indications about the heating of the surfaces by the arc, the temperature measurements have been used to find a best fit of the data with a temperature function according to eq. (1) and starting at an initial temperature T_i. Using A and T_i as fit parameters, eq. (1) has been solved numerically. Because of the low acquisition frequency of the thermography camera the delay between end of arc interaction and acquisition time has been chosen as an additional fit parameter. Examples of measurements and fitted temperature courses are shown in Fig. 12. Unfortunately, the measured values show a considerable scattering with a large overlapping between the different materials. However, significant differences in the extrapolated initial temperatures T_i can be obtained. From these results it can be roughly estimated that higher initial temperatures (above 1300 K) occur in the case of $CaCO_3$ layers and lower temperatures (around 1300 K) occur at the surfaces of $MgCO_3$ and $Mg(OH)_2$ layers.

In final experiments the glass substrates have been exchanged by thin ceramic plates of 1 mm thickness. In these experiments, a significantly different heating of the backside of the substrates could be found. The low thermal conductivity of the ceramics causes a slow temperature increase at the backside surface only after arc impact. The maximum temperature is reached typically after 4 to 7 seconds. Again, samples with different layer materials and at different positions with respect to the electrodes (but in the same distance from the arc axis) have been studied. The backside surfaces of samples coated with $Mg(OH)_2$ have been heated up in average by about 20 K (scattering up to ± 5 K), that with $MgCO_3$ by about 15 K whereas about 10 K difference has been obtained for $CaCO_3$ and plaster. In contrast, the backside of an uncoated ceramic sample shows a heating by 70 to 80 K. The much lower heating of the coated samples may result from the low thermal conduction in the layer and in addition from the cooling effect due to evaporation and dissociation.

5. Summary and outlook

The study confirms that an open setup with copper electrodes can be used to generate a free burning arc with currents in the kA range under almost stable and reproducible conditions and to study arc interaction with protective layers. Copper line radiation can be used for a rough estimation of arc temperature and the corresponding heat load on the layer surfaces. Layers made by ceramic powders can be used for an effective protection against heating and erosion of walls. However, coating techniques and the use of binders should be improved.

Next studies will focus on a more accurate analysis of the material evaporation and the heat balance of coated substrates. Furthermore, the more detailed investigation of high-current arc radiation and self-absorption effects in copper vapour is aimed.

References

[1] Lee A, Heberlein J V R and Meyer T N 1985 High-current arc gap with ablative wall: dielectric recovery and wall-contact interaction *IEEE Trans. Components, Hybrids, and Manufacturing Technol.* **8** 129-134

[2] Seeger M, Tepper J, Christen T and Abrahamson J 2006 Experimental study on PTFE ablation in high voltage circuit-breakers *J. Phys. D: Appl. Phys.* **39** 5016-5024

[3] Rieder W, Veit Ch and Gauster E 1992 Interaction of magnetically blown break arcs with insulating walls in the contact region of interrupters *IEEE Trans. Components, Hybrids, and Manufacturing Technol.* **15** 1123-1137

[4] Doméjean E, Chévrier P, Fiévet C and Petit P 1997 Arcwall interaction modelling in a low-voltage circuit breaker *J. Phys. D: Appl. Phys.* **30** 2132-2142

[5] Rodriguez P, Didier J, Bernard G and Rowe S 1998 Arc-contact-insulating wall interactions in low voltage circuit-breakers *IEEE Trans. Power Delivery* **13** 480-488

[6] Seeger M, Niemeyer L, Christen T, Schwinne M and Dommerque R 2006 An integral arc model for ablation controlled arcs based on CFD simulations *J. Phys. D: Appl. Phys.* **39** 21802191

[7] Franke St, Methling R, Uhrlandt D, Bianchetti R, Gati R and Schwinne M 2013 Temperature determination in copper-dominated free-burning arcs *J. Phys. D: Appl. Phys.* **47** 015202

[8] Cressault Y and Gleizes A 2013 Thermal plasma properties for ArAl, ArFe and ArCu mixtures used in welding plasmas processes: I. Net emission coefficients at atmospheric pressure *J. Phys. D: Appl. Phys.* **46** 415206

Time Response of Arc Driven by Alternating Magnetic Field

K Takeda, H Okubo and M Sugimoto

Department of Machine Intelligence and System Engineering, Akita Prefectural University, Yuri-Honjyo, Akita, 015-0055, Japan

E-mail: ktkd@ruby.ocn.ne.jp

Abstract. The expanded arc has been developed by imposing an alternating magnetic field to the arc. The time response of the arc motion is investigated in the present work. The alternating magnetic field in the form of B=Bo cos($\omega t+\phi$) is assumed to be imposed perpendicularly to the arc. Physical/mathematical models on the oscillatory motion of the plasma gas are described. Solving a set of three 3rd order linear ordinary differential equations, the trajectories of the plasma gas from the torch to the anode are obtained for various frequencies of the alternating magnetic field. Numerical analyses reveal that the amplitude of the oscillatory arc motion remains constant for small ω but it reduces with the increase of the frequency for large ω. Theoretical predictions are compared with experimental results. It is confirmed that their agreements are good.

1. Introduction

Transferred arcs have been widely used in various industrial fields such as welding and cutting metals, because of the intense energy concentrated into a small arc-root. However, conventional arcs remain inconvenient for heating or melting over a wide area. Several attempts were made to expand arc profiles using an external magnetic field [1]. Takeda et al developed a magnetically driven arc by imposing an alternating magnetic field perpendicular to a transferred arc [2]. The high speed oscillating arc can be regarded as a broad heat source with the width of the oscillation amplitude as shown in figure 1.

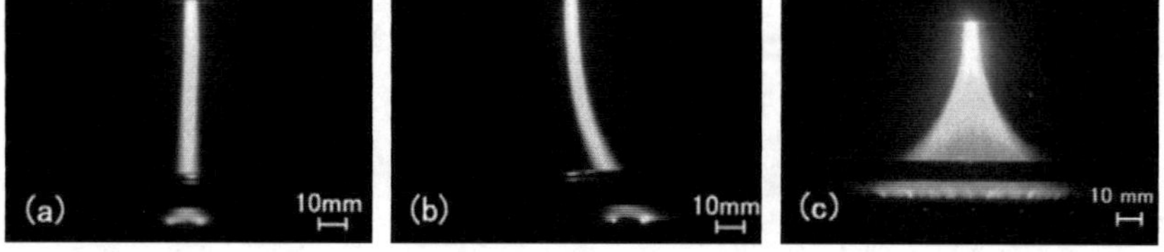

Figure 1. Typical arc profiles in various magnetic fields; (a) no magnetic field, (b) steady magnetic field and (c) alternating magnetic field.

A magnetically driven arc presents the following advantages: the oscillation amplitude can be varied easily by adjusting the magnetic flux density and the space distribution of the heat flux can be controlled by varying the wave form of the alternating field. For example, if the waveform is rectangular, then the heat flows are localized at both ends of the oscillation. Arc profiles and heat flux distributions have been investigated under the alternating magnetic field imposed perpendicularly to the arc [3-6]. Recently arc motions not only in the perpendicular magnetic field but also in the oblique field have been investigated [7]. However in these analyses, the magnetic field is restricted to a low frequency so that the plasma gas may travel in a quasi-static magnetic flux density during its flight from an arc torch to an anode. In the present work, we focus our attention to the time response of the arc motion. The variation of the arc profiles with the field frequency is investigated.

2. Theoretical consideration of the arc driven by alternating magnetic field

2.1. Description of the model

The schematic illustration of a magnetically driven arc is shown in figure 2. A transferred arc is generated between an arc torch and an anode plate. The origin of the coordinate system (x, y, z) used in the theoretical model is located at the torch exit. Plasma forming gas is introduced to the torch at a certain flow rate. And it is ionized in the torch. An alternating magnetic field is imposed parallel to x-axis. It is uniform in space but its flux density and direction vary with time at a certain frequency. The ionized gas (or partially ionized plasma) is ejected from the orifice of the torch nozzle and travels toward the anode.

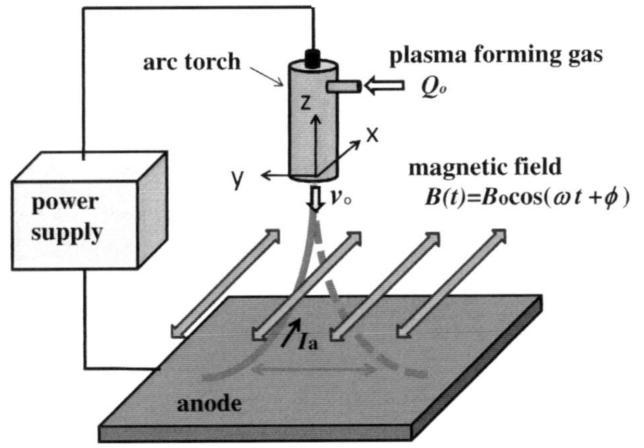

Figure 2. Schematic illustration of magnetically driven arc.

The fundamental concept of the present modeling is based on the idea that the streamline of the plasma gas flow after the torch represents the arc profile. This is because electric current cannot exist in a space without plasma gas. Therefore the pass of the plasma gas gives the pass of arc current. The interaction between the arc current and the imposed magnetic field produces the electro-magnetic force, which drives the plasma in the y-direction. The authors consider A small element of the plasma ejected from the torch is considered for calculations; it is called "fluid parcel". The fluid parcel travels from the torch to the anode under the influence of the electromagnetic force. In order to understand the trajectory of the fluid parcel, the momentum equation governing the parcel motion is analyzed by a Lagrangian method. Physical parameters and their nomenclature are listed in table 1.

The momentum equation for the fluid parcel is generally expressed as

$$\rho\frac{d\vec{v}}{dt} = \vec{F} - \nabla p + \mu\Delta\vec{v} + \frac{\mu}{3}\nabla(\nabla\cdot\vec{v}), \qquad (1)$$

where $\vec{F}$ represents the external force. To simplify the consideration, the following assumptions are used.

(1) Pressure is constant, as the plasma moves in an open atmosphere.
(2) Compressible and viscous effects are negligible.
(3) External force consists of the electromagnetic force ($\vec{F} = \vec{j}\times\vec{B}$) only.
(4) The arc current is anti-parallel to the motion of the plasma gas. The relation of

$\vec{j} = -\alpha\vec{v}$ is satisfied, where α is a positive proportional constant. This is because the arc current is along the plasma gas flow as mentioned above. However, it should be noticed that the direction of the arc current is opposite to that of plasma gas. The arc current runs from the anode to the torch, while the plasma gas moves from the torch to the anode.

The validity of these assumptions was confirmed experimentally in the previous work [5].

Table 1. List of main parameters used in the present model.

symbol	nomenclature
$\vec{B}$	magnetic flux density, [T], $\vec{B} = (B_x, B_y, B_z) = (B_x, 0, 0)$
B_x	x-component of $\vec{B}$, [T], $B_x = B_0 \cos(\omega t + \phi)$
B_0	amplitude of B_x, [T]
I_a	arc current, [A]
Q_0	mass flow rate of plasma forming gas, [kg/s]
T_{field}	characteristic changing time of the magnetic field, [s]
T_{travel}	characteristic traveling time of the parcel, [s]
(X_a, Y_a, Z_a)	position of the arc root on the anode, [m]
f	frequency of the alternating magnetic field, [Hz]
f^*	critical frequency
$\vec{j}$	arc current density, [A/m^2], $\lvert \vec{j} \rvert = j$
$\vec{v}$	velocity of the plasma gas, [m/s], $\vec{v} = (v_x, v_y, v_z) = (0, v_y, v_z)$
v_0	initial velocity of the plasma gas at t=0, [m/s]
α	proportional constant, [A s/m], $\alpha = j / \lvert \vec{v} \rvert$
ϕ	phase shift, [rad]
λ	$\lambda = I_a B_0 / Q_0$, [A T s/kg]
ρ	density of the plasma gas, [kg/m^3]
ω	angular velocity of the alternating magnetic field, [rad/s], $\omega = 2\pi / f$

Taking these assumptions into account, equation (1) is simplified as

$$\rho\frac{d\vec{v}}{dt} = -\alpha(\vec{v} \times \vec{B}). \tag{2}$$

Considering that the imposed magnetic field has no y-component or z-component as $\vec{B} = (B_x,\ 0,\ 0)$, equation (1) expressed in vector form is decomposed into the following three equations in scalar form.

$$\rho\frac{dv_x}{dt} = 0 \tag{3}$$

$$\rho\frac{dv_y}{dt} = -\alpha(v_z B_x) \tag{4}$$

$$\rho\frac{dv_z}{dt} = \alpha(v_y B_x) \tag{5}$$

Differentiating equation (5) with respect to t and then substituting equation (4), the following equation is obtained.

$$\frac{d^2 v_y}{dt^2} - \frac{1}{B_x}\frac{dB_x}{dt}\frac{dv_y}{dt} + \left(\frac{\alpha}{\rho}\right)^2 B_x^{\,2} v_y = 0 \tag{6}$$

In a similar manner, equation (7) is also obtained.

$$\frac{d^2 v_z}{dt^2} - \frac{1}{B_x}\frac{dB_x}{dt}\frac{dv}{dt} + \left(\frac{\alpha}{\rho}\right)^2 B_x^{\,2} v_z = 0 \tag{7}$$

Then, substituting the following relations of equation (8) into equations (3), (6) and (7),

$$v_x = \frac{dx}{dt},\quad v_y = \frac{dy}{dt},\quad v_z = \frac{dz}{dt}, \tag{8}$$

the governing equations for the movement of the fluid parcel are obtained;

$$\frac{d^2x}{dt^2} = 0, \tag{9}$$

$$\frac{d^3y}{dt^3} - \frac{1}{B_x}\frac{dB_x}{dt}\frac{dy^2}{dt^2} + (\frac{\alpha B_0}{\rho})^2\frac{dy}{dt} = 0, \tag{10}$$

and

$$\frac{d^3z}{dt^3} - \frac{1}{B_x}\frac{dB_x}{dt}\frac{dz^2}{dt^2} + (\frac{\alpha B_0}{\rho})^2\frac{dz}{dt} = 0. \tag{11}$$

The imposed magnetic field is assumed as

$$B_x = B_0\cos(\omega t + \phi). \tag{12}$$

Then, equations (11) and (12) are written respectively as

$$\frac{d^3y}{dt^3} - \omega\tan(\omega t + \phi)\frac{dy^2}{dt^2} + \lambda^2\cos(\omega t + \phi)\frac{dy}{dt} = 0, \tag{13}$$

and

$$\frac{d^3z}{dt^3} - \omega\tan(\omega t + \phi)\frac{dz^2}{dt^2} + \lambda^2\cos(\omega t + \phi)\frac{dz}{dt} = 0. \tag{14}$$

Where, λ is defined as

$$\lambda = \frac{\alpha B_0}{\rho}. \tag{15}$$

If the cross sectional area of the plasma flow tube (or the arc current) is represented as S, the following relations are obtained; $jS = I_a$ and $\rho|v|S = Q_0$. Substituting these relations and $\alpha = j/|\vec{v}|$ into equation (15), the new parameter λ can be expressed by familiar experimental parameters such as the arc current (I_a), the flow rate of plasma forming gas (Q_0) and the imposed magnetic flux density (B_0).

$$\lambda = \frac{I_a B_0}{Q_0}. \tag{16}$$

2.2. *Initial conditions*

In order to solve the differential equations represented by equations (9), (12) and (13), it is necessary to give the initial conditions for each equation. As the fluid parcel starts from the exit of the arc torch at *t=0*, the initial condition for the position is;

$$(x, y, z) = (0,0,0) \qquad at\ \ t = 0. \tag{17}$$

And as the plasma gas is ejected vertically from the torch exit, the initial condition for the velocity is;

$$(\frac{dx}{dt},\frac{dy}{dt},\frac{dz}{dt}) = (v_x, v_y, v_z) = (0,0,v_0) \qquad at\ \ t = 0 \tag{18}$$

Substituting the above conditions to equations (4) and (5), the third initial condition is obtained.

$$(\frac{d^2x}{dt^2},\frac{d^2y}{dt^2},\frac{d^2z}{dt^2}) = (\frac{dv_x}{dt},\frac{dv_y}{dt},\frac{dv_z}{dt}) = (0,-\lambda v_0,0), \qquad at\ \ t = 0 \tag{19}$$

2.3. *Determination of the initial velocity*

The plasma forming gas goes into the arc torch at the flow rate of Q_0 and is ionized there. After drastic expansion of its volume, it is ejected from the torch. Experimentally it is not easy to measure the gas velocity. Theoretically, it is also difficult to predict its value, since the velocity depends on the torch configuration, the magnitude of the arc current and many other experimental parameters.

Apart from the magnetically driven arc, the movement of electrons under a stationary and uniform magnetic field is considered. The momentum equation for the electron is expressed in a similar form to equation (2). Suppose that the magnetic flux density is presented as $\vec{B} = (B_x, B_y, B_z) = (B_0,0,0)$ and the initial velocity of the electron at $(x, y, z) = (0,0,0)$ is denoted as $v = (v_x, v_y v_z) = (0,0,v_0)$. It is known that the trajectory of the electron is a circle with a Larmor radius described as $R_L=mv_0/qB_0$, where m and q represent mass of electron and its charge respectively.

It is apparent that the fluid parcel in a steady and uniform magnetic field also travels along a circle with a radius R and its position (X,Y,Z) is expressed in the similar manner to the electron as;

$$Y=-R+\left(R^2-Z^2\right)^{1/2}, \tag{20}$$

where R is defined as

$$R=\frac{v_0}{\lambda}=\frac{Q_0v_0}{I_aB_0}. \tag{21}$$

If an arc is generated under a stationary magnetic field in the arrangement shown in Figure 2, the position on the anode, (X_a, Y_a, Z_a) satisfies the following relations;

$$X_a=0, \tag{22}$$

and

$$Y_a=-\frac{Q_0v_0}{I_aB_0}+\left\{\left(\frac{Q_0v_0}{I_aB_0}\right)^2-Z_a^{\ 2}\right\}^{1/2}. \tag{23}$$

Rearranging equation (23), the initial velocity is represented as

$$v_0=\frac{I_aB_0}{Q_0}\frac{Y_a^{\ 2}+Z_a^{\ 2}}{2Y_a}. \tag{24}$$

Therefore if the position of the arc root on the anode can be measured by experiment in the conditions of I_a, Q_0 and B_0, the unknown value v_0 can be evaluated by these experimental parameters.

2.4. *Critical frequency*

Then, we consider how much change of the magnetic flux density the fluid parcel experiences during the travel from the torch to the anode. As the alternating magnetic field changes with time in the form of cos(ωt), the characteristic time for the magnetic field change (T_{field}) can be estimated as

$$T_{field}\approx\frac{1}{\omega}=\frac{1}{2\pi f}. \tag{25}$$

Where, f represents the frequency of the alternating field. On the other hand, the characteristic time for the fluid parcel to travel from the torch to the anode (T_{travel}) is approximated as

$$T_{travel}\approx\left|\frac{Z_a}{v_0}\right|. \tag{26}$$

Here, we define the critical frequency f^* as

$$f^*=\frac{1}{2\pi}\left|\frac{v_0}{Z_a}\right|. \tag{27}$$

If the frequency of the alternating magnetic field is less than the critical frequency ($f < f^*$), then the characteristic time for the magnetic field change is larger than that for the fluid parcel to travel ($T_{\text{field}} > T_{\text{travel}}$). In this case, the fluid parcel travels during its flight under almost constant magnetic field. The frequency change of the magnetic field has small effect on the trajectory of the parcel. The amplitude of the oscillatory arc motion is maintained constant in the frequency range from 0 to f^*. On the contrary, if the frequency of the magnetic field is larger than f^*, the parcel experiences different electromagnetic forces from place to place. In such conditions, the trajectory of the parcel becomes rather complicated.

3. Numerical results

Numerical calculations were conducted. Commercial software (Mathematica) [8] was used to solve the set of differential equations presented by equations (9), (13) and (14) under the initial conditions of equations (17), (18) and (19). The fluid parcel starts from the torch exit at a time t=0 and travels toward the anode located at $z=Z_a$. It reaches there at a time $t=T_{\text{travel}}$. At $t=0$, the magnetic flux density is $B=B_0\cos\phi$. The location of the parcel at time t is given as the set of $x(t)$, $y(t)$ and $z(t)$. The trajectory of the parcel is described by the variation of (x,y,z) with time $(0\le t\le T_{travel})$. The initial velocity of the fluid parcel is assumed v_0 = -60 [m/s]. The anode plate locates at Z_a=6.0 x 10^{-2} [m]. The magnitude of λ is assumed 800 [s^{-1}]. Under these conditions, the critical frequency, f^* is estimated to be 160 [Hz].

The trajectories for various frequencies are shown in figure 3, where ϕ is fixed at zero. As mentioned in 2.4, for $f < f^*$ the profile of the trajectory is almost the same as that calculated for $f=0$, and for $f > f^*$ its profile varies with the change of the frequency.

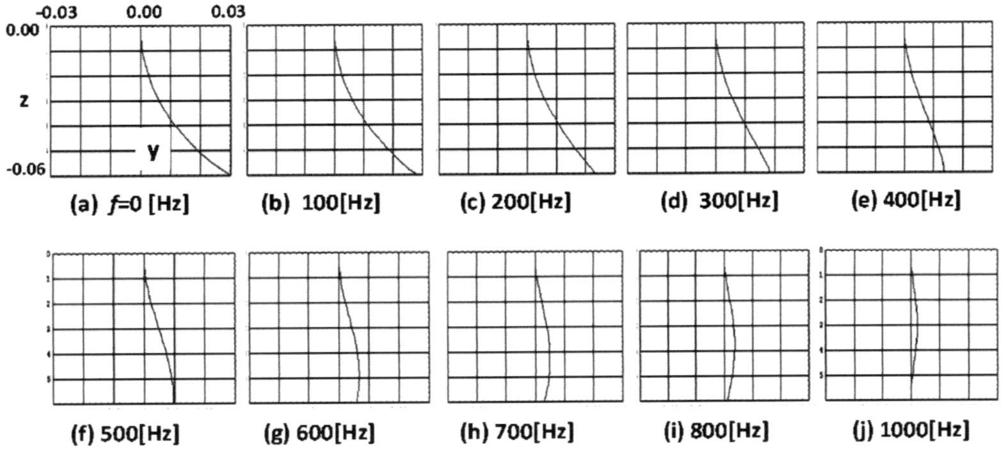

Figure 3. Variation of the trajectories with the frequency under the condition of ϕ=0. The abscissa and the ordinate in each figure represent respectively y-position and z-position in [m].

The trajectories with various ϕ at f =10Hz are illustrated in figures 4 (a)-(k). Whole profile of the oscillatory motion is depicted in figure 4-(l). As the frequency is much smaller than f^*, the fluid parcel travels from the torch exit to the anode in the quasi-stationary magnetic field. The magnitude of the stationary field strength is approximated as $B_0 \cos\phi$. Uniform and steady electro-magnetic force acts on the fluid parcel during the flight and therefore the curvature of the trajectory does not change. Because of the periodic property of the alternating field, the profiles of the trajectory are repeated at the interval of $\phi = 2\pi$.

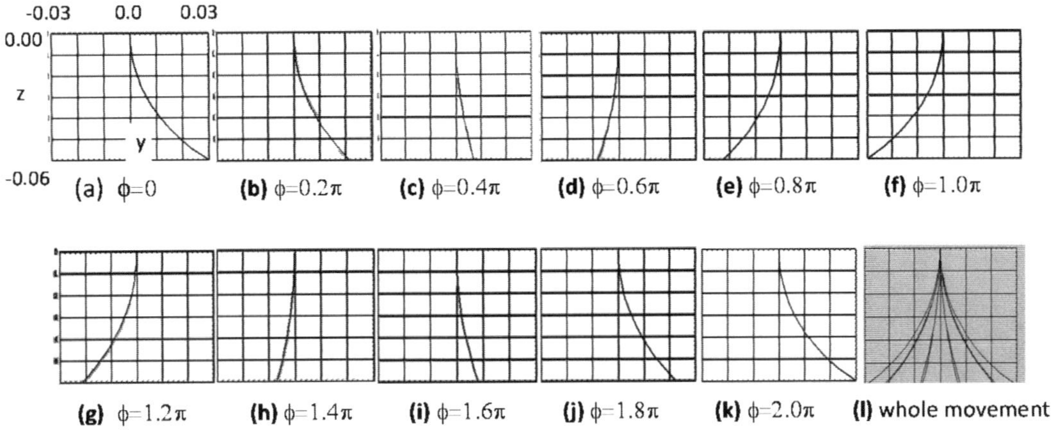

Figure 4. Variations of the trajectory with ϕ (shown in (a) – (k)) and the whole oscillating movement (shown in (l)); the frequency of the imposed magnetic field is at f=10Hz.

And as shown in figure 5, similar calculations are carried out for f=500Hz. In this case, the change of the magnetic field is so fast that the direction of the electro-magnetic force may vary not only in its magnitude but also its direction during the flight of the parcel. The curvature of the trajectory changes from place to place.

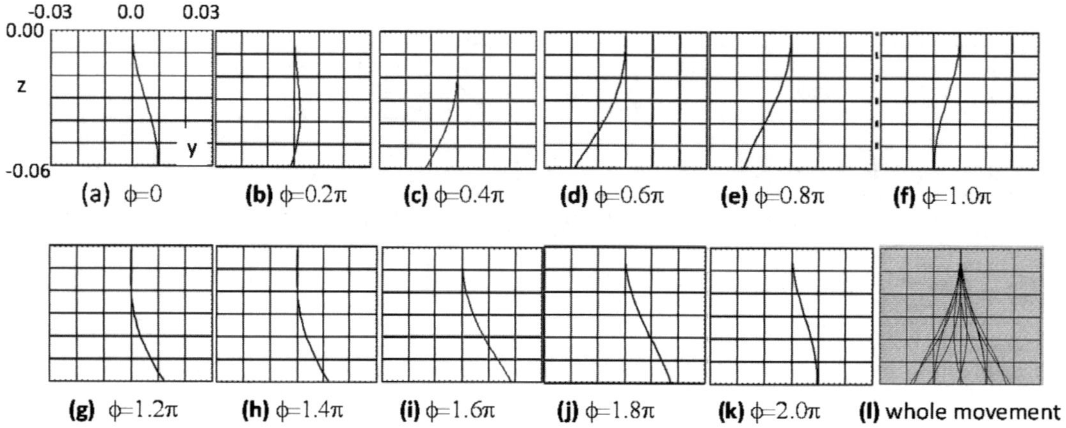

Figure 5. Variations of the trajectory with ϕ (shown in (a) – (k)) and the whole oscillating movement (shown in (l)); the frequency of the imposed magnetic field is f=500Hz.

The variation of the amplitude in the oscillatory motion with frequency is depicted in figure 6. The solid line represents the numerical result and the critical frequency, f^* is also shown in this figure.

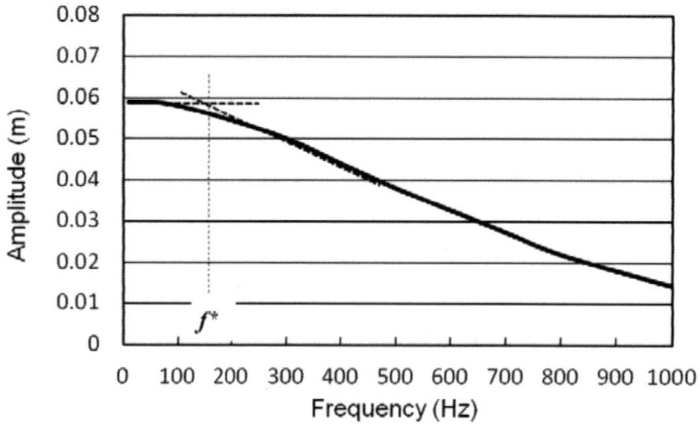

Figure 6. Variation of the amplitude in the oscillatory motion with frequency. Critical frequency is represented by f^*.

4. Experiment

4.1. Experimental procedure

Experiments were carried out to assess the theoretical results. Experimental arrangement is illustrated in figure 7. A tungsten cathode in the arc torch and an anode copper tube were connected to a dc power supply which was operated in a constant current mode. An arc current was fixed at $I_a = 100$ [A]. Argon gas was fed into the arc torch at the flow rate of $Q_0 = 1.8\times10^{-4}$ [kg/s]. Standoff distance between the torch exit and the anode was adjusted to be $Z_a = 6\times10^{-2}$ [m]. A magnetic field was generated by a water cooled rectangular coil though which electric currents were supplied with various frequencies ($f = 0-1000$ [Hz]). All the parameters used in the experiment were adjusted to be same as those in the numerical analyses.

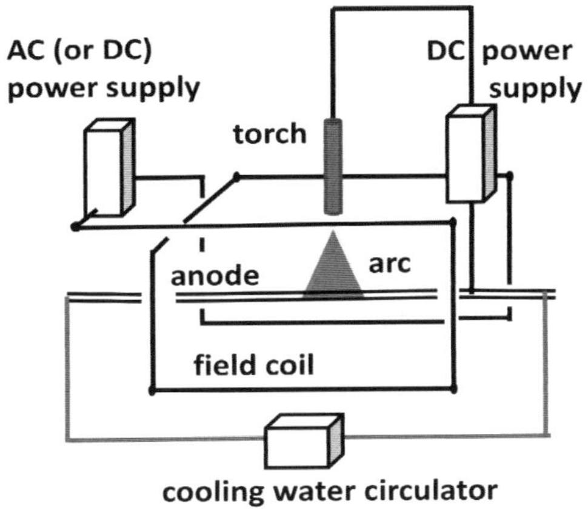

Figure 7. Schematic illustration of the experimental arrangement.

4.2 . Determination of the initial velocity of the plasma gas

In order to determine the initial velocity of the plasma gas ejected from the torch, the arc was produced under the stationary magnetic field at $B_0 = 1.8\times10^{-3}$ [T]. Other operating conditions were same as those described in the subsection *4.1*. The arc profile obtained under these conditions is shown in figure 8. From the measurement of the anode root position, $Y_a = -4.4\times10^{-2}$ [m] was obtained. Substituting Y_a and other known experimental parameters into equation (24), the initial velocity of the fluid parcel (or plasma gas) was evaluated as $v_0 = -60$ [m].

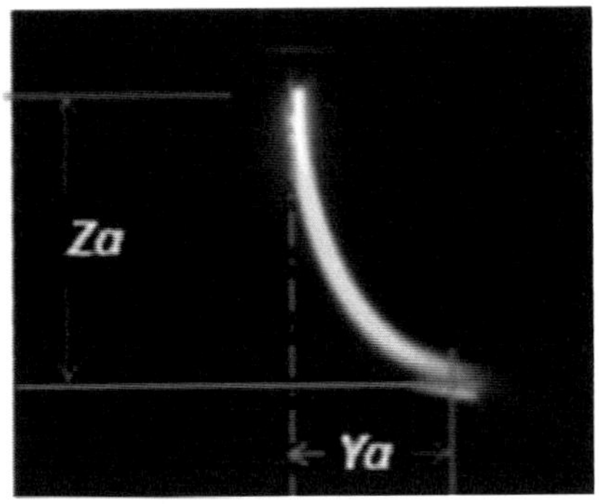

Figure 8. Deformation of arc by dc magnetic field. Experimental conditions; B_o=1.8x10^{-3} [T], Z_a=6x10^{-2}[m], Q_o=1.8x10^{-4}[kg/s] and I_a=100[A].

4.3. Experimental results

Then the arc was driven by alternating magnetic fields with various frequencies. The magnitude of the magnetic flux density was adjusted at $B_0 = 1.4\times10^{-3}$ [T]. The variations of the arc profiles with the frequency are shown in figure 9. As was predicted in the theoretical consideration, the arc oscillated in the Y-Z plane two dimensionally. The reduction of the oscillation amplitude was observed with the increase of the frequency.

The amplitudes of the oscillatory motion were measured for various frequencies of the magnetic field. In figure 10, the experimental results on the variation of the amplitude were compared with theoretical ones. Closed circles represent the experimental results and a solid curve depicts the theoretical variation. In spite of various assumptions in the modeling, experimental results are in fairly good agreement with numerical predictions. The critical frequency observed experimentally is in reasonable agreement with that obtained theoretically.

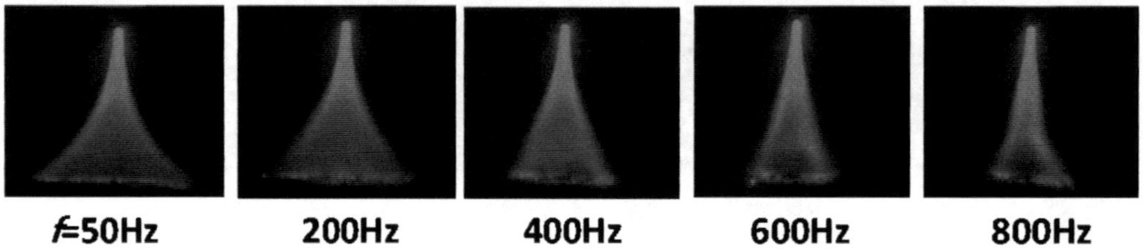

Figure 9. Variation of the arc profiles with the frequency of the alternating magnetic field.

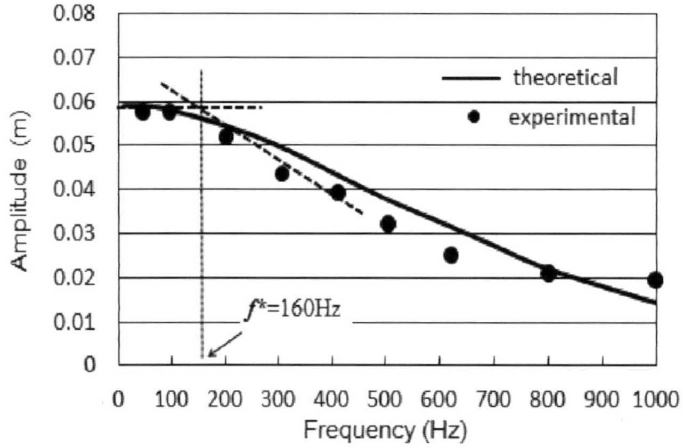

Figure 10. Comparison of the amplitude variation with frequency between theoretical results and experimental measurements.

5. Conclusion

The time response of the arc motion driven by alternating magnetic field was investigated. A theoretical model was constructed on the idea that the trajectory of the plasma gas from the arc torch to the anode represented the arc profile. Numerical analyses revealed that the amplitude of the oscillatory arc motion decreased with the increase of the alternating frequency. The critical frequency f* was defined to characterize the amplitude reduction. Theoretical consideration predicted that the variation of the amplitude was small if $f < f^*$ and then it became large if $f > f^*$. Experiment was conducted to assess the theoretical results. Good agreement was obtained between theoretical predictions and experimental observations.

References

[1] Harry J E and Goodwin D 1978, Surface heat treatment using a plasma torch with a magnetically traversed arc, *Proceedings of 4th Intl. Conf. on Advances in Welding Processes,* Welding Institute , Cambridge, 181

[2] Takeda K 1990, Oscillating plasma arc by external AC magnetic field, *Proceedings Intl. Workshop, on Plasma Jet in the Development of New Materials Technology,* Frunze,485

[3] Toh T, Tanaka J, Muraki J, Yamamoto Y and Takeda K 2005, Magneto-hydrodynamic simulation of DC arc plasma under AC magnetic field, *ISIJ International*, **45**, 947

[4] Gonzalez J J, Lago F, Freton P, Masquere M and Franceries X 2005, Numerical modelling of an electrical arc and its interaction with the anode: part 2 Three dimensional model- influence of external force on the arc column, *J. Phys. D: Appl. Phys.* **36**, 306

[5] Yamamto T 2010, Fundamental study of the arc control for heating over large surface, *Doctor Thesis, Akita Prefectural University*, Japan

[6] Yamamoto T, Matsumoto N, Sugimoto M, Sudo S, Toh T and Takeda K 2009, Theoretical consideration of heat flux distribution of arc driven by ac magnetic field, *ISIJ Inetnational*, **49** , 1845

[7] Akiho R, Takeda K and Sugimoto M 2012, Arc motion in an obliquely imposed alternating magnetic field, *Journal of Physics, Conference Series*, **406**, 012036

[8] Wolfram S 1998, The Mathematica book, 3rd ed., WolramResearch Inc., Champaign

Spectroscopic study and high speed imaging of a transient arc

P Ratovoson[1], F Valensi[1], M Razafinimanana[1] and T Tmenova[2]
[1]Université de Toulouse; UPS, INPT; LAPLACE (Laboratoire Plasma et Conversion d'Energie); 118 route de Narbonne, F-31062 Toulouse Cedex 9, France
[2]Taras Shevchenko Kiev Natl Univ, Radio Phys Fac, Kiev, Ukraine

Abstract. An experimental investigation of a transient arc discharge is reported in this work. The following parameters affecting the discharge properties and electrode erosion were investigated: current, time constant, electrode types. Plasma temperature and electron density were characterized by spatial and time resolved emission spectroscopy. Results are correlated to electrode erosion and show the influence of the resulting metal vapours on the plasma characteristics. Their repartitions were also visualised by high speed imaging using interference filters.

1. Introduction

Transient arcs conform to high current (several hundred or thousand amps) and short duration (a few milliseconds to a few hundred milliseconds).They occur during the separation of two electric contacts for instance in the case of losing contact between pantograph and catenary in railway electric traction system [1] or with breaking arcs in circuit breakers [2]. Lightning impacts on structures such as aircrafts also fall in this category [3]. These arcs are undesired but unavoidable; they can cause damages due to the high energy released to the surrounding. Since energy from the arc leads to electrodes heating, melting and vaporization, the evaporated material will contaminate the plasma and modify its characteristics [4], [5]. Better understanding of transient arc and its interaction with electrode is the key issue to improve systems performances. In order to reproduce and to study this kind of discharge a specific experimental device was conceived. It is composed of a power supply with adjustable current and time constant. The arc established between a graphite cathode (which vaporization can be neglected) and graphite or copper anode was characterized by mean of optical emission spectroscopy. The distribution in the plasma of vapor resulting from anode erosion was analyzed qualitatively by high speed imaging using appropriate interference filter.

2. Experimental setup

2.1. The power supply and electrode configuration

The power supply consists of a capacitor bank with capacity up to 144 mF (12mF to 144mF by steps of 12mF). Arc ignition is performed using a high voltage source (HV) connected in parallel to the capacitor bank. The discharge circuit is composed of a ballast resistor and the two electrodes arc setup (figure 1). The maximal current is set by choosing the load voltage (up to 430 V) and the ballast resistor (down to 0.04 Ω) leading to a maximal current of about 10 kA. The time constant can be modified independently (from 1 to several hundred milliseconds) by adjusting the number of connected capacitors.

Electrodes are placed vertically with a gap of 4 mm. The cathode in the upper position is constituted by a 6mm in diameter graphite rod with sharp end. Anode in lower position consists of 6mm diameter with flat surface made of graphite or copper. All experiments are carried out at atmospheric pressure in air.

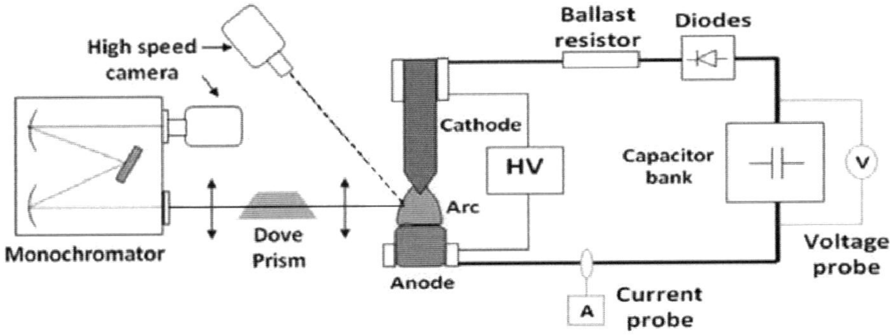

Figure 1. Power supply and spectroscopic acquisition system

Arc voltage and current were measured respectively with voltage and current probes. In figure 2, we present the temporal variation of the current for Imax = 500A for three different time constants (τ=24ms, 41ms, 91ms) and for Imax = 1kA for two different time constants (τ =24ms, 41ms). For both studied cases, maximum of current value is reached after 0.1ms as shown in figure 2.

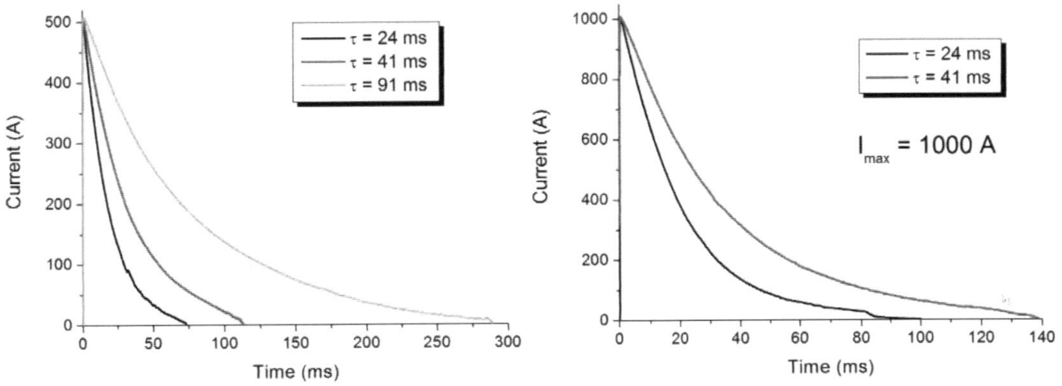

Figure 2. Time variation of the arc current

2.2. *Spectroscopic and high speed imaging observation system*

Phantom V9.1 high speed camera is used for spectra acquisition and was adapted on the exit slit of the monochromator type Jobin Yvon with 1m of focal distance and a 1200 grooves/mm grating or placed to view directly the arc. In this case the camera gives images of the plasma with high temporal and spatial resolution allowing study of arc displacement and electrode erosion during the discharge extinction. When connected to the monochromator an horizontal plasma slice is observed at a chosen position above the anode, using a dove prism and a two lenses setup with a magnification of 1(figure 1). For spectra acquisition, frame rate was 1000 fps and resolution was 800 x 800 pixels and for images of plasma, frame rate was 4000fps and resolution was 480 x 400 pixels.

3. Spectroscopic data processing

3.1. *Temperature determination*

Plasma is composed by atom and ion from metallic vapours from anode erosion and air. Plasma temperature was deduced from the Boltzmann plot method. Due to the strong presence of copper vapours in the plasma when using copper anodes, copper lines are self-absorbed and cannot be used.

Beside, the observed neutral nitrogen and oxygen lines do not present significant gap between the upper energy transition levels. Thus, we used ionized nitrogen lines (NII) (λ=444.7 , 460.1 , 463.05 , 566.6 , 567.9 , 571.07 , 592.7 and 661,05nm). They are largely separated in wavelength but unfortunately cannot be obtained in a single acquisition which requires calibration of these lines by the use of tungsten ribbon lamp. For all spectroscopic results presented in this paper, acquisition was made at an intermediary vertical position z=2 mm above the anode.

3.2. Electron density

Electron density was determined by using two different methods:

- Saha's law applied to N I neutral nitrogen line at λ = 744,2nm and N II ionized nitrogen lines at λ= 661,05nm
- Stark broadening of Hα hydrogen line at λ = 656,2nm and of N I neutral nitrogen line at λ=744,2nm

3.3. Reproducibility

It is important to verify the reproducibility of the lines used because they can't be obtained in a single acquisition. We have done several acquisitions for each line in order to have a mean value. Figure 3 shows four measurements of relative intensity of an N II ionized nitrogen line situated at λ= 661,05nm at 4ms and for I_{max} = 1kA and τ = 24ms as a function of radial distance x. Maximal deviation from the average value is estimated at 8%.

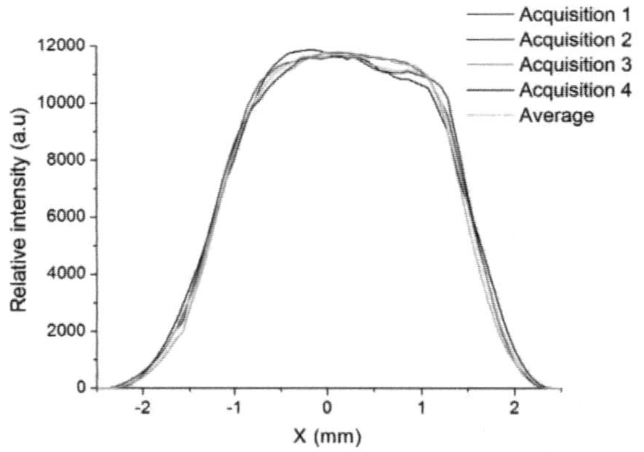

Figure 3. Measured relative intensity of ionized nitrogen line (λ=461nm) along radial distance x

4. Results and discussion

4.1. Imaging

In the case of copper anode, we compared images obtained using two different interference filters, one centered on Cu I copper lines (λ = 515 nm, FWHM of 10 nm) (figure 4) and the second one centered on N I nitrogen lines (λ = 750 nm, FWHM of 10 nm) (figure 5). The aim is to evaluate qualitatively the temporal evolution of anode erosion, copper vapours and nitrogen distribution in the plasma. These acquisitions were made with a maximal current of 1kA.

Journal of Physics: Conference Series **550** (2014) 012012 doi:10.1088/1742-6596/550/1/012012

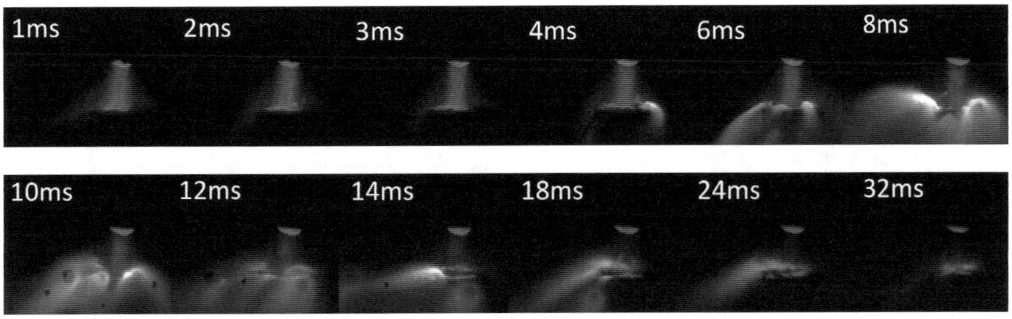

Figure 4. Evolution of copper vapour distribution (filter centered on copper lines: λ=515nm)

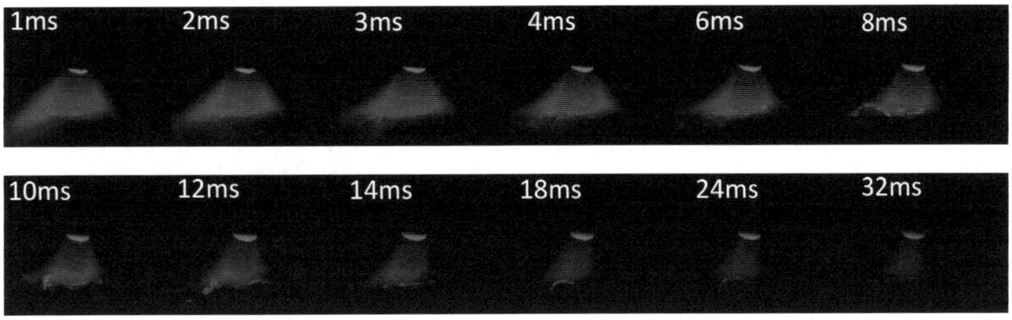

Figure 5. Evolution of nitrogen distribution (filter centered on nitrogen lines: λ=750nm)

In figure 4, the brighter zones correspond to high copper vapours concentration. One can notice that erosion is very weak and no anode deformation is observable until 3 ms. Then erosion increases as a function of time. It reaches a maximum between 8ms and 10ms and then decreases gradually until the arc extinction. Copper vapours distribution are not uniform in the plasma. On the other hand, the repartition of brighter zones corresponding to high emission of nitrogen lines in figure 5 indicates that nitrogen is distributed uniformly in the plasma. Indeed the lines emission changes seems mostly due to neutral nitrogen concentration; conclusions regarding power density cannot be drawn yet just on the basis of these pictures. We also see the cathode which radiate like a blackbody, especially within the range of transmitted wavelength by the filter. The arc presents as a bell-shape that shrinks gradually until the arc extinction.

4.2. *Influence of Metal vapour on plasma properties*

Ablated masses for the two different maximal currents, three different time constants and two different anode materials are presented in table 1. These are average values obtained by weighting the anode before and after the arc for several experiments. According to the ablated mass difference, we suppose that plasma obtained with graphite anode can be considered as pure air plasma for I_{max}=500A. Thus, comparison between graphite and copper anodes reports the influence of metallic vapours on the plasma characteristics. It is important to mention that for the case of I_{max}=1kA, droplets of copper forming on the anode and ejected outside were observed during the arc discharge (figure 4). They were taken into account in the ablated mass values presented in table1 but do not contribute to plasma contamination.

Table1. Ablated mass for copper and graphite anode

Maximal current I_{max}(A)	Time constant τ (ms)	Mass loss m (mg)	
		Copper	graphite
500	24	1.35	<0.1
	41	1.92	<0.1
	91	2.78	<0.1
1000	24	14.81	1.23
	41	53.03	2.46

4.2.1. Influence of time constant on plasma temperature

a) Case of I_{max} = 500A

We have reported in figure 6 a) and figure 6 b) the temporal variation of plasma temperature on the axis (x=0) at z=2mm respectively for the case of graphite anode and cooper anode and for three different time constants (τ = 24ms, 41ms, 91ms). One can note that temperature decreases for the two different anodes from a maximal temperature around 14kK. In the case of graphite anode and considering that erosion is negligible, temporal evolution of the temperature depends only on the variation of current. In the case of cooper anode, the temperatures drop more rapidly than with graphite. We have reported in table 2 copper anode erosion and the time during which the temperature is above 10kK for the two anode materials and the three time constants.

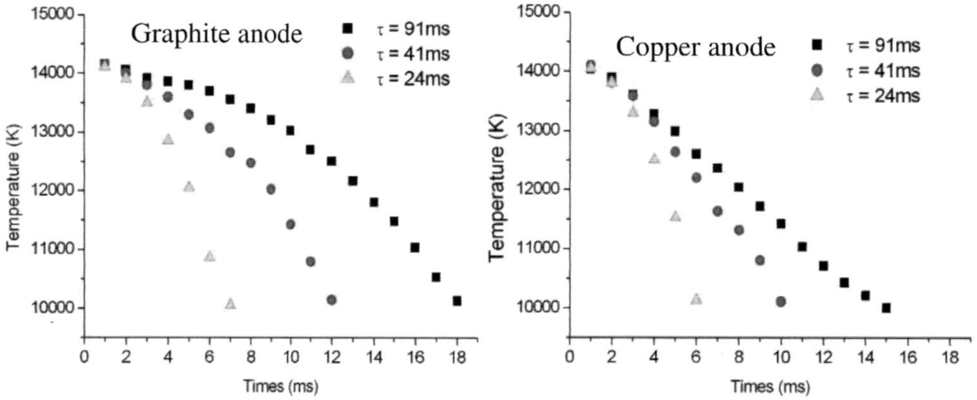

a) b)

Figure 6. Temporal variation of temperature for I_{max}=500A and for different time constant: a)graphite anode b) copper anode

Table 2. Comparison of time during which temperature is above 10kK

Time constant (ms)	Time for T>10kK		Copper anode erosion(mg)
	graphite anode (ms)	copper anode (ms)	
24	7	6	1.35
41	12	10	1.92
91	18	15	2.78

In both cases, temperatures above 10kK are observed for a longer time as time constant. For the copper anode, we noticed that erosion increases the same way as the time during which T>10kK. This is consistent to the stronger temperature decrease with cooper anode.

To highlight these observations, we reported in figure 7 the difference of the temperature between graphite and copper anodes for the three time constants.

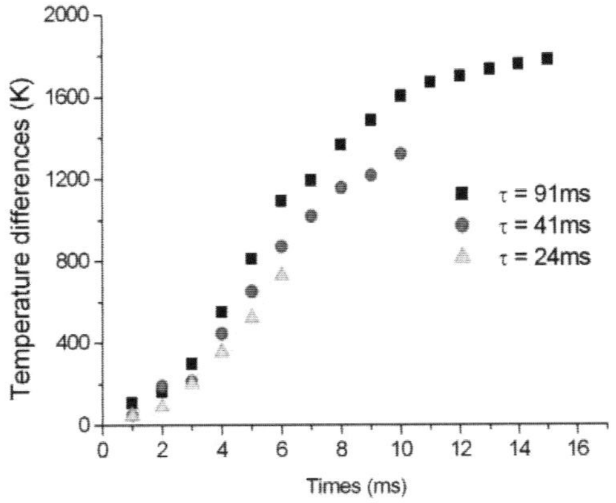

Figure 7. Temporal variation of temperature differential for three time constants

During the first 3 ms after arc ignition, the temperature differences are weak for the three considered time constants. Significant difference is observed from 4 ms and the larger the time constant, the higher the temperature difference.

b) Case of I_{max} = 1000A

Results presented in this paragraph concern the case of maximal current fixed at 1kA. Two different time constants were considered (τ = 24ms and 41ms). In figures 8 a) and b), we reported temporal variation of temperature on the discharge axis (x=0) at z=2mm for the two different time constants and for cooper and graphite anodes.

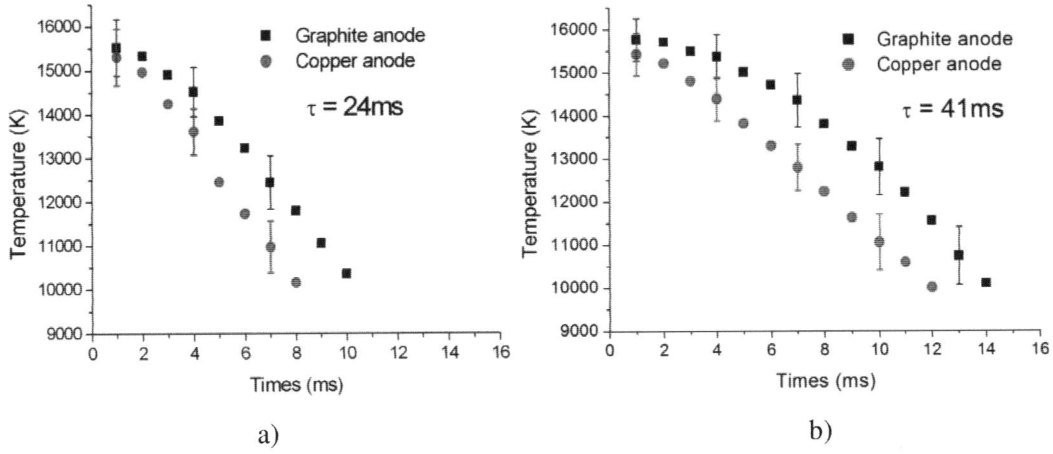

Figure 8. Temporal variation of temperature for different anode materials: a) τ = 24ms b) τ = 41ms

For the first 3ms the two curves overlap. This corresponds to the delayed erosion noticed by high-speed imaging (figure 4). Difference occurs from 4 ms when erosion begins to appear for the copper anode, with temperature significantly lower in this case. This difference is about 900K at 4 ms and increases up to 1600K at 8ms for the two time constants (τ = 24ms and τ=41ms). Considering the ablated mass in both cases (table1) and with the hypothesis that in the case of copper only half of it was distributed into the plasma, copper vapour rate is five times higher than that of graphite for τ =24ms and about ten times higher for τ = 41ms. This explains the fact that temperature is lower in the case of copper. Moreover, graphite anode erosion is not negligible, so plasma formed is composed by air and carbon vapours mixtures. Thus, even with lower erosion, plasma in this case undergoes to a cooling effect since carbon like copper is highly emissive.

These results can be explained by the following reasons [6]:

- the presence of metallic vapours leads to an increase of the plasma electrical conductivity. The resulting widening of the conduction area causes plasma cooling.
- the metallic vapours increase the energy loss by radiation leading also to the plasma cooling.

4.3. Electron density number n_e

For the electron density number, we consider the case of the maximal current I_{max} = 1kA and the time constant τ = 24ms. Figure 10 shows the temporal variation of electron density on the axis (x=0) at z=2mm for the two different anode materials (cooper and graphite).

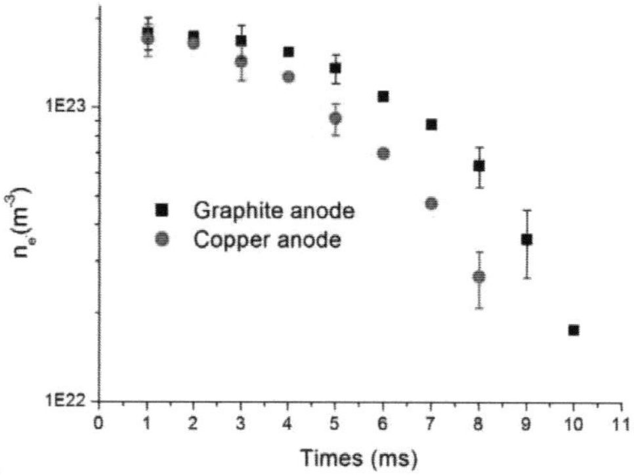

Figure 10. Comparison of the temporal variation of electron density for different copper and graphite anode for I_{max}=1000A and τ = 24ms.

One can see that electron density presents a plateau during the first 4 ms for the graphite and the first 2 ms for the copper. These instants correspond to temperature higher than 15kK (figure8). Let us note that at atmospheric pressure, electron density maximal value is about $2 * 10^{23}\ m^{-3}$ corresponding to a temperature value of about 15kK. At 4 ms electron density presents a difference with lower value in the case of copper anode. This difference increases until 8ms. This observation can be explained by the fact that the electron density in the plasma results mainly from ionization of the air (nitrogen and oxygen) beyond temperatures of 10kK [7]. We presented in figure 11, the theoretical value of electron density calculated from the equilibrium composition assuming LTE at atmospheric pressure, in function of mole fraction of copper and for different temperatures. At high temperature (T>10kK) electron density does not vary much whatever the proportion of copper vapour in the plasma between 0.1 %mol and 10%mol.

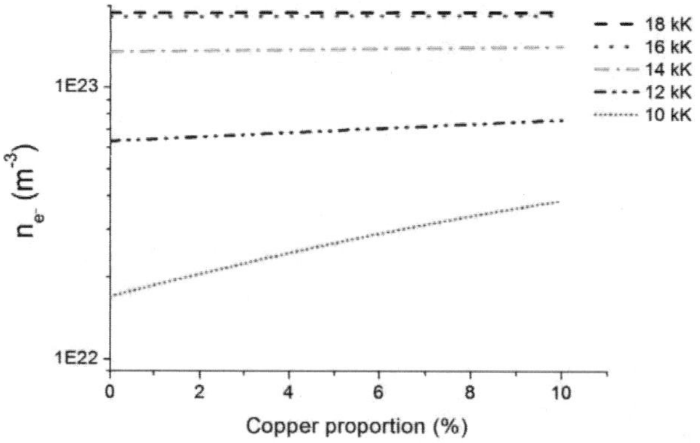

Figure 11. Theoretical values of electron density depending on copper proportion

The variation of electron density in function of the position deduced from the two different methods are presented in figure 12 under the following conditions: I_{max} = 1kA, τ = 24ms and copper anode.

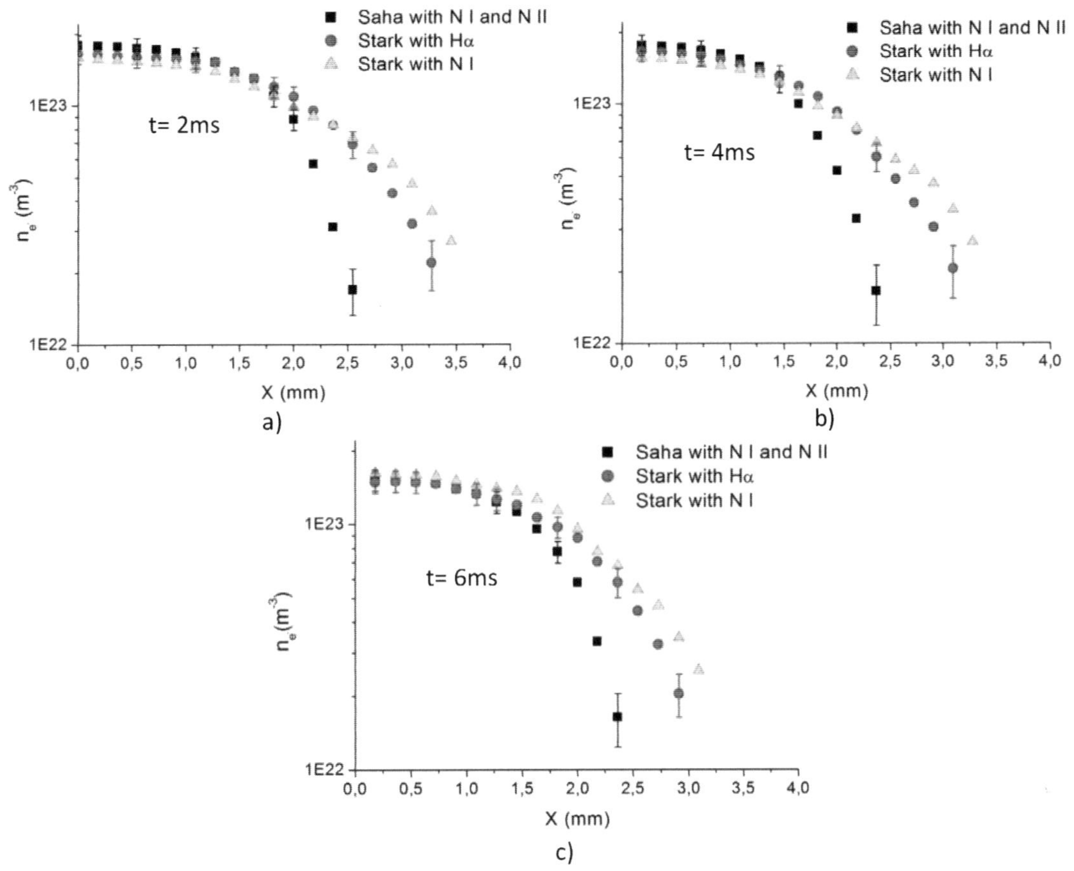

Figure 12. Radial profile of electron number density from Stark broadening and Saha's law: a) t = 2ms b) t = 4ms c) t = 6ms

- Electron density deduced from Stark broadening of Hα and N I are in good agreement along the profile on a distance of about 2.5mm from the discharge axis.
- Good agreement is observed between electron number densities deduced from Stark broadening measurements and from Saha's law which assume LTE on a distance from the discharge axis. This distance in which LTE assumption is valid decreases with the time: x=2mm at t = 2ms, x = 1,5mm at t = 4ms and x=1mm at t=6ms. Deviation from LTE is observed at the edge of the plasma.

5. Conclusion

In this paper, experimental setup simulating transient arc was described. A qualitative analysis of anode erosion, metal vapour and nitrogen atom distributions in the plasma was made by the mean of high speed imaging. A delay of about 3ms is observed for the anode erosion. Results allow expecting modification of plasma characteristics. Plasma temperature and electron density obtained by spectroscopic study are presented for various value of current, time constant and for different anode materials. Results show that presence of copper vapours tends to cool down the plasma. Comparison between electron densities obtained using Saha's law and Stark broadening measurement allows us to confirm that LTE is valid within a distance from the discharge axis which decreases with the time.

6. References

[1] Bormann D, Midya S, and Thottappillil R Oct. 2007 DC components in pantograph arcing: Mechanisms and influence of various parameters *Proceedings of 18th International Zurich Symposium on Electromagnetic Compatibility, Munich, Germany* pp 369–72

[2] McBride J W and Weaver P M January 2001 Review of arcing phenomena in low voltage current limiting circuit breakers *Proc IEE* Vol. **148** pp 1-7

[3] Chemartin L, Lalande P, Peyrou B, Chazottes A, Elias P Q, Delalondre C, Cheron B G, Lago 2012 Direct effects of lightning on aircraft structure: analysis of the thermal, electrical and mechanical constraints *Journal Aerospacelab*, Issue 5

[4] Qiang M a, Rong Mingzhe, Wu Yi, Xu Tiejun, Sun Zhiqiang Jun. 2008 Influence of copper vapour on low-voltage circuit breaker arcs during stationary and moving states *Plasma Science and Technology*, Vol.**10**, No.3

[5] Vacquie S 1996 Influence of metal vapours on arc properties *Pure &Appl. Chem.*, Vol. **68** No. 5 pp 1133-6

[6] Cressault Y, Hannachi R, Teulet Ph, Gleizes A, Gonnet J P and Battandier J Y 2008 Influence of metallic vapours on the properties of air thermal plasmas *Plasma Sources Sci. Technol.* 17 035016

[7] Ouajji H, Cheminat B and Andans P 1986 Composition and conductivity of copper-air plasma *J.Phys. D Appl. Phys.* **19** pp 1903-16

Journal of Physics: Conference Series **550** (2014) 012013 doi:10.1088/1742-6596/550/1/012013

Strange collective behaviors of cathode spots in low vacuum arc

M Sugimoto and S Takasugi
Faculty of Systems Science and Technology, Akita Prefectural University,
84-4 Tsuchiya-Ebinokuchi, Yurihonjo, Akita 015-0055, JAPAN

E-mail: sugimoto@akita-pu.ac.jp

Abstract. Removal of an oxide layer from all surfaces of a metal plate using an arc in vacuum is conducted for this study. Results confirmed that cathode spots that appeared on the side faced to anode were able to exist on another side opposite from the anode as well. Although some features of the cathode spots are not unusual, some characteristics disagree with conventional explanations for low-vacuum arc behavior.

1. Introduction

Metal surface processing is playing an increasingly important role in metal manufacturing industries because life span extension of iron steel infrastructures or metal products is necessary to realize a self-sustaining society. As a preliminary treatment, surface cleaning such as descaling or decontamination is necessary to establish sufficient performance of surface processing. Although chemical and mechanical methods have been used conventionally for the metal surface cleaning, more efficient and eco-friendly methods are desired. The authors are studying and developing a method using arc in low vacuum as a metal surface cleaning innovation [1-2]. The method depends on characteristics of the "cathode spot", which is a small, bright region that appears and moves on the cathode surface. Although the cathode spot is a well-known phenomenon, its behavior is mysterious. Many studies have been conducted both theoretically and experimentally for objects of various kinds and shapes, and not only objects of metal but of other electrically conductive materials such as carbon fiber [3-9]. Moreover, the improvement of energetic efficiency has been studied by changing the surface conditions [10-11].

One shortcoming of this method is that it requires a low pressure state such as several tens Pascals, even though it must not be high vacuum. The cathode spot activity is influenced by pressure [12-13]. The time necessary to produce the low-vacuum conditions occupies most part of one cycle of the processing. Therefore, reduction of this preparation duration is strongly desired for practical use. The authors attempt to satisfy that demand and to improve the efficiency by decreasing the number of times that waste gases are exhausted from the system. When this method was applied for descaling of both sides of a metal plate, for example, two exhaust operations were conducted because it was expected in general that the cathode spots exist only on the side facing the anode [14].

In the present research, the same treatment is attempted with a room that is opposite from the anode, which can enable the cathode spots to enter into that space. Using this system, the moving cathode spot behavior is investigated. Some strange features were found.

2. Experimental setup

Figure 1(a) presents a schematic diagram of the experimental arrangement used for this study. Descaling experiments in a low vacuum arc are conducted with an anode and test piece as a cathode set in a cylindrical stainless-steel vacuum chamber. The 300-mm-long vacuum chamber has 250 mm inner diameter. A cylindrical anode of 100 mm diameter is made of copper and is inserted through the upper side of the vacuum chamber. Facing this anode, a test piece support stage is inserted through the lower side of the vacuum chamber. Both the anode and the test piece support stage are water-cooled. Test pieces are a SS400 steel plate, which are 100 mm square and 3 mm thick. The test piece surface is covered with an approximately 5-μm-thick oxide layer, which makes the test piece dark, as presented in figure 1(b). Two stainless steel rods extend from the stage through two 7-mm-diameter holes in the test piece, which keeps the test piece about 60 mm separated from the stage surface. Actually, this room differs greatly from the previous experimental arrangement. The gap separating the anode and the test piece surface is 40-50 mm. In this test piece support device, a few ohms resistance is confirmed between the test piece and the negative electrode of the power supply. A 1-mm-diameter zinc wire is set in that electrode gap to short out the circuit and thereby initiate discharge. Before initiation of the discharge, the vacuum chamber gases are exhausted to about 10 Pa using a mechanical booster pump with a rotary pump. No gas is fed into the chamber. Therefore an arc is generated in residual air. The arc in low vacuum is started and sustained with a DC power supply (AR-SC120P; Osaka Electric Co. Ltd.) using constant current operation mode. Although the pressure in the vacuum chamber becomes higher than the initial value of 10 Pa, it does not exceed 100 Pa during the discharge. The arc current is controllable between about 50 A and 120 A. The current in the circuit and inter-electrode voltage of the DC power supply during the discharge are monitored respectively using a current probe and a data scanner. Photographs of the test sample during the discharge are taken using a camera (EOS X5; Cannon Inc.) and a high-speed digital camera (Fastcam MC2.1 10K-M2; Photron Ltd.).

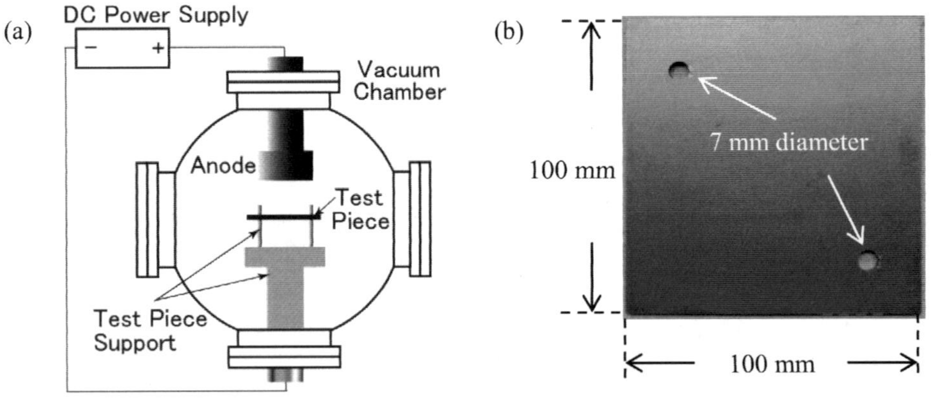

Figure 1. (a) Schematic of experimental apparatus and (b) SS400 steel plate as a test piece.

3. Experimental results and discussion

After initiation of the discharge, some small bright spots resemling those presented in figure 2(a), which are called cathode spots, appear on the test piece. In the experiments, the arc is initiated in low vacuum using a large current flowing into a zinc wire that spans the electrodes gap. A part of the wire is melted in an extremely short time; an electric spark leads to arc discharge. Therefore, all cathode spots are on the upper side of the test piece at the beginning of the discharge. Each spot moves independently and in a random manner. There is apparently no regular pattern to their behavior, as shown also in our previous experiments [1-2]. High-density energy that is confined within the cathode spot causes rapid heating and vaporization of the oxide layer on the test piece surface, which results in

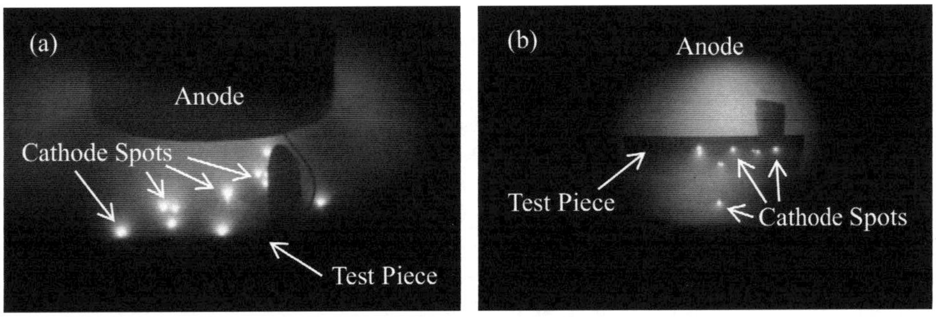

Figure 2. Cathode spots on (a) upper side and (b) lower side of test piece, respectively. Discharge current is 100 A.

descaling. As descaling progresses and the residual oxide layer diminishes, the cathode spot motions gradually become quicker.

As arranged for this study, there is some room under the test piece. After most of the oxide layer on the upper side of the test piece has been removed, all cathode spots move into the lower side of the test piece, which is opposite from the anode. Figure 2(b) shows the lower side immediately after all spots have moved there. This feature agrees with previously described characteristics: the cathode spot preferably stays on the oxide layer [1, 8]. It is particularly interesting that their side change occurs in almost the same amount of time, going through the test piece periphery. The number of the cathode spots remains almost constant, although a slight change can be observed at every moment. Subsequently they continue removal of the oxide layer on the lower side toward the center. Their motions are slow as long as the oxide layer remains sufficient.

During oxide layer removal of the lower side, all cathode spots suddenly go back to the upper side and remain there for several seconds, as depicted in figure 3(a). Compared with the spots in figure 2(a), their sizes are apparently decreased. Their velocities are greatly increased. This phenomenon sometimes occurs. It is not appreciated because it prevents the oxide layer removal. However, figure 3(b) also shows the upper side of the test piece, where all cathode spots are on the lower side. In contrast to figure 3(a), the electrode gap is filled with an extremely bright cloud even though no cathode spot is confirmed on the upper side. These results seem strange because the cathode material vapor is believed to be ionized. Thereby, generated plasma is supplied by cathode spots [7, 15]. As the oxide layer remaining in the lower side decreases, and because it is restricted mainly around the center of the lower side, it becomes difficult for the spots to go to the central region. For that reason, the descaling efficiency is worse.

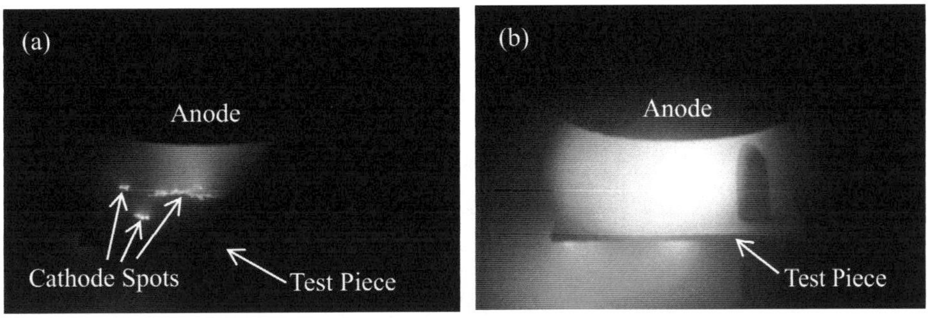

Figure 3. Photographs of the inter-electrode gap when (a) cathode spots come back and (b) no cathode spots are on the upper side.

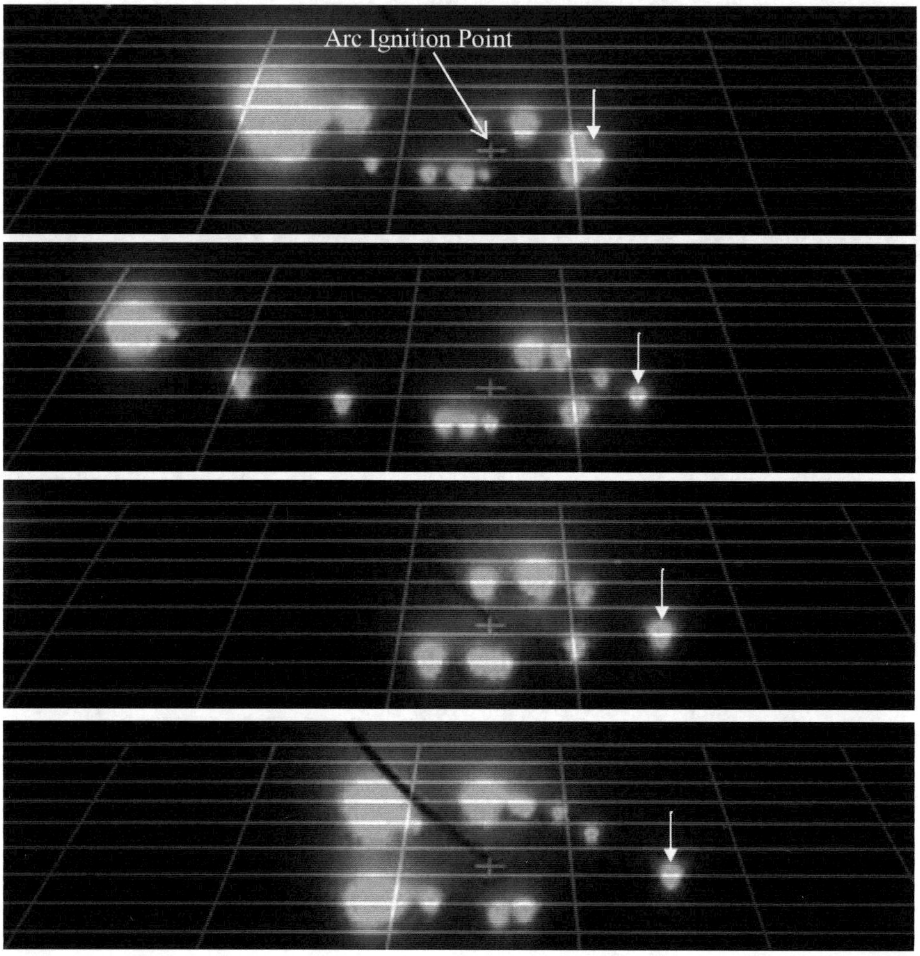

Figure 4. Cathode spots motion on oxide layer as well as figure 2(a). Time interval between the pictures is 12.5 ms.

To investigate differences of the cathode spots between figures 2(a) and 3(a), their motion can be recorded using a high-speed digital camera. However, it is extremely difficult to catch the moving cathode spots in the field of view. Instead, the cathode spots images at the beginning of discharge are taken because the point of arc ignition can be decided easily by the zinc wire position. Figure 4 depicts a series of photographs, captured from motion images, in which the test piece upper side is covered with an oxide layer as well as figure 2(a). The time interval between the pictures is 12.5 ms. A picture of crossed lines with every 10 mm gap is overlapped to ascertain the cathode spot velocities. The arc starts at the position shown by a cross mark. The cathode spots spread out immediately. One cathode spot indicated by an arrow in the pictures, for example, is traveling about 5 mm along the lateral line for 37.5 ms. Therefore its average velocity is estimated as about 0.13 m/s.

Figure 5 also displays a set of photographs of the cathode spot motion, in which the oxide layer on the upper side of the test piece is almost removed in advance, just as shown in figure 3(a). The number of observed spots decreases to fewer than half of those shown in figure 2(a), which means that electric power in a spot increases to more than twice as large. The time interval between the pictures is 0.125

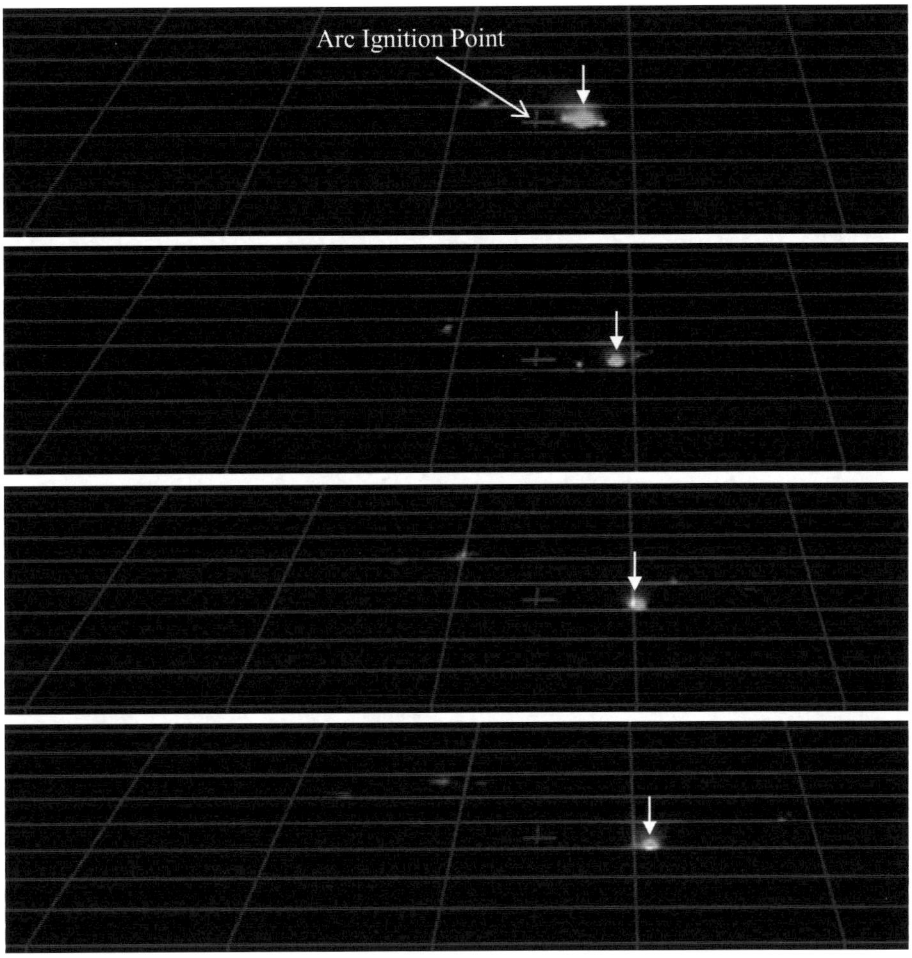

Figure 5. Cathode spots motion not on oxide layer as well as figure 3(a). Time interval between the pictures is 0.125 ms.

ms, which is a hundredth of that of figure 2(a). Because the cathode spot indicated by an arrow in the pictures seems to move about 4 mm laterally and about 5 mm perpendicularly for 0.375 ms, the average velocity is about 17.1 m/s, which is about 130 times higher than that of the cathode spot in figure 2(a).

Although these results were obtained at the beginning of arc discharge, they suggest that the behaviors and characteristics of cathode spots depend strongly on the cathode surface state, which is similar to the results reported in some literatures [10, 16]. Therefore, a drastic and sudden change of cathode spot behavior is apparent when some small differences exist in a test piece, as they do in this research case.

4. Conclusions

Descaling experiments for all surfaces of a plate type test piece covered with an oxide layer were conducted. Cathode spot behavior was investigated in a low-vacuum arc. The cathode spots, which appear on the side facing the anode, remove the entire oxide layer on that side and move into the opposite side from the anode together, continuing the descaling. However, they sometimes go back to the original side and remain there for a few seconds moving with extremely high velocity. The result obtained using high-speed digital camera observation reveals that the average velocity differs by two orders of magnitude with and without the oxide layer on the surface.

Acknowledgement
This work was supported by the Japan Society for Promotion of Science KAKENHI Grant Number 25400536.

References

[1] Takeda K and Sugimoto M 2003 *IEEE Trans. Plasma Sci.* **31** 983
[2] Sugimoto M *et al* 2004 *Thin Solid Films* **457** 168
[3] Beilis I I 2002 *Appl. Phys. Lett.* **81** 3936
[4] Anders A, Oks E M and Gera Y Y 2005 *Appl. Phys. Lett.* **86** 211503-1-3
[5] Rosen J and Anders A 2005 *J. Phys. D: Appl. Phys.* **38** 4184
[6] Anders A 2008 *Cathodic Arcs* (New York, Springer Science + Business Media, LLC) chapters 2 and 3 pp 7-174
[7] Beilis I I 2013 *IEEE Trans. Plasma Sci.* **41** 1979
[8] Shi Z *et al* 2013 *IEEE Trans. Plasma Sci.* **41** 2068
[9] Zhang C *et al* 2013 *Vacuum* **93** 45
[10] Arai Y, Shimizu Y, Sugimoto M and Takeda K 2011 *ISIJ Int.* **51** 1890
[11] Arustamov V N *et al* 2013 *J. Surf. Investigation. X-ray, Synchrotron and Neutron Techniques* **7** 1104
[12] Takeda K *et al* 2006 *ISIJ Int.* **46** 335
[13] Shi Z *et al* 2011 *IEEE Trans. Plasma Sci.* **39** 1585
[14] Boxman R L, Martin P J and Sanders D M 1995 *Handbook of Vacuum Arc Science and Technology* (New Jersey, Noyes Publications) chapter 3 pp 73-281
[15] Anders A 2008 *Cathodic Arcs* (New York, Springer Science + Business Media, LLC) chapter 4 pp 175-225
[16] Szente R N, Munz R J and Drouet M G 1990 *J. Phys. D: Appl. Phys.* **23** 1193

Investigation of plasma flow in vacuum arc with hot cathode

R Amirov[1], N Vorona[1,2], A Gavrikov[1,2], G Lizyakin[1], V Polistchook[1], I Samoylov[1], V Smirnov[1], R Usmanov[1,2], I Yartsev[1]

[1]Joint Institute for High Temperature of RAS, Moscow, Russia
[2]Moscow Institute of Physics and Technology (State University), Moscow, Russia

E-mail: glizyakin@gmail.com

Abstract. One of the crucial problems which appear under development of plasma technology processing of spent nuclear fuel (SNF) is the design of plasma source. The plasma source must use solid SNF as a raw material. This article is devoted to experimental study of vacuum arc with hot cathode made of gadolinium that may consider as the simple model of SNF. This vacuum discharge was investigated in wide range of parameters. During the experiments arc current and voltage, cathode temperature, and heat flux to the cathode were measured. The data on plasma spectrum and electron temperature were obtained. It was shown that external heating of the cathode allows change significantly the main parameters of plasma. It was established by spectral and probe methods that plasma jet in studied discharge may completely consist of single charged ions.

1. Introduction

Nowadays actual problem of atomic energy is recycling of SNF. In order to better involve resource fuel in reactors it is necessary to provide the closed fuel cycle. Existing chemical methods produce lots of liquid radioactive waste. It should also be noted that any civil technology refining of SNF must comply with requirements of the non-proliferation of nuclear weapons i.e. required to prevent the separation of plutonium. In general, the plasma technology of separation fulfills the above requirements [1-3].

The recycling technology involves the production of the low-temperature plasma of SNF and subsequent spatial separation of ions by mass. In contrast to the electromagnetic methods, in plasma separator ions move under compensated space charge condition [4] that removes the restrictions on the value of the ion current and significantly increases productivity.

The ability to use plasma separation for controlled thermonuclear fusion program has been featured for more than 30 years ago [5]. The centrifugal trap, which was created in magnetized plasma by system of annular electrodes, was offered in work [6]. Ohkawa's mass filter [7, 8] and the recent works on the analysis of asymmetric centrifugal trap with respect to the processing of SNF became the development of this concept [9]. Another approach to separation was proposed by A. Morozov [4, 10] who suggested plasma-optical focus of ions of different masses departing from the plasma accelerator. Currently studies in this direction are focused on improving the parameters of plasma-optical systems and achieving an acceptable technology for spatial separation of flows of matter [11, 12]. It should mention other plasma-optical schemes [13, 14] which may be useful for the separation of components of SNF. Separation of plasma ions may be also achieved by various resonance techniques including ion-cyclotron resonance and electron-beam discharge [15, 16].

Three main tasks need to be solved for creating plasma separation technology: to create the plasma source in which the conversion of solid SNF in plasma SNF will take place: to create separation device where the separation of the components of SNF in electric and magnetic fields will occur and finally to construct the unit of collectors to collect separated substances. Currently in our research team the field configuration has been developed in which the separation of the components of SNF can be obtained, the creation of the experimental model of separator has come to the end, the configuration of collectors has been designed and series of experiments of the deposition of substances have been carried out [17, 18]. The question about the source of plasma remains open. This work is devoted to the study of this problem.

Almost completely ionized plasma must come in to the working volume for effective separation. As the experience of American colleagues shows these seemingly contradictory requirements can be satisfied simultaneously [1].

The analysis of the whole number of ways of transferring condensed matter in the plasma state was conducted: magnetron sputtering, crucible heating with subsequent ionization of vapor jet, evaporation of solid substance with the help of electron beam or laser radiation and finally vacuum arc. Under requirements for SNF plasma (degree of ionization is about 100%, identical multiplicity of ions) the vacuum arc discharge with cathode diffuse spot (CDS) was elected as the most perspective method. Such form of arc can exist if the cathode temperature is above 2 kK [19]. The current is evenly distributed over the cathode surface with the density j ~10–100 A/cm^2. This discharge generates the supersonic plasma jet in which degree of ionization can reach 100%. The temperature of electrons is several electron-volt, the electron distribution function is close to Maxwellian and their concentration — 10^{13} cm^{-3} [19–21]. The important distinctive feature of vacuum arc with cathode diffuse spot is the absence of micro droplets fraction (the part of which in usual vacuum arc on a cold cathode can exceed 80%).

For the experimental study vacuum arc with CDS as a source for the development of technologies for SNF plasma separation we have proposed the following research program:

- The study of metal analogue of uranium (from the point of view of a vacuum arc) and the answer to the fundamental question – is it possible to satisfy the requirements to the plasma source performance with the help of vacuum arc, the extent and degree of ionization?
- The study of vacuum arc on composite electrodes which contains ceramic components. This step is necessary because uranium dioxide presents more than 95% of SNF [22].

This work was focused on the first point of the problem. For the second point there were analyzed oxides like uranium dioxide depending on the electrical conductivity from temperature, saturated vapor pressure and relation to the oxide melting temperature and pure substance. The choice was made in a favor of niobium oxides (V), cobalt (II) and titanium (IV). Generally there could be the vacuum arc burning on the dielectric substance however, this issue is currently not fully covered in the scientific literature and requires additional study.

Gadolinium was elected as a substance that imitates uranium in arc discharge. The first three ionization potentials of gadolinium and uranium are practically identical (U: 6.19 – 11.9 – 20.0 eV; Gd: 6.15 – 12.1 – 20.6 eV) [23]. Furthermore, the properties of vacuum arc are determined by the ratio g of the flows of thermally evaporated atoms and electrons [24]. For gadolinium this ratio is $g = 0.05$, and for uranium - $g = 0.01$, for both elements it is much less than unit, so the characteristics of vacuum arc on gadolinium and uranium cathodes must be similar [24].

2. Experimental setup and method of experiments

During the experiments we measured the cathode temperatures, the heat flux to the cathode, and the average rate of evaporation of gadolinium in the discharge. We also studied such plasma parameters as the temperature of electrons, intensity of atoms ions lines and mean charge of ions depending on arc voltage. The regime of discharge burning was established in which the plasma has a maximum degree of ionization, and multiple-charged ions are absent.

Figure. 1 shows the scheme of the experiment. Discharge was burnt in vacuum chamber with a volume of 100 litre evacuated to a pressure of residual gases less than 10 mPa. Gadolinium was placed in the molybdenum crucible with an outer diameter of 25 mm. Under the crucible there was electron-beam heater (EBH) consisting of wolfram-rhenium wire (emitter) and an electrostatic lens. EBH serves to heat the crucible and its contents to a temperature at which the vacuum arc with CDS arises. During the experiment, changing the power of EBH $N=IU$ (0-1000 W) allows to vary the cathode temperature and discharge voltage under constant arc current. Under arc operation gadolinium was in the molten state (T_m = 1586 K) with the surface of melting area of about 5 cm^2. Water- cooling steel disk was used as the anode. The anode had the diagnostic hole with diameter of 30 mm. Interelectrode gap was about 30 mm. The photo of experiment is shown on figure 2.

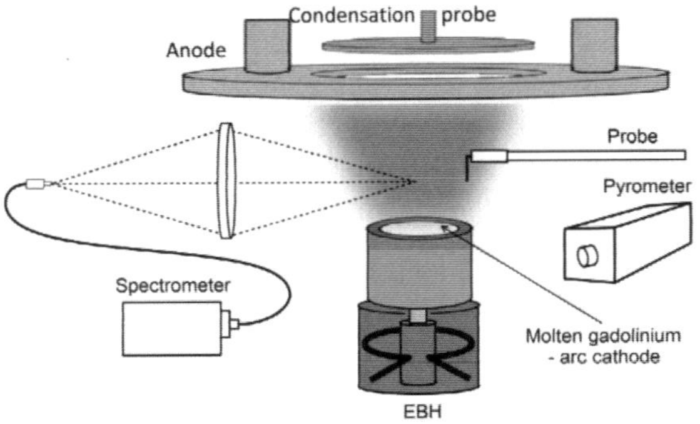

Figure 1. Experiment scheme.

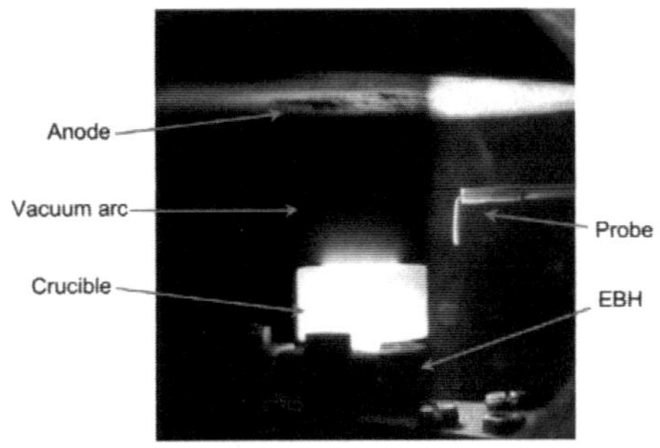

Figure 2. Photo of the vacuum arc discharge with CDS during the experiment.

Electric schemes of the experimental stand are shown in figure 3. The rectifier with an output voltage of 380 V served as a power source of arc, discharge current was regulated by ballast resistance which was a rheostat with water cooling. Consistently with the arc the shunt (R=0.001 Ohms) was turned on to measure the discharge current. Discharge voltage was registered by voltmeter.

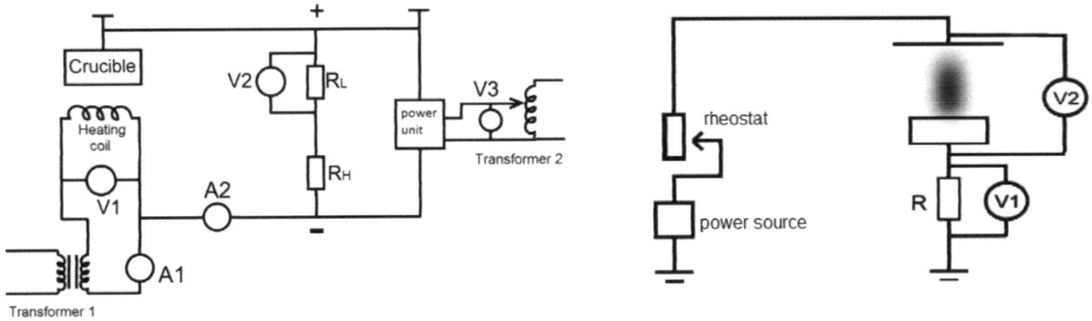

(a) (b)

Figure 3. Electric schemes: (a), the scheme of EBH; (b), the scheme of power supply of the arc.

The crucible temperature T_c was measured by photoelectric pyrometer. By estimations the difference between measured temperature and the average temperature of gadolinium surface was less than 3 %. During the experiment this temperature reached 2.1 kK. Plasma radiation was recorded by using spectrometer with a resolution of 0.28 nm in the spectral range 280-600 nm. The radiation was derived from the vacuum chamber through an extension cord with a quartz glass. This measure was necessary to protect the diagnostic windows from pollution by arc products. The image of discharge with the help of quartz lens was focused on the entrance of spectrometer optical fiber.

The isolated Langmuir's probe was applied to determine electron temperature and estimate the concentration of plasma ions. The tungsten wire with diameter of 0.5 mm was used as this probe. It was bent in such way that its collecting surface was parallel to the lines of the electric field in the discharge. This requirement ensures that the charges were fallen on the probe only due to their thermal motion. The probe had an active length of 10 mm, the other part was isolated with a thin ceramic tube with a metallic screen (figure 2). This screen was designed to protect insulation end from deposition, which can increase an effective area of the current collecting surface. During the experiment, the probe can be moved along the radius of the chamber within 14-60 mm from the axis of the discharge.

The condensation probe used to determine the average particle charge of gadolinium was set after anode. This parameter was determined from the ratio of charge coming to the probe and increase of its mass. Condensation probe with surface of 6 cm^2 was made of molybdenum foil. The distance between probe and rear surface of anode was 2 cm.

Before the experiment, the crucible with gadolinium was weighed. Then the assembly with EBH and crucible was placed into the chamber. After evacuating the chamber cathode was heated by EBH to the temperature about 1.9 kK at which saturated vapor pressure of gadolinium is about 2 Pa. Then the voltage from rectifier was applied between electrodes, and arc discharge was arisen. After experiment the crucible was weighed again.

There were several series of experiments on the stand:

- With the constant current I = 44 A the emission spectra at different distances from the surface of the crucible (2.2-8.2 mm) at arc voltage (U = 3-23 V) were recorded.
- Probe current–voltage characteristics was measured at distances of 14-60 mm from the discharge axis at a fixed current I = 44 A and arc voltage U = 3-23 V.
- The dependence of the temperature of the crucible on power of EBH during discharge (I = 44 A) and without (I = 0 A) was registered.
- In every experiment the average rate of gadolinium evaporation by the mass difference of the crucible was measured before every experiment and after it. For different series of experiments this evaporation rate was 1.5-2.2 mg/s.
- The mean charge of gadolinium particles after anode was measured by the condensation probe at arc current of 50 A in the range of arc voltage 3-8 V.

3. Experimental results and discussion

Figure 4 presents the crucible temperature in the absence of an arc (I = 0 A) and at current I = 44 A depending on the power of EBH. The figure shows that with the EBH power N less than 500 W the temperature T_c with the arc is greater than at zero current, i.e. the arc heats the cathode. With greater power of EBH the opposite relation between these temperatures takes place, and the discharge starts to cool the crucible. This fact is explained by the cathode cooling due to the heat losses on electron emission [20].

The arc voltage depends significantly on EBH power. At N above 600 W the arc voltage seeks to 3 V. Decreasing power N down to 400 W causes increasing of arc voltages to 20 V , and at zero heating this voltage would be about 70 V (arc current was 44 A).

The plasma radiation spectra were analyzed on lines of atomic gadolinium matching and its two first ions [25]. The example of spectrum where the appearance of lines Gd III traces with the transition from a voltage of 4.9 V to 7.9 is shown in figure 5. This spectrum was recorded at 5 mm from the cathode crucible. The radiation intensity of the individual lines depends on factors such as electron energy, excitation cross section of displayed levels, transition probabilities, scattering of radiation, passing through the optical system, etc. [26].

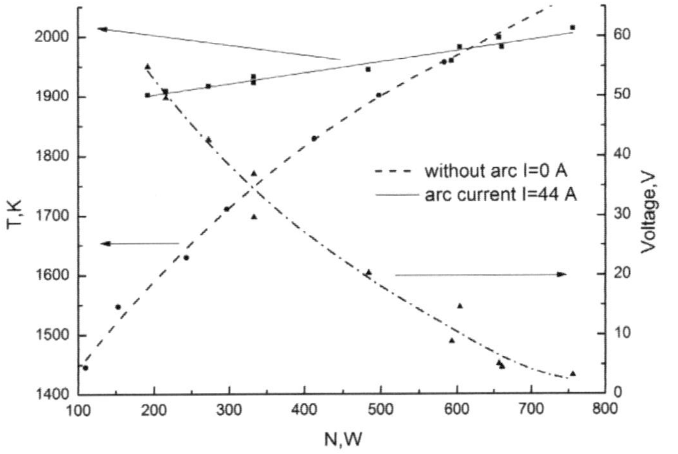

Figure 4. Crucible temperature and interelectrode voltage depended on EBH power.

Thus the intensity of individual lines between themselves can not be correctly compared in the interelectrode gap. The overall picture of the behaviour of the line intensities of ions and atoms together is shown in figure 6. The intensity of each line on this figure is normalized to its maximum. This spectrum was recorded at 5 mm from the cathode crucible. This figure shows that the maximum intensity of the Gd II occurs at approximately 5 V of arc voltage when the ions with the charge +2 are practically absent. The maximum intensity of double charged gadolinium ions takes place at arc voltage about 12 V. The figure 6 also shows the density of saturated gadolinium vapor N_0 determined from the measured crucible temperatures.

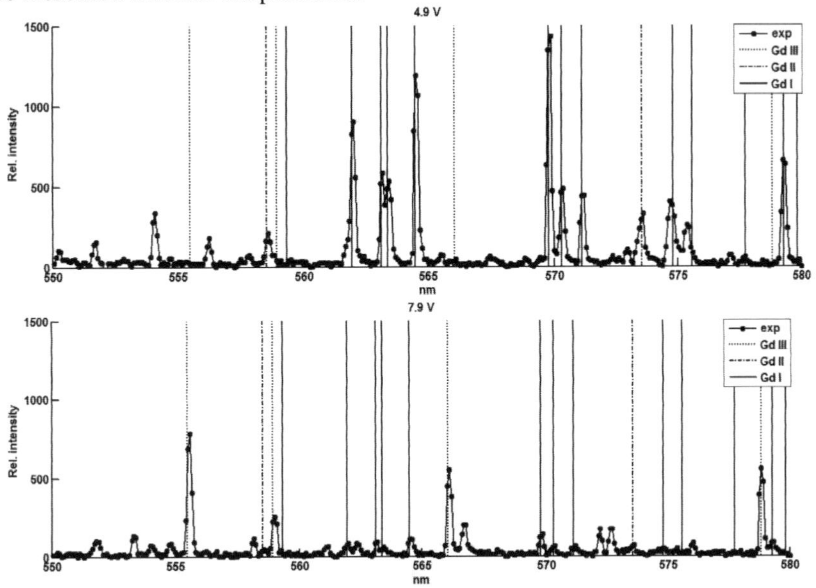

(a)

(b)

Figure 5. Plasma emission spectrum at a voltage of: (a), 4.9 V; (b), 7.9 V (Gd III, Gd II - tabular spectrum of double and single ionized gadolinium, Gd I - atomic spectrum, exp – experimental results).

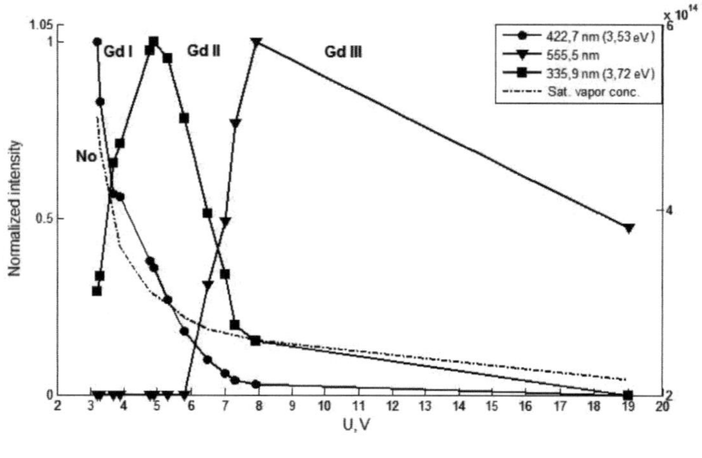

Figure 6. Line intensities depending on the arc voltage at current $I = 44$ A, N_0 – is the density of saturated gadolinium vapor

The behavior of the line intensities of atomic gadolinium and ion at distances from 2.2 mm to 8.2 mm from the crucible surface is shown in figure 7.

When processing probe current-voltage characteristics there was especially verified the lack of thermionic emission from the probe. By estimations during the experiment probe was heated up to 1 kK, but thermionic current from its surface which can distort current-voltage characteristics appears at temperatures above 2 kK [23].

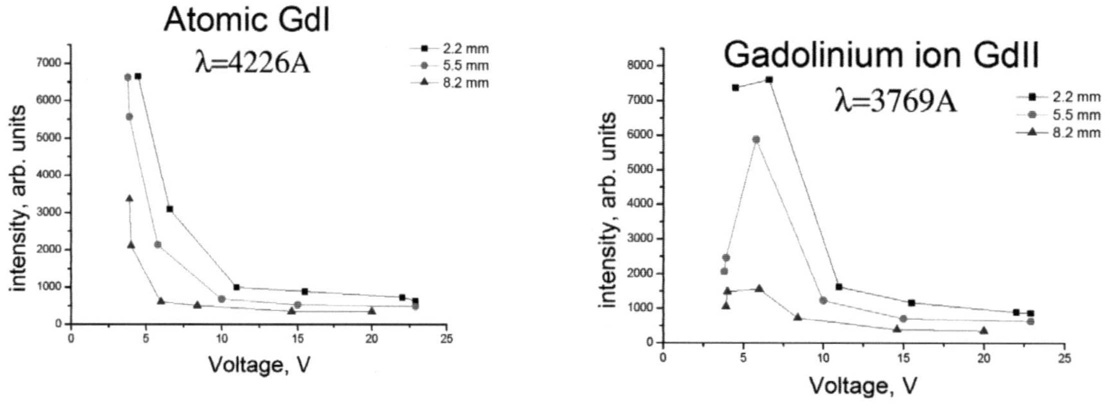

(a) (b)

Figure 7. - Intensity of lines (a) Gd I and (b) Gd II depending on the arc voltage at different distance from the crucible surface.

A typical current-voltage characteristic is shown in figure 8. The dependence of the electron temperature of the arc voltage at distances from 14 mm to 51 mm from the axis of the discharge measured by probe is shown in figure 9. Thus the electron temperature increases with the voltage and practically does not depend on the distance to the axis of the arc.

The probe measurements allowed estimating the density of plasma ions; it was (0.3-2.4) $\times 10^{12}$ cm^{-3} depending on the point of registration and electron temperature.

Figure 10 presents the charge of gadolinium particles in plasma after anode. It should be noted that data on this figure in general corresponds to results of spectral diagnostic and electron temperatures described above.

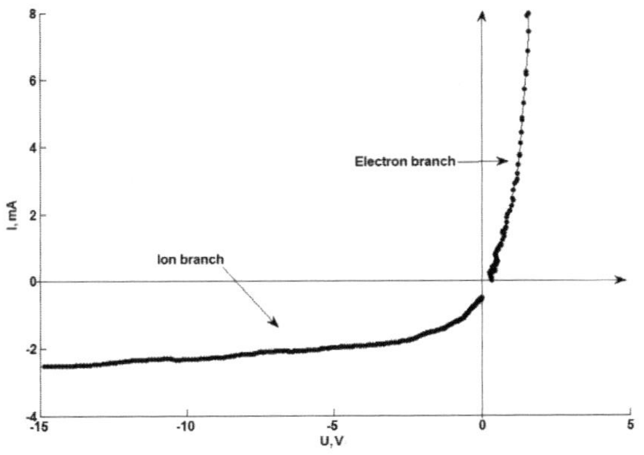

Figure 8. Current-voltage characteristic of the probe.

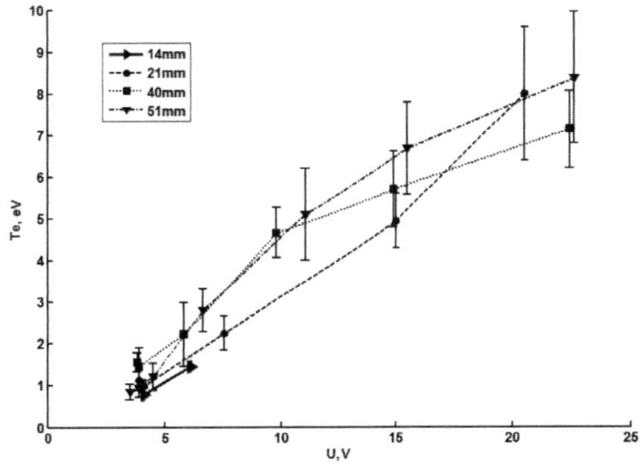

Figure 9. Electron temperature depending on the arc voltage.

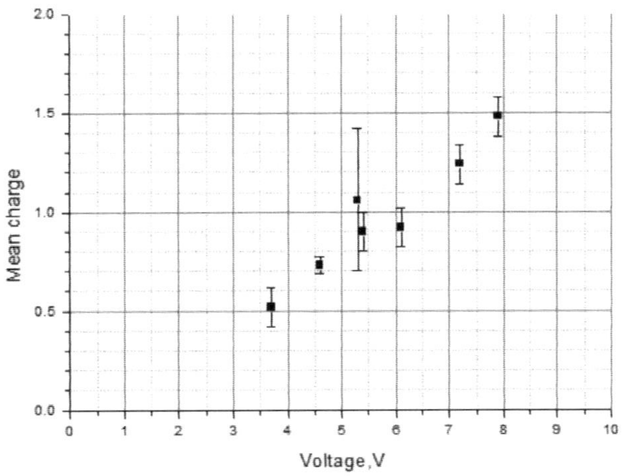

Figure 10. Mean charge of the gadolinium particles, arc current 50 A.

4. Conclusion

To conclude let us summarize the obtained results once again. Measured average evaporation rate of gadolinium in a vacuum arc with cathode diffuse spot was 1.5-2.2 mg/s (5.4-8 g/hr). Note that the nuclear power plant with the power 1 GW annually accumulating 5-10 tons of SNF (1 kg/hr) and for the experimental plasma separator it was decided to focus on the performance of 100 g/hr (or 27 mg/s).

The productivity obtained during the experiment hasn't met the requirements of SNF plasma separator yet but it can be easily increased by changing of the crucible size. Modes were determined in experiments in which there is maximum emission intensity of singly ionized gadolinium lines where the double charged ions are practically absent. It is determined that the appearance of plasma with high ionization degree is expected when the arc voltage is about 5 V. There was also explored the dependence of the electron temperature of the arc from voltage (Fig. 8) which has shown a weak dependence on the distance from the discharge axis. The received data on the cathode temperatures and heat flow of plasma at the cathode allow constructing the plasma source with the desired settings for experimental plasma separation.

Generally, the results indicate that the vacuum arc with cathode diffuse spot as a source of plasma is perspective for the technology of plasma separation. This study will be continued, and at the next step the experiments will be carried out on the substances which simulate nonconducting SNF (oxides of cobalt, niobium and titanium) and multi-component mixtures.

References

[1] Litvak A., Agnew S., Anderegg F. et al. 7-11 July 2003 Archimedes Plasma Mass Filter // 30th EPS Conference on Contr. Fusion and Plasma Phys. St. Petersburg, ECAVol. 27A, O-1.6A

[2] Zhil'tsov V, Kulygin V, Semashko N, Skovoroda A, Smirnov V, Timofeev A, Kudryavtsev E, Rachkov V and Orlov V 2006 *Atomic Energy* **101** p. 755-759

[3] Smirnov V,Samohin A,Vorona N and Gavrikov A 2013 *Fizika Plozmi* (Physics of plasma) **39 6** 523-533

[4] Morosov A 2008 *Introduce in plasmadynamic* (Moscow: Science)

[5] Bekhtenev A and Volosov V 1980 *Nuclear Fusion* **20 5** 579

[6] Volosov V and Pekker M 1981 *Nuclear Fusion* **21 10** 1275

[7] Ohkawa T and Miller R 2002 *Phys. Plasmas.* **9** 5116.

[8] Freeman R., Agnew S., Anderegg F. et al. 2003 Archimedes Plasma Mass Filter // *Radio Frequency Power in Plasmas: 15th Topical Conference. AIP Conference Proceedings* **694** 403-410

[9] Fetterman A and Fisch N 2011 *Phys. of Plasmas* **18** 094503.

[10] Morozov A and Saveliev V 2005 *Fizika Plozmi (Physics of plasma)* **31 5** 458–465.

[11] Brdakov V,Kichigin G and Strokin N 2010 *Pis'ma v ZTF (letter to JTPh)* **36 4** 75.

[12] Brdakov V, Vo Ny Zan Kichigin G et al 2011 *Izvestiya Vuzov. Yadernaya energetika (nuclear energy)* **2** 3

[13] Korobkin Y, Lebedev N and Paaperny 2012 *Pis'ma v ZTF (letter to JTPh)* **38 6** 1

[14] Egorov A, Yuferov V and Shariy 2009 *VANT. Seriya: Fizika Plozmi (Series Physics of plasma)* **1 (59)** 122–124

[15] Timofeev A 2007 *Fizika Plozmi (Physics of plasma)* **33 11** 971–987.

[16] Skibenko E and Kovtun Y 2009 *VANT. Seriya* Fizika radiatsionnyh povrezdeniy i radiazionnoe materialovedenie **4 (62)** 286–292

[17] Vorona N, Smirnov V, Antonov N, Gavrikov A, Samokhin A and Zhabin S 2013 *28th International Conference on Interaction of Intense Energy Fluxes with Matter, Book of Abstracs* 175

[18] Antonov N, Vorona N, Gavrikov A, Zhabin S and Smirnov V 2012 *Trudy 55 nauchnoy konferentsii MFTI "Problemy sovremennoy fiziki"* 132

[19] Vasin A, Dorodnov A and Petrosov V 1979 *Pis'ma v ZTF (letter to JTPh)* **5 24** 1499–1505

[20] Polistchook V, Sichev P, Shabashov V and Yartsev I 1986 ZTF (JTPh) **56 11** 2233–2235

[21] Bronin S, Polistchook V, Sichev P, Shabashov V and Yartsev I 1993 *TVT (THT)* **31 1** 29

[22] Beckman I 2005 *Yadernaya industriya (Nucliet industry)* (Moscow: Moscow State University)

[23] Grigoreva I and Meylihova E 1991 *Fizicheskie velechiny (Physicul value*) (Moscow Energoizdat)

[24] Polistchook V, Serdukova O and Yartsev I 1993 ZTF (JTPh) **63 3** 66–74.

[25] Meggers W, Corliss C and Scribner B 1975 *Tables of Spectral-Line Intensities, Part I — Arranged by Elements, Monograph 145 — U.S.: Natl. Bur. Stand.,. NIST compilation*

[26] Lohte-Holtgreven V *1971 Metody issledovaniya plazmi (The methods of plasma investigation)* (Moscow: Mir)

Comparison of CO_2 and oxygen DC submerged thermal plasmas for decomposition of carboxylic acid in aqueous solution

S Safa[1], A Hekmat-Ardakan[1] and G Soucy[1]
[1]Department of chemical and biotechnology engineering, Université de Sherbrooke, Sherbrooke, Canada, J1K 2R1

E-mail: Gervais.Soucy@USherbrooke.ca

Abstract. The feasibility of the carboxylic acid decomposition with two different direct current (DC) thermal plasma torches was investigated. An oxygen DC submerged thermal plasma torch and a newly designed submerged DC plasma torch operating with a mixture of carbon dioxide and methane (CO_2/CH_4) were used. Sebacic acid was selected as a representative of pollutants in the most wastewater produced by chemical process industries. The effect of different operational conditions including treatment time, the reactor pressure as well as the role of oxidizing agents such as (H_2O_2) were investigated on the decomposition rate of sebacic acid. Concentration of sebacic acid was quantified by Ion Chromatography/Mass Spectrometry (IC/MS). The oxygen plasma showed higher decomposition rate in basic medium. Adding H_2O_2 into aqueous solution enhanced the sebacic acid decomposition rate with the CO_2/CH_4 plasma up to the same decomposition rate of the oxygen plasma. Increasing the pressure also increased the decomposition rate for both plasmas with an increase twice higher for the CO_2/CH_4 plasma than that of the oxygen plasma. This work therefore presents the conditions in which these plasmas can provide the same decomposition rate for contaminants in aqueous solution.

1. Introduction

The increase in the world's population and at the same time, changes in our lifestyles along with industrialization, especially in developing countries, are leading to increase the amounts of wastewaters and contaminated aqueous solutions. This ongoing increase in the amounts of wastewater raises the risk of discharging the polluted water into the natural environment. The nature of the contaminants is different. A significant amount of pollutants in wastewater of chemical process industries, such as dyes, pigments, paints, and petrochemicals, are carboxylic acids [1]. They also form a significant part of organic contaminants in process liquors in some industries, such as the aluminum industry (Bayer process) [2]. A large amount of solubilised and built up carboxylic acids in process liquors and wastewaters has negative effects on the environment and the production efficiency of these industries. Therefore, demands for effective and environmentally friendly technologies for aqueous solution treatment have increased [2]. Among different types of thermal plasma, direct current (DC) arc plasma is a common type of thermal plasma for treating liquids and solutions [2]. DC submerged thermal plasma reactor is a specific type of DC thermal plasma reactors because it can offer a direct contact between the plasma and solution [3].The advantages of the submerged thermal plasma also include the presence of steep thermal gradients, the photolysis reaction as a result of UV radiation of plasma, ability to treat high volumes of wastewater, and proper mixing between solution and plasma [4]. In the case of DC plasma torches, different plasma gases can lead to different decomposition rates of the treated material depending on their reactive species, enthalpy and thermal conductivity. They also have a direct effect on the electrode life-time, which is an important issue for DC plasma torches [2]. Air, oxygen, argon, and nitrogen are considered as common plasma gases [5]. Carbon-containing molecular gases, such as CO_2-based gas mixtures, are recently presented by Mostaghimi et al. [6] as plasma gases for a newly designed DC plasma torch. This new DC plasma torch shows some advantages compared to DC plasma torches with common plasma gases. It can efficiently transfer heat to the treated waste due to its high plasma enthalpy and high thermal conductivity. Moreover, it produces a positive carbon ion current, which can be deposited on the surface of a negatively charged cathode. This carbon deposition protects the electrode from erosion and increases the electrode life-time. Furthermore, CO_2 can be found in almost pure form as a waste gas in certain chemical industries [7]. Therefore, using easily available plasma gas like CO_2 along with a long-life electrode makes this torch valuable. However, the performance of this new DC thermal plasma torch has not been yet compared with common DC thermal plasma torches for treatment of organic contaminants in solutions and liquids.

Therefore, the scope of this study is to investigate the comparison between the performances of CO_2-based gas mixture thermal plasma torch and oxygen thermal plasma torch in a submerged mode at different operational conditions for decomposition of carboxylic acid in aqueous solution. Sebacic acid, a typical organic contaminant in Bayer liquor, was selected as a representative of organic contaminant. The effectiveness of the process was therefore defined by decomposition rate of sebacic acid. The operational parameters such as treatment time, and reactor pressure as well as the role of hydrogen peroxide as an oxidizing agent were investigated on the decomposition of sebacic acid for both plasmas.

2. Material and methods

Figure 1 shows a photo of the apparatus used in this work. This apparatus was initially designed by Bernier et al. [3] based on a submerged thermal plasma system. In this study, two different direct current (DC) plasma torches using CO_2-CH_4 and O_2 gases were separately used to generate thermal plasma at the bottom of the reactor. A DC plasma torch designed and built in the Centre for Advanced Coating Technology at University of Toronto with a 39 kW power (60% thermal efficiency), and a graphite cathode was used while the plasma gas was CO_2 and CH_4 in a volume ratio of (CO_2/CH_4= 1.875). The presence of CH_4 in plasma gas composition produces a protective graphite deposition on the surface of the negatively charged graphite cathode protecting the cathode form erosion. Therefore, the cathode will be operational as long as the CH_4 is present in the plasma gas composition. Another 39 kW DC plasma torch with a copper cathode, around 60% thermal efficiency, and oxygen as plasma gas designed and built in the thermal plasma laboratory at Université de Sherbrooke was also used for generation of thermal plasma. The operating procedure for the decomposition of sebacic acid by submerged thermal plasma was as follows: 20 L of distilled water and 4.5g/L of sebacic acid (supplied by Alfa Aesar®) and known amount of sodium hydroxide (97%, from Alfa Aesar®) for changing the solution pH were placed in the solution tank. The solution was stirred before injecting into the reactor for making homogenous solution. The solution was injected into the reactor via the solution pump after preheating to 50 °C. Totally dry CH_4 (99.99% pure) and CO_2 (99.99% pure) in a volume ratio of (CO_2/CH_4= 1.875), and dry oxygen (99.99 % pure) were used as plasma gases for the experiments. Plasma gas (CO_2/CH_4 or oxygen) were then injected into the discharge region of the torch. The beginning of the treatment process (t=0) was set with the starting of the plasma ignition. The solution samples were then taken periodically from the sampling line at the middle of the reactor for analytical characterization. IC/MS (A Dionex ICS-3000 ion chromatography system equipped with a Thermo Scientific MSQ Plus Mass Detector) was used to quantify sebacic acid present in the sample. More details about the analytical characterization can be found in [5].

The emission of UV radiation (UVB and partially UVC) from plasma in the submerged mode was quantified by means of an IL-1700 research radiometer with SED240 narrow-band germicidal light probe (International Light, Inc., Newburyport, Mass.) as can be seen in figure1 b. The zero point for the detector of radiometer was set at the room light. The organic compounds in the solution absorb the major amount of UV radiation [8]. Therefore, to measure and compare the intensity of UV radiation from the CO_2-based gas mixture plasma and the oxygen plasma in the submerged mode, the blank experiments were performed with distilled water, without adding sebacic acid.

(a)

(b)

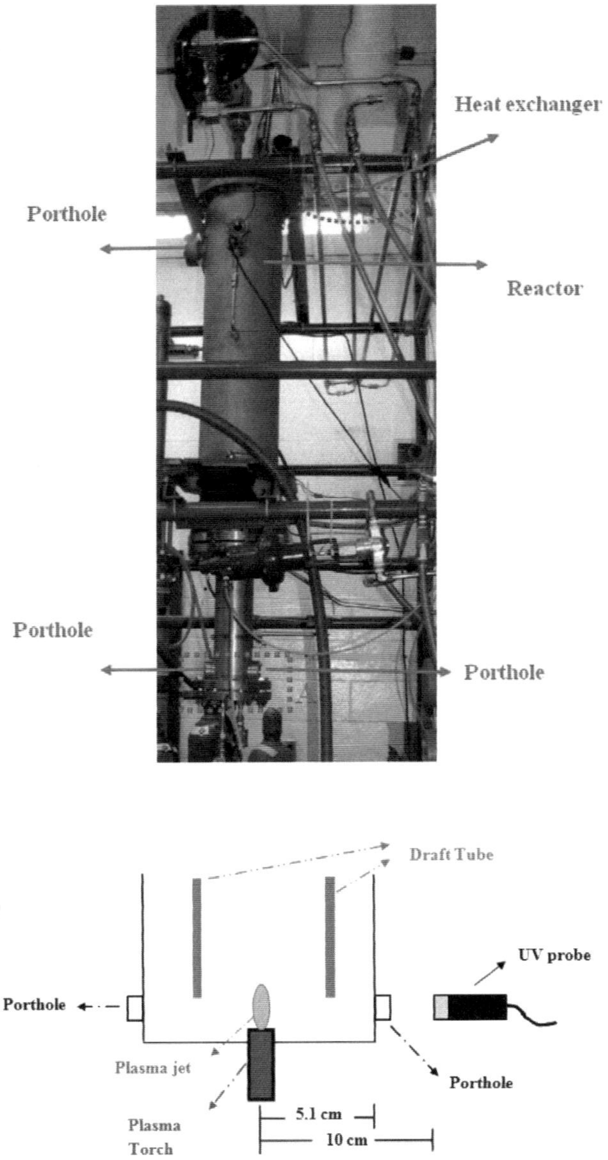

Figure 1. (a) The experimental set-up (b) Magnified zone of location A shown in figure 1 (a) with UV probe for plasma radiation measurement.

3. Result and discussion

3.1. Decomposition vs. treatment time

Decomposition rate of sebacic acid with the oxygen and the CO_2/CH_4 plasmas in basic medium is shown in figure 2. It is observed that the decomposition of sebacic acid is feasible with both plasmas. However, the decomposition rate of sebacic acid with the CO_2/CH_4 plasma is lower than that of the oxygen plasma in basic medium. The higher decomposition rate of sebacic acid via the oxygen plasma can be attributed to the higher amount of contributing plasma oxidant species in the decomposition of sebacic acid. The presence of CO_2 in the CO_2/CH_4 plasma leads to the production of dissolved bicarbonate and carbonate species upon contact with the solution, without any sebacic acid decomposition. These species can act as scavenger of oxidant species, which decrease the amount of oxidant species contributing in the decomposition of sebacic acid with the CO_2/CH_4 plasma. Therefore, lower decomposition rate of sebacic acid with the CO_2/CH_4 plasma compared to the oxygen plasma in basic medium can be due to high amount of bicarbonate and carbonate species in solution as well as low amount of plasma oxidant species.

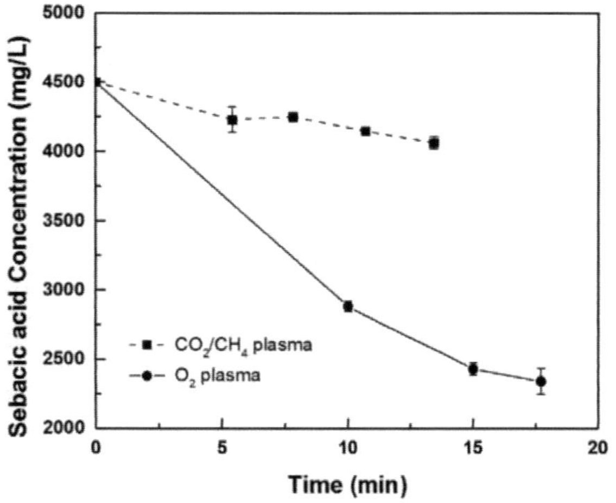

Figure 2. Sebacic acid concentration versus treatment time, solution pH=12.5, $C_{\text{Sebacic acid}}$ = 4.5g/l, CO_2/CH_4=1.875, reactor pressure = 1 atm (108 kPa).

3.2. Decomposition rate in the presence of hydrogen peroxide

The role of hydrogen peroxide as an oxidizing agent on decomposition of sebacic acid is illustrated in figure 3 for the oxygen, and the CO_2/CH_4 submerged thermal plasmas. Hydrogen peroxide is a common industrial oxidizing agent and is easy to provide. Hydrogen peroxide was added through the make-up water (5% volume of make-up water) to the solution during the treatment. As observed in figure 3, adding H_2O_2 has a significant positive effect on further decomposition of sebacic acid with the CO_2/CH_4 plasma. It can be attributed to the high amount of UV radiation of the CO_2/CH_4 plasma.

However, adding H_2O_2 does not make any difference on decomposition rate of sebacic acid with the oxygen plasma. The UV emission detected by UV detector from the oxygen plasma in submerged mode was (4.26E-06 (W/cm^2)), which is much lower than that of the CO_2/CH_4 plasma 5.63E-04 (W/cm^2). The UV radiation leads to the production of hydroxyl radicals via cleavage of the hydrogen peroxide molecule (equation (1)).

$$H_2O_2 \xrightarrow{hv} 2HO^{\bullet} \tag{1}$$

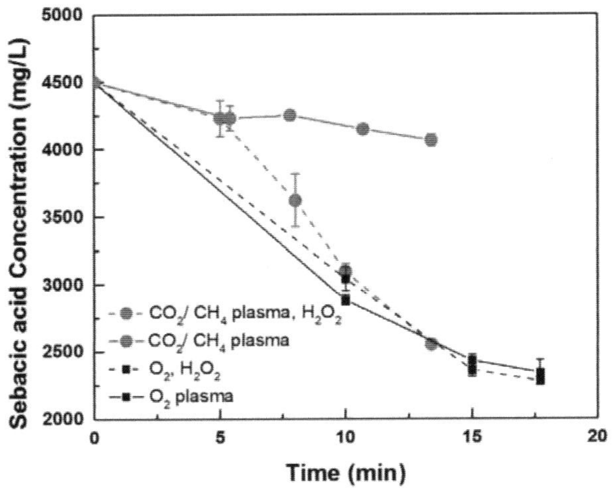

Figure 3. Effect of adding H_2O_2 as oxidizing agent on sebacic acid concentration vs. time, $C_{Sebacic\ acid} = 4.5$g/l, pH=12.5, CO_2/CH_4=1.875, reactor pressure = 1 atm (108 kPa).

The positive effect of adding hydrogen peroxide on decomposition rate of sebacic acid with the CO_2/CH_4 plasma is therefore attributed to the formation of hydroxyl radicals, which is a highly oxidant specie, through the photolysis of H_2O_2. As seen in figure 3, the addition of H_2O_2 in the solution treated by the CO_2/CH_4 plasma increases the sebacic acid decomposition rate up to the same decomposition rate with the oxygen plasma. The high intensity of UV radiation of CO_2 plasma, which leads to photolysis of hydrogen peroxide and formation of hydroxyl radical, has positive effect on sebacic acid decomposition. Therefore, the CO_2/CH_4 plasma in the presence of hydrogen peroxide can be suitable choice as the oxygen plasma for the treatment of contaminated solution containing sebacic acid.

3.3. Decomposition rate at non-atmospheric pressure

The effect of non-atmospheric pressure on decomposition of sebacic acid with both plasmas was investigated by performing the experiments at atmospheric and non-atmospheric (10 psig equivalent to 1.7 atm) pressure. The comparison of sebacic acid decomposition rate at atmospheric and non-atmospheric pressure for both plasmas is shown in figure 4. It can be seen that higher pressure leads to higher decomposition rate of sebacic acid regardless of the plasma gas type. However, the increase in sebacic acid decomposition rate is lower for the oxygen plasma than that of the CO_2/CH_4 plasma. It has been shown that the intensity of UV radiation from plasma could be increased by increasing pressure [4, 8, 9]. The higher UV radiation resulting from higher reactor pressure leads to producing higher amount of hydroxyl radicals (equation (2)), and subsequently higher sebacic acid decomposition rate.

$$H_2O + OH^- \xrightarrow{h\nu} H^+ + HO^\bullet + e^- \qquad (2)$$

Furthermore, as shown in figure 4, the treatment time for the same amount of sebacic acid decomposition at non-atmospheric pressure with the CO_2/CH_4 plasma is higher compared to that with the oxygen plasma. This again indicates the lack of oxidizing hydroxyl radicals present in the solution treated by CO_2/CH_4 plasma due to the formation of bicarbonate and carbonate species.

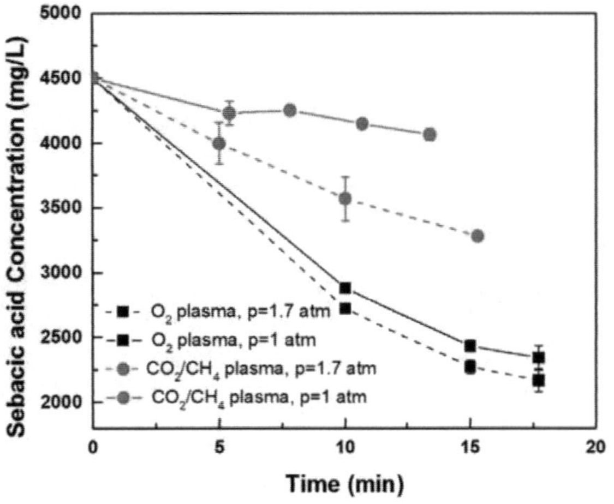

Figure 4. Effect of reactor pressure on sebacic acid concentration, $C_{Sebacic\ acid}$ = 4.5g/l, CO_2/CH_4=1.875, solution pH=12.5.

4. Conclusion

The performance of oxygen direct current (DC) thermal plasma torch and a newly designed carbon dioxide based gas mixture DC plasma torch has been studied for treatment of an aqueous solution containing carboxylic acid. Sebacic acid was selected as a representative of high molecular weight carboxylic acid contaminant in wastewater. The feasibility study confirmed the decomposition of sebacic acid in aqueous solution with both plasma torches. However, the oxygen plasma showed higher conversion rate in basic medium. The results also revealed the positive role of H_2O_2 in the sebacic acid decomposition rate with the CO_2/CH_4 plasma due to synergetic effect between H_2O_2 and UV radiation. It was observed that, by adding H_2O_2 to the solution treated by the CO_2/CH_4 plasma, decomposition rate can attain to the same decomposition rate of the oxygen plasma. Increasing the pressure also enhanced the decomposition of sebacic acid for both plasmas. However, the increasing tendency is lower for the oxygen plasma than that of the CO_2/CH_4 plasma. This work therefore not only shows the difference between the performance of two plasma torches at different operational conditions, but also presents the condition in which these plasmas can provide the same decomposition rate for contaminants in aqueous solution.

Acknowledgments

This study was funded by NSERC (Natural Sciences and Engineering Research Council of Canada) discovery grant. We are grateful for the support of NSERC for funding this research.

References

[1] Gandhi V, Mishra M and Joshi P A 2012 *Mater. Sci. Forum.* 175-89
[2] Safa S and Soucy G 2013 *Int. J. Environ. Sci. Technol.* DOI: 10.1007/s13762-013-0356-3
[3] Bernier J L, Fortin L, Boulos M I, Kasireddy V, Jonquiere C and Soucy G 2001 *US patent* 6,187,206
[4] Fortin L, Soucy G, Kasireddy V K and Bernier J L 2000 *Can. J. Chem. Eng.* **78** 643-649
[5] Safa S and Soucy G 2014 Chem. Eng. J. **244** 178-187
[6] Chen L, Pershin L and Mostaghimi J 2008 IEEE Trans. Plasma Sci. **4** 1068-1069
[7] Kobayashi A, Osaki K and Yamabe C 2002 Vacuum **3-4** 475-479
[8] Poirier G 2001 Traitement de liqueur Bayer synthétique par plasma submergé, M.Sc thesis, Université de Sherbrooke, Sherbrooke, QC, Canada
[9] Kawakami R, Niibe M, Takeuchi H, Konishi M, Mori Y, Shirahama T, Yamada T and Tominaga K 2013 Nucl. Instrum. Meth. B **315** 213-217

Large-Eddy Simulation of Subsonic-Supersonic Flow and Heat Transfer in a Hybrid Gas–Water Stabilized Arc

J Jeništa[1], H Takana[2], H Nishiyama[2], M Bartlová[3], V Aubrecht[3] and P Křenek[1]
[1]Institute of Plasma Physics AS CR, v.v.i., Za Slovankou 3, 182 00 Praha 8, Czech Republic
[2]Institute of Fluid Science, Tohoku University, 2-1-1 Katahira, Aoba-ku, Sendai, Miyagi, 980-8577, Japan
[3]Brno University of Technology, Technická 8, 616 00 Brno, Czech Republic

E-mail: jenista@ipp.cas.cz

Abstract. The paper presents numerical simulations of the turbulence effect in the discharge and near-outlet regions of the hybrid-stabilized argon-water electric arc. Calculations were carried out for the assumption of laminar and turbulent plasma flow models, respectively. Results of simulation for currents 300-600 A on a fine numerical grid show that the influence of turbulence is weak and the maximum difference for all the monitored physical quantities is less than 2% in the discharge region and less than 6% in the near-outlet region. Comparison with experimental temperature profiles exhibits good agreement.

1. Introduction

The so-called hybrid-stabilized electric arc, which was developed a decade ago at the Institute of Plasma Physics AS CR, v.v.i. (IPP AS CR, v.v.i.) in Prague, utilizes a combination of gas and vortex stabilization. A picture and its principle is shown in figure 1. In the hybrid argon–water plasma torch, the arc chamber is divided into the short cathode part, where the arc is stabilized by tangential argon flow, and the longer part, stabilized by water vortex created by tangential injection of water in the chamber. The arc burns between the cathode, made of a small piece of zirconium pressed into a copper rod, and the external water-cooled rotating disc anode, made of a copper and placed a few millimeters downstream of the torch nozzle exit. This arrangement provides not only the additional stabilization of the arc in the cathode region and the protection of the cathode tip, but also offers the possibility of controlling argon–water plasma jet characteristics in a wider range than that of pure gas- or liquid-stabilized arcs [1]. The experiments made on this type of torch [2] showed that the plasma mass flow rate, velocity and momentum flux in the jet can be controlled by changing the mass flow rate in the gas-stabilized section, whereas thermal characteristics are determined by the processes in the water-stabilized section.

The hybrid arc has been used at IPP AS CR, v.v.i., in the plasma spraying torch WSP®H (160 kW) for spraying metallic or ceramic powders (TiO_2, Al_2O_3, $ZrSiO_4$, W-based, Ni-based alloys, Al, steel) injected into the plasma jet [3]. Several years ago, an experimental plasmachemical reactor PLASGAS equipped with the spraying torch WSP®H was utilized for the innovative and environmentally friendly plasma gasification and pyrolysis of wastes with a perspective to their sustainable energetic and chemical valorization and to a reduction of the emission of greenhouse gases [4]. Pyrolysis of biomass was experimentally studied in a reactor using crushed wood, sunflower seeds or rubber oil as model substances. A high content of a combustible mixture of hydrogen and carbon monoxide was produced (i.e., syngas).

Some of our previous calculations analyzed flow regimes, thermal, electrical characteristics and power losses from the hybrid-stabilized electric arc [5-6]. Images of plasma jet taken with the fast shutter camera show the laminar structure of the plasma flowing out of the discharge chamber downstream of the nozzle exit for currents up to 600 A, i.e., the operational regime of our torch. Our previous calculation demonstrates that the Reynolds number based on the 6 mm outlet diameter reaches 13,000 at maximum in the axial region just behind the exit nozzle and decreases to 300 in arc fringes.

The type of flow inside the discharge chamber has been already studied in [7] and it was concluded that the influence of turbulence inside the discharge region is small. Here we study this phenomenon in more detail. First, since the difference between laminar and turbulent flow regimes is expected to be small, it is necessary to adjust numerical grid to reduce at the minimum grid-dependence of the calculated physical quantities. Second, the character of flow was investigated here not only inside the discharge region as in [7] but also in the nozzle and near-outlet regions of the arc torch.

Sections 2 and 3 provide information about the model assumptions, plasma properties, boundary conditions and the numerical scheme. Section 4 reveals the most important findings and a comparison of our calculated results with experiments.

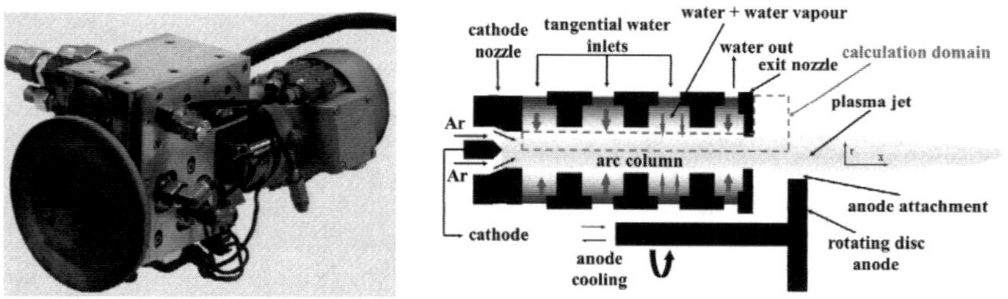

Figure 1: The plasma spraying torch WSP®H with hybrid stabilization (left), i.e., the combined stabilization of the arc by axial gas flow (Ar or N_2) and water vortex. The principle of the hybrid plasma torch WSP®H is shown at right. The calculation domain is shown by a red-dashed line.

2. Physical model

The following assumptions for the model were applied:

1) The numerical model is two-dimensional with the discharge axis being the axis of symmetry (axisymmetric);
2) Plasma flow is laminar/turbulent and compressible in the state of local thermodynamic equilibrium (LTE);
3) Argon and water create a uniform mixture in the arc chamber (single gas phase);
4) Only a self-generated magnetic field by the arc itself is considered;
5) Gravity effects are negligible;
6) The partial characteristics method for radiation losses from the arc is employed.

The complete set of conservation equations representing the mass, electric charge, momentum and energy transport of such plasma can be written in the vector notation as follows:

continuity equation:

$$\frac{\partial}{\partial t}\rho + \nabla\cdot(\rho\vec{u}) = 0 \quad , \tag{1}$$

momentum equations:

$$\frac{\partial}{\partial t}(\rho\vec{u})+\nabla\cdot(\rho\vec{u}\vec{u})=-\nabla p+\nabla\cdot\vec{\tau}+\vec{j}\times\vec{B} \quad , \qquad \tau_{ij}=\eta\left(\frac{\partial u_i}{\partial x_j}+\frac{\partial u_j}{\partial x_i}-\frac{2}{3}\delta_{ij}\frac{\partial u_l}{\partial x_l}\right) \quad , \tag{2}$$

energy equation:

$$\frac{\partial e}{\partial t}+\nabla\cdot\left[(e+p)\vec{u}-\lambda\nabla T\right]=\vec{j}\cdot\vec{E}+\Phi_{diss}-\dot{R} \quad , \tag{3}$$

charge continuity equation:

$$\nabla\cdot(\sigma\nabla\Phi)=0 \quad , \tag{4}$$

equation of state:

$$p=\rho R_g(p,T)T \quad . \tag{5}$$

Here ρ is the mass density, $\vec{u}$ the velocity vector containing the axial (u) and radial (v) velocity components, p the pressure, $\vec{\tau}$ the stress tensor, $\vec{j}$ the current density, $\vec{B}$ the magnetic field, e the density of internal and kinetic energy produced or dissipated in the unit volume, T the temperature, $\vec{E}$ the electric field strength, Φ_{diss} the viscous dissipation term, $\dot{R}$ the divergence of radiation flux, Φ the electric potential, λ and σ are the thermal and electrical conductivities, respectively, and R_g is the specific gas constant dependent on pressure and temperature.

The assumption 3) means that argon- and water-plasma species create single gas (or plasma) phase with a constant argon molar fraction everywhere in the discharge. For example, for 500 A and 40 slm of argon, the argon molar fraction in the water-argon plasma is 43%. No species conservation equation is thus needed to be solved. In the simulation, the appropriate tables of the transport and thermodynamic properties for this argon molar content must be loaded. We have used the set of tables with the argon molar fraction from 0 to 1 with the step of 0.05. The thermodynamic properties and the transport coefficients of the argon-water plasma mixture were calculated according to the Chapman–Enskog method in the 4th approximation [8] for temperatures of 400–50,000 K [9] and pressures of 0.1-0.3 MPa in the local thermodynamic equilibrium. For the temperature range 400 – 20 000 K the total number of 35 chemical species was considered [9]. The calculations of plasma composition were performed using the modified Newton method for the solution of the nonlinear equations system, which is composed of equations of Saha and mass action law type expressing individual complex components by means of basic ones (e, H, O, Ar). The system was completed by the usual particle and charge balance assuming quasineutrality and equilibrium. The use of perfect gas law in the present model (equation of state) is justified by the fact that the Debye–Hűckel pressure correction at the given thermodynamic conditions was found to be of the order 10^{-3} which cannot influence the relations between molar fractions [9].

Turbulence was modeled by Large-Eddy Simulation using the Smagorinsky sub-grid scale model [10-13] with the constant values of the Smagorinsky coefficient (C_S = 0.1) and the turbulent Prandtl number (Pr_t = 0.9). To suppress turbulence near the walls, we applied a Van Driest damping function near the walls [10] in the form

$$l_S=C_S\Delta\left[1-exp\left(-y^+/A^+\right)\right] \quad , \tag{6}$$

where l_S is the Smagorinsky lengthscale, C_S is the Smagorinsky coefficient, Δ is the filter width, y^+ is the distance from the wall normalized by the viscous lengthscale, and A^+ = 25. The eddy viscosity μ_t and the turbulent thermal conductivity λ_t are modeled as [10, 11]

$$\mu_t = l_S^2 \overline{S} = (C_S \Delta)^2 \overline{S} \qquad , \qquad \lambda_t = c_p \mu_t / Pr_t \qquad , \tag{7}$$

$\overline{S}$ is the characteristic filtered rate of strain and c_p is the specific heat at constant pressure. The filter width Δ is proportional to the grid size with generally nonequidistant spacing Δx_i and Δy_j, defined here as $\Delta \equiv \Delta_k = \sqrt{\Delta x_i \Delta y_j}$.

Radiation losses from the argon-water arc were calculated by the partial characteristics method for different molar fractions of argon and water plasma species as a function of temperature and pressure. The principle of the method is that the calculation of integrals over radiation frequencies in equations for radiation quantities is performed in advance to form functions Som and Δ Sim, called partial characteristics [14] which include all time demanding integrations over frequencies. These functions are stored in data tables according to several parameters, such as plasma pressure, temperature and geometrical dimensions. The tables are then used in numerical simulations of plasma flow with spatial integrations only. The advantage of this method is that it directly calculates the reabsorption of radiation in the low-temperature regions of the arc. Continuous radiation due to photo-recombination and *"bremsstrahlung"* processes, discrete radiation consisting of thousands of spectral lines, molecular bands of O_2, H_2, OH and H_2O have been included in the calculation of partial characteristics [15]. Broadening mechanisms of atomic and ionic spectral lines due to Doppler, resonance and Stark effects have been considered.

3. Boundary conditions and the numerical method

The calculation region and the corresponding boundary conditions are presented in figure 2. This configuration agrees with the hybrid torch experimental setup. The only difference in boundary conditions between the laminar and turbulent models is a Van Driest damping function, which is applied only in the turbulent model.

(a) Inlet boundary (AB) is represented by the nozzle exit for argon. Along this boundary we assume the zero radial velocity component, $v = 0$. Because of the lack of experimental data, the temperature profile $T(r, z = 0)$ and the electric field strength $E_z = -\partial\Phi/\partial z = const.$ for a given current are calculated at this boundary before the start of the fluid-dynamic calculation itself iteratively from the Elenbaas-Heller equation including the radiation losses from the arc. The inlet velocity profile $u(r)$ for argon plasma for the obtained temperature profile $T(r, z = 0)$ is pre-calculated from the axial momentum equation under the assumption of a fully developed flow.

(b) Axis of symmetry (BC). The zero radial velocity and symmetry conditions for the temperature, axial velocity and electric potential are specified here, i.e., $\partial T/\partial r = \partial u/\partial r = \partial\Phi/\partial r = 0$, $v = 0$.

(c) Arc gas outlet plane (CD). The zero electric potential $\Phi = 0$ (the reference value) and zero axial derivatives of the temperature and velocity components are defined at CD, i.e., $\partial T/\partial z = \partial u/\partial z = \partial v/\partial z = 0$.

(d) Arc gas outlet plane (DE). The zero radial derivatives of the temperature, velocity and electric potential are defined here, $\partial T/\partial r = \partial u/\partial r = \partial v/\partial r = \partial\Phi/\partial r = 0$. The total pressure is fixed at 1 atmosphere, p = 1 atm.

(e) Outlet wall and the nozzle (EF). We specify no slip conditions for velocities, $u = v = 0$, constant values of E_r and E_z ($\partial\Phi/\partial z = \partial\Phi/\partial r = 0$) and $T(r, z) = 773$ K (500° C) for the temperature of the nozzle.

(f) Water vapor boundary (FA). Along this line we specify the so-called "effective water vapor boundary," designated in Fig. 2 as the "water vapor boundary" with a prescribed temperature of water vapor $T(R = 3.3mm, z) = 773$ K. This is a numerical simplification of a more complex physical reality assumed to be near the phase transition water-vapor in the discharge chamber. Since we are not

Journal of Physics: Conference Series **550** (2014) 012016 doi:10.1088/1742-6596/550/1/012016

able to see inside the discharge chamber, the shape of the phase transition between water and vapor in the discharge chamber is not experimentally known and it is unclear as yet if the structure of the transition is simple or very complicated, for example, with a time-dependent form. In Ref. [16] the iteration procedure for determination of the mass flow rate $\dot{m}$ and the radius of the "water vapor boundary" for each current was proposed. The final corresponding values of water mass flow rates $\dot{m}$ are: 0.228 g s^{-1} (300 A), 0.315 g s^{-1} (400 A), 0.329 g s^{-1} (500 A), 0.363 g s^{-1} (600 A). The magnitude of the radial inflow velocity is calculated from the definition of mass flow rate

$$v(R) = \frac{\dot{m}}{2\pi R \sum_{\Delta z} \rho(R,z)\Delta z} , \tag{8}$$

where, $\rho(R,z)$ is a function of pressure and thus dependent on the axial position z, Δz being the distance between the neighboring grid points.

Due to practically zero current density in the cold vapor region (no current goes outside of the lateral domain edges), the radial component of the electric field strength is put at zero, i.e., $E_r = -\partial\Phi/\partial r = 0$. The axial velocity component is set to zero, $u = 0$.

LU-SGS (Lower-Upper Symmetric Gauss-Seidel) algorithm [17], belonging to the group of the density-based methods coupled with the Newtonian iteration method, are used for the integration of discretized Eqs. (1)~(4) in time and space. To resolve compressible phenomena accurately, the Roe Flux Differential Scheme coupled with the third-order MUSCL-type (Monotone Upstream-centered Schemes for Conservation Laws) TVD (Total Variation Diminishing) scheme are used for a convective term. The electric potential from equation (4) is solved in a separate subroutine by the TDMA (Tri-Diagonal Matrix Algorithm) line-by-line method, including the block correction technique to speed up the convergence. Under-relaxation is employed to avoid divergence.

The computer program was written in the Fortran language. The task was solved on oblique structured grids with nonequidistant spacing. We chose three different grids in order to make numerical tests on grid-independence and appropriate resolution of the LES model. The coarsest grid contains totally 38,553 grid points (Grid1) with 543 and 71 points in the axial and radial directions respectively. The two finer grids have 115,018 (Grid2) and 193,914 (Grid3) grid points with 878 x 131 (Grid2) and 1134 x 171 (Grid3) points in the axial and radial directions respectively.

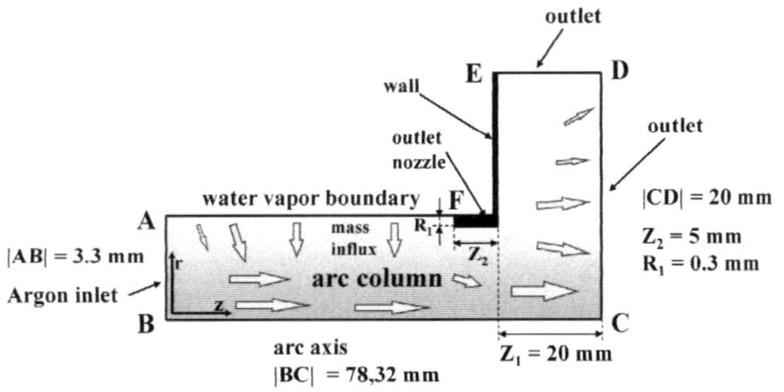

Figure 2: Discharge area geometry. The origin of the r-z coordinate system is set at the point B. The dimensions of the domain are 3.3 mm for the radius of the discharge region (AB), 20 mm for the outlet region (Z_1) and 78.32 mm for the total length (BC). Inlet boundary (AB) is represented by the nozzle exit for argon. Along the line FE the outlet nozzle and the wall of the torch equipment is specified.

4. Results of Calculation

Calculations for the assumptions of turbulent and laminar flows have been carried out for currents 300, 400, 500 and 600 A. The mass flow rate for the water-stabilized section of the discharge was taken for each current between 300 and 600 A from our previously published work [16]. The values of argon mass flow rates were chosen 22.5, 27.5, 32.5 and 40 slm (standard liters per minute) in accordance with experimental conditions. It was found from the mass and energy balances at the torch exit nozzle [18] that part of argon is taken away before it reaches the torch exit because argon is mixed with vapor steam and removed to the water system of the torch. The amount of argon transferred in such a way from the discharge is around 50 % for the currents studied [18]. Since the present model does not treat the mechanism of argon removal in the model, we consider in the calculations that argon mass flow rate present in the discharge equals one-half of argon mass flow rate at the torch inlet (i.e., for 22.5 slm we take 50% of this amount, etc.).

4.1 Grid-independence study

It is known that a computational grid density can influence the stability of a solution, convergence of a solution and the final distribution of physical quantities. It usually holds that an increasing number of grid points changes the distribution of calculated quantities. There is however a limit number of points behind of which the results become nearly independent of the grid density. Here we employed Grids1-3 mentioned above to study this phenomenon.

Some of the results presented in this paper are evaluated as an average quantity in the discharge region or in the nozzle and near-outlet regions. For this purpose we defined two subdomains in the calculation domain - Domainl and Domain2, displayed in figure 3. Textured Domain 1 includes the whole discharge region and the cylindrical outer region 2 mm thick and 3.3 mm in radius. The yellow Domain2 covers the nozzle wall region and a shear layer between hot plasma and cold surrounding gas with the axial coordinates from 50 to 78 mm and the radial ones between 0 and 5 mm.

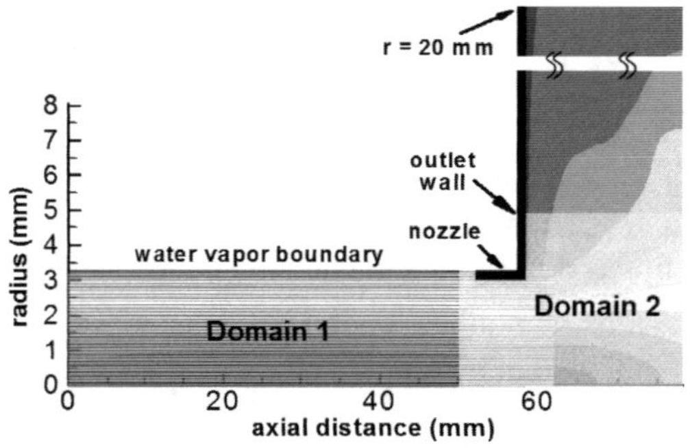

Figure 3. Detail of subdomains (Domainl, Domain2) in the whole calculation domain.

Figure 4 shows the relative difference of solutions between two sets of grids – Gridl and Grid2 (left) and Grid2 and Grid3 (right). Temperature, Velocity and the Mach number are taken at the axial position 2 mm downstrean of the nozzle exit, overpressure and the electric potential are taken within the length of the discharge, 0-78 mm (see figure 2). It is obvious that the difference becomes the smallest for the case of finer grids, Grid2 and Grid3, nevertheless the difference even for coarser Gridl and Grid2 is in this case less than only 0.6%. The same tendency occurs for all studied currents and mass flow rates.

Figure 5 presents the so called space-average relative difference between turbulent and laminar models in Domainl and Domain2 (see figure 3) for velocity, Reynolds number and temperature. The space-average relative difference for a given physical quantity X is calculated as

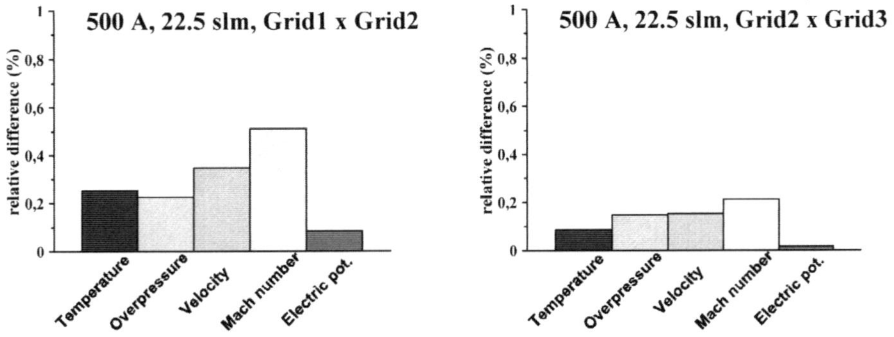

Figure 4. Relative difference of solutions between two sets of grids – Grid1 and Grid2 (left) and Grid2 and Grid3 (right) for the assumption of turbulent flow.

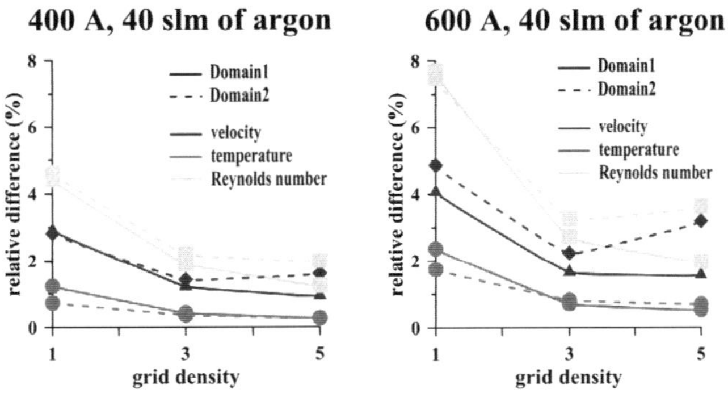

Figure 5. The space-average relative difference between turbulent and laminar models in Domain1 and Domain2.

$$\Delta_{av}^{X} = \frac{100}{N} \cdot \sum_{i=1}^{N} abs\left(X_{turb}^{i} - X_{lam}^{i}\right) / \left[0.5 * \left(\left|X_{turb}^{i}\right| + \left|X_{lam}^{i}\right|\right)\right] \quad , \tag{9}$$

where N is the total number of grid points in Domain1 or Domain2; "*turb*" and "*lam*" stand for the turbulent and laminar values of *X*, respectively. Grid density is defined as the ratio of the number of grid points to the Grid1 number of points (Grid2 ~ grid density = 3, Grid3 ~ grid density = 5). The results shown here demonstrate that, except for temperature, the space-average relative difference is higher in Domain2 due to the increased eddy viscosity in the shear layer (a transition region between the hot arc and the surrounding atmosphere with steep radial gradients). The maximum difference reaches several per cent. One can also see that space-average relative differences for Grid2 and Grid3 are closer to each other than in the case of Grid1 and Grid2.

From the presented numerical tests it comes out that the most accurate results were obtained using Grid3 so that it was employed in our calculations in the next Section.

4.2 Results of calculation for the finest Grid3

Figure 6 shows contours of axial and radial velocities, pressure and temperature in the nozzle and outlet regions for 500 A and 27.5 slm of argon for the finest Grid3. It was confirmed in simulations that results for the studied range of parameters are slightly fluctuating with time. The contours displayed here are static, calculated as arithmetic means of the values in 10-15 consecutive time layers with the time step of about 2-4 ms. There is some visible difference between laminar and turbulent flow regimes for the radial velocity and pressure, while a negligible difference for the axial velocity and temperature. The ratio of axial velocity to radial velocity is 10-30 so the bigger difference in the radial velocity practically does not influence the flow field. The displayed case is the transonic flow with the Mach number ~ 1 at the arc axis.

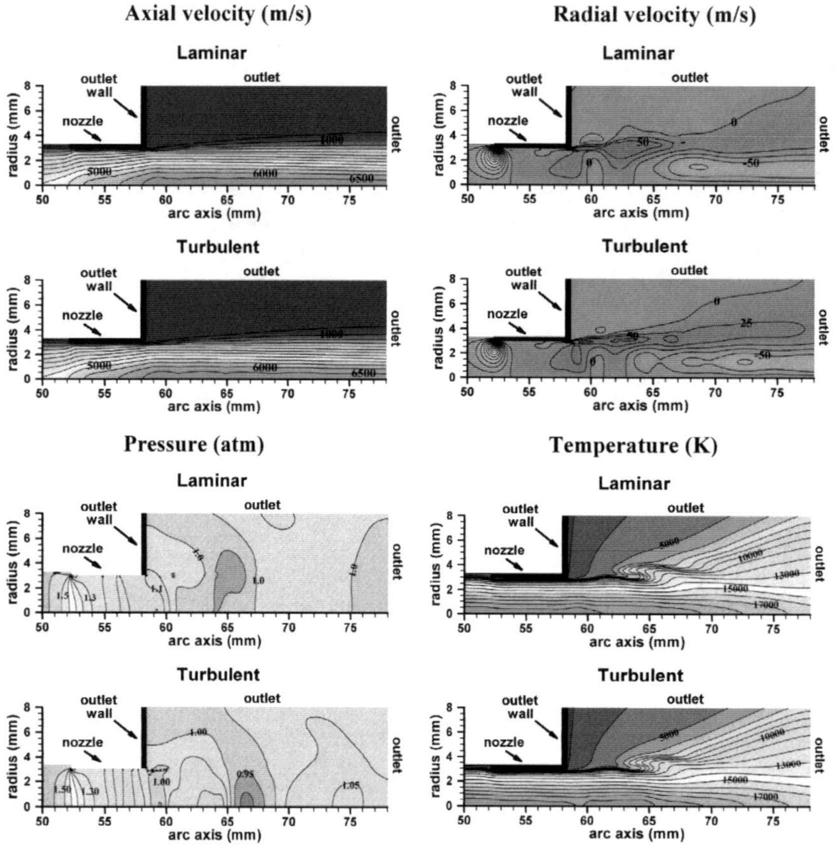

Figure 6. Contours of axial and radial velocities, pressure and temperature in Domain2 (extended here for the radial coordinate up to 8 mm) for 500 A and 27.5 slm of argon for the finest Grid3.

The most comprehensive result about turbulent effects for the all range of studied currents and mass flow rates is shown in figure 7, displaying the contours of the space-average relative differences (equation (9)) between laminar and turbulent flow regimes of several physical quantities calculated in Domains1, 2. To understand these plots, one must realize that calculations for laminar and turbulent flow regimes have been carried out for 16 different combinations of currents and argon mass flow rates. Each calculation provided one value of the space-average relative difference, i.e. one point in the plot. The final contours shown in figure 7 were obtained as the result of interpolation through the matrix of the discrete set of points. It can be seen that the maximum difference in Domain1 is below 2% and occurs for the Reynolds number, derived from the nozzle diameter. Domain2 exhibits higher difference up to 6% for axial velocity and therefore also for the Reynolds number. The reason for increased difference consists in steep radial gradients of temperature and velocity in the shear layer where the eddy viscosity starts to be significant. The results shown here clearly demonstrate that the plasma flow inside the discharge chamber and in the near-outlet can be regarded as quasi-laminar with negligible turbulent effects.

Finally, we compare measured and calculated radial temperature profiles 2 mm downstream of the nozzle exit for 500 A and 600 A for 22.5 slm of argon (figure 8). In our experiment, the radial profiles of the temperature near the nozzle exit were calculated from optical emission spectroscopy measurements. The procedure is based on the ratio of emission coefficients of the hydrogen H_β line and four argon ionic lines using calculated LTE composition of the plasma for various argon mole fractions as a function of temperature [9]. It is obvious that the calculated profiles for Grids1-3 nearly overlap. Very good agreement is demonstrated between experiment and calculation for 500 A, and excellent agreement for 600 A where the measured profile nearly coincides with our calculation.

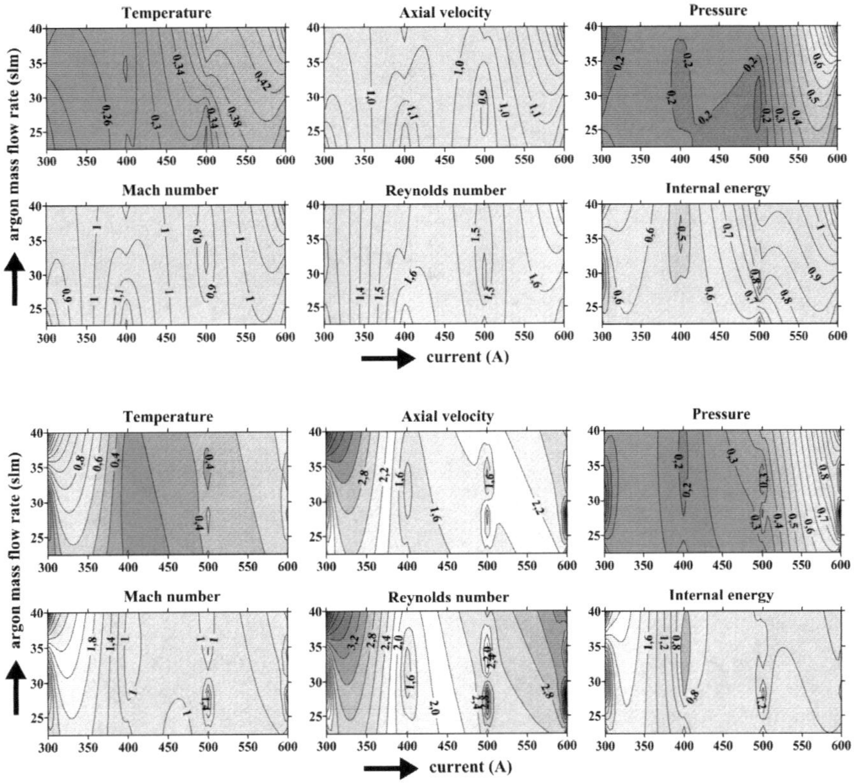

Figure 7. The space-average relative difference between turbulent and laminar models for several quantities in Domain1 (top) and Domain2 (bottom).

5. Conclusions

The results obtained for currents 300-600 A and for argon mass flow rates 22.5-40 slm show that:

(a) The plasma flow in the hybrid-stabilized argon-water electric arc can be regarded as quasi-laminar: The maximum space-averaged relative difference between laminar and turbulent models for each of the monitored physical quantities, calculated for the finest grid, is less than 2% within the volume of the discharge, and less that 6% for the outlet region.

(b) The proper resolution of the LES model has been reached for the finest grid (Grid3).

(c) Turbulent effects are stronger in the regions with high radial temperature and velocity gradients - near sharp edges of the outlet nozzle and in the shear layer.

(d) Comparison with available experimental temperature profiles demonstrates very good agreement for temperature radial profiles near the outlet nozzle.

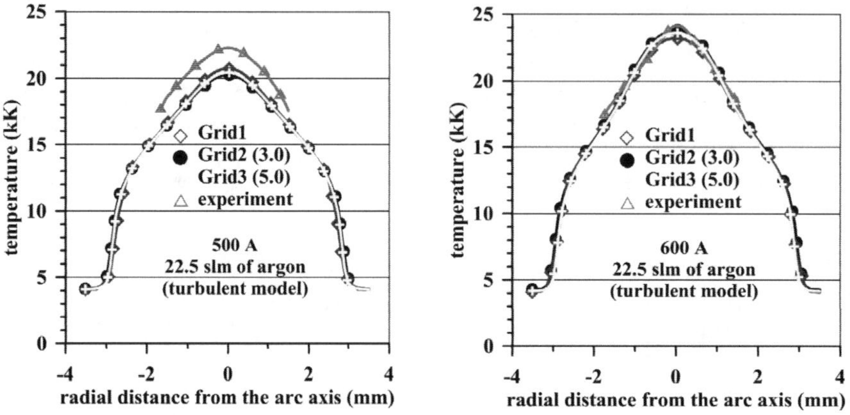

Figure 8. Comparison between measured and calculated radial temperature profiles 2 mm downstream of the nozzle exit for 500 A and 600 A for 22.5 slm of argon. Turbulent flow assumption is adopted.

Acknowledgments
This work was supported by the Collaborative Research Project of the Institute of Fluid Science, Tohoku University, Sendai, Japan; by the Grant Agency of the Czech Republic (GA CR) 205/11/2070; and by the Center CERIT Scientific Cloud, reg. no. CZ. 1.05/3.2.00/08.0144.

References

[1] Hrabovský M, Kopecký V and Sember V *Proc. 16th Int. Symp. on Plasma Chemistry (Taormina)* Italy, on CD-ROM
[2] Hrabovský M, Kopecký V, Sember V, Kavka T, Chumak O and Konrád M 2006 *IEEE Trans. Plasma Sci.* **34** 1566
[3] Chráska T, Neufuss K, Dubský J, Ctibor P and Klementová M 2008 *J. Thermal Spray Technology* **17** 872
[4] Van Oost G, Hrabovský M, Kopecký V, Konrád M, Hlína M and Kavka T 2008 *Vacuum* **83** 209
[5] Jeništa J 2004 *IEEE Trans. Plasma Sci.* **32** 464
[6] Jeništa J, Takana H, Nishiyama H, Bartlová M, Aubrecht V, Křenek P, Hrabovský M, Kavka T, Sember V and Mašláni A 2011 *J. Phys. D: Appl. Phys.* **44** 435204
[7] Jenista J, Takana H, Nishiyama H, Bartlová M, Aubrecht V and Křenek P 2013 *J. Thermal Sci. Technol.* **8** 435
[8] Křenek P and Něnička V 1984 *Acta Technica CSAV* **28** 420
[9] Křenek P 2008 *Plasma Chem. Plasma Processes* **28** 107
[10] Pope S B 2000 *Turbulent Flows* (Cambridge University Press)
[11] Lesieur M, Metais O and Comte P 2005 *Large-Eddy Simulations of Turbulence* (Cambridge University Press)
[12] Schiestel R 2008 *Modeling and Simulation of Turbulent Flows* (John Wiley & Sons, Inc.)
[13] Fox O 2003 *Computational Models for Turbulent Reacting Flows* (Cambridge University Press)
[14] Sevast'yanenko V G 1979 *J. Eng. Phys.* **36** 138
[15] Bartlová M and Aubrecht V 2006 *Czech. J. Phys.* **56** Suppl. B632
[16] Jeništa J 2003 *J. Phys. D: Appl. Phys.* **36** 2995
[17] Yoon S and Jameson A 1988 *AIAA J.* **26** 1025
[18] Kavka T, Chumak O, Sember V and Hrabovský M 2007 *Proc. 28th Int. Conf. on Phenomena in Ionized Gases (Prague)* Czech Republic, p 1819

Journal of Physics: Conference Series **550** (2014) 012017 doi:10.1088/1742-6596/550/1/012017

Advanced oxide powders processing based on cascade plasma

O P Solonenko and A V Smirnov
Khristianovich Institute of Theoretical and Applied Mechanics, Siberian Branch of RAS, Novosibirsk, Russia

E-mail: solo@itam.nsc.ru

Abstract. Analysis of the potential advantages offered to thermal spraying and powder processing by the implementation of plasma torches with inter-electrode insert (IEI) or, in other words, cascade plasma torches (CPTs) is presented. The paper provides evidence that the modular designed single cathode CPT helps eliminate the following major disadvantages of conventional plasma torches: plasma parameters drifting, 1-5 kHz pulsing of plasma flow, as well as excessive erosion of electrodes. More stable plasma results in higher quality, homogeneity and reproducibility of plasma sprayed coatings and powders treated. In addition, CPT offers an extremely wide operating window, which allows better control of plasma parameters, particle dwell time and, consequently, particle temperature and velocity within a wide range by generating high enthalpy quasi-laminar plasmas, medium enthalpy transient plasmas, as well as relatively low enthalpy turbulent plasmas. Stable operation, flexibility with plasma gases as well as wide operating window of CPT should help significantly improve the existing plasma spraying processes and coatings, and also help develop new advanced technologies.

1. Introduction

Knowledge of the fundamental physical processes, taking place in the linear direct current (DC) plasma torches, allowed to suggest a simple classification and to reduce multiple constructive solutions to three principal classes of torches [1, 2] (see figure 1).

The first torch class is widely used in the industry and in scientific works. This torch is equipped with tube exit electrode and has a self-installing length of the arc [1, 2]. Volt-Ampere characteristics (VACs) of such plasma torches have a falling nature (figure 1, curve 1). The mean arc length l_a is a function of arc current, chamber diameter, gas flow rate and pressure. It further depends on the type of working gas and on polarity of the exit electrode. A large-scale shunting, taking place in the transient zone of the discharge channel, forms the mechanism for arc length changes.

The second class of the plasma guns is characterised by the fact that the average arc length l_a is constant in sufficiently wide range of current changes ($I_a<I<I_l$) while other above-mentioned parameters are constant. The arc length l_a is always less than the length of the self-installing arc $l_a<l_{si}$ within the first torch class. VAC has a U-shape form (figure 1, curve 2). Without taking special measures value U is limited to the value where VAC curves 1 and 2 intercept. There are several technical solutions, providing the constant average length of the arc. One of them is the use of a backward-facing stepped anode [1. 2]. The peculiarities of flow gasdynamic structure inside the backward-facing stepped electrode, that consists of two cylinders with different diameters, the diameter of the exit part of the electrode d_3 being larger than d_2, have been analyzed in [2, 3]. It was mentioned that in this case VAC of the arc is situated below VAC of the arc with self-installing mean length. The plasma torches of second type are working stable without additional resistance in the circuit on an upward going branch of arc VAC.

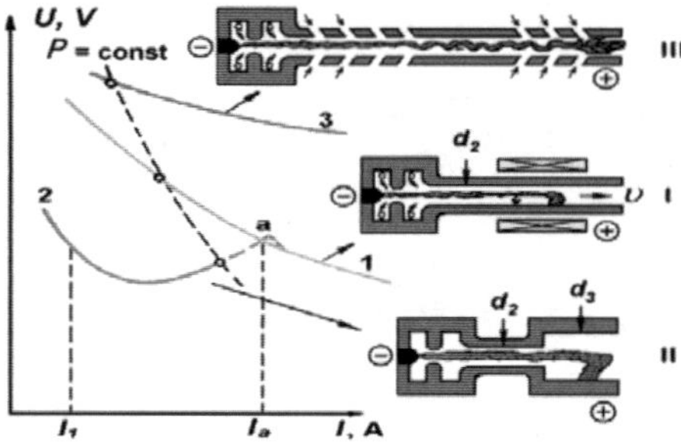

Figure 1 Classification of the linear type DC torches. VACs correspond to three classes of torches. I – torches with self-installing length of arc, II – torches with length of arc fixed by a backward-facing stepped anode, III – cascade plasma torches. Dash line corresponds to fixed electric power of torches (*P*=const).

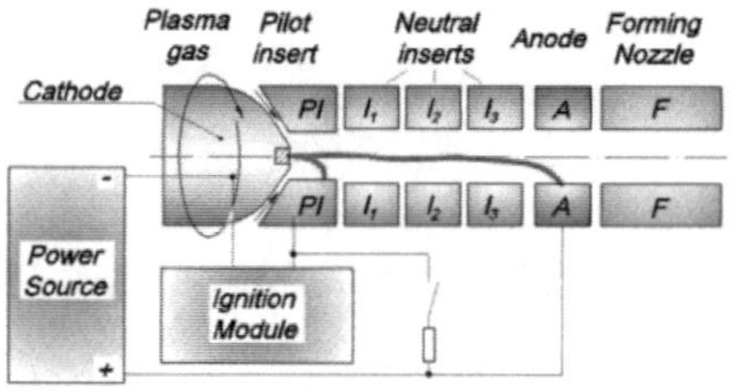

Figure 2 Schematic diagram of the cascade torch.

During the last decades a great attention is paid to the third type of the non-transferred arc torches, in which the averaged arc length in non-changeable, but longer than that of the self-installing length of arc [1, 2]. VAC of this arc (figure 1, curve 3) is situated higher than the two previously mentioned cases. The increase of the average length of the arc is achieved by IEI placed between the anode and the cathode, the length of which is larger than l_{si}. The insert may be solid, porous, with any gas injection through porous structure or sectional with/without gas injection into intersection gaps. This type of torch will be reffered to as cascade torch/gun.

Below is a brief overview of the research and development, carried out in the Khristianovich Institute of Theoretical and Applied Mechanics SB RAS, illustrating the possibilities of cascade torches in context of elaboration of high-tech processes of plasma spraying and powder treatment.

2. Demands on torches for plasma spraying and powders processing

Requirements to a hypothetical "ideal" torch could be used as a set of criteria to compare different possible designs of plasma guns and to make a decision regarding the future of a plasma product line.

In this connection, the following main requirements could be formulated:

- high quality consistent and reproducible sprayed coatings and powders, in particular, hollow microspheres;
- high efficiency and low cost process;
- wide operational window for spraying of different materials and low dependence on some unavoidable changes of parameters (for example, independent of spraying distance changes in a case of parts with a complicated shape);
- possibility to get a high productivity, for instance, deposition rate;
- low maintenance, high reliability, user-friendly;
- controllable within a wide plasma jet temperature and velocity range (to control dwell time, particle temperature and aggregate state) with possibility to provide different regimes of plasma jet flow (quasi-laminar, transient and turbulent);
- axially symmetric with the uniform distributions of plasma velocity and temperature;
- minimized pulsations of arc voltage and plasma jet thermal power;
- possibility to use relatively large inner diameter of anode and forming nozzle to allow the following: (i) easier arrangement of several powder injectors to control a foot print and/or increase homogeneity of powder treatment due to decrease of the velocity and temperature gradients in the cross-sections of the plasma jet, (ii) low dependence on spray distance

changes, carrier gas flow variations and powder injection conditions due to decreased sensitivity to particle trajectory;
- long lifetime of the electrodes (no contamination of coatings and powders by electrode material: W, Cu etc.);
- possibility to scale up the design within a wide range, if necessary (different applications and production rates need different power level);
- possibility to use different gases and their mixtures, in particular inexpensive gases, as plasma-forming gases.

3. Background
The following main designs are available now on the marketplace:
- Conventional Metco torches (9MB, F4,...);
- Mettech 100 kW Axial III gun with an axial powder injection;
- Water-stabilized plasma guns;
- PlazJet;
- Triplex 50 kW (Metco).

The analysis showed that none of the designs could satisfy all requirements to the "ideal" gun. The analysis also showed that a "cascade" approach is the most promising from a theoretical standpoint to satisfy the requirements.

The disadvantages of the torches with self-installing length of arc are studied during last two decades. The main of them are the following:

1) the low-frequency pulsations of the arc voltage, i.e. fluctuation of the thermal power of flowing out plasma jet, due to arc length variation and the movement of anode spot attachment conditioned by magnetic and gas dynamic interaction of a radial arm of the arc with a cold wall boundary layer [1, 3, 4]. This results in a considerable inhomogeneity of powder material processing [4, 5];

2) absence of axial symmetry of velocity and temperature fields in flowing out plasma jet and, as consequence, non-uniform heating and acceleration of particles in the cross-section of the jet; the latter causes the absence of axial symmetry of particles' parameter distributions within a spraying spot.

Discussion regarding the other disadvantages of the conventional DC plasma torches can be found in the publications [4, 5], etc.

4. Features and advantages of the "Cascade" design
A major contribution to study and developing the cascade plasma torches has been made by Zhukov and colleagues [1-3], etc. Results of the investigations and developments of the cascade plasma torches as applied to plasma spraying and powder materials processing have been published in [6-9]. Without going into details, below a brief overview of principal results will be given with the focus on the above-mentioned requirements for the "ideal" torch.

Electrically insulated inter-electrode inserts are applied in the cascade torch (figure 2) to extend and stabilize the arc length over a wide range of gas mass flow rates. Inter-electrode inserts or the neutral electrodes are intermediate electrodes between the cathode and the anode.

A long, high-voltage arc column is stabilized by these neutral inserts rather than by high gas flow rate resulting in more stable plasma jet with minimized pulsation, higher thermal efficiency and, consequently, more efficient particle heating. The concept of low-current and relatively high-voltage design also provides lower cathode and anode erosion [1, 2], longer service life and minimized or no contamination of coating and powder by products of electrodes' erosion.

The following main advantages for the cascade approach could be formulated more specifically.

1. Significantly lower level of electric power low-frequency pulsations because of the "neutral" electrode with a fixed length (see figures 3*a,b*). It should result in better coating quality as there is less unmelted particles, better coating homogeneity and deposition efficiency;

2. Very low or no voltage drifting (figure 3*c*). It results in consistent coating quality and deposition efficiency.

Journal of Physics: Conference Series **550** (2014) 012017 doi:10.1088/1742-6596/550/1/012017

3. Relatively low arc current at fixed electric power (see figure 1) results in a low level of electrodes' erosion because of the actual heat fluxes into electrodes mainly depend on the arc current. Therefore, the gun should have low maintenance and coatings should not be contaminated by products of erosion (Cu, W, etc.). Cleaner coating results in better quality.

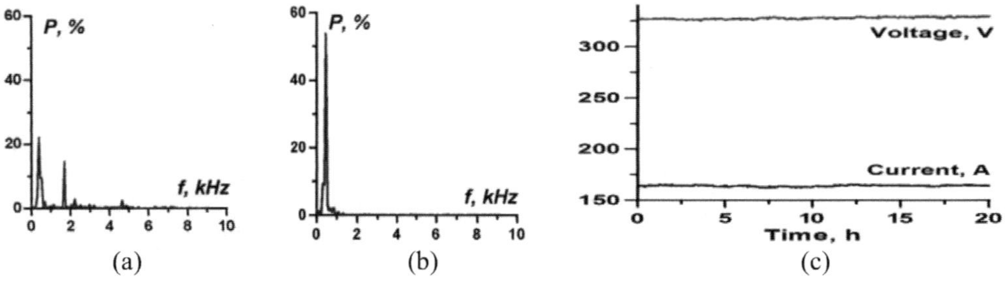

Figure 3 Power spectral density of electric power oscillation for conventional torch – (a), and the cascade torch – (b), other conditions are equal: diameter of the channel – $8 \cdot 10^{-3}$ m, its length – $8 \cdot 10^{-2}$ m, mass flow rate of the working gas (air) – $2 \cdot 10^{-3}$ kg/s, arc current - 250 A. Frequency 300 Hz corresponds to the frequency of the power supply used. (c) - illustration of absence of arc voltage drifting for 100 kW cascade gun, working gas – nitrogen, mass flow rate – $1.5 \cdot 10^{-3}$ kg/s.

4. The "cascade" approach allows changing plasma gases mass flow rate (i.e. Reynolds number, $\mathrm{Re} = 4G / \pi d_a \overline{\mu}$, where G is the mass flow rate of the working gas, d_a is the inner diameter of the outlet electrode (anode), and $\overline{\mu}_{f0}$ is the mean-mass dynamic viscosity of the working gas at the exit cross-section of the outlet electrode) within relatively wide range without significant plasma voltage changes (see figure 4). Therefore, plasma temperature, specific enthalpy and velocity could be changed within a wide range, thus increasing the operational window (figure 4). It could create some new technological opportunities.

5. The "cascade" designs have higher thermal efficiency [1]: 60-70% and more vs. 40-60%. Therefore, for the same power level temperature and specific enthalpy of the plasma jet should be higher.

6. The "cascade" approach allows to use more than one swirl (for example, double swirl, first one – between cathode and pilot section, and second one – between pilot section and inter-electrode insert; the second swirl flow can be concurrent or cross-current to the first one. Therefore, relatively wide range of electric force, arc voltage, turbulence number and more spatially uniform plasma temperature, specific enthalpy and velocity distributions could be achieved [2, 3] (see figures 5 and 6, taken from [1, 3]). The uniform plasma plume will result in more reproducible powder treatment, higher coating quality and higher deposition efficiency. In accordance with [1,3], the following designations are used in figures 5 and 6: d – diameter of channel, m; I – arc current, A; G – total mass flow rate of plasma-forming gas, kg/s; $m_s = g_2/g_1$ – ratio of the mass flow rate of gas supplied between the cathode and pilot section to mass flow rate of gas supplied between the pilot section and inter-electrode insert (position $\overline{z}_s = z_s / d$ is marked by arrow); g_i – mass flow rate of gas supplied between sections of the inter-electrode insert, kg/s; p – pressure inside the channel, Pa; $\overline{l} = l / d$ – relative length of the electric-discharge channel; $\overline{a} = a / d$ – relative length of the inter-electrode insert; $\overline{z} = z / d$ - non-dimensional axial coordinate.

7. Low current gives possibility to use low cost, high enthalpy gas mixtures like air + natural gas or CO_2 + natural gas or other combustible gases. These options could be attractive for the following: 1) low cost manufacturing of micro-spherical, in particular HOSP, powders or other powders where plasma treatment has to be involved, 2) implementation of plasma technologies in countries or areas where high purity inert gases are not available or extremely expensive, 3) air or CO_2 mixtures with

different amounts of combustible gases allow to create oxidizing, reducing or neutral atmospheres that could result in some technological advantages. For example, chrome oxide coatings sprayed by Ar/H_2 or N_2/H_2 mixtures could have some inclusions of metallic Cr. This disadvantage could be avoided by application of "oxidizing" plasma.

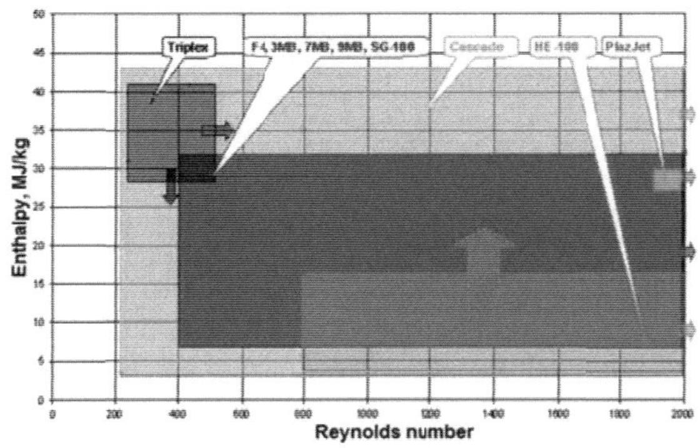

Figure 4 Operation window of 50-kW cascade torch vs. some commercial torches.

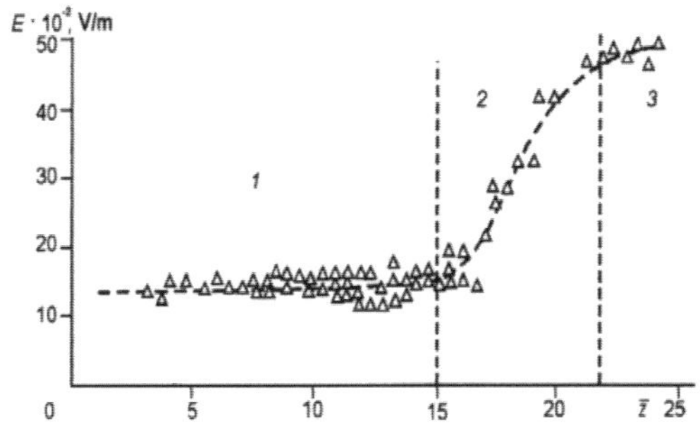

(a)

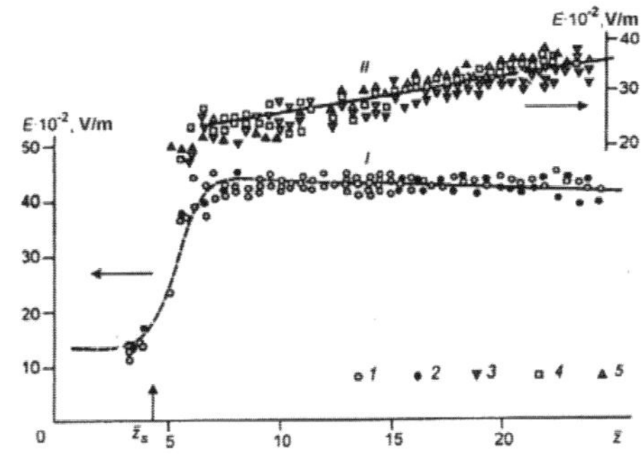

(b)

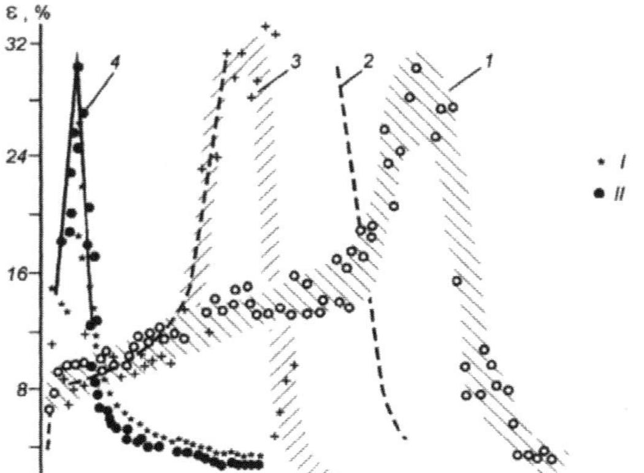

(c)

Figure 5 Distributions of the electric field strength (E) – (a), (b), and turbulent number (ε) – (c), along the discharge chamber axis at different conditions of plasma gas (air) supply. (a) – single supply of swirl working gas between the cathode and pilot section; 1, 2 and 3 correspond to initial, transient and turbulent section of the arc, respectively. (b) – distributed supply of air along the channel. d=2·10^{-2} m; I=100 A; I): g_i=0, G=24.6·10^{-3} kg/s; 1 – m_s=1; 2 – m_s=1.1; II): g_i=0.4·10^{-3} kg/s, G=28·10^{-3} kg/s; 3÷5 – m_s=1.1; 1.43; 1.65. Dashed line was computed from the following formula, generalizing the experiments

$$E_T d = 115(I/d)^{-0.23}(G/d)^{0.47}(pd)^{0.2}.$$

(c) – distribution of turbulence scale ε along the channel centerline (d=10^{-2} m; G=5·10^{-3} kg/s). 1 – channel with smooth wall, $\bar{l} = l/d$ =77; 2-4 – divided channel, $\bar{a} = a/d$ = 32 (2 – g_i=0; 3 – g_i=0.5·10^{-3} kg/s; 4 – m_s=1 in the cross-section $\bar{z}_s$ =4.3); I – g_i=0; II – g_i = 1·10^{-3} kg/s.

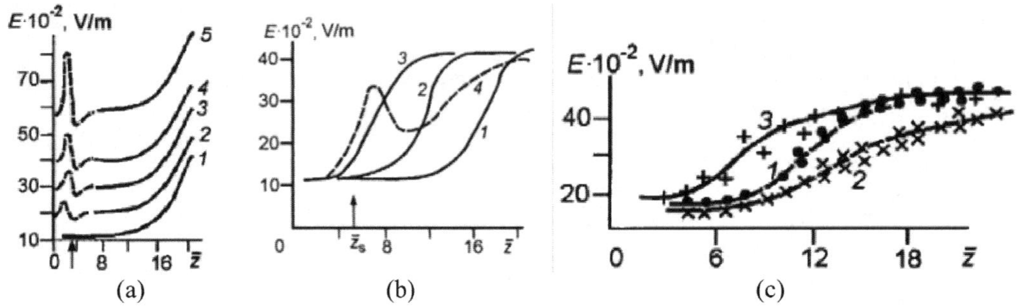

Figure 6 Distributions of the electric field strength (E) along the axis of the electric-discharge chamber at different conditions of plasma gas (air) supply. (a) – cocurrent swirl of gas supplied between pilot and first inter-electrode sections. d=2·10^{-2} m, $\bar{a}$ =21.5; $\bar{z}_s$=3.2, I=120 A, G=30·10^{-3} kg/s, g_i=0.5·10^{-3} kg/s, g_1+g_2=const=15·10^{-3} kg/s, 1÷5 – m_s=0.08; 0.18; 0.39; 0.62 and 1.2, respectively. (b) – cross-current swirl of gas supplied between pilot and first inter-electrode sections. 1 ÷ 4: m_s = 0.08; 0.37; 1.1 and 2.1; $\bar{z}_s$=5. (c) – with and without swirl of gas supplied between pilot and first inter-electrode sections. d=15·10^{-3} m; g_1=1.5·10^{-3} kg/s; G=17.9·10^{-3} kg/s, I=120 A. 1, 2 – concurrent swirl; 3 – without swirl.

8. High enthalpy plasma gas mixtures result in a longer plasma plume in comparison with Ar based mixtures. Longer plume results in higher "useful" dwell time and possibility to improve heat treatment of sprayed or treated particles. Besides, "long plume" plasma spraying is more forgivable to spraying distance changes that should be important for the product line.

9. The "cascade" design could be scaled up for a wide range of power levels without big problems [9]. It means that the same approach could be used, for example, for 10-30, 30-75, 75-120 kW, etc.

10. The "cascade" design allows realization of both options: sub-sonic and supersonic plasma velocities.

11. Possibility of quasi-laminar and transition plasma flows. These options allow increasing the particles dwell time and, consequently, efficiency of heat transfer from plasma to particles and treated surface [7, 8] (see figures 7, 8 and 9).

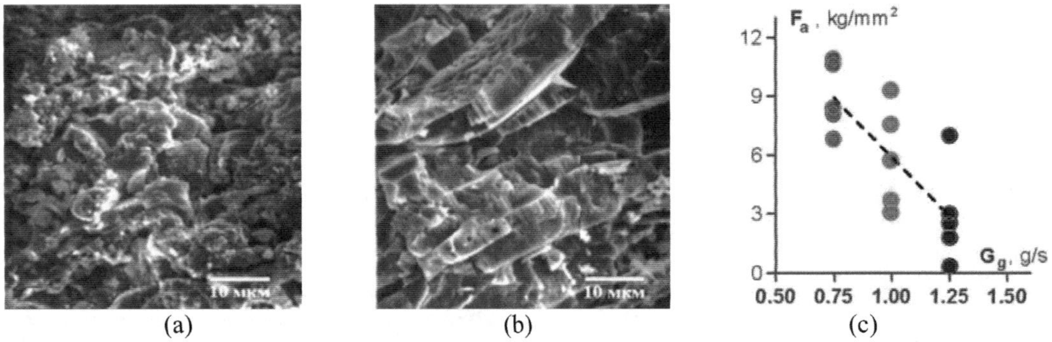

Figure 7 Fracture surface of alumina coatings sprayed in turbulent – (a), and quasi-laminar – (b), plasma jet outflow; (c) – experimental data characterizing a bond strength of α-Al_2O_3 coatings sprayed on the stainless steel substrates at different air plasma jet outflow: quasi-laminar (0.75 g/s), transient (1 g/s) and turbulent (1.25 g/s).

Journal of Physics: Conference Series **550** (2014) 012017 doi:10.1088/1742-6596/550/1/012017

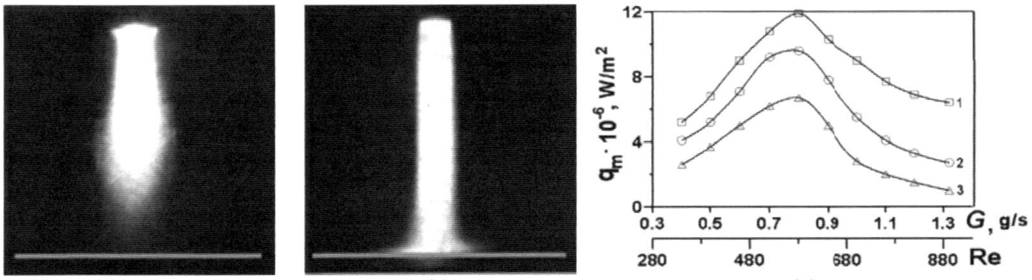

(a) (b) (c)

Figure 8 Photos of plasma jets impinging onto flat substrate at different regimes of outflow: (a) – turbulent (Re=820); (b) – quasi-laminar (Re=580). (c) –heat flux density from the nitrogen plasma jet to the flat substrate vs. mass flow rate of nitrogen and Reynolds number. curve 1 – $\bar{z}$=6, curve 2 – $\bar{z}$ =8, curve 3 – $\bar{z}$=10, $\bar{z}=z/d$ is non-dimensional distance from the plasma torch nozzle to obstacle.

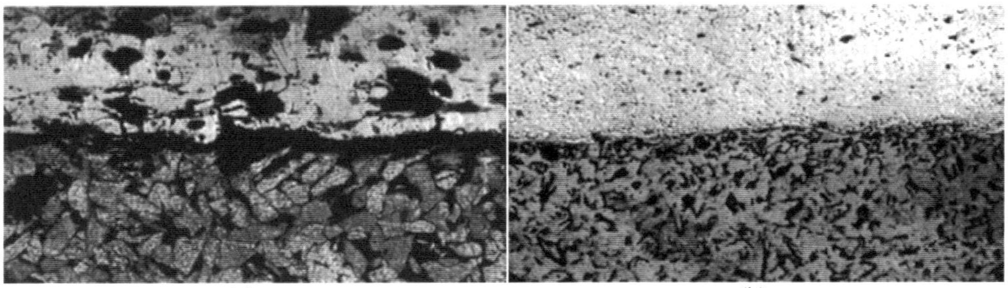

(a) (b)

Figure 9 Cross cut of Ni-Cr-B-Si-C coating sprayed in turbulent regime of plasma jet outflow – (a), and fused by quasi-laminar plasma jet – (b).

5. Applications of cascade torches for powders processing

In the present-day industry of structural and protective materials, powders made up by hollow spheres, or hollow powders, gain increasing acceptance. Hollow ceramic powders are used in the production of plasma-sprayed thermal barrier coatings, composite heat- and sound-insulating materials, light construction fillers and backfills, buoyancy materials, and explosive mixtures. Such powders also form basis for catalysts, adsorbents, filter elements, encapsulating media, etc. HOSP powders of metals, metal alloys and oxide ceramics are of great interest for powder metallurgy, material science and thermal spraying. HOSP powders can be produced by different methods [10].

Great interest has been recently expressed in studying the characteristics of thermal barrier coatings sprayed from HOSP YSZ powders [11-13], etc. Figure 10 shows photographs of the initial (*a*) and plasma treated (*b* and *c*) spray-dried YSZ powders. Particles were injected radially into the plasma torch exit region to undergo spheroidization in quasi-laminar jet. Working gas – air+CH_4, electric power – 75 kW, four-point powder injection, productivity of process - 20 kg/h.

Another example illustrates the synthesis of α-Al_2O_3 hollow powders (figure 11) under plasma treatment of aluminum hydroxide particles from 40 to 120 μm in size, containing from 10 to 15 wt.% of chemically bound water, in air plasma jet flowing out in quasi-laminar mode of 50-kW cascade gun [14].

At the same time, in our publication [14], the experimental data regarding the novel class of corundum powder particles is presented. They have a size of 40-250 μm, a specific surface area of 150-200 m^2/g and characterized by multi-level inner structure (nano-, submicron- and micrometer) produced by plasma treatment of initial aluminum hydroxide powder under transient and turbulent plasma jet outflow. Figure 12 shows the morphology and texture of the obtained powder with particle sizes from 90 to 120 μm.

As it can be seen, the α-Al_2O_3 powder particles with rounded surface are formed by crystallites from nano- to micrometer in size. In accordance with derivatographic data, the obtained alumina sample, exhibiting a polycrystalline surface, contained 2.25 wt.% of γ-Al_2O_3 and 2.27 wt.% of more high-temperature phases. Thus, the synthesized alumina is 95 wt.% of α-Al_2O_3 (corundum). This fact was also confirmed by x-ray data. This powder and articles sintered herefrom (tablets, hollow cylinders) are of considerable interest as thermostable catalysts supports.

(a)

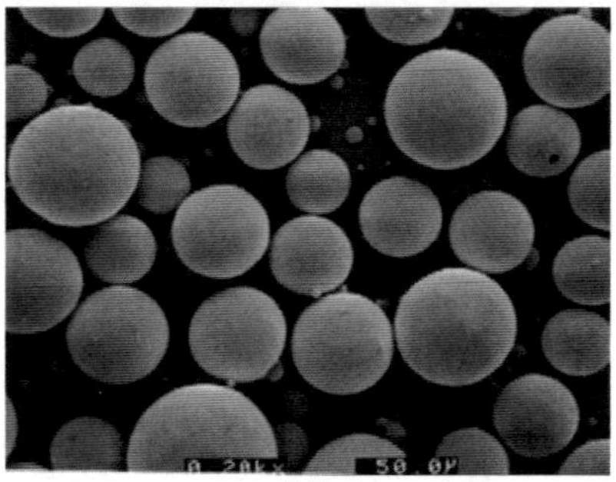

(b)

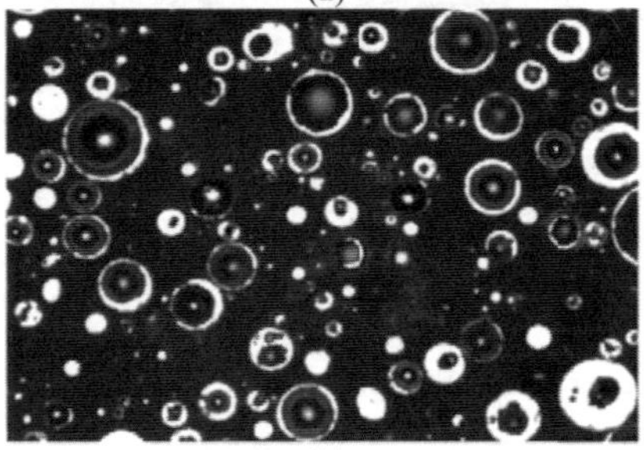

(c)

Figure 10 SEM photos of initial spray-dried YSZ powder - (a), and HOSP powder made using the cascade torch: (b), (c) – particles' general view and their cross cut, correspondingly.

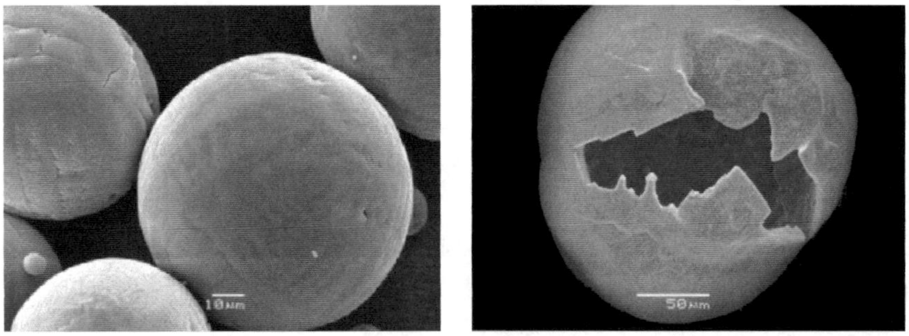

(a) (b)

Figure 11 SEM photos of the hollow α-Al_2O_3 spheres - (a), and broken single hollow α-Al_2O_3 microsphere illustrating the shell's thickness - (b).

Finally, with the example of SiO_2 powder we have shown the possibility to synthesize the hollow particles several tens of micrometers in size (see figure 13) using the cascade torch treatment of the suspended nano- and submicrosized powder [10]. On injection into the plasma, the liquid suspension was atomized into droplets that contained quartz particles (mean particle size 110 nm). The droplets were several tens of micrometers in size. On droplets evaporation, the plasma conditions led to the formation of agglomerates of the initial SiO_2 particles, whose subsequent heating and melting resulted in the formation of hollow spheres.

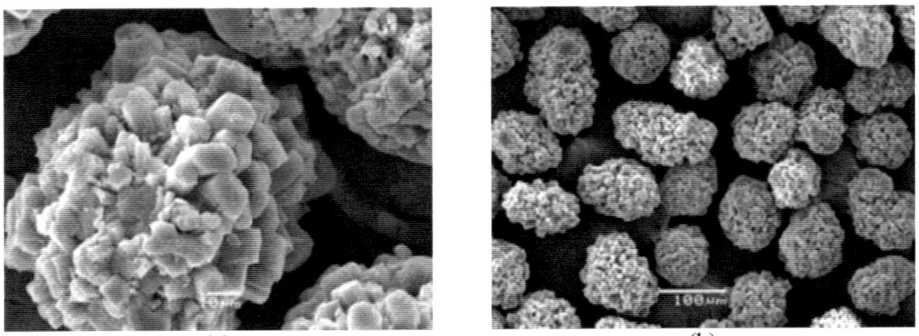

(a) (b)

Figure 12 SEM photos of single α-Al_2O_3 particle characterized by multi-level (nano- submicron- and micrometer) structure - (a), general view of powder - (b).

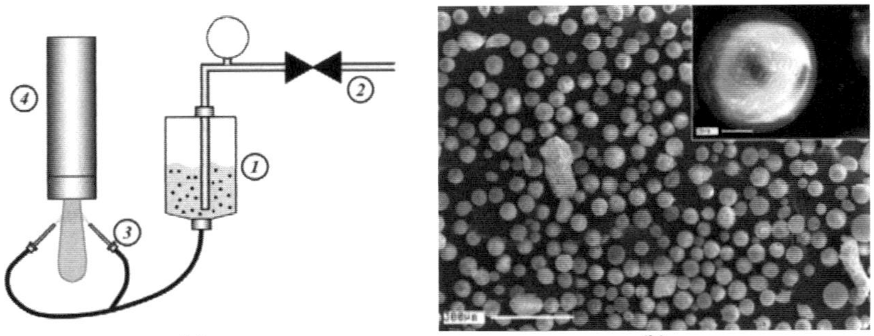

(a) (b)

Figure 13 Suspension powder plasma treatment (a): 1 – pressure vessel filled with suspension, 2 – high-pressure gas pipeline, 3 – injectors, 4 – plasma gun; appearance of the plasma-synthesized HOSP SiO_2 powder (b).

6. Conclusions

Unique dominating platform based on "cascade" approach allows the satisfaction of practically all customer existing and future needs:

1) Cascade family of plasma torches is easy scaled up and possesses an extremely wide operating window, allowing reproducing plasma parameters and related coatings of practically all torches presently available on market place.

2) Cascade design has capability of generating laminar, transient and turbulent plasmas using different plasma gases; it could allow to develop new coatings and powders.

3) Cascade process is capable of using inexpensive gases to generate plasmas; latter should allow expanding business in developing countries where high purity Ar, He, H_2 are expensive and have some availability problems.

Acknowledgement. This work was supported in part in the framework of Interdisciplinary Integration Projects No. 2 and 98 of Siberian Branch of the Russian Academy of Sciences for 2012-2014. Authors thank Dr. V.E. Belashchenko for fruitful discussions regarding the technological aspects of thermal plasma torches application.

References

[1] Zhukov M F, An'shakov A S, Zasypkin I M, Kurochkin Yu V, Lukashov V P, Desyatkov G A, Pustogarov A V 1981, *Electric-Arc Generators With Interelectrode Inserts*, Novosibirsk, Nauka, Siberian Branch 221 p (in Russian)

[2] Zhukov M F 1994 *Thermal Plasma and New Materials Technology Vol.1 Investigation and Design of Thermal Plasma Generators*, ed O P Solonenko and M F Zhukov (England: Cambridge Interscience Publishing) pp 5-43

[3] *Electric Arc Generators of Thermal Plasma* ed M F Zhukov and I M Zasypkin (Novosibirsk: Nauka SB RAS) 1999 712 p (in Russian)

[4] Fauchais P 2004 Understanding plasma spraying *J. Phys. D: Appl. Phys.* **37** 86

[5] Leblanc L and Moreau C 2000 Study on the long-term stability of plasma spraying *Proc. of ITSC'00*, May 2000, Montreal, Canada

[6] Donskoi A V and Klubnikin V S 1979 *Electro-Plasma Processes and Apparatuses in Mechanical Engineering* (Leningrad: Mechanical Engineering Publishing) 221 p, in Russian

[7] Solonenko O P, Alkhymov A P, Marusin V V, Orishich A M, Rakhimyanov Kh M, Salimov R A, Shchukin V G and Kosarev V F 2000 *High-Energy Processes of Materials Treatment* (Novosibirsk: Nauka SB RAS) 425 p (in Russian)

[8] Solonenko O P 1995 State-of-the art of thermophysical fundamentals of plasma spraying *Thermal Plasma and New Materials Technology, Vol.2, Investigations and Design of Thermal Plasma Technologies* ed O P Solonenko and M F Zhukov (England: Cambridge Interscience Publishing) pp 7-96

[9] Solonenko O P, Takeda K, Kuz'min V I, Smirnov A V, Sakashita M and Nakamura M 2001 Advanced approach to developing equipment for plasma assisted oil free starting systems for pulverized coal power station *Proc. of 15th Intern. Symp. on Plasma Chemistry*, 8-13 July 2001, Orleans, France **5** 2079

[10] Solonenko O P, Gulyaev I P and Smirnov A V 2011 Thermal plasma processes for production of hollow spherical powders: theory and experiment *J. of Therm. Sci. and Technol.* **6(2)** 219

[11] Tan Y, Longtin J P and Sampath S 2006 Modeling thermal conductivity of thermal spray coatings: comparing predictions to experiments *Proc. of ITSC'06*, Seattle, Washington, USA, (Electronic publication)

[12] Ladru F, Reymann H, Mensing M, Zimmermann S and Fischer J 2006 Tailoring of YSZ powders to meet fixed process windows, parameters and properties for land based turbine applications *Proc. of ITSC'06*, Seattle, Washington, USA, (Electronic publication)

[13] Markocsan N, Nylen P, Wigren J and Li X-H 2006 Low thermal conductivity coatings for gas turbine applications *Proc. of ITSC'06*, Seattle, Washington, USA, (Electronic publication)

[14] Solonenko O P, Leonchikov A M, Efimova E V and Nedosekin F V 2010 Powdery α-alumina, method of its production and article from it, *Patent RU #2386589* Priority on September 17, 2008

Development of a 100 kW plasma torch for plasma assisted combustion of low heating value fuels

S Takali, F Fabry, V Rohani, F Cauneau and L Fulcheri[1]
MINES ParisTech, PSL-Research University, PERSEE Centre procédés, énergies renouvelables et systèmes énergétiques, CS 10207 rue Claude Daunesse 06904 Sophia Antipolis Cedex, France, Phone:+33.493.957.540.

E-mail: sabri.takali@mines-paristech.fr.

Abstract. Most thermal power plants need an auxiliary power source to (i) heat-up the boiler during start up phases before reaching autonomy power and (ii) sustain combustion at low load. This supplementary power is commonly provided with high LHV fossil fuel burners which increases operational expenses and disables the use of anti-pollutant filters. A Promising alternative is under development and consists in high temperature plasma assisted AC electro-burners. In this paper, the development of a new 100 kW three phase plasma torch with graphite electrodes is detailed. This plasma torch is working at atmospheric pressure with air as plasma gas and has three-phase power supply and working at 680 Hz. The nominal air flow rate is 60 $Nm^3.h^{-1}$ and the outlet gas temperature is above 2 500 K. At the beginning, graphite electrodes erosion by oxidizing medium was studied and controlling parameters were identified through parametric set of experiments and tuned for optimal electrodes life time. Then, a new 3-phase plasma torch design was modelled and simulated on ANSYS platform. The characteristics of the plasma flow and its interaction with the environing elements of the torch are detailed hereafter.

1. Introduction

The distribution of worldwide electrical power production shows that the main part of our electricity is produced by thermal power plants. Whether these plants are using fossil fuels or renewable combustibles, they all face start-up difficulties and the majority requires heavy fuel burner for several hours and consumes tons of high LHV fossil fuel to heat up the boilers and reach autonomy power. Some alternatives to these burners are actually under development. For instance, oxy-combustion [1], co-combustion [2] and plasma assisted combustion are three different alternatives[3][4][5]. The principal of the last alternative is mainly based on thermochemical activation of the combustion by thermal plasma (Figure 1). At chemical level, the plasma accelerates reaction kinetics and enhances the production of highly reactive radicals and ionised species such as atomic oxygen. At a thermal level, plasma torches can deliver high density of energy in a large volume of gas with a very high efficiency. Actually, the majority of thermal plasma torches is based on DC technology which suffers from reliability problems when operating with air, needs high maintenance costs and uses non consumable tungsten or copper electrodes. At centre PERSEE a new plasma torch is being developed based on years

[1] To whom any correspondence should be addressed

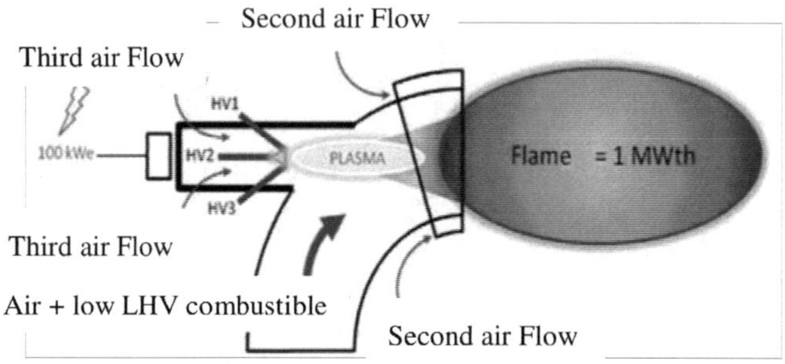

Figure 1. Schematic principle of Plasma electro-burner.

of know-how in the domain of three phase plasma torches for different application domains like syngas production or black carbon synthesis.

For the development of a more powerful three phase plasma torch working at atmospheric pressure with air and with graphite electrodes, preliminary tests are performed on previous generation of plasma torches to identify the main improvement axes. Actually, the existing plasma torches are mainly working with non-oxidizing gas such as hydrogen, nitrogen, CO, argon, helium or methane. So, the first step consists on studying the graphite electrodes erosion with air plasma gas in order to find the key parameters that can lead to erosion reduction. Furthermore, since we are aiming to increase the total power, arc discharge phenomena are also investigated in order to optimise the electro-thermal conversion efficiency at 100 kWe.

2. Preliminary study

2.1. Graphite electrode erosion

The first test campaign is conducted to evaluate the electrode erosion at different operating conditions: with non-oxidizing gas, with air or with a mixture of air and carbon black as plasma gas. The idea of injecting carbon powder with air consists in providing an alternative reactive medium for oxygen and protecting the graphite electrodes from oxygen. The test results are summarized in Table 1. These tests were performed at 200 A, the electric power reaches approximately 30 kW, with a total plasma gas flow rate of 5.6 Nm3.h-1. The first test is completed with nitrogen. If we neglect the formation of cyanide molecules (C-N) that occurs only at temperatures exceeding 3000 K [6], nitrogen can be considered as chemically neutral in presence of atomic carbon. Consequently, the eroded mass with nitrogen is low and mainly due to the electric discharge. In fact, under the effect of the temperature, the graphite sublimate at the cathode and anode spots [7]. Arc roots are also under the effect of ionic and electronic bombardment [8]. In presence of air, erosion by chemical reaction between oxygen and graphite is added to the previous erosion. Figure 2 shows the impact of both types of erosion: arc discharge effect is visible through impacts on the electrode tips with shape of craters and oxygen erosion is noticed on the lateral surface of each electrode.

The simulation of thermodynamic equilibrium of C, O_2, and N_2 mixtures between 273 K and 5000 K gives the proportions of all species that can coexist in the plasma zone as a function of the temperature. When oxygen is in excess, it reacts with all carbon atoms to form CO_2 until 2000 K which dissociates then to atomic oxygen and CO. Yet, when carbon is more preponderate, it reacts with oxygen to produce CO_2 until 700 K and then dissociates to give place to CO without residual atomic oxygen. It is this situation that we want to reproduce in the inter-electrode zone by injecting carbon powder at very small granulometry in order to enhance CO formation and avoid the reaction between oxygen and graphite electrodes. Test results with CO as plasma gas show the pertinence of the idea. However, Table 1shows that contrarily to what is expected, increasing the C/O ratio leads to an increase of electrode erosion. A previous study of carbon black radiation in a plasma torch [9] shows that these particles behave like a thermal shield near the walls by absorbing the heat in the plasma zone leading to a decrease in wall heat losses. It is then possible that increasing the quantity of carbon particle for the same amount of energy, and for a very short residence time, concentrates more the energy in the discharge zone without reaching sublimation temperature to react with air. Consequently, the temperature rises and the impacts of air and discharge erosions are higher.

Table 1. Equivalent electrode life time in hours for a 0.12 m length and 0.017 m diameter electrode as a function of injected gas. The equivalent electrode life time is deduced from the equivalent eroded length calculated from the actual eroded mass as if the erosion takes place homogenously in the transversal section of the electrode.

Inlet plasma gas	Equivalent electrode life time in hours
Nitrogen	8.07
Air	2.80
Carbon monoxide	17.64
Mixture of air and solid carbon black (C/O=1)	2.81
Mixture of air and solid carbon black (C/O=1.15)	1.64
Mixture of air and solid carbon black (C/O=1.38)	1.60
Mixture of air and solid carbon black (C/O=1.75)	1.28

Figure 2. Top view of the discharge zone, the three graphite electrodes are in planar position.

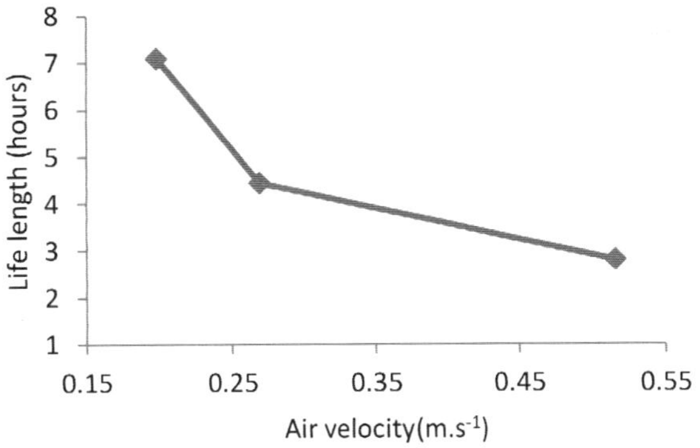

Figure 3. Graphite electrode life time as a function of air axial velocity at electrode tips in the planar three phase plasma torch geometry.

Another critical parameter for electrode erosion consists of air velocity impacting the lateral surface of the electrodes. In Figure 3 is plotted the electrode life time from different tests conducted at different air velocities in the inter-electrode zone. It can be noticed that by simply reducing the air flow velocity in this location, the electrode erosion decreases drastically. In addition, a minimum amount of nitrogen gas can be injected as protective layer around the electrode tips to prevent their erosion by air.

2.2. Power enhancement

In order to increase the electrical power of the new plasma torch design, the present study is based on the work of C. Rehmet [9] about electric arc behaviours in different geometries of three phase plasma torches. It is concluded that the impact of hydrodynamic electrode jet force is so significant that it changes the amount of power that could be provided by the arc for the same imposed current. For instance, in a planar electrode configuration, Lorentz force has a low impact on the arc column and the direction of the hydrodynamic electrode jets force -normal to the electrode tips- keeps the arc discharge confined in the inter-electrode gap (Figure 4). However, in an angular configuration, this hydrodynamic force is not directed toward the electrodes and the arc length increases (Figure 5). Besides, the interaction between the current lines in the arc column induces an increase of Lorentz force which leads to a global centrifugal arc motion. Consequently, the arc elongation due to the spatial position of the electrodes rises the arc column resistance and increases the electric discharge power for the same value of arc current. The choice of an electrode angle of 20° from the axis of the plasma torch is the result of these observations.

For technical reasons, mainly linked to plasma ignition and electrode gap, it is very difficult to have parallel electrodes although simulations show it is the best configuration for maximum arc power.

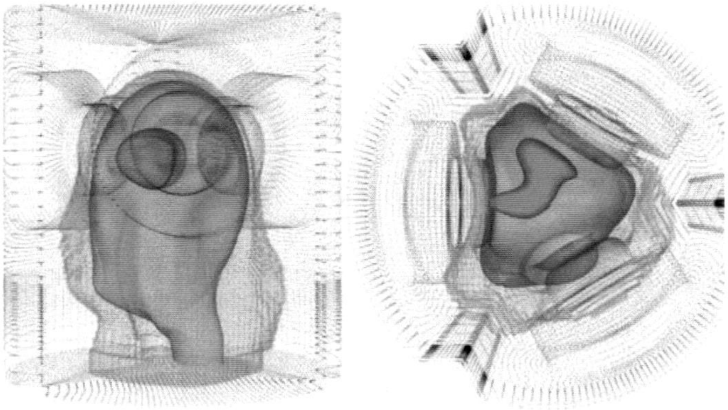

Figure 4. [Image courtesy of C. Rehmet]. Arc represented with temperature isosurfaces 15 000 K, 10 000 K and 5 000 K in coplanar electrodes configuration issued from MHD modelling.

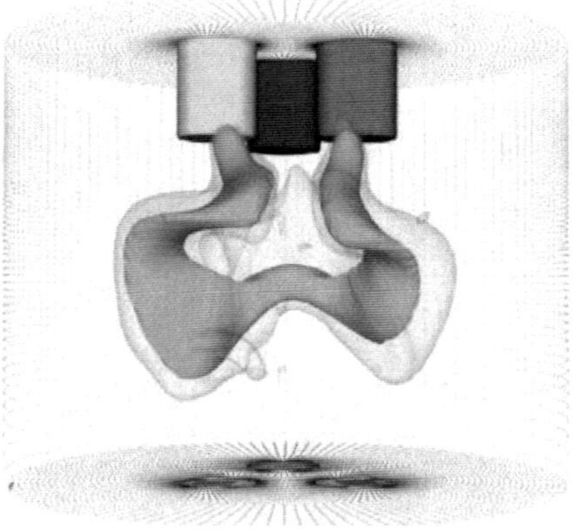

Figure 5. [Image courtesy of C. Rehmet].Arc discharge represented with temperature isosurfaces at 5 000 K and 8 000 K in parallel electrodes configuration issued from MHD modelling.

For better combustion assistance of the plasma, the gas at the outlet of the torch should reach a temperature high enough to provide radicals for the burner. A temperature range between 2 500 and 3 000 K is a compromise between the lower limit of oxygen dissociation and the upper limit of material resistance. For a 100 kWe plasma torch with a 70% thermal efficiency, the inlet flow rate needed to obtain a minimum of 2 500 K is about 0.02 kg.s-1 which gives an energy density of 1.11 kWh.Nm-3. In this case, the exhaust diameter has to be less than 0.15 m to obtain an outlet gas velocity of 10 m.s-1.

3. Simulation

3.1. Model, assumptions and boundary conditions

The objective of the modelling is to predict the plasma flow behaviour in regions where thermo-mechanical effects are preponderant (thermal and velocity fields), the quantification of heat losses through the walls and the validation of thermal reliability of the plasma torch components. In order to obtain coherent results yet inexpensive calculation cost, some assumptions are made and simplifications are suggested.

For simplification purposes only, one third of the plasma torch is modelled. This assumption is very rough since it implies neglecting the hydrodynamic force effect of the two opposite inlets but allows to considerably reduce the calculation cost. The second assumption consists on assuming that the plasma flow tends to a stationary state. However, the periodicity of the current and the nature of three phase plasma discharge -with all what it implies as electromagnetic and hydrodynamic forces- induce continuous perturbations. Yet, these forces are far of being periodic as shown in [10] and have some random aspects. Thus, adding the periodicity of the discharge to the simulation would not be a fair reproduction of the reality. In addition, the turbulence generated by these forces is located around the electric arc [10]. It is assumed that the unsteady phenomena due to the nature of three phase discharge are confined in a zone called the source volume. In the other hand, it is assumed that the local thermodynamic equilibrium (LTE) is verified which allows to define a unique temperature at each point of the plasma zone for all the species. The gas is considered incompressible i.e. apart from thermal dilatation effects, since the Mach number is considered always less than 0.3. The electromagnetic phenomenon is neglected here, since it is estimated that its effect is located in the power source volume.

The local balance equation for any intensive parameter g contained in a volume V with a delimiting surface S and containing the extensive parameter G could be expressed as; the variation of G inside V as a function of the time t is the sum of net entering flux of G through S and the creation of G inside V. With analytic terms this could be written as follows:

$$\frac{\partial \rho g}{\partial t} + \mathrm{div}\, \rho \vec{V} g + \mathrm{div}\, \overrightarrow{\varphi_{\mathrm{DG}}} - \rho g^{*} = 0 \qquad (1)$$

Where $\vec{V}$ and g^* are successively the velocity vector, the production term of g. Here, the flux is divided into convective and diffusive terms: $div\ \rho\vec{V}g$, $div\ \overrightarrow{\varphi_{DG}}$.

For mass balance, one should consider G as the mass and 1 for the intensive parameter g. Since there is no mass production, mass balance equation is directly:

$$\frac{\partial \rho}{\partial t} + \mathrm{div}(\rho \vec{V}) = 0 \qquad (2)$$

For the momentum balance equation, the velocity vector should be considered as the intensive parameter and the surface density of diffusive flux is equal to the stress tensor which could be written as: $\overrightarrow{\varphi_{DG}} = p\,\bar{\bar{I}} - \bar{\bar{\tau}}$ where p, $\bar{\bar{I}}$ and $\bar{\bar{\tau}}$ are successively the pressure, the unity tensor and the viscous stress tensor. The simplified equation becomes:

$$\rho\frac{\partial \vec{V}}{\partial t} + \mathrm{div}\, \rho\vec{V} \times \vec{V} + \overrightarrow{\mathrm{grad}}\, p - \overrightarrow{\mathrm{div}}\, \bar{\bar{\tau}} - \rho g^{*} = 0 \qquad (3)$$

To include turbulent phenomena in the simulation, the velocity is divided into average and fluctuating terms u_i. The k-ε turbulence model is the most common model used for plasma flow modelling [11], and adds two additional transport equations to equilibrate the additional two fluctuation parameters. The transport variables are the turbulent kinetic energy k, and the turbulent dissipation ε. In the particular case of RNG k-ε model, these equations are as follows:

$$\frac{\partial(\rho k)}{\partial t}+\frac{\partial(\rho k u_i)}{\partial x_i}=\frac{\partial}{\partial x_j}\left(\alpha_k \mu_{eff}\frac{\partial k}{\partial x_j}\right)+\mathrm{G_k}+\mathrm{G_b}-\rho\varepsilon \tag{3.1}$$

$$\frac{\partial(\rho\varepsilon)}{\partial t}+\frac{\partial(\rho\varepsilon u_i)}{\partial x_i}=\frac{\partial}{\partial x_j}\left(\alpha_\varepsilon \mu_{eff}\frac{\partial \varepsilon}{\partial x_j}\right)+\mathrm{C}_{1\varepsilon}\frac{\varepsilon}{k}(\mathrm{G_k}+\mathrm{G_b})-\mathrm{C}_{2\varepsilon}\rho\frac{\varepsilon^2}{k}-R_\varepsilon \tag{3.2}$$

When replacing the intensive parameter with the gas enthalpy h we obtain the energy balance equation. The diffusion term consists of the heat flux by means of molecular conductivity which gives:

$$\frac{\partial \rho h}{\partial t}+\operatorname{div}\rho\vec{V}h+\operatorname{div}\left(\frac{\lambda}{C_p}\overrightarrow{\operatorname{grad}}\,h\right)-\rho g^*=0 \tag{4}$$

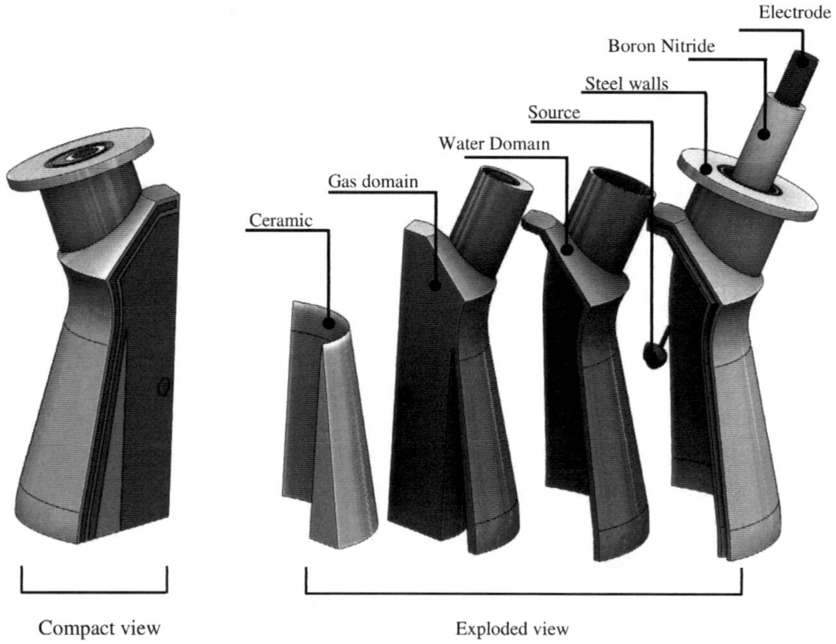

Figure 6. Compact and exploded views of the calculation domain.

These Navier-Stokes equations are the transport equation being solved numerically in Fluent. In this particular case, the integration of the water cooling circuit is necessary because the increase of the water temperature is a critical data at the laboratory scale were the external cooling system has a limited capacity for power extraction. Since the cooling system is part of the model, adjustment of the boundary conditions has minimal effect and is limited to outer walls where only convection heat transfer is considered, at a rate of 20 $W.k^{-1}.m^{-2}$ corresponding to forced air convection. Water inlet flow rate is at 0.27 $kg.s^{-1}$. For gas inlets, 18.33 $Nm^3.h^{-1}$ and 1.66 $Nm^3.h^{-1}$ are defined respectively for air and nitrogen inlet flow rates. The third of the power source contains 33.33 kW. These flow values coupled with the power injected and an exhaust diameter of 0.15m, give an estimation for the gas enthalpy of about 1.11

Journal of Physics: Conference Series **550** (2014) 012018 doi:10.1088/1742-6596/550/1/012018

kWh.Nm^{-3}, an average temperature of around 2400 K for the outlet gas with an average velocity of around 7 m.s^{-1}.

The solid materials used for the model are stainless steel for the walls, ceramic for thermal sheeting, boron nitride for electric insulation and graphite for electrodes. Concerning air characteristics, the data used are defined up to 20 000 K and generated using TT-Winner software [6]. The geometry is edited and meshed using ANSYS geometry and meshing standard tools. Figure 6 show the different subdomains included in the simulation. For CPU consumption reason, the triangular meshing is enhanced with layer inflation at each interference and wall contact without predefined squeezed nodes. The maximum distance between nodes is set to 0.005 m and the total number of elements is 375000.

3.2. Simulation results

Figure 7 shows the gas expansion from the power source into the surrounding volume. The temperature inside the power source reaches almost 19 000 K according to the simulation, representing arc column and then decreases outsides with very high gradient to 1000 K in almost 0.03m. The gas accelerates inside the hot zone to a maximum of 50 m.s^{-1} due to gas expansion. Nearby the power volume, we notice that the high energy content creates an important pressure gradient that aspires the air into the volume and forms vortices around the arc column. The ceramic element plays its role as a temperature shield and protects the steal walls from overheating by confining the gas expansion into the outlet and its maximum temperature is about 2000 K; 400 K less than its top operating limit. The heat losses to the walls are negligible and this is mainly due to the absence of radiation. This leads to higher outlet averaged temperature of about 4000 K. The electric insulation element is not also overheating although the proximity of the power source. The air layer passing between the ceramic and steel walls plays also its role in cooling the ceramic and avoiding the conduction of heat if they would be in contact. The current lines from the gas inlets show no backflow in the exhaust but some recirculation cells are created at the vicinity of the walls which may disturb arc discharge. The mass weighted average of molecular viscosity, density and velocity at the outlet are successively: 0.000115 kg.m^{-1}.s^{-1}, 12.2 m.s^{-1} and 0.155 kg.m^{-3}, and knowing that the outlet diameter is 0.15 m, the Reynolds number is 2466. This means that the flow is not laminar and starts passing to turbulent state. At the frontier between cold and hot gas where the temperature gradient is high, Prandtl number reaches the value of 0.8 showing that thermal diffusion governed by the gas conduction and its specific heat capacity becomes as strong as the viscous diffusion characterized by the cinematic viscosity, whereas, inside and outside the plasma, thermal diffusion dominates. This means that this gas layer with high temperature gradient and high Prandtl number behaves more like a barrier that slows down the cold gas. The penetration of inlet air into the plasma is more occurring because of the depression. It is expected that hot and cold gas would not mix easily.

For graphite electrode erosion control, 5 Nm^{3}.h^{-1} of nitrogen are injected between the electrodes and the electric insulation element surrounding the electrodes. In Figure 8, we see how the nitrogen is lowering the concentration of air in the inter-electrode zone which means less oxygen for electrode erosion. The nitrogen sheathing is more efficient near the torch axis. In fact, the torch is designed for lower velocity inside the discharge zone in order to prevent arc extinguishing. Figure 9, shows the temperature at different height levels. Above the arc column, we see how the air has higher velocity at the wall side of the electrode. The temperature is becoming more homogenous when moving down to the outlet. In Figure 10, the gas recirculation at the tip of the electrode shows how important is the shape of the source volume for a fair reproduction of the flow.

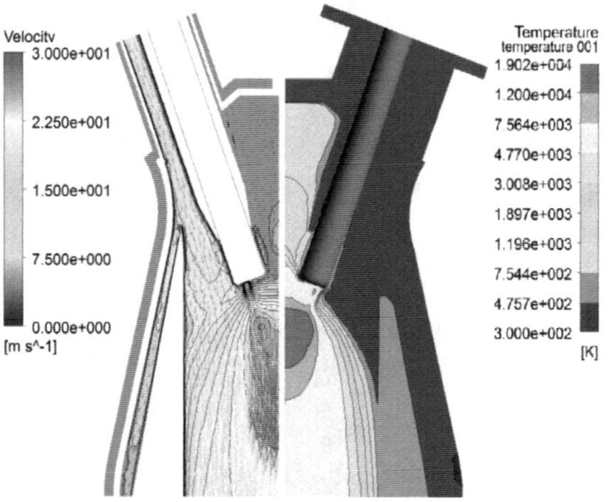

Figure 7. Velocity field (left), temperature field (right).

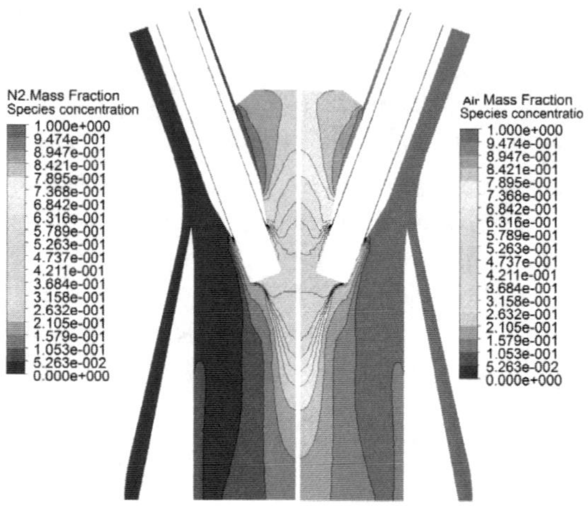

Figure 8. N2 mass fraction (left), Air mass fraction (right).

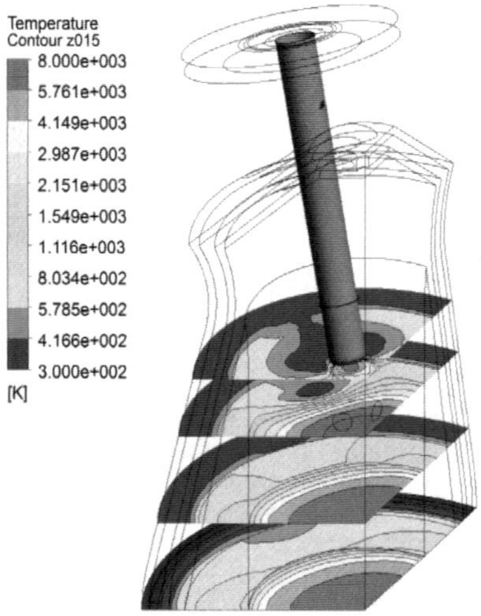

Figure 9. Temperature field at different level from the source volume (Temperature in the arc region exceeds 8 000 K).

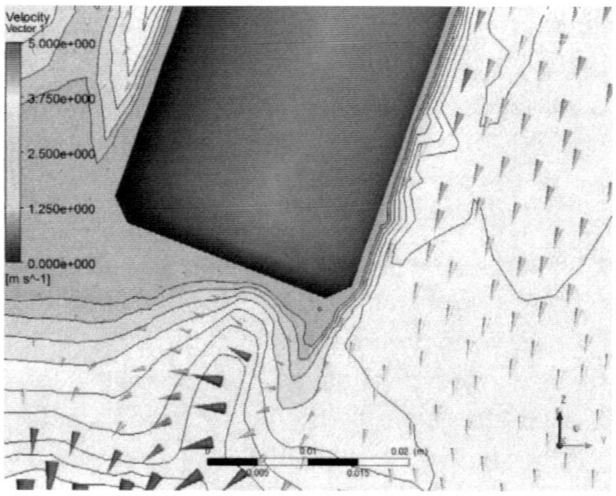

Figure 10. Velocity field around the electrode tip.

A parametric study is performed to investigate the effect of the power source volume on the plasma flow. Four different volumes are tested: 2 cm^3, 4 cm^3, 8 cm^3 and 16 cm^3. Figure 11 shows the gas temperature along radial position at different heights inside the torch: at 0.04m, 0.07m, 0.10 and 0.16m from the tip of the electrode. The mass averaged temperature of the power source volume are successively 17 011 K, 15 645 K, 14 538 K and 12 069 K starting from the smallest volume. Thus, when increasing the volume, the power density decreases and the mass averaged temperature decreases also. Nevertheless, the temperature outside the volume at $z = 0.04$ m is higher when the volume is bigger and this is mainly due to the non-linearity of gas conductivity at high temperature. Besides, the volume change implies some differences in gas recirculation and thus the temperature field distribution is not exactly the same in the 4 cases as we see in Figure 12. These temperature variations become negligible at 0.06 m down to the source (0.10 m from the electrode tip) and the difference is also not noticeable above the source volume which means that for a global study of the plasma flow the source volume would not have big impact on the results.

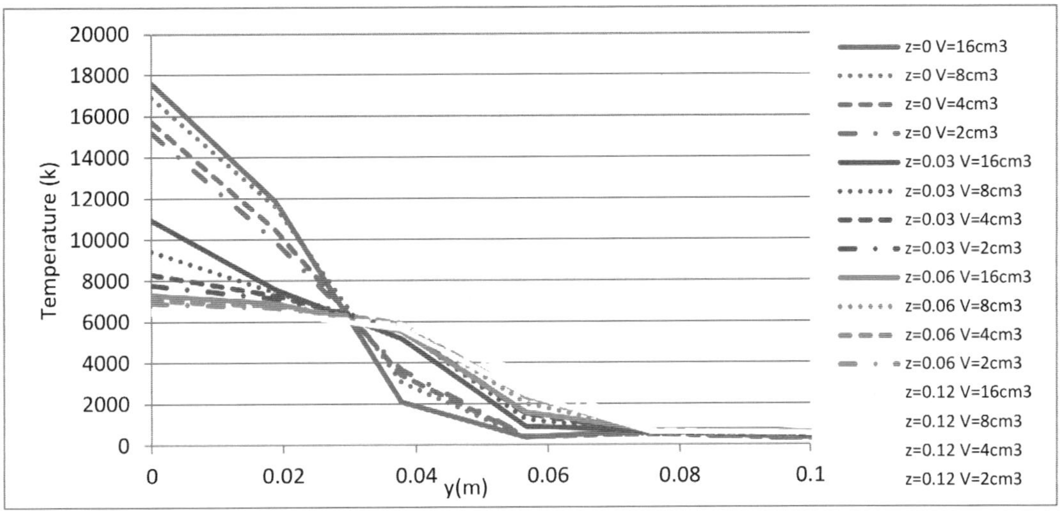

Figure 11. Gas temperature along the radial position (different source volume and vertical positions). (The plan defined by z = 0 is at 0.04m down to the electrode tips)

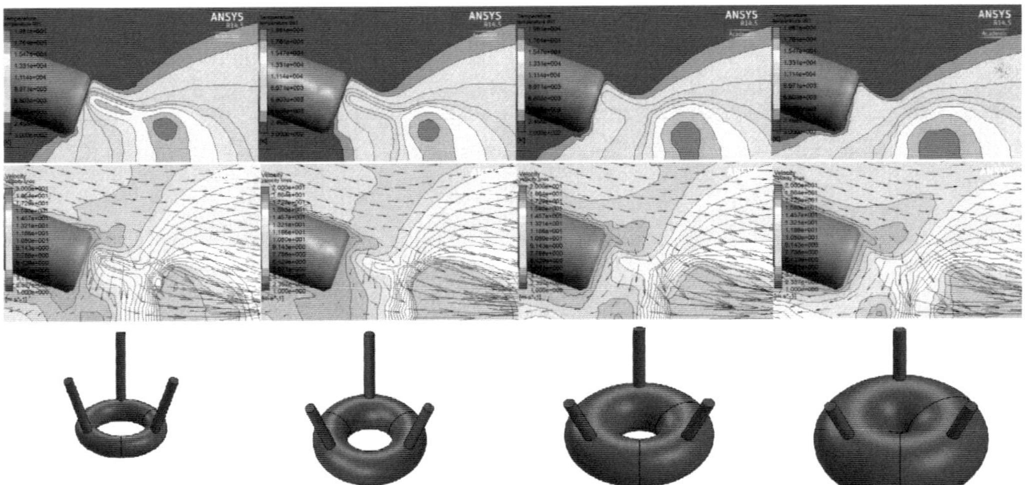

Figure 12. Temperature and velocity fields nearby the electrode tip (top) and arc column simplified shape (bottom) in 4 different volumes: 2 cm^3, 4 cm^3, 8 cm^3 and 16 cm^3.

The validation of the plasma flow and thermal behaviour of the different components of the 100 kW torch led to the manufacturing of the new torch which has been mounted on test bench and tested. The first start up test was successful and showed up to 80% heat efficiency. A series of test campaigns is programmed for reliability validation and characterization of this torch. In addition, for better understanding of the plasma flow, more accurate simulations that take into account radiation will be performed.

4. Conclusion

The complexity of the graphite electrode erosion near the discharge zone in oxidizing atmosphere makes chemical control of the erosion very difficult and for this reason the aerodynamic control of air flow around the electrodes is kept as a solution by decreasing air speed at the tips. A shield of nitrogen is also injected around each electrode to provide a neutral boundary layer. The injection of methane could also be a solution for graphite erosion by depositing carbon atoms on the electrode surface. This alternative will be tested by replacing the nitrogen flow with methane. At the simulation level, we can deduce that, although the complexity of the phenomena taking place nearby the arc column, for global simulation it is possible to represent it with a source volume with the adequate shape.

5. References

[1] I. Pfaff and A. Kather, "Comparative thermodynamic analysis and integration issues of CCS steam power plants based on oxy-combustion with cryogenic or membrane based air separation," *Energy Procedia*, vol. **1**, no. 1, pp. 495–502, Feb. 2009.

[2] S. G. Sahu, N. Chakraborty, and P. Sarkar, "Coal–biomass co-combustion: An overview," *Renew. Sustain. Energy Rev.*, vol. **39**, pp. 575–586, Nov. 2014.

[3] M. A. GOROKHOVSKI, Z. JANKOSKI, F. C. LOCKWOOD, E. I. KARPENKO, V. E. MESSERLE, and A. B. USTIMENKO, "Enhancement of Pulverized Coal Combustion by Plasma Technology," *Combust. Sci. Technol.*, vol. **179**, no. 10, pp. 2065–2090, 2007.

[4] A. Askarova, E. . Karpenko, Y. . Lavrishcheva, V. E. Messerle, and A. Ustimenko, "Plasma-Supported Coal Combustion in Boiler Furnace," *IEEE Trans. Plasma Sci.*, vol. **35**, no. 6, pp. 1607–1616, Dec. 2007.

[5] T. Matsuura, K. Taniguchi, and T. Watanabe, "A new type of arc plasma reactor with 12-phase alternating current discharge for synthesis of carbon nanotubes," *Thin Solid Films*, vol. **515**, no. 9, pp. 4240–4246, Mar. 2007.

[6] G. D. Bernard PATEYRON, *T&Twinner, calculation and data bases of thermodynamics and transportation properties*. Direction des bibliothèques, des Musées et de l'Information Scientifique et technique, 1999.

[7] K. Katou, T. Asou, Y. Kurauchi, and R. Sameshima, "Melting municipal solid waste incineration residue by plasma melting furnace with a graphite electrode," *Thin Solid Films*, vol. **386**, no. 2, pp. 183–188, May 2001.

[8] Club E.D.F / Arc Electrique, Ed., *L'Arc électrique et ses applications*. Paris: Cnrs, 1984.

[9] C. Rehmet, V. Rohani, F. Cauneau, and L. Fulcheri, "3D Unsteady State MHD Modeling of a 3-Phase AC Hot Graphite Electrodes Plasma Torch," *Plasma Chem. Plasma Process.*, vol. **33**, no. 2, pp. 491–515, Apr. 2013.

[10] C. Rehmet, F. Fabry, V. Rohani, F. Cauneau, and L. Fulcheri, "High Speed Video Camera and Electrical Signal Analyses of Arcs Behavior in a 3-Phase AC Arc Plasma Torch," *Plasma Chem. Plasma Process.*, vol. **33**, no. 4, pp. 779–796, Aug. 2013.

[11] Q. Zhou, H. Li, X. Xu, F. Liu, S. Guo, X. Chang, W. Guo, and P. Xu, "Comparative study of turbulence models on highly constricted plasma cutting arc," *J. Phys. Appl. Phys.*, vol. **42**, no. 1, p. 015210, Jan. 2009.

Journal of Physics: Conference Series **550** (2014) 012019 doi:10.1088/1742-6596/550/1/012019

Effect of suspension characteristics on in-flight particle properties and coating microstructures achieved by suspension plasma spray

E Aubignat[1,2], M P Planche[1], A Allimant[2], D Billières[3], L Girardot[4], Y Bailly[4] and G Montavon[1,5]
[1]IRTES-LERMPS, UTBM, 90010 Belfort cedex, France
[2]Saint-Gobain C.R.E.E., 84300 Cavaillon, France
[3]Saint-Gobain Coating Solutions, 84000 Avignon, France
[4]Institut FEMTO-ST, UMR 6174, CNRS, Université de Franche-Comté, 90000 Belfort, France

Abstract. This paper focuses on the influence of suspension properties on the manufacturing of coatings by suspension plasma spraying (SPS). For this purpose, alumina suspensions were formulated with two different liquid phases: water and ethanol. Suspensions were atomized with a twin-fluid nozzle and injected in an atmospheric plasma jet. Suspension injection was optimized thanks to shadowgraphy observations and drop size distribution measurements performed by laser diffraction. In-flight particle velocities were evaluated by particle image velocimetry. In addition, splats were collected on glass substrates, with the same conditions as the ones used during the spray process. Scanning electron microscopy (SEM) and profilometry analyses were then performed to observe the splat morphology and thus to get information on plasma / suspension interactions, such as particle agglomeration. Finally, coatings were manufactured, characterized by SEM and compared to each other.

1. Introduction

Suspension plasma spraying (SPS) is a coating manufacturing process that aims at injecting a suspension (solid particles of about 1 µm or less, dispersed in a liquid phase) in a high-energy plasma flow. Particles are heated, accelerated towards a substrate, flattened and subjected to a rapid solidification. Layer after layer, a coating develops to provide functional properties to the substrate on top of which it is built [1][2][3]. Figure 1 depicts the principle of SPS. This variation of the conventional plasma spray process that injects, with a carrier gas, coarser particles (order of magnitude: 30 µm in diameter) is able to achieve coatings with finer thickness (from 10 to 100 µm instead of hundreds of µm) and microstructure. The particle size reduction leads to a size reduction of pores embedded in the coating and created mostly by particle stacking defects [4]. Moreover, submicrometer-sized particles having lower kinetic and thermal inertia, their velocity and temperature decrease rapidly and thus, their spreading on the substrate is lower than larger particles (*i.e.*, their flattening ratio is smaller). This results in a granular microstructure [5] rather than a lamellar one encountered with conventional plasma spraying.

[5] ghislain.montavon@utbm.fr

Though this emerging process has been studying since the mid-1990's and achieving a fast-growing interest, it has no industrial applications yet and still needs some developments due to the complex phenomena, such as suspension fragmentation and liquid phase evaporation, that occur in the plasma which are not totally understood and controlled. The suspension injection is a particularly challenging point of the process because it is necessary that particles follow an optimal trajectory inside the plasma in order to be correctly treated. Besides, due to low momentum quantity, suspension droplets and then submicrometer-sized particles are extremely sensitive to plasma instabilities.

Aiming for an optimized injection, various strategies have been considered by the scientific community:

- Axial injection within the plasma torch [6] should be "ideal" for a complete thermal and kinetic treatment of particles, such as the one encountered in the High Velocity Oxy Fuel (HVOF) process [7]. However, this approach requires an adapted torch geometry that allows internal injection. Such configurations are generally more complicated, and costly, than classical plasma torches. Moreover, internal injection increases the risk of clogging.
- In the same manner as conventional powder, suspension can also be radially injected. Suspension droplets can only penetrate the plasma if their momentum is superior to the plasma one [8]. With this type of injection, a part of the suspension, more or less important, may not be injected in the hot central zone of the plasma core and a distribution of trajectories is observed for penetrated particles. This is why injection conditions must be adjusted in order to fulfill at best this requirement.

Since radial suspension injection can be easily implemented on most existing plasma spray equipments, it is the object of numerous studies. Among them, two main techniques are met:

- Mechanical injection [9][10][11]: most of the time, the suspension is stored under pressure in a reservoir and is forced through a capillary injector of a given internal diameter. The suspension is injected in the form of a liquid jet. This type of injector is easy to handle and allows a good penetration of the suspension thanks to the important momentum quantity of the liquid jet. However, as all the input parameters (internal diameter, pressure and suspension flow rate) are correlated, for given plasma flow and injector, the suspension flow rate cannot be varied. The injection is also much localized and there is a risk of clogging at the injector exit.
- Atomization [12][13]: a coaxial twin-fluid nozzle (as shown on Figure 2) is usually used to generate a stream of fine droplets. The co-current gas fragments once the suspension and serves as a carrier so that droplets penetrate the plasma flow before being breaking up a second time. Atomizer geometry is more complex, its control more difficult but the mass flow is independent from the chosen plasma flow. Freedom in adjusting the injection conditions allows exploring a broader spectrum of coating microstructures achievable with this technique, which has brought the choice on this kind of injector for this study.

The behavior of atomizers adapted to SPS process, and phenomena that arise in the plasma, are still not fully understood. This is why two suspensions, an aqueous and an ethanol-based one, whose properties were characterized, were selected to understand how droplets are generated by a home-made twin-fluid atomizer. Based on previous studies, shadowgraphy [11] and laser diffraction [14] were used to characterize the nozzle and optimize the suspension injection. In-flight diagnosis was also carried out to get information on particle velocities and molten state, just before impact on the substrate. SEM analyses of coatings obtained with the two selected suspensions are finally presented and hypotheses are developed, based on the diagnosis observations, in order to give a possible explanation to the difference of noticed morphology.

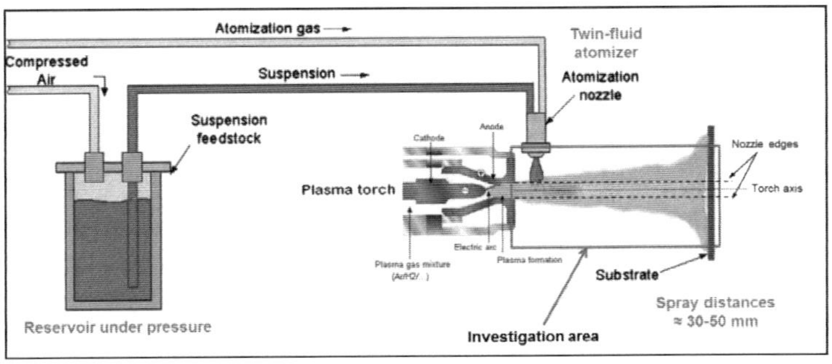

Figure 1. Scheme of the SPS process.

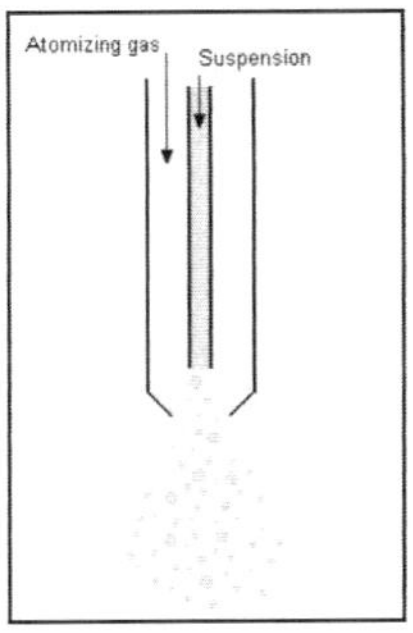

Figure 2. Scheme of a twin-fluid atomizer.

2. Experimental setups and methods

2.1. Suspension plasma spray conditions

An α-alumina feedstock powder (P152SB, Aluminium Péchiney, France, $D_v(50)$ = 1.6 µm, SSA = 2.4 $m^2.g^{-1}$) was added either to distilled water or to ethanol. A solid mass load of 20 wt.% was chosen, corresponding to a volume fraction of 0.06. An ammonium salt of polyacrylic acid (noted PAA, from Coatex, France, Mw = 2000-2500 $g.mol^{-1}$, pH = 7-8) and a phosphoric ester (Beycostat C213, CECA, France) were added respectively to aqueous and alcoholic suspensions to achieve good dispersion, checked by zeta potential measurements and sedimentation tests. Rheology measurements were performed and showed that suspensions with such low volume fraction could be assimilated to the liquid phase: viscosity does not go over a few mPa.s whatever the liquid phase. Surface tensions of liquid phases were also measured. Suspension characteristics are given in Table 1.

Table 1. Alumina suspension main characteristics.

Liquid phase	Dispersant (wt.%[a])	Mass density ($kg.m^{-3}$)	Surface tension ($x10^3$ $N.m^{-1}$)
Water	PAA (0.2%)	1176	82.5 ± 0.3
Ethanol	Beycostat (0.5%)	940	24.0 ± 0.1

[a] Weight percentage of dispersant is indicated in relation to mass of powder.

During the spray process, suspension was stored in a pressurized tank, under magnetic stirring to avoid sedimentation issues, and was delivered through a home-made twin-fluid atomizer, supplied with argon as atomizing gas. Properties of the spray can be adjusted by varying the input liquid and gas pressures that directly act on flow rates and velocities. The spray unit consists in an atmospheric plasma ProPlasma torch (Saint-Gobain Coating Solutions, Avignon, France), used in a "high performance" mode and equipped with a 6.5 mm diameter anode. The torch was attached to a 6-axis robotic arm. The atomizer is positioned perpendicularly to the torch axis, at 10 mm from the torch axis and 6 mm downstream the anode face.

Table 2. Plasma operating conditions.

Plasma	Intensity (A)	Ar/H_2:Total gas flow rate (slpm)	Enthalpy ($\times10^6$ $J.kg^{-1}$)	Effective power[a] (kW)
1	600	48/12:60	13.0	19
2	600	50/10:60	11.4	17

[a] Effective power is defined as the total electric power supplied to the plasma torch minus the thermal losses due the torch cooling.

Two sets of plasma operating spray parameters, given in Table 2, were used in conjunction with the different liquid phase used for the suspension formulation: the ethanol-based suspension, easier to evaporate, was injected in the plasma whose enthalpy is the lowest.

2.2. Characterization of the atomizer and the injection

A shadowgraphy technique (SprayCam, Control Vision Inc, USA) was used to observe on the one hand, the atomized jet generated by the twin-fluid nozzle without the plasma, and on the other hand, the suspension injection in the plasma and the droplet trajectory. The area of investigation was (22 × 16.5) mm² allowing the detection of drops and particles > 30 µm. The system is able to acquire up to 10 images per second. The software ImageJ (www.scion.org) was used to process images, to quantify the spray width and to superpose some images in order to highlight the quality degree of the injection in the plasma by revealing droplets, which did not penetrate the plasma. Droplet size distributions were measured using an optical instrument based on laser diffraction (Spraytec, Malvern Instruments Ltd, Malvern, UK). This measurement rests on the Mie scattering and uses a 10 mm diameter laser beam whose wavelength is 670 nm. Distributions with median size $D_v(50)$ from 2.5 to 125 µm are measurable.

2.3. In-flight diagnosis

In addition to suspension injection observations via shadowgraphy, some in-flight diagnoses were carried out to qualify by other means the quality of injection and to understand what happens between the moment that drops are injected into the plasma and the moment they impact the substrate.

2.3.1. Particle Image Velocimetry (PIV). PIV [15] is a technique able to measure the velocity of features (droplets and particles) and to give simultaneously a visual overview of the injection inside the plasma. It consists in acquiring with a CCD camera two successive images of a cross section of the plasma flow containing particles, by illuminating them with extremely short laser pulses (7.3 ns @532 nm in this case), turned into light sheets. The light sheets are positioned in order to include both the torch and atomizer axes. According to the particle velocities and displacements, a time of 200 ns between the two laser pulses, and so images, was taken on. The velocity of a small set of particles can be estimated measuring its displacement observed from the first image to the second one. This is done thanks to an algorithm analyses based on cross-correlation. The plane chosen for this experiment was including both plasma torch and atomizer axes. The spatial resolution of this experiment is 23 µm per pixel. Insight 3G software (TSI Inc., USA) was used for image acquisitions, instantaneous displacement field determinations and instantaneous velocity field derivation. The average velocity vector fields were calculated and plotted using two software TECPLOT (Tecplot Inc., Bellevue, WA, USA) and Mathematica (Wolfram, USA). More details on diagnosis were given in a previous work [16].

2.3.2. In-flight particle collection. Particles were collected in the same injection conditions as for coating manufacturing; in particular, suspension flow rate was kept identical. Glass substrates at ambient temperature were selected for their easy implementation of the tests. They passed only once in front of the plasma torch, at the spray distance and at a relative speed of about 15 $m.s^{-1}$ in order to get separated splats and to allow further analyses. Samples were then observed by scanning electron microscopy (SEM) and a zone of 180 × 130 µm² per sample was characterized with an interferometric profilometer to measure particle dimensions (surface and thickness). To eliminate noise, splats whose diameter was below 1.6 µm (corresponding to 5 pixels on images) were not taken account for the analysis.

2.4. Coatings

Some coatings were manufactured using the alumina suspensions described in Table 1, plasma parameters of Table 2 and optimized suspension injection conditions. They were sprayed on rotating aluminum plates (5 × 30 × 40 mm^3), preheated with the plasma flow at 150°C, at least. Coatings were sprayed at a distance of 35 mm from the plasma torch exit, moving at a relative speed of 2 $m.s^{-1}$. Coating morphology was finally observed by SEM.

3. Results and discussions

3.1. Optimization of the suspension injection

Before the coating manufacturing by SPS process, the selected home-made atomizer must be characterized in order to check whether it could be convenient for this process. Shadowgraphy observations were first performed, shown on Figure 3 and Figure 4.

For a large range of gas and liquid pressures, observations revealed that the home-made nozzle generates atomized jet whose spray cone angles are sufficiently fine so that the whole spray pattern can penetrate the plasma flow, symbolized by pink circles. It can also be noticed the spray is a little bit finer in case of ethanol compared to water.

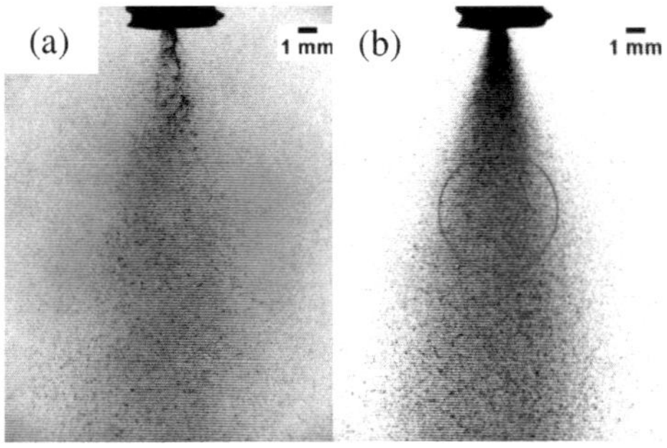

Figure 3. (a) Single image and (b) superposition of 100 images of a spray obtained with a home-made atomizer and water.

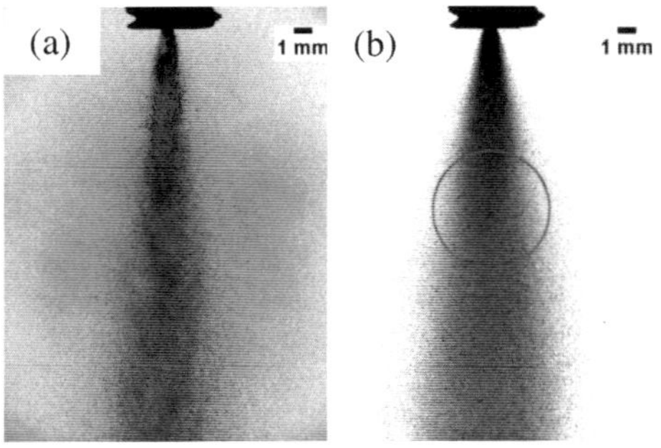

Figure 4. (a) Single image and (b) superposition of 100 images of a spray obtained with a home-made atomizer and ethanol.

Since the sprays observed in free atmosphere seem to be adequate for SPS process, that is to say sprays with fine cone angle and well-defined geometry without outlying secondary droplets, some injection tests were then carried out with the two alumina suspensions. Figure 5 and Figure 6 present the results, obtained with the same gas and liquid pressures for both suspensions. In both cases, image overviews bring out that all the droplets penetrate the plasma flow. Once again, the ethanol-based jet appears to be a little finer. Besides, few millimeters after the liquid injection, droplets are too small to be detected on images, which is consistent with a good treatment leading to fragmentation of droplets and liquid evaporation phase. In parallel, spray beads were also manufactured to confirm the quality of the injection and neat lines of molten alumina were obtained and very little overspray was observed on either side far from the spray bead.

To finalize the spray characterization, droplet size distributions were measured by laser diffraction with the same conditions as for injection tests. The results, presented in Figure 7, show that droplets are smaller in case of ethanol with a median diameter in number of 28 μm compared to 44 μm for water, containing respectively 260 and 1260 alumina particles. This is concordant with its lower surface tension that requires less energy to break out drops.

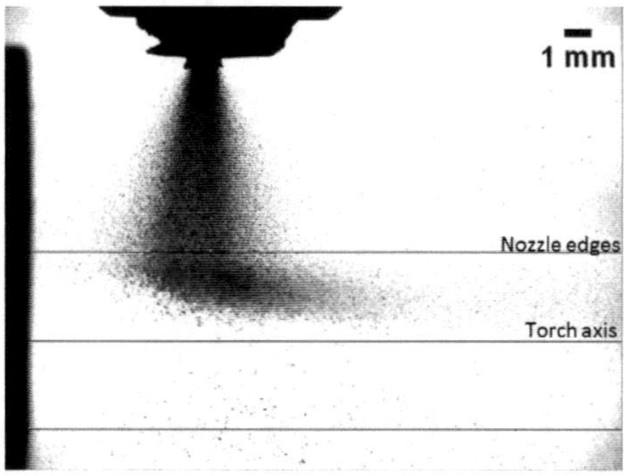

Figure 5. Shadowgraphy observation (superposition of 100 images) of the injection of the aqueous alumina suspension with a home-made atomizer.

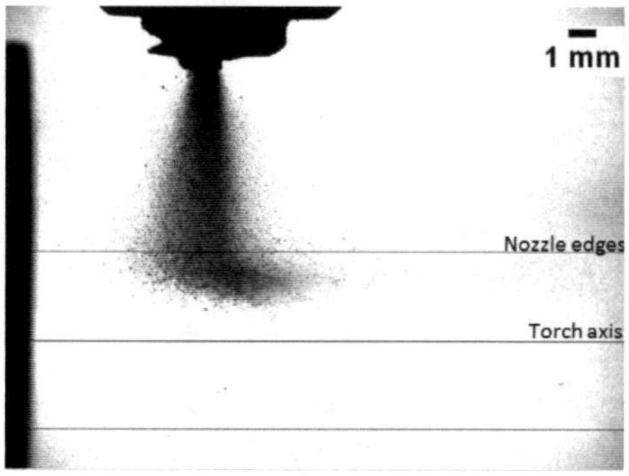

Figure 6. Shadowgraphy observation (superposition of 100 images) of the injection of the alcoholic alumina suspension with a home-made atomizer.

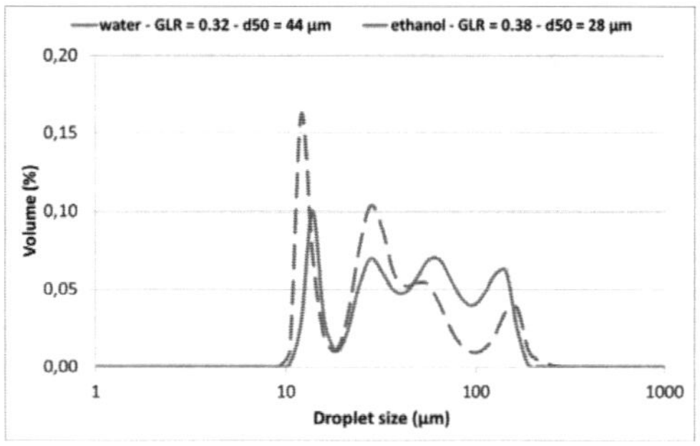

Figure 7. Droplet size distributions of sprays obtained with a home-made atomizer with the injection conditions.

3.2. Contribution to the understanding of phenomena occurring into the plasma

Now that optimal conditions were found for a proper injection of suspensions, attention was focused on particle treatment achieved by the plasma. Particle velocities were measured for the two suspension conditions. Figure 8 and Figure 9 present the velocity vector profiles obtained. It was not possible to calculate vectors in the plasma due to too high emittance, whence a lack of high velocity vectors in this zone.

Considering the important velocities, it appears that droplets and then particles seem to have penetrated in the plasma in both cases. For the alcoholic suspension, the field seems more homogeneous with vectors all over the area of investigation. Given that alcoholic droplets are initially smaller, less dense and evaporate faster, alumina particles get quickly rid of the solvent and then, are carried away by the plasma flow. On the contrary, aqueous droplets penetrate the plasma and follow a trajectory close to the torch axis, thanks to their higher momentum. Concerning the particle velocity value, it is equivalent for both cases (a maximum around 500 $m.s^{-1}$ with an uncertainty of the measurement of about 10%). However, it should be noted that these vector fields are not correlated with a number of particles: particles with highest velocities are not representative of particles that contribute most to the coating building.

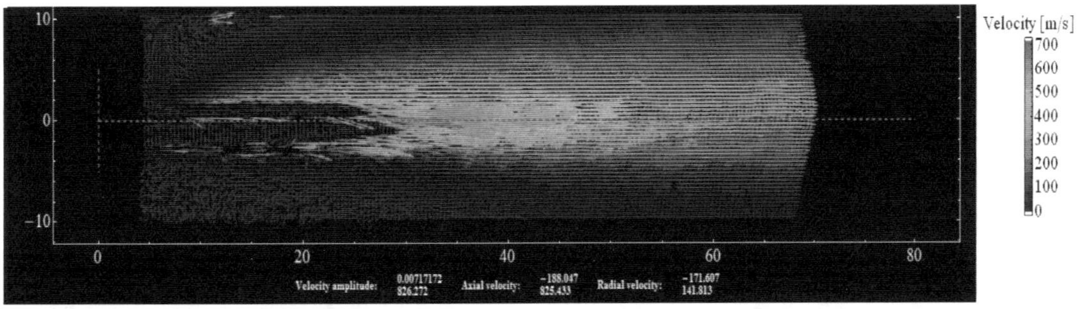

Figure 8. Velocity vector fields for injection of the alcoholic alumina suspension into plasma.

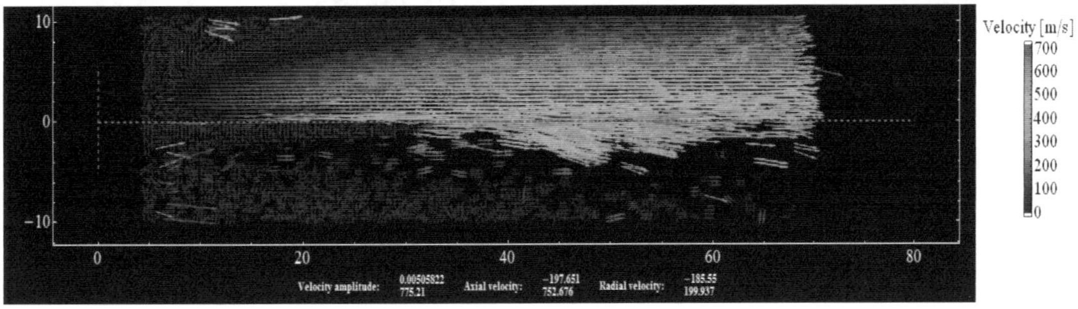

Figure 9. Velocity vector fields for injection of the aqueous alumina suspension into plasma.

PIV measurements give information on particle velocities and trajectories but data on particle temperatures, or at least, their melting state, are also very useful to understand the phenomena occurring inside the plasma. This can be achieved thanks to particle collection onto a substrate: a pancake-like splat is sign of a totally molten particle whereas if the particle is not correctly flattened, it is an indication that the particle was not treated by the plasma or already solidified before the impact. The size of the collected particles could also give some explanation to what happened to particles in the plasma, such as agglomeration. Figure 10 gives the SEM observations performed on particle collection with the two suspensions.

These observations revealed the presence of some splats correctly molten and splashed for both cases. It also showed the presence of numerous fine and spherical particles. These particles may have already cooled down before the impact on substrate. In any case, it is difficult for such fine particles to spread due to their lack of momentum quantity. On higher magnification, it is noticed that some particles have agglomerated during their stay in the plasma as grain boundaries are still visible. Moreover, this particle agglomeration led to porosity creation.

Splats surface were measured by profilometry and an equivalent diameter was calculated for each surface measured. A histogram of diameters was plotted on Figure 11 and Figure 12 shows the average diameter and the number of particles analyzed.

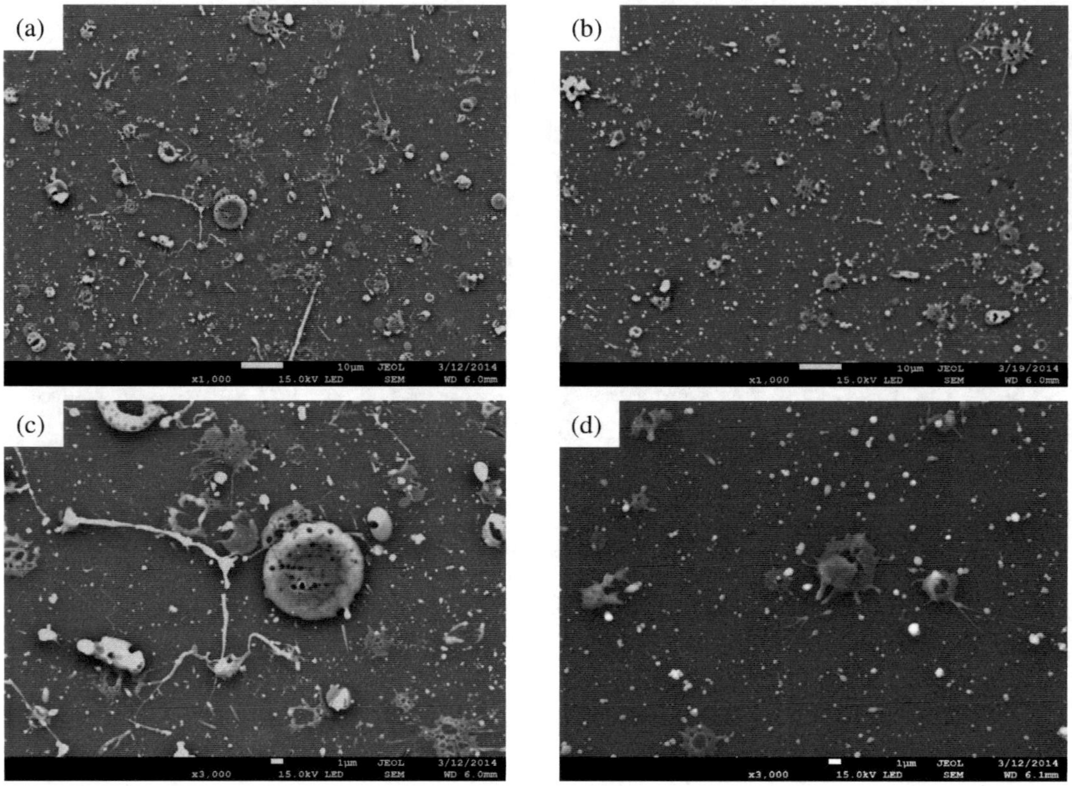

Figure 10. SEM observations of particle collection: (a) and (c) injection with aqueous suspension; (b) and (d) injection with alcoholic suspension.

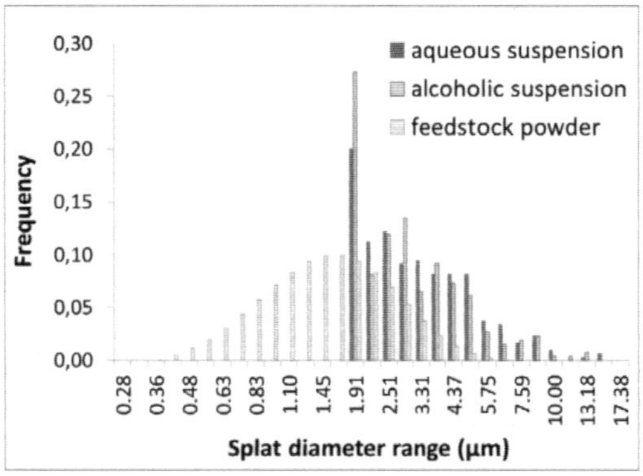

Figure 11. Histogram of splat diameter.

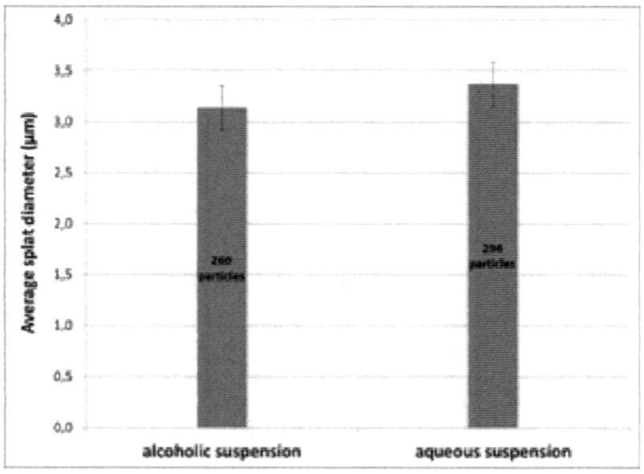

Figure 12. Average splat diameter.

These results highlight that particles arriving on the substrate have same dimensions whether the suspension is water or ethanol-based with an average diameter of 3 µm, which supposes that these two suspensions are fragmented in the same manner by the plasma. Assuming a squashing rate of about 2 [17], most particles have not agglomerated. However, a such size distributions of splats could lead to voids as it is explained in [18].

Finally, coating microstructures were manufactured with the same injection conditions as for injection tests and were observed by SEM (cf. Figure 13 and Figure 14).

Figure 13. Coating microstructure obtained with the alumina aqueous suspension.

Figure 14. Coating microstructure obtained with the alumina alcoholic suspension.

It is found that the coating with the aqueous suspension is denser than the one with ethanol. Moreover, the deposition efficiency is lower for water. As said previously, aqueous droplets follow a trajectory close to the torch axis, which has for consequence a better stacking with less porosity. On the other hand, because of lower vaporization enthalpy and boiling point of ethanol and despite the difference of plasma enthalpy (cf. Table 2), surface coating temperature is higher in case of alcoholic suspension injection. This could enhance the sticking of particles, which explains the higher deposition efficiency.

Conclusions

This study presents some phenomena occurring inside plasma during SPS coating manufacturing using an aqueous suspension and an alcoholic one. Atomized jets generated by a home-made twin-fluid nozzle were observed by shadowgraphy. Their fine spray cone angle led to a proper injection into the plasma. Droplet size measurements showed that ethanol droplets are slightly smaller than water ones, consistent with the difference of surface tension. Velocity vector fields were obtained by PIV. They showed that in case of alcoholic suspension, particles were carried away by the plasma flow all over the area of investigation, whereas aqueous droplets travel close to the torch axis. They also revealed particle velocities of more than 500 $m.s^{-1}$ at spray distance, equivalent for both suspensions. In-flight particle collections were also performed, which showed that particles arriving on the substrate had approximately the same dimensions whatever the suspension. The average splat diameter of about 3 µm suggested that most particles did not agglomerate, although agglomeration is not totally excluded according the SEM observations. Finally, the morphology of resulting coatings was observed by SEM. The deposition rate of the coating produced with the aqueous suspension is much lower. However, it appeared denser than the alcoholic one. The difference of porosity rate could be explained by a difference of particle trajectories inside the plasma, whereas the difference of deposition efficiency comes from the surface coating temperature, higher in case of ethanol, which can favor the sticking of particles. Nevertheless, further work should be carried on to confirm this hypothesis, such as PIV measurements on a plane perpendicular to the twin-fluid atomizer and also at its exit.This work demonstrates that different microstructures, more or less porous, can be obtained by changing the suspension liquid phase. Thus, coatings satisfying requirements of several applications can be manufactured just by adapting the suspension composition.

References

[1] Fazilleau J, Delbos C, Rat V, Coudert J F, Fauchais P and Pateyron B 2006 Phenomena Involved in Suspension Plasma Spraying Part 1: Suspension Injection and Behavior *Plasma Chem. Plasma Process.* **26** n° 4 371 - 391

[2] Delbos C, Fazilleau J, Rat V, Coudert J F, Fauchais P and Pateyron B 2006 Phenomena Involved in Suspension Plasma Spraying Part 2: Zirconia Particle Treatment and Coating Formation *Plasma Chem. Plasma Process.* **26** n° 4 393–414

[3] Fauchais P, Etchart-Salas R, Rat V, Coudert J F, Caron N and Wittmann-Ténèze K 2008 Parameters Controlling Liquid Plasma Spraying: Solutions, Sols, or Suspensions *J. Therm. Spray Technol.* **17** n° 1 31–59

[4] Pawlowski L 2008 Finely grained nanometric and submicrometric coatings by thermal spraying: A review *Surf. Coat. Technol.* **202** n° 18 4318–4328

[5] Fauchais P, Rat V, Delbos C, Coudert J F, Chartier T and Bianchi L 2005 Understanding of suspension DC plasma spraying of finely structured coatings for SOFC *IEEE Trans. On Plasma Sci.* **33** n° 2 920–930

[6] Waldbillig D and Kesler O 2009 Characterization of metal-supported axial injection plasma sprayed solid oxide fuel cells with aqueous suspension plasma sprayed electrolyte layers *J. Power Sources* **191** n° 2 320–329

[7] Gadow R, Killinger A and Rauch J 2008 New results in High Velocity Suspension Flame Spraying (HVSFS) *Surf. Coat. Technol.* **202** n° 18 4329–36

[8] Fauchais P, Joulia A, Goutier S, Chazelas C, Vardelle M, Vardelle A and Rossignol S 2013 Suspension and solution plasma spraying *J. Phys. -Appl. Phys.* **46** n° 22 224015

[9] Delbos C 2004 Contribution à la compréhension de l'injection par voie liquide de céramiques (Y.S.Z., Pérovskite,...) ou métaux (Ni,...) dans un plasma d'arc soufflé afin d'élaborer des dépôts finement structurés pour S.O.F.Cs. Université de Limoges thesis

[10] Vert R, Chicot D, Dublanche-Tixier C, Meillot E, Vardelle A and Mariaux G 2010 Adhesion of YSZ suspension plasma-sprayed coating on smooth and thin substrates *Surf. Coat. Technol.* **205** n° 4 999–1003

[11] Soysal D and Ansar A 2013 A new approach to understand liquid injection into atmospheric plasma jets *Surf. Coat. Technol.* **220** 187–190

[12] Marchand O 2010 Etude du procédé de projection plasma de suspensions pour l'élaboration du coeur de pile à combustible SOFC Université de Franche-Comté, Université de Technologie de Belfort-Montbéliard thesis

[13] Karthikeyan J, Berndt C, Tikkanen J, Reddy S and Herman H 1997 Plasma spray synthesis of nanomaterial powders and deposits *Mater. Sci. Eng. A* **238** n° 2 275–286

[14] Rampon R, Filiatre C and Bertrand G 2007 Suspension Plasma Spraying of YPSZ Coatings: Suspension Atomization and Injection *J. Therm. Spray Technol.* **17** n° 1 105-114

[15] Raffel M, Willert C E and Kompenhans J 1998 *Particle Image Velocimetry - A Practical Guide* (Berlin: Springer, 1st edition)

[16] Aubignat E, Planche M P, Allimant A, Billières D, Girardot L, Bailly Y and Montavon G, A three-step characterization of a twin-fluid atomizer for suspension plasma spray process *16th Int. Symp. on Flow Visualization* (Okinawa, Japan, 24-27 June 2014)

[17] Tingaud O, Bacciochini A, Montavon G, Denoirjean A and Fauchais P 2009 Suspension DC plasma spraying of thick finely-structured ceramic coatings: Process manufacturing mechanisms *Surf. Coat. Technol.* **203** n° 15 2157-61

[18] Joulia A, Duarte W, Goutier S, Vardelle M, Vardelle A and Rossignol S Tailoring the spray conditions for suspension plasma spraying *Int. Therm. Spray Conf.* (Barcelona, Spain, 21-23 May 2014)

Pulsed laminar arc jet with synchronized suspension injection-spectroscopic studies

J Krowka[1], V Rat[1], C Chazelas[1] and J- F Coudert[1]
CNRS, University of Limoges, SPCTS UMR7315, European Ceramic Centre, Limoges Cedex, 87068, France

E-mail: joanna.krowka@unilim.fr

Abstract. The uncontrolled arc plasma instabilities are one of the difficulties encountered in in suspension plasma spraying. The improvement of this process is usually attempted by means of the reduction of arc fluctuations. The following paper presents a new approach to overcome these arc instabilities. The principle is to produce a pulsed laminar plasma jet combined with phased injection of liquid droplets. This is achieved by the particular design of the plasma torch which allows coupling two modes of the plasma oscillations, restrike and Helmholtz. The plasma produced in a new mode is laminar and pulsed, characterized by a modulated enthalpy, what permits to make the synchronization with the suspension injection. The droplets are injected using a piezoelectric device, based on drop-on-demand method, triggered by the voltage signal. The results are evaluated by time-resolved imaging technique and Time-Resolved Optical Emission Spectroscopy.

1. Introduction

Suspension plasma spraying is a well-established process which allows producing finely structured and dense coatings. However, in spite of the fact that the plasma torch is supplied by a direct current (DC) power source, the plasma jet presents unsteady characteristics. The research studies have verified that this time-fluctuating momentum of the plasma have a strong influence on the droplet fragmentation and vaporization process [1, 2]. It results in a poor reproducibility and reliability of the method, what causes the limited applications of suspension plasma spraying in advanced processing. Therefore, for many years, the special efforts have been devoted to understand the arc behavior in dc plasma torch. One of the sources of the plasma instabilities is the 'stick and slip' motion of the arc inside the nozzle [3-4]. This kind of oscillations, known as 'restrike' mode, gives an irregular voltage signal approximately saw-tooth in shape. The other major source of the plasma fluctuations has been more recently identified in conventional plasma torches. Delair et al. have suggested that Helmholtz oscillation in the arc chamber can be the reason for high frequency fluctuations of the arc voltage [5]. Rat and Coudert have shown that the plasma instabilities are related to compressibility effects of plasma forming gas in the cathode cavity [6, 7]. These instabilities appear in power spectrum of the arc voltage as a strong, sharp peak (2–5 kHz), and have been referred to Helmholtz mode. The following work highlights thc possibility to control the instabilities of plasma. The arc jet working in a pulsed laminar regime, obtained by coupling Helmholtz and restrike modes in a mechanism of phase-locked loop, is presented. This pulsed plasma, provided the use of synchronous injection, is expected

[1] To whom any correspondence should be addressed.

to improve the control of dynamic and thermal interaction between the plasma and injected material. The synchronization of suspension injection with the plasma is possible by the use an innovative injection system based on drop-on-demand (DOD) technique. The results obtained by this new suspension plasma spraying system are evaluated by time-resolved imaging technique and Time-Resolved Optical Emission Spectroscopy.

2. Experimental setup

As has been highlighted above, in the plasma jet generated by dc torch the oscillations modes, restrike and Helmholtz, have been identified. The following paper shows the possibility of coupling these two modes to obtain pulsed laminar arc plasma jet [8, 9]. It is possible to obtain by a special torch characterized by a larger cathode cavity (V_g=17.8 cm^3). The study of Helmholtz mode shows that the resonance frequency, f_H, linked to this oscillation, depends on: $f_H = 1/2\pi(\gamma_g.p_g/\rho_p).(S/L_p.V_g)$, where: γ_g, p_g are, respectively, the isentropic coefficient of the cold gas, the mean pressure in cathode cavity, ρ_p is the plasma density, S, L_p and V_g are, successively, the cross section area, the length of the nozzle channel and the volume of the cathode cavity. According to the above equation, the enlargement of the cathode cavity reinforces the Q factor, which is linked to the band pass of resonator, Δf, and to the damping factor, ξ, as follows: $Q = f_H/\Delta f = 1/2\xi$, what decreases the Helmholtz mode frequency. Nitrogen is used as plasma gas with mean mass flow rates, $\overline{\dot{m}}$, between 0.042 and 0.104 g.s^{-1} and different nozzles are tested with channel diameters, d, of 2.5, 3, 3.5 and 4 mm. The regulation of the above mentioned parameters and the arc current results in a new oscillation mode, called mosquito mode because of the sound produced by a torch that recalled the flight of a mosquito, characterized by a very regular voltage signal, presented in figure. 1.

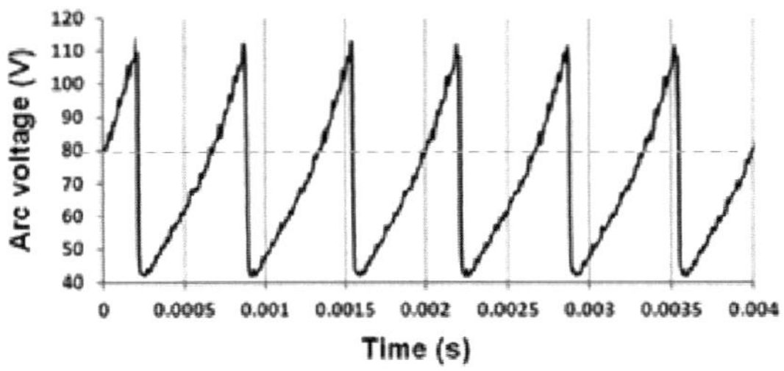

Figure 1. Arc voltage signal for: d = 4 mm, I=14.9 A, N_2: 0.042 g.s^{-1}, $\overline{V}$ = 74 V.

The measurements of the heat losses, Q_{loss}, to the electrodes and the electric power supplied to the torch, P_{elec}, permitted to determine the specific enthalpy of the plasma by the equation: $h=(P_{elec}-Q_{loss})/\overline{\dot{m}}$. The thermal losses are proportional to the arc current I, so that the ratio (Q_{loss}/I), that is homogeneous to a voltage, V_{th}, represented about 35 V. That means that the recorded voltage, V(t) must be reduced by 35 V before the conversion into time dependent specific enthalpy, following the relationship, $h(t)=(V(t)-V_{th}).I/\overline{\dot{m}}$. Then it can be predicted that the specific enthalpy in the mosquito mode is modulated in the proportion, h_{max}/h_{min} up to 16, with a mean value of 13.3 MJ.kg^{-1}. That last value is comparable to what is currently obtained for conventional plasma spraying torches. Figure 2 presents the suspension plasma spraying system used in the experiment. It consists of a new home-made torch, described above, synchronized with an external injector based on the drop-on-demand (DOD) method (Ceradrop, Limoges, France).

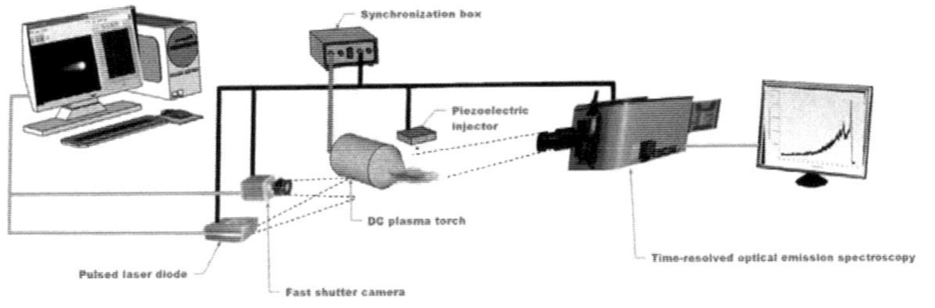

Figure 2. Experimental setup.

The aqueous suspension, stored in a buffer tank and composed of TiO_2 crystallites (42 wt.%), is injected into the plasma jet using a piezoelectric head actuated by voltage pulses. This piezoelectric injector allows obtaining a single calibrated droplet with a diameter of 50 µm and a velocity between 2 and 10 $m.s^{-1}$. The principle of synchronous injection with the pulsed plasma jet consists in the insertion of the suspension droplet at the chosen moment of the periodic plasma jet oscillations, providing that the injection frequency is equal to that of the Helmholtz resonant mode. The emission of each droplet is triggered from sampling of the torch voltage after an adjustable time delay, τ. The time-resolved imaging system allows to observe the interaction between the plasma and the suspension droplet. It is composed of a fast shutter camera (PCO, Kelheim, Germany), characterized by high resolution (1392 x 1040 pixels), coupled with a laser (HiWatch, Oseir, Tampere, Finland). The pulsed laminar arc jet is diagnosed by the Time-Resolved Optical Emission Spectroscopy (TROES) which consists of an IsoPlane spectrograph (Princeton Instruments, Trenton, New Jersey) equipped with 1200 g/mm gratings. The spectra acquisition is realized by using a PI-MAX4 ICCD camera (Princeton Instruments) connected to the PC and controlled by LigthField software. Both the time-resolved imaging system and TROES are synchronized with the injection by the synchronization box.

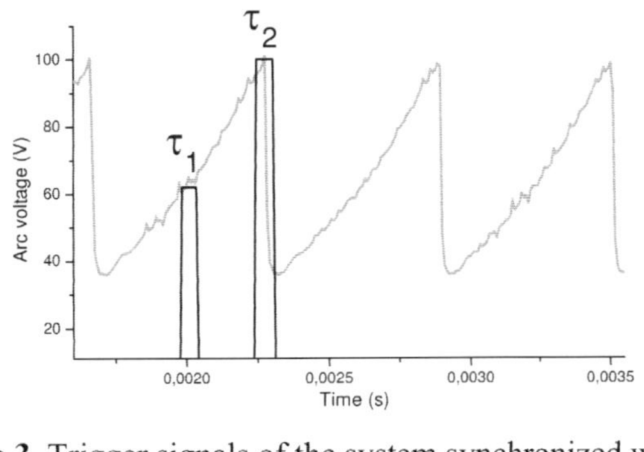

Figure 3. Trigger signals of the system synchronized with the arc voltage.

Two different moments, τ_1 and τ_2 presented in figure 3, of the periodic arc voltage have been chosen to trig the suspension injection, the camera, laser and spectroscopy recording.

3. Results

The synchronization of the time-resolved imaging system with the arc voltage signal allows to observe the plasma jet in two triggering moments, shown in figure 3. Figure 4 presents the photo of a pulsed plasma jet taken in: a) τ_1 moment at the arc voltage level of 60 V, characterized by a low specific enthalpy, b) τ_2 moment characterized by the highest level of the enthalpy, corresponded to the maximum voltage level of 100 V. These following results show the effect of the coupling restrike and Helmholtz modes- periodic modulated plasma jet.

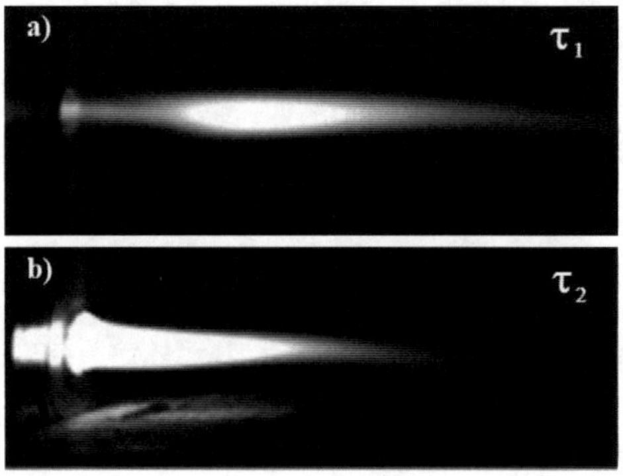

Figure 4. Pulsed arc jet with: a) low level of local specific enthalpy, corresponded to the arc voltage level of 60 V, b) high level, corresponded to 100 V.

To better analysis the periodicity of this new pulsed plasma jet and the modulation of the specific enthalpy the spectroscopic measurements have been carried out for the two moments of the plasma jet presented in figure 4. The temperature of the plasma flow is commonly determined by the rotational lines of the first negative system $B^2\Sigma_u^+ \rightarrow X^2\Sigma_g^+$ [10, 11]. Therefore, the emission spectrum of the first negative system of the N_2^+ molecule have been measured at a distance of 1 mm from the nozzle. Figure 5 presents the spectrum of the nitrogen plasma obtained in the triggering moment τ_2. The spectroscopy system has been calibrated, both in the intensity and wavelength axis.

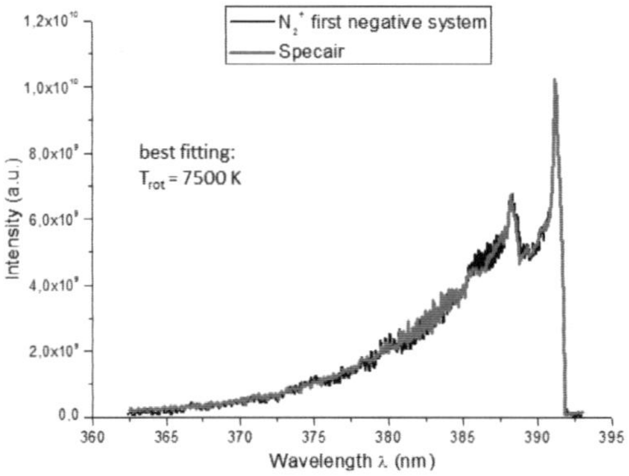

Figure 5. Comparison of the measured and simulated by Specair software (N_2^+) first negative spectra.

The rotational temperature, T_{rot}, has been calculated by using Specair software, which allows comparing the experimental and simulated spectra. Bruggeman et al. have shown that the rotational temperature is strongly determined by the slit function [12]. Therefore, to ameliorate the accuracy, this instrument function has been carefully chosen. It has been determined by recording the line profile of the emission from a low-pressure Hg lamp at the wavelength of 404.7 nm.

The measured spectrum has been fitted to the theoretical one, what is presented in figure 5. Well-fitted rotational structures permit to determine the rotational temperature with good precision.

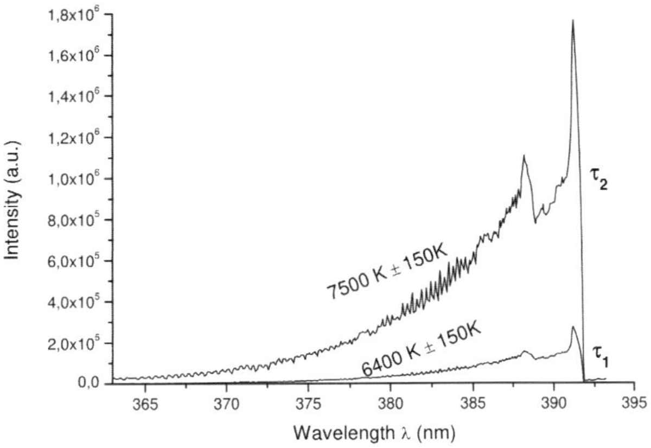

Figure 6. (N_2^+) first negative spectra for two different moments, τ_1 and τ_2 of the periodic arc voltage.

Figure 6 presents the molecular emission spectra of the first negative system of the N_2^+ molecule of the nitrogen plasma obtained at the moments, τ_1 and τ_2, of the periodic arc voltage. Analyzing the N_2^+ molecule emission at 391 nm, it can be stated that the spectrum measured at the moment τ_2, corresponded to the maximum voltage level of 100 V, is characterized by the higher value of the intensity than the spectrum detected in the moment τ_1. The rotational temperature determined for the plasma measurement at the moment τ_2 is much higher and equals to 7500 ± 150 K. The spectroscopic estimation of the rotational temperature of the nitrogen plasma permits to obtain the values of the specific enthalpy of the plasma jet measured at different moments, τ_1 and τ_2.

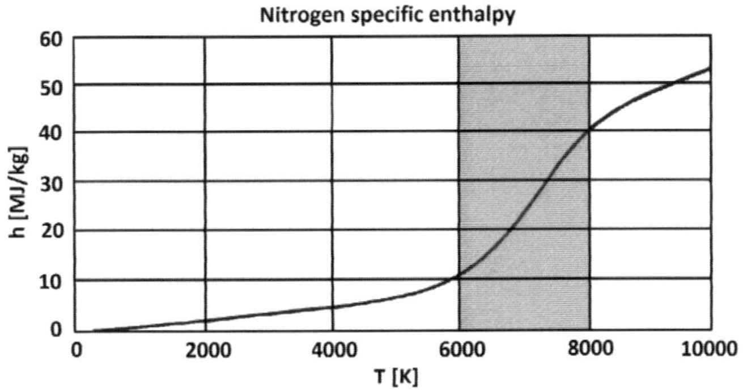

Figure 7. The nitrogen specific enthalpy [13].

The comparison of the obtained results with the nitrogen enthalpy graph, presented in figure 7, shows that the plasma jet analyzed at the moment τ_2 is characterized by the enthalpy value of around 35 ± 3 MJ/kg and at the moment τ_1 by the enthalpy of 15 ± 2 MJ/kg. Figure 8 presents the material injection to the pulsed plasma jet.

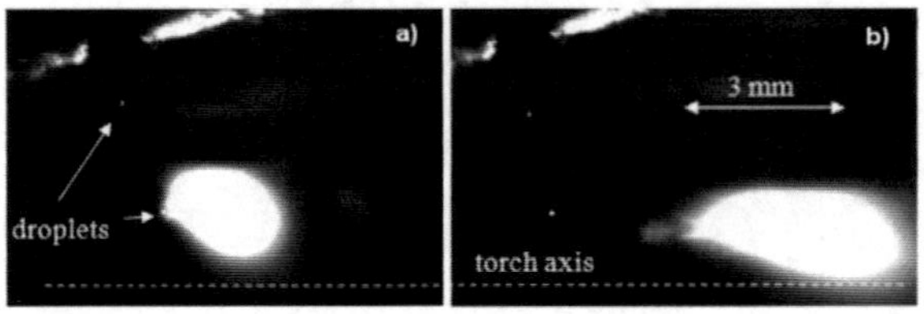

Figure 8. Influence of the local instantaneous specific enthalpy on droplet thermal treatment: a) high level of local specific enthalpy, b) low level.

In figure 8 a) the droplet enters the plasma 4 mm downstream of the nozzle exit, at the moment τ_2 characterized by a high level of specific enthalpy, around 35 MJ/kg. The almost immediate vaporization process of the droplet has been observed. Another situation is shown in figure 8 b). The droplet is injected into the plasma with a medium level of specific enthalpy. At that time the vaporization-seeding process does not concern this droplet but that one which was injected one period earlier. These two pictures are observed through a narrow band-pass filter centered on the laser wavelength. The emission wavelength of the laser is in the range of 801 ± 2 nm. Therefore, the interferential filter of 801 nm has been applied. This configuration permits to eliminate on the image the light coming from the pure nitrogen plasma, compared to figure 4. Only the particles that reflected the laser light are seen, together with the light ball that resulted of the interaction of the nitrogen plasma with the material contained in the droplet that produced a strong increase of the brightness and an enrichment of the spectrum.

4. Conclusions

A new approach to overcome the plasma instabilities has been presented, obtained by a phase locking Helmholtz oscillation and rearcing events in the nozzle. The torch works in a new resonant mode, called mosquito mode, characterized by very regular, saw-tooth shaped arc voltage signal. The plasma produced by this new mode is periodic, pulsed with the enthalpy highly modulated (h_{max}/h_{min}=13). This periodicity has been investigated by using Time-Resolved Optical Emission Spectroscopy synchronized with the arc voltage signal. The spectroscopic diagnostics allow to determine the rotational temperature of the pulsed nitrogen plasma and demonstrate the significant modulation of the plasma specific enthalpy. The modulation of the arc voltage has been used to synchronize the plasma with the injection system, based on drop-on-demand method, which enables to obtain the suspension injection in a chosen moment. The interaction between the periodic plasma flow and the droplets has been examined by the time resolved imaging triggered by arc voltage signal and synchronized with the suspension injection. The thermal treatment of injected material is shown to be very sensitive on the moment at which the droplet penetrates the plasma. This type of synchronous injection with a periodic pulsed arc jet is supposed to control the dynamic and thermal interaction between plasma and material.

5. Acknowledgments

The French National Research Agency is acknowledged for financial support in the frame of PLASMAT program (ANR-12-JCJC-JS09-0006-01).

6. References

[1] Vardelle A, Chazelas C, Marchand C and Mariaux G 2008 *Pure Appl. Chem.* **80** 1981-1991
[2] Etchart-Salas R, Rat V, Coudert J-F, Fauchais P, Caron N, Wittman K and Alexandre S 2007 *J. Therm. Spray Techn.* **16** 857-865
[3] Duan Z and Heberlein H 2002 *J. Therm. Spray Techn.* **11** 45-51
[4] Rat V and Coudert J-F 2008 *J. Phys. D: Appl. Phys.* **4** 205208
[5] Delair L, Tu X, Bultel A and Cheron B G 2005 *High Temp. Mater. Processes* **9** 583-597

Journal of Physics: Conference Series **550** (2014) 012020 doi:10.1088/1742-6596/550/1/012020

[6] Rat V and Coudert J-F 2010 *J. Phys. D: Appl. Phys.* **108** 043304
[7] Rat V and Coudert J-F 2010 *Appl. Phys. Lett.* **96** 101503
[8] Krowka J, Rat V and Coudert J F *2013 J. Phys. D: Appl. Phys.* **46** 224018
[9] Krowka J, Rat V, Goutier S and Coudert J F 2014 *J. Therm. Spray Techn.* **23** 786-794
[10] Laux C O, Gessman R J, Kruger C H, Roux F, Michaud F and Davis S.P. 2001 *J. Quant. Spectrosc. Radiat. Transfer* **68** 473-482
[11] Parigger C, Plemmons D H, Hornkohl J O and Lewis J W L 1995 *Appl. Opt.* **34** 3331-5
[12] Bruggeman P J, Sadeghi N, Schram D C and Linss V 2014 *Plasma Sources Sci. Technol.* **23** 023001
[13] Boulos M I, Fauchais P and Pfender E 1994 *Thermal plasmas: fundamentals and applications, Volume 1* (New York: Plenum Press)

Analysis of processes in DC arc plasma torches for spraying that use air as plasma forming gas

V Frolov, D Ivanov, A Toropchin
Saint Petersburg State Polytechnical University, Saint-Petersburg, Russia

E-mail: frolov.eed@gmail.com

Abstract. Developed in Saint Petersburg State Polytechnical University technological processes of air-plasma spraying of wear-resistant, regenerating, hardening and decorative coatings used in number of industrial areas are described. The article contains examples of applications of air plasma spraying of coatings as well as results of mathematical modelling of processes in air plasma torches for spraying.

1. Introduction

Application of DC arc plasma torches for spraying of coating is well known [1–3]. For that purpose they usually use a DC arc plasma torch with a variable arc length in which they use nitrogen, inert gases (argon, helium) and their mixtures with hydrogen as the plasma forming gas [2].
Studies being performed for many years in the Department of Electrical Power Engineering and Equipment of St. Petersburg State Polytechnical University (SPbSPU) were a basis for a development of DC arc plasma torches for spraying that have the following main features: a design is based on an interelectrode insert and an application of air as plasma forming gas. There were developed a large number of plasma spraying technologies for various purposes (heat-resistant coating, wear-resistant one, corrosion-resistant one, protective one, decorative one, etc.) using such plasma torches [3, 4].
The article presents examples of application of used plasma equipment as well as an analysis of processes in the DC arc plasma torch for spraying using a mathematical modelling.

2. Examples of applications

Department of Electrical Power Engineering and Equipment of Saint Petersburg State Polytechnical University and Science and Educational Technological Centre "Electrotechnology" perform theoretical and experimental researches in the area of plasma techniques and technology over 50 years [3].
As a result of performed researches there were developed plasma torches and realized technologies of spraying of different protective coatings on new manufactured parts and technologies of recovery of out-of-repair items [4]. There were realized technologies of recovery of out-of-repair parts of industrial machines (spindles, seats for bearings), plain bearings of turbocharging compressor shafts for internal-combustion engines as well as technologies of air-plasma spraying of wear-resistant coatings on automotive crankshaft journal (freezer compressor) [4].
Fig. 1 shows the process of plasma spraying of aluminium coatings onto carbon fiber to improve the properties of supercapacitors [5].

There were developed technologies of surface hardening of stop equipment of gas main lines. There were worked through conditions of spraying of wear-resistant and corrosion-resistant coatings on surfaces of plungers of autocrane elevators, face seals and rods of water pumps.
Air-plasma spraying provides to create coatings from metals that are usually exposed to considerable oxidation, for example, copper coating [4].
Fig. 2 presents parts of gas turbine GTN-25 with a heat-shielding zirconium dioxide coating created by air-plasma spraying.

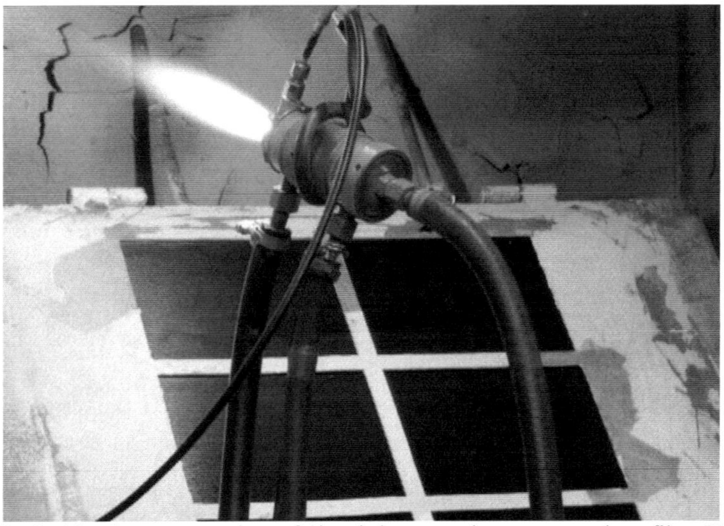

Fig.1. Air plasma spraying of aluminium coatings onto carbon fiber [5]

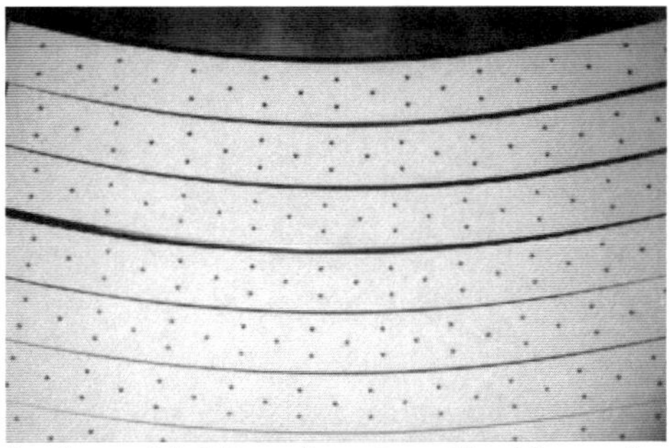

Fig. 2. Parts of gas turbine GTN-25 with a heat-shielding zirconium dioxide coating

Modern industry requires increasing of energetic branches, particularly in oil industry and power production. One of the problems, which were solved with air-plasma spraying, was a problem of wearing of threaded connection of pump-compressor pipes that used in oil industry. That problem was very important because during operations the threaded connection of pipe breaks down in the first place.
Coatings were created from powder materials of different compositions with size of particles less than 50 μm. Optimal conditions of spraying were determined versus variation of following operational conditions: arc current, arc voltage, plasma forming air flow rate, spraying distance, torch velocity relative to pipe, angle between particles flow and pipe axis [6, 7].
Example of pipes with threaded connection that was hardening by coatings were exposed to wear-resistant tests using a stand imitating pipe screwing. Coating wear-resistance as well as pipe breaking strength in an area of the threaded connection was estimated. Results of tests have shown that a wear

resistance of thread with coating is increased in several times in comparison with an initial version without coating.

A new important trend is a coating for restoring of metal sculptures and monuments. In St. Petersburg, as in other historical centres of the world, there are a lot of monuments which have different damages because of aggressive influence of city's atmosphere.

There was elaborated an air-plasma technology of corrosion-resistant and decorative coatings for copper alloys. A preoxidised copper powder was chosen as a coating material. Previously, it was investigated a level of powder's oxidation and regimes of the spraying. One can choose the colour of the coating by varying the level of powder's preliminary oxidation (see. Fig. 3) [8].

Total technology of coatings includes several stages [9]:

- stream-vortex cleaning of surface;
- air-plasma spraying;
- impregnation an inhibitor of corrosion into the coating's porosity;
- surface treatment with natural wax.

Preliminary test was realized on bronze, brass and copper plates with size 150×150 mm. Thickness of the formed coatings was about 100 μm. The covering samples were processed by solution of inhibitor and fitted for test in climatic camera. As a result based on accelerated test on corrosion stability it was determined an optimum regime of the air-plasma spraying. Test in climatic camera have shown that such covering can stand more than 70 years in our climatic condition without any trace of the corrosion.

The elaborated technology was used on practice in Saint Petersburg during last restorations of sculptures group "Tamer of horses" by P.Klodt (the Anichkov bridge), metal parts of "Aleksander's column" (the Palace square) and of sculptures group on the Senate and Synod Building (see Fig. 4).

Fig. 3. The color palette of protective and decorative coatings based on copper powder sprayed by the air-plasma arc plasma torch [8]

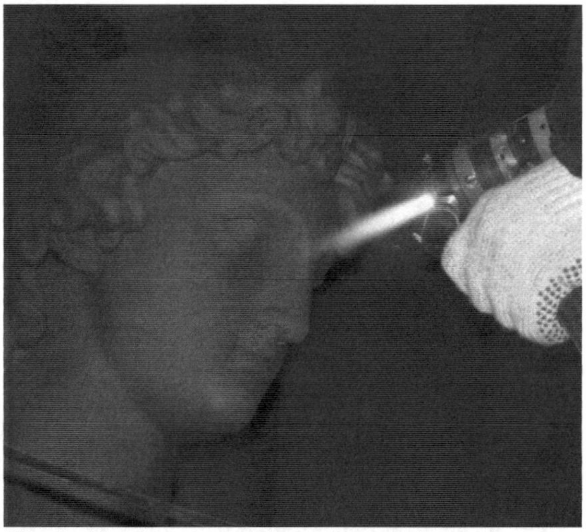

Fig. 4. Air plasma spraying of protective and decorative coatings onto sculpture of the Senate and Synod Building (Saint Petersburg)

There is a possibility of applying of such covering not only for restoring purposes but as anti-corrosion covering in different branches of industry.

3. Mathematical modelling

In conditions of widespread applications of air-plasma spraying it is very important to know the qualitative and quantitative relationship between technological efficiency of the spraying process, on the one hand, and the geometry of the plasma torch and its mode of operation (arc current, gas flow rate etc.) on the other hand. An effective way of obtaining this information is the mathematical modelling of plasma processes in the arc plasma torch for spraying.

Used mathematical model of plasma processes in the arc plasma torch for spraying is based on the following assumptions: plasma is in the state of local thermodynamic equilibrium; plasma is laminar and optically thin. Those assumptions let us to consider plasma as a continuous medium.

Equations included in the model express fundamental conservation laws (energy, momentum, mass) and are given in [10]. The region inside the plasma torch and a region of plasma jet were taken as the computational domain as can be seen in Figs. 5 and 6.

Plasma properties are also included in the mathematical model. Their dependencies on temperature (typically at atmospheric pressure) are given in the literature for main gases used in arc plasma torches for spraying, for example in [11].

A series of calculations of plasma processes in the air-plasma DC arc plasma torch PN-V1 was carried out with the following operational parameters: the arc current was varied between 150 and 200 A, the gas flow rate was varied between 0.6 and 1.2 g/s. Distributions of plasma temperature, velocity, pressure and electromagnetic functions were obtained as results of calculations. Examples of obtained distributions of plasma temperature and axial velocity are presented in Fig. 5 and 6.

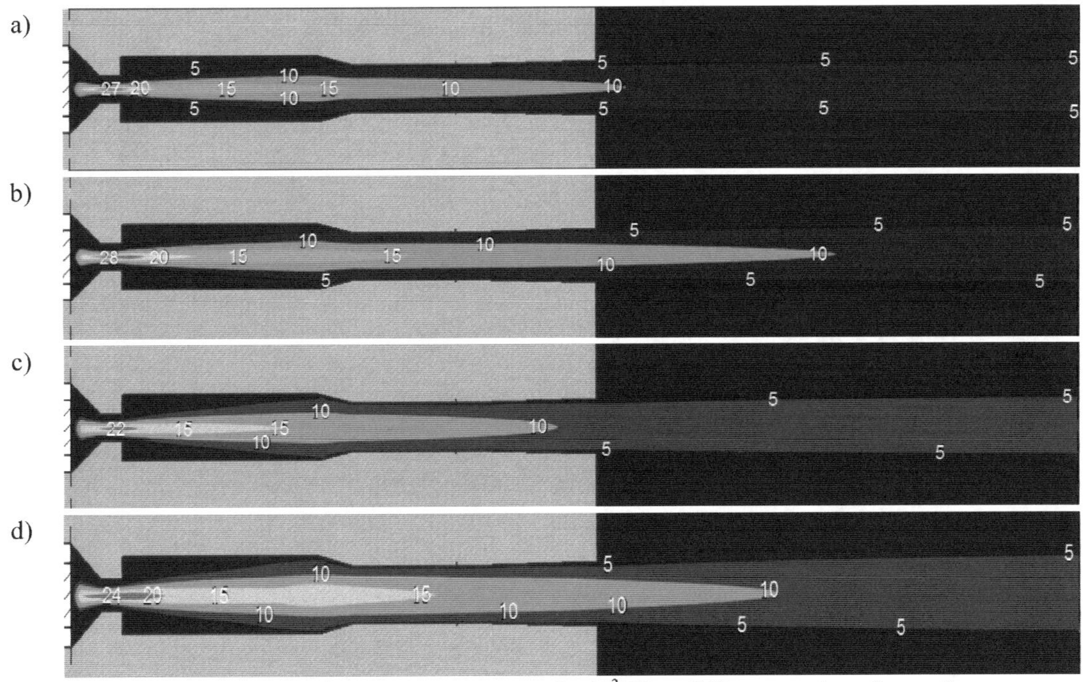

Fig. 5. Distributions of plasma temperature (in 10^3 K) at different modes of operations:
a – I=150 A, G=1.2 g/s; b – I=200 A, G=1.2 g/s; c – I=150 A, G=0.6 g/s; d – I=200 A, G=0.6 g/s

Comparing the results of calculations one can draw the following conclusions:

- temperature at the outlet of the plasma torch increases when the arc current increases (at a constant flow of gas) and when the gas flow rate increases (at a constant arc current);
- velocity at the outlet of the plasma torch (1256 m / sec) at the arc current of 200 A and the gas flow rate of 1.2 g/s is higher than for other operational conditions. Comparing the results at the same arc current one can see that the plasma velocity depends on the gas flow rate namely the plasma velocity is higher when the gas flow rate is higher.

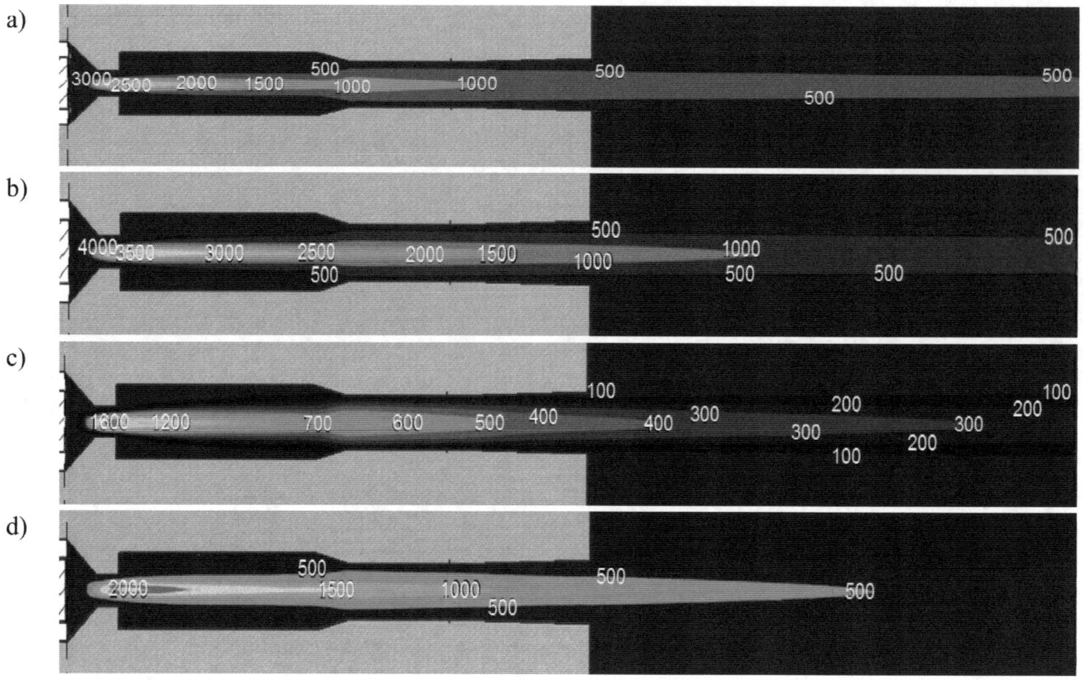

Fig. 6. Distributions of axial plasma velocity (in m/s) at different modes of operations:
a – I=150 A, G=1.2 g/s; b – I=200 A, G=1.2 g/s; c – I=150 A, G=0.6 g/s; d – I=200 A, G=0.6 g/s

4. Conclusion

It was shown that an application of air as the plasma forming gas in plasma torches for spraying has certain advantages in a number of cases. These include not only the low cost and high availability of air as the plasma forming gas but also an ability to control the properties of the sprayed coating, namely, the degree of oxidation.

Developed equipment, techniques of experimental investigations, mathematical models of plasma processes and results of investigations are the basis for the implementation of the technology of air-plasma spraying of wear-resistant, corrosion-resistant, recovering and thermal barrier coatings as well as of protective and decorative coatings on the monuments of different materials.

References

[1] Pfender E., 1999 Thermal plasma technology: where do we stand and where are we going? *Plasma Chem. Plasma Process* **19** 1.

[2] Fauchais P. 2004 Understanding plasma spraying *J. Phys. D: Appl. Phys.* **37** 9 R86.

[3] Frolov V. Ya., Klubnikin V. S., Petrov G. K., Yushin B. A. 2008 *Technique and technology of coatings* (St. Petersburg Polytechnical Univ., in Russian).

[4] Frolov V., Petrov G., Yushin B., Dubov M., Churkin I., Ivanov D. 2009 Research and development of plasma technologies of spraying of coatings *Proc. 18th Symposium on Physics of Switching Arc (Nové Mĕsto na Moravĕ, Czech Republic, Sept. 7.-11., 2009)* eds V.Aubrecht, M.Bartlova pp 162–165.

[5] Isaeva E.M. 2014 *Investigation of the process of plasma spraying of aluminum coating onto the carbon fiber*: Master Thesis (SPbSPU, St. Petersburg, in Russian).

[6] Petrov G.K. et al. 2006 Equipment and technology of spraying of air plasma coatings: basic scientific and technical results and achievements *SPbSTU Proceedings* **501** pp 139-150 (in Russian).

[7] Baeva M., Zalach J., Petrov G., Ushin B., Ivanov D., Frolov V., Uhrlandt D. 2011 Study of plasma torches used for air plasma spraying of protective and decorative coatings *Proc. XIXth Symposium on Physics of Switching Arc (Nové Mĕsto na Moravĕ, Czech Republic, Sept. 5.-9., 2011)* eds V.Aubrecht, M.Bartlova pp 105–108.

[8] Ushin B.A. 2010 *Development of air-plasma technology of spraying of protective and decorative coatings*: PhD Thesis (SPbSPU, St. Petersburg)

[9] Petrov G.K., Frolov V.Ya., Ushin B.A. 2013 Method of forming of protective and decorative coating on a metal surface *RF Patent 2486276*, filed February 29, 2012, and issued June 27, 2013.

[10] Dresvin S.V., Ivanov D.V., Frolov V.Ya. 2012 Method of calculation of thermal plasma processes *Induction heating* 4(22) 25 (in Russian).

[11] Boulos M. I., Fauchais P., Pfender E. 1994 *Thermal Plasmas: Fundamentals and Applications, Vol. 1* (New York: Plenum Press)

Mechanisms of heat transfer between a droplet and a plasma jet in Suspension Plasma Spraying.

F. Girard(1), E. Meillot(1), S. Vincent(2), L. Bianchi(1) and J.P. Caltagirone(2)

(1) CEA, DAM, Le Ripault, F−37260, Monts, France

(2) University of BORDEAUX 1, I2M Laboratory, UMR CNRS 5295, 16 avenue Pey Berland, F-33607 Pessac, France

Corresponding author : Erick Meillot. erick.meillot@cea.fr

Abstract. This article describes a small-scale modeling investigation of the suspension plasma spraying process. The heat transfer between a droplet of pure water and an Ar/H_2 plasma jet was analyzed. The low dwell time of the droplet in the flow before impacting the substrate leads to consider radiation as not the main mechanism while convection and conduction were enhanced as the droplet became deformed.

Key words: suspension plasma spraying, heat transfer, conduction, radiation, deformation, vaporization, modeling, small scales

1. Introduction

Obtaining higher efficiency rates or increasing the lifetimes of functional industrial parts requires the development of new materials such as ceramic coatings. For many years now, the industry has relied upon surface treatment processes by deposition of various materials to improve these parameters. Since the end of the 1990's, the processing of nanostructured materials has been performed using thermal spraying processes **[1-4]**. The characteristic length of the various physical processes (electronic, mechanical, optical…) is included in this scale field, in which the structural organization plays a main role. Moreover, it confers enhanced properties to the specific coatings.

Among the various processes, plasma spraying is well adapted to ceramic coating deposition. Conventional plasma spraying uses a micrometer-sized powder, injected with a carrier gas into the plasma plume, to produce thick layers, generally from 100 µm to close to or greater than a millimeter. Nanomaterials can be deposited by such adapted plasma spraying, but their small particle mass requires a liquid media, like water or ethanol, to increase the particle momentum and ensure penetration into the plasma jet. The nanoparticles are put in suspension and pushed into the plasma jet by pressure via an injector located close to the torch exit. After vaporization of the media, the particles are treated in the plasma plume, impact the substrate and thus generate a coating through piling of lamellae.

This process, represented in **Figure 1,** is a concentration of complexity through a succession of physical phenomena (liquid jet fragmentation, vaporization, heating, agglomeration…) detailed in [**5**].

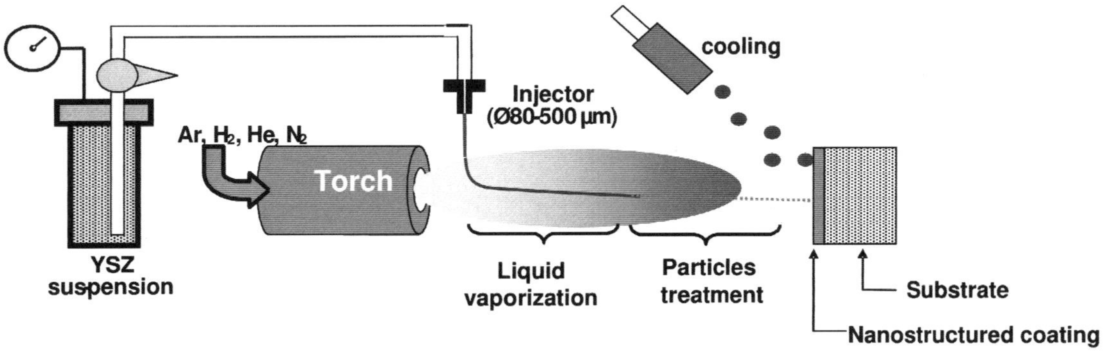

Figure 1: Representation of the suspension plasma spraying process

All these multiphase phenomena play important roles both in a reduced space scale (where the short stand-off distance ranges from 30 to 50 mm) and the associated time scale (from a few microseconds for the primary fragmentation to a few milliseconds for the vaporization phenomenon).

The in-situ analysis of the succession of phenomena is particularly difficult because of the high plasma jet velocity with important gradients, the significant radiation from the plasma due to the high temperature, again with high gradients which lead to composition, thermodynamic properties and transport coefficient gradients, and the liquid which rapidly changes into gas at low temperature.

For these reasons, modeling can be one way of investigating this process. To this end, previous papers have put forward numerical investigations in the first steps of the interactions (penetration and fragmentation) between an argon/hydrogen plasma flow and a continuous jet of pure water [5-8]. These studies have shown the possibility of modeling the hydrodynamic interactions between the two fluids leading to the liquid fragmentation without use of an analytical model, merely solving the NAVIER-STOKES equations involving VOF (Volume of Fluids) methods to track the interface [**9**] [**10**]. At each step (in space and time), the equations, described in [**5**], are resolved which causes the interfaces to move between liquid and gas and then to a scattered liquid phase in case of breakup.

Based on this, the next step, i.e., heating of the droplet, is described here. The aim of the present work was to determine the mechanisms of heat transfer, especially its activation, between the as-fragmented droplet and the surrounding gas. We start by focusing our interest on previous investigations concerning heat transfer in suspension plasma spraying.

Suspension plasma spraying has been intensively studied generally through shadowgraph imaging analysis [11-14] or infrared imaging [15] but also by numerical approaches [16-23]. These investigations deal with the breakup of the droplets in particular. In fact, few studies concern the heat transfer between liquid droplets and a plasma jet where low temperature phase transformation can appear rapidly.

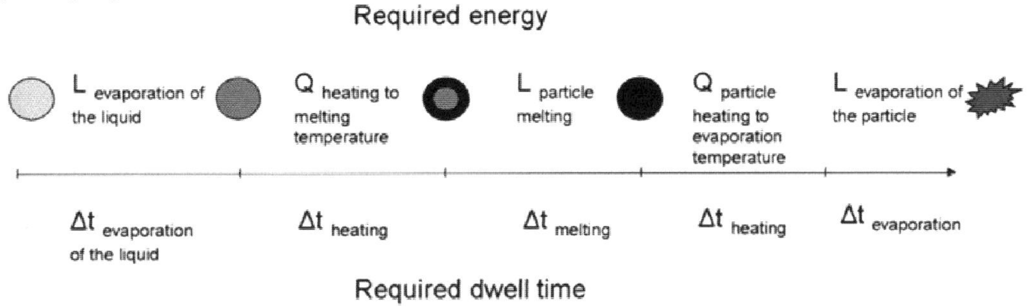

Figure 2: Required dwell time and energy

Concerning particle-loaded droplet heat transfer, KASSNER et al **[24]** proposed the mechanism depicted in **Figure 2** with the following assumed stages: evaporation of the liquid and particle agglomeration, heating of the solid ceramic particles to the melting temperature, particle melting, heating of the liquid particle to the evaporation temperature, and finally, evaporation of the ceramic. For the sake of simplification, the droplet break-up at the beginning of the injection as well as during the dwell time in the plasma until evaporation of the fluid has been left out. The single steps from one stage to another are thus assumed to be separate. To reach the next stage, a defined energy amount for the phase transformation L is needed, or alternatively to reach the temperature for the next phase transformation Q. This energy comes from the plasma by radiation and convection during a short required time for the energy transfer.

The following parts of this paper describe the investigations concerning the first stage of this model in an attempt to understand how the heating of the droplet leads to its evaporation. First, a description of the global experimental conditions is proposed. From these parameters, we extracted local conditions concerning on the one hand the plasma plume as the surrounding atmosphere of one droplet and on the other hand the size and position in the plasma flow of one characteristic water droplet. They were the result of previous studies referenced in the section "Local experimental conditions".

After that, modeling investigations concerning different physical phenomena were carried out and were analyzed separately to describe their influence upon the droplet behavior. The evaporation was not taken into account due to the short calculation time.

2. Experiments and modeling

2.1. Global experimental conditions

The following analyses were based upon the injection of a water jet into a plasma plume. The operating conditions of the plasma are summarized in **Table 1.** An explanation of these choices can be found in [5] **[25] [26]** and are not detailed here. Using a mixture with hydrogen implied that the downstream plasma flow fluctuated and presented stretching and shrinking with hot gas puffs and a modification of the gas velocity. These instabilities led to different liquid treatments depending on the instantaneous state of the plasma **[25] [27]** with various degrees of penetration of the liquid in the jet and different trajectories. The suspension parameters are listed in Table 2. In the first approximation, the liquid was not loaded with the particle.

2.2. Assumptions and model

The investigations presented here require determining the characteristics of the plasma flow, such as its composition, velocity and temperature, in addition to the droplet size and velocity distributions. Such calculations have been carried out by CARUYER et al **[7]** using a Computational Fluid Dynamic (C.F.D.) library solving the compressible and multiphase Navier-Stokes equations. For the sake of simplicity, all details concerning how to write and solve the fluid mechanics equations can be found in **[6]**, **[7]** and **[9]**. The basic assumptions of the calculations both under steady-state or time-dependent conditions are detailed in **[8]**, **[28]** and **[29]**. As explained in these articles, the arc root motion was simulated by the 3-D simple Joule effect model. In addition, the flow scale was considered to be larger than the size of the mesh grid, in order to perform resolved scale multiphase flow simulation by a volume-of-fluid interface tracking approach.

In **section 4.3.** the CFD code was also used with the following calculation conditions. Computations were carried out in 2D configuration for a shorter calculation time and because of agreements with 3D test case results. This increased the accuracy of the mesh without having a high number of cells. The mesh size around the droplet was 50 nm, which means that there were 400 cells in the diameter of the droplet. In order to have a Courant number lower than 0.3 (required for stability reasons by the interface tracking algorithm), the time step was 2.5 10^{-11}s. The domain size was 200 μm long and 100 μm large (**Figure 3**).

Journal of Physics: Conference Series **550** (2014) 012022 doi:10.1088/1742-6596/550/1/012022

Table 1: Spraying operating conditions

Plasma torch	**SULZER METCO F4-VB**
Nozzle diameter (mm)	**6**
Plasma gas mixture	**Argon / Hydrogen**
Plasma gas flow rate (slpm)	**45 / 15**
Arc current intensity (A)	**500**
Mean voltage (V)	**76**
Thermal efficiency (%))	**52**
Standoff distance (mm)	**40**

Table 2: YSZ suspension parameters

Media	**water**
Injection pressure (Pa)	**$2.5\ 10^5$**
Diameter of the injector (µm)	**200**

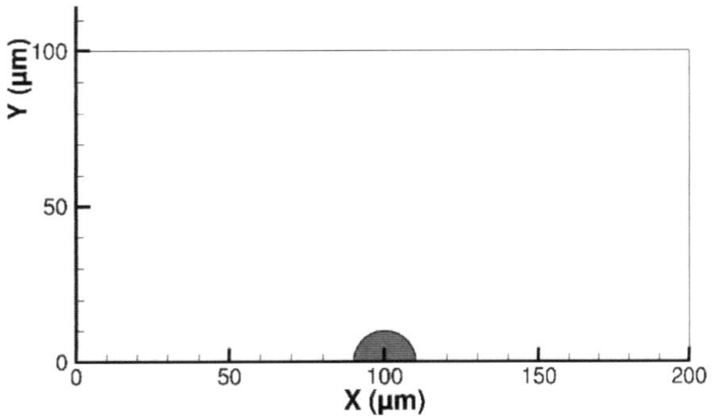

Figure 3: Definition sketch of the calculation domain with the initial droplet position.

The low dimension of meshes was a high assumption regarding the mean free path of the plasma which certainly implied an important error percentage for the calculation of the gas characteristics. Nevertheless, close to the droplet and inside it, the mean free path of the molecules was larger because of a lower temperature leading to validity of the continuity equation.

After 20 µm in the y-direction, the mesh became exponential, to reduce the computational time. The left side was the input side, with a 500 $m.s^{-1}$ condition for the gas velocity and the right side was the output side. The top boundary had a sliding condition, while the bottom boundary presented a planar symmetry in order to be able to divide the number of cells by 2. In order to properly calculate the Laplace pressure and capillary effects, an accurate representation of the interface properties was implemented for the local normal and curvature by **[30]**. Indeed, without this method and even with a quite thin mesh, spurious oscillations of the interface could be induced, which would disturb the interface in an artificial and non-physical way.

3. Local experimental conditions

Based on the previous calculations, the following figures show the characteristics of the plasma flow. These characteristics were the results of long time calculations since they correspond to mean values integrating the fluctuations of the plasma (due to the arc root motion). First of all, **Figure 4** shows that the as-fragmented droplets were completely immerged in an atmosphere of plasma at a high temperature, more than 9000 K close to the centerline (**Figure 5)** while the surrounding gas

Journal of Physics: Conference Series **550** (2014) 012022 doi:10.1088/1742-6596/550/1/012022

flowed at more than 500 m/s (**Figure 6**) if the fragmentation was assumed to end 10 mm from the torch exit and close to the centerline.

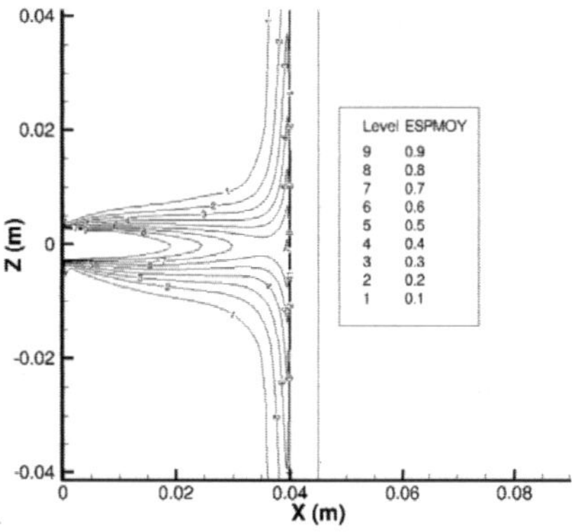

Figure 4: Average weight percentage of plasma gas versus surrounding air.

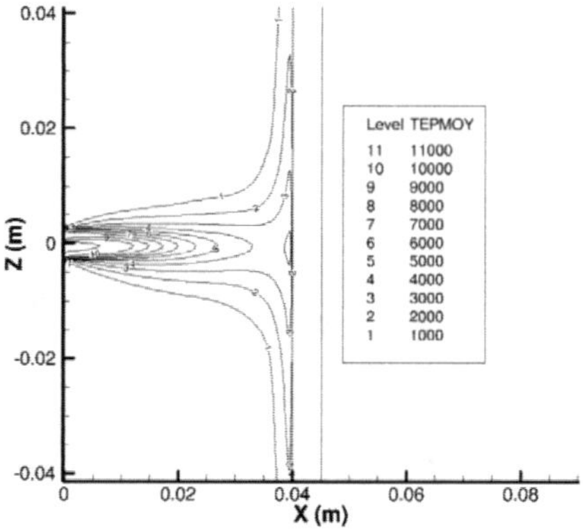

Figure 5: Average calculated temperature field (K)

At this position in this environment, the liquid shape was assumed to be completely fragmented into small droplets following the size distribution in **Figure 7**. This figure shows that the main population of the fragmented droplets (defined by their near-spherical shape) had a diameter close to 20 µm regardless of the time **[31]**. Still-existing ligaments were not incorporated in these distributions.

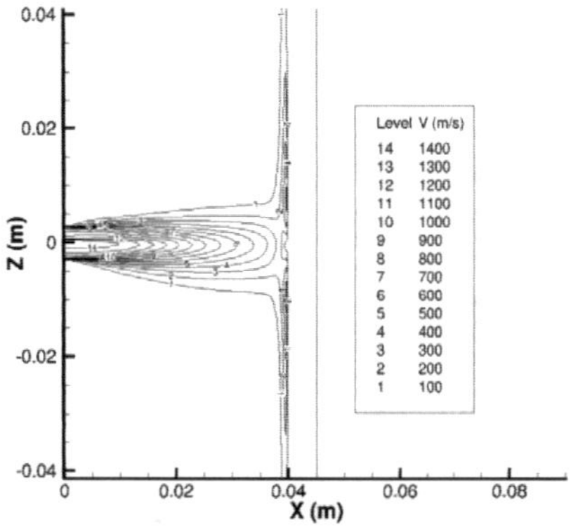

Figure 6: Average calculated velocity field (K)

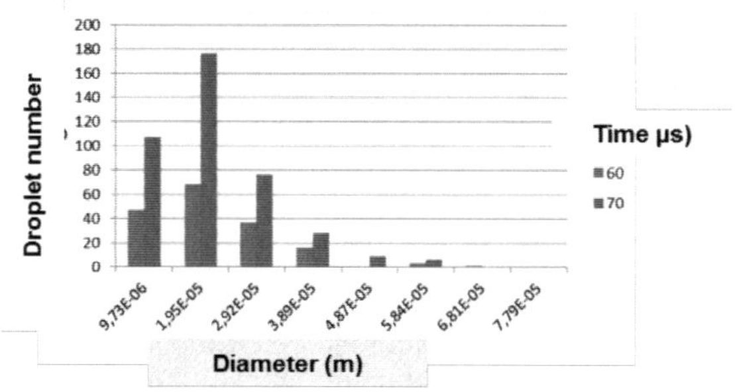

Figure 7: Size distribution of the droplets at different times of the unsteady plasma flow.

4. Analysis of the heat transfer mechanism

Evaporation of the liquid requires the activation of heat transfer. In this section, the evaporation was not taken into account and the investigations concerned the different mechanisms of heat transfer, that of the radiation, and that of convection assuming a spherical droplet. Finally, the mechanism of non-spherical droplet deformation was investigated.

4.1. Radiation

The highest black body emission intensity is in the wavelength range centered around 0.25 µm. Considering the spectral absorption coefficient of the different plasma gas species **[32]**, **Figure 8** indicates that the radiation was free to heat droplets in the frequency range 0.5 10^{15} – 3. 10^{15} Hz (wavelength from 0.6 to 0.1 µm) and moreover, **Figure 9** shows at which wavelength the liquid water could absorb the radiation when its environment was at different black body temperatures **[33]**. The heating was efficient in two wavelength ranges, one from around 0.06 µm to 0.2 µm, the second from 1 µm to 10 µm. So the common range for efficient heating was between 0.1 and 0.2 µm. From this, **Table 3** shows the integrated power upon the entire wavelength range considering the plasma as a black body radiating at the different temperatures. In **[34]**, GIRARD et al. calculated the dwell time it takes for a mean droplet to reach the substrate: it was less than 100 µs. So, **Table 3** also indicates the energy absorbed by a 20-µm water droplet flowing in the plasma jet during this dwell time. We must add that plasma radiation is lower than that of a black body at the same temperature so that the values of absorbed energy in table 3 are overestimated.

Table 3: Integrated available power and absorbed energy by a 20 µm sized water droplet in 100 µs.

Temperature (K)	**10 000**	**11 000**	**12 000**
Integrated power (W)	**0.028**	**0.063**	**0.129**
Absorbed energy (J)	**2.8 10^{-6}**	**6.3 10^{-6}**	**1.29 10^{-5}**

From the model previously proposed by KASSNER **[24]**, the energy required for heating and evaporating a 20-µm sized water droplet was around 1.10^{-5} J. When comparing these two energies, it appeared that the available radiation energy of the plasma was not high enough to completely evaporate the droplet. Radiation is only one activation mode to start this process. But as Basu et al. explained in **[35]**, it takes more than 200 µs to start the significant vaporization of the fuel droplet. In SPS, the dwell time is only 100 µs for the 20-µm water droplet before reaching the substrate. Consequently, radiation is not the main mechanism for heating the droplet.

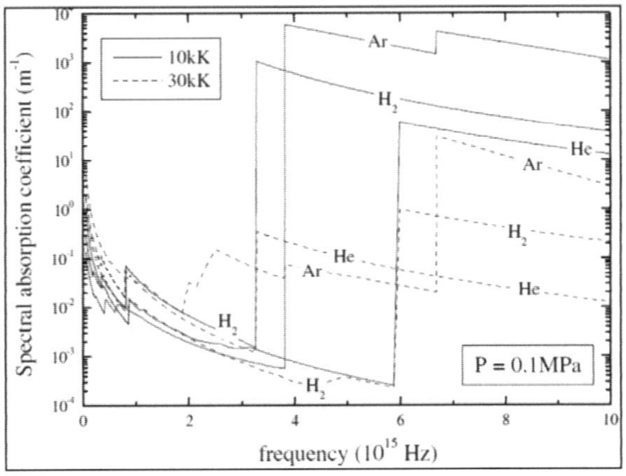

Figure 8: Spectral absorption coefficient of argon, helium and hydrogen versus frequency for two temperatures (10 000 and 30 000 K).

4.2. Convection for a spherical droplet

This section describes how to determine whether the convection mechanism is efficient during the flight of the droplet between the fragmentation position and the impact onto the substrate. This mechanism has been completely detailed in **[34]**. Here, we only mention the main conclusions. Due to heat transfer, each droplet receives energy from the plasma which leads to vaporization. Assuming that the droplet is spherical, the interface phase transformation velocity is more than 0.1 m/s. Due to the phase transformation above the boiling temperature, the steam might expand with a speed of around 160 m/s. This is a lower limit of the boiling phenomenon.

On the contrary, the water vaporization is not a supersonic phenomenon if the heating occurs by convection. So the vapor speed is less than the sound speed in steam (evaluated to 440 m/s at 100°C) and so the evaporation time is more than 38 µs. During this lapse of time, the plasma jet and steam are in interaction with each other. Calculations of their volume momentum show that the momentum of steam is higher than that of plasma (respectively 90 and 15 kg m s^{-1} m^{-3} for steam and plasma). This implies that if the boiling is uniform, the steam pushes the plasma, and prevents it from touching the droplet. Consequently, heat transfer does not occur between the hot plasma and the cold droplet, but through a layer of steam, from a hot source to a cold target. This is also one of the conclusions made by Basu et al [**35**]. Then, considering the position where the momentums of the plasma jet and of the steam sphere are equal, it can be deduced that the steam radius (including the liquid droplet) is close to 18.3 µm leading to a steam thickness of 8.3 µm (**Figure 10**).

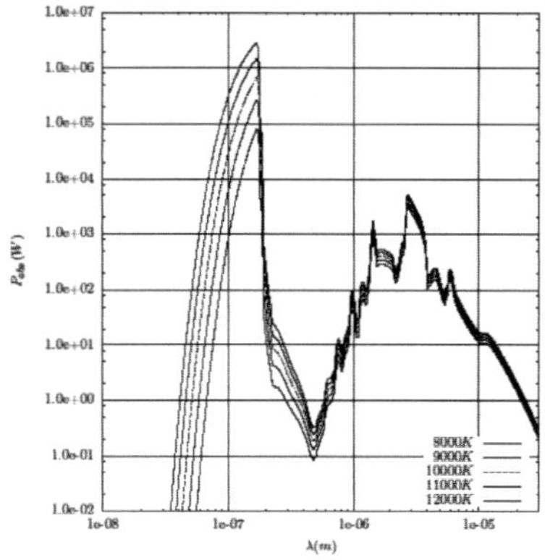

Figure 9: Water absorption power versus wavelength at different black body temperatures.

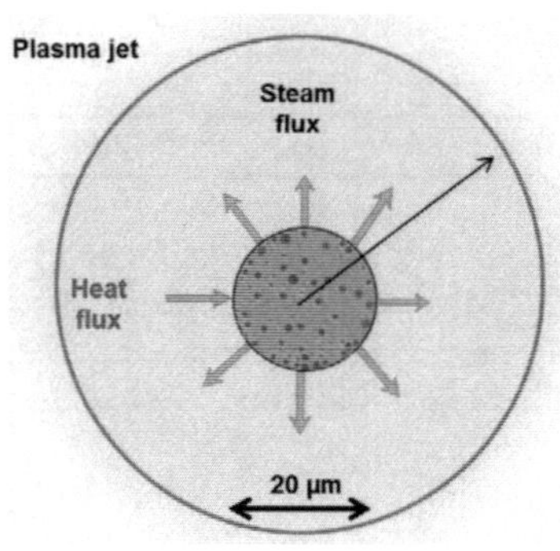

Figure 10: Droplet with its surrounding sphere of steam.

In conclusion, if the droplet remains spherical during its flight, heating by convection cannot be considered to be the main mechanism due to the steam sphere around it. Based on this, the next section is dedicated to the deformation of the in-flight droplet.

4.3. Deformation of a droplet.

The following figures present different parameters for a droplet immerged in a plasma flow at different calculation times (t= 0.25 µs, 1 µs, 2.25 µs and 4.25 µs). The initial time corresponds to the beginning of the plasma flow appearing at the left side of the domain. Of course in this case, no thermal effect is taken into account due to the low calculation time.

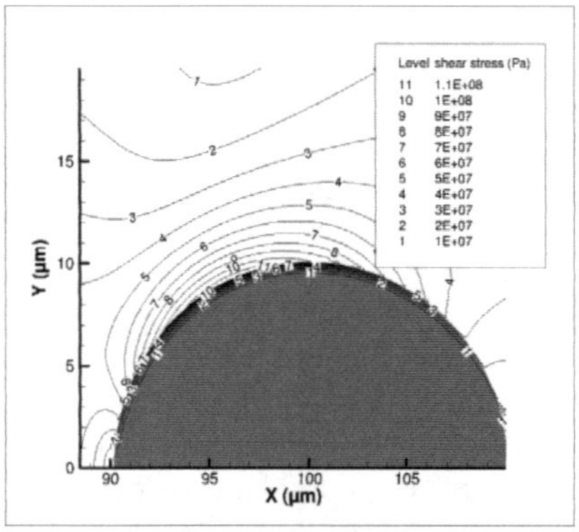

Figure 11: Shear stress around the droplet immerged in the plasma flow at 0.25 µs.

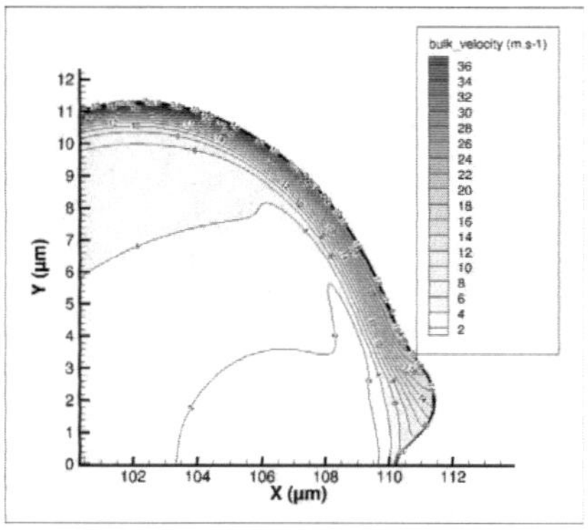

Figure 12: Internal velocity of the droplet immerged in the plasma flow at 1 µs.

Figure 11 shows the shear stress beginning to appear around the droplet with an asymmetry between its left side, its upper side and its right side. As a consequence, the shear stress on the interface accelerates the external layer of the liquid, which rotates over the center of the droplet. As the time increases, the internal velocity grows (**Figure 12**) and leads to an accumulation of the fluid on the right side (downstream) of the droplet with the beginning of the interface geometry deformation, as shown in **Figure 12, Figure 13 and Figure 14**. While this appendix appears, the external layers continue to be accelerated and transform the spherical droplet into a kind of medusa shape, much thinner and with a higher surface contact with the plasma. This higher surface contact leads to a greater drag force, and thus to a greater dynamic pressure from the upstream side inside the droplet (**Figure 15**).

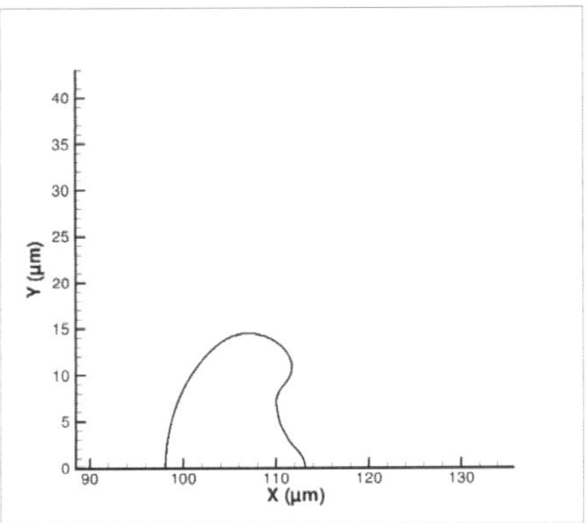

Figure 13: Droplet shape immerged in the plasma flow at 2.25 μs.

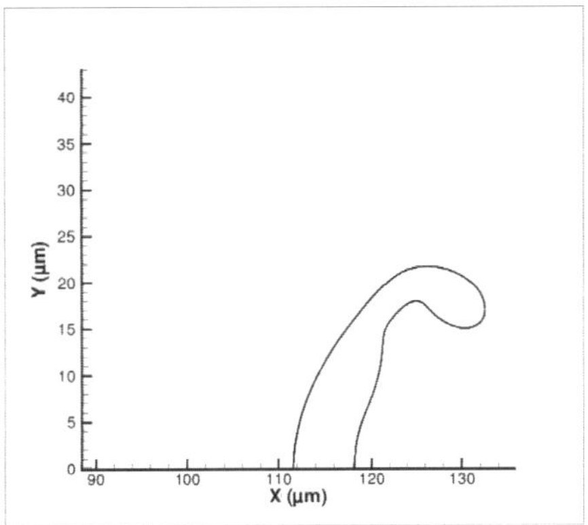

Figure 14: Droplet shape immerged in the plasma flow at 4.25 μs.

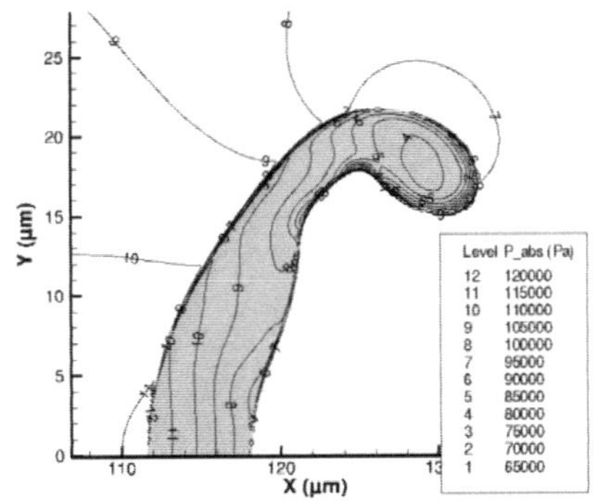

Figure **15**: Internal pressure of a droplet immerged in the plasma flow at 4.25 μs.

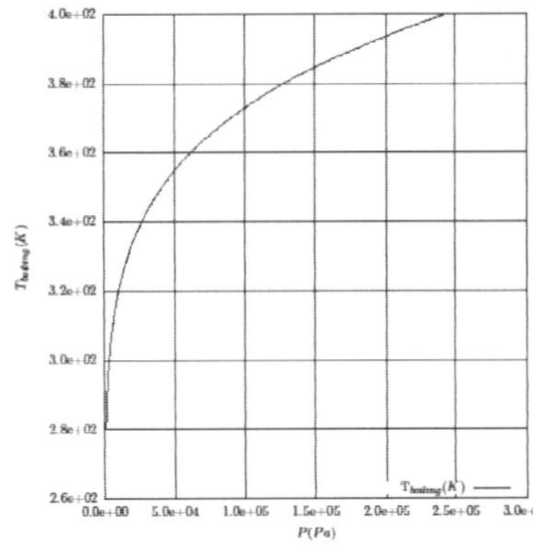

Figure 16: Boiling temperature of water versus pressure.

According to **Figure 16 [35]**, the boiling temperature of the water varied depending on the pressure. Consequently, the droplet that immerged in the plasma flow could boil from 363 K to 383 K. This pressure distortion accelerated the boiling phenomenon, and this temperature difference led to an energy saving of approx. 4.2 10^{-7} J for a 20-μm diameter droplet. This value is low, representing around 3% of the latent heat but it is enough to accelerate the process of the droplet heating.

5. Conclusions

The present article has been dedicated to investigating the heat transfer between a water droplet and the plasma jet surrounding it. The objective of these studies was to understand in which way the boiling phenomenon appeared in order for only the nanoparticles to reach the substrate.

Different mechanisms have been considered: radiation, convection around a spherical droplet and deformation of the droplet due to shear stress. Radiation could have an influence on the heating but is not enough efficient to boil completely a droplet while convection to a spherical droplet had no influence at all due to a thick layer of steam encompassing the liquid droplet.

Another mechanism was identified due to the deformation of the droplet before boiling because of the shear stress from the plasma flow. This shear stress also led to an internal flow inside the droplet, mixing the droplet temperature. Moreover, the deformation increased the surface exchange resulting in a more efficient heating. As only water droplets were considered, further studies will integrate the influence of ceramic particles inside the liquid jet via modification of the chemical and physical properties of the liquid. Finally, we will also consider how change of the liquid phase influences the plasma flow.

References

[1] GITZHOFER F., BOUYER E., and BOULOS M., 1997, Suspension Plasma Spraying, *U.S. Patents*, 5, 609, 921

[2] KARTHIKEYAN J.J., BERNDT C.C., TIKKANEN J., REDDY S., HERMAN H., 1997, Plasma Spray Synthesis of Nanomaterial Powders and Deposits, *Surf. Coat. Technol.*, **238** (2), pp. 275-286.

[3] KARTHIKEYAN J.J., BERNDT C.C., REDDY S., WANG J.Y., KING A.H., HERMAN H., 1998, *J. American Ceram. Soc.,* **81** (1), pp. 121-128.

[4] BOUYER E., GITZHOFER F., BOULOS M., 1996, “Parametric Study of Suspension Plasma Sprayed Hydroxyapatite”, *Proceedings of the ITSC 1996*, (Ed.) C.C. Berndt, (Pub.) ASM International, pp 683-691.

[5] MEILLOT E., VINCENT S., CARUYER C., DAMIANI D., CALTAGIRONE J.P., 2013, *J. Phys. D: Appl. Phys.,* **46** (22).

[6] MEILLOT E., DAMIANI D., VINCENT S., CARUYER C., CALTAGIRONE J.P., 2013, Analysis by Modelling of the Plasma Flow Interactions with a Liquid Injection, *Surf. Coat. Technol.*, **220**, pp. 149-156.

[7] CARUYER C., VINCENT S., MEILLOT E., DAMIANI D., CALTAGIRONE J.P., 2011, Effects of plasma conditions on the fragmentation and the atomization of a liquid jet, ILASS – Europe 2011, *Proceedings of 24th European Conference on Liquid Atomization and Spray Systems,* ESTORIL, PORTUGAL, September 2011

[8] CARUYER C., VINCENT S., MEILLOT E., CALTAGIRONE J.P., 2010, Modeling the First Instant of the Interaction Between a Liquid and a Plasma Jet with a Compressible Approach, *Surf. Coat. Technol.,* **205**, pp. 974–979.

[9] VINCENT S., BALMIGERE G., CALTAGIRONE J.P., MEILLOT E., 2010, Eulerian-Lagrangian multi-scale Methods for Solving Scalar Equations - Application to Incompressible Two-phase Flows. *J. Comput. Phys.,* **229**, pp. 73–106.

[10] LOPEZ J., HERNANDEZ J., GOMEZ P., FAURA F., 2005 *J. Comput. Phys.,* **208** pp. 51–74.

[11] DAMIANI D., TARLET D., MEILLOT E., 2014, A Particle-Tracking-Velocimetry (PTV) Investigation of the Injection of Liquid Solution in a Thermal Plasma Jet, *J. Thermal Spray Technol.*, **23** (3).

[12] DAMIANI D., MEILLOT E., CARUYER C., VERT R., 2011, Méthodes optiques pour la visualisation des écoulements dans les plasmas, *14ème Congrès Français de Visualisation et de Traitement d'Images en Mécanique des Fluides*, Lille, France,21 - 25 novembre 2011.

[13] MEILLOT E., DAMIANI D., CARUYER C., VINCENT S., CALTAGIRONE J. P., 2011, *IEEE Trans. Plasma Sci.*, **39** (11)

[14] ETCHART-SALAS R., 2007, Projection par plasma d'arc de particules submicroniques en suspension. Approche expérimentale et analytique des phénomènes impliqués dans la reproductibilité et la qualité des dépôts. *Ph. D. Thesis (in French), Limoges University*, 2007.
[15] SOYSAL D., ANSAR A., 2013 *Surf. Coat. Technol.,* **220**, pp. 187-190.
[16] MARCHAND C., VARDELLE A., MARIAUX G., LEFORT P., 2008 *Surf. Coat. Technol.,* **202**, 4458–4464.
[17] MARCHAND C., CHAZELAS C., MARIAUX G., VARDELLE A., 2007 *J. Thermal Spray Technol.* **16** (5-6) pp. 705-712.
[18] SHAN Y., COYLE T.W., MOSTAGHIMI J., 2007 *J. Thermal Spray Technol.,* **16** (5-6), pp 698-704.
[19] SHAN Y., COYLE T.W., MOSTAGHIMI J., 2007 *J. Thermal Spray Technol.*, **16** (5-6), pp 736 – 743.
[20] QIAN L.J., LIN J.Z., H.B. XIONG, 2011 *Int. J. of Thermal Sci.,* **50**, 1417-1427.
[21] XIONG H.B., QIAN L.J., J.Z. LIN, 2012 *J. Thermal Spray Technol.,* **21** (2) pp. 226-239.
[22] CHEN X., PFENDER E., 1982 *Plasma Chem. and Plasma Process.* **2**, 2, pp 185-212.
[23] PFENDER E., LEE Y.C., 1985 *Plasma Chem. and Plasma Process.*, **5**, 3, pp 211- 237.
[24] KASSNER H., VASSEN R., STÖVER D., 2008, Study on Instant Droplet and Particle Stages During Suspension Plasma Spraying (SPS), *Surf. Coat. Technol.*, **202**, pp. 4355 - 4361.
[25] COUDERT J.F., RAT V., 2010, The role of torch instabilities in the suspension plasma spraying process, *Surf. Coat. Technol.*, **205,** pp. 949-953.
[26] JANISSON S., VARDELLE A., COUDERT J.F., FAUCHAIS P., MEILLOT E., 1999, Analysis of The Stability of DC Plasma Gun Operating With Ar–He–H2 Gas Mixtures, in: *Annals of the New York Academy of Sciences*, **891**, pp. 407–416.
[27] FAZILLEAU J., DELBOS C., RAT V., COUDERT J.F., FAUCHAIS P., PATEYRON B., 2006, Phenomena Involved in Suspension Plasma Spraying. Part 1: Suspension Injection and Behavior, *Plasma Chem. Plasma Process*, **26** pp. 371-391.
[28] MEILLOT E., VINCENT S., CARUYER C., CALTAGIRONE J.P., DAMIANI D., 2009, From D.C. time-dependent thermal plasma generation to suspension plasma spraying interactions, *J. Thermal Spray Technol*, **18** pp. 857–86.
[29] CARUYER C., VINCENT S., MEILLOT E, CALTAGIRONE J. P., DAMIANI D., 2010, Analysis of the unsteadiness of a plasma jet and the related turbulence, *Surf. Coat. Technol.,* **205** pp. 1165–70
[30] PIANET G., VINCENT S., LEBOI J., CALTAGIRONE J.P., ANDERHUBER M., 2010, Simulating compressible gas bubbles with a smooth volume tracking 1-fluid method, *Int. J. of Multiphase Flow,* **36** pp. 273-283.
[31] CARUYER C., 2011, Modélisation de nanomatériaux injectés par voie liquide dans un jet de plasma pour la fabrication de nanostructures, *Ph. D. Thesis (in French), Université de Bordeaux 1*, 4329, 2011.
[32] CRESSAULT Y., ROUFFET M.E., GLEIZES A., MEILLOT E., 2010, Net Emission of Ar-H_2-He Thermal Plasmas at Atmospheric Pressure, *J. Phys. D: Applied Phys.*, **43**.
[33] MESENBRINK M., 1996, Complex indices of refraction for water and ice from visible to long wavelengths, *Ph. D. Thesis, Florida State University*, March 1996.
[34] GIRARD F., MEILLOT E., VINCENT S., CALTAGIRONE J.P., BIANCHI L., Understanding of heat and mass transfers between plasma jet and colloidal droplets in Suspension Plasma Spraying process, *Surf. Coat. Technol.*, submitted, in progress.
[35] BASU S., SAHA A., KUMAR R., 2012, Thermally Induced Secondary Atomization of Droplet in an Acoustic Field, *Applied Physics Letters* 100, 054101.

Journal of Physics: Conference Series **550** (2014) 012023 doi:10.1088/1742-6596/550/1/012023

Synthesis of carbon nanotubes by high current divergent anode-channel plasma torch

R H Amirov[1], E Kh Isakaev[1], M B Shavelkina[1] and T B Shatalova[2]
[1] Laboratory of Plasma, Joint Institute for High Temperatures of Russian Academy of Sciences, Izhorskaya Street 13/2, 125412 Moscow, Russia
[2] Department of Chemistry, Lomonosov Moscow State University, Lenin Mountains 1, 119991 Moscow, Russia

E-mail: amirovravil@yahoo.com

Abstract. In this study we propose the high-performance technology to produce carbon nanotubes (CNT) in plasma jet reactor by means of a direct current plasma torch. This technology provides excellent opportunities to investigate a direct evaporation of materials and their subsequent condensation on the carbon surface. Experiments were carried out at the electric power of a plasma torch up to 30 kW. Helium and argon served as plasma gases. CNT synthesis at pyrolysis of soot was catalyzed by the metal disperse powders of Ni, Co, Y_2O_3. We applied x-ray diffraction and electronic microscopy to investigate the structure of obtained products. Also we utilize the thermogravimetric analysis to determine the phase structure of carbon nanomaterials. Using available experimental data we were able to sequentially scale the production process of CNT of desirable space structure. Finally we established that structural and morphological properties of CNT produced at evaporation of soot in the presence of high-percentage combined catalysts depend upon the catalyst structure.

1. Introduction

Although the mechanisms of synthesis for carbon nanotubes are not fully understood nowadays, the key methods of their production were already described. So the question is which of these methods are most suitable for commercial manufacturing. In this respect it is known that the methods using vacuum or high pressure are less productive than the methods of synthesis at atmospheric pressure. Evaporation of graphite by an electron or ion beam and also usage of diamond anvils restrict the productivity of the methods. Also damaging effects are defects of the electrolysis or the explosive synthesis. It is concluded [1-3] that the most widespread for laboratory usage are the arc and laser methods.

During CNT synthesis in the electric arc the direct current is generated between the anode and the cathode situated at the several millimeters from each other. This process generates the plasma arc with a small distance between the electrodes. On the carbon surface (for example, graphite) the carbon evaporates from the anode and subsequently condenses on the cathode forming carbon nanotubes. In the arc method [4] any external magnetic field generated by constant magnets or electromagnets was not used and there was no recirculation of the evaporated carbon. With the help of this method one adds the various chemical catalysts into the anode. This increases the production of CNT and affects the diameter of nanotubes. In order to improve an arc method the cathode and the anode were made in the form of a plate on a tip to create the interplanar micro-discharge. Also the plasma volume was slightly increased [5]. An idea for the future development is to increase the synthesis productivity by increase of arc current or electrode diameter. The later is considered impracticable because the precipitated carbon nanotubes evaporates again from the cathode. To sum up the standard arc method is discrete and unstable process and by this method it is not possible to reach mass production of high-quality carbon nanotubes [5].

The much bigger quantities of carbon nanotubes (CNT) can be obtained in a continuous manner by plasma torch method and other chemical vapour deposition methods [6, 7]. Using plasma torch method in particular and other similar methods in general one can significantly increase productivity and synthesis time by transition to volume synthesis. Since 2001 several scientific centers [6-16] were investigating the synthesis of carbon nanotubes in the plasma jet reactors with the help of plasma torch method. In the study [7] the serial plasma torch of induction type in the plasma jet reactor was described as a perspective method for future development. The presented data indicated the high productivity of CNT synthesis, which considerably exceed the analogues results reached with the help of arc and laser methods of synthesis. The quality of synthetized CNT remains as high as in the arc and lasers methods.

Among existing variety of designs and schemes of direct current plasma torch with self-established length of an arc the common feature for all of them is the cylindrical channel. The technological disadvantages of such plasma torch are faults in volt-ampere characteristic, the low efficiency and an unstable operating mode. The arc discharge in such plasma torch is characterized by strong heterogeneity of electro-physical and heat-physical properties. Using the extending channel of the output electrode in the plasma torch one can significantly increase the velocity of cold gas at the entrance to the channel and intensify heat exchange between an arc and plasma-forming gas. This process aims to reduce length of the arc discharge and obtain uniform distribution of thermal streams in electrode walls. The arc column can occupy the considerable volume of a discharge gap, and its anode short circuit is achieved by micro-arcs that considerably reduce the average density of current and the thermal stream in the anode. Thus uniform magnetic field of current is established and as a result stability of arc column is achieved by expansion of an output electrode [17]. Another feature of the plasma torch with the extending channel of the output electrode is that the length of the arc discharge is not fixed and strongly depends on the gas type, gas flow and intensity of the current. It allows effective usage of this type plasma torch in various plasma technologies, especially in those where the considerable change of parameters of a mode is required during technological process [17, 18].

The advantage of this type plasma torch system was demonstrated by comparison to several developed technologies. Among these technologies were plasma hardening and nitriding of railway wheels, plasma coating for recovering and manufacture of railway frogs and plasma coating of copper with wear resistant refractory layer [17, 18]. On the basis of these technologies the plasma jet reactor was developed and the experimental investigation of synthesis of carbon nanotubes was started to investigate the mechanisms of their synthesis, structure assembly and yield increase.

The main aim of the present study is to obtain experimental data on synthesis of carbon nanotubes from soot in the presence of catalysts by means of the direct current plasma torch with the extending channel of an output electrode. The synthesis depends on a sort of plasma-forming gas, its pressure, the consumption of soot and structure of catalysts.

2. Experimental setup and procedure

The detail description of experimental setup was given in the study [19]. Shown in the figure 1 is the scheme of the experiment. Photo of plasma torch and reactor is listed in figure 2. The main difference of our approach from the arc method is that the original materials for synthesis of nanotubes (carbon and disperse metal catalysts) are put together in plasma torch with used gas (argon, helium) under the electric power equalled 10-30 kW. The evaporation of these materials takes place in plasma jet with the further rapid cooling process during which from the formed carbon vapour and catalysts the synthesis of carbon nanotubes occurs in gas volume as well as on a surface of reactor. Evaporation of carbon and metal catalysts occurs not only in plasma jet but also in the vicinity of the arc discharge by processing the original disperse materials. The flow rate of carbon mass, quantities of catalysts substances, used gas and powers of plasma torch are controlled independently from each other. By changing geometry of a reactor, pressure and velocity of a plasma jet it is possible to widely manipulate the rate range of used cooling carbon vapour. Time of continuous work of plasma jet is limited by the lifetime of the tungsten cathode. The pressure of gas is within the range from 10 to 760 Torr.

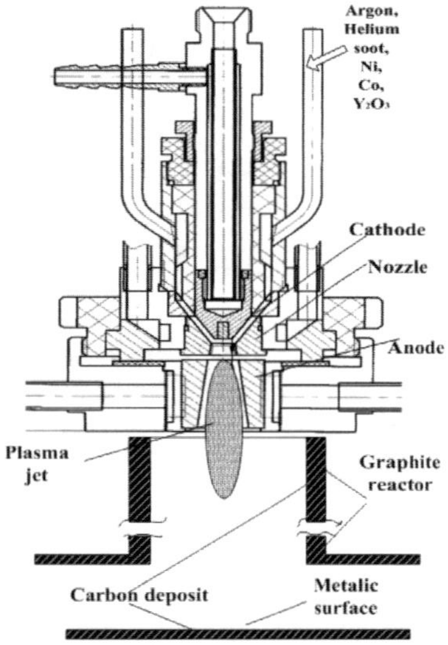

Figure 1. Construction of plasma reactor.

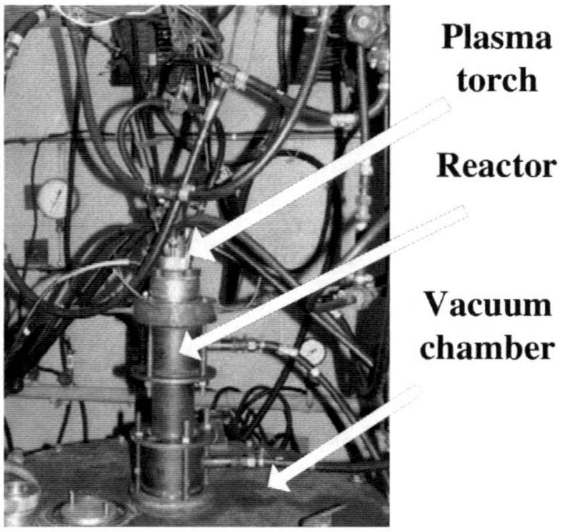

Figure 2. Photo of plasma torch and reactor.

Soot was used in the form of powder or small granules. Synthesis of CNT was conducted by pyrolysis of mechanical mixture of soot and the metals Ni, Co and Y_2O_3 in the powder form. The concentration of a pair Ni + Co was fixed. For a combination of Ni+Co+Y_2O_3 catalysts the wide range of concentrations were used. Substance combinations were made in such a way that there was a variation in each component concentration from minimum to maximum allowed value. In the arc methods a mixture of two catalysts is more commonly used rather then a mixture of three and more substances [20].

In each experiment the mixture was injected only after stabilization of a temperature distribution in the graphite reactor. Stabilization usually takes 5-10 min and was indicated by measuring the water temperature in cooling system. Table 1 lists the typical operating conditions.

The methods of x-ray diffraction and electronic microscopy were used to investigate the structure of original materials and their synthesized products on a scanning electronic microscope of MIRA 3 TESCAN and a raster electronic microscope LEO SUPRA 50 VP (Germany) with the powerful dispersive X-ray Oxford X-MAX micro spectral analyzer (Great Britain).

The method of thermo gravitation measurements was used to measure the efficiency of synthesis and describe the phase structure of carbon products (STA 449 the Netzsch firm F3 Jupiter platform). Also using this method the data was obtained on thermal stability, purity, structural perfection and uniformity of samples with UNT depending on catalysts substances quantitates in mixture with soot. For this technology the large amount of various materials was used. The sample weights for the analysis ranges from 0,58 to 6,8 mg. As a gas an atmospheric air was used. Temperature was increased with the rate 10 degrees a minute starting from room temperature to 1000^0 C. The method takes into account a range of stability conditions for various carbons forms and also considers impurity of metals during their heating in the oxidative environment. The efficiency of nanotube synthesis and its discharge characteristics (voltage, current, pressure, composition of gas) can be controlled by the velocity of a gas stream, the sizes of the reactor and its configuration, the nature and purity of materials of electrodes and other parameters. The main parameter affecting the nanotube synthesis is the growth rate, which in turn depends on many parameters. The growth rate is used for scaling and currently there is a lack of theoretical models for exact quantification of the growth rate. Nowadays there is no quantitative model to explain influence of catalyst type on CNT synthesis. To make progress in modeling the experimental measurements should to be obtained for the broad range of system parameters.

Table 1. Technological conditions.

Parameter	Value
Power	10-30 kW
Current	100-400 A
Voltage	35-80 V
Plasma gas	Argon, Helium
Gas flow rate	1-3.6 g/sec
Pressure	50-760 Torr
Flow rate of solid carbon	0-2.2 g/min
Carbon precursors: soot	< 10 μm
Catalyst <15% on weight	Ni, Co, Y_2O_3
Size of catalyst	< 10 μm
Time of continuous work	1 hour

3. Results of experiment

For the experimental conditions the electric power of a plasma torch was set up to 30 kW. Helium and argon were used as a plasma gases. Helium was taken under pressure of 350-710 Torr with gas flow rate from 0.5 to 1 g/sec. The argon flow rate ranged from 1.5 to 3 g/sec with pressure interval from 450 to 600 Torr. CNT synthesis at pyrolysis of soot was conducted at the presence of catalysts. The consumption of soot with catalysts set as 0.3-2.1 g/min. For the analysis the products of carbon condensation were taken from different part of the metal target [3] under different concentrations of catalysts, pressure and rate of the gas flow. The soot was analyzed in advance. According to the thermogravimetric analysis the synthesis product contains about 4 % of amorphous carbon, 48.37 % of a graphitic phase of soot and 38.21 % of polycrystalline graphite.

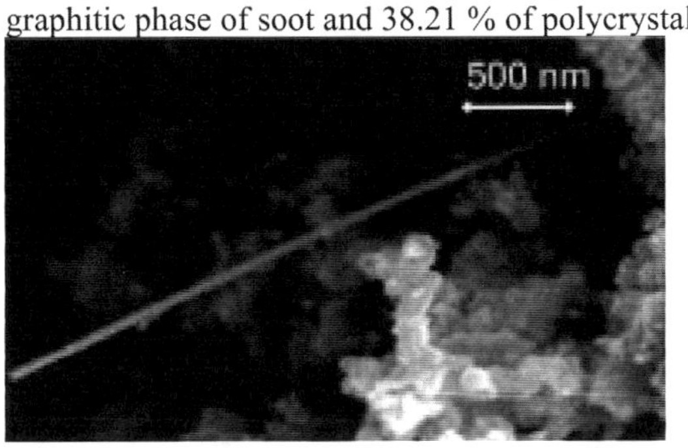

Figure 3. Structure of synthesis products in the presence of catalyst ratios Ni/Co/Y_2O_3 - 1.06/1.06/0.94 at.%.

Figure 4. The SEM image of synthesis products with catalyst ratios: Ni/Co/Y_2O_3 - 1.46/1.47/1.29 at.%, the pressure set as 500 Torr, current is 400 A and consumption of soot with catalysts is 0.45 g/min.

Journal of Physics: Conference Series **550** (2014) 012023 doi:10.1088/1742-6596/550/1/012023

The formation of the quasiamorphous structure of plastically deformed material with the coating of Ni and Co particles was detected under experimental conditions helium pressure 600 Torr and the soot content of catalysts Ni/Co - 0.64/0.64 at.%. The quasiamorphous state fills boundary areas connecting randomly situated nanoelements forming a dendrite structure.

The single extended cylindrical structures of a fixed diameter were formed from a quasiamorphous matrix with the following experimental conditions: mixture Y_2O_3 with the soot was done according to the ratio Ni/Co/Y_2O_3 = 1.06/1.06/0.94 at. %, helium pressure was decreased to 500 Torr, the current set as 400 A and the consumption of soot was fixed as 1.25 g/min. Shown in the figure 3 is the SEM image of nanotube at above experimental conditions.

The key difference of the created cylindrical nanostructure from known CNT [1] is the formation of clusters of the ordered framework with the open end and rather pure walls. Diameter of cylindrical structures is 60 nanometers. One can see in figure 3 the structural features of nanoelements of soot in figure 3. They have more deformed surface but smaller degree of disorder then previosly described CNT [1]. Also if Y_2O_3 concentration is less than Ni concentration and relates to the concentration of Co when the single-wall nanotubes with characteristical defective structure are formed with the consumption of soot with catalysts reduced twice.

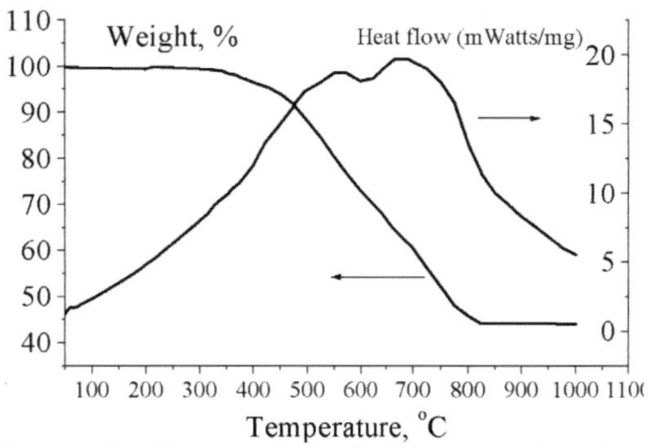

Figure 5. The results of thermogravimetric analyses of synthesis products with catalysts taken with the ratios Ni/Co/Y_2O_3 1.46/1.47/1.29 at.%; the pressure was 500 Torr, current was 400 A and consumption of soot with catalysts was 0.45 g/min.

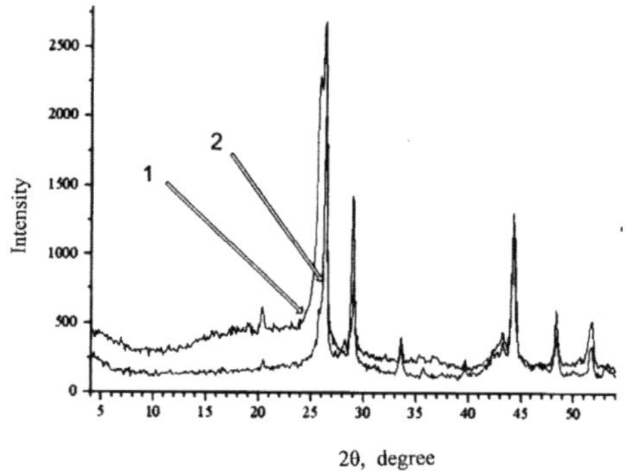

Figure 6. Results of X-ray analyses of synthesis products with CNT (1) produced in presence of catalysts and (2) produced without catalysts.

The CNT production was increaed by decrease of a soot consumption with catalysts taking rations Ni/Co/Y_2O_3 1.05/1.05/0.73 at.% to 0.32 g/min and with pressure of He 500 Torr. Thus the distribution of CNT diameter was observed. Many CNT situated on a quasiamorphous surface of a matrix of soot but not perpendicular to it was with the mishaped struture.

The difference in the structure of formed products was detected on the center of the target and on the periphery. On the periphery polyhedral particles formed nanostructures. We noted that the formation of CNT with helium is more effective then with argon. A large amount of amorphous carbon and spheroidal nanoparticles is formed using argon.

Diameter of carbon nanotubes ranged from 16 to 74 nanometers depending on experimental conditions. The maximum production of carbon nanotubes was observed in helium with the following experimental conditions: ratio Ni/Co/Y_2O_3 1.46/1.47/1.29 at.%, the pressure 500 Torr, current of arc 400 A and a consumption of mix 0.45 g/min. Shown in the figure 4 is the structure of the synthesis products at optimized conditions. SEM images corresponded to the thermogravimetric data. In the figure 5 there are results of thermogravimetric analyses of synthesis products at optimized conditions. According to the thermogravimetric analysis the maximum weight ratio of nanotubes was 13%. At temperatures ranges from 310 to 380°C and from 380 to 473°C the oxidation of black particles to amorphous and quasiamorphous structure on the air took place. Further increase of temperature to ranges 450-660°C and 660-820°C was accompanied by appearance of two peaks on DSC curve related to production of CO and CO_2. This result was confirmed by comparison of distribution curves for intensity of dispersion of X-rays in experimental data for different combinations of catalysts in soot mixture and literature data. Figure 6 illustrates the example of X-ray analyses.

To sum up the experimental data allowed sequential scaling process of CNT production to obtain required structures. The structural and of the morphological properties of CNT produced at soot evaporation at the set of highly concentrated catalysts depended upon the catalyst composition.

4. Conclusions

We investigated of synthesis of carbon nanotubes at soot pyrolysis in the presence of metal $Ni+Co+Y_2O_3$ catalysts in the plasma jet reactor using the direct current plasma torch with the extending channel anode in plasma-forming gas helium and argon.

It was possible to influence the production of the carbon nanotubes and their spatial structure by changing the gas flow rate, pressure and type of plasma-forming gas. The spheroidal particles have been formed in the presence a large amount of amorphous carbon in argon. The decreased arc current increased the impurity amount in the different forms of nanoparticles and amorphous carbon in products of synthesis.

The catalyst structure influenced structural and morphological properties of the carbon nanostructures produced at pyrolysis of soot in the presence of highly concentrated catalysts containing nickel, cobalt and yttrium.

To sum up the experimental data allowed sequential scaling process to produce CNT of required spatial structure.

5. Acknowledgements

The authors gratefully acknowledge Russian Fond of Fundamental Research for the support by grant No 12-08-00695.

6. References

[1] Rakov E G 2001 *Russian Chemical Reviews* **70** 827
[2] Rakov E G 2000 *Russian Chemical Reviews* **69** 35
[3] Eletskii A V 1997 *Physics Uspechi* **40** 899
[4] Ebbesen T W and Ajayan P M 1992 *Nature* **358** 220
[5] Lee S J, Baik H K, Yoo J and Han J H 2002 *Diamond and Related Materials* **11** 914
[6] Lim S Y, Mohamad Amran bin Mohd Salleh, Hamdan b. Mohamed Yusoff, Suraya Binti Abdul Rashid and Jeefferie b. Abd. Raza 2011 *J. of Industrial and Eng. Chemistry* **17** 367
[7] Kim K S, Cota-Sanchez G, Kingston C T, Imris M, Simard B and Soucy G 2007 *J. Phys. D: Appl.* **40** 2375
[8] Arabzadeh Esfarjani S, Dworkin S B, J Mostaghimi J, Kim K S, Kingston C T, Simard B and Soucy G 2012 *J. of Phys.: Conference Series* **406** 012011
[9] Smiljanic O, Stansfield B L, Dodelet J-P, Serventi A and Desilets S 2002 *Chem. Phys. Lett.* **356** 189
[10] Bystrzejewski M, Huczko A, Plotczyk W W, Pichler T, Gemming T and Ruemmeli M H 2008 *Appl. Phys.* **A91** 223
[11] Fulcheri L, Flamant G, Fabry F, Gruenberger T M, Gonzalez-Aguilar J, Probst N and Grivei E 2003 *Proc. 16th Int. Symp. on Plasma Chemistry (Taormina, Italy, June 22-27)* p 522.
[12] Chen C-K, Perry W L, Xu H, Jiang Y and Phillips J 2003 *Carbon* **41** 2555
[13] Harbec D, Muenier J L and Gaivin R 2003 *Proc. 16th Int. Symp. on Plasma Chemistry (Taormina, Italy, June 22-27)* p 540
[14] Fulcheri L, Flamant G, Gruenberger T, Gonzalez-Aguilar J, Fabry F, Probst N and Grivei E 2004 *Abstract Amer. Chem. Soc.* **227** 808

[15] Fulcheri L, Gruenberger T M, Aguilar J G, Fabry F, Grivei E, Probst N; Flamant G, Okuno H and Charlier J C 2004 *High Temperature Material Processes* **8** 119
[16] Okuno H, Grivei E, Fabry F, Gruenberger T M, Gonzalez-Aguilar J, Palnichenko A, Fulcheri L and Probst N 2004 *Carbon* **42** 2543
[17] Isakaev E Kh, Sinkevich O A, Tyuftyaev A S and Chinnov V F 2010 *High Temperature* **48** 97
[18] Isakaev E Kh, Sinkevich O A, Spektor N O, Tazikova T F, Tyuftyaev A S and Khachaturova A G 2010 *High Temperature* **48** 777
[19] Amirov R H, Asinovsky E I, Isakaev E Kh and Kiselev V I 2006 *J. of High Temp. Material Process* **10** 197
[20] Tarasov B P, Muradyan V E, Shul'ga Y M, Krinichnaya E P, Kuyunko N S, Efimov O N, Obraztsova E D, Schur D V, Maehlen J P, Yartys V A and Lai H J 2003 *Carbon* **41** 1357

Journal of Physics: Conference Series **550** (2014) 012024 doi:10.1088/1742-6596/550/1/012024

UV Fe I absorption in the carbon-iron arc periphery under formation of carbon-encapsulated iron nanoparticles

H Lange, O Łabędź and M Bystrzejewski
Faculty of Chemistry, University of Warsaw, Warsaw, Poland

E-mail: lanhub@chem.uw.edu.pl

Abstract. The UV (384.5 – 393.1 nm) absorption spectra of Fe I atoms (21 lines) originating from a $^5D_{0\text{-}4}$, a $^5F_{1\text{-}4}$ and a $^3F_{2\text{-}3}$ electronic levels were used to determine the Fe I column density in function of carbon-iron anode composition and the distance from the arc gap, beneath and above the horizontally oriented electrodes. The results revealed a wide variation between the gradients of Fe I column density above and below the arc zone. Also the electronic excitation temperature of the Fe I levels was evaluated.

1. Introduction

The carbon arc route is frequently used for fullerene [1], carbon nanotube [2,3] and carbon-encapsulated magnetic nanoparticles (CE) [4] synthesis. Magnetic CE are promising hybrid nanomaterials with unique potential applications, e.g. in environmental protection [5,6,7]. The systematic studies on the synthesis of CEs by carbon arc route have already been performed elsewhere [8]. The ability to control the process of CE growth is of primary importance. The formation of its precursor species, their temporal evolution within the various plasma zones and transport through the specific temperature gradient certainly plays an important role on the diameter, quality and yield of the final product [8]. Because of high temperature, several kK, a variety of carbon species is present: from atoms, ions and simple molecules in the interelectrode gap to the complex carbon clusters at the arc periphery. Obviously, neutral and ionized Fe species, which form the core of CE (e.g Fe@C), as well as atoms of the buffer gas, are also the arc plasma constituents. The detailed diagnostics of the arc plasma was the subject of a previous work [9], in which combined optical emission and absorption spectroscopy was applied to determine the temperature and C_2 radical content in function of carbon-iron anode composition. Also, the average density of Fe I atoms in the inter-electrode space was evaluated on the basis of the integrated absorbance of 385.99 nm line.

The aim of this work was to explore the so-called dark zone where the nucleation of iron nanoparticles and coating by graphene layers, thus forming the CEs, takes place, and to find how far iron atom vapor spread out in the reactor volume. To our knowledge such studies have never been carried out. However, some results on the iron content in the arc plasma zone are described elsewhere [9].

2. Experimental

The homogenous anodes used for experiments (0 – 22.5 Fe at. %) were prepared using a fine graphite powder (1 - 2 μm), blended with Fe powder (6 - 9 μm) and a tar pitch, which was added as a binding agent. The as-prepared mixtures were then pressed under 2 MPa to form rods of 8 mm in diameter and 30 mm in length. Next, the rods were calcinated under Ar atmosphere at 1000° C for 10 hours.

The experimental system for arc generation and spectroscopic studies is described in detail elsewhere [10]. All experiments were conducted under $Ar:H_2$ (1 : 1) atmosphere and at constant pressure of 60 kPa. The discharge current was 40 A and the voltage drop varied within the range between 30 and 38 V. The light of a 200 Watt Xe arc lamp was used as a background for absorption spectra, which were recorded with the use of a two-dimensional CCD detector coupled with a spectrograph of 3 m in focal length and 1.6 nm/mm in reciprocal dispersion. The spectral resolution on the CCD detector was 0.0056 nm/pixel. Four dark zones were explored. Locations of the zones are marked by the dashed lines in figure 1. The CCD detector covered about 9 mm of coordinate y and 8.6 nm of the spectral interval.

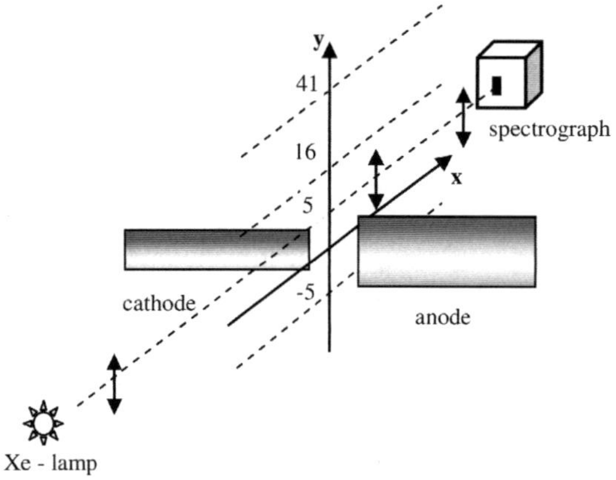

Figure 1. Pictorial representation of electrode configuration (side view) and arrangement for absorption measurement. Dashed lines – absorption paths.

The measurements in each zone were associated with the exchange of the electrodes and new positioning of the electrodes against the optical axis of the lamp-spectrograph system.
The absorption measurements were time limited during the discharge, because of generation of a dust. The incident light passing through the reactor undergoes the Mie scattering on the micro-carbon dust particles formed during the discharge. Thus, this- effect reduces the reactor transparency. Figure 2 illustrates the time evolution of the reactor transparency in the UV spectral window. The transparency rapidly decreases with the increasing of Fe content in the anode. The reliable absorption spectra could be obtained within a short time interval, about 20 seconds, in case of the highest Fe contents in the anode. Obviously in this case the absorption measurements could be influenced by the largest error.

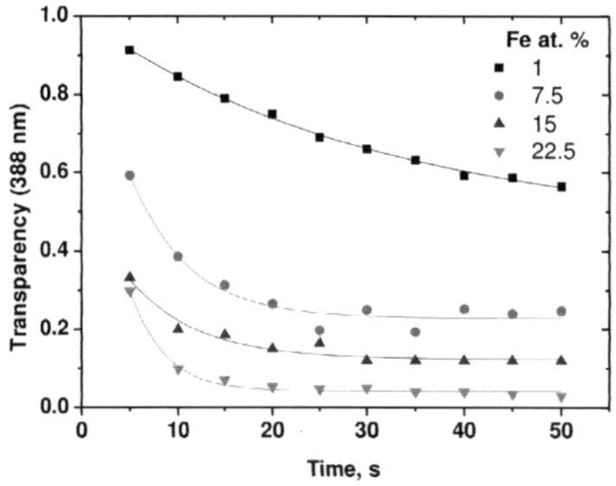

Figure 2. Time dependence of reactor transparency

for anodes with different Fe content.

3. General remarks on Fe I absorption

The UV (384.5 – 393.1 nm) absorption spectra of Fe I atoms (21 lines) pertaining to the transition from the $^5D_{0\text{-}4}$, a $^5F_{1\text{-}4}$ and a $^3F_{2\text{-}3}$ electronic levels were used to determine the Fe I column density. The column density (in units cm^{-2}) is defined as the product of atom concentration and column length along the line of sight. The examples of the absorption spectra recorded at different positions, below and above the arc discharge are shown in figure 3. The spectrum C contains only lines associated with the absorption from $^5D_{0\text{-}4}$ levels. It should also be noted that as with the distance from the anode the width of absorption lines decreases. It is demonstrated in figure 4, in which the full width at half maximum (FWHM) of some $^5D_{0\text{-}4}$ lines is plotted against the observation coordinate y, which is perpendicular to the electrode axis (figure 1). The FWHM follows the population of the levels, hence the most broadened line is that one which originates from the lowest, and thereby the most populated, $^5D_4$ sublevel. One can assume that the main factor which influences the line broadening is connected with the resonance interaction between atoms of the same kind [11]. Thus the width depends on the local Fe I concentration. Therefore such a line broadening becomes much weaker further from the arc (figure 3, spectrum C). In this zone all the recorded spectral lines are of similar width, which approximately is equal to the width of the apparatus function (≈ 0.023 nm).

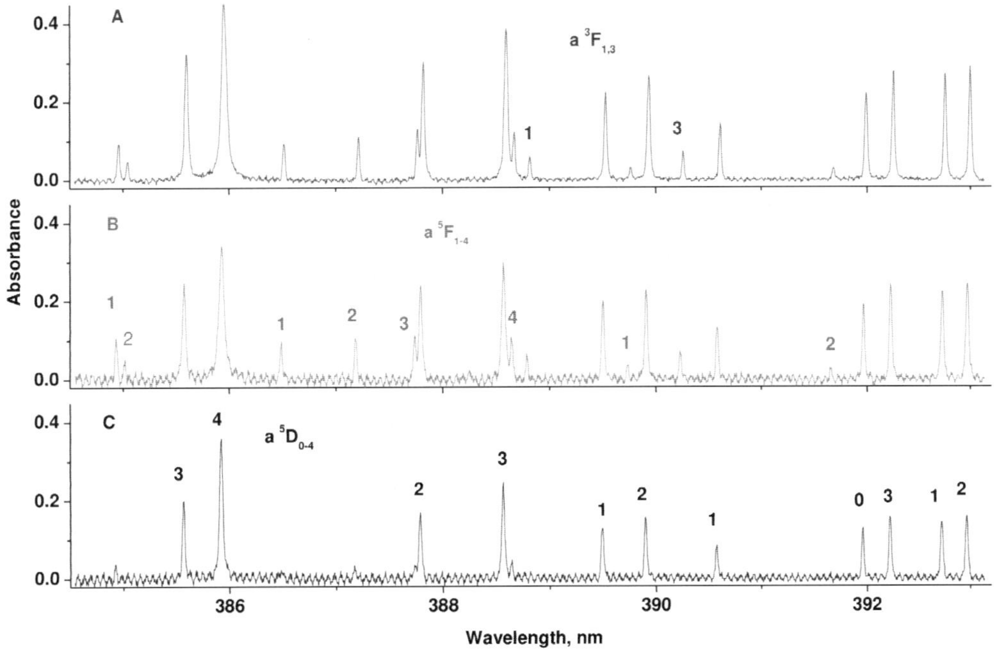

Figure 3. Examples of Fe I absorption spectra recorded below and above electrodes at y = - 5 mm (A), y = 5 mm (B) and y = 15 mm (C). Anode composition: 7.5 at. Fe%.

The Fe I column density was determined taking into account both the FWHM and the value of the absorbance at the line center (peak absorbance). At first, a set of absorption spectra covering all lines under consideration were computed. Three parameters were taken into account, namely Fe I column density, temperature and width of spectral absorption coefficient. The latter one was described by Voigt function (P_V) :

$$\kappa_\lambda = \kappa_i P_V^i(\lambda - \lambda_i)$$

Journal of Physics: Conference Series **550** (2014) 012024 doi:10.1088/1742-6596/550/1/012024

(1)

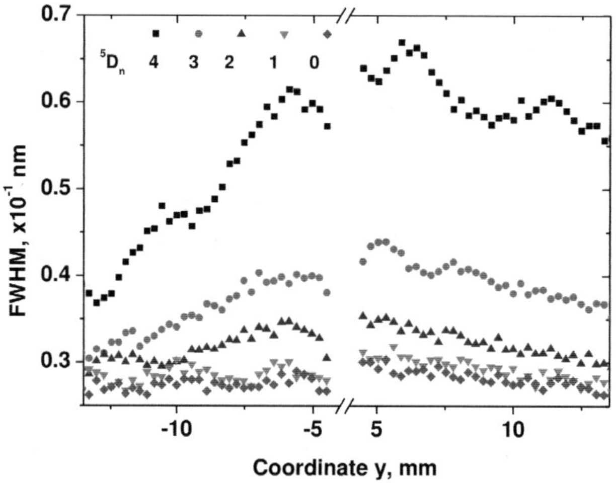

Figure 4. Examples of FWHMs of selected absorption lines vs. coordinate y (see figure 1).

The width of Voigt function depends on the damping constant α ($\alpha = \Delta\lambda_D/\Delta\lambda_L$), i.e. the ratio of Doppler and collision broadening width, which in turn depends on temperature. The as-computed spectra were then convoluted with the apparatus function to be Gaussian with the FWHM equal to 0.023 nm. Next the absorption spectra were transformed into absorbance spectra and a three dimensional data table (three parameters) of peak absorbance of all lines was created. Prior to transformation of the recorded peak absorbance into column density the parameter α was evaluated by comparing the experimental FWHM values with a set of computed FWHM as a function of α and temperature. All the calculations were performed assuming constant temperature and broadening of absorption lines within a column. Three temperature values were considered: 3000, 4000 and 5000 K. Neither the temperature nor the Fe I atom concentration is constant along any column. Therefore the Fe column density output can be converted only to mean concentration.

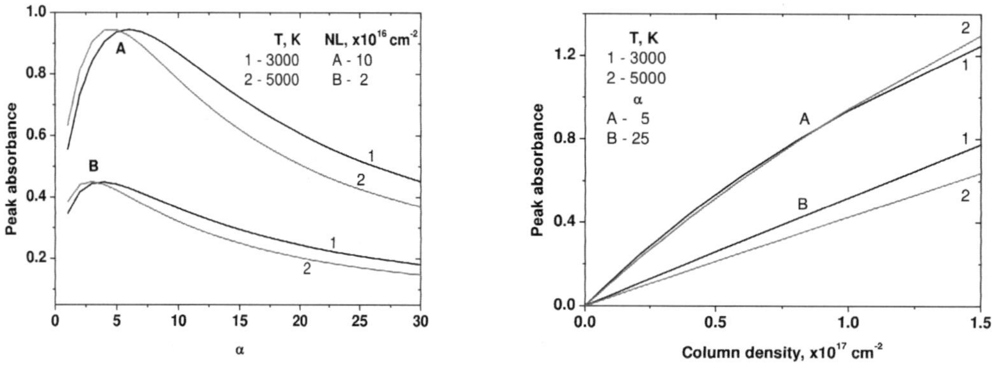

Figure 5. Influence of parameter α and Fe I column density on peak absorbance of $^5D_4$ line (left and right panel, respectively).

The influence of parameter α on the recorded peak absorbance of the $^5D_4$ line for two distinct temperatures and column densities is illustrated in figure 5 (left panel). The example of peak absorbance growth curves for two different temperatures (3000 and 5000 K) and parameters- α (5 and 25) are also presented in figure 5 (right panel). Let us note that strong line broadening leads to lowering the peak absorbance and hence to linear dependence of peak absorbance on column density.

Taking into account that the average population of a particular level is proportional to the column density the mean temperature within a column was also evaluated using the Boltzmann plot method. The examples of Boltzmann plots for two distinct column coordinates, 5 and 40 mm, are shown in figure 6. Please note that the points belonging to $^5D_{0\text{-}4}$ levels are on a straight line, but with a positive slope coefficient, as it is marked by red lines. Such a graph appearance was characteristic for all absorption spectra regardless the zone and discharge conditions. Therefore to be able to at least evaluate the temperature also in the zones where $^5D_{0\text{-}4}$ lines are only present (e.g. figure 3 C) the oscillator strength values had to be modified, i.e. multiplied by appropriate factors.

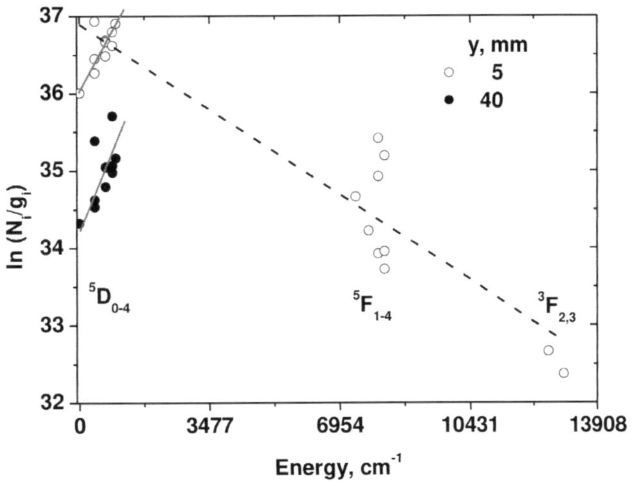

Figure 6. Example of Boltzmann plots for temperature determination.
Column coordinate y: 5 mm (open markers) and 40 mm (solid markers).

Both, literature and modified oscillator strengths are listed in table 1. One has to highlight here that the plasma inhomogeneity surely had an influence on the adopted oscillator strengths. Therefore the new values require further verification because in terms of accuracy all the literature oscillator strengths are assigned to class A [12].

Table 1. Literature [13] and new oscillator strengths ($f_{ik}*10^3$).

λ [nm]	386.0	388.6	392.3	385.6	387.9	390.0	393.0	389.6	390.7	392.8	392.0
lit.	21.4	11.9	3.2	7.3	8.8	5.8	6.4	7.0	1.9	9.9	17.7
new	8.7	6.7	4.0	5.5	6.5	5.8	5.8	8.6	4.7	9.4	23.7

4. Results and discussion

The Fe I column density distributions determined in different zones above and below the arc gap are shown in figure 7. The panels A, B, C, and D correspond to different anode composition 1, 7.5, 15 and 22.5 at. %, respectively. The dashed lines indicate the middle of the electrode gap. In each panel the distributions are determined from the set of absorption spectra which were recorded every five seconds after the arc ignition.

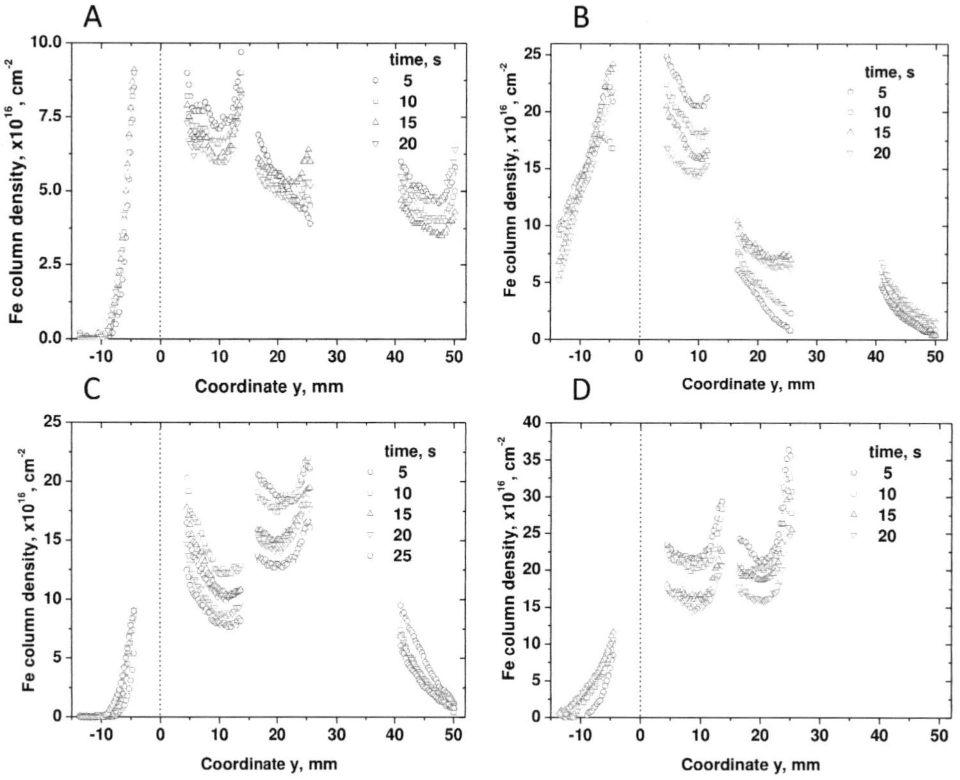

Figure 7. Fe I column density distributions. Content of iron in anode in unit at.%: A – 1, B – 7.5, C – 15 and D – 22.5.

From figure 7 it follows that atomic iron distribution in the arc reactor is not symmetrical. Moreover, the Fe species are mostly concentrated above the arc discharge zone. In the space below the electrodes one could detect atomic iron only within the distance less than 10 mm, whilst above the electrodes iron is present even at a distance exceeding 50 mm (e.g. panel A). It is a result of the vertical motion of the carbon-iron vapour caused by convection among other things. As it was mentioned above, the measurements performed for the each zone were carried out with new electrodes and their space adjustment with respect to the optical axis. Therefore a very accurate reproduction of measurement conditions was not possible. Additionally, in each experimental run the arc spot randomly attached itself to different places on the anode head surface causing thereby the non-symmetrical erosion of the anode. This all explains the shifts between the sub-distributions seen in figure 7. Nevertheless the overall trend of the atomic iron distributions seems to be correct. One can also conclude that the amount of free iron atoms in the reactor increases with iron content in the anode. It should also be mentioned that in the case of the anode containing 22.5 at.% Fe (figure 7, panel D) the measurements between 40 and 50 mm from the electrodes were not possible because of rapid dust formation. Nevertheless, one can expect that the trend in this zone should be similar to those shown in panels B and C.

The results of the mean temperature determination are shown in figure 8. The temperature values at two sides close to the arc show the same decreasing trend, i.e. from about 5000 ±500 K to about 3500 K. The spread of the temperature values in more distant areas is quite significant. Probably this is due to a large fluctuation of dust moving by convection towards the reactor lid. Thus, the dust formed more intensively in the zones located further from the arc plasma may randomly change the local gas temperature. Of course, one cannot ignore the fact that the temperatures in these zones were determined only on the basis of the relative population of a $^5D(0\text{-}4)$ sublevels covering a small energy interval between 0 and 0.12 eV, while the energy of a ^5F and a ^3F levels is about 1 and 1.5 eV, respectively. For the sake of comparison the rotational temperature distributions in the arc discharge zone of C_2 radical in both, excited (d $^3\Pi_g$) and non-excited (a $^3\Pi_u$) states are also shown in figure 8. [9]. Obviously, the rotational temperature of C_2 in the ground state (a $^3\Pi_u$) can be interpreted as the gas temperature. Therefore one can conclude that close to the arc zone the population of higher iron levels (a ^5F and a ^3F) are in non-equilibrium with the gas temperature.

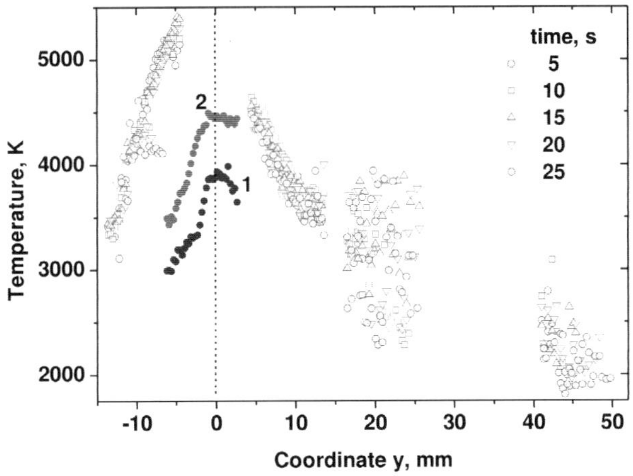

Figure 8. Mean temperature distributions. Anode composition: 7.5 Fe at. %. Full markers: rotational temperature of C_2 in a $^3\Pi_u(1)$ and d $^3\Pi_g$ (2) states.

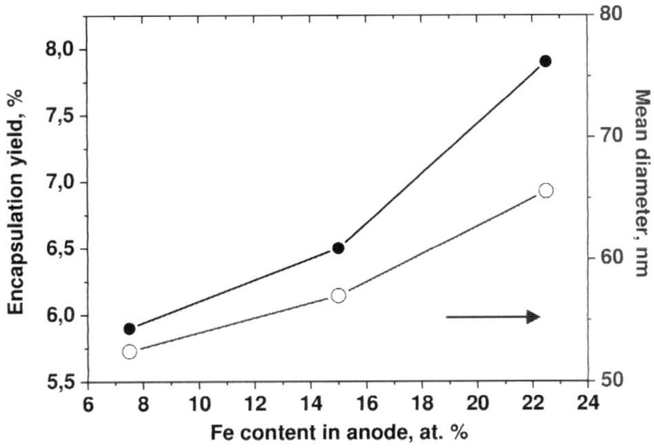

Figure 9. Encapsulation yield and mean diameter of carbon-encapsulated iron nanoparticles vs. anode composition.

Since this work is related to carbon-encapsulated magnetic nanoparticles, some data showing the influence of anode composition on the CEs formation are presented in figure 9 [8]. The data are not shown for 1 at. % Fe, because at this condition the main product consists of single-wall carbon nanotubes [14]. It can be seen that the encapsulation yield gradually increases with Fe content in the anode. The same trend concerns the mean diameter of CEs. These findings are consistent with the above spectroscopic results, especially the data shown in figure 7. The increase of atomic iron density in the gas phase increases the collision rate between iron species and thereby enhances the formation of larger Fe clusters. Consequently it leads to the higher encapsulation yield and larger mean diameter of carbon-encapsulated iron nanoparticles.

5. Conclusions

The diagnostics of the dark zones above and below the carbon-iron arc discharge during synthesis of carbon-encapsulated iron nanoparticles was performed by absorption spectroscopy using UV transitions in Fe atoms. The Fe I column density and temperature distributions were determined. The results revealed different column density gradients. A much higher gradient was found beneath the arc. In this zone the content of atomic iron rapidly decreases and nearly totally disappears ca. 10 mm from the anode edge, while in the upper part of the arc reactor it extends to at least 50 mm. The obtained results clearly indicate that the growth of carbon-encapsulated iron nanoparticles occurs mainly beyond 50 mm from the arc.

Acknowledgements

This work was supported by the National Research Center under Grant N N204 248140.

References

[1] Krätschmer W, Lamb L D, Fostiropoulos K and Huffman D R 1990 *Nature* **347** 354-8
[2] Iijima S 1991 Helical *Nature* **354** 56-8
[3] Zhao X, Inoue S, Jinno M, Suzuki T and Ando Y 2003 *Chem. Phys. Lett.* **373** 266-71
[4] Rodney S, Ruoff RS, Lorents DC, Chan B, Malhotra R and Subramoney S 1993 *Science* **259** 346-8
[5] Niu H, Wang Y, Zhang X, Meng Z and Cai Y *ACS Appl. Mater. Interfaces* **4**(1) 286-95
[6] Bystrzejewski M, Pyrzyńska K, Huczko A and Lange H 2009 *Carbon* **47(4)** 1201-4
[7] Bystrzejewski M, Cudziło S, Huczko A, Cota-Sanchez G and Kaszuwara W 2007 *Biomolecular Engineering* **24 (5)** 555-558
[8] Bystrzejewski M, Łabędź O, Kaszuwara W, Huczko A and Lange H 2013 *Powder Technology* **246** 7
[9] Bystrzejewski M, Łabędź O and Lange H 2013 *J. Phys. D: Appl. Phys.* **46** 355501
[10] Lange H, Baranowski P and Huczko A 1997 *Rev Sci Instrum* **68** 3723-7
[11] Ali AW and Griem HR 1965 *Physical Review* **140** 4A A1044-A1049
[12] physics.nist.gov
[13] Blackwell DE, Ibbetson PA, Petford AD and Shallis MJ 1979 *Mon Not R astr Soc* 186 633-650
[14] Łabędź O, Lange H, Huczko A, Borysiuk J, Szybowicz M and Bystrzejewski M 2009 *Carbon* **47** 2847-2854

Journal of Physics: Conference Series **550** (2014) 012025 doi:10.1088/1742-6596/550/1/012025

Experimental tests for carbon nanomaterial synthesis using DC plasma jet

H Lange, O Łabędź, I Tylska, A Huczko and M Bystrzejewski
Faculty of Chemistry, University of Warsaw, Pasteur str. 1, 02093 Warsaw, Poland

E-mail: lanhub@chem.uw.edu.pl

Abstract. In the frame of this work some experimental tests were performed in the plasma jet. Pure ethanol vapour alone or with the addition of fine iron powder were used to synthesize few-layer graphene or carbon-encapsulated iron nanoparticles, respectively.

1. Introduction

Carbon-encapsulated magnetic nanoparticles (CEINPs) were discovered in 1993 during the intensive studies on fullerene and nanotube formation using the carbon arc process. Recently, carbon encapsulates containing ferromagnetic phases have attracted significant interest. This is primarily due to their great potential for applications in biomedicine (e.g. contrast agents in magnetic resonance imaging [1]), drug delivery [2], hyperthermia [3], catalysis, and environmental protection (removal of heavy metal ions from aqueous solutions [4]). Hence, the development of efficient and selective synthesis routes for carbon-encapsulated magnetic nanoparticles is very important.

The flowing plasma jet is a powerful alternative to the conventional synthesis route of carbon nanomaterials using carbon arc discharge. These two methods of CEINPs synthesis have a very similar approach. The high temperature causes vaporization of carbon and iron precursors, and nucleation of Fe-C nanoparticles. When they leave the elevated temperature zone, the solidification of Fe–C nanoparticles with simultaneous expulsion of the crystalline carbon coating takes place. The major advantage of a flowing plasma jet over its predecessor stems from the fact, that it can be operated in a continuous regime. Additionally, the starting materials can be introduced into the reactor in any state, i.e. solid, liquid or gaseous. As for example carbon nanotubes have already been synthesized in a thermal plasma jet from ethanol [5], coal particles [6] or via ablation of a catalyst doped graphite rod [7].

This work has an exploratory character, as it was aimed at establishing the optimal conditions for the nanomaterial synthesis via dc plasma jet and carrying out several preliminary tests involving starting materials required for the formation of CEINPs.

2. Experimental

The experimental tests were carried out in a flow-through dc plasma jet reactor (figure 1). The pictured setup consists of two main sections: a plasma torch, where the reactants are introduced, and a water-cooled chamber, where the products condense. The starting materials were in this case ethanol and fine iron powder (10 μm). Previous studies have shown that ethanol is an optimal carbon precursor, since its processing in high temperature plasma environment results in formation of highly graphitic materials [8]. Importantly, the products formed from ethanol do not contain amorphous carbon. The alcohol was delivered into the reaction zone in a form of vapor. The iron powder was delivered by a powder feeder using Ar as a carrier gas. The reactant flow rates ranged between 8 and 35 g min^{-1} for ethanol, and between 2 to 7 g min^{-1} for iron powder. According to the presented flow rate values the experiments were grouped into two categories: low (m_{Fe}/m_{EtOH} < 0.1) and high (m_{Fe}/m_{EtOH} > 0.1) Fe content.

Argon was the main plasma gas with the mean flow rate of 30 (or 21 - 35) slpm. The dc arc current was 600 A, and the voltage drop ca. 22 V. During the tests, the emission spectra of hydrogen atoms H_α (656.2 nm) and H_β (486.13 nm), and C_2 (d-a, Δv=0, 516.56 nm) Swan band were recorded (figure 2) through an optical fiber inserted into one of two small view ports, close to the nozzle and 6 cm below (figure 1). On the basis of Stark broadening of hydrogen lines the mean electron density and temperature was evaluated. Obviously lines of Ar I and O I were also detected (figure 2).

The products were collected from different parts of the reactor system and subjected to a two-step purification procedure consisting of boiling for 5 h in 3M HNO_3 and afterwards for 2.5 h in 30% H_2O_2, in order to remove the non-encapsulated iron particles and some insufficiently graphitized carbon (if any).

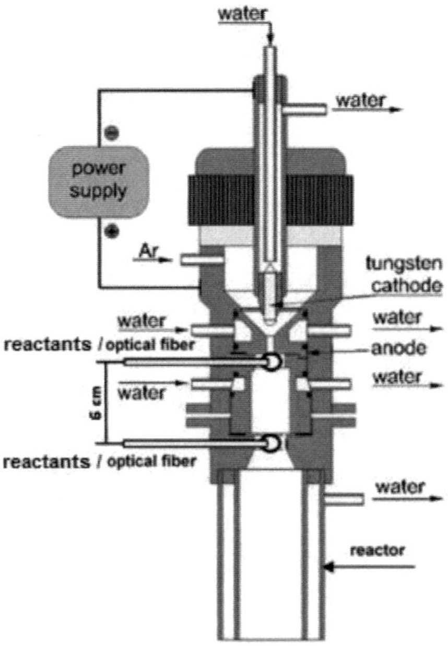

Figure 1. Reactor set-up.

3. Results and discussion

3.1. Plasma spectroscopy

Figure 2 shows emission spectra of hydrogen lines, C_2, Ar I and O I. The Stark broadening of H_α and H_β lines was used to evaluate the mean electron density along the observation path (i.e. the optical axis). To obtain information on temperature the thermodynamic calculations of plasma composition (including electron density) were performed assuming the local thermodynamic equilibrium. The dependence of electron density on temperature is shown in figure 3. Using the data from figure 3 and the experimental mean electron density the mean temperature was drawn out. The electron density and mean temperature for various Ar flow rates are shown in figure 4. The mean electron density and temperature at the nozzle was about 6x10^{16} cm^{-3} and 12000 K. These values describe the plasma jet at its middle zone. The mean temperature evaluated from the emission spectra of C_2 radical was found to be between 5800 and 6200 K. The atomic spectra recorded 6 cm below the nozzle were very weak. However, at this site in the reactor the molecular emission band of C_2 was acquired and used for estimation of the mean gas temperature, which was 3800 ± 500 K. It means that at 6 cm below the nozzle the vaporization of iron powder is still possible.

It follows from the thermodynamic calculations that C_2 radicals can only be present in the periphery of the jet, where the temperature is substantially lower in comparison to the middle. To summarize, the higher and lower temperatures describe different plasma zones.

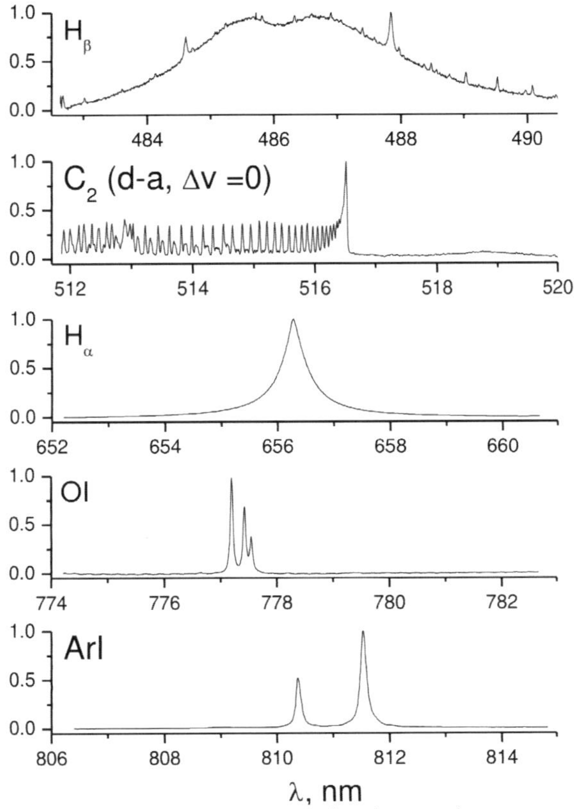

Figure 2. Emission spectra recorded at the nozzle.

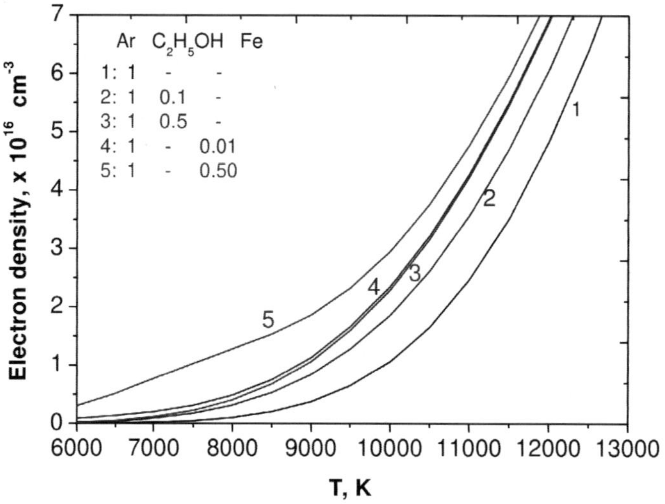

Figure 3. Influence of reactant composition and temperature on electron density.

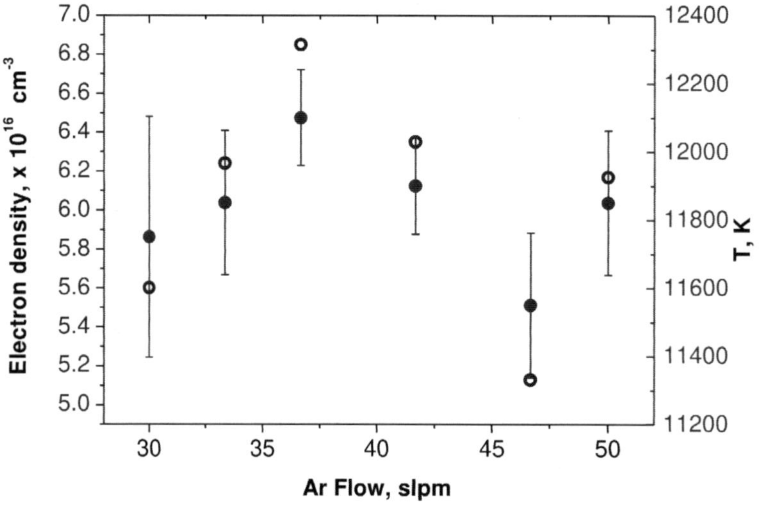

Figure 4. Influence of Ar flow rate on electron density and temperature (empty and full circles, respectively).

3.2 Plasma processing of pure ethanol

The plasma processing of pure ethanol resulted in a black powder, which was recovered from the inner wall of the cooling chamber. The TEM investigations (figure 5) show that the products are composed of few-layer graphene flakes. Their lateral size is between 100 and 500 nm. The thickness (evaluated from the bending zones) is between 2 and 10 nm. This finding informs that the graphitic flakes are comprised of ca. 5 to 30 individual graphene layers. Further structural details were derived from

Raman spectra (figure 6). The G band (located at. 1590 cm^{-1}) has substantially larger intensity in comparison to the D band (1360 cm^{-1}). This observation proves that the obtained product has a relatively high graphitization degree.

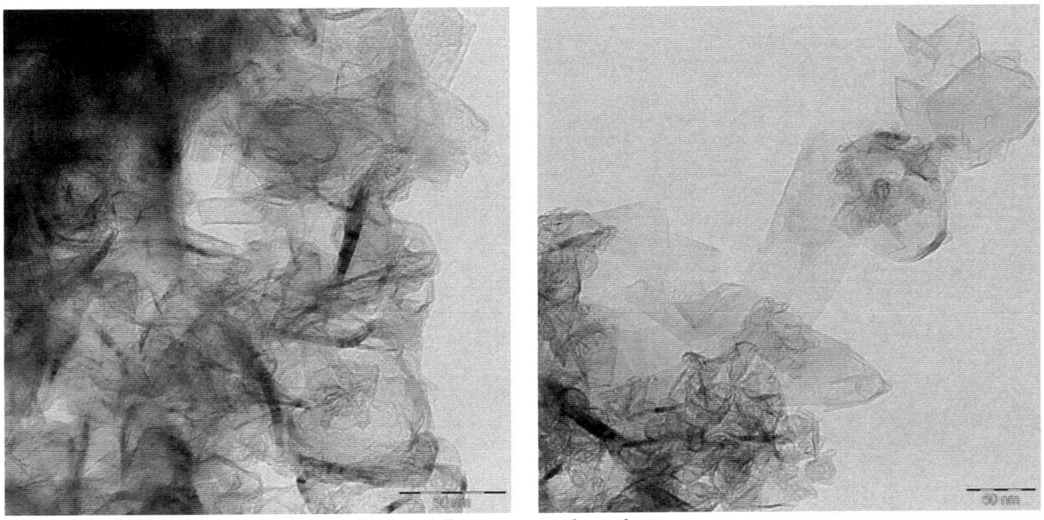

Figure 5. TEM of raw products obtained from pure ethanol.

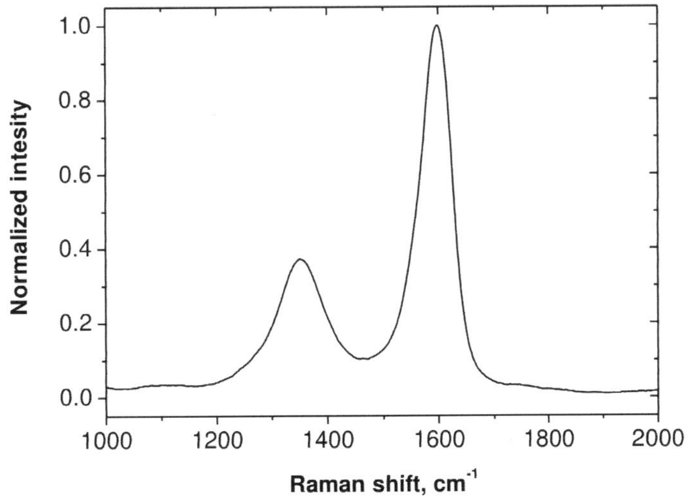

Figure 6. Raman spectrum of products obtained from pure ethanol.

Multi-layer graphene sheets have already been fabricated by decomposition of ethanol in thermal plasma jet [9].

3.3 Plasma processing of ethanol and iron

The TEM investigation of products obtained from all tests which involved both ethanol and iron as the starting materials revealed that all of them contain mainly agglomerates of nanosized Fe particles surrounded by a layer of carbon – carbon encapsulated iron nanoparticles (CEINPs, figure 7). As a result of the purification procedure, the insufficiently protected iron was removed from the samples leaving behind some empty shells (data not shown). Importantly, there is a clear noticeable difference in the diameter distribution of CEINPs. The encapsulates with narrower diameter distribution are primarily synthesized at higher Fe content in the initial reactant mixture.

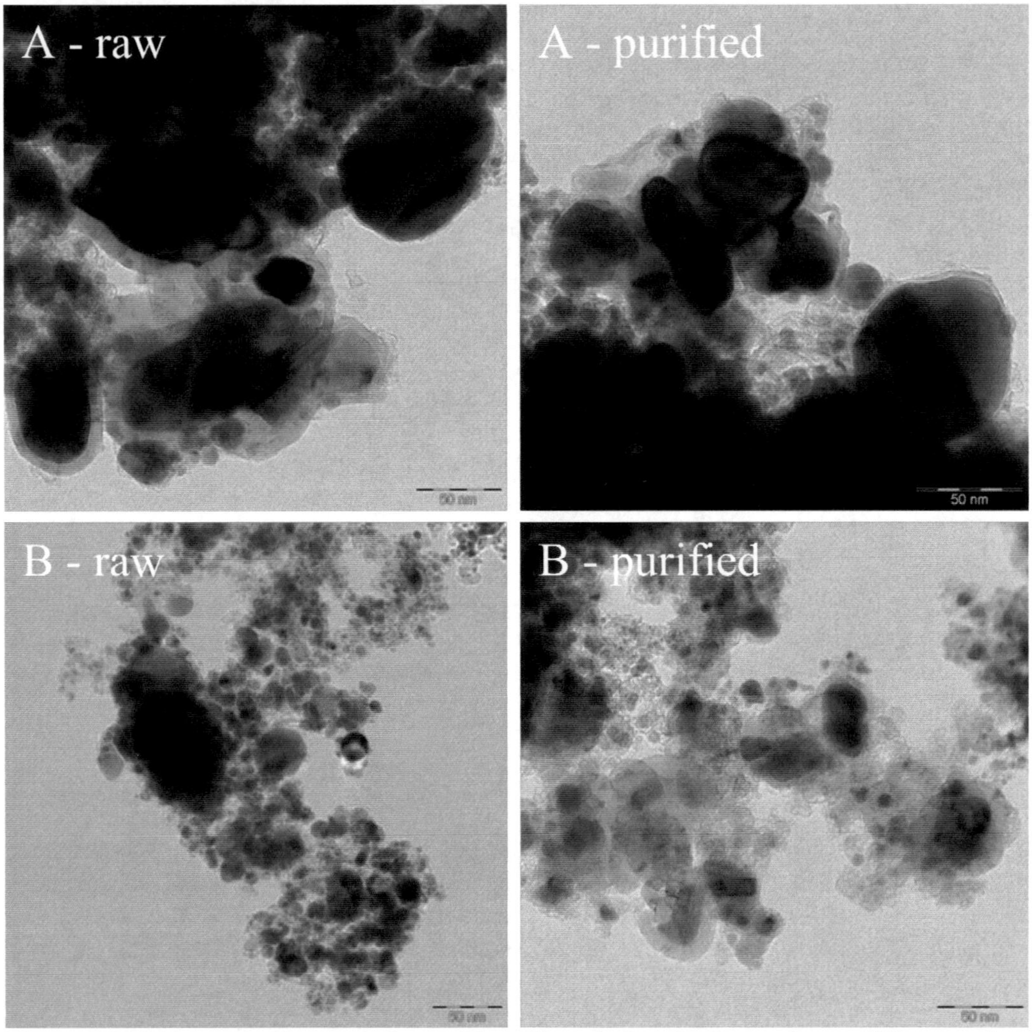

Figure 7. TEM images of products obtained from Fe and ethanol mixtures in two different ratios: A – $m_{Fe}/m_{EtOH} < 0.1$ and B - $m_{Fe}/m_{EtOH} > 0.1$.

The Fe content in both raw and purified products was evaluated by thermogravimetric analysis (TGA, figure 8). The analysis is performed under oxygen and during the TGA measurement the solid carbon phase is gasified, whilst Fe-bearing phases are transformed to the solid iron oxide. The ratio between the solid residue collected after the measurement and the initial mass is an indirect indicator of Fe content. As it stems from the TGA curves the Fe content in the raw products and the products after each stage of the purification procedure do not depend on the Fe/EtOH ratio. The Fe content in the samples obtained after nitric acid treatment is ca 1.5-fold lower in comparison to the raw product. This observation is a consequence of irreversible dissolution of non-encapsulated Fe particles which are present in the raw product. Noticeably, the Fe content increases after H_2O_2 treatment. In this case it is resulted from the removal of some amount of carbon.

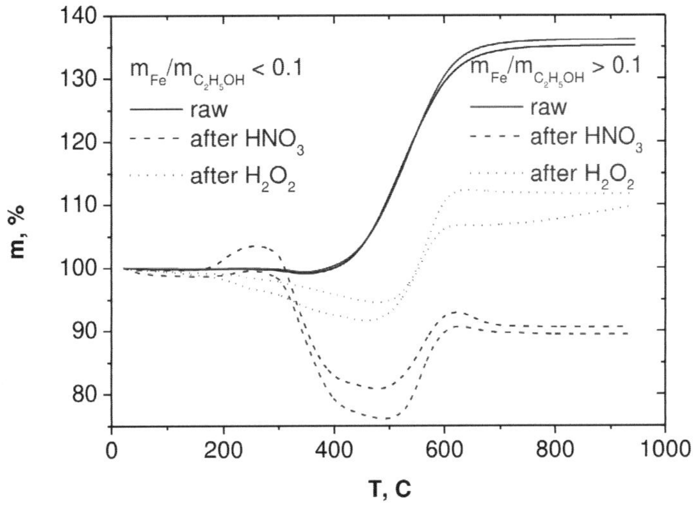

Figure 8. TGA of products on every stage of the purification process.

The Raman spectra of CEINPs are shown in figure 9. There is no remarkable difference in the graphitization degree for the products synthesis at low and high Fe/EtOH ratio (please note the G and D bands are of similar intensity). However, the graphitization degree of CEINPs is lower in comparison to the products obtained from pure ethanol (figure 6). This is an expected result, because carbon coatings in carbon encapsulates contain more defects than the relatively flat few-layer graphene structures.

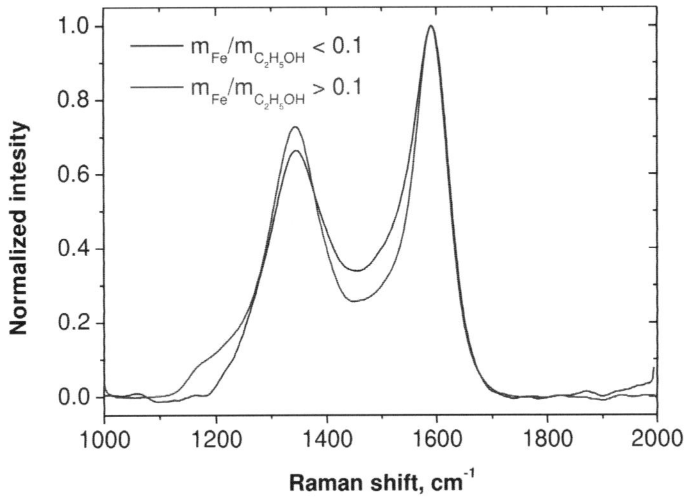

Figure 9. Raman spectra of purified products.

The overall synthesis yield of carbon-encapsulated iron nanoparticles has been evaluated. The yield is defined as the ratio between the mass of products and the mass of the introduced starting material. Then, the yield is multiplied by the purification yield (this factor illustrates the fraction of a material which was recovered after purification). The overall yield was found to be ca. 1% and 5% for low ($m_{Fe}/m_{EtOH} < 0.1$) and high ($m_{Fe}/m_{EtOH} > 0.1$) Fe content in the initial mixture of starting reactants.

4. Conclusions

The plasma processing of pure ethanol and its mixture with fine Fe powder was studied. It was found that the mean temperature in the port where the initial reactants were introduced was ca. 2 times greater than the sublimation point of Fe and high enough for decomposition of the carbon precursor. The performed tests have shown that the plasma processing of pure ethanol results in formation of few-layer graphene flakes with the lateral dimension between 100 and 500 nm and the corresponding thickness up to 10 nm. The inclusion of Fe in the reactant system causes the production of carbon-encapsulated iron nanoparticles.

5. Acknowledgements

The tests were supported by the National Research Center under Grant UMO-2012/05/B/ST5/00709 (graphene) and N N204 248140 (CEINPs).

References

[1] Frey NA, Peng S, Cheng K and Sun S 2009 *Chem. Soc. Rev.* **38** 9 2532-2542

[2] Wozniak MJ, Wozniak P, Bystrzejewski M, Cudzilo S, Huczko A, Jelen P, Kaszuwara W, Kozubowski JA, Lange H, Leonowicz M and Lewandowska-Szumiel M 2006 *J. Alloys Compd.* **423** 87-91

[3] Lukanov P, Anuganti VK, Krupskaya Y, Galiber A-M, Soula B, Tilmaciu C, Velders AH Klingeler R, Büchner B and Flahaut E 2011 *Adv. Funct. Mater.* **21** 18 3583-3588

[4] Bystrzejewski M, Pyrzyńska K, Huczko A and Lange H 2009 *Carbon* **47** 4 1201-1204

[5] Hahn J, Heo SB and Suh JS 2005 *Carbon* **43** 2618–2641

[6] Tian Y, Zhang Y, Wang B, Ji W, Zhang Y and Xie K 2004 *Carbon* **42** 2597–2601

[7] Bystrzejewski M, Huczko A, Lange H, Płotczyk W, Stankiewicz R, Pichler T, Gemming T and Rümmeli MH 2008 *Appl. Phys. A* **91** 223-228

[8] Bystrzejewski M, Huczko A, Byszewski P, Domańska M, Rümmeli MH, Gemming T and Lange H 2009 *Carbon* **17** 298-307

[9] Kim J, Heo SB, Gu GH and Suh JS 2010 *Nanotechnology* **21** 095601

Two-dimensional spectroscopic observation of a pulse-modulated induction thermal plasma torch for nanopowder synthesis

N Kodama, K Kita, Y Tanaka, Y Uesugi, T Ishijima, S Watanabe†, K Nakamura†

Faculty of Electrical and Computer Engineering, Kanazawa University, Kakuma, Kanazawa 920-1192, JAPAN

†Research Center for Production & Technology, Nisshin Seifun Group Inc., 5-3-1 Tsurugaoka, Fujimino 356-8511, JAPAN

E-mail: n_kodama@stu.kanazawa-u.ac.jp, tanaka@ec.t.kanazawa-u.ac.jp

Abstract. The two-dimensional distributions of spectral radiation intensities in the plasma torch were observed for the pulse modulated induction thermal plasmas (PMITP) with continuous or intermittent feedstock feeding for TiO_2 nanopowder synthesis. For this observation, an imaging spectrophotometer with a high speed video camera were adopted. The evaporation of feedstock Ti powder, the formation of TiO and TiO transportation were investigated from the observation results of a Ti atomic spectral line and TiO molecule spectra as well as those of Ar and O atomic lines. An interpretation was suggested from the observation results for Ti feedstock evaporation and TiO formation in nanoparticle synthesis using a PMITP with intermittent feedstock feeding.

1. Introduction

Nanoparticles/nanopowder are expected as promising next generation material used in various applications such as in electronics, energy and environmental fields. Titanium dioxide (TiO_2) nanopowder is receiving attention as photocatalyst material, photonic crystals, photovoltaic cells, and gas sensors [1]–[4]. However, TiO_2 works as photocatalyst only under ultraviolet light because of its wide energy band gap. Attention has also been paid to TiO_2 nanopowder with metallic-ion doping as photocatalyst material. This is because the impurity energy levels in the band gap improve their visible light absorption efficiency [5]. Another application of metallic-ion doped TiO_2 is in the biomedical field [6]. It has been recently reported that Al-doped TiO_2 nanopowder has a protein-adsorption ability, which is effective to skincare for atopic dermatitis [7]. For biomedical applications, no contamination in TiO_2 or metallic-ion doped TiO_2 nanopowder is greatly desired. However, effective mass production methods have not yet been developed for such TiO_2 and metallic-ion doped TiO_2 nanopowder without contamination.

Various kinds of nanopowder synthesis methods have been developed so far. Among them, the inductively-coupled thermal plasma (ICTP) method offers great advantages [8]. For example, it can provide nanopowder with short processing time, and with simple one-step and continuous process. It can offer nanopowder in non-equilibrium or metastable phase. The most important feature of the ICTP method is that it can essentially synthesize nanopowder without contamination due to the absence of electrodes. Many researchers have studied various kinds of nanopowder synthesis method using a steady-type of ICTP. However, nanopowder synthesis methods using steady-type ICTPs still have some issues such as the difficulty in controlling the synthesized particles size, and in the scale-up of the process for higher production rate [9]–[12].

The authors have developed a method to synthesize large amounts of nanopowder using the pulse modulated induction thermal plasma (PMITP) with time-controlled feeding of feedstock (TCFF), i.e. PMITP-TCFF method. The PMITP was developed to be controlled the time evolution of the temperature and reactive species fields in thermal plasmas using the coil current modulation. In addition to this, a method was also developed for feeding of feedstock powder. In this method, feedstock powder is intermittently and synchronously supplied with the coil current modulation of the PMITP. In our previous work, the PMITP-TCFF method was applied to synthesize pure TiO_2 and Al-doped TiO_2 nanopowder with large-scale production. Results indicated that this method using a 20 kW PMITP offered a high production rate of 500 g h^{-1} for pure TiO_2 nanopowder with a mean particle diameter around 43 nm, and 400 g h^{-1} for Al-doped TiO_2 nanopowder with a mean particle diameter around 67 nm [13, 14].

For a detailed understanding of this PMITP-TCFF method, it is important to study the evaporation process of the supplied feedstock powder and precursor production in gas phase for precise control of TiO_2 nanoparticle synthesis. This paper describes the two-dimensional spectroscopic observation of the PMITP-TCFF for TiO_2 nanopowder synthesis. For this purpose and experimental simplification, the feedstock feeding rate was intentionally adopted to be several g min^{-1} instead of several tens of g min^{-1}. The temporal-spatial variation for radiation intensities from some atomic spectral lines and molecular emission bands were measured during the coil current modulation and the intermittent feedstock powder injection. Finally, the evaporation of the feedstock powder and precursor formation processes are discussed from the obtained results.

2. PMITP-TCFF approach for large-scale nanopowder synthesis

The PMITP was developed to control the temperature and reactive species in the thermal plasma jet. The PMITP is sustained by the coil current modulated in a rectangular waveform. The PMITP has four control parameters: on-time, off-time, HCL and LCL. The 'on-time' is the duration of the higher current level (HCL) phase where a higher input power is supplied to the PMITP. In this 'on-time' temperatures in the plasma torch are higher. The 'off-time' represents the duration of the lower current level (LCL) phase where the input power to the PMITP is lower. During this phase, the plasma jet temperature is lower. In addition to these four parameters, we also defined a shimmer current level (SCL) and duty factor (DF). The SCL is the ratio of LCL to HCL, and the DF is the ratio of 'on-time' to a modulation cycle ('on-time'+ 'off-time'). The condition of 100%SCL or 100%DF corresponds to the non-modulation condition.

Figure 1 illustrates the comparison between two methods for nanopowder synthesis: One is the conventional method in which a steady type of ICTP (non-modulated ICTP) is used with continuous feedstock feeding, and the other is the PMITP-TCFF method. In the PMITP-TCFF method, a solenoid valve with a response time of 2 ms is installed on the feedstock feeding tube between the powder feeder and the PMITP torch. By switching the solenoid valve on and off, the feedstock is fed intermittently into the plasma torch intermittently and synchronously with the coil-current modulation. The feedstock is supplied into the plasma torch only during the high-temperature period. This intermittently and synchronously feedstock feeding can be executed by controlling the delay time t_{d} for opening the valve in reference to the pulse modulation signal of the PMITP coil-current. In the actual experiment, the feedstock reaches the PMITP torch in a finite time of t_{adt}. The t_{adt} was measured in our previous work as about 6-8 ms. Therefore, the actual delay time to the coil-current modulation signal is $(t_{\mathrm{d}} + t_{\mathrm{adt}}) \simeq (t_{\mathrm{d}} + 7)$ ms.

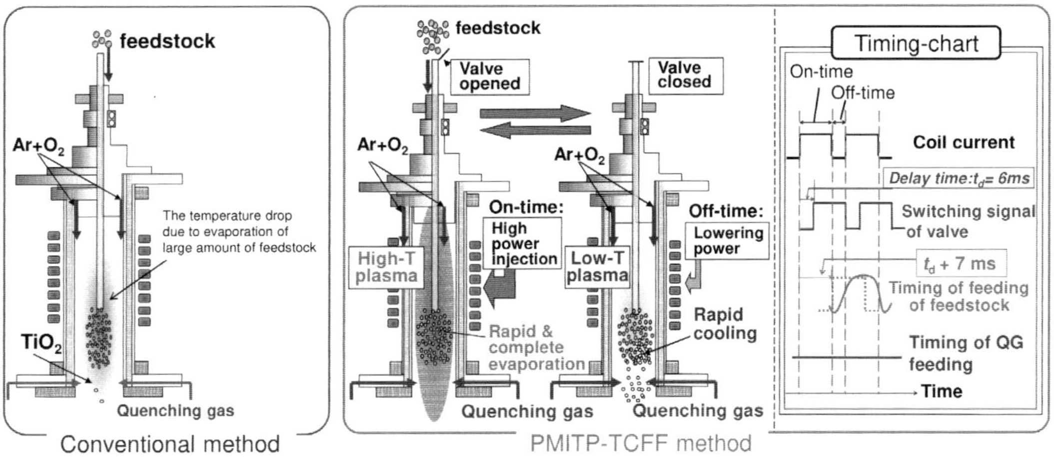

Figure 1. Method for large-scale nanopowder synthesis using induction thermal plasma. Left-hand side: conventional non-modulated ICTP with continuous feedstock feeding; Right-hand side: PMITP with intermittent feedstock feeding.

The injected feedstock is rapidly and completely evaporated during the high-temperature phase of the PMITP. The feedstock feeding is stopped by closing the solenoid valve during successive off-time. In the off-time, the evaporated feedstock material is rapidly and efficiently cooled down because the thermal plasma temperature decreases by the decreased input power to the PMITP. The evaporated Ti atom is mixed with Ar and O atoms from the thermal plasmas. The rapid cooling of the above atomic mixture promotes the nucleation of nanoparticles under the saturated vapor conditions. The nucleated nanoparticles are transported downstream of the PMITP while particle growth take place. The quenching gas (QG) injection downstream of the torch nozzle further cools the evaporated material to promote the nucleation of nanoparticles, thus restraining the particle growth. In this way, the PMITP-TCFF method enables the synthesis of nanopowder with a high production rate.

3. Experimental

3.1. Experimental conditions

The evaporation behavior of the powder injected in the PMITP-TCFF plasma torch is investigated. For comparison, we also measured the behavior for conventional non-modulated ICTP with continuous powder feeding, and for the PMITP with continuous powder feeding. Table 1 summarizes the experimental conditions. The plasma torch used is the same we used in our previous work[13, 14]. The total sheath gas flow rate was fixed at 100 l min^{-1}. The O_2/Ar gas admixture ratio was set at 10mol% in the sheath gas. The pressure in the chamber was regulated at 300 torr (=40 kPa). The quenching gas (QG) was not supplied in this work. The time-averaged input power was fixed at 20 kW. It is noted that instantaneous input power was changed periodically with time according to the coil current modulation. The modulation cycle was fixed at 15 ms. The DF of the modulated coil current was fixed at 80%. The SCL was set to two conditions: 100% (non-modulation condition) and 80%. The feedstock was Ti powder with a mean diameter of 27 μm (TILOP-45; Osaka Titanium Technologies Co.Ltd). The feedstock was fed into the thermal plasma using a rotary powder feeder with Ar carrier gas flow. The Ar carrier gas flow rate was fixed at 4 l min^{-1}. In the present experiments, the feeding rate of the feedstock was regulated at 4-7 g min^{-1}. This feed rate is much lower than what is

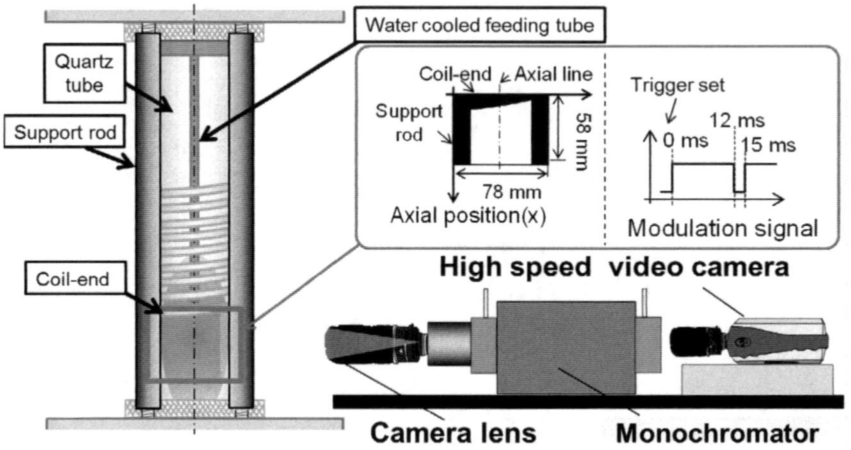

Figure 2. Spectroscopic observation system and observation area for an induction thermal plasma torch.

typically used for synthesis of large amounts of nanopowder [13, 14]. This small feed rate was adopted to carefully investigate the evaporation process. We set two feedstock feeding methods: the conventional continuous feeding and the intermittent feeding methods. The intermittent feedstock feeding was synchronized with the coil current modulation with a delay time t_d of 6 ms. The open and close time of the valve for the intermittent feedstock feeding were 12 ms and 3 ms, respectively. The powder was actually fed from the tip of the water-cooled feeding tube directly to the thermal plasma region. The tip of the water-cooled feeding tube was located between just above the coil end on the axis of the torch to allow spectroscopic observation.

3.2. Spectroscopic observation condition

Spectroscopic observation was carried out using a two-dimensional imaging spectrometry system(Anfi Inc., AN-IMC-DD) shown in figure 2. This system contains an objective lens (Nikon, AF Zoom-Nikkor 80-200 mm f/2.8D ED), a two-dimensional imaging spectrophotometer (Princeton Inst., Acton Spectra Pro 2300i), and a high speed video camera (NAC Imaging Tech., GX-8). The observation area was set to a 78×58 mm^2 region below the coil end. It is noted that the feedstock powder is supplied just above the coil end from the water-cooled feeding tube. The 1200 grooves/mm of diffraction grating was used for the spectrophotometer. In this case, the wavelength resolution of the spectrometer was 0.4 nm. This system can resolve time varying two-dimensional images at a selected wavelength using a high speed video camera. The atomic spectral lines of Ar I at a wavelength of 811.53 nm, O I at 777.54 nm were observed in the thermal plasma itself, and an atomic line of Ti I at 453.32 nm from the evaporated material. We also measured the molecular emission of TiO around 615.91 nm , which specie is formed by the chemical reaction of Ti + O. The frame rate of the high speed camera was set at 3000 fps.

4. Results and discussion

4.1. Synthesized nanoparticles

Figure 3 shows FE-SEM images of nanoparticles synthesized (a) by non-modulated ICTP (100%SCL) with continuous feedstock feeding , and (b) by PMITP (80%SCL) with intermittent feedstock feeding. As seen in figure 3, in both cases (a) and (b), many nanoparticles with diameters less than 100 nm were produced. The adopted feedstock feed rates being much

Table 1. Experimental conditions and spectroscopic observation conditions.

Time-averaged input power	20 kW
Fundamental frequency of the coil current	450 kHz
Modulation cycle	15 ms
Shimmer current level, SCL	100% (Non-modulation) or 80%
Duty factor, DF	80% (On-time / Off-time=12 ms / 3 ms)
Pressure	300 torr (= 40 kPa)
Gas composition	Ar:90 l min^{-1},O_2:10 l min^{-1}
Carrier gas flow rate	Ar:4 l min^{-1}
Quenching gas flow rate	None
Delay time of solenoid valve open t_d	6 ms
Feedstock feeding rate	4-7 g min^{-1}
Raw powder and its maximum size	Ti, 45 μm
Observation area	78×58 mm^2 region below the coil end
Wavelength resolution	0.4 nm
Spectral lines observed	Ar I (811.53 nm), O I (777.54 nm), Ti I (453.32 nm) and TiO (615.91 nm)
Frame rate for high speed video camera	3000 fps

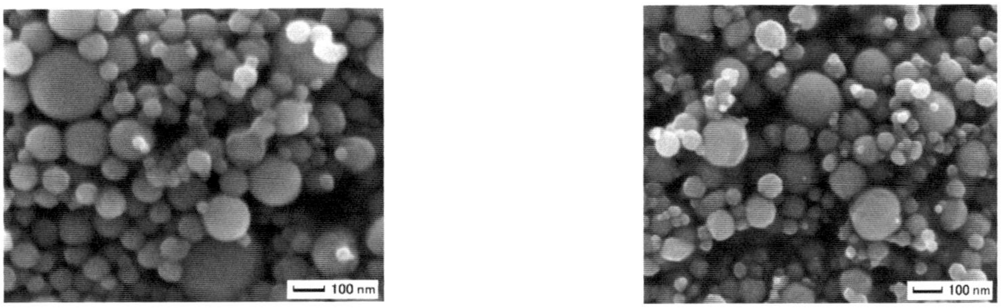

(a) (b)

Figure 3. FE-SEM images of nanoparticles synthesized (a) by non-modulated ICTP, and (b) by PMITP with intermittent feedstock feeding (80%SCL)[15].

lower typical value used in synthesis of large-scale of nanoparticles [13, 14] , the feedstock was completely evaporated.

From randomly selected 200 particles in each of these FE-SEM images, the particle size distribution of synthesized particles were evaluated for different conditions. Figure 4 depicts the particle size distribution of nanoparticles synthesized (a) by non-modulated ICTP with continuous feedstock feeding, and (b) by PMITP with intermittent feedstock feeding (80%SCL). The mean particle diameter, and the standard deviation are also represented in figure 4. Nanoparticles with mean particle diameters around 55-65 nm were synthesized by the both methods even without quenching gas injection. In the next section, we focus an attention on

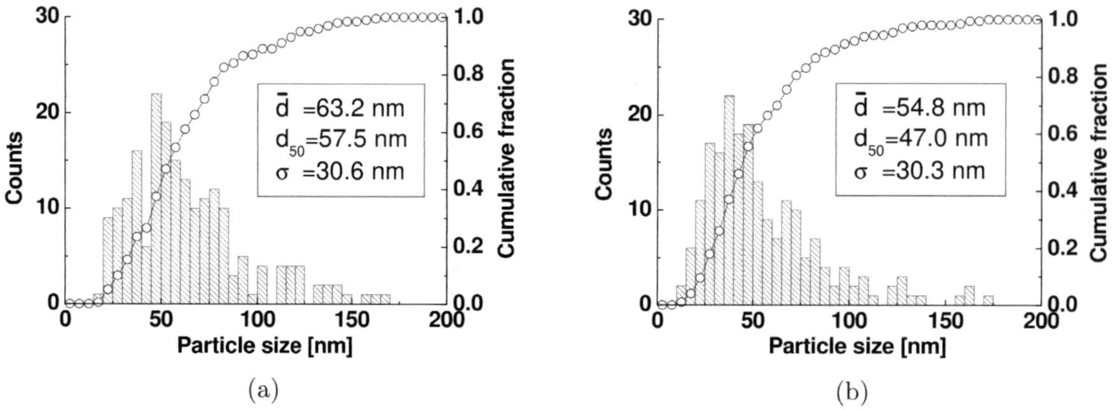

Figure 4. Particle size distribution of nanoparticles synthesized (a) by non-modulated ICTP and (b) by PMITP with intermittent feedstock feeding (80%SCL)[15].

the influence of the coil-current modulation and intermittent feedstock feeding on the feedstock evaporation process as well as the vapor cooling process.

4.2. Two-dimensional distribution of radiation intensities

Figures 5– 7 show temporal variations of the two-dimensional distribution of radiation intensities for (*a*) Ar I and (*b*) O I lines from the thermal plasma itself, that of (*c*) Ti I line from the evaporated material, and those of (*d*) TiO from by-product of Ti and O. In these figures, the timing t was also denoted at the bottom-left of each image. The coil current amplitude rises up at $t=$ 0 ms, and falls down at $t=$ 12 ms.

In case of the non-modulated ICTP with continuous feedstock feeding, the Ar I and O I radiation intensities were almost unchanged with time as shown in figures 5 (*a*) and (*b*). The Ar I and O I radiation intensities were almost uniform in the radial direction just below the coil end. However, the Ar I and O I radiation intensities decreased sharply downstream around the bottom of these figures, clearly indicating that the plasma tail of the ICTP is rapidly quenched. On the other hand, Ti atoms emit strongly around the axis, and especially in off-axis region below the coil end, as depicted in figure 5 (*c*). This implies that the injected Ti powder was evaporated by the thermal plasma to produce a high-density of Ti atoms, which were transported downstream. The Ti radiation intensity observed on the axis is lower than that in the fringe. This may be because the induction heating is present in the fringe, and the Ar carrier gas cools down the temperature on the axis. Figure 5 (*d*) presents the TiO spectral intensity distribution. The TiO is one important precursor for TiO_2 nanoparticle formation. The radiation intensity of TiO was detected locally on the axis. This means that TiO was formed in gas phase around the axis, and it transferred downstream by the Ar carrier gas flow.

Results for the PMITP with continuous feedstock feeding are presented in figure 6. The Ar I and O I lines had strong radiation intensities detectable from t=3–12 ms during the on-time, whereas these intensities became weaker in the off-time as shown in figures 6 (*a*) and (*b*). These results indicate that high and low temperature fields in the PMITP torch were repeatedly created with the coil-current modulation. Panel (*c*) indicates that Ti I radiation intensity from evaporated material was also varied with the coil-current modulation. The TiO radiation intensity were also detected on the axis with a peculiar time variation according to the coil-current modulation.

The most important result is presented in figure 7 for the PMITP with intermittent feedstock

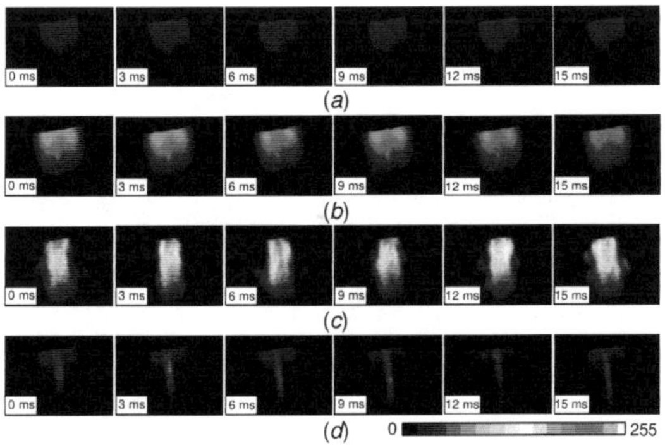

Figure 5. Two-dimensional distribution of radiation intensities for the non-modulated plasma (100%SCL) with continuous feedstock feeding. (*a*) Ar I: 811.53 nm, (*b*) O I: 777.54 nm, (*c*) Ti I: 453.32 nm, and (*d*) TiO: 615.91 nm.

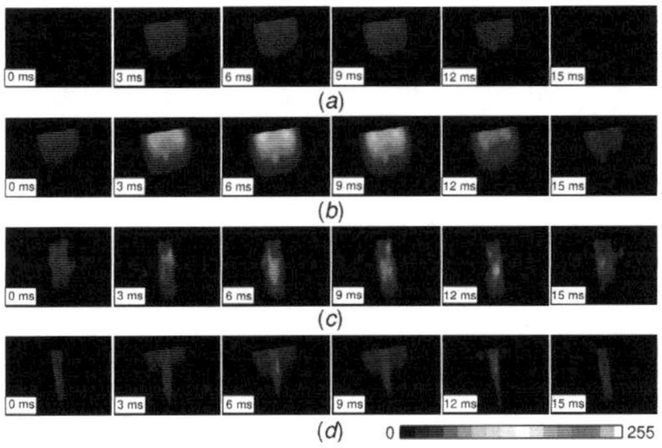

Figure 6. Two-dimensional distribution of radiation intensities from the PMITP for 80%SCL with continuous feedstock feeding. (*a*) Ar I: 811.53 nm, (*b*) O I: 777.54 nm, (*c*) Ti I: 453.32 nm, and (*d*) TiO: 615.91 nm.

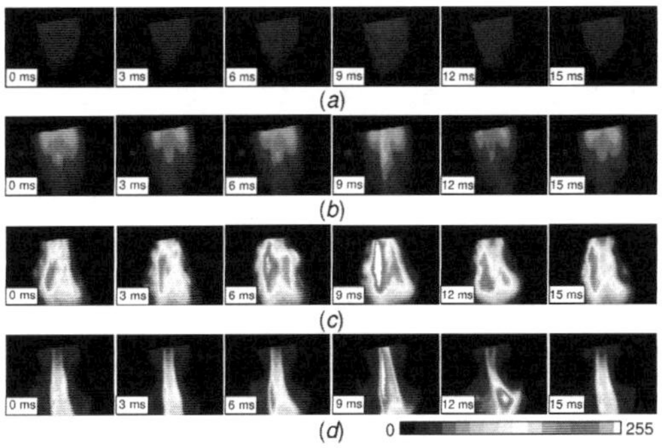

Figure 7. Two-dimensional distribution of radiation intensities from the PMITP for 80%SCL with intermittent feedstock feeding. (*a*) Ar I: 811.53 nm, (*b*) O I: 777.54 nm, (*c*) Ti I: 453.32 nm, and (*d*) TiO: 615.91 nm.

feeding. In this case, Ar I and O I radiation intensities similarly changed with the coil-current modulation. However, a large difference appears in Ti and TiO radiation intensities from those observed with continuous feedstock feeding. The radiation intensity of Ti I in intermittent feedstock feeding became much stronger than with continuous feedstock feeding. This is attributed to the fact that intermittent feedstock feeding has a higher instantaneous feeding rate during the on-time, and that the fed feedstock is efficiently evaporated by during the PMITP on-time. The Ti I radiation intensity depends on both Ti excitation temperature and Ti atomic density determined by evaporation amount of Ti feedstock. The result of the high Ti I intensity in panel (*c*) indicates that high-density Ti is present almost without a temperature drop even from feedstock loading because of high input-power to the PMITP during the on-time. In addition, the TiO intensity was strong just under the coil end. This result suggests that larger amounts of TiO was formed under the coil end region in this condition, and that it was transported downstream of the PMITP torch.

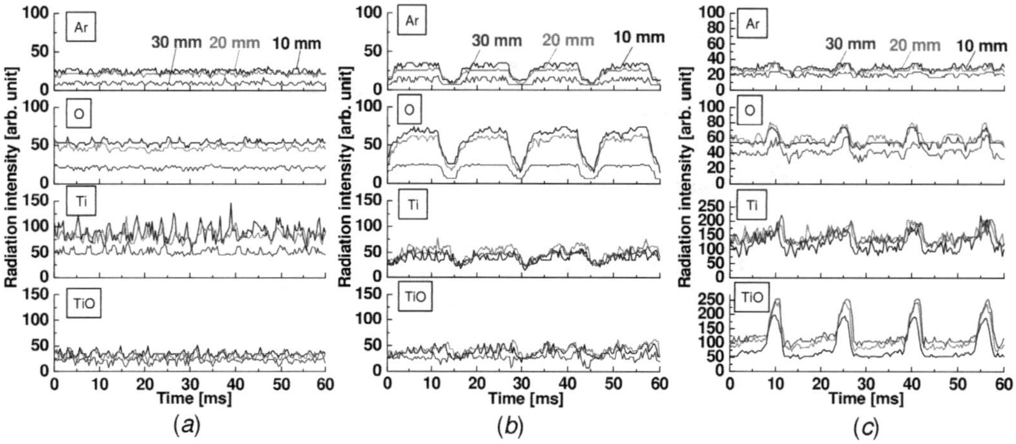

Figure 8. Temporal variation of radiation intensities of different spectral lines observed below the coil end. (*a*) Non-modulated coil current with continuous feedstock feeding, (*b*) modulated coil current of 80%SCL with continuous feedstock feeding, (*c*) modulated coil current of 80%SCL with intermittent feedstock feeding.

4.3. Temporal variation in radiation intensities for different spectral lines

Figure 8 illustrates the temporal variations in the radiation intensities of Ar I, O I, Ti I and TiO observed 10 mm, 20 mm and 30 mm below the coil end on the axis of the plasma torch. Panel (*a*) is the result for non-modulated coil current condition with continuous feedstock feeding; panel (*b*) is that for modulated coil current condition with continuous feedstock feeding, and panel (*c*) corresponds to the result for modulated coil current with intermittent feedstock feeding.

In case of non-modulation with continuous feedstock feeding, the Ar I and O I have almost unchanged radiation intensities, indicating stable sustainability of the thermal plasma torch. The radiation intensity of Ti I fluctuate due to the fluctuations in feedstock powder feeding and its evaporation. The TiO intensity is also hardly changed with time.

For modulated coil current with continuous feedstock feeding, the Ar I and O I radiation intensities periodically varied following the coil-current modulation. This result clearly shows that high-temperature and low-temperature fields in the PMITP torch were generated repeatedly. The radiation intensities from Ti I and TiO also change with time similarly.

A significantly different behavior of the time-varying radiation intensities for modulated coil current with intermittent feedstock feeding is observed. First, Ar I in the panel (*c*) has almost unchanged radiation intensity although the coil current was modulated. The O emission line intensity at 777.54 nm is almost constant similar to Ar I intensity except just before transition to the off-time. The radiation intensity of Ti I changes according to the pulse modulation but in almost triangular waveform. On the other hand, the TiO spectral intensity changes with time in periodical waveform, having a drastic increase just before transition to the off-time. We infer from those observations that the Ar I has almost unchanged intensity, which shows the unchanged plasma temperature in spite of the pulse modulation. This is inferred due to the energy balance between the increased input power by the coil current modulation and the increased loss power for evaporation of large amounts of feedstock during the on-time. The O emission line intensity increase just before transition to the off-time may be caused only by overlapping intensities from TiO band spectra around this wavelength because TiO band spectra has a wide wavelength region and high intensity in this case. The drastic increase in

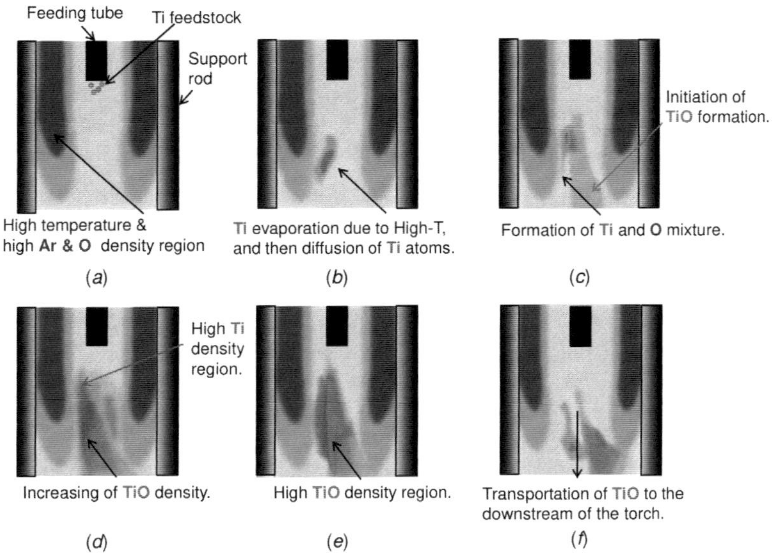

Figure 9. An interpretation on the evaporation process of Ti feedstock and the formation of TiO for pulse-modulation with intermittent feedstock feeding. (*a*) Input powder increase and initiation of feedstock feeding in on-time; (*b*) The evaporation of Ti feedstock, and then diffusion of Ti atoms in on-time (t= 0 ms); (*c*) Mixing of Ti and O atoms, and initiation of TiO formation around the axis (t= 3 ms); (*d*) Increasing of TiO density (t= 6 ms); (*e*) Formation of high TiO density region (t= 9 ms); (*f*) Transportation of TiO to the downstream of the PMITP torch (t= 12 ms).

TiO emission intensity may arise from a remarkable increase in TiO number density just before transition to the off-time. The remarkable increase in the TiO number density originates from promotion of formation reaction Ti + O $\rightarrow$ TiO because of Ti density increase from feedstock evaporation.

4.4. Discussion on Ti evaporation and TiO formation

The observations presented in the preceding sections lead to the following understanding of the evaporation process of Ti feedstock, formation of TiO, and the transport of TiO(precursor of TiO_2 nanoparticles). Figure 9 illustrates the schematic interpretation for phenomena happening in the plasma torch. The steps described as follows, where t is the time in figure 7:

(*a*) The Ti feedstock is supplied into the high-temperature thermal plasma just before the 'on-time' through the feeding tube. High Ar and O atomic density regions are present in off-axis region.

(*b*) At t= 0 ms, the coil current amplitude is increased to establish the high temperature thermal plasma. The Ti feedstock starts being evaporated to generate a high atomic density Ti region. Ti atoms in gas phase then diffuse in all directions.

(*c*) From t= 3 ms, a region containing Ti and O atoms is formed between the center-axis and the high-temperature region, and then TiO formation is initiated there.

(*d*) Around t= 6ms, TiO density increases with time around the center axis.

(*e*) Around t= 9 ms, further higher TiO density region is formed around the center axis.

(*f*) At t= 15 ms, TiO is transported downstream of the PMITP torch between off-time.

5. Conclusions

The two-dimensional distributions of spectral radiation intensities from pulse modulated induction thermal plasmas (PMITP) were investigated with continuous/intermittent feedstock feeding for TiO_2 nanopowder synthesis. For this measurement, an imaging spectrophotometer and a high speed video camera were adopted. The tempo-spatial distribution of radiation intensities for Ar I, O I, Ti I and TiO spectra were observed selectively. The observations provided spatial information on the evaporation of the Ti feedstock, mixing of Ti and O atoms, formation of TiO and the transport of TiO, i.e. the precursor of TiO_2 nanoparticles. This study dearly highlights the advantage of using the pulse modulation of the RF power and synchronous injection of feedstock over the conventional continuous process.

Acknowledgments

This study was supported in part by Grant-in-Aid for Scientific Research (KAKENHI) (A) (23246050), the Japan Society for the Promotion of Science (JSPS), and by Adaptable and Seamless Technology Transfer Program through Target-driven (A-STEP) R&D (AS242Z01379M), Japan Science and Technology Agency (JST), Japan

References

[1] Malato S, Blanco J, Alarcon D C, Maldonado M I, Fernandez-Ibanez P, and Gernjak W 2007 Photocatalytic decontamination and disinfection of water with solar collectors *Catalysis Today* **122** pp 137–49
[2] Li J, Zhao X, Wei H, Gu Z Z, and Lu Z 2008 Macroporous ordered titanium dioxide (TiO_2) inverse opal as a new label-free immunosensor. *Analytica Chemica Acta* **625** pp 63–9
[3] Gratzel M 2001 Photoelectrochemical cells. *Nature* **414** pp 338–44
[4] Biju K P and Jain M K 2008 Effect of crystallization on humidity sensing properties of sol-gel derived nanocrystalline TiO_2 thin films. *Thin Solid Films* **516** pp 2175–80
[5] Tsai Y C, Hsi C H, Bai H, Fan S K and Sun D H 2012 Single-step synthesis of Al-doped TiO_2 nanoparticles using non-transferred thermal plasma torch. *Jpn. J. Appl. Phys.* **51** 01AL01
[6] Lee K N, Kim Y, Lee C W and Jai-Sung Lee J S 2011 Simultaneous amination of TiO_2 nanoparticles in the gas phase synthesis for bio-medical applications *Mater. Sci. Eng.* **18** 082021
[7] Mio M, Kogoma M, Fukui H, and Kogoma A 2010 Control of inflammation using adsorption of inflammatory by nano-particles *Chem. Eng.* **55** pp 603-8
[8] Girshick S L, Chiu C P, Muno R, Wu C Y, Yang L, Singh S K, and McMurry P H 1993 Thermal plasma synthesis of ultrafine iron particles. *J. Aerosol Sci.* **24** pp 367–82
[9] Oh S M and Ishigaki T 2004 Preparation of pure rutile and anatase TiO_2 nanopowders using RF thermal plasma. *Thin Solid Films* **457** pp 186–91
[10] Tong L and Reddy R G 2006 Thermal plasma synthesis of SiC nano-powders/nano-fibers. *Materials Research Bulletin* **41** pp 2303–10
[11] Li J G, Ikeda M, Ye R, Moriyoshi Y, and Ishigaki T 2007 Control of particle size and phase formation of TiO_2 nanoparticles synthesized in RF induction plasma. *J. Phys. D: Appl. Phys.* **40** pp 2348–53
[12] Ishigaki T and Li J G 2007 Synthesis of functional nanocrystallites through reactive thermal plasma processing. *Sci. and Technol. Advanced Mater.* **8** pp 617–23
[13] Tanaka Y, Tsuke T, Guo W, Uesugi Y, Ishijima T, Watanabe S, and Nakamura K 2012 A large amount synthesis of nanopowder using modulated induction thermal plasmas synchronized with intermittent feeding of raw materials. *J. Phys. D: Conf. Ser.* **406** 012001
[14] Kodama N, Kita K, Tanaka Y, Uesugi Y, Ishijima T, Watanabe S, and Nakamura K 2014 A method for large-scale synthesis of Al-doped TiO_2 nanopowder using pulse-modulated induction thermal plasmas with time-controlled feedstock feeding. *J. Phys. D: Appl. Phys.* **47** 195304
[15] Tsuke T, Guo W, Tanaka Y, Uesugi Y, Ishijima T, Sakai Y, and Nakamura K 2012 The paper of Technical Meeting on Plasma Sci. & Technolo. IEE of Japan, PST-10-122

Thermal plasma source for processing of MoO_3 crystals

A Lebid[1], A Veklich[1], V Boretskij[1], S Savenok[2] and O Andreev[2]

[1] Taras Shevchenko National University of Kyiv, Faculty of Radio Physics, Electronics and Computer Systems, Kyiv, Ukraine
[2] Taras Shevchenko National University of Kyiv, Faculty of Geology, Kyiv, Ukraine

E-mail: tgctg@yandex.ru

Abstract. Simple original technique for fabrication of molybdenum trioxide MoO_3 crystals during electric arc discharge directly from metallic molybdenum was proposed. The vertically oriented free-burning arc was ignited between the end surfaces of metallic molybdenum non-cooled electrodes. Molybdenum oxide crystals were deposited on side surface of bottom electrode (anode). Observation of crystals formation zone was used for determination of main process stages. The resulting products were studied by X-rays diffraction and optical microscopy methods. Investigations indicate, that self-organizing vapor-deposition process of MoO_3 crystals formation has place. The resulting product is colorless sparkling prisms and platelets, which mainly consist of orthorhombic α-MoO_3 phase. Optical microscopy indicates formation of closely packed feather-like pins structures by vapor-solid process.

1. Introduction

Molybdenum trioxide MoO_3 attracts scientific interests due to it's unique physicochemical properties. There are useful applications of this material:

- as perspective material for lithium-ion batteries [1,2];
- as highly field emission cathode for display devices [3];
- as catalyst for hydrocarbons transformation reactions [4];
- as material for thin film gas sensors [5].

Fabrication of MoO_3 usually is performed by chemical or physical methods. The physical methods allow to obtain micro- and nano- structured materials, particularly crystals. These methods are mainly based on vapor-deposition processes. These processes are organized by evaporation of molybdenum or molybdenum oxide powders in special furnace [1,7], evaporation of molybdenum foil by infra-red heating [3], atmospheric plasma processes based on UHF discharge [2,6].

In the work [1] α-MoO_3 obtained by vapor transportation method for application in lithium-ion rechargeable batteries was studied. Molybdenum trioxide powder was evaporated inside tube furnace in argon flow. Microstructured material was deposited on quartz substrate near the open side of furnace. It takes 1 hour to produce the final material as white α-MoO_3 microrods at temperature 780-800° C. Lithium-ion battery cells, fabricated with using the microstructured α-MoO_3 had higher stability and capacity in comparison with non-structured MoO_3 and other materials. These advantages are explained by tolerance of orthorhombic α-MoO_3 structure to intercalation/deintercalation of lithium ions during battery charging/discharging processes.

Atmospheric microplasma process was used in work [2] for preparation of monoclinic β-MoO_3 nanosheets. Molybdenum wire (Ø 0.1 mm) was sputtered by plasma of ultra high frequency (UHF) antenna discharge. Duration of nanomaterial deposition on Si-Pt substrate at atmosphere conditions was 45 s. Obtained material β-MoO_3 is very attractive due to its specific crystal structure, electrochemical properties and morphology for lithium battery and catalytic applications.

Other configuration of atmospheric microplasma process for molybdenum oxide was studied in work [6]. Molybdenum wire was closed inside quartz capillary with 60-70 μm nozzle. UHF discharge was initiated inside capillary by coil antenna. Reaction products transported by argon-oxygen mixture with various flow rates. Resulting nanoparticles are injected from nozzle on the silicon substrate. It was found that percentage of different oxides MoO_2 and MoO_3 in resulting particles was drastically affected by gas mixture flow rate. The changing of particle generation mechanism explains this phenomenon.

This paper deals with alternative technique, which allows formation of MoO_3 crystals directly from metallic molybdenum in electric arc plasma source at atmospheric pressure. Advantages of the technique are simplicity of equipment and conditions, self-organization of process and high reproducibility of resulting products.

2. Setup description

The vertically oriented free-burning arc was ignited in air between the end surfaces of metallic molybdenum non-cooled electrodes (figure 1, a). The diameter of the rod electrodes was 6 mm, the discharge gap was 8 mm and DC current was 3.5 A. Molybdenum oxide crystals were deposited on side surface of anode (figure 1, b). It must be noted, that zone of crystals formation has place at 3-5 mm below the end surface of electrode.

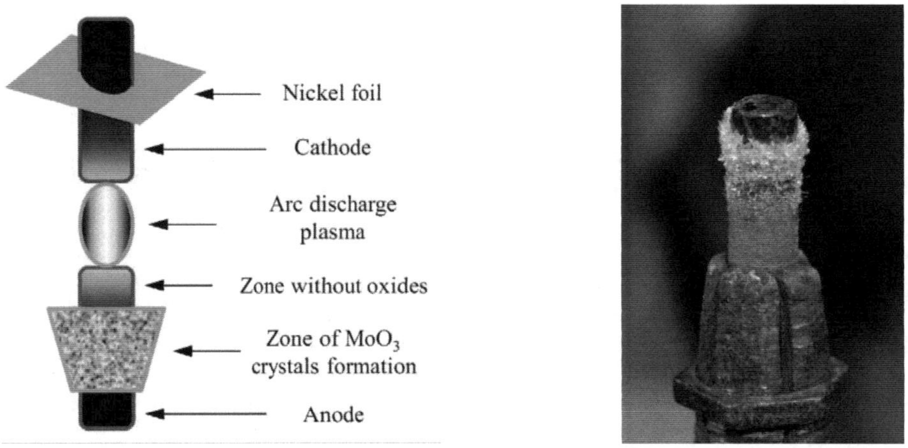

a b

Figure 1. Experimental scheme (a) and common view of anode with deposited MoO_3 crystals (b).

Video registration of crystal formation zone was used for determination of main process stages.

Chemical composition and structure of obtained crystals were determined by X-rays diffraction method, performed by DRON-UM1 setup. Peculiarities of crystals formation were studied by optical microscopy method with using of MBI-1 microscope.

3. Formation of crystals

The process of crystals formation during arc discharge can be separated on specific sequential stages. Immediately after arc ignition there wasn't evaporation due to relatively low temperature of anode surface (figure 2, a).

After few seconds a white fume was observed around the electrode (figure 2, b). This stage was explained by oxidation of metallic molybdenum and volatilization of oxides at increasing electrode temperature. Really, oxidation of metallic molybdenum surface and volatilization of oxide layer were observed during heating in the furnance [1,7]. It was mentioned in work [7] that oxide layer completely evaporates from molybdenum surface at temperature above 1150° C. So, this assumption explains the absence of crystals near the end surface of electrode (figure 1) where surface temperature was obviously more higher.

The MoO_3 crystallites on electrode surface appeared at the next process phase (figure 2, c). Crystals start growing from white fume evaporations, which are transported by convectional air flow. Probably, initial crystallization starts on surface defects or on greyish-black particles, which can be Mo_2O_3. Friable layer around electrode (figure 1, b), which consists of variously oriented transparent prisms and platelets, was formed by vapor deposition.

Re-evaporation of deposited crystals is avoided by two reasons. The first one is low thermal conductivity between electrode and crystallites due to their irregular orientation. The crystallites are weakly connected to electrode surface but have numerous connection with others. The second reason is cooling of crystallites by convectional flows.

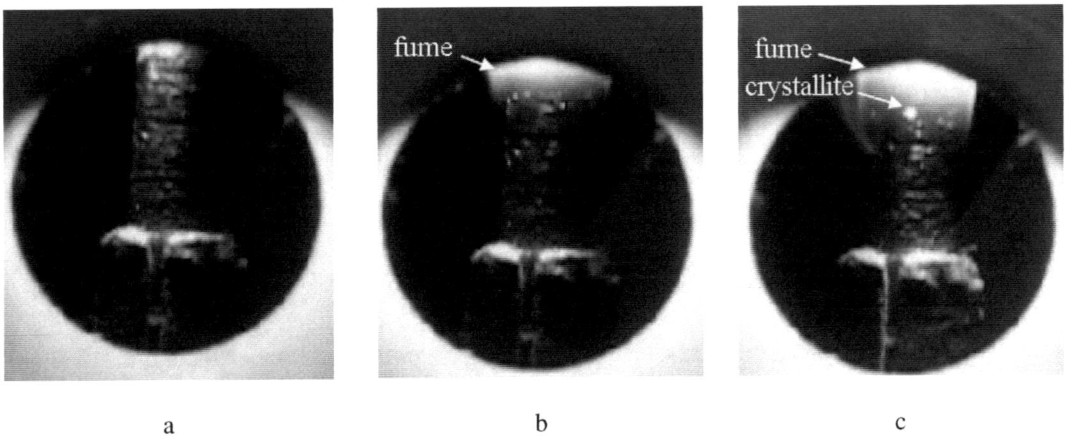

Figure 2. View of anode at different process stages. Initial view - 7 s after arc ignition (a), appearance of white fume at 17 s (b), formation of MoO_3 crystallites at 25 s (c).

Therefore, in proposed plasma source self-organizing vapor-deposition process of MoO_3 crystals formation has place. The process consists of molybdenum surface oxidation, evaporation of oxide layers, vapour transportation by convectional air flow and crystal growth.

Usually formation of crystals is terminated after 2 min after arc ignition. It can be explained by overlapping of molybdenum surface by crystals, which complicates following evaporation and transportation of building material.

4. Chemical composition and structure

Chemical composition and structure of produced crystals were determined by X-rays diffraction method (XRD). Obtained crystals were detached from electrode surface and milled before the investigations. X-ray diffraction study indicates that resulting crystals consist of orthorhombic α-MoO_3 phase (figure 3). Positions of diffraction peaks are in good agreement with reference data [8]. High intensity of some diffraction peaks can be explained by orientation effects in structure of obtained crystalline material. Only slight admixture of monoclinic β-MoO_3 phase was detected.

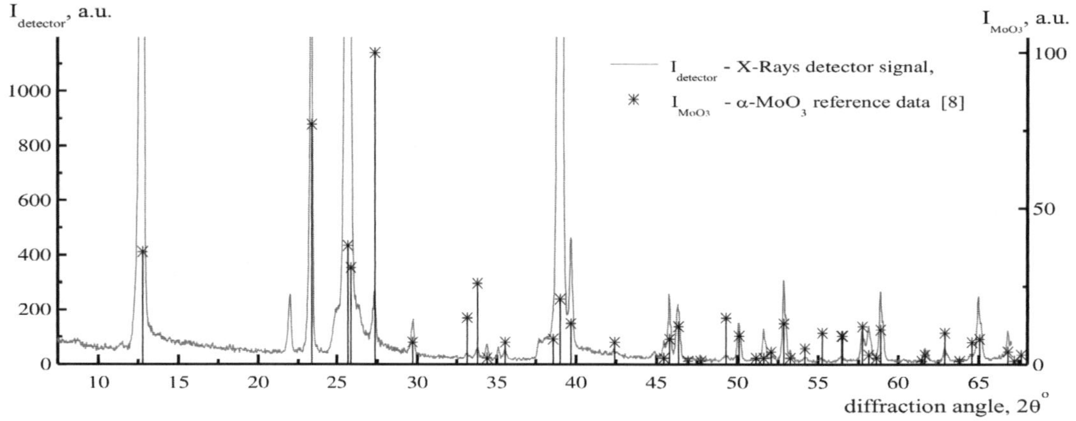

Figure 3. X-rays diffraction analysis of obtained crystals.

Two forms of deposited crystals has place. The prismatic transparent crystallites with longitudinal dimension up to 3 mm and flat platelets with dimensions up to 3x3 mm are obtained.

Optical microscopy indicates formation of closely packed feather-like structures (figure 4). Every “feather” consists of closely packed parallel needles, which are unfinished structures of directional crystal growth. These needles are resembled to microrods observed in work [1], but in contrast, these microrods were oriented in a random way. Probably, attaching of building material on these pins has place during vapor deposition process and supports further translation of crystal structure.

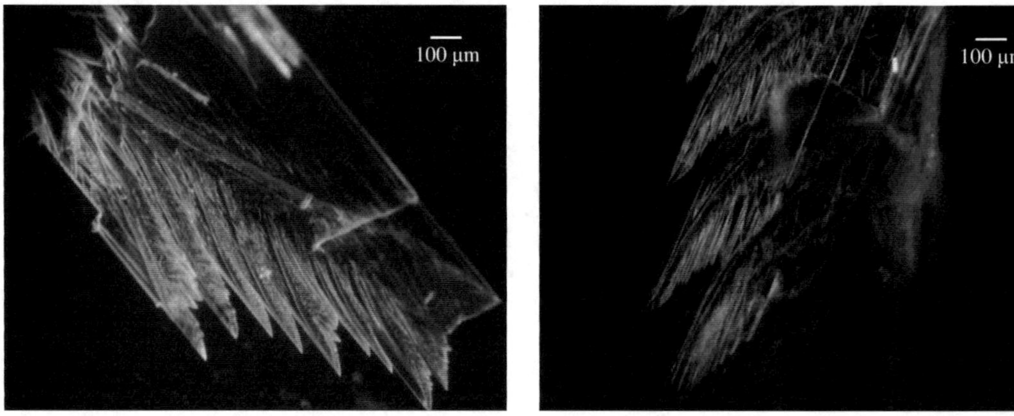

a b

Figure 4. Optical microscopy of crystals growth structures.

Conclusions

Plasma source for processing of MoO_3 crystals directly from metallic molybdenum is proposed. XRD analysis indicates that resulting crystals mainly consist of orthorhombic α-MoO_3 phase and only slight admixture of monoclinic β-MoO_3 phase was detected. The crystals are prismatic transparent prisms and flat platelets with dimensions up to 3 mm.

Formation of closely packed feather-like structures was observed. Every “feather” consists of parallel needles – unfinished structures of directional crystal growth.

So, self-organizing vapor-deposition process of MoO_3 crystals formation has place in proposed plasma source. Observation of crystal formation zone allows to conclude the further process mechanisms: molybdenum surface oxidation, evaporation of oxide layer, vapour transportation by convectional air flow and crystal growth.

Acknowledgement

The authors would like to thank to Professor Sergiy A. Vyzhva and Doctor Ida L. Babich for the help in organization of experiment and preparing of the paper.

References

[1] Li W, Cheng F, Tao Z, and Chen J 2006 Vapor-transportation preparation and reversible lithium intercalation/deintercalation of α-MoO_3 microrods *J. Phys. Chem. B* **110** 119-124

[2] Mariotti D, Lindstrom H, Chandra Bose A and Ostrikov K 2008 Monoclinic β-MoO_3 nanosheets produced by atmospheric microplasma: application to lithium-ion batteries *Nanotechnology* **19** 495302

[3] Li Y B, Bando Y, Golberg D and Kurashima K 2002 Field emission from MoO_3 nanobelts *Appl. Phys. Lett.* **81** 5048-50

[4] Queeney K T and Friend C M 2000 Site-Selective Surface Reactions: Hydrocarbon Oxidation Processes on Oxidized Mo(110) *J. Phys. Chem. B* **104** 409-415

[5] Comini E, Faglia G, Sberveglieri G, Cantlini C, Passacantando M, Santucci S, Li Y, Wlodarski W and Qu W 2000 Carbon monoxide response of molybdenum oxide thin films deposited by different techniques *Sensors and Actuators B* **68** 168-174

[6] Chandra Bose A, Shimizu Y, Mariotti D, Sasaki T, Terashima K and Koshizaki N 2006 Flow rate effect on the structure and morphology of molybdenum oxide nanoparticles deposited by atmospheric-pressure microplasma processing *Nanotechnology* **17** 5976-82

[7] Vahldiek F W 1968 Growth and microstructure of molybdenum oxide *Journal of the Less-Common Metals* **16** 351-359

[8] Powder diffraction files. Pdf-2 2013 *The International Centre for Diffraction Data (*Available from: http://www.icdd.com/products/pdf2.htm)

Journal of Physics: Conference Series **550** (2014) 012028 doi:10.1088/1742-6596/550/1/012028

Similarity criteria of HOSP formation at plasma processing of agglomerated particles

O P Solonenko
Khristianovich Institute of Theoretical and Applied Mechanics, Siberian Branch of RAS, Novosibirsk, 630090, Russia

E-mail: solo@itam.nsc.ru

Abstract. The theoretical analysis of basic phenomena at plasma processing of the agglomerated powders, in context of hollow microspherical particles production, was fulfilled. With the use of the obtained theoretical solutions for the first time the key similarity criteria were derived that allow to formulate the requirements to plasma flow, characteristics of agglomerate and its material as well as condition of quenching of the formed hollow microdroplet which realization provides the formation of hollow microsphere with necessary characteristics, in particular, of maximal diameter and minimum thickness of shell.

1. Introduction

In the present-day industry of structural and protective materials, powders made up by hollow spheres (HOSP), or hollow powders, gain increasing acceptance. Hollow particles can be obtained by various methods [1-17], etc. Ceramic powders that are used in the production of plasma-sprayed thermal barrier coatings, composite heat- and sound-insulating materials, light construction fillers and backfills, buoyancy materials, and explosive mixtures, deserve special attention. Such powders are also used as catalysts carriers, adsorbents, filter elements, encapsulating media, etc. Application of hollow powders has entered a stage at which the range of scientific and engineering problems to be solved widens permanently.

Over the recent decade, in connection with permanently increasing interest in hollow powders and widening application fields of such powders, considerable efforts of researchers have been focused on the development of new approaches to the problem of formation of hollow microspheres in thermal plasma. In this connection, guaranteed plasma synthesis of hollow microspheres with desired morphology, chemical composition, and mechanical properties becomes an urgent matter.

As it was noted in [18], the simplest production method of hollow powders implies processing of void-free, or dense, particles in plasma. For instance, for a number of materials such as oxides, their densities $\rho_{\rm p}^{(s)}$ and $\rho_{\rm p}^{(l)}$ in the solid (*s*) and liquid (*l*) state may differ substantially. As a result, the particle with $\rho_{\rm p}^{(l)} < \rho_{\rm p}^{(s)}$, as it melts, grows in volume, so that its effective size increases. The radius of the completely melted particle increases by $\Delta R_{\rm p} = R_{\rm p0}\left(\sqrt[3]{\rho_{\rm p}^{(s)} / \rho_{\rm p}^{(l)}} - 1\right)$, where $R_{\rm p0}$ is the initial particle radius. If overheated melt droplet did undergo subsequent fast quenching, its solidification starts from the surface, so that the particle size turns out to be fixed at $R_{{\rm p},end} = R_{\rm p0} + \Delta R_{\rm p} = R_{p0}\sqrt[3]{\rho_{\rm p}^{(s)} / \rho_{\rm p}^{(l)}}$.

Following solidification-induced volume shrinkage, a spheroidal cavity forms inside the particle, the corresponding radius of the cavity being $R_{\rm h} = R_{\rm p0} + \Delta R_{\rm p} - \Delta_{\rm p}$, and the shell thickness $\Delta_{\rm p} = R_{p0}\left(1 - \sqrt[3]{1 - \rho_{\rm p}^{(l)} / \rho_{\rm p}^{(s)}}\right) \cdot \sqrt[3]{\rho_{\rm p}^{(s)} / \rho_{\rm p}^{(l)}}$.

Using these dependences at hand, we can obtain numerical estimates, for instance, for α-Al_2O_3 (corundum) particles. Analyzing available experimental data, the authors of [19] proposed the following dependence for α-Al_2O_3 density in liquid state: $\rho_p^{(l)}(T) = 3030 - 1.08 \cdot (T - 2327)$, where the dimension of density is kg/m^3, and the dimension of temperature T is K. For the density of α-Al_2O_3 in solid state, we adopt the value $\rho_p^{(s)} = 3960$ kg/m^3. Then, for the droplet of radius 50μm at the melting point T_{pm}=2327K we have: $\rho_p^{(l)}(2327) = 3030$ kg/m^3, $\Delta R_p \approx 4.7$ μm, $R_{p,end} \approx 54.7$ μm, $\Delta_p \approx 20.9$ μm, and $R_h \approx 33.8$ μm. For the same droplet overheated over the melting point by 400K, we similarly obtain: $\rho_p^{(l)}(2727) = 2598$ kg/m^3, $\Delta R_p \approx 7.5$ μm, $R_{p,end} \approx 57.5$ μm, $\Delta_p \approx 17.2$ μm, and $R_h \approx 40.3$ μm.

Hence, the characteristics of hollow spheres solidified from overheated molten particles are fully defined by the densities of the particle material in the solid and liquid state, and also by the overheating value of the droplets. Provided that the particle shell is impermeable, a spheroidal vacuum cavity is formed inside each particle.

In paper [17] technological capabilities of thermal plasma are shown when processing various agglomerated powders for the purpose of obtaining the hollow microspheres of metals, alloys and oxide ceramics. Despite great practical and scientific interest to such powders [1, 2, 12, 20-22], etc., so far there are no physical and theoretical bases allowing *a priori* to optimize the mode of plasma processing and for certain to predict diameter and thickness of shell of obtained hollow microspheres.

2. Analysis of basic processes

In the present paper as the initial particles are considered the agglomerated particles of decamicrometer sizes obtained by means of spray drying. Such particles are formed of large number of randomly packed ultra-fine particles (UFPs) and are characterized by open porosity. Their processing in plasma jet can lead to formation, both hollow microspheres, and microspheres with distributed gas inclusions. Implementation of this or that scenario depends on the intensity of interphase heat transfer "agglomerated particle - plasma" and agglomerate residence time in plasma jet. As it was noted in [17], irrespective of method of obtaining the agglomerated particles (spray drying, mechanoactivation in high-energy planetary-type mills, etc.) formation of hollow microspheres has much in common. This process consists of following stages (see figure 1):

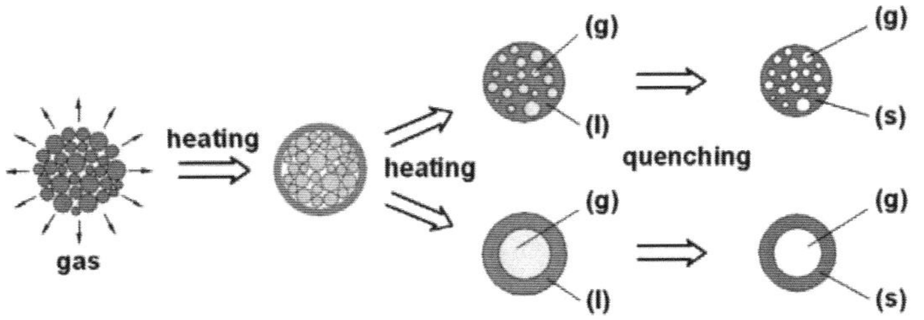

Figure 1 Phenomenological diagram of the stages succession of plasma processing of the agglomerated particle with hollow microsphere formation

1) heating the agglomerate until its surface reaches the melting point; at the same time, owing to the thermal expansion, there is the free outflow of gas from porous frame of particle in environment;

2) formation of stable surface film of the melt encapsulating remained gas in porous frame of particle; as it will be shown further, mass of the gas $m_{g0}^{(c)}$ which has remained in volume of agglomerated particle in many respects determines the final diameter of hollow microsphere and thickness of its shell;

3) further heating of particle and melting of UFPs directed from its surface to the center, their merge and involvement in growing layer of melt, owing to ideal wetting and minimization of surface

of the front of melting; at the same time there is gas displacement in solid porous frame by moving front or capture of gas inclusions by melt;

4) completion of melting of agglomerate, formation of either internal gas cavity and liquid spherical shell, or spherical droplet with distributed gas bubbles, further heating of droplet-balloon;

5) the subsequent quenching at which the hollow microspherical droplet solidifies from surface.

The purpose of the present work is obtaining the key similarity criteria allowing the formulation of the requirements to heat flux density $q^{(+)}$ (W/m^2) from plasma to particle, taking into account the properties of heterogeneous material of the agglomerate, which realization is necessary for formation of hollow microdroplet of controlled diameter and thickness of shell.

Besides, the criterion condition which realization provides preservation of size and thickness of hollow droplet shell as a result of obtained hollow droplet quenching is derived.

3. Similarity criteria characterizing the stages of HOSP production

At first, we will consider the processes of heating and melting of spherical agglomerated particle of size D_{p0} of several tens microns, formed of solid spherical UFPs with characteristic size $d_p << D_{p0}$.

Let's assume that P is its volume open porosity, and T_{p0} is initial temperature at the moment of input in plasma flow characterized by the heat flux density $q^{(+)}$. Let's take into consideration the effective properties of heterogeneous material: $(\rho c)_p^{(s)} = (1-P)(\rho c)_1^{(s)} + P(\rho c)_2$ is specific volumetric heat capacity, where $(\rho c)_1^{(s)}$ and $(\rho c)_2$ are specific heat capacities of UFPs material and gas; ρ and c are density and heat capacity; inferior indexes of i=1,2 correspond to UFPs material and gas, respectively; superscripts "s" and "l" characterize the values of parameters for solid and liquid state of material. We will define thermal conductivity $\lambda_p^{(s)}$ of particle porous material in accordance with [23], i.e. $\lambda_p^{(s)} = \lambda_1^{(s)} / [1 + P(\lambda_1^{(s)} / \lambda_2 - 1)]$. We consider that effective thermophysical properties $\overline{(\rho c)}_p^{(s)}$, $\overline{\lambda}_p^{(s)}$, etc. of porous material do not depend on temperature and are obtained by averaging in interval of temperatures [T_{p0}; T_{1m}], T_{1m} is melting point of UFPs; $\overline{a}_p^{(s)} = \overline{\lambda}_p^{(s)} / \overline{(\rho c)}_p^{(s)}$ is thermal diffusivity. Under $L_p(P) = (1-P)L_{1m}$ we understand efficient melting heat of heterogeneous material (L_{1m} is heat of melting of the UFPs material). Thus, we have evaluated all required thermophysical characteristics of the heterogeneous material. Those characteristics parametrically depend on the porosity P, which is the parameter to be set *a priori*.

Below, in figures, we will omit the upper line under variables. As a model ceramic material we will use practically interesting to many technologies α-Al_2O_3 which properties are rather well studied both for solid, and for liquid states. Let's consider that the gas filling pore space of agglomerates is air and ambient pressure is equal to atmosphere pressure.

3.1. Heating of agglomerated particle before melting

Apparently, under other equal conditions, final diameter D_p and thickness of shell Δ_p of hollow droplet will depend on the mass of gas $m_{g0}^{(c)}$, remained (encapsulated) in porous volume of agglomerate by the time t_f of formation of the surface film of melt, and also on the degree of the subsequent overheat of hollow droplet formed after complete melting of agglomerate. It is evident, that overheating of hollow droplet is limited by boiling point T_{1b} of UFPs material. On one hand, the value $m_{g0}^{(c)} < m_{g,\max} = \pi \rho_g (T_{p0}) D_{p0}^3 P / 6$, i.e. is limited above the initial mass of gas, and, on the other hand $m_{g0}^{(c)} \geq m_{g,\min} = \pi \rho_g (T_{1m}) D_{p0}^3 P / 6$ that answers the case when by the time t_f agglomerated particle is uniformly heated, and its temperature is equal to T_{1m}. The actual mass of gas depends on temperature distribution on particle radius at the time $t = t_{\mathrm{f}}$, i.e. $m_{\mathrm{g0}}^{(\mathrm{c})} = 4\pi P \int_0^{R_{\mathrm{p0}}} \rho_{\mathrm{g}}(T) r^2 \, \mathrm{d}r$, $R_{\mathrm{p0}} = D_{\mathrm{p0}}/2$. Therefore, value $m_{\mathrm{g0}}^{(\mathrm{c})}$ is defined by thickness $\delta(t_{\mathrm{f}}) = \sqrt{12\bar{a}_{\mathrm{p}}^{(\mathrm{s})} t_{\mathrm{f}}}$ of the thermal boundary layer in particle which at the moment $t_{\mathrm{a}} = R_{\mathrm{p0}}^2 / 12 a_{\mathrm{p}}^{(\mathrm{s})}$ is equal to R_{p0}, i.e. at $t = t_{\mathrm{a}}$ thermal disturbance will reach the particle center.

For consideration of the irregular stage of heating of agglomerated particle, that we are interested in, we will formulate the following boundary value problem:

$$\frac{\partial T}{\partial t} = \frac{a_{\mathrm{p}}^{(\mathrm{s})}}{r^2} \cdot \frac{\partial}{\partial r} r^2 \frac{\partial T}{\partial r}, \quad T(r,0) = T_{\mathrm{p0}}, \quad r \in [0; R_{\mathrm{p0}}]; \quad \left.\frac{\partial T}{\partial r}\right|_{r = R_{\mathrm{p0}} - 0} = \frac{q^{(+)}}{\bar{\lambda}_{\mathrm{p}}^{(\mathrm{s})}}, \ 0 < t \le t_{\mathrm{a}}.$$

Taking into consideration the dimensionless variables $\bar{r} = r / R_{\mathrm{p0}}$, $\mathrm{Fo} = \bar{a}_{\mathrm{p}}^{(\mathrm{s})} t / R_{\mathrm{p0}}^2$, $\vartheta^{(\mathrm{s})} = [T(\mathrm{Fo}, \bar{r}) - T_{\mathrm{p0}}]/[T_{1\mathrm{m}} - T_{\mathrm{p0}}]$, we will formulate the boundary value problem characterizing heating of solid agglomerated particle:

$$\frac{\partial \vartheta^{(\mathrm{s})}}{\partial \mathrm{Fo}} = \frac{1}{\bar{r}^2} \cdot \frac{\partial}{\partial \bar{r}} \bar{r}^2 \frac{\partial \vartheta^{(\mathrm{s})}}{\partial \bar{r}}; \ \vartheta^{(\mathrm{s})}(0, \bar{r}) = 0, \ \bar{r} \in [0; 1], \tag{1}$$

$$\left.\frac{\partial \vartheta^{(\mathrm{s})}}{\partial \bar{r}}\right|_{\bar{r} = +0} = 0, \ \left.\frac{\partial \vartheta^{(\mathrm{s})}}{\partial \bar{r}}\right|_{\bar{r} = 1 - 0} = \mathrm{Ki}_{\mathrm{p}}^{(\mathrm{s})}, \ 0 < \mathrm{Fo} \le \mathrm{Fo}_{\mathrm{a}}, \tag{2}$$

where $\mathrm{Ki}_{\mathrm{p}}^{(\mathrm{s})} = q^{(+)} R_{\mathrm{p0}} / \lambda_{\mathrm{p}}^{(\mathrm{s})} (T_{1\mathrm{m}} - T_{\mathrm{p0}})$ is Kirpichyev's similarity criterion [24]. By analogy with work [25], the approximate solution of the boundary value problem (1), (2) characterizing dynamics of surface temperature of spherical particle at the stage of its irregular heating ($\mathrm{Fo} \le \mathrm{Fo}_{\mathrm{a}} = 1/12$), looks like:

$$\vartheta^{(\mathrm{s})}(\mathrm{Fo}, 1) = \mathrm{Ki}_{\mathrm{p}}^{(\mathrm{s})} \cdot (\mathrm{Fo} + 2\sqrt{\mathrm{Fo}/3}), \tag{3}$$

Having used (3), we will find the value of time $\mathrm{Fo}_{\mathrm{f}} = (\sqrt{1 + 3/\mathrm{Ki}_{\mathrm{p}}^{(\mathrm{s})}} - 1)^2 / 3$ necessary for heating of particle surface to the melting point, solving the equation (3) at $\vartheta^{(\mathrm{s})}(\mathrm{Fo}_{\mathrm{f}}, 1) = 1$. Let's demand the realization of the condition $\mathrm{Fo}_{\mathrm{f}} / \mathrm{Fo}_{\bar{\delta}} = 1$, where $\mathrm{Fo}_{\bar{\delta}} = \bar{\delta}^2 / 12$ is an instant of time for which the dimensionless thickness of thermal boundary layer in agglomerated particle is equal to $\bar{\delta} = \delta / R_{p0}$, $0 < \bar{\delta} \le 1$. As a result we will obtain the condition in similarity criterion and dimensional forms

$$\mathrm{Ki}_{\mathrm{p}}^{(\mathrm{s})}(\bar{\delta}) = 3 / [(1 + \bar{\delta}/2)^2 - 1], \tag{4}$$

$$q^{(+)}(\bar{\delta}) = 3\bar{\lambda}_{\mathrm{p}}^{(\mathrm{s})} (T_{1\mathrm{m}} - T_{\mathrm{p0}}) / [(1 + \bar{\delta}/2)^2 - 1] R_{\mathrm{p0}}. \tag{5}$$

In figure 2*a,b* are presented the generalized dependences for Kirpichyev's similarity criterion $\mathrm{Ki}_{\mathrm{p}}^{(\mathrm{s})}$ and the relative mass of gas $\overline{m}_{\mathrm{g0}}^{(\mathrm{c})} = m_{\mathrm{g0}}^{(\mathrm{c})} / m_{\mathrm{g,max}}$, which has encapsulated in porous volume of agglomerated particle by the instant of time t_{f}, versus dimensionless thickness $\bar{\delta}$. The corresponding generalized correlation between the relative mass of the gas $\overline{m}_{\mathrm{g0}}^{(\mathrm{c})}$, which has remained in particle by the instant of time t_{f}, and criterion $\mathrm{Ki}_{\mathrm{p}}^{(\mathrm{s})}$ is presented in figure 2*c*.

As appears from the presented data, parameter $\overline{m}_{\mathrm{g0}}^{(\mathrm{c})}$, depending on intensity of particle heating, undergoes essential variation. It should be noted that, other things being equal, the relative mass of gas $\overline{m}_{\mathrm{g0}}^{(\mathrm{c})}$, encapsulated in volume of agglomerated particle in an instant of time t_{f}, does not depend on porosity P of agglomerate, and is defined by temperature distribution in it. But, as appears from (3), the heat flux density $q^{(+)}(\bar{\delta})$ demanded for ensuring the given thickness $\bar{\delta}$ of thermal boundary layer in an instant of time t_f, depends in direct proportion on effective thermal conductivity $\bar{\lambda}_p^{(s)}$ which, in turn, depends on porosity of agglomerate.

In accordance with [26], for random dense packing arrangements of spherical UFPs of the set size in the agglomerated particle their volume concentration 1-P ranges from 0.59 to 0.65. Therefore, when studying the influence of porosity P of agglomerates on dynamics of their heating and phase changes, as minimum value of porosity it is possible to assume the value P =0.35.

The values of the heat flux density $q^{(+)}$ are given in Figure 3, demanded for providing stated above conditions of heating of agglomerated particles α-Al_2O_3 of sizes D_{p0}=50, 100 and 150 μm at their porosity of P=0.35. We consider that in initial instant of time the porous volume of agglomerates is filled with air at temperature T_{p0}=300 K. It is clear that requirements to heat flux density $q^{(+)}$ significantly increase with decrease of size D_{p0} of agglomerated particle and mass of gas $\overline{m}_{g0}^{(c)}$.

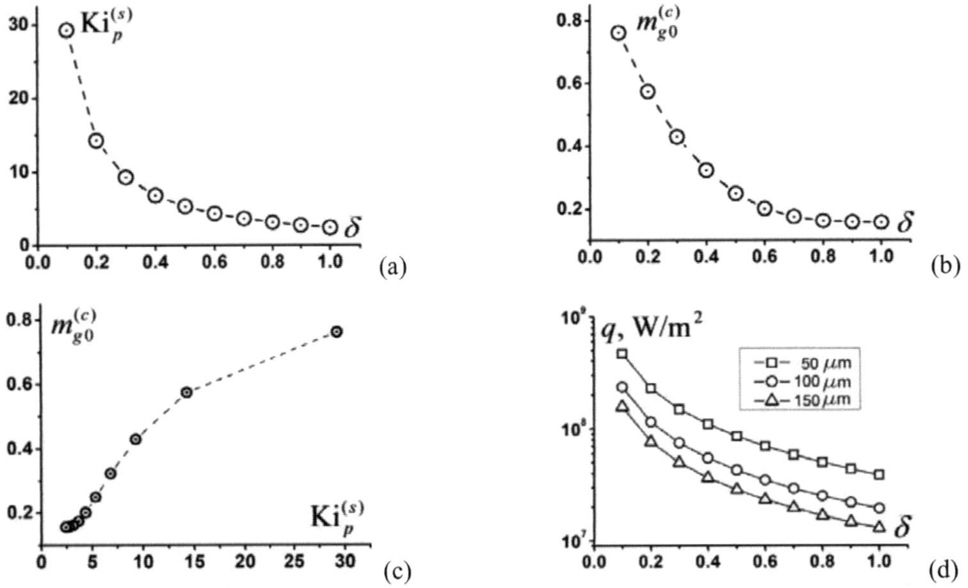

Figure 2 Generalized dependences of Kirpichyev's similarity criterion (a) and relative mass of encapsulated gas (b) vs. the relative thickness of thermal boundary layer in agglomerated particles; generalized dependence of relative mass of encapsulated gas vs. Kirpichyev's criterion (c); required values of the heat flux density when processing agglomerates α-Al_2O_3 of sizes D_{p0}=50, 100 and 150 μm (porosity P=0.35) vs. the relative thickness of thermal boundary layer in agglomerated particles.

3.2. Melting of agglomerate and hollow droplet formation

Let's make an estimation of the relative values of diameter $\bar{D}_{p,m} = D_{p,m} / D_{p0}$ and thickness of shell $\bar{\Delta}_{p,m} = \Delta_{p,m} / D_{p,m}$ of hollow microspheres, having assumed that agglomerates α-Al_2O_3, obtained at $t = t_f$, then are completely melted, and their temperature of T_p is equal to melting point of α-Al_2O_3 (T_{1m} =2327 K). According to [27], the equilibrium value of hollow droplet radius R_p can be defined from the equation

$$p_{g\infty}+\frac{2\sigma_{1m}^{(l)}}{R_p}\cdot\left(1+\frac{1}{1-\Delta_p/R_p}\right)-\frac{3m_{g0}^{(c)}RT_g}{4\pi R_p^3 M_g(T_g)}\left(1-\Delta_p/R_p\right)^{-3}=0,\ 1-\Delta_p/R_p=\sqrt[3]{1-3m_{p0}/4\pi\rho_p^{(l)}R_p^3}, \quad (6)$$

derived from the equality of initial agglomerated particle mass and mass of hollow droplet formed.

Here $m_{p0}=4\pi\rho_1^{(s)}(1-P)R_{p0}^3/3$ is the mass of UFPs in agglomerate, $p_{g\infty}$ is the pressure of environment, R=8.314·10^3 J/(kmole·K) is the universal gas constant, $M_g(T_g)$ is the relative molar mass of gas at temperature $T_g=T_p$. For efficient solution of the non-linear equation (6) the Newton method is used.

As the calculations showed, searching the solution of the equation (6) with relative error $\varepsilon=10^{-6}$ at given ambient pressure and temperature of encapsulated gas equal to temperature of the hollow droplet shell, usually requires 5-6 iterations if as an initial approximation to required diameter of hollow droplet the initial size of the agglomerated particle is assumed.

From the data presented in figure 3, it is visible that the mass $\overline{m}_{g0}^{(c)}$ of gas encapsulated in the particle at the moment $t=t_f$, considerably influences the final characteristics of hollow droplet. So, at temperatures T_{1m} their maximal diameter surpasses initial diameter of agglomerated particle by more than 40%. It should be noted that the condition (4) is necessary, but, generally speaking, not sufficient. For obtaining the hollow droplet it is necessary to melt agglomerate completely and if necessary to overheat it up to temperature T_{1b}. Let's find the time $\mathrm{Fo_m}$ of complete melting of agglomerate at given value $q^{(+)}$. Having written down for particle the equation of heat balance and having brought it to dimensionless form, we will obtain demanded condition in the similarity criterion form

$$\mathrm{Fo_m}=\frac{2}{3(1+\overline{D}_{p,m}^2)\mathrm{Ki}_p^{(s)}}\left\{1+\mathrm{Ste}_p^{(s)}+\overline{D}_{p,m}^3\cdot\frac{(\rho c)_1^{(l)}[1-(1-\overline{\Delta}_{p,m})^3]}{(\rho c)_p^{(s)}}\cdot\frac{1+\mathrm{Ki}_p^{(l)}/2}{1-T_{p0}/T_{1m}}\right\}, \quad (7)$$

where $\mathrm{Ste}_p^{(s)}=(1-P)\rho_1^{(s)}L_{1m}/(\rho c)_p^{(s)}(T_{1m}-T_{p0})$ is the Stefan similarity criterion of phase transition, $\mathrm{Ki}_{pb}^{(l)}=q^{(+)}\Delta_{p,b}/\lambda_1^{(l)}T_{1b}$ is the Kirpichyev's similarity criterion characterizing heating the liquid shell of formed hollow droplet in an instant of time $\mathrm{Fo_m}$, $\lambda_1^{(l)}$ is the effective thermal conductivity of melt; absolute values of diameter $D_{p,m}$ and thickness $\Delta_{p,m}$ of hollow droplet shell are found from (6).

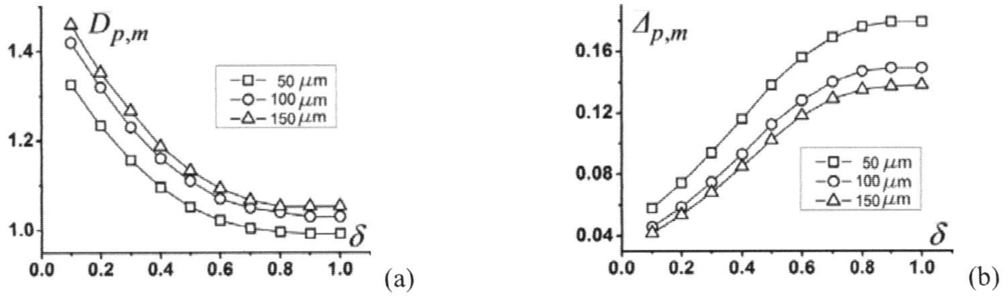

Figure 3 The relative diameter $\overline{D}_{p,m}=D_{p,m}/D_{p0}$ - (a), and the relative thickness of shell $\overline{\Delta}_{p,m}=\Delta_{p,m}/D_{p,m}$ - (b), of hollow droplets α-Al_2O_3, obtained when processing agglomerates of sizes D_{p0}=50, 100 and 150 μm (porosity P=0.35), at their heating to melting point (T_{1m}) vs. relative thickness $\overline{\delta}$ of thermal boundary layer at the instant of time of the first stage completion.

At deriving (7) it was supposed that temperature distribution in the cross-section of melted shell is quasi-stationary. Therefore, the temperature of its outer surface at the moment $t_{\mathrm{m}} = R_{\mathrm{p0}}^2 \, \mathrm{Fo_m} / \overline{a}_{\mathrm{p}}^{(\mathrm{s})}$ is equal to $T_{\mathrm{surf}}^{(\mathrm{l})}(t_{\mathrm{m}}) = T_{\mathrm{lm}}(1 + \mathrm{Ki}_{\mathrm{p}}^{(\mathrm{l})})$.

In figure 4, as an example, are provided, calculated according to (6), the value of time t_{m} for agglomerated particles α-Al_2O_3, characterized by porosity *P*=0.35, and generalized processing conditions given in figure 2*a*. Analyzing the data presented in figure 4, it is possible to conclude that by means of the thermal plasma generated by DC and RF plasma torches, at the corresponding choice of power, working gas, processing conditions and subsequent quenching of hollow microdroplets it is possible to obtain the hollow droplets with demanded characteristics.\

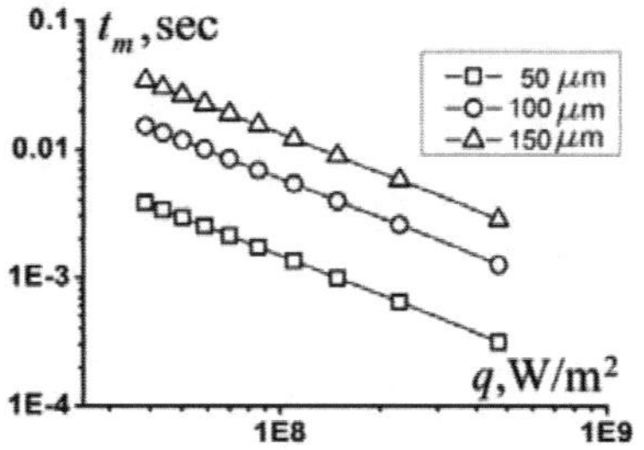

(a)

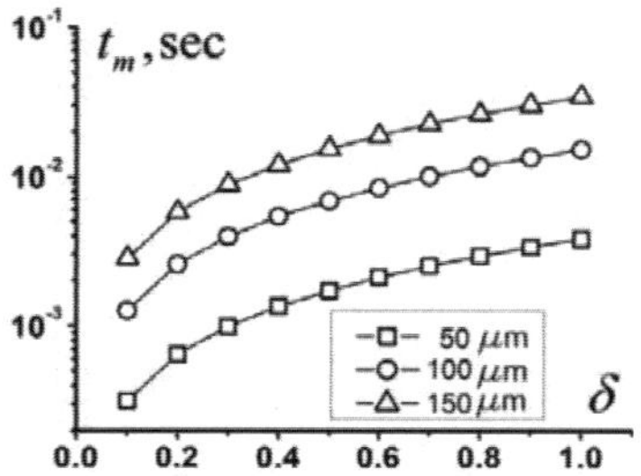

(b)

Figure 4 Dependences of time t_{m} of the complete melting of agglomerated particles α-Al_2O_3 (sizes D_{p0}=50, 100 and 150 μm, porosity *P*=0.35): (*a*) - vs. heat flux density $q^{(+)} = q^{(+)}(\overline{\delta})$ from plasma to particle required for completion of the first stage at relative thickness $\overline{\delta}$ of thermal boundary layer (see Figure 3); (*b*) – vs. relative thickness $\overline{\delta}$ of thermal boundary layer at the instant of time of the first stage completion.

At directed melting of agglomerated particle (from its surface to the center) the current surface of the melting front, generally speaking, is not spherical, and is piecewise both envelope surfaces of completely melted UFPs, and dividing solid and liquid phases in separate, not completely melted UFPs. Owing to merge of the neighbor liquid ultrafine particles randomly distributed in the layer adjacent to the melting front, and to their absorption by the front, space volume in the layer (with initial porosity of *P*), filled with gas, decreases. The characteristic time t_{j} of the complete coalescence of two liquid UFPs of diameter d_{p} corresponds to the closing-time of pore of the same diameter $t_{\mathrm{j}} = 2\mu_{\mathrm{p}} d_{\mathrm{p}} / 3\sigma_{\mathrm{p}}$, where μ_{p}, σ_{p} is dynamic viscosity and the surface intention of the melt. For the melt of alumina μ_{p}=0.0573 Pa·s, σ_{p}=0.69 N/m, at d_{p}=10^{-6} and $5 \cdot 10^{-6}$ m we will have, respectively, t_{j}= $5.5 \cdot 10^{-8}$ and $2.8 \cdot 10^{-7}$ s that is, at least, three-four orders less than the characteristic time of the complete melting of agglomerate and is one-two orders less than the characteristic time of passing of the melting front through one layer of UFPs. It leads to replacement of the gas having temperature of T_g~T_{lm}, by porous connected pinholes inside the particle. Depending on UFPs packing in agglomerate, gas is forced out as to the particle center, and crosswise, owing to gas pressure equalization in the porous framework before the melting front.

3.3. Overheating of hollow droplet to boiling point

Let's make an estimation of the relative values of the diameter $\overline{D}_{\mathrm{p,b}} = D_{\mathrm{p,b}} / D_{\mathrm{p0}}$, and also the thickness of shell $\overline{\varDelta}_{\mathrm{p,b}} = \varDelta_{\mathrm{p,b}} / D_{\mathrm{p,b}}$ of hollow microdroplet formed at the melting point and then overheated to the boiling points of α-Al_2O_3.

Absolute values of diameter $D_{\mathrm{p,b}}$ and thickness $\varDelta_{\mathrm{p,b}}$ of shell of hollow droplet are obtained from equation (6) at temperature of encapsulated gas $T_g = T_{1b}$.

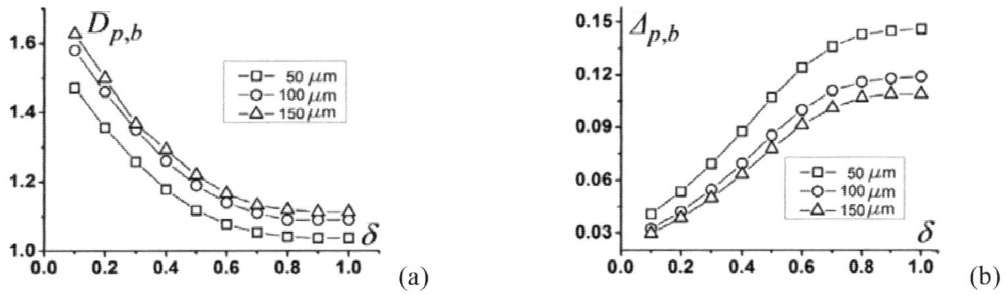

Figure 5 Relative diameter $\overline{D}_{\rm p,b} = D_{\rm p,b} / D_{\rm p0}$ - (a), and relative thickness of shell $\overline{\Delta}_{\rm p,b} = \Delta_{\rm p,b} / D_{\rm p,b}$ - (b), of hollow microdroplets α-Al_2O_3, obtained when processing the agglomerates of sizes D_{p0}=50, 100 and 150 μm (porosity P=0.35), at their heating to boiling point (T_{1b}) of α-Al_2O_3, vs. relative thickness $\overline{\delta}$ of thermal boundary layer at the instant of time of first stages completion.

From the data presented in the figure 5 it is visible that the mass $\overline{m}_{\rm g0}^{(c)}$ of gas encapsulated in particle at the moment $t = t_{\rm f}$, considerably influences the final characteristics of hollow droplet. So, at temperatures T_{1b} its maximal diameter surpasses initial diameter of agglomerated particle more, than for 60%.

3.4. Quenching of hollow droplet to produce HOSP

Let's assume that the formed hollow droplet having diameter $D_{\rm p,\mathit{end}}^{(l)}$ and shell thickness $\Delta_{\rm p,\mathit{end}}^{(l)}$, is evenly heated to temperature $T_{p,end}$, $T_{pm} < T_{p,end} \le T_{1b}$ and is exposed to cooling until achievement by its external surface of the melting point of UFPs material and to the subsequent solidification. Similar to the first stage of the irregular heating, in this case we need to investigate the problem of irregular cooling of liquid shell, i.e. to consider the following boundary value problem:

$$\frac{\partial T}{\partial t} = \frac{\overline{a}_{\rm p}^{(l)}}{r^2} \cdot \frac{\partial}{\partial r} r^2 \frac{\partial T}{\partial r}, \quad T(r,0) = T_{{\rm p},end}, r \in [R_{{\rm p},end}^{(l)} - \Delta_{{\rm p},end}^{(l)}; R_{{\rm p},end}^{(l)}]; \quad \left.\frac{\partial T}{\partial r}\right|_{r=R_{{\rm p},end}^{(l)}-0} = -\frac{q^{(-)}}{\overline{\lambda}_{\rm p}^{(l)}}, \ 0 < t \le t_{\Delta}.$$

Taking into consideration the dimensionless variables

$$\overline{r} = [r - R_{{\rm p},end}^{(l)} + \Delta_{{\rm p},end}^{(l)}] / \Delta_{{\rm p},end}^{(l)}, \quad r \in [R_{{\rm p},end}^{(l)} - \Delta_{{\rm p},end}^{(l)}; R_{{\rm p},end}^{(l)}],$$

$$\mathrm{Fo} = \overline{a}_{\rm p}^{(l)} t / \Delta_{{\rm p},end}^{(l)2}, \quad \vartheta^{(l)}(\mathrm{Fo}, \overline{r}) = [T_{{\rm p},end} - T(\mathrm{Fo}, \overline{r})] / [T_{{\rm p},end} - T_{1\rm m}],$$

we can formulate the boundary value problem characterizing the initial stage of cooling under formation of final hollow solid microsphere:

$$\frac{\partial \vartheta^{(l)}}{\partial \mathrm{Fo}} = \frac{1}{\overline{r}^2} \cdot \frac{\partial}{\partial \overline{r}} \overline{r}^2 \frac{\partial \vartheta^{(l)}}{\partial \overline{r}}; \ \vartheta^{(l)}(0, \overline{r}) = 0, \ \overline{r} \in [0;1]; \ \left.\frac{\partial \vartheta^{(l)}}{\partial \overline{r}}\right|_{\overline{r}=+0} = 0, \ \left.\frac{\partial \vartheta^{(l)}}{\partial \overline{r}}\right|_{\overline{r}=1-0} = \mathrm{Ki}_{{\rm p},end}^{(l)}, \ 0 < \mathrm{Fo} \le \mathrm{Fo}_{\Delta}, \quad (8)$$

where $\mathrm{Ki}_{{\rm p},end}^{(l)} = q^{(-)} \Delta_{{\rm p},end}^{(l)} / \overline{\lambda}_{\rm p}^{(l)} (T_{{\rm p},end} - T_{1\rm m})$ is the Kirpichyev's similarity criterion, characterizing cooling of liquid shell.

The approximate solution of the boundary value problem (8) characterizing dynamics of the external surface temperature of shell at the stage of its irregular cooling ($\mathrm{Fo} \le \mathrm{Fo}_{\Delta} = 1/12$), looks like:

$$\vartheta^{(l)}(\mathrm{Fo}, 1) = \mathrm{Ki}_{{\rm p},end}^{(l)} \cdot (\mathrm{Fo} + 2\sqrt{\mathrm{Fo}/3}), \quad (9)$$

Having used (9), we will find the value of time $\mathrm{Fo}_{\rm s} = (\sqrt{1 + 3/\mathrm{Ki}_{{\rm p},end}^{(l)}} - 1)^2 / 3$ necessary for the external shell surface cooling to the melting point, solving the equation (9) at $\vartheta^{(l)}(\mathrm{Fo},1)=1$. Let's demand the realization of the condition $\mathrm{Fo_s}/\mathrm{Fo}_{\Delta}\leq 1$, where $\mathrm{Fo}_{\Delta}=1/12$ is an instant of non-dimensional time for which the dimensionless thickness of thermal boundary layer in shell is equal to $\Delta^{(l)}_{\mathrm{p},end}$. As a result we will obtain the condition in similarity criterion and dimensional forms

$$\mathrm{Ki}^{(l)}_{\mathrm{p},end}\geq 2.4\,,\; q^{(-)}\geq 2.4\bar{\lambda}^{(l)}_{\mathrm{p}}(T_{\mathrm{p},end}-T_{\mathrm{1m}})/\Delta^{(l)}_{\mathrm{p},end}\,. \quad (10)$$

Realization of condition (10) means that at the time of the beginning of hollow drops solidification temperature of encapsulated gas does not undergo any changes, and, therefore, the size of solidified hollow particle will also not undergo any changes. At less intensive surface cooling of hollow droplet at the time of achievement of the melting point by external surface of the shell temperature in its cross-section will be distributed more uniformly. It will lead to cooling of encapsulated gas, and, therefore, to decrease in the size of the hollow droplet and increase in thickness of its shell. So, energy consumptions on overheat of hollow droplet will not give the necessary effect as at low rate of cooling the hollow droplet restores its parameters (diameter and thickness of shell) which it had at the moment of formation at temperature close to melting point. As a result of its solidification the hollow microsphere with practically same parameters will be obtained.

Estimates show that characteristic time of heating/cooling of air cavity with a radius ≤ 50 μm at temperature of shell equal to the melting point of alumina is less by at least an order of characteristic time of thermal relaxation of 5 μm thick shell. Therefore, at heating/cooling of hollow droplet temperature of encapsulated gas "follows up" the temperature of shell of hollow microdroplet.

Thus, the final size of hollow microsphere depends not only on intensity of heating of initial agglomerate, its melting and overheating the formed hollow droplet up to the boiling point of UFPs material, but also on the rate of cooling (quenching) of the final hollow droplet formed in plasma flow.

If solidification of hollow drop began at radius $R_{\mathrm{p},end}$ and shells thickness of $\Delta_{\mathrm{p},end}$, then after solidification and cooling up to temperature $T_{\mathrm{g}0}$ of obtained hollow microsphere the internal gas pressure is defined according to dependence

$$p_{\mathrm{i},end}=\frac{3m^{(\mathrm{c})}_{\mathrm{g}0}RT_{\mathrm{g}0}}{4\pi(R_{\mathrm{p},end}-\Delta_{\mathrm{p},end})^3M_{\mathrm{g}}(T_{\mathrm{g}0})},$$

which is obtained from (6). Therefore, if hollow droplet of alumina was heated to boiling point and as a result of fast quenching kept the size, after cooling up to the temperature $T_{\mathrm{g}0}$=300 K then internal gas pressure will be $p_{\mathrm{i},end}\approx 0.85\cdot 10^4$ Pa. The estimate is obtained at following conditions: $R_{\mathrm{p}0}$=50 μm, P=0.35, $p_{\mathrm{g}\infty}=10^5$ Pa, $q^{(+)}=3\cdot 10^8$ W/m^2, $\bar{\delta}$=0.1, $q^{(-)}=6.9\cdot 10^8$ W/m^2, $m^{(\mathrm{c})}_{\mathrm{g}0}=1.85\cdot 10^{-13}$ kg, $M_{\mathrm{g}}(T_{\mathrm{g}0})$ =28.97 kg/kmole, $R_{\mathrm{p},end}$=76.5 μm, $\Delta_{\mathrm{p},end}$=6.6 μm. As it can be seen, at the given conditions of obtaining the hollow microspheres with impermeable shell, inner gas pressure is lower by at least one order than ambient pressure in which agglomerated particle was treated.

4. Conclusions

1. The engineering physical model which has allowed for the first time to formulate key similarity criteria, responsible for formation of hollow ceramic particles at plasma processing of the agglomerated powders obtained at spray drying is developed.

2. On the example of alumina it is shown that, other things being equal, varying the heat flux density from plasma to the particles of agglomerated powder, and also varying the rate of quenching of the produced hollow droplets, it is possible to control diameter and thickness of shell of obtained hollow microspheres in the wide range.

3. The obtained results allow to formulate requirements to choice of plasma processing mode of any agglomerated powders with decamicron particles size for production of hollow microspherical powders with given characteristics.

Acknowledgement. This work was supported in part in the framework of Interdisciplinary Integration Projects No. 2 and 98 of Siberian Branch of the Russian Academy of Sciences for 2012-2014.

References

[1] Karoly Z and Szepvolgyi J 2003 *Powder Technology*.**132** 211
[2] Roy P, Bertrand G and Coddet C 2003 *Thermal Spray 2003: Advancing the Science & Applying the Technology* ed C Moreau and B Marple, Published by ASM International, Materials Park, Ohio, USA 1617
[3] Komeya K, Tatami K J, Hotta M, Zhang C, Meguro T and Cheng Y-B 2003 *Key Engineering Materials* **237** 87
[4] Bica I 2005 *Plasma Chemistry and Plasma Processing* **25** 121
[5] Sodeyama K and Sakka Y 2005 *Key Engineering Materials* **280-283** 701
[6] Maeda H, Miura N, Kasuga T and Nogami M 2005 *Key Engineering Materials* **284-286** 301
[7] Han Y S, Tarutani Y, Fuji M and Takahashi M 2006 *Advanced Materials Research* **11-12** 673
[8] Hadiko G, Han Y S, Fuji M and Takahashi M 2006 *Advanced Materials Research* **11-12** 677
[9] Itatani K, Itokazu H, Umeda T, Davies I J, Musha Y, Mizutani K and Koda S 2006 *Key Engineering Materials* **309-311** 129
[10] Lee C W, Kim S G and Lee J S 2006 *Key Engineering Materials* **317-318** 219
[11] Nakamura R, Tokozakura D, Lee J-G, Mori H and Nakajima H 2007 *Materials Science Forum* **561-565** 1703
[12] Xu J L, Khor K A and Kumar R 2007 *Thermal Spray 2007: Global Coating Solutions* ed B R Marple, M M Hyland, Y-C Lau, C-J Li, R S Lima and G Montavon, Published by ASM International®, Materials Park, Ohio, USA pp. 890-893
[13] Liu N, Cao H, Chen C and Zhang W 2007 *Solid State Phenomena* **121-123** 219
[14] Wu X F, Chen Y F, Li Q Y and Wei L Q 2007 *Solid State Phenomena* **121-123** 187
[15] Solonenko O P, Gulyaev I P and Smirnov A V 2008 *Technical Physics Letters* **34** 22
[16] Márquez F, Campo T, Cotto M, Polanco R, Roque R, Fierro P, Sanz J M, Elizalde E and Morant C 2011 *Soft Nanoscience Letters* **1** 25
[17] Solonenko O P, Gulyaev I P and Smirnov A V 2011 *Journal of Therm. Sci. and Techn.* **6(2)** 219
[18] Zhukov M F and Solonenko O P 1990 *High-Temperature Dusted Jets in the Powder Materials Processing*, ed V E. Nakoryakov (Novosibirsk: Institute of Thermophysics, Siberian Branch of USSR Academy of Sciences) 516 p (in Russian)
[19] Maurakh M A and Mitin B S 1979 *Liquid Refractory Oxides* (Moscow: Metallurgy Publishing) 288 p (in Russian)
[20] Hurevich V, Pawlowski L and Smurov I 2002 *Proc. of Intern. Thermal Spray Conf.* March 4-6, 2002, Essen, Germany p 32
[21] Bertrand G, Bertrand P, Roy P, Rio C and Mevrel R 2008 *Surface & Coatings Technology* **202** 1994
[22] Kumar A and Gu S 2012 *Intern. Journal of Heat and Fluid Flow* **37** 189
[23] *Physico-Chemical Processes of Materials Treatment by Concentrated Energy Fluxes* ed A A Uglov (Moscow: Nauka Publishing) 1989 268 p (in Russia)
[24] Lykov A V 1978 *Heat Transfer (Handbook)* (Moscow: Energiya Publishing) 480 p (in Russian)
[25] Deineko V V 1992 *Methods of Approximate Solution of Ordinary Differential Equations* (Novosibirsk: Publisher) 240 p. (in Russian)
[26] Ur'evb N B and Potanin A A 1992 *Fluidity of Suspensions and Powders* (Moscow: Khimiya Publishing) 256 p (in Russian)
[27] Solonenko O P, Smirnov A V and Gulyaev I P 2008 *AIP Conference Proceedings* **982** 561

Development of super-hydrophobic PTFE and PET surfaces by means of plasma processes

S Zanini[1], R Barni[1], R Della Pergola[2], C Riccardi[1]

[1] *Dipartimento di Fisica G. Occhialini, Università degli Studi di Milano-Bicocca, piazza della Scienza 3, I-20126 Milano, Italy*
[2] *Dipartimento di Scienze dell'Ambiente e del Territorio e di Scienze della Terra, Università degli Studi di Milano-Bicocca, piazza della Scienza 1, I-20126 Milano, Italy*

Abstract. In this work, Poly(tetrafluoroethylene) and Poly(ethylene terephtalate) substrates were modified by means of plasma techniques for the creation of super-hydrophobic surfaces. Both the materials were etched with an O_2 plasma, thus increasing their surface roughness which was investigated by means of Atomic Force Microscopy analysis. Plasma etching of PTFE surfaces under appropriate conditions results in the creation of super-hydrophobic surfaces, as assessed by measurements of dynamic contact angles and sliding angles. Chemical modifications of the PTFE surfaces was investigated with Attenuated Total Reflectance Fourier Transform Infrared spectroscopy and X-ray Photoelectron Spectroscopy analysis. The realization of super-hydrophobic PET surfaces needs the deposition of a hydrophobic top coating, which was realized through an hexamethyldisiloxane (HMDSO) plasma. The thickness of this top layer was varied by changing the plasma deposition time and the effects on the hydrophobic performances of the modified PET were investigated. Micro-nano structures created by plasma on PTFE and PET surfaces were characterized and correlated with the wettability.

1. Introduction

The wettability is usually determined by measuring the contact angle (CA) of a water droplet on a solid surface. When the CA of water is larger than 150° and the sliding angle of the drop is lower than 10°, the surface is called super-hydrophobic. Due to their ample variety of possible applications, super-hydrophobic surfaces have gained a lot of interest in the past few decades. Some practical applications of super-hydrophobic surfaces include biomedical devices, liquid separation, drag reduction, anti-icing, self-cleaning, anti-fouling [1]. In general, in order to create a super-hydrophobic surface, two factors are needed. First, the surface should have low surface energy (it should be initially hydrophobic). Second, the surface should exhibit micro- or nano-sized roughness, as well as hierarchical structure. Roughness combined with hydrophobicity often results in air pockets being trapped between the solid and liquid (composite solid–liquid–air interface), thus reducing the contact area between the surface and the water drop. This composite interface is needed for super-hydrophobicity.

In the past years, many approaches and techniques have been used to develop super-hydrophobic surfaces. For example, literature reports the creation of such surfaces by phase separation, electrochemical deposition, crystallization control, chemical vapor deposition, wet chemical reaction, layer-by-layer deposition, spin coating, sol–gel processing, lithography, electrospinning [1]. Plasma-based techniques are one of the most promising approach to the development of super-hydrophobic surfaces on polymeric substrates [2-4]. In particular, plasma technology is of great interest for the creation of super-hydrophobic surfaces because plasma processing can be used for the modification of both the chemical and the morphological characteristics of a polymeric surface. For example, low surface energy coatings can be deposited by means of plasma polymerization [5]. On the other hand, plasma etching (which is the process of selective removal of materials from a surface by reactive plasma radicals) increases the surface roughness [3,6-8].

In general, etching can be used as pretreatment before coating the surface with a low surface energy layer, or it can be used to generate roughness on a low surface energy material. In this work, O_2 plasma etching was used to create super-hydrophobic Poly(tetrafluoroethylene) (PTFE) surfaces. Chemical modifications of the PTFE surfaces were investigated with Attenuated Total Reflectance Fourier Transform Infrared (ATR/FTIR) spectroscopy and X-ray Photoelectron Spectroscopy (XPS) analysis. Instead, super-hydrophobic Poly(ethylene terephtalate) (PET) surfaces were realized by means of an O_2 plasma etching followed by the deposition of an hydrophobic top coating, which was realized through a HMDSO plasma. The thickness of this top layer was varied by changing the plasma deposition time and the effects on the hydrophobic performances of the modified PET were investigated. Micro-nano structures created by plasma on PTFE and PET surfaces were characterized by means of Atomic Force Microscopy (AFM) analysis and correlated with the wettability.

2. Experimental

2.1 Plasma treatments

The O_2 plasma was produced inside a cylindrical stainless steel vacuum chamber (diameter 30 cm) with a symmetric parallel plate configuration (two parallel aluminum dishes, diameter 20 cm). The PTFE sheets (1 cm^2, thickness= 2 mm) and the PET films (thickness= 36 μm) were positioned on the centre of the lower grounded electrode. The upper electrode is externally connected, through a semi-automated matching network, to a 13.56 MHz RF power supplier (Cesar 133, Dressler) which provides a RF voltage with respect to the lower electrode and to the grounded chamber. O_2 was uniformly distributed in the reactor by the upper showerhead electrode (with pinholes diameter of 2 mm). Before operating the discharge the device is evacuated to 10^{-3} Pa by means of a rotary pump, combined with a turbo-molecular pump. During the plasma treatment, the chamber is evacuated only by means of the rotary pump.

Plasma treatments were performed at a fixed pressure of 9 Pa. The RF power input was varied between 20 and 300 W, while the duration of the plasma treatment was varied between 2 minutes and 40 minutes. HMDSO plasma polymerization was performed in the same reactor, only on PET films. Only deposition time was varied (between 2 and 20 minutes), while the other relevant plasma parameters were kept constant (HMDSO flow rate= 0.6 sccm, HMDSO pressure= 3 Pa, power input = 30 W).

2.2 Samples characterization

Contact angles (CAs) were measured with 3 μl deionized water droplet on a Dataphysics OCA 20 (Dataphysics) instrument at room temperature. All the contact angles were determined by averaging the values obtained at 5–6 different points on each sample surface. Roll-off angles were determined with 10 μl and 50 μl water droplets, using a dedicated apparatus equipped with a goniometer.

Chemical modifications owing to the O_2 plasma treatment and HMDSO film structure were determined by means of a Fourier transform infrared (FT-IR) spectrometer (Nicolet iS10, Thermo Scientific) equipped with a ATR sampling accessory (Smart iTR). For each spectrum 32 scans, with a spectral resolution of 2 cm^{-1}, were recorded.

X-ray photoelectron spectra were recorded using a 5500-PHI (Physical Electronics) apparatus, with a monochromatic Al-anode K_α source and an electron take-off angle of 45°. The analyzed circular area had a diameter of 0.8 mm. The pressure in the chamber was around 10^{-6} Pa. The spectrometer was calibrated by using the Ag 3d5/2 peak and the resulting energetic resolution was 0.46 eV.

AFM measurements were carried out using a Solver P47-PRO (NT-MDT, Moscow, Russia), in semi-contact (tapping) mode on dry samples using HA_NC (High Accuracy Non Contact) silicon tips

(NT-MDT) of typical spring constant 5.8 ± 20% N/m and resonance frequency 200 ± 10% kHz. Square images (5×5 μm^2) were collected at a typical frequency of 1.5 Hz and with 256 points per line.

3. Results and discussion

3.1 Creation of super-hydrophobic PTFE surfaces

Chemical characterization of O_2 plasma treated PTFE surfaces was performed by means of Attenuated Total Reflectance Fourier Transform Infrared (ATR/FTIR) spectroscopy and X-ray Photoelectron spectroscopy (XPS). Inset in Figure 1a shows the ATR/FTIR spectra of the untreated and of a plasma treated PTFE (power input= 200 W; O_2 pressure = 9 Pa; treatment time = 10 minutes). The spectrum of the untreated PTFE displays two main bands, at 1150 cm^{-1} and 1204 cm^{-1}, which can be attributed respectively to the asymmetric and the symmetric stretching of the CF_2 groups. The spectrum of the plasma treated sample displays an additional band at 3000-3700 cm^{-1}, typical of the stretching of OH groups. A semi-quantitative evaluation of the extent of the plasma functionalization has been performed by integration of the OH and CF_2 bands, followed by the calculation of the OH/CF_2 band ratio. Results are shown in Figure 1a and Figure 1b.

By increasing the power input from 20 W to 200 W (Figure 1a), the OH/CF_2 band ratio increases, while a further increase to 300 W leads to a decrease in the insertion of OH groups. The influence of the treatment time on the OH functionalization is displayed in Figure 1b. It can be noticed that the plasma treatment at 200 W always leads to a higher insertion of OH groups on the PTFE surface and that, for both the power inputs, the OH/CF_2 ratio increases with the treatment time. These trends are in agreement with literature results: by increasing the power input plasma etching becomes prevailing over surface functionalization [3].

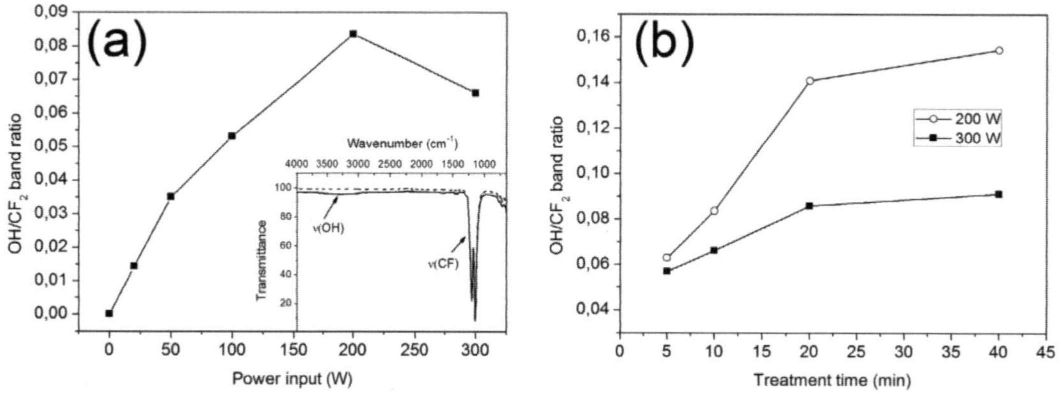

Figure 1. *(a)* Variation of the OH/CF_2 ATR/FTIR band ratio with the power input. *Inset* - ATR/FTIR spectra of the untreated (dash line) and the plasma treated (full black line) PTFE (power input= 200W; O_2 pressure= 9 Pa; treatment time= 10 minutes). *(b)* Variation of the OH/CF_2 ATR/FTIR band ratio with the treatment time for samples prepared with power inputs of 200 W and 300 W.

In order to better investigate the chemical changes owing to the plasma treatments at different power levels, we performed XPS analyses on sample treated between 20 W and 200 W. Results are displayed in Figure 2. The oxygen surface concentration (expressed as O/C ratio) is higher for samples prepared at power input ≤50 W. XPS analysis reveals also an increase in the oxygen content of the plasma treated samples with the ageing time. This post-plasma oxidation is a well documented phenomenon, which derives from the reaction between remaining radicals on the plasma treated polymeric material and atmospheric oxygen [9]. The trend in the oxygen surface concentration detected by XPS is not in agreement with that of the OH/CF_2 ratio obtained from the ATR/FTIR spectra. This result can be easily explained taking into account the different penetration depth of the two techniques. While XPS analysis probes polymeric material for a depth of few nanometers, the penetration of ATR/FTIR analysis is of the order of few microns. Taken together, XPS and ATR/FTIR results indicate that for power levels ≤50 W there is a higher degree of OH functionalization of the outermost layers of polymeric chains, but a low penetration of the OH functionalization into the PTFE bulk. By increasing the power level, the penetration of the OH functionalization increases too, while the degree of OH functionalization of the polymeric chains near the surface decreases.

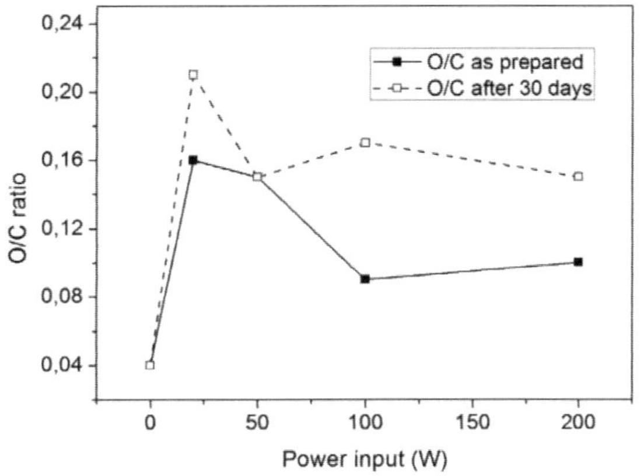

Figure 2. O/C ratio, detected by XPS, on samples prepared at different power inputs.

Morphological characterization of the plasma treated PTFE samples was performed by means of AFM analysis. Figure 3 displays the trend of the root mean square (rms) of the heights (for 5 × 5 μm^2 AFM images) of the surfaces of PTFE samples treated at different power inputs. The values which are reported are the average of the r.m.s. of at least five images, acquired in different zones of the samples. Plasma treatments at low power inputs (20 W, 50 W) decrease the surface roughness of PTFE. Instead, treatments at higher power inputs (100 W, 200 W, 300 W) result in the formation of bigger globular structures on the PTFE surface, with a strong increase of the surface roughness. The formation of the globular structures observed for power levels ≥100 W is the consequence of a differential etching of the PTFE surface. This is a common phenomenon, which has been observed also for other polymeric materials. Some authors have hypothesized the deposition of etch inhibitors (unetchable material) to be the most important mechanism which can be related to the formation of the surface micro/nano structures [4,8]. In our case, XPS analysis gives important information with regard to the changes in the surface chemistry that could be connected with the roughness formation. For very low power input (20 W), no aluminum was detected on the surface of the plasma treated PTFE sample (Figure 3). Instead, all the samples prepared at power input ≥ 50 W are contaminated by aluminum. Since both the electrodes of our plasma reactor are made of aluminum, it can be hypothesized that the energetic and intense ion bombardment releases in the gas phase Al particles. In particular the upper RF electrode gets heated when high powers and long treatment times are used. Due to the RF sheath formation, ion bombardment happens causing some sputtering. Nonvolatile Al particles will redeposit on the nearby surface, including the PTFE surface. There, they can act as micromasks, causing local differential etching rates and promoting the development of micro/nano structures. The increasing amount of aluminum with the power input can be related to the increasing surface roughness. A more quantitative analysis, for instance by changing the electrode materials, although interesting, is beyond the scope of the paper.

Figure 4a shows the variation of the advancing and the receding contact angles with the RF power input, keeping constant the other relevant plasma parameters (O_2 pressure = 9 Pa; treatment time = 10 minutes).

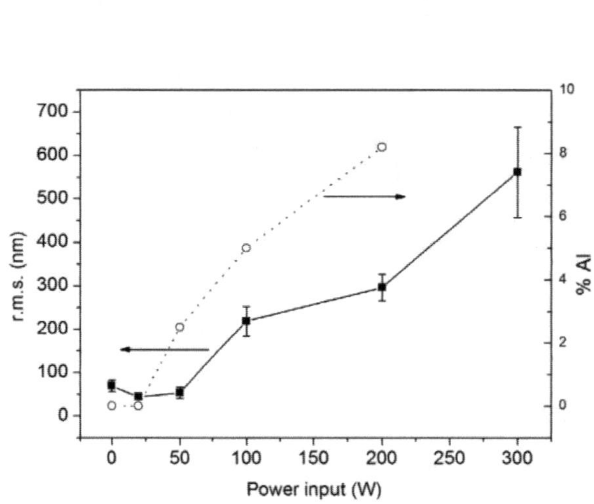

Figure 3. Variation of the root mean square (rms) of the heights of the surfaces (for 5 μm x 5 μm AFM images) with the power input and Al percentage detected by XPS.

The contact angles were measured immediately after the oxygen plasma treatment, and the values for power input= 0 refer to the untreated PTFE. It can be noticed that the advancing contact angle remains substantially unchanged up to 50 W, then it increases with the power input. Instead, the receding contact angle strongly decreases (from 70° to a value around 20°) for power input ≤100 W, then it gradually increases. When the plasma treatment was performed at 300 W, a super-hydrophobic PTFE surface was obtained. This trend with the RF power input is in agreement with previous literature results [3]. ATR/FTIR data do not explain these results, since the measured OH/CF_2 ratio increases with the power input and the water contact angle should be expected to decrease. Instead, the trend of contact angle is better in agreement with the XPS analysis, which probes a lower depth into the surface and displays a decrease of the oxygen atomic concentration by increasing the power input. Moreover, the trend of the advancing and receding contact angle can be explained taking into account the increase in surface roughness which occurs by increasing the power input. In fact, surface roughness strongly influences the wettability, and contact angles measured on a rough surface can be interpreted by means of two main models. The Wenzel model predicts that the water penetrates the holes on the surface and therefore the contact area between water and solid surface increases with roughness. The contact angle on the rough surface can be calculated by means of the equation:

$$\cos \theta^{W} = r_s \cos \theta_e$$

where θ^W is the contact angle measured on the rough surface, θ_e is the equilibrium contact angle (measured on the flat solid surface) and r_s is the surface roughness (the ratio of the actual surface area to the projected area of the surface). Since r_s is always >1, for θ_e>90° (as for our PTFE surfaces) the model predicts an increase of contact angle by increasing the surface roughness.

Instead, Cassie and Baxter proposed a model in which the liquid does not fill the grooves of a rough solid and sits on a composite surface made of solid and air [24]. They suggested that:

$$\cos \theta^{CB} = f_1 \cos \theta_e - f_2$$

where θ^{CB} is the contact angle measured on the composite rough surface, θ_e is the equilibrium contact angle (measured on the flat solid surface), f_1 is the surface area of the liquid in contact with the solid (in our case PTFE) divided by the projected area and f_2 is the surface area of the liquid in contact with air divided by the projected area ($f_1 + f_2 = 1$).

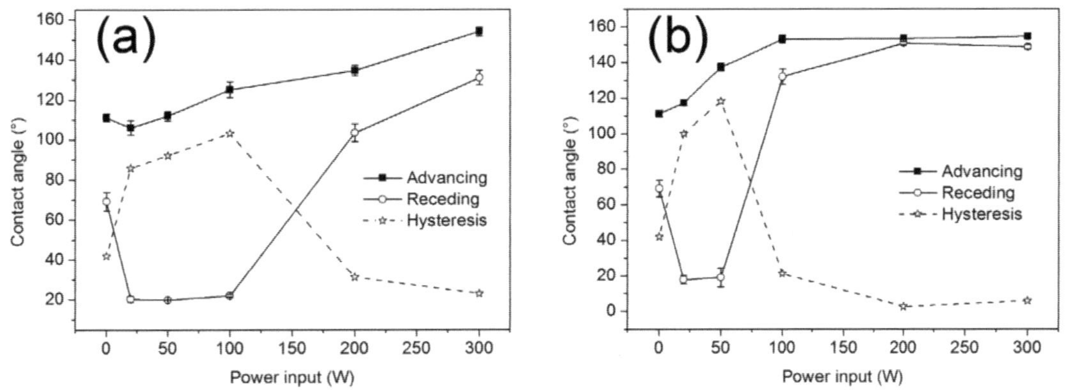

Figure 4. Variation of the advancing and receding contact angles with the power input: *(a)* angles measured immediately after the treatment; *(b)* angles measured after 30 days.

On the basis of the chemical and morphological characterization already discussed, the trend of contact angle displayed in Figure 4a can be interpreted in terms of a competition between surface functionalization (insertion of polar oxygenated groups) and chemical etching of the PTFE. At low RF power inputs the surface functionalization prevails, and the wettability increases. By increasing the power input, etching of PTFE surface becomes prevailing (Figure 3) and wettability decreases.

Figure 4b shows the contact angles measured 30 days after the plasma treatment, as a function of the power input. In contrast with XPS analysis results (which showed an increase of the surface oxygen content after 30 days), the advancing contact angles are greater than those measured immediately after treatment and, for power inputs ≥100 W, they are also greater than 150°, thus fulfilling the first condition for super-hydrophobic surfaces. The receding contact angles don't change with ageing time for power input ≤50 W, while they increase for power inputs ≥100 W. For power inputs of 200 W and 300 W the hysteresis angle significantly decreases with ageing time. This leads to super-hydrophobic surfaces, with very low roll-off angle (below 5°). Particularly interesting is the ageing behavior of the sample prepared at 100 W, with the receding contact angle that increases from 30° to over 130°. Noteworthy, receding contact angles measured after 30 days display a trend which strongly resemble that of the surface roughness of the samples, with a strong increase for power input higher than 50 W. The contact angle increase detected on aged samples can be easily explained in terms of surface adaptation. As a consequence of the reptation motion (that move some of the polymer chains from the surface to the bulk) and of chain rotation, the density of hydroxyl groups exposed to air decreases with the ageing time [9].

Figure 5 shows the variation of the advancing and the receding contact angles with the treatment time, for samples prepared at 200 W.

Full lines, which report contact angles measured immediately after the oxygen plasma treatment, display a maximum of hydrophobicity for 20 minutes of duration. Hysteresis angles are quite large, always greater than 20°, as a consequence of the introduction of some hydrophilic OH groups which was detected by ATR/FTIR spectroscopy (data not shown). The OH/CF_2 band ratio of these sample increases with the treatment time. Since it is known from literature that also surface roughness increases with treatment time [10], the trends displayed by full lines in Figure 5 (with a maximum in hydrophobicity at intermediate treatment time) can be interpreted in terms of competition between these two effects of the oxygen plasma treatment, with the highest contact angles which are obtained by reaching an optimal compromise.

Both advancing and receding contact angles increase with the ageing time, and hysteresis decreases. All the aged samples were super-hydrophobic, with roll-off angle lower than 5°.

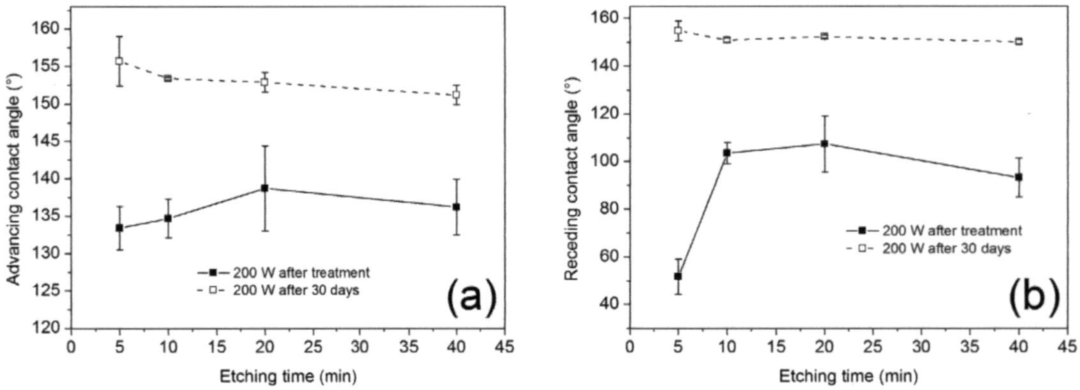

Figure 5. Variation of the advancing and receding contact angles with the treatment time: *(a)* advancing contact angle after treatment (full line) and after 30 days (dashed line); *(b)* receding contact angle after treatment (full line) and after 30 days (dashed line). Samples were prepared at a power level of 200 W.

3.2 Creation of super-hydrophobic PET surfaces

Plasma etched PET films were super-hydrophilic (contact angle <5°), and the realization of a super-hydrophobic PET surfaces needed the deposition of a hydrophobic top coating, which was realized through an HMDSO plasma [5]. The composition of this organosilicon coating, measured by XPS analysis and expressed as atomic percentage of the elements, was: C= 60%; O= 14%; Si= 26%. Figure 6 shows the variation of the advancing and the receding contact angles measured on PET films with the O_2 etching duration. The samples were etched at a power input of 200 W and then coated with a plasma-polymerized HMDSO layer, keeping constant the relevant operating parameters (HMDSO flow rate= 0.6 sccm, HMDSO pressure= 3 Pa, power input= 30 W, deposition time= 2 minutes, coating thickness= 50 nm). The sample at etching time= 0 refer to the PET film which was only coated with the plasma-polymerized HMDSO layer. Advancing contact angles increase with the etching time up to 10 minutes, then they remain almost constant. Instead, receding contact angles slightly decrease for short etching time (5 min), and display a big jump for longer etching times. Samples prepared with etching times of 10 minutes and 20 minutes were super-hydrophobic (dynamic contact angle >150°, hysteresis angle <5°). $5 \times 5\ \mu m^2$ AFM images in Figure 6 refer to samples before HMDSO deposition (only etched samples). It can be noticed the difference in surface roughness of samples etched for 5 and 20 minutes, which results in different hydrophobic performances after the deposition of the HMDSO coating.

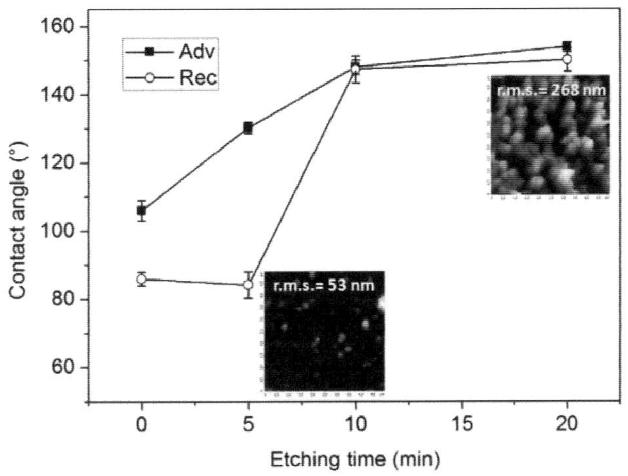

Figure 6. Variation of the advancing and receding contact angles of PET films with the etching time. Samples were etched at a power level of 200 W.

In the subsequent experiments, the thickness of the hydrophobic top layer was varied by changing the plasma deposition time. Before deposition, all the samples were etched with the oxygen plasma at 200 W for 20 minutes. The thickness of the coating (measured on Si wafers by means of a profilometer) was 50 nm for deposition time of 2 minutes, 220 nm for 10 minutes and 440 nm for 20 minutes. Results are displayed in Figure 7. It can be noticed that by increasing the deposition time, hydrophobic performances of PET samples decreases (decrease of dynamic contact angles, increase of roll-off angles). AFM analysis explains this result: by increasing the deposition time, surface structures seems to merge due to a partial filling of the voids between them, with a consequent decrease of the surface roughness.

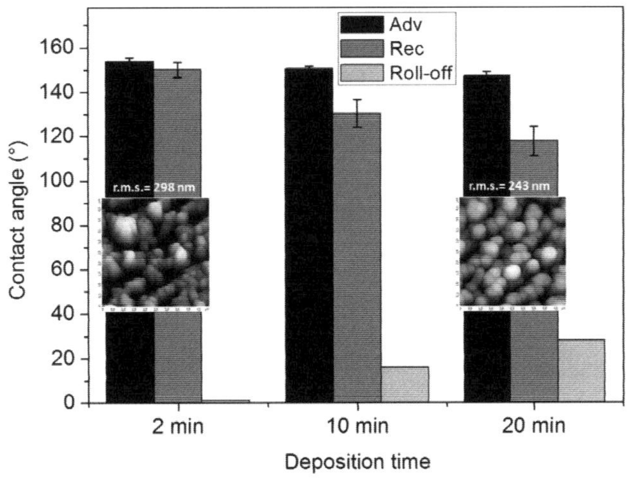

Figure 7. Variation of the advancing and receding contact angles of PET films with the HMDSO deposition time.

Images in Figure 8a and 8b show the AFM Z-Height profiles of plasma etched PTFE and PET samples, measured on a line extracted from a 5×5 μm^2 image. Both the surfaces present columnar structures, although with different aspect ratio. Structures observed on PTFE sample are 500 to 600 nm high, and have a width at the base of about 1.5 μm. Moreover, a superimposed nano-roughness has been detected. Instead, structures on PTFE sample are 700 to 800 nm high, and have a width at the base of about 700 nm. These differences can derive from the different etching rates of the two materials and/or from the different duration of the plasma etching. Images in Figure 8c and 8d show the AFM Z-Height profiles of two PET samples which were plasma etched and then coated with HMDSO layers of different thicknesses. It can be noticed that the HMDSO coating reduces the height of the structures (probably by filling the voids between them), which are only 400 nm high for the sample in Figure 4d. This "snowfall effect" has detrimental results on the hydrophobic performances (Figure 7).

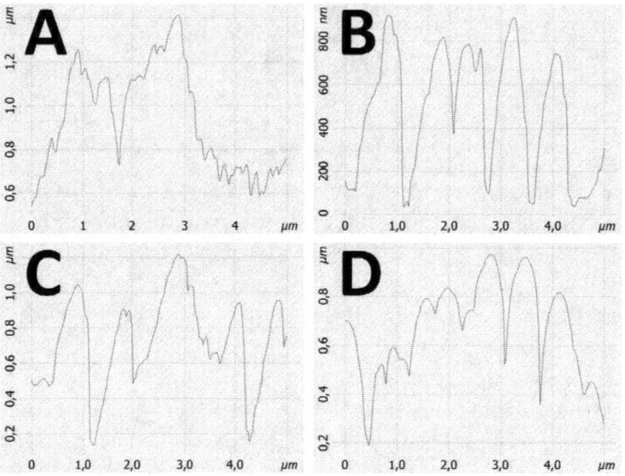

Figure 8. AFM Z-Height profiles of: (A) Plasma etched PTFE (200 W, 20 min); (B) Plasma etched PET (200 W, 20 min); (C) Plasma etched PET with a HMDSO coating of 50 nm; (D) Plasma etched PET with a HMDSO coating of 440 nm.

4. Conclusions

PTFE sheets were modified with a pure oxygen plasma varying the RF power input and the treatment time. We found that both these operating parameters influence the chemical/morphological characteristics and the wettability of the plasma treated samples. PTFE samples treated at low power input (up to 50 W) are more hydrophilic than the untreated one, but dynamic contact angles gradually increase with the power input. XPS analysis suggests that for low power levels there is a higher degree of OH functionalization of the outermost layers of polymeric chains, which leads to an increased wettability. AFM analysis reveals that plasma treatments at high power inputs result in the formation of globular structures as a consequence of the differential etching of the PTFE surface, with a strong increase of the surface roughness. Although XPS analyses of the aged samples reveal a post-plasma oxidation, the dynamic contact angles measured after 30 days are greater than those measured immediately after treatment and, for power input ≥100 W, the plasma etched PTFE surfaces are super-hydrophobic. This result has been interpreted in terms of surface adaptation, which leads to a decreased density of the exposed hydroxyl groups.

Plasma etched PET films need to be coated with a HMDSO top layer in order to create super-hydrophobic surfaces. The thickness of this layer has been varied by changing the plasma deposition time. Increase of the HMDSO layer thickness result in a decrease of the hydrophobic performances of PET samples (decrease of dynamic contact angles, increase of roll-off angles). This has been interpreted in terms of a "snowfall effect": by increasing the layer thickness, AFM reveals that the surface structures merge together, with a consequent decrease of the surface roughness.

References

[1] Guo Z, Liu W and Su B-L 2011 *J. Colloid Interface Sci.* **353** 335
[2] Jafari R, Asadollahi S and Farzaneh M 2013 *Plasma Chem. Plasma Process.* **33** 177
[3] Vandencasteele N, Nisol B, Viville P, Lazzaroni R, Castner D G and Reniers F 2008 *Plasma Processes Polym.* **5** 661
[4] Vourdas N, Tserepi A and Gogolides E 2007 *Nanotechnology* **18** 125304
[5] Zanini S, Massini P, Mietta M, Grimoldi E and Riccardi C 2008 *Journal of Colloid and Interface Science* **322** 566
[6] Morra M, Occhiello E and Garbassi F 1989 *Langmuir* **5** 872
[7] Ryan M E and Badyal J P S 1995 *Macromolecules* **28** 1377
[8] Tsougeni K, Vourdas N, Tserepi A, Gogolides E and Cardinaud C 2009 *Langmuir* **25** 11748
[9] Siow K S, Britcher L, Kumar S and Griesser H J 2006 *Plasma Processes Polym.* **3** 392
[10] Salapare III H S, Guittard F, Noblin X, Taffin de Givenchy E, Celestini F and Ramos H J 2013 *J. Colloid Interface Sci.* **396** 287

Journal of Physics: Conference Series **550** (2014) 012030 doi:10.1088/1742-6596/550/1/012030

Numerical Simulation of Bosch Processing for Deep Silicon Plasma Etching

P Moroz[1] and D J Moroz[2]

[1] Technology Development Center, Tokyo Electron U.S. Holdings Inc, Billerica, MA 01821, USA

[2] Department of Computer Science, University of Pennsylvania, Philadelphia, PA 19104, USA

E-mail: paul.moroz@us.tel.com

Abstract. We present a simulation of the Bosch process using the feature-scale modeling software FPS3D. FPS3D is a generic simulator that can be applied to any set of materials, plasmas, reactive gases, and reactions for both 2D and 3D simulations of etching and deposition. FPS3D can simulate multi-time-step processes for which the fluxes, species, reactions, ion energies, angular distributions, and other parameters can change with each time-step; it is thus well-suited for Bosch process simulations. The polymer deposition and etching time-steps of the Bosch process are modeled and discussed in more detail than was previously attainable.

1. Introduction

Numerical simulations for plasma materials processing have become more important as materials and chemistries have increased in complexity and process requirements have become more stringent. Multi-step processing has often been adopted both for IC fabrication and for micromachining in cases for which deep etching was required.

The two most well-known high-etch-rate techniques for deep silicon etching are cryogenic etching (see, for example, [1–4]) which requires temperatures to be in the range of -110°C during etching, and the so-called Bosch etching [5–8]. Bosch etching requires high density inductively coupled plasmas and uses alternating steps of isotropic etching by SF_6 and wall passivation (polymer deposition) by c-C_4F_8 to obtain vertical profiles.

Many other schemes for efficient deep silicon etching were proposed and investigated. These sometimes used gases other than the usual SF_6 and c-C_4F_8, or used mixtures of gases, and did not necessarily require high density inductively coupled plasmas (see, for example, [9–15] and references there). A similar technique to that used in Bosch processing is sometimes called gas chopping [9–12] or time multiplexed deep etching [6, 8, 13, 15].

The goal of Bosch and other deep-etch processes is to provide high-speed etching (in the range of 1000 Å/s, and higher) and to obtain deep profiles (up to 100s of microns). These high etch rates require high fluxes of active radicals to the wafer, which in turn require high RF power, high pumping speed, and high gas flows.

Using fluorine chemistry is advantageous as fluorine etches silicon at a high rate. Sulphur hexafluoride, SF_6, is commonly used for this purpose as it is largely comprised of fluorine, F. In plasma, SF_6 breaks into the highly reactive radicals F and SF_x and into the ions $SF_x{}^+$, where x is an integer 1,2, etc. These ions provide energy to break through the bottom of the polymer layer and to etch anisotropically in the vertical direction. Ion energy is also needed to accelerate etching of Si by F and by SF_x radicals.

Fluorine and other flourine-rich radicals etch Si isotropically. To ensure unidirectional vertical etching, the passivation or protection of sidewalls is needed. In the Bosch process, this is accomplished by alternating the etching step with the polymer deposition step. Separating these two steps allows each to be done optimally.

The deposition step often uses octofluorocyclobutane, c-C_4F_8, chemistry to protect the sidewalls with the polymer. In plasma, c-C_4F_8 gas breaks into a wide spectrum of radicals and ions. The plasma then contains such fluorocarbon radicals as CF, CF_2, CF_3, and C_2F_4 which actively deposit on the sidewalls forming a Teflon-type polymer. The polymer is stable at temperatures below about 50°C but disintegrates at higher temperatures. Since energy delivered by fluxes of energetic species from the plasma heats the substrate and the etching reactions are exothermic, the substrate requires active cooling. If the polymer layer on the sidewalls is not thick enough by the beginning of the etching time-step, it cannot prevent lateral etching of Si.

The Bosch process, if operated at conditions in which the etch rate is maximally high, leaves significant roughness and apparent scalloping on the sidewalls. Shortening the deposition and etching time-steps improves that negative phenomenon but at the cost of a reduced overall etch rate. Further smoothing of the sidewalls requires a change of chemistry (see, for example, [11] in which smooth sidewalls were obtained by using CHF_3 gas instead of SF_6).

For typical reactive ion etching (RIE) processes, the vertical anisotropy of obtained profiles depends on the ion energy level. The higher the ion energy, the higher the aspect ratio and degree of vertical anisotropy that can be obtained. The Bosch process also requires energetic ions in order to anisotropically break the polymer layer at the bottom and to provide energy which significantly increases the etch rate there. However, the Bosch process primarily relies on sidewall protection to ensure vertical profiles and the excessive ion energy may damage the profiles. Thus, reliable control of ion fluxes and ion energy is important for deep etch processing. The Bosch process operates in regimes in which ion fluxes are significantly lower than the fluxes of active radicals.

The etch rate of the Bosch process depends on the aspect ratios of etched features (aspect ratio dependent etch, ARDE). Sidewall reactions play a significant role in reducing radical fluxes coming to the bottom of the etched features [16]. This causes gradual reduction of etch rates at the bottom as the aspect ratios of etched features grow with time. This effect becomes strong at high aspect ratios and could lead in some cases to etch stop.

2. Surface Reactions

Setting the reactions between incoming species and solid materials is likely the most difficult part of simulating the Bosch process. Plasma experiments often do not provide sufficient information on the probabilities or yields of reactions for particular species, especially for ions for which reaction rates also depend on energy and angle of incidence. Multiple radicals and ions come to the surface simultaneously during a typical plasma experiment which makes it difficult to distinguish between reactions of different species and determine the reaction coefficients for each particular species. Specially designed beam experiments could help resolve that uncertainty because the species, energy, and angle of incidence could be selected and accurately controlled.

Our goal here is to demonstrate the capabilities of the feature-scale simulator FPS3D, which allow it to simulate such complex processes as Bosch etching or similar deep Si processing with alternating time-steps. There are a number of publications which provide specific information on particular experiments which use the same chemistry as does the Bosch process; they suggest a set of main reactions (see, for example, [17, 18]). We have used some of this data but only to put the reaction parameters in the right range. Proper analysis of the Bosch process and the determination of a complete set of reactions and main reaction pathways with determined reaction rates is a difficult task that, to our knowledge, has not yet been accomplished. We did not try here to build such a set of reactions by fitting experimental data. Instead, we have used a simplified set of reactions, sufficient to capture the main phenomena of the Bosch process.

During the deposition time-step, we consider only the polymer depositing species, and select the primary CF_2 to represent all others such as CF, CF_2, CF_3 and C_2F_4. During the etching step, the SF_6 plasma produces many radicals and ions. We consider reactions with the main etching species, F radicals, to represent all reactions with radicals, and consider reactions with ${CF_3}^+$ to represent all reactions with ions.

A simplified set of reactions between gaseous species and solid materials was built. It includes reactions of neutral species, ions and implants with solid materials. Reactions with neutrals and implants have temperature dependencies, while reactions with ions have energy and angular dependencies. Solid species are denoted by a concluding (s), implants are denoted by a concluding (i), and gaseous species have no concluding symbol. The polymer is presented as Poly. Species with a concluding E represent molecules with partially broken bonds due to interaction with energetic ions. These are typically more likely to react with gases than are the non-energized species.

The following reactions of neutrals with solids were considered:

$$\begin{aligned}
\mathrm{Si(s)} + \mathrm{CF_2} &\longrightarrow \mathrm{Si(s)} + \mathrm{Poly(s)} \\
\mathrm{SiF_2(s)} + \mathrm{CF_2} &\longrightarrow \mathrm{SiF_2(s)} + \mathrm{Poly(s)} \\
\mathrm{Poly(s)} + \mathrm{CF_2} &\longrightarrow 2\,\mathrm{Poly(s)} \\
\mathrm{Mask(s)} + \mathrm{CF_2} &\longrightarrow \mathrm{Mask(s)} + \mathrm{Poly(s)} \\
\mathrm{Si(s)} + 2\,\mathrm{F} &\longrightarrow \mathrm{SiF_2(s)} \\
\mathrm{SiF_2(s)} + 2\,\mathrm{F} &\longrightarrow \mathrm{SiF_4} \\
\mathrm{Poly(s)} + 2\,\mathrm{F} &\longrightarrow \mathrm{CF_4} \\
\mathrm{Mask(s)} + \mathrm{F} &\longrightarrow \mathrm{Mask} + \mathrm{F} \\
\mathrm{SiE(s)} + \mathrm{CF_2} &\longrightarrow \mathrm{Si(s)} + \mathrm{Poly(s)} \\
\mathrm{SiF_2E(s)} + \mathrm{CF_2} &\longrightarrow \mathrm{SiF_2(s)} + \mathrm{Poly(s)} \\
\mathrm{PolyE(s)} + \mathrm{CF_2} &\longrightarrow 2\,\mathrm{Poly(s)} \\
\mathrm{MaskE(s)} + \mathrm{CF_2} &\longrightarrow \mathrm{Mask(s)} + \mathrm{Poly(s)} \\
\mathrm{SiE(s)} + 2\,\mathrm{F} &\longrightarrow \mathrm{SiF_2(s)} \\
\mathrm{SiF_2E(s)} + 2\,\mathrm{F} &\longrightarrow \mathrm{SiF_4} \\
\mathrm{PolyE(s)} + 2\,\mathrm{F} &\longrightarrow \mathrm{CF_4} \\
\mathrm{MaskE(s)} + \mathrm{F} &\longrightarrow \mathrm{Mask} + \mathrm{F}
\end{aligned}$$

Journal of Physics: Conference Series **550** (2014) 012030 doi:10.1088/1742-6596/550/1/012030

The following reactions of ions with solids were taken into account:

$$
\begin{aligned}
\mathrm{Si(s)} + \mathrm{SF_3}^+ &\longrightarrow \mathrm{SiE(s)} + \mathrm{S(i)} + 3\,\mathrm{F(i)}\\
\mathrm{Poly(s)} + \mathrm{SF_3}^+ &\longrightarrow \mathrm{PolyE(s)} + \mathrm{S(i)} + 3\,\mathrm{F(i)}\\
\mathrm{SiF_2(s)} + \mathrm{SF_3}^+ &\longrightarrow \mathrm{SiF_2E(s)} + \mathrm{S(i)} + 3\,\mathrm{F(i)}\\
\mathrm{Mask(s)} + \mathrm{SF_3}^+ &\longrightarrow \mathrm{MaskE(s)} + \mathrm{S(i)} + 3\,\mathrm{F(i)}\\
\mathrm{SiE(s)} + \mathrm{SF_3}^+ &\longrightarrow \mathrm{Si} + \mathrm{S(i)} + 3\,\mathrm{F(i)}\\
\mathrm{PolyE(s)} + \mathrm{SF_3}^+ &\longrightarrow \mathrm{CF_2} + \mathrm{S(i)} + 3\,\mathrm{F(i)}\\
\mathrm{SiF_2E(s)} + \mathrm{SF_3}^+ &\longrightarrow \mathrm{Si} + 2\,\mathrm{F} + \mathrm{S(i)} + 3\,\mathrm{F(i)}\\
\mathrm{MaskE(s)} + \mathrm{SF_3}^+ &\longrightarrow \mathrm{Mask} + \mathrm{S(i)} + 3\,\mathrm{F(i)}
\end{aligned}
$$

While the S(i) implant was considered non-reactive, the following reactions for the F(i) implant were considered:

$$
\begin{aligned}
\mathrm{Si(s)} + 4\,\mathrm{F(i)} &\longrightarrow \mathrm{SiF_4}\\
\mathrm{SiF_2(s)} + 2\,\mathrm{F(i)} &\longrightarrow \mathrm{SiF_4}\\
\mathrm{Poly(s)} + 2\,\mathrm{F(i)} &\longrightarrow \mathrm{CF_4}\\
\mathrm{Mask(s)} + \mathrm{F(i)} &\longrightarrow \mathrm{Mask} + \mathrm{F}\\
\mathrm{SiE(s)} + 4\,\mathrm{F(i)} &\longrightarrow \mathrm{SiF_4}\\
\mathrm{SiF_2E(s)} + 2\,\mathrm{F(i)} &\longrightarrow \mathrm{SiF_4}\\
\mathrm{PolyE(s)} + 2\,\mathrm{F(i)} &\longrightarrow \mathrm{CF_4}\\
\mathrm{MaskE(s)} + \mathrm{F(i)} &\longrightarrow \mathrm{Mask} + \mathrm{F}
\end{aligned}
$$

This set of reactions is significantly simplified compared to what we believe actually occurs, but it is enough to capture the main characteristics of Bosch processing. In FPS3D, if a reaction between a gaseous species and a solid species is not listed, the default is a reflection of the gaseous species.

3. Incoming Fluxes

The FPS3D simulator requires fluxes for all incoming reactive species to be set. The Bosch process typically utilizes high plasma density resulting in high fluxes of radicals, while the ion fluxes are limited to a small fraction of the radical fluxes. In simulations presented in this article, the following parameters were used. During the polymer deposition time-step, the flux of the main polymer depositing species, CF_2, was set to be $3 * 10^{18}$ $cm^{-2}s^{-1}$ for a duration of 4s. An intermediate transitional time-step of 1s was set, during which the fluxes gradually changed from the deposition regime to the etching regime. For the etching time-step, the fluxes of the main species, F radicals and SF_3^+ ions, were set at the levels of $1.2 * 10^{19}$ and $1.8 * 10^{16}$ $cm^{-2}s^{-1}$, respectively. After the etching step, an intermediate transitional time-step was set again during which the fluxes changed back to the polymer deposition regime. These four time-steps represent a single cycle. During Bosch processing, this cycle of fluxes is repeated many times (demonstrated in Figure 1).

4. The FPS3D Feature Scale Simulator

Our goal is to numerically model the Bosch process using the feature-scale simulator FPS3D [19, 20]. Bosch processing is challenging to simulate because it includes both etching and

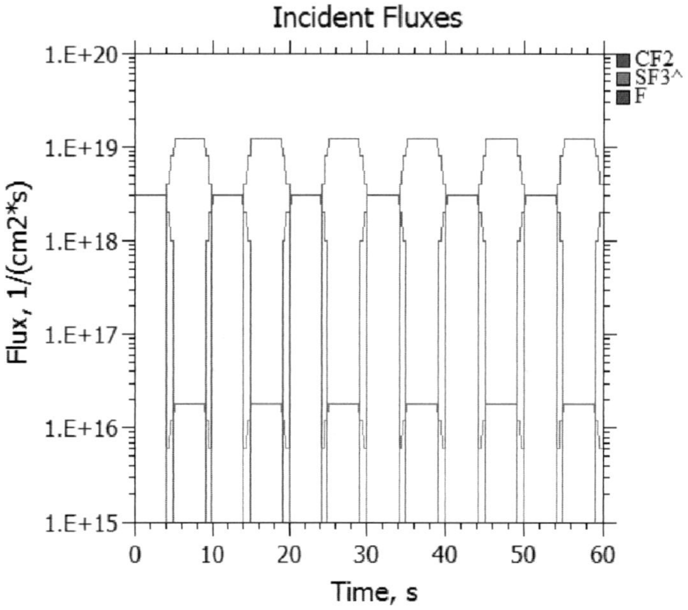

Figure 1. Time variation of species fluxes. The cycle duration is 10s.

deposition. Additional complications appear because the fluxes of species are large and change frequently, and because silicon and polymer have highly disparate densities (meaning that the approximation that all full cells contains the same number of molecules cannot be used). Large fluxes can pose complications to Monte Carlo simulation software if it includes the assumption that each gas particle contains a single molecule, but that assumption is not used in FPS3D.

In FPS3D, all solid materials are represented via rectangular cells of equal dimensions, and the code presents all incoming fluxes of species as Monte Carlo particles. The number of molecules in a full cell depends on the density of the corresponding materials. Each particle and each cell can include any number of molecules, ranging from one to an arbitrarily large number. When possible, to improve statistics and reduce artificial numerical roughness of profiles, the number of molecules in a particle is chosen to be significantly less than that in a cell. FPS3D allows consideration of the large fluxes and large feature dimensions of typical Bosch processing, both in 2D and 3D.

5. Results of Simulations

The incident fluxes of reactive species to the wafer change over time in cycles, as shown in Figure 1. The CF_2 flux strikes the wafer at a constant level during the deposition time-step of 4s, and the F radical and SF_3^+ ion fluxes strike at constant levels during the etching time-step of 4s. These two main steps overlap during the intermediate step of 1s, in which the fluxes gradually change one into the other. This approach allows simulation of the realistic condition of gas chopping. It is impossible to change fluxes instantaneously, although researchers strive to make the corresponding gas changes as quick as possible.

Note: the following images are taken from the FPS3D software, which uses a concluding carat symbol (^) to represent an ion, a concluding underscore (_) symbol to represent a solid material, uses A/s to mean Angstroms per second, and does not use subscripts in chemical notation.

Journal of Physics: Conference Series **550** (2014) 012030 doi:10.1088/1742-6596/550/1/012030

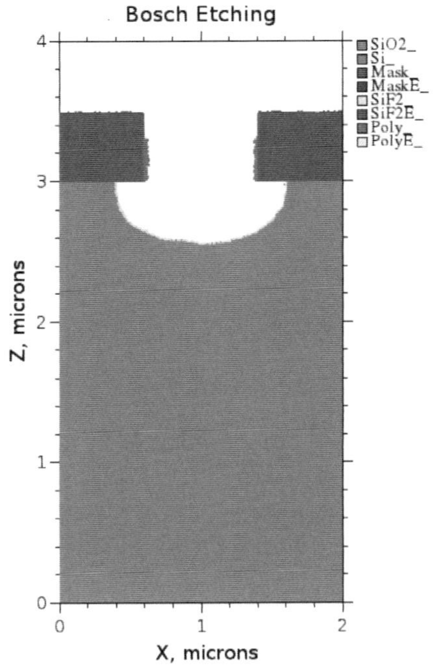

Figure 2. The first etching step. Si is removed isotropically.

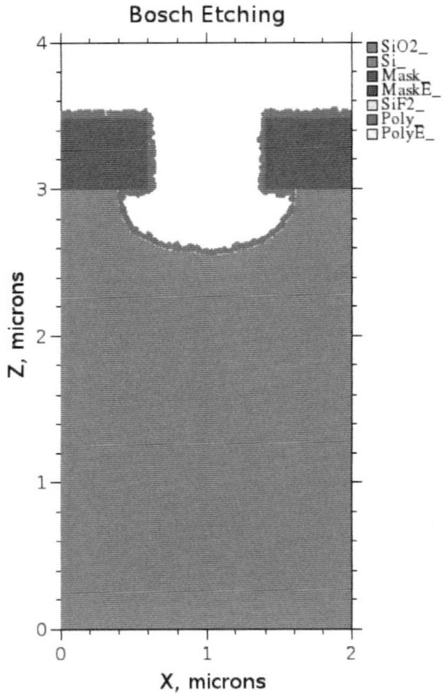

Figure 3. The following polymer deposition step.

During the etching time-step, fluorine removes silicon mostly isotropically, as shown in Figure 2 for the first etching time-step. The following deposition time-step results in the polymer deposited everywhere on Si and on mask, as shown in Figure 3.

Figure 4 and Figure 5 show the deposition step and the following etching step after a few cycles have finished. Strong roughness (scalloping) has developed along the sidewalls, corresponding to the number of cycles passed. Etching during a single cycle is significant, but the etch rate gradually decreases when the trench becomes deeper.

FPS3D allows the user to record parameters at any number of selected locations. We selected three x-locations of 0.7, 1, and 1.3 microns to record such parameters as the height (6), the etch rate (Figure 7 – instantaneous etch rate, Figure 8 – average etch rate) and etch yield (Figure 9 – instantaneous etch yield, Figure 10 – average etch yield). Etch yield shows the number of Si atoms removed per single ion.

The height changes in a step-like fashion in Figure 6, corresponding to the cycles. The height increases slightly during the deposition time-step due to polymer deposition but then quickly decreases during the etching time-step due to removal of polymer and Si.

The instantaneous etch rate and instantaneous etch yield in Figure 7 and Figure 9 are slightly negative during the polymer deposition time-steps but then are strongly positive during the etching time-steps. The instantaneous etch rate initially reaches about 1000 Å/s but then gradually decreases to about 500 Å/s, after 100s of etching.

Similar results were obtained during 3D simulations for the case of etching trenches. Sample 3D profiles are shown in Figure 11 and Figure 12 for the deposition time-step during the second cycle and for the following etching time-step, respectively. The cell size was chosen to be larger than that in the 2D simulations to accelerate calculations.

The results of simulations for more complex 3D structures, such as high aspect ratio vias (not

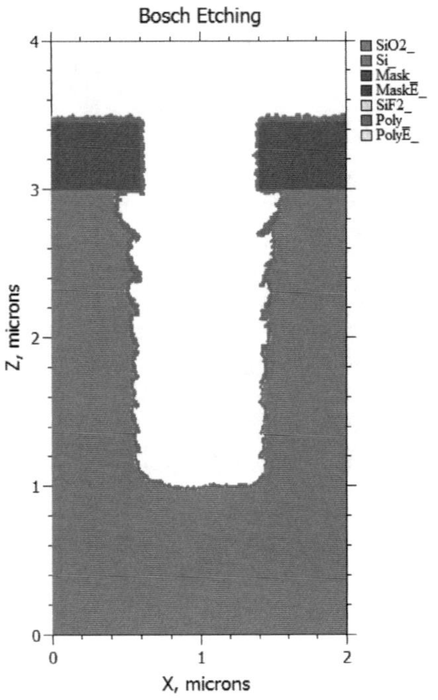

Figure 4. The polymer deposition step.

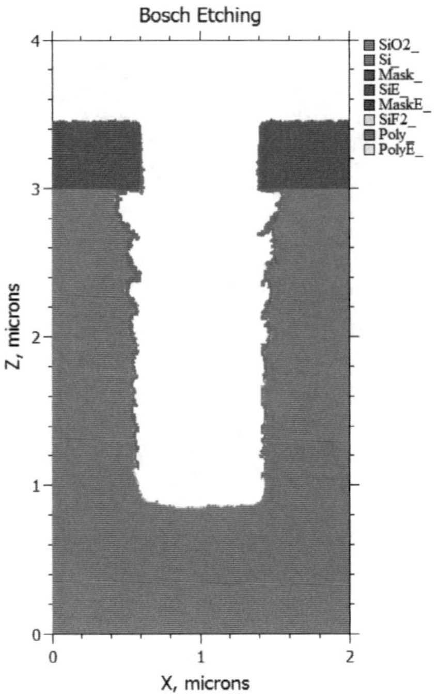

Figure 5. The following etching step.

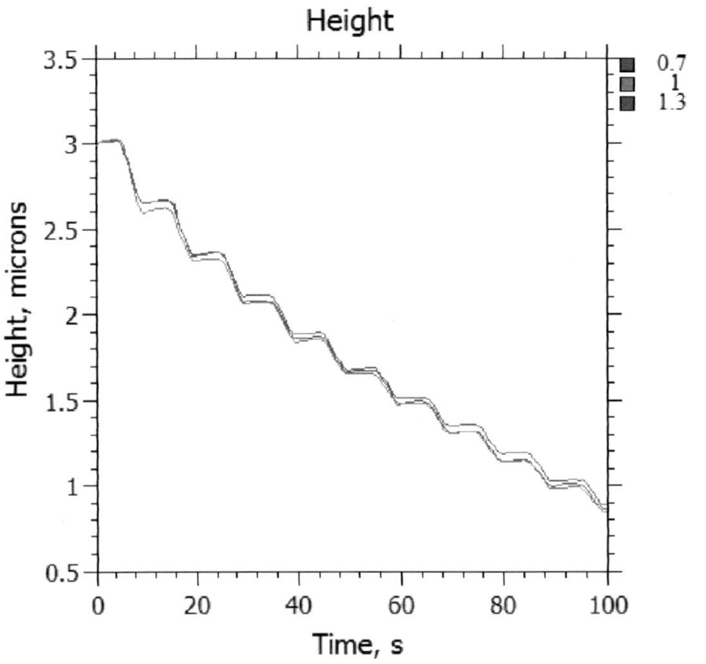

Figure 6. Height change over time.

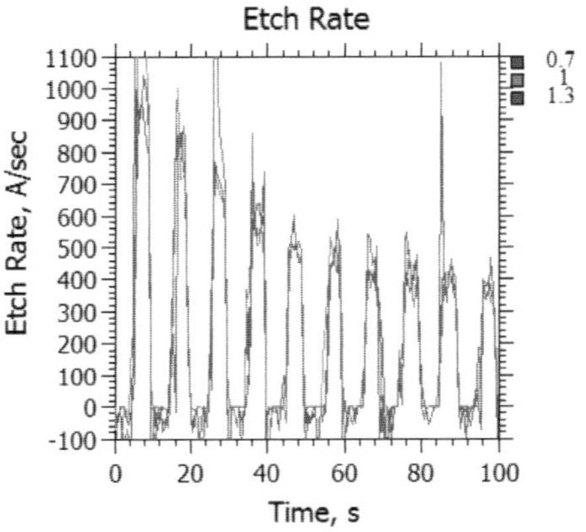

Figure 7. Instantaneous etch rate.

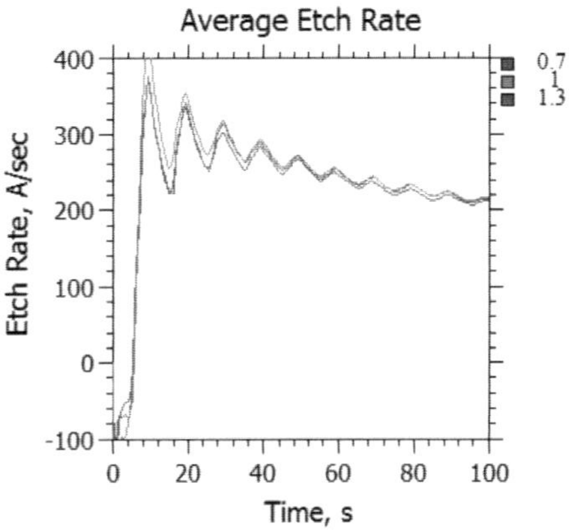

Figure 8. Average etch rate.

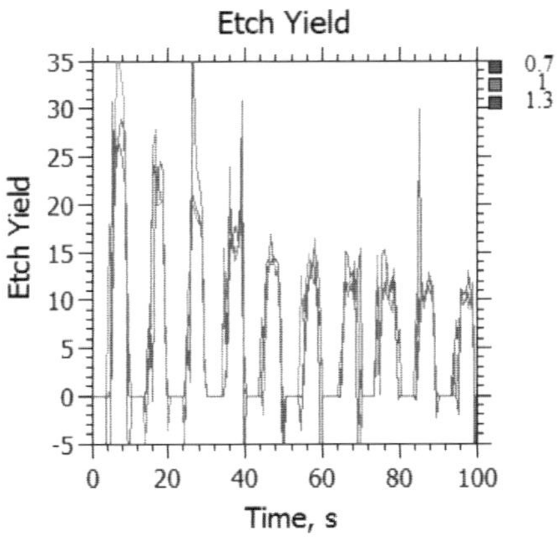

Figure 9. Instantaneous etch yield.

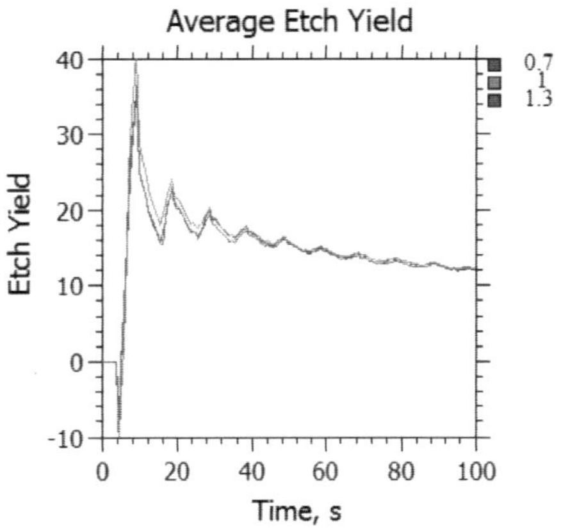

Figure 10. Average etch yield.

presented here), show significant differences between 2D and 3D calculations – the etch rate is always lower in 3D calculations than it is in 2D. For these cases, 2D calculations cannot properly capture the flow limitations and the magnitude of wall interactions, thus 3D calculations should be used.

6. Discussion and Conclusion

We have presented simulation results for Bosch-type deep Si etching. Bosch processing is designed for high etch-rate etching and uses alternating time-steps of etching and polymer deposition in cycles.

In our simulations, the duration of one cycle was chosen to be 10s, with the deposition and etching steps each 4s long, and with two intermediate time-steps simulating gas switching between etching and deposition, each having the duration of 1s.

We have used the advanced feature-scale simulator FPS3D which is designed to simulate etching and deposition during materials processing. The simulator can consider any set of materials, any set of incoming fluxes of species, and any set of reactions happening at the

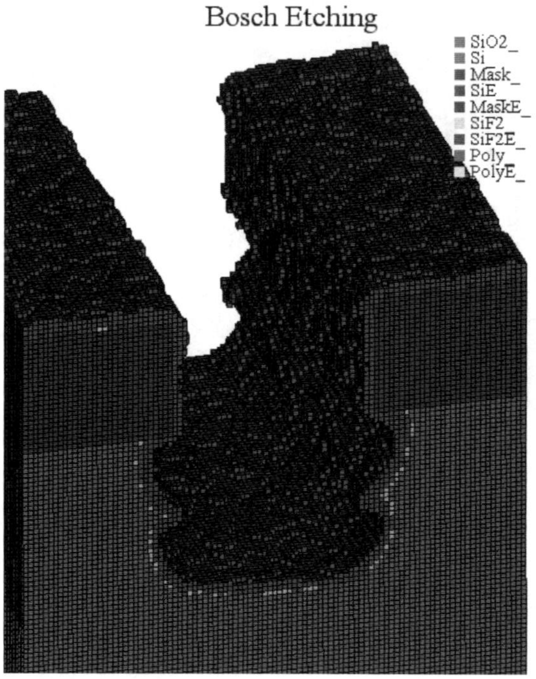

Figure 11. Polymer deposition step.

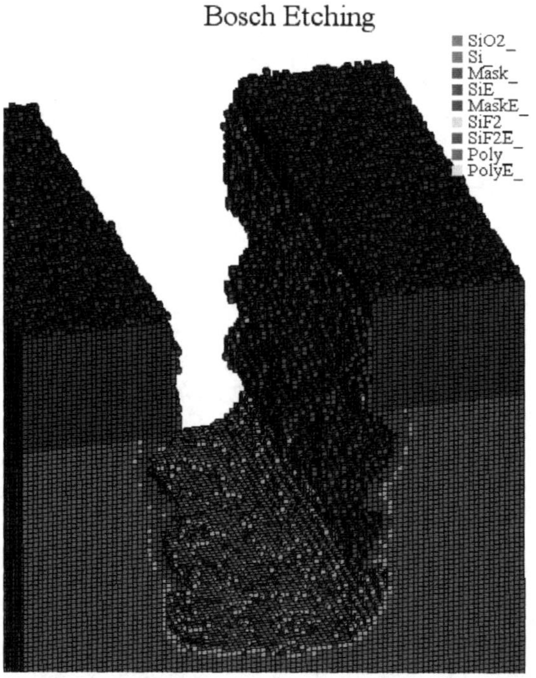

Figure 12. The following etching step.

material surfaces. In this article, we used a simplified set of reactions which was nevertheless capable of capturing the main characteristics of the Bosch process. The deposition and etching time-steps were simulated with details not previously attainable.

Acknowledgements

The authors are thankful to many people at Tokyo Electron Inc. and Tokyo Electron U.S. Holdings for their interest, support, and advice. Special thanks go to Dr. Masaaki Matsukuma of Japan's Technology Development Center for valuable discussions.

References

[1] Aachboun S, Ranson P 1999 *J. Vac. Sci. Technol.* A **17** 2270
[2] Aachboun S, Ranson P, Hilbert C, Boufnichel M 2000 *J. Vac. Sci. Technol.* A **18** 1848
[3] Mellhaoui X Dussart R, Tillocher T, Lefaucheux P, Ranson P, Boufnichel M, Overzet L J 2005 *J. Appl. Phys.* **98** 104901
[4] Kamto A, Divan R, Sumant A V, Burkett S L 2010 *J. Vac. Sci. Technol.* A **28** 719
[5] Labelle C B, Donnelly V M, Bogart G R, Opila R L, Komblit A 2004 *J. Vac. Sci. Technol.* A **22** 2500
[6] Blauw M A, Zijlstra T, van der Drift E 2001 *J. Vac. Sci. Technol.* B **19** 2930
[7] Kokkoris G., Panagiotopoulos A, Goodyear A, Cooke M, Gogolides E 2009 *J. Phys. D: Appl. Phys.* **42** 055209
[8] Ayon A A, Braff R, Lin C C, Sawin H H, Schmidt M A 1999 *J. Electrochem. Soc.* **146** 339
[9] Rangelow I W 1990 *Proc. SPIE* **1392** 180
[10] Rangelow I W, Biehl St 2001 *J. Vac. Sci. Technol.* B **19** 916
[11] Rangelow I W 2003 *J. Vac. Sci. Technol.* A **21** 1550
[12] Volland B, Shi F, Hudek P, Heerlein H, Rangelow I W 1999 *J. Vac. Sci. Technol.* B **17** 2768
[13] Blauw M A, Craciun G, Sloof W G, French P J, van der Drift E 2002 *J. Vac. Sci. Technol.* B **20** 3106
[14] Sammak A, Azimi S, Izadi N, Hosseinieh B K, Mohajerzadeh S 2007 *J. Microelectromech. Syst.* **16** 912
[15] Azimi S, Mehran M, Amini A, Vali A, Mohajerzadeh S, Fathipour M 2010 *J. Vac. Sci. Technol.* B **28** 1125
[16] Kiihamaki J 2000 *J. Vac. Sci. Technol.* A **18** 1385
[17] Rauf S, Dauksher W J, Clemens S B, Smith K H 2002 *J. Vac. Sci. Technol.* A **20** 1177
[18] Kokkoris G, Tserepi A, Boudouvis A G, Gogolides E 2004 *J. Vac. Sci. Technol.* A **22** 1896
[19] Moroz P 2011 *IEEE Trans. on Plasma Science* **39** 2804
[20] Moroz P, Moroz D J 2013 *ECS Transactions* **50** 61

Journal of Physics: Conference Series **550** (2014) 012031 doi:10.1088/1742-6596/550/1/012031

ECWR plasma enhanced chemical vapour deposition of microcrystalline silicon thin films.

E Farsari, A G Kalampounias, E Amanatides, D Mataras
Department of Chemical Engineering, University of Patras
P.O. Box 1407, 26504 Patras Greece

E-mail: lef@plasmatech.gr

Abstract. An ECWR plasma source was used in order to deposit microcrystalline silicon thin films. The effect of input power and silane content on the deposition rate and the materials properties was investigated. Deposition rates, up to 25Å/sec, and high crystallinity degree were achieved using high silane content which is in contrast to the conventional CCP method. The amorphous to crystalline transition zone was determined revealing that in ECWR discharges microcrystalline silicon growth is favoured in a wide range of experimental conditions.

1. Introduction

Hydrogenated microcrystalline silicon thin films are widely applied in a variety of optoelectronic devices such as thin film solar cells and thin film transistors. Industrially, the material is produced by silane/ hydrogen discharges in capacitevely coupled plasma (CCP) reactors, although the deposition rate achieved by this method does not exceed 1 nm/sec. In order to increase the deposition rate and the gas conversion to film, several deposition techniques have been proposed such as high pressure depletion method (HPD), very high frequency method (VHF) and use of alternative precursors [1-3]. Moreover, high density methods such as Inductively Coupled Plasmas, microwave plasmas, Electron Cyclotron Reactors and Hollow cathode discharges have been suggested [4-7]. The aim of this work is to use an Electron Cyclotron Wave Resonance (ECWR) plasma reactor in order to deposit microcrystalline silicon thin films and determine the process parameters leading to high deposition rate and device grade materials.

ECWR plasma sources are inductively coupled sources in a weak superimposed magnetic field. The rf magnetic field is usually produced from a single sector coil whose axis is perpendicular to the stationary magnetic field and its frequency is lower than the electron cyclotron frequency (ω_c) corresponding to the static magnetic field. Depending on the value of the static magnetic field, plasma dimensions and the radiofrequency, ECWR sources present a resonant-like increase of plasma density. In order to describe the ECWR effect, B. Pfeiffer proposed a theory according to which when the magnetic field enters the plasma region, it splits into a right-hand polarized wave and a left-hand polarized wave [8, 9]. When the wave frequency is higher than the cut-off frequency (which depends on the plasma frequency and the ω_c) both waves penetrate the plasma region. On the other hand for wave frequency lower than the cut- off frequency, the refractive index of the left- hand polarized wave is imaginary and the wave does not propagate into the plasma region. However, when the wave frequency is lower than the electron cyclotron frequency (ω_c), the refractive number of the right hand polarized wave becomes higher than unity and this wave penetrates the plasma region. These right hand polarized waves, which can propagate in the plasma, are called Electron Cyclotron Waves, or whistler waves. Resonant plasma excitation by whistler waves is possible when an odd number of the wavelength of the propagating waves fits the plasma thickness in the direction of the wave vector. Experimentally this can be achieved by varying the static magnetic field for a specific radiofrequency. Thus the electron density can reach values more than one order of magnitude higher than that of conventional ICP and CCP discharges [8-11]. The enhancement of the electron density allows the ECWR sources to be used for ionization processes and for plasma enhanced chemical vapour deposition of various films such as hydrogenated Si and Ge, boron and carbon nitride films and amorphous carbon films [12-15].

In this work the deposition of hydrogenated microcrystalline silicon thin films by ECWR discharges is examined. More precisely, the effect of process parameters, such as silane content and input power on the deposition rate and the materials properties is investigated. Laser Raman Spectroscopy was used in order to calculate the crystallinity of the materials and to define the transition zone from amorphous to crystalline phase, while Fourier Transform Infrared Spectroscopy measurements were performed in order to estimate the hydrogen to silicon bonding. Atomic force microscopy measurements were also performed to monitor the morphology of the films and to calculate the materials roughness.

2. Experimental

The experimental set-up is illustrated in figure 1. The ECWR source (COPRA DN250 by CCR Technology) is placed vertically in a high vacuum chamber. The source consists of a single turn copper coil which surrounds the plasma region. The coil is powered via a matching network by RF power of 13.56 MHz. Helmholtz coils are placed around the source in order to create the necessary static magnetic field, for the ECWR effect. The substrate holder is grounded and it is placed at a distance of 10 cm from the plasma source. An external heating system allows the sample heating up to 200°C, which is the typical temperature for μc-Si depositions. A system of flow meter controllers and valves is used in order to feed the reactor with the precursor gases. The hydrogen is fed directly in the plasma source while silane enters the reactor in the region between the substrate holder and the source. The pumping system, which consists of a rotary pump, a roots pump and a turbomolecular pump, is capable to maintain low pressure, in the range of 10^{-2} mbar, even for high flow rates. All experiments presented in this work, were performed by keeping the process pressure equal to 10^{-2} mbar, while the input power and the silane content varied. Corning glass and c-Si wafers were used as substrates. Just after the depositions, all materials were characterized by Laser Raman Spectroscopy, Fourier Transform Infrared Spectroscopy and Atomic Force Microscopy while the deposition rate was estimated by optical methods.

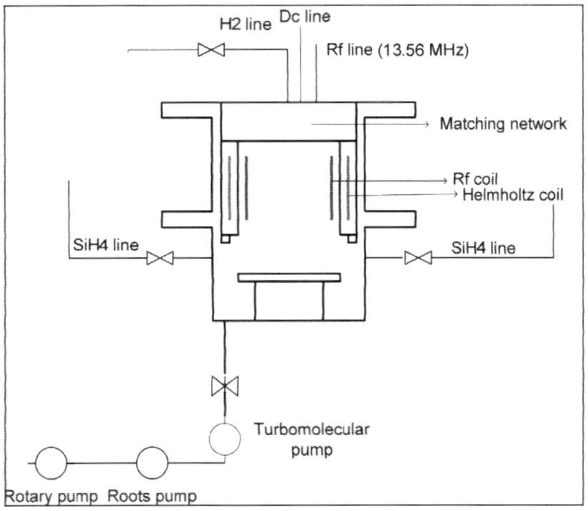

Figure 1. Experimental set-up.

Raman spectra, excited with the 441.6 nm line of an air-cooled HeCd laser (Kimmon Electric Co.), were measured with the UV-Vis Labram HR-800 spectrograph (Jobin Yvon). A narrow-bandpass interference filter was used for the elimination of the laser plasma lines. The excitation beam was directed to the sample compartment of a confocal open microscope attachment for punctual analysis. The microscope was used for the delivery of the excitation laser beam on the sample and the collection of the backscattered light through a notch-filter and the objective lens adapted to the aperture of the microscope. The focusing objective was a Long Working distance (8 mm) 50x/0.55 Olympus lens. The spectra were obtained using fixed 0.07 mW laser power on the specimen for a total integration time of 200 s. The Raman scattered radiation was focused on a confocal entrance slit (200 μm width) of an achromatic flat field spectrograph with 800 nm focal length. A 1800-grooves/mm grating and a two-dimensional CCD detector cooled at 140K were used for the dispersion and detection of the Raman photons, respectively.

Infrared spectra of films deposited on Si-wafers were measured in the transmission mode on a Bruker Equinox 55 Fourier transformed mid-IR spectrometer with sealed and desiccated optics. All spectra were measured at room temperature and represent the average of 50 scans at a resolution of 2 cm^{-1}.

A NTEGRA Prima by NTMDT, Atomic Force Microscope was used for imaging the materials surface. All films were investigated at semi-contact mode and the RMS roughness was calculated using appropriate software.

3. Results and discussion

3.1 Deposition Rate

The effect of the input power on the deposition rate was examined for silane content equal to 20 and 43%. In order to keep the process pressure constant at 10^{-2} mbar both silane and hydrogen flow rates were adjusted. The experimental results are presented in figure 2. The deposition rate in all cases studied here is higher than the rates observed for conventional capacitively coupled reactors. A maximum value ~25 Å/sec is achieved for the highest silane content which is 5 times higher than the industrial rate.

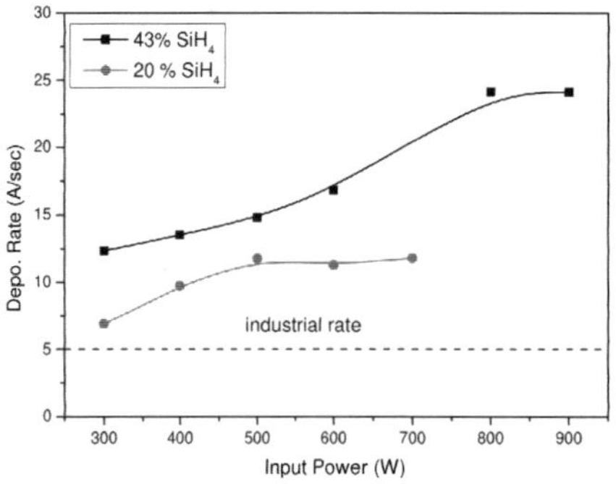

Figure 2. Deposition Rate as a function of the input power for silane content equal to 20 and 43%.

As the input power increases, the deposition rate also increases for both mixtures. Within ECWR discharges the electron density varies linearly with the input power, thus the silane dissociation and the SiH_x (x<4) radicals production are promoted [16]. Moreover the pressure dependent reactions which lead to radicals consumption, silane polymerization and powder formation in the gas phase are suppressed as the process pressure is very low and the secondary gas phase reactions are not favored. Even highly reactive radicals can travel long distances and reach the surface rather than consumed in the gas phase. The enhancement of radicals production and the limitation of secondary reactions establish an efficient silane conversion to Si-H thin film.

3.2 Materials Characterization

The Raman spectra of films deposited on glass substrates at input power ranging from 300 to 700 W and constant silane content are illustrated in figure 3. The spectrum at the lowest power is dominated by a broad band around ~480 cm^{-1} attributed to amorphous silicon. At higher input power a band near ~517 cm^{-1} appears assigned to crystalline phase of the material. The c-Si phonon peak position in all materials is shifted from its position at 520 cm^{-1} to lower wavenumbers which is related to the crystallite size and the film stress [17- 19].

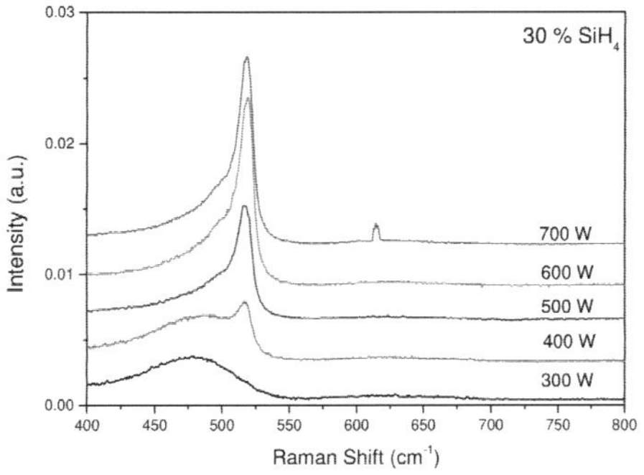

Figure 3. Raman spectra of Si-H thin films deposited at different input power.

Crystallinity calculations were performed using spectrum deconvolution in three Gaussian peaks: one at 480 cm^{-1} assigned to a-Si, one at 517-520 assigned to c-Si and one at 505 assigned to grain boundaries and crystallite sizes below 5nm [20]. The results are presented in figure 4 for materials deposited on glass substrates by varying the silane content and the power input. All materials deposited at the highest silane content (66%) appeared amorphous while an increase of Raman crystallinity with power is observed for silane content 30 and 43%. This is a significant difference compared to conventional CCP method where the microcrystalline silicon deposition is usually achieved when silane mole fraction in the mixture with H_2 is lower than 10% [21].

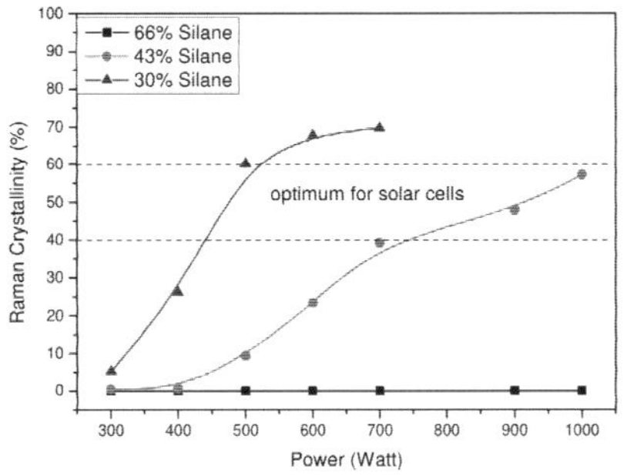

Figure 4. Raman crystallinity as a function of the input power on glass substrates for silane content 30, 43 and 66% and pressure 10^{-2} mbar.

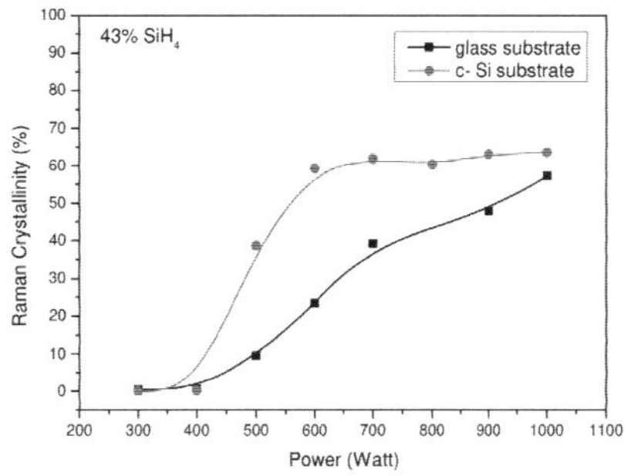

Figure 5. Raman crystallinity as a function of the input power on c-Si and glass substrates for silane content 43% and pressure 10^{-2} mbar.

The deposition of highly crystalline materials in ECWR discharges, even for high silane content, is achieved due to the enhancement of silane and hydrogen dissociation and the increase of atomic hydrogen density in the reactor. Atomic hydrogen reaching the film surface has an important role on microcrystalline silicon growth. The models that have been proposed in order to explain the beneficial role of atomic hydrogen on the crystallinity of Si: H thin films are summarized in ref [22]. In silane – hydrogen discharges, atomic hydrogen can be produced by both gases. Thus as the input power increases and the dissociation of gases is enhanced, a discharge rich in H is established which in turn leads to higher film crystallinity. As expected, the crystallinity enhancement is more pronounced for the lowest silane content and it reaches a value up to 60% at power 500W. On the other hand, the transition zone from amorphous to crystalline materials is more extended for the intermediate silane content (43%) making possible the microcrystalline silicon deposition in a wider range of input power.

Figure 5 illustrates the Raman crystallinity as a function of input power for materials deposited on glass and c-Si wafers for silane content 43%. Raman crystallinity shows the same trend for both substrates but the transition zone for materials deposited on c-Si wafers is shifted to lower values of the input power. Crystalline silicon wafers offer higher nucleation sites density [23], allowing the crystallization at lower power.

Materials deposited on c-Si wafers were used in order to examine the hydrogen to silicon bonding by means of Fourier Transform Infrared Spectroscopy. A typical spectrum of hydrogenated silicon films consists of three absorption regions: 600-700 cm^{-1}, 850-910 cm^{-1} and 1850-2250 cm^{-1} assigned to Si- H_x wagging, Si-H_2 bending and Si-H_x stretching vibrations [24]. In this work the analysis is focused on the range of hydrides stretching mode (SM). Figure 6 illustrates the FTIR absorbance spectra of materials deposited by varying the power and for constant silane content 20%. The deposition rate was estimated between 10 and 12 Å/sec and the crystallinity for all samples was calculated between 50 and 60%. The thickness of the films was kept at ~600 nm.

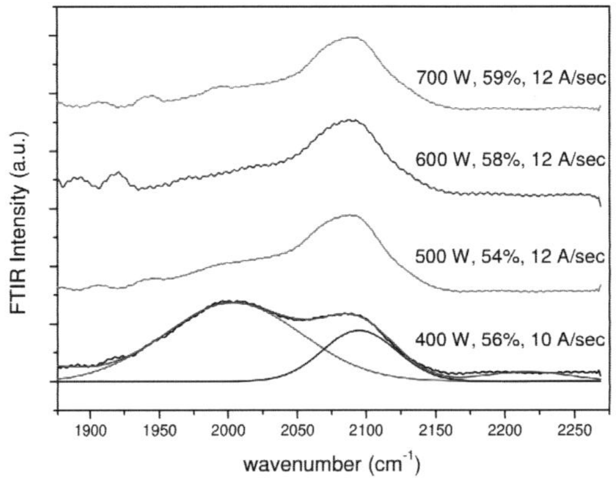

Figure 6. FTIR absorbance spectra of materials deposited by varying the input power.

In the SM region of the spectrum two bands at 2000 and 2090 cm^{-1} can be observed for all materials. The Low Stretching Mode (2000 cm^{-1}) is attributed to hydrogen bonded in vacancies while the High Stretching Mode (2090 cm^{-1}) to hydrogen bonded on voids surfaces [25]. However the relative intensities of LSM and HSM bands vary significantly with the deposition conditions. The LSM is dominant for the film deposited at the lowest power, while the HSM intensity increases as the input power and the deposition rate increase, regardless the crystalline volume of the material. The enhancement of HSM intensity is indicative for porous materials and it has been correlated to post oxidation phenomena [26].

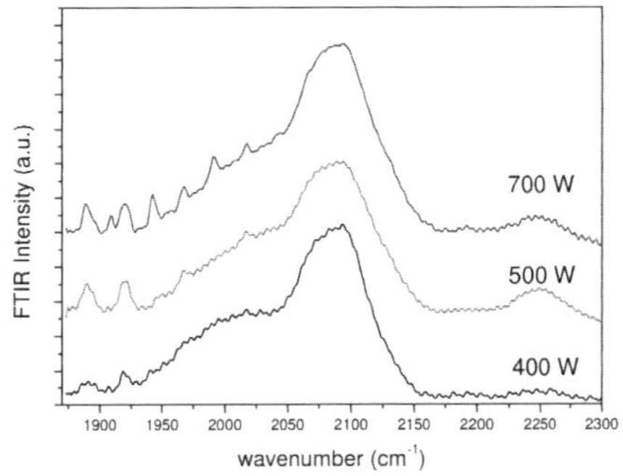

Figure 7. SM region of materials after the exposure to ambient conditions for 1 month.

In order to examine if post oxidation phenomena take place, the materials were exposed in ambient conditions for 1 month and FTIR measurement were repeated. Figure 7 illustrates the sm region of the spectrum for films after the exposure in the atmosphere. For samples deposited at power range 500-700 W two bands arise in the spectrum at 2190 and 2250 cm^{-1}, assigned to $O_2Si\text{-}H_x$ and $O_3Si\text{-}H_x$ [27], indicating materials oxidation. However, $O_ySi\text{-}H_x$ bands are barely indentified for the film deposited at the lowest input power suggesting that high quality crystalline materials can be deposited at high deposition rates in ECWR discharges.

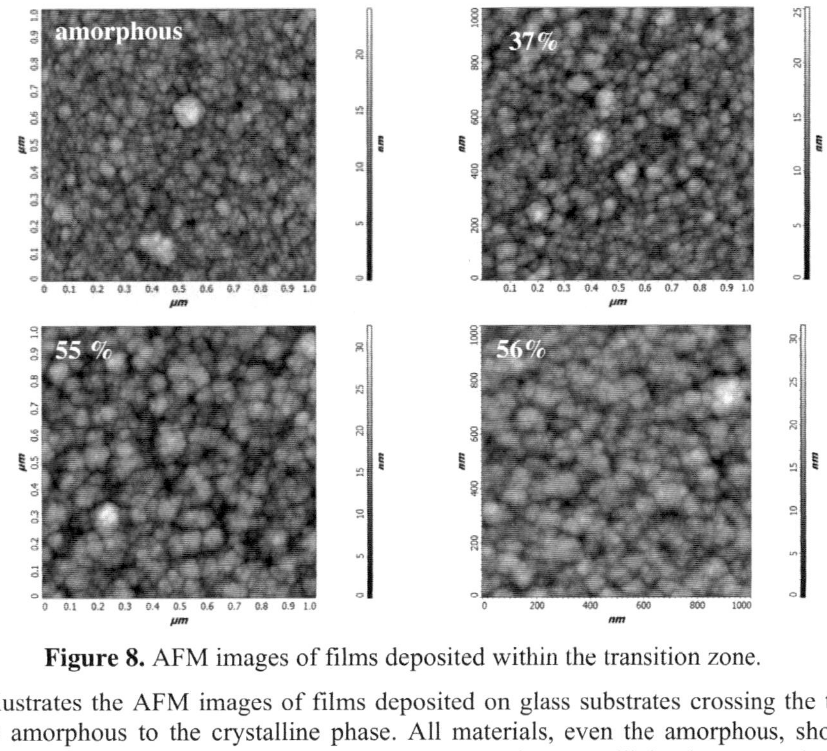

Figure 8. AFM images of films deposited within the transition zone.

Figure 8 illustrates the AFM images of films deposited on glass substrates crossing the transition zone from the amorphous to the crystalline phase. All materials, even the amorphous, show a fine structure and the grain size is of several nm. However as the crystallinity increases the material becomes compact due to grains coalescence. This is also illustrated in figure 9, where the Rms roughness and Raman Crystallinity are plotted. The Rms roughness increases significantly in the transition zone of the material, while the roughness drops when conditions of crystalline materials deposition are established.

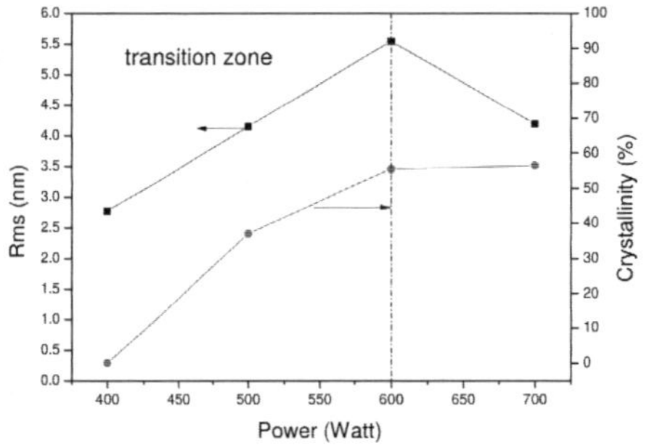

Figure 9. Rms roughness and Raman crystallinity as a function of the input power.

4. Conclusions

Hydrogenated silicon thin films have been deposited by Electron Cyclotron Wave Resonance discharges. The effect of the silane content and input power on deposition rate and materials properties was investigated.

The deposition rate shows a significant increase as silane content and input power increases and can reach a value up to 25Å/sec for silane content 43%. The enhancement of the deposition rate is related to the high dissociation degree of silane. The production of SiH_x ($x<4$) radicals, which are the building blocks of the film, is more pronounced in ECWR discharges. Moreover, secondary reactions, which promote the silane consumption in the gas phase, are suppressed at the process pressure (10^{-2} mbar). The combination of the above mentioned phenomena leads to effective utilization of the precursor gas.

At constant silane content, Raman Crystallinity increases with the input power due to high flux of atomic hydrogen which is produced by silane and hydrogen dissociation. The high dissociation degree of silane provides sufficient atomic hydrogen, so microcrystalline materials can be deposited for silane content up to 43%. It is worth noticing that for ECWR discharges the experimental conditions which lead to crystalline materials are wider than for conventional ccp discharges.

Furthmore, FTIR analysis revealed that highly crystalline materials deposited at deposition rates ~12 Å/sec are porous and oxidation can take place as the material is exposed in ambient conditions. On the contrary, crystalline materials deposited at intermediate rates (~10 Å/sec) do not show significant post oxidation phenomena.

Acknowledgement
The authors wish to thank Pr. G.N. Papatheodorou and Dr. G. A. Voyiatzis, Foundation for Research and Technology Hellas – Institute of Chemical Engineering and High Temperature Chemical Processes, for providing spectroscopic facilities and Dr. F. Farmakis, Department of Electrical and Computer Engineering-Democritus University of Thrace for the film thickness measurements.

Part of this Research has been co-financed by the European Union (European Social Fund – ESF) and Greek national funds through the Operational Program "Educational and Lifelong Learning" of the National Strategic Reference Framework (NSRF) – Research Funding Program: Herakleitus II. Investing in knowledge society through the European Social Fund. Another part of the research has also been co-financed by OC Oerlikon Balzers Ltd through the UPAT – OER contract entitled

"Plasma enhanced Chemical Vapor Deposition of microcrystalline silicon from high density ECWR discharges"

References

[1] Roschek T, Repmann T, Muller J, Rech B and Wagner H 2002 *J. Vac. Sci. Technol.* A **20** 492
[2] Fukawa M, Suzuki S, Guo L, Kondo M and Matsuda A 2001 *Sol. Energy Mater. Sol. Cells* **66** 217
[3] Hammad A, Amanatides E, Mataras D and Rapakoulias D 2004 *Thin Solid Films* **451-452** 255
[4] Kosku N and Miyazaki S 2006 *Thin Solid Films* **511** 265
[5] Jia H, Saha JK, Ohse N and Shirai H 2007 *Thin Solid Films* **515** 6713
[6] Roca i Cabarrocas P. Bulkin P, Daineka D, Dao TH, Leempoel P, Descamps P, Kervyn de Meerendre T and Charliac J 2008 *Thin Solid Films* **516** 6834
[7] Dimitrakellis P, Amanatides E, Mataras D and Rapakoulias DE 2011 *J. Phys. : Conf. Ser.* **275** 012014
[8] Pfeiffer B 1965 *J. Appl. Phys.* **37** 4 1624
[9] Pfeiffer B 1965 *J. Appl. Phys.* **37** 4 1628
[10] Oechsner H 1974 *Plasma Physics* **16** 835
[11] Oechsner H, Scheib M and Goebel H 1999 *Thin Solid Films* **341** 101
[12] Oechsner H 2009 *Vacuum* **83** 727
[13] Scheib M, Schroder B and Oechner H 1996 *J. Non-Cryst. Sol* **198-200** 895
[14] Rodil S, Morrison NA, Milne WI, Robertson J, Stolojan V and Jayawardane DN 2000 *Diamond and Related Materials* **9** 524
[15] Weiller M, Lang K, Li E and Robertson J 1998 *Appl. Phys. Lett.* **72** 1314
[16] Schreiber SJ 2001 *Plasma Deposition Of Microcrystalline Silicon For Thin Film Transistors* (Cambridge University)
[17] Iqbal Z and Veprek S 1982 *J. Phys. C* **15** 377
[18] Viera G, Huet S and Boufendi L 2001 *J. Appl. Phys.* **90** 3276
[19] Paillard V, Puech P, Sirvin R, Hamma S and Roca i Cabarrocas P 2001 *J. Appl. Phys.*,**90** 3276.
[20] Veprek S, Sarott FA and Iqbal Z 1987 *Phys. Rev. B* **36** 3344
[21] Matsuda A 2004 *J. Non-Cryst. Solids* **338-340** 1
[22] Matsuda A 1999 *J. Non-Cryst. Solids* **337** 1
[23] Collins RW, Ferlauto AS, Ferreira GM, Chen C, Koh J, Koval RJ, Lee Y, Pearce JM and Wronski CR 2003 *Sol. Energy Mater. Sol. Cells* **78** 143
[24] Brodski MH, Cardona M and Cuorno JJ 1977 *Phys. Rev. B* **16** 3556
[25] Smets AHM, Matsui T and Kondo M 2008 *Appl. Phys. Lett.* **92** 033506
[26] Bronneberg AC, Smets AHM, Creatore M and Van de Sanden MCM 2011 *J. Non-Cryst. Solids* **357** 884
[27] Niwano M, Kageyama J, Kurita K, Kinashi K, Takahashi I and Miyamoto N 1994 *J. Appl. Phys.* **76** 2175

Journal of Physics: Conference Series **550** (2014) 012032 doi:10.1088/1742-6596/550/1/012032

Morphological and chemical analysis of PP film treated by Dielectric Barrier Discharge

E C Dell'Orto[1], A Vaccaro[2], C Riccardi[1]
[1] Dipartimento di Fisica, Università degli Studi di Milano-Bicocca, Piazza della Scienza 3, Italy
[2]Manucor S.p.A., 81037 Sessa Aurunca (Caserta, Italy)

Abstract. PP films have been activated by DBD, with different gases (air, N_2, H_2, CO_2) and at different energy treatments. Then film surfaces were analysed by AFM, contact-angle technique, ink test measurements and XPS, to explore modifications in surface topography, wettability, and oxidation state. After the treatments, it is possible to observe the formation of water-soluble low-molecular-weight oxidized materials (LMWOMs), which dimensions and spatial density are related to treatment parameters. Using AFM data it is possible to analyse the statistical distribution of the bubbles sizes. The analysis of XPS spectra shows the formation of oxidized polar groups on surface (e.g. C-O, C=O, COH) and allows to evaluate the extent and the nature of film oxidation, and how it influences surfaces properties. The relationship between surface chemical and morphological modifications and treatment parameters is pointed out. Then PP sample have been monitored in the time, to determine the ageing effects.

1. Introduction

Many industrial applications of polypropylene (PP) films, like printing or coating, request good adhesion properties, that are a high surface energy [1]. This characteristic is related both to the polar groups present on PP surface and to film morphology [2].

Dielectric Barrier Discharge (DBD) treatments could be used to modify surface properties, increasing wettability [3-4]. Plasma surface treatment of polymers primarily results in the oxidation of the surface, by the generation of functional groups such as C–OH (alcohol), H–C=O (aldehyde), C=O (carbonyl) and C–O–C=O (ester). Furthermore surface activation causes both a cross-linking of the polymer chains and a oxidative polymer chain degradation, for example the creation of more or less mobile chemical degradation products on surface areas of the polymer [5-6]. Thus it is possible to observe the formation of low molecular weight oxidized materials (LMWOMs). They appear as "droplets" on the treated surface that could be removed by wiping the film surface [7]. The size of this droplets results to be related to treatment dose.

Chemical modifications occurred on PP surface could be monitored by XPS measurements and FTIR/ATR measurements [8-9]. Using AFM technique it is possible to study surface topography of treated Biaxially Oriented PP (BOPP) films.

Even if in literature the effects of dose treatments on chemical and morphological modifications on PP are widely studied [10-12], here we propose a systematic study on DBD treatment of BOPP. We have analysed changes in surface energy, surface chemistry and surface topography as a function of treatment dose, and in relationship with gas used during treatments. Then aging of treated samples are monitored, to evaluate the causes of surface energy change after a storage time.

2. Experimental

The BOPP film used in this study is about 30 μm thick and has a total surface energy lower than 34 dynes/cm.
The DBD experimental setup is showed in figure 1. It consists of a stainless steel plate electrode coated by Polytetrafluoroethylene (PTFE), where samples to treat are placed on. Two rod electrodes (220 mm long, 8 mm diameter) coated with pure (>99.7%) Al_2O_3 synthesized ceramic dielectric (thickness 2 mm) are displaced in a mechanical arm, distant 20 mm from each other. The two rods are connected through an HV cable to the secondary coil of a transformer, whose primary circuit is connected to a power generator, providing the driving high voltage for the discharge. Distance between plate electrode and rod electrode could be varied between 0.5 mm to few millimetres. In our experiments it is fixed at 1 mm. During treatments the mechanical arm can move horizontally with a speed between 0.7 to 7 m/min. The electrical power deposited into the discharge could be varied until 450 W. For each treatment, the energy density, or dose, could be estimated by: $D=Pn/vl$, where P is the electrical power, v is the speed of rod electrodes, l is their length and n is the numbers of treatments.
A system of 2 mass flow controllers and a controlled evaporator and mixer allows mixing gas and vapours. The inlet fluxes are injected between the high voltage electrodes through an injection nozzle that ensures uniform fluxes on the whole width of the electrodes up to 20 l/min.

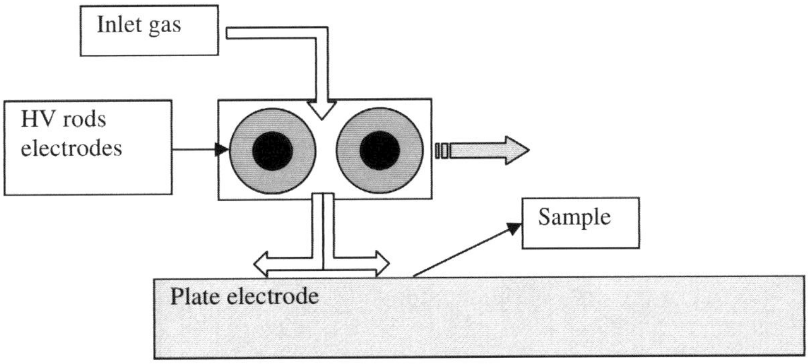

Figure 1. Schematic view of DBD treatment apparatus

Before treatments the samples are washed using acetone and then they are dried in air. After treatments PP films are characterized and then stored in air to evaluate aging affects.
The analysis of discharge properties is not a purpose of this research, chemical and physical BOPP films modifications were studied in detail.
X-ray photoelectron spectra (XPS) were recorded using a 5500-PHI (Physical Electronics) apparatus, with a monochromatic Al-anode Kα source and an electron take-off angle of 45°. The analysed circular area had a diameter of 0.8 mm. The pressure in the chamber was around 10^{-6} Pa. The spectrometer was calibrated by using the Ag 3d5/2 peak and the resulting energetic resolution was 0.46 eV.
AFM characterizations were made using a Solver P47-PRO NT-MDT, operating in semi-contact mode. Mono-crystal silicon tip with a radius of curvature lower than 10 nm was employed.
Contact angle measurements were performed using two different probe liquids: distilled water and 1-Bromonaphthalene 97% Sigma Aldrich ($C_{10}H_7Br$). Wettability was estimated using the standard test method described in literature [13], equivalent to ISO 8296.
The treatments were realized with different doses and different gas fluxes.

Journal of Physics: Conference Series **550** (2014) 012032 doi:10.1088/1742-6596/550/1/012032

3. Results

3.1. Surface energy

First we have performed air treatments. Different doses have been tested, to evaluate the effect of energy density on surface energy. In table 1 results are reported.

Table 1. Measurements of surface energy and contact angles on BOPP at different doses treatments. The first column reports the data obtained on BOPP without any treatments.

Dose [W min/m^2]	Surface energy [Dyne/cm]	Contact angle water	Contact angle $C_{10}H_7Br$
BOPP	<34	109±1.5	49±0.8
90	36	74.1±2.5	32.9±1.8
180	40	74±0.7	35.1±0.5
270	40	79.3±1.4	34.1±2.1
540	40	70±0.7	34.2±1.2
1100	42	71.8±0.9	35.1±1.1
1400	44	70.6±1.2	32.6±2

It is possible to observe how wettability raises increasing treatment dose. Anyway there is a range of dose, from 180 W min/m^2 to 540 W min/m^2 where surface energy seems to remain the same. This could be related to the existence of threshold energy in activation process. The higher value of wettability is 44 dyne/cm, reached at 1400 W min/m^2. After treatments contact angles decrease, according to surface energy data. A clear decrease occurs just after the treatment at 90 W min/m^2, while for greater doses contact angle changes are less relevant, or not appreciable at all for $C_{10}H_7Br$.
To evaluate the effect of using different gases during treatments, we have fixed a dose of 1100 W min/m^2, and have performed treatments using a mixture of nitrogen and hydrogen (95:5), and carbon dioxide.
Results are reported in table 2. Results show as using a different feed gas during treatments, surface energy increases. This improvement is more evident in CO_2 treatment. In fact wettability raises until 46 dyne/cm, while using N_2-H_2 it reaches 44 dyne/cm. Observed modifications could be related both to chemical and morphological modifications, in the next sections we examine in details the reasons of wettability change.

Table 2. Surface energy and contacts angles measured on BOPP treated using H_2-N_2 and CO_2, with a dose of 1100 Wmin/m^2

Gas	Surface energy [Dyne/cm]	Contact angle water	Contact angle $C_{10}H_7Br$
H_2-N_2	44	67.4±1.1	34.8±1.5
CO_2	46	68.2±2.1	34.5±1.4

3.2. Surface AFM

It is well known in literature how plasma treatments on BOPP cause surface modifications through surface oxidation and scission, which combine and create water-soluble low-molecular-weight oxidized materials (LMWOMs). The presence of these materials could affect wettability, stability and adhesion properties of treated films.

The formation of LMWOMs is observable by morphological modifications on film surface. They create a typical "drop-like" structure, which could be detectable by microscopy techniques, like AFM. In figure 2 and 3 the images obtained analyzing BOPP samples treated in air different doses are reported. We have stated scan images obtained using different scan size, 10 μm and 2 μm. Observing the scans performed at 10 μm, it is possible to observe how the drops are uniformly distributed on the surface, and that drops diameter increase until 1100 Wmin/m^2, then it seems to remain constant. Then the scans executed at 2 μm, allow examining in details LMWOM structures. Starting from a dose of 540 Wmin/m^2, large spread bubbles are superimposed to join smaller ones. In table 3 the height and

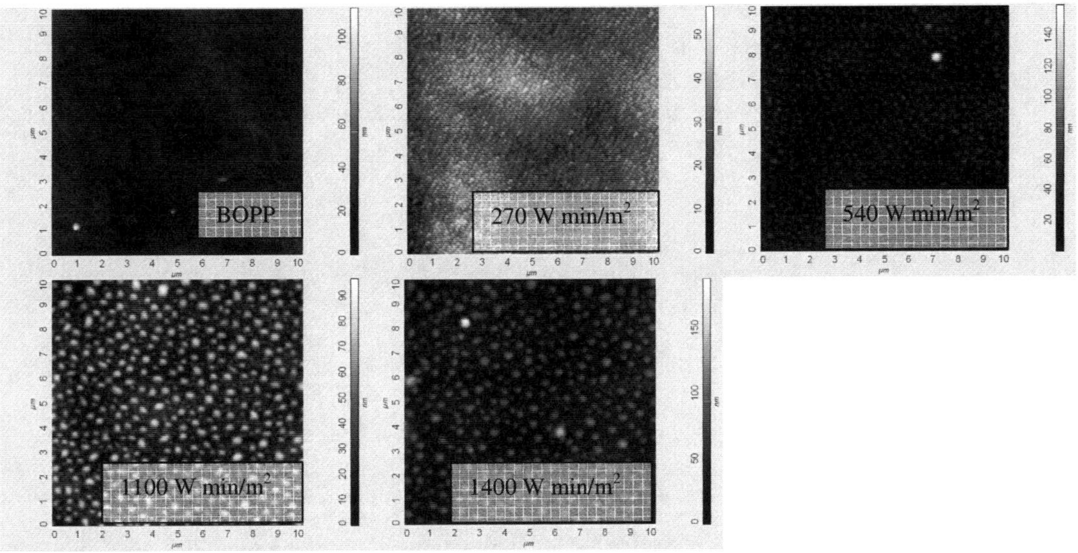

Figure 2. AFM images obtained on BOPP films treated in air at different doses, with a scan size of 10 μm. The formations of LMWOM are pointed out from the presence of "bubble" on the surface.

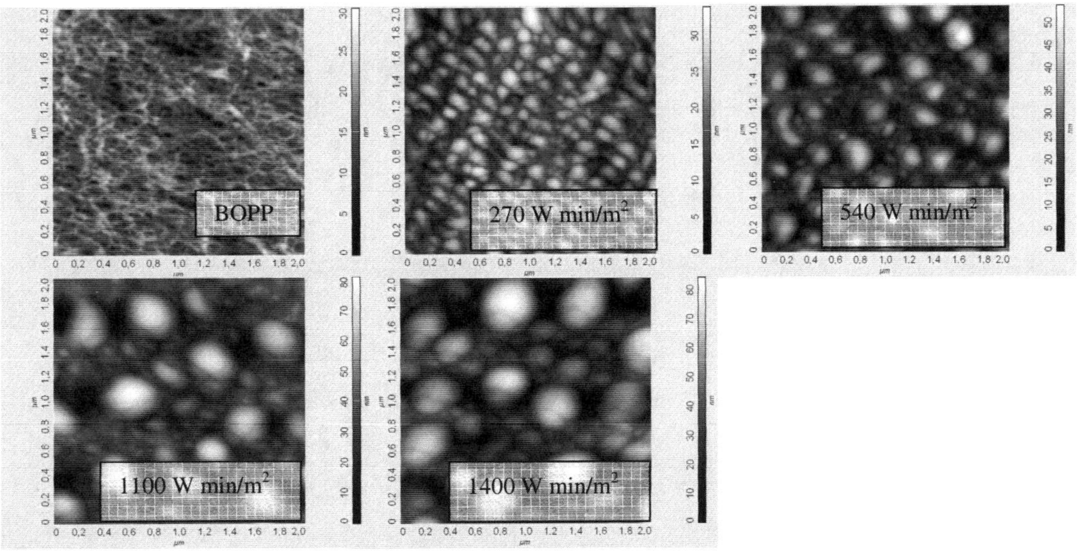

Figure 3. AFM images of BOPP films treated in air at different doses, with a scan size of 2 μm.

Table 3. LMWOM structures parameters. The "drop-like" structures are separated in two groups, depending on size and spatial distribution.

Dose	Bigger structures		Secondary structures	
[W min/m^2]	Diameter [nm]	Height [nm]	Diameter [nm]	Height [nm]
270	100-200	15-20	-	-
540	200-300	30-40	100	10-15
1100	300-500	50-70	100-200	10-30
1400	300-500	50-70	150-200	15-30

the width of the bubbles are reported, discerning, where present, the bigger structure from smaller ones. Observing data reported in table 3, it is possible to note as the bubbles are low and wide, the ratio between the diameter and the height is about 10 for all samples. This characteristic will be useful to analyze samples profile.

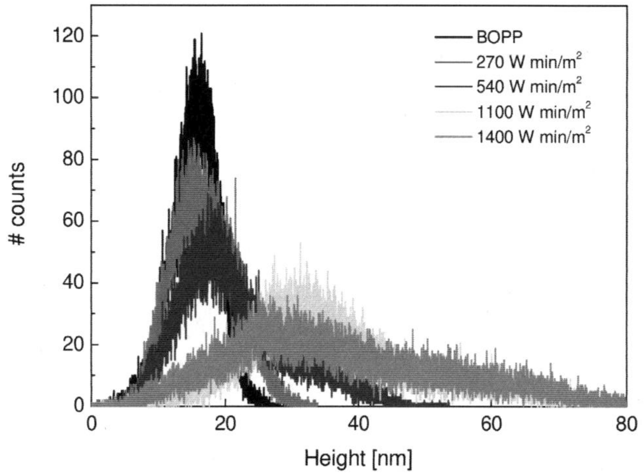

Dose [Wmin/m^2]	rms[nm]
0	3.54
270	4.78
540	10.39
1100	13.5
1400	17.83

Figure 4. Histograms of the surfaces of the scans showed in figure 3, obtained with a scan size of 2 μm. In the table on the right, roughness values are reported.

Focusing on the analysis of distributions shape, it is possible to examine the contributions of different LMWOMs structures. In figure 5 we have reported the height distributions relative to treatments

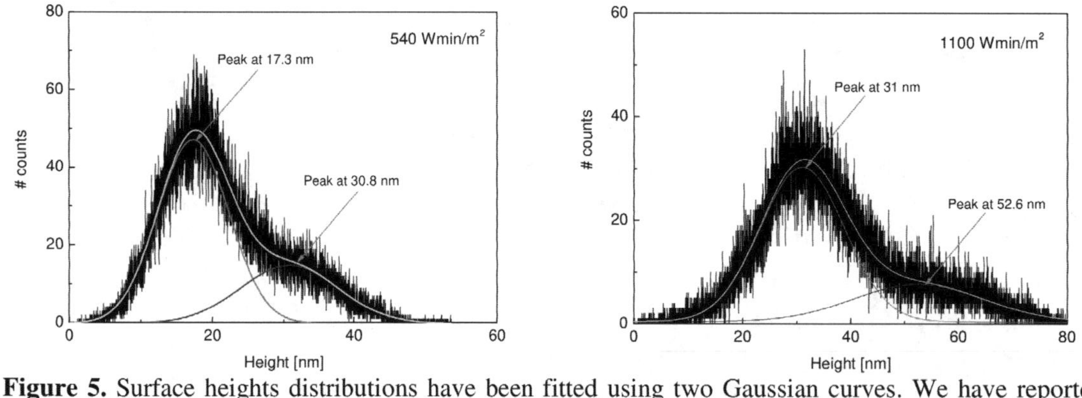

Figure 5. Surface heights distributions have been fitted using two Gaussian curves. We have reported results obtained on samples treated at 540 and 1100 Wmin/m^2.

performed at 540 and 1100 Wmin/m^2, and the numerical fits obtained using two Gaussian distributions. The peaks of the Gaussian distributions corresponds roughly to the values reported in table 3, those were extrapolated from profile scans. Then peaks intensity allows to estimate the relative presence of "drop-like" structures with different heights on the same surface.

After the analysis of the effect of dose on morphology, it is interesting to evaluate the effect of using gas during treatments. In figure 6 AFM images performed on CO_2 treated BOPP (D = 1100 Wmin/m^2), are reported. Also in this case, it is possible to observe the typical "drop-like" structure, even if in this case the smaller drops are not so well defined. The drops heights are about 20-40 nm, the diameters are about 200-400 nm.

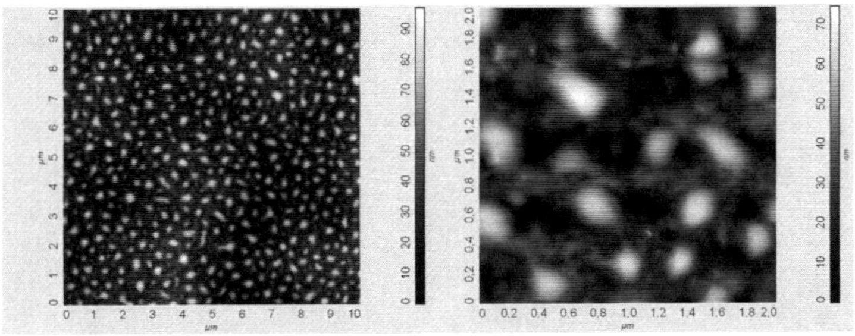

Figure 6. AFM images performed on CO_2 BOPP treated at 1100 Wmin/m^2.

LMWOMs are soluble in polar liquids, such as water used in contact angles measurements. We have tested the effect of water on film morphology. The results have showed that, during contact angle measurements, the amount of water is to small and the time interval is to short, to cause morphological modification, and then contact angle values.

In this paragraph we have analysed morphological modifications that occur on BOPP surface after DBD treatments. These modifications turn out to be related to treatment dose. We have pointed out in previous sections how also energy surface modifications are related to treatment dose. However the relationship between surface morphology and wettability properties is not so simple to understand. Observing data it is possible to note that, even if wettability doesn't change performing a treatment with a dose of 270 or 540 Wmin/m^2, samples surface appears different. At the same time, the change in wettability that occurs between 1100 and 1400 Wmin/m^2 is not reflected in a change of morphological structure. It emerges how more than one cause occur in alter wettability.

The next paragraph analyse chemical variations that take place on the surface as a consequence of DBD treatments.

3.3. Surface chemistry

XPS measurements have been performed on the following samples: untreated BOPP, air treated BOPP at 270 and 1100 Wmin/m^2, to evaluate the effect of dose on chemical modifications, and CO_2 treated BOPP at 1100 Wmin/m^2, to evaluate the effect of feed gas. Table 4 shows the atomic percentage of elements determined on these sample. After all treatments the amount of oxygen increases, and this increase is more relevant, as treatment dose rises. So it is possible to state a relationship between dose intensity and the presence of oxygenated functional groups on sample surface. In the CO_2 treated sample, a small amount of nitrogen has been detected.

The results of the high resolution C1s and O1s peak deconvolutions are summarized in table 5. In the table the assignments are reported too [14]. Plasma treatments on BOPP have the effect to increase the quantity of carbon atoms that are singly or doubly-bonded to oxygen. In detail, the creation of C-O groups is favored over the formation of C=O groups. The small amount of nitrogen, detected in case (c), could be due to chemical reaction between air and BOPP surface, immediately after the treatment.

Table 4. Atomic percentage of elements, for studied samples: untreated BOPP; (a) air treated BOPP (D = 270-1100 Wmin/m^2); (b) air treated BOPP (D = 1100 Wmin/m^2); (c) CO_2 treated BOPP (D = 1100 Wmin/m^2).

Sample	O1s	C1s	N1s
BOPP	2,5	96,9	
(a)	13,1	86,9	
(b)	18,1	81,9	
(c)	17,0	82,5	0,5

Table 5. High resolution XPS C1s and O1s peak deconvolutions for studied samples: untreated BOPP; (a) air treated BOPP (D = 270-1100 Wmin/m^2); (b) air treated BOPP (D = 1100 Wmin/m^2); (c) CO_2 treated BOPP (D = 1100 Wmin/m^2).

				% Area			
	Center [eV]	ΔE [eV]	Assignment	BOPP	(a)	(b)	(c)
C1s	285.0	-	C-H, C-C	92,5	80,6	76,5	76,6
	286.7	1.7	C-O	6,2	12,0	13,5	14,4
	288.3	3.3	C=O	1,2	5,4	6,8	6,6
	289.7	4.7	O-C=O	-	2,0	3,1	2,4
O1s	532.7	-	C-O, C=O				

3.4. Ageing

All the characterizations that have been discussed before, have been monitored during a storage time of up to more than one month, on BOPP treated in air at 270 and 1100 Wmin/m^2 and BOPP treated in CO_2 at 1100 Wmin/m^2. Measurements of surface energy and contact angles are plotted in figure 7.

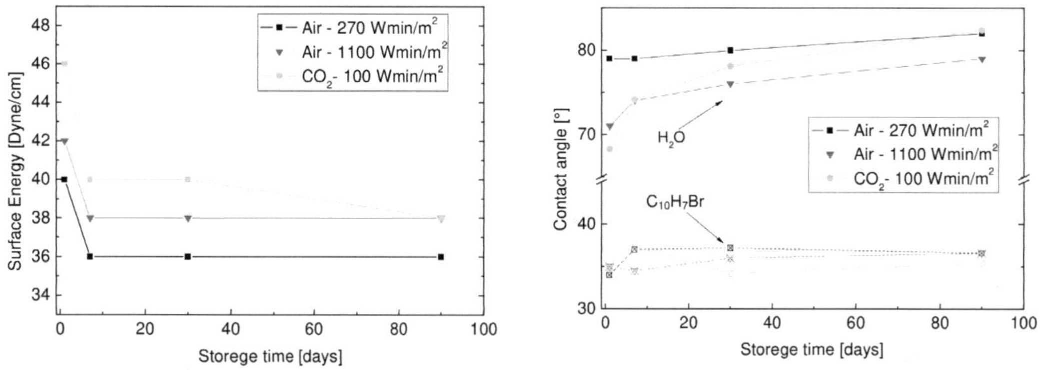

Figure 7. Surface energy and contact angles measurements on treated BOPP, as a function of storage time. Straight and dotted lines are guides for eyes.

The observed decrease of hydrophilic properties on treated samples could be attributed to several factors, such as diffusion, agglomeration or sublimation of LMWOM, modification in polymer chains and migration of additives from the bulk to the surface [15-16]. To better understand the origin of detected "hydrophobic recovery", morphological and chemical properties of treated films were examined during the storage period.

AFM images obtained on stored samples (figure 8) don't show relevant modifications after three month from the treatment. So morphology of treated BOPP gives information about the amount and the size of LMWOMs, and these properties affect wettability. However the lack of hydrophilic properties is not related to morphology changes.

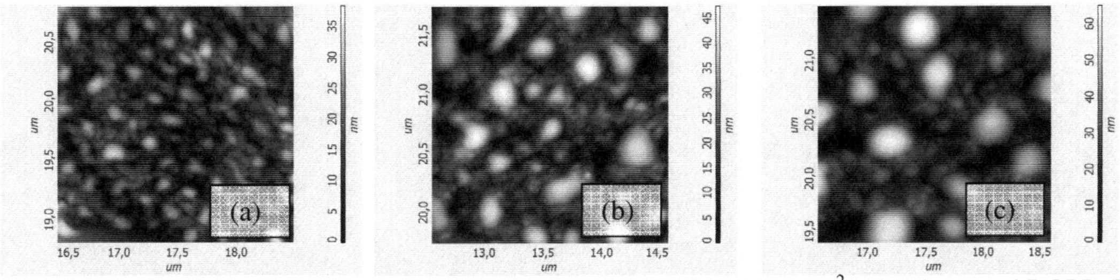

Figure 8. AFM images of: (a) air treated BOPP (D = 270-1100 Wmin/m^2); (b) air treated BOPP (D = 1100 Wmin/m^2); (c) CO_2 treated BOPP (D = 1100 Wmin/m^2). Images are obtained with a scan size of 2 μm.

Table 6. Atomic percentage of elements for studied samples: untreated BOPP; (a) air treated BOPP (D = 270-1100 Wmin/m^2); (b) air treated BOPP (D = 1100 Wmin/m^2); (c) CO_2 treated BOPP (D = 1100 Wmin/m^2). Data are reported at different storage time.

Sample	O1s			C1s			N1s		
	0 day	1 week	1 month	0 day	1 week	1 month	0 day	1 week	1 month
BOPP	2,5			96,9					
(a)	13,1	12,3	11,6	86,9	86,8	88,3			
(b)	18,1	13,8	13,0	81,9	86,2	86,2			0,5
(c)	17,0	16,4	13,9	82,5	81,6	85,3	0,5	0,8	0,6

Table 6 reports XPS data collected at different storage time. All samples show a decrease in oxygen percentage. This decrease however is not enough to explain the change in wettability. For example CO_2 treated sample shows, after 1 month, an oxygen percentage of about 13.9%, greater respect to air treated BOPP at D= 270 Wmin/m^2, however its wettability is lower. The explanation of this result could be the phenomena of the polymer chain migration toward the film bulk and chain reorientation. In this way, the functional groups can be still detected by the XPS analysis, since its penetration depth is of the order of few nanometers, but they don't influence the surface energy, since they are not exposed to air.

4. Conclusion

We have performed a systematic study to compare the effects of treatment dose and used gas on BOPP properties. We have pointed out how these parameters affect wettability, and how it is possible to relate the increase in energy surface to chemical and morphological modifications. Then we have monitored surface properties during time, to observe the hydrophobic recovery. We have point out how this recovery is related to variations in polymer chains, not to sublimation or reorganization of LMWOMs.

5. References

[1] Uehara T., 1999. *Corona discharge treatment of polymers,* MATERIALS ENGINEERING-NEW YORK, 14, 191-204.

[2] Overney R M, Guntherodt H-J, Hild S, 1994, *Corona-treated isotactic polypropylene films investigated by friction force microscopy*, Journal of Applied Physics 75, 1401.

[3] De Geyter N, Morent R, Leys C, Gengembre L, Payen E, 2007, *Surface and Coatings Technology*, 201(16), 7066-7075.

[4] Strobel M, Jones V, Lyons C S, Ulsh M, Kushner M J, Dorai R, Branch M C, 2003, *A comparison of corona-treated and flame-treated polypropylene films*, Plasmas and Polymers, 8, 1, 61-95.

[5] Grossman R F, Beasley W A, 1959, *Effects of corona discharge upon polyethylene*, J. Appl. Polym. Sci. 2, 163.

[6] Kim C Y, Evans J, Goring A I, 1971, *Corona-induced autohesion of polyethylene*, J. Appl. Polym. Sci. 15, 1357.
[7] Overney R M, Luthi R, Haefke H, Frommer J, Meyer E, Gu¨ntherodt H-J, Hild S, Fuhrmann J, 1993, *An atomic force microscopy study of corona-treated polypropylene films*, Appl. Surf. Sci., 64, 197.
[8] Suzer S, Argun A, O. Vatansever O, Aral O, 1999, *XPS and Water Contact Angle Measurements on Aged and Corona-Treated PP*, J. Appl. Polym. Sci., 74: 1846–1850.
[9] Massines F., Gouda G, Gherardi N, Duran M, Croquesel E, 2001, *The Role of Dielectric Barrier Discharge Atmosphere and Physics on Polypropylene Surface Treatment*, 6, 35-49.
[10] Kostov K G, Hamia Y A , Mota R P, dos Santos A L R, Nascente P A P, 2014, *Treatment of polycarbonate by dielectric barrier discharge (DBD) at atmospheric pressure*, Journal of Physics Conference Series 05/2014, 511, 012075.
[11] Jones, V., Strobel, M., & Prokosch, M. J. (2005). *Development of poly (propylene) surface topography during corona treatment,* Plasma Processes and Polymers, 2(7), 547-553.
[12] Mix R, Friedrich J, Rau A, 2009, Polymer surface modification by aerosol based DBD treatment of foils, Plasma Processes and Polymers, 6, 9, 566-574.
[13] ASTM International, Standard Test Method for Wetting tension of Polyethylene and Polypropylene Films.
[14] Beamson G, Briggs D, 1992, High resolution XPS of Organic Polymers: The Scienta Acta ESCA 300 Database, John Wiley, Chichester.
[15] Strobel J M, Strobel M, Lyons C S, Dunatov C, Perron S J, 1991, Aging of air-corona-treated polypropylene film. Journal of Adhesion Science and Technology, 5, 119-130.
[16] Morent R, De Geyter N, Leys C, Gengembre L, Payen E, 2007, Study of the ageing behavior of polymer films treated with a dielectric barrier discharge in air, helium and argon at medium pressure, Surface and Coatings Technology, 201(18), 7847-7854.

Journal of Physics: Conference Series **550** (2014) 012033 doi:10.1088/1742-6596/550/1/012033

Comparison of ICP and DBD plasma process for the degreasing of metal reinforcements

F Judee[1,2], **E Duffour**[3,4], **F Perisse**[3,4], **S Menecier**[3,4] **and D Vacher**[3,4]
[1] Université de Toulouse, LAPLACE-UMR 5213, F-31062 Toulouse, France
[2] CNRS, LAPLACE-UMR 5213, F-31062 Toulouse, France
[3]Clermont Université, Institut Pascal, BP 10448, F-63000 CLERMONT-FERRAND
[4] CNRS, UMR 6602, IP, F-63171 Aubière, FRANCE

E-mail: judee@laplace.univ-tlse.fr, sebastien.menecier@univ-bpclermont.fr, emmanuel.duffour@univ-bpclermont.fr, frederic.perisse@univ-bpclermont.fr, damien.vacher@univ-bpclermont.fr

Abstract. This paper presents two different ways of degreasing metallic wires based on plasma technology. An ICP torch and a DBD with different electrodes shapes were compared using both conventional method and design of experiments. Results show that with both methods a good degreasing can be obtained but the process optimization is uneasy because of the large number of impacting parameters. For the ICP, significant temperature gradients also make it difficult to find a good repeatability while in the case of the DBD, the filamentary discharge does not affect the regularity of treatment.

1. Introduction

Metal reinforcements are used in many industries to confer rigidity to a soft structure. In that purpose, plasma technologies were tested, especially ICP and DBD technologies. Metal wires were treated using both technologies in wide variations of different parameters such as plasma power or wire displacement velocity in the active plasma. Degreasing efficiency was then evaluated using Scanning Electron Microscope and mechanical properties were tested. Designs of experiments were used instead of conventional ones. Relevant parameters of the process were deduced from the designs and optimization was achieved for a better degreasing.

2. Methodology

As each analysis requires a relatively long processing time, it is crucial to determine the right number of tests to carry out to study the effects of our different plasma reactors on the wire while maintaining quality analysis data. In order to obtain the best degreasing while keeping the initial mechanical properties of wire, a design of experiments was used. The principle of design of experiments is to link measurable physical quantities, called "response" Y, to varying experimental elements, the "factors" x_i, by a mathematical model, typically a polynomial equation. This type of packaging actually allows to achieve a minimum of experiments to determine all the model parameters. Indeed if there is n unknowns in this model, n equations thus n measurements are required. For instance, a response Y can be modelled using the first degree polynomial with interactions:

$$Y = a_0 + \sum_{i=1}^{n} a_i x_i + \sum_{i=1}^{n} \sum_{j=1;\, j \neq i}^{n} a_{ij} x_i x_j \qquad (1)$$

Based on this principle, the n factors requires $2n$ tests illustrated in a minimum space of n dimensions by points located at the apex of an n-dimensional hypercube to minimize errors. This type of model can be easily implemented in the screening phase. But once obtained ideas about the type of global influence of each factor it is necessary to pass into more faithful model called response surface. In these models we add the terms of the second degree:

$$Y = a_0 + \sum_{i=1}^{n} a_{ii} x_i^2 + \sum_{i=1}^{n} \sum_{j=1;\, j \neq i}^{n} a_{ij} x_i x_j \qquad (2)$$

The optimization of the position of points in the measurement space allows different configurations among which we mention Doelhert plans where measurements points are uniformly distributed on a circle. For our measurements we use for targeting phase a composite plan. This type of plan has the advantage of being part of a plan consisting in first degree and change of a model to another is easier and it allows a gain in measure. To obtain model as accurate as possible, it is important to keep all other factors constant potentially influenced during all testing (hydrometric of the room, for example). In addition to this points, a measurement point with average values of all the factors is repeated at least three times. So, we obtained an information about the repeatability of the process.

Experimental design identify quickly the important experimental parameters on each study area defines but also to find local optimum if they exist. The results presented in this paper are the product of many experimental design and their successive optimizations.

3. Experimental setup and analytical devices

Here are presented the plasma technologies used as well as the characterization devices.

3.1. Plasma facilities

Two different plasma technologies were used:

- Inductive Coupled Plasma torch (ICP);
- Dielectric Barrier Discharges (DBD).

3.1.1. ICP torch. The inductively coupled plasma torch ICP-T64 can operate with different kinds of plasma gas (air, argon, CO2, N2 and gas mixtures). A seven-turn induction coil, made in inconel, cooled by an inner air flux is used to sustain the plasma. Figure 1b shows the experimental setup. The operating conditions of the experimental set-up are given in table 1.
The plasma is generated through the induction coil by a radio frequency (RF) of 64 MHz at an electrical applied power of up to 4 kW. The plasma is confined within a 28 mm wide cylindrical, vertically mounted quartz tube. The plasma gas, with a swirl injection, has a fixed total mass flow rate of 0.2 *g/s*.

Table 1. Operating conditions of the plasma torch set-up. Pressure was atmospheric pressure.

Applied voltage (kV)	2.83	3.05	3.30
Anode current (mA)	537.5	576.0	620.5
Power supply (kW)	1.52	1.76	2.05
Plasma gas flow rate (L/min)	15	18	21

3.1.2. DBD devices. In this study, two different electrodes were used. One is called DBD strait figure 2 and the second DBD comb figure 3 in reference to their geometrical forms.

For both of them, planar dielectric is made of 1.6 mm thick polyepoxides and 35 μm thick copper electrodes. The power supply is done by a single phase angle controller, Celduc SG4, with Step-up transformer 230/10 000 V, allows to generate high voltage required for application. Power change is attained through an input setpoint 0-10 V on the phase angle controller. The experimental conditions prensented in table 2 correspond at the optimal power value for each generator.

Table 2. Operating conditions of the dielectric barrier discharge set-up.

Applied voltage (kV)	10
Frequency (Hz)	50
Phase angle controller (°)	98
Operating gas :	network air
Operating pressure :	atmospheric pressure

.

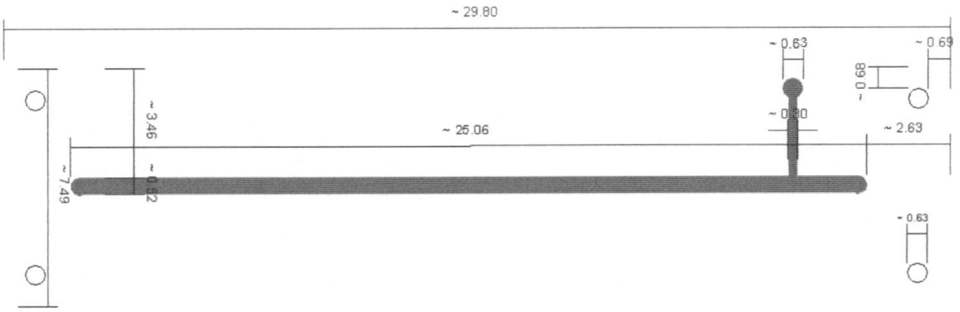

Figure 2. Geometric characteristics of DBD strait, length are expressed in cm.

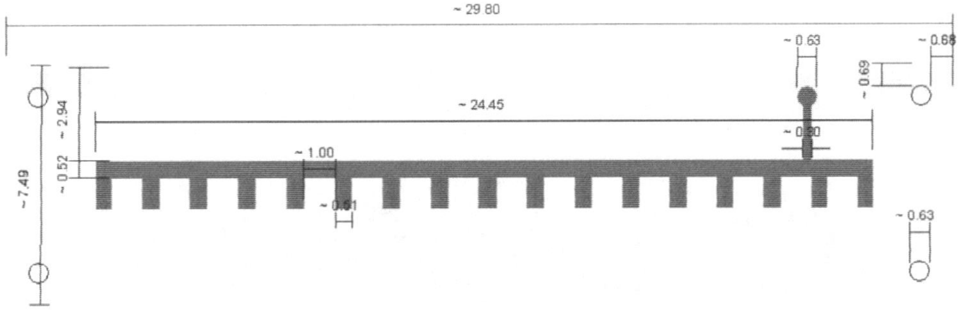

Figure 3. Geometric characteristics of DBD comb, length are expressed in cm.

3.2. Gas

Gas used in experiments were argon and network air. The air composition of the laboratory network summarized in table 3 is supposed to be identical to that of dry air under standard conditions of temperature and pressure (molar proportion). The air moisture is neglected.

Table 3. Composition of network air.

N_2	78.08 %
O2	20.95 %
CO2	0.03%
Ar	0.93 %

3.3. Analyses
Three methods of analysis are detailed below, one optical and two mechanical.

3.3.1. SEM. Wires visual studies were made with a scanning electron microscope (Phillips XL30), allowing a very low acceleration voltage, and therefore very surface sensitive. Treatment efficiency is evaluated with a mark between -10 and +10:

- 0 means that the process was useless for the wire degreasing (identical to the reference sample);
- +10 corresponds to a perfect degreasing;
- -10 to a complete damage of the wire surface.

Figure 4 shows for instance two different results with two different degreasing efficiency. Figure 4(a) is a typical view of untreated sample with the presence of grease at the surface identified by its dark tint and more random distribution. Figure 4 (b) shows the result in a particular treatment condition by ICP torch described in Part 4.1. The brighter areas at high and low ends of the pictures are caused by the fact that the sample is cylindrical.

3.3.2. Mechanical measurements. Plasma treatment should change anything in the intrinsic mechanical properties of the metal wire, which is why control tests are performed by measuring the resistance to mechanical forces that are applied by a standard method of pulling out. The comparison of the maximum permissible load (maximum load accepted before passing the deformation zone at break) with control wires can guarantee a certain quality of the finished product.

To study the impact of degreasing on adhesion properties of the wire, we have made wire eight test pieces (treated or untreated) embedded in a nylon base melted after heating in an incubator of beads of PA6 nylon. The study of these in traction specimens provides adhesion contact materials and also the support of our wire with this material. The multiplicity of wire into the sample provides information on the repeatability of the process.

Figure 4. Visual inspection of wire surface of 500μm by 90μm and marking. (a) Untreated sample (mark 0), (b) sample treated with the ICP torch with a mark of reference estimated at +10.

3.3.3. Contact angle. Such as tensile adhesion tests are lengthy process and requires a significant length of wire, we want to evaluate another parameter that could learn about the trends in this traction grip. Measuring the contact angle can actually perform this function since it shares many concepts with common physical adhesion phenomena such as roughness and surface tension.

To perform this measurement, it is necessary to immerse the wire in a liquid and measure the angle between the meniscus ant the wire at the triple point of contact [7] (contact line between solid, liquid and gas phases). Usually done when the wire is upright and tense, inclining the wire (see figure 5) facilitates measures and makes it more accurate. Indeed it is sufficient to tilt the wire until a side of the wire meniscus disappears. In this condition the contact angle is precisely the angle formed between the wire and the free surface of the water.

Figure 5. Contact angle measurement by the method of the surface level.

4. Results

4.1. ICP
Experiments were achieved using ICP torch whose parameters are described in table 4.

4.1.1. SEM. For this testing phase the distribution of scores obtained after SEM analysis is relatively diffuse. These results are exposed in figure 6. Eight studies have negative scores symbol of degradation upper layer of wire, two are semblabe to the control and was therefore not affected by the plasma treatment. Finally eleven samples get a positive mark with perfectly treated sample, marked +10. For these tests a significant gain, for fives of them, is noted when the decrease of fat on the surface. With regard to the +10 test evaluated, the SEM image show an almost total disappearance of the latter figure 4.

From these results, it turned out to be very difficult to point out the most relevants parameters of the process. The configuration that allows this measure is tested when the perceived energy of the wire is the most intense that is to say when it is near the plasma, at maximum power and flow and a speed of scrolling weak. Repeatability is it more questionable indeed for all four grades obtained, based on an analysis of the same wire treated in the same way, the notes were +7, +2, +1 and 0.

Table 4. Limits study of factors for ICP torch.

Factors	Maximum value	*Minimum value*
Distance $_{\text{plasma wire}}$ (cm)	12	20
Electrical power (kW)	1.5	3.5
Argon flow (L/min)	10	15
Displacement velocity (m/s)	0.50	2.51

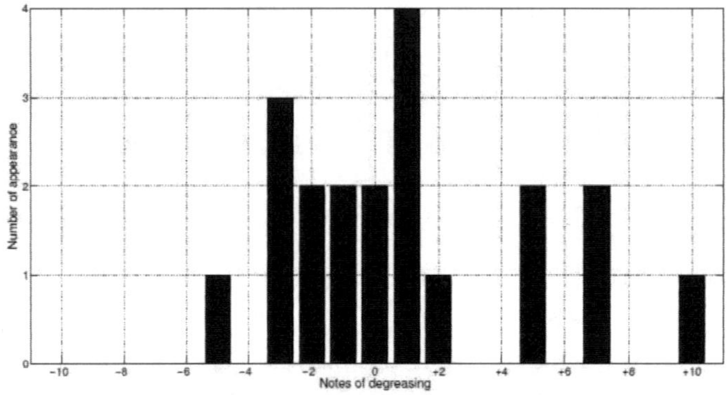

Figure 6. Efficiency of degreasing distribution achieved through optimised design of experiments to ICP torch.

4.1.2. Stress tests. Behavioral studies after treatment of cables with respect to force traction do not show any major degeneration with an average maximum load and maximum displacement relatively close to the control (Table 5). For testing traction adhesion, the mechanism is more variable. Indeed, only three points allow us to gain adhesion, while all other trials are losing adhesion significantly, up to 14 % (Figure 7). So even if the generator provides a relatively good degreasing without damaging the mechanical properties it is not possible to maximize adherence with PA6 nylon.

Table 5. *Stress tests ICP performed on 20 tests of the plan.*

	Reference	*Metal wire after treatment*	
		Average	99.73 % confidence interval (Student law)
Maximum load (N)	248.2	248.1	±7.9
Maximum move of strength (mm)	3	2.9	±0.8

4.1.3. Comparative study of the contact angle and adhesion. The comparative study of the contact angle and adhesion does not identify a real dependency between the two factors, one can only note that all samples submitted possess a contact angle greater than our samples control. It is possible to note that the maximum adhesion gain is observed for test 3 which is also the test number for which the angle of contact is the closest to that of the control.

4.1.4. Analyse by optical microscopy of the wire's surface. The best score is obtained for higher thermal and chemical energies received by the wire, tests are carried out getting closer and closer over the plasma plume. In this way we can observe in figure 8 the evolution of the outer layer of the wire at the optical microscope according to this energy. Whereas the wire has a pale color and sometimes iridescent elements when it is not treated, it then goes through different stages. For medium energy, so it takes a relatively diffuse bright orange and pink.

Such measures tend to confirm a chemical oxidation at the surface of the elements forming the outer layer of the wire during the process. In addition, oxidation distribution shows the holistic character of the treatment. Finally when an energy threshold has been exceeded, the oxidation process is such that the surface of the wire disappear and reveals the wire core. Finally, these data allow us to think that the problem of repeatability of the phenomenon is mainly due to the system of movement of the wire and not to the plasma.

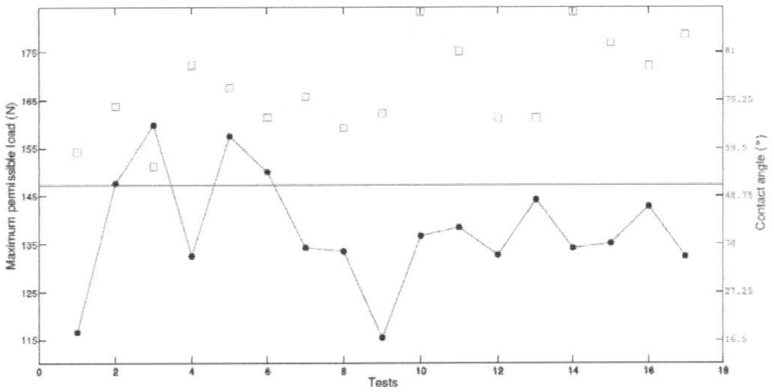

Figure 7. Comparison between maximum permissible load in tensil-adherence test and contact angle during a series of 17 tests with ICP devices. The red line represents the measured value for the same tests on a control untreated wires. Blue line with • markers is for the maximum load and green line with markers is associated with the contact angle.

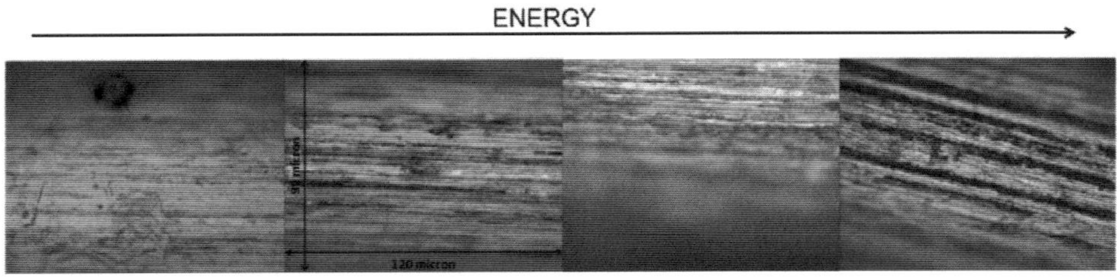

Figure 8. Surface texture of the wire (90 μm-120 μm) for different energy (thermal and chemical) levels applied. Left picture shows the untreated wire, while right picture was taken after a treatment with low displacement velocity into the plasma plume.

4.2. DBD

Experiments were also achieved using DBD devices whose parameters are described table 6.

Table 6. Limits study of factors for DBD devices.

Factors	Maximum value	Minimum value
Distance inter-electrode (mm)	3.2	4.8
Electrical power (W)	2	3.5
Air flow (L/min)	0	3
Displacement velocity (m/s)	0.42	0.84

4.2.1. SEM. While distributions notes of torch ICP showed a lot of damage, the DBD system itself have only slightly damaged two trials, three are similar to untreated and eleven test with a degreasing efficiency of our wire. Results of degreasing efficiency is presented in figure 9. The maximum score in is only +6 which is lower than for the ICP torch but is still quite interesting especially that repeatability is greatly improved because its four test offers notes -2, -2, +1 and +2. This method of treatment is able to degrease and is repeatable. It has especially the advantage of not causing degradation of the wire if the conditions are respected. This conditions for the device are low power and low distance between electrodes without air gas flow. As regards the crolling system, must privilege a fast speed of displacement to protect the wire. The application is therefore all the more interesting that she has the advantage of a possible industrial adjustment with higher speed. Disparate treatment due to filamentary DBD seems relatively well avoided by the movement of the wire.

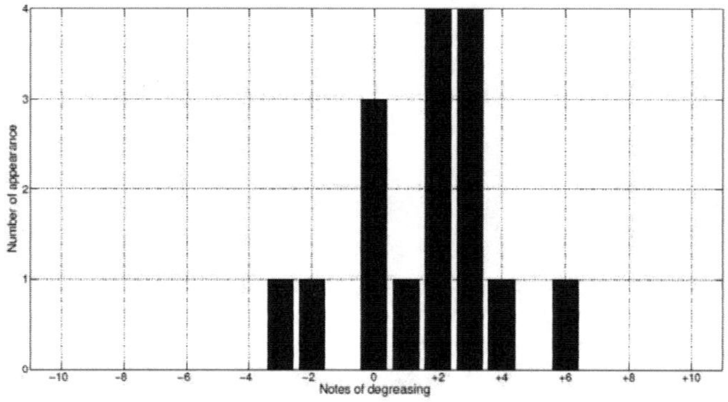

Figure 9. Efficiency of degreasing obtained for several time exposition.

4.2.2. Stress tests. Similarly, the results of traction tests shown in table 7 are relatively close to the values of untreated cables. There can be a slight drop in quality though. A supplementary study of the rotating bending wire (test applies cyclically contractions and compressions efforts in the extreme surface of the wire) have been achieved to demonstrate that the mechanical characteristic of the wire are well respected even after treatment. Furthermore, the traction adhesion measurements are also consistent with those found for the ICP torch is no significant increase in membership with losses that may reach up to 20 % in adhesion.

4.2.3. Comparative study of the contact angle and adhesion. With the DBD system, the behavior of the contact angle and adhesion seems more connected (See figure 10). We can find the same overall changes depending on test. However, and considering the test with the ICP torch it is difficult to clearly identify a correlation between the two variables. Thus the contact angle measurement cannot allow in any case, any deduction of the evolution of the adhesion of wire.

Table 7. Stress tests DBD performed on 16 tests of the plan.

	Reference	*Metal wire after treatment*	
		Average	99.73 % confidence interval (Student law)
Maximum load (N)	248.9	246.8	±9.2
Maximum move of strength (mm)	3.6	3	±0.4

4.2.4. Analyse by optical microscopy of the wire's surface. To compare the different geometry of the generator DBD, we first observed the optimal conditions previous planes. Unfortunately no significant difference appeared. In order to observed changes, some measurements were carried out under "extrem" conditions. To do this, we fix the wire speed at its minimum i.e. 0.25 m/s (corresponding to the passage of wire during one second in the plasma), some tests in static complemented these measures. Results are presented in figure 11. The static tests provide results of degreasing of 0 to 6 s for the DBD strait and comb DBD respectively: (0, 0, 0, +2, +1 and +5) and (0, +1, +3, +7, and two negatives values). Logically the DBD comb focuses energy on narrower field having, in addition to tip effect, who thus encourages a more rapid degreasing of the wire for comb DBD. As regards the two dynamic tests (one second scroll processing), they are of +4 for DBD strait and of +3 for the DBD comb. We can clearly identify the fact that scrolling promotes degreasing without explanation at the

Journal of Physics: Conference Series **550** (2014) 012033 doi:10.1088/1742-6596/550/1/012033

moment. Red elements appearing on the DBD comb pictures are actually oxide glasses, which relates the important local heating thus allowing the merger of oxide but also equally abrupt cooling to freeze the elements in this state.

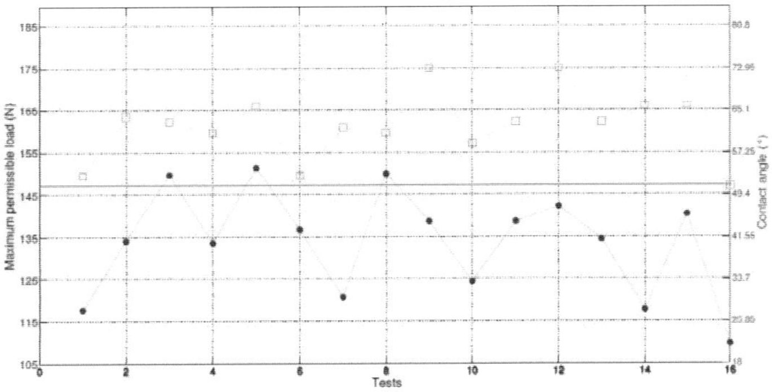

Figure 10. Comparison between maximum permissible load in Tensil-adherence test and contact angle during a series of 16 test with DBD devices. The red line represents the measured value for the same tests on a control untreated wires. Blue line with point markers is for the maximum load and green line with square markers is associated with the contact angle.

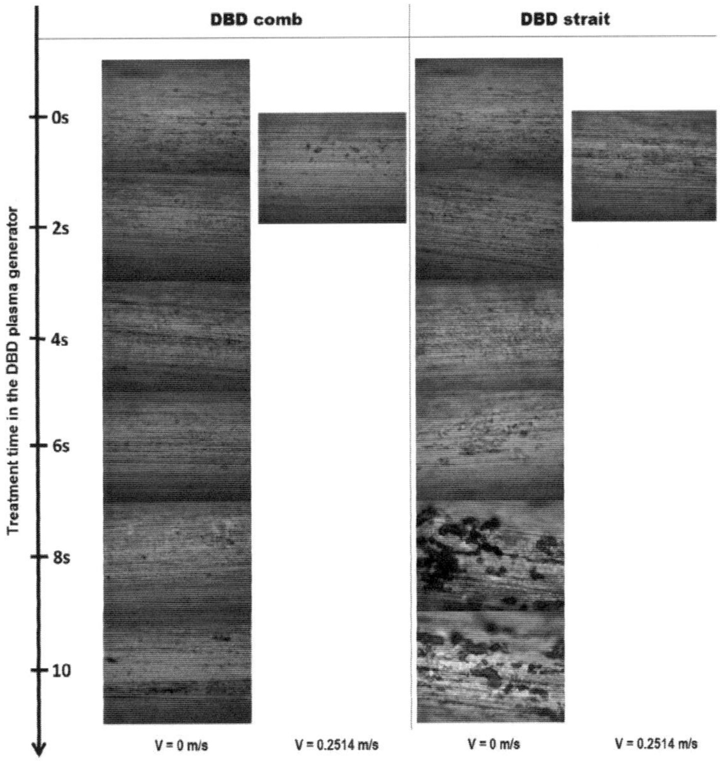

Figure 11. Surface texture of the wire (90 μm - 120 μm) for different treatment times and geometrical DBD devices.

5. Conclusion

From this study, we show that wire degreasing was possible using plasma technologies like DBD and ICP torch. The use of design of experiments method was not so fruitful since we were not able to point out the most relevant parameters in each process. The possible reasons are many:

- some important parameters maybe were not taken in account like room temperature or moisture rate;
- some response measurements were only qualitative like degreasing efficiency.

Anyway, from all the results we can claim that:

- ICP process seems to be a workable process for degreasing, but degreasing efficiency is very sensible to parameters variations. Indeed near the plasma, temperature gradients are very important, leading to other side-effects like for instance turbulence. Therefore, it turned out to be very difficult to find the good parameters to put just energy enough for a good degreasing without oxidizing or deteriorating the wire. It also have an impact onto the measures repeatability;
- DBD results are very promising even if we did not get a +10 degreasing efficiency. Further studies will focus on the DBD improving, especially the electrodes shapes, the power supply and signal shape and frequency.

6. References

[1] Ruimi M 1996 Solvants chlorés et environnement *GalvanoOrgano : Traitements de surface* **668** 589-94.

[2] Hellweg S, Demou E, Scheringer M, McKone T, Hungerbuhlerk K 2005 Confronting Workplace Exposure to Chemicals with LCA: Examples of Trichloroethylene and Perchloroethylene in Metal Degreasing and Dry Cleaning *Environ. Sci. Technol.* **39** 7741-48

[3] Tendero C, Torche plasma microonde à la pression atmosphérique : application au traitement de surfaces métalliques, thèse de l'Université de Limoges, 2005

[4] Goossens O, Dekempeneer E, Vangeneugden D, Van de leest R, Leys C 2001 Application of atmospheric pressure dielectric barrier discharge in deposition, cleaning and activation *Surface & Coatings Technology* **142-144** 474-81

[5] Thiebaut JM, Belmote T, Chaleix D, Choquet P, Baravian G, Puech V, Michel H 2003 Comparison of surface cleaning by two atmospheric pressure discharges *Surface & Coatings Technology* **169-170** 181-85

[6] Herrmann HW, Henins I, Park J, Selwyn GS 1999 Decontamination of chemical and biological war-fare (CBW) agents using an atmospheric pressure plasma jet (APPJ) *Physics of Plasmas* **6** 2284-91

[7] Yuehua Y, T.Randall L 2013 *Contact Angle and Wetting Properties (Surface Science Techniques* vol 51*)* ed Gianangelo B, Bodil H (Berlin: Springer) chapter 1 pp 3-34

Journal of Physics: Conference Series **550** (2014) 012034 doi:10.1088/1742-6596/550/1/012034

Photoinduced superhydrophilicity of amorphous TiO_x-like thin films by a simple room temperature sol-gel deposition and atmospheric plasma jet treatment.

V E Vrakatseli, E Pagonis, E Amanatides and D Mataras
University of Patras, Dept. of Chemical Engineering, Laboratory of Plasma Technology

E-mail: lef@plasmatech.gr (Dr E. Amanatides)

Abstract. A room temperature sol gel process of TTIP / iPrOH / H_2O /HNO_3 sol was applied for the deposition of functional Ti alkoxide thin films on glass and polymeric substrates (PEEK). The unheated - amorphous films become superhydrophilic after 7 minutes of UV exposure which deteriorates after one day of storage in dark, exhibiting stable amphiphilic behavior. Superhydrophilicity is also obtained after 5 min of atmospheric pressure Ar - O_2 plasma jet treatment. As the plasma power and the oxygen content of the mixture of the treatment increase (70W, 3.2-5% O_2) the films high hydrophilicity is maintained for many days even in dark atmospheric conditions providing long term hydrophilic coatings.

1. Introduction

Titania (TiO_2) thin films have been extensively studied because of the unique properties of TiO_2 such as its photocatalytic activity and the photoinduced hydrophilicity. Since 1997 Wang et al.[1] discovered the ability of TiO_2 surfaces to alternate their wetting characteristics from hydrophobic or relatively hydrophilic to super hydrophilic under UV irradiation. Most of the research focused on the deposition of superhydrophilic TiO2 thin films because of the great prospects for many new applications such as self-cleaning, antifogging and anti-ice surfaces. These applications are limited by the fact that TiO_2 hydrophilicity is not persistent in dark or under visible light illumination. An interesting related state of the art application for TiO_2 thin films is their use in microfluidic devices on polymeric substrates, where they could control fluid delivery without the need of external pumps or valves due to their aphiphilicity [2] ,[3] or simply act as hydrophilic fluid delivery channels [4] depending on the stability of hydrophilic films.

Much effort has been done on the in situ formation of the anatase structure since it is believed that the anatase crystal form is dominating in TiO_2 films exhibiting the best photocatalytic and hydrophilic functionality [5],[6]. However, anatase deposition and/or densification processes demand high temperatures (> 300°C) [7], thus restricting their application to high temperature stable substrates. Plasma sputtering has been reported as a method for depositing TiO_2 thin films having high crystallinity in the anatase form also at low temperatures. [8], [9]. Nevertheless, plasma sputtering is an equipment demanding process with usually low deposition rates. On the other hand sol gel processes may be a cost effective tool for metal oxides thin film deposition at low temperatures although the as prepared films are amorphous and therefore a post deposition sintering process is needed to form anatase [10]. Yet another approach for the low temperature crystallization of amorphous TiO_2 is plasma treatment [11]. In this approach, Ar, N_2 and O_2 plasma treatments can be applied aiming to the enhancement of photocatalytic or hydrophilic properties of TiO_2 films [12]

Taking into account the limitations and opportunities described above, this work aims on room temperature deposition of TiOx-like thin films for application on glass and polymeric substrates, using a simple sol gel process, with no post deposition thermal annealing of the films. The thin films thus obtained are examined in terms of their photo induced hydrophilicity and stability before and after atmospheric Ar – O_2 plasma treatment, while the persistency of their wetting characteristics as they age in the dark is also recorded.

2. Experimental

2.1 Preparation of the solution and thin film deposition.

Titanium alkoxide thin films were deposited using the sol–gel spin coating technique at room temperature on microscope glass slides and on PEEK substrates (Goodfellow 0.038mm), using a Laurell WS-650S spin coater. Before the deposition process all the substrates were ultrasonically cleaned, then rinsed with distilled water and ethanol and dried with N_2 gas. Ti-alkoxide solutions were prepared by hydrolysis and condensation of Titanium tetraisopropoxide, $Ti(OiPr)_4$, (Sigma Aldrich 99.999%) while isopropanol (PENTA,99.8%) was used as the solvent. Titanium isopropoxide is a highly reactive precursor thus it was first diluted in isopropanol while a second solution of 0.5 ml of 3d water and isopropanol was added rapidly to TTIP/iPrOH solution. After 10 min of magnetically stirring, 0.25 ml HNO_3 (Fisher Scientific, 70%) was added to the solution thus reaching a pH = 1 and then the solution was covered and magnetically stirred at 300 rpm for 1h at room temperature. Another 3 ml of H_2O was then added to the solution which was stirred for 10 minutes before deposition. The volume ratio of the final solution is iPrOH : TTIP : H_2O : HNO_3 = 140/9/3.5/0.25

The deposition of the films was performed using the prepared Ti- alkoxide transparent sol before complete gelation occurred, by applying 1 ml of the solution for every deposited layer for a total of 10 layers, with a two-step spinning at 400 rpm followed by 900 rpm. All the films were left to dry at atmospheric conditions at least for 24h and stored in dark until examined. No heat treatment was applied on the films.

2.2 Thin film characterization.

In order to evaluate photo induced hydrophilicity of the films, water contact angle measurements were taken as the samples were exposed to UV irradiation (Puritec UVC lamp G8T5/OF) with a wavelength centered at 254 nm and 0.2 W/m^2 (@1 m). The samples were placed 5 cm under the UV source, so the corresponding intensity of the UV irradiance which is calculated by Newton's inverse square law equals 8 mW/cm^2. Water contact angle measurements were carried out using the Kruss DSA 100 goniometer with 2μl drops of 2D water carefully placed on the sample. WCA measurements were performed 24 h hours after the deposition, after the UV or plasma treatments and as the films were aging in the dark both for the as deposited and treated films. Ambient temperature and humidity may affect the hydrophilicity and the contact angle conversion rate of the films [13]. The storage conditions remained stable during the experiments and aging with a mean temperature $\bar{T}$=25° C and relative humidity $\overline{RH}$=55%, so that the results can be comparable. The XRD spectra of as deposited and aged films were recorded using a Bruker, D8 Advance X-ray diffractometer, in the range of 20° to 70°. SEM images and EDX analysis were taken using a JEOL JSM-6300 microscope.

2.3 Plasma Treatment

All plasma treatments of the coatings were performed at atmospheric pressure using an RF plasma jet source of 1 mm in diameter operating at 13.56 MHz provided by the Low Temperature Plasma Laboratory, National Institute for Laser, Bucharest, Romania [14]. Atmospheric plasma treatments were applied for 5 minutes with a distance between plasma jet source to the sample of 2.2 cm and the Ar gas flow at 8.5 slm. The parameters examined were the oxygen content of the Ar-O_2 mixture and the plasma generator power (40, 50 and 70W).

3. Results

As confirmed by the X-ray diffraction pattern (fig. 1), the as deposited films are as it is expected, in an amorphous state since no thermal treatment was used. The sample was stored in dark atmospheric conditions ($\bar{T}$ =25°C, $\overline{RH}$=55%) for 4 months before the XRD spectra measurement and this ensures that possible hydrolysis due to the atmosphere's water and/or solvent evaporation has been completed.

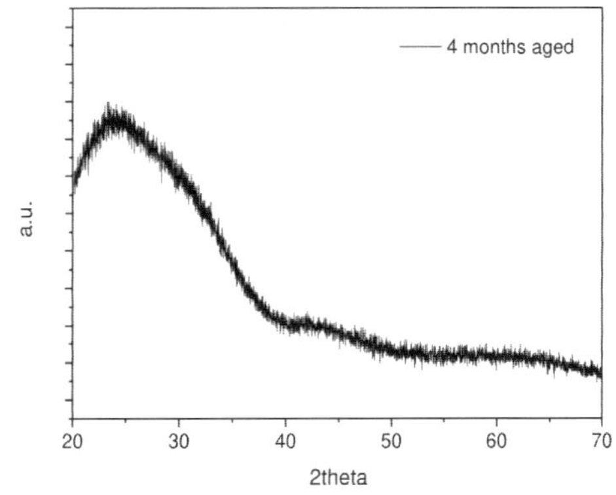

Figure 1. XRD spectra of 4 months aged Ti alkoxide film

3.1 Surface morphology

Macroscopically, the as prepared films show good adherence and transparency, they are smooth and crack free, however SEM images revealed a few cracks on the films at the scale of some μm (Figure 2a). Figure 2b shows that some aggregates are entrapped in or refrained from the network of the TiOx film. These solid particles can be attributed to the vigorous hydrolysis of $Ti(iOR)_4$ and formation of amorphous TiO_2 or $Ti(OH)_4$ despite the low pH value of the solution [15-17]. The EDX analysis indicates a high concentration of Ti and O on the agglomerates and a lower concentration of Ti and O in the bulk of the film with a corresponding higher concentration in carbon. The presence of carbon is attributed to the terminal –OR groups of the Ti alkoxide network. Even though, the results of the present work indicate that the films performance and durability is not influenced neither by the carbon content nor by the aggregates formed on the surface.

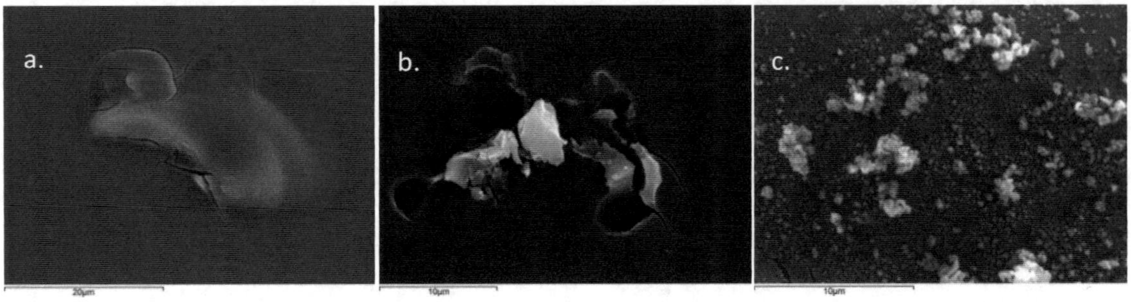

Figure 2. SEM images of the as deposited film a) 20μm, b) 10μm and c) Ar-O_2 plasma treated film(10μm)

3.2 UV induced superhydrophilicity.

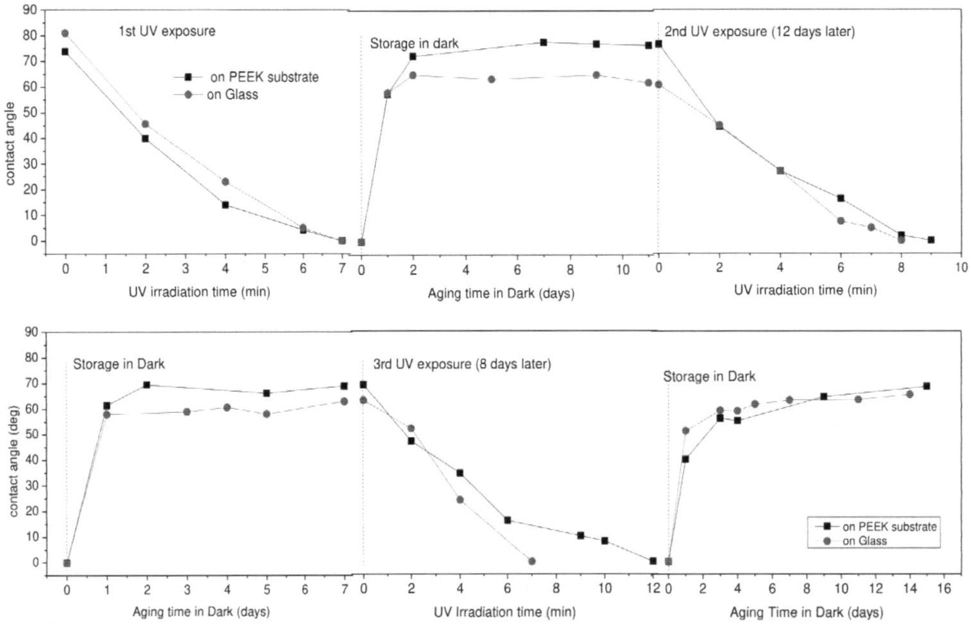

Figure 3. Contact angles of Ti alkoxide films deposited on PEEK and glass substrates after multiple UV exposures and aging in dark and atmospheric conditions (25°C, RH=55%). First UV exposure referred as day 0, 2nd is 12 days later and 3rd is 20 days after the first one.

Thus, figure 3 show the dependence of the contact angle on UV irradiation time as well as the recovery of the contact angle when the untreated samples are stored in the dark in atmospheric conditions. In order to evaluate the stability of the films, we repeated the exposures on the same sample during 20 days. After every UV exposure the sample was stored in the dark and the recovery of the contact angle was measured.

Before the first UV exposure the contact angle of the films were 70° to 80° depending on the type of the substrate and the ageing of the materials (days after deposition). All TiOx films, both on glass or on PEEK substrates become superhydrophillic when exposed to UV irradiance, reaching a 0° contact angle within 7 minutes of exposure, no matter what was their initial CA. The superhydrophilicity is lost if the films are stored in dark and the water contact angle almost reaches its initial value 24h after the UV exposure. It is then shown that the photo induced hydrophilicity is not a property of crystalline TiO_2 exclusively, but UV induced changes occur on the amorphous TiOx network as well. Gao et al. [18] claimed that the photoinduced hydrophilicity of the amorphous TiO_2 is associated with the shift of Ti-OH bonds to Ti=O dangling bonds on the surface resulting in chemisorption of H_2O molecules. Other authors attribute the good UV hydrophilicity of amorphous porous TiO_2 in the higher specific area compared to the crystalline, resulting in a high density of oxygen bridging sites on the surface [19].

In our case, the unheated Ti alkoxide films were deposited from the Ti alkoxide sol in which the molecular hydrolysis ratio is $H_2O/TTIP = 6.3$. The network formed derives from hydrolytic and poly-condensation reactions:

$$\equiv \mathrm{Ti} - \mathrm{OR} + \mathrm{H_2O} \rightarrow \equiv \mathrm{Ti} - \mathrm{OH} + \mathrm{ROH} \quad (1)$$

$$\equiv \mathrm{Ti} - \mathrm{OH} + \mathrm{RO} - \mathrm{Ti} \equiv \rightarrow \equiv \mathrm{Ti} - \mathrm{O} - \mathrm{Ti} \equiv + \mathrm{ROH} \quad (2)$$

The hydrolysis ratio in this experiment does not permit the formation of a very high percentage of the oxide network. The terminating bonds of this expanding oxide polymer contain -OH and -OR groups. [20]. On the other hand, the morphology of the material (fig. 2 a,b) does not imply a porous structure. Therefore, the mechanism proposed by Gao et al. may be valid for this case as well. The photogenerated electrons break the O-H bond of the terminal Ti-OH groups resulting in the formation of Ti=O bonds which is unstable and is more likely to accept the H_2O molecules.

3.3 Atmospheric Pressure Plasma Treatments.

Furthermore, a superhydrophilic surface can also be obtained after the application of 5 min Ar–O_2 atmospheric plasma treatment on the films. All the treated films obtain a zero contact angle after the treatment at any plasma power level. Figure 4 shows the measured contact angles as the films are aging in dark after the treatments at different plasma power levels. All the films were treated with 3.2 % volume fraction O_2 in Ar for 5 min at a distance of 2.2 cm from the plasma source. Moderate plasma power of 40 and 50 W does not severely influence the films aging. When higher plasma power (70 W) is applied the hydrophilic character of the films is prolonged reaching a 42° contact angle only, after 35 days stored in the dark.

Figures 5a and 5b present the influence of the oxygen content of the Ar-O_2 mixture of plasma treatments at 70 W on the persistence of the hydrophilicity of TiOx films deposited on glass or on peek substrates. For this evaluation, mixtures with 1%, 3.2% and 5% O_2 were tested. When a higher content of oxygen (3.2%) and a relatively high plasma power (70 W) is applied, the films exhibit a considerably prolonged hydrophilicity under atmospheric dark conditions. Increase of oxygen content to 5% does not induce any further changes. The Ar treatment (fig. 5a) affects also the persistence of the films hydrophilicity. Ar ion bombardment may induce a higher roughness on the surface which could lead to an enhancement of the hydrophilicity according to wenzel's law [21] and it will induce changes on the crosslinks of the polymeric part of the material. On the other hand atmospheric plasma contains highly reactive species of atmosphere's oxygen which could partially oxidize the surface. The SEM image of 3.2% O_2 in Ar plasma treated surface (fig.2c) reveals also a significant change in the film morphology and more precisely an increase of roughness and the formation of grains. The formation of grains on the film surface may have a strong effect on the film surface properties, due to the larger real area and the density of defects than on the smooth surface [22].

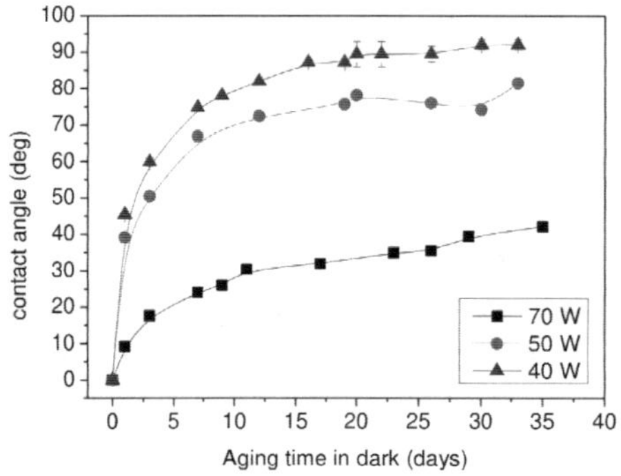

Figure 4. Change of the contact angle of atmopspheric Ar – 3.2% O_2 plasma treated (5 min, @ 22 mm) Ti alkoxide films on PEEK substrate with different plasma power (40 W -70 W), as they age in dark atmospheric conditions of the laboratory

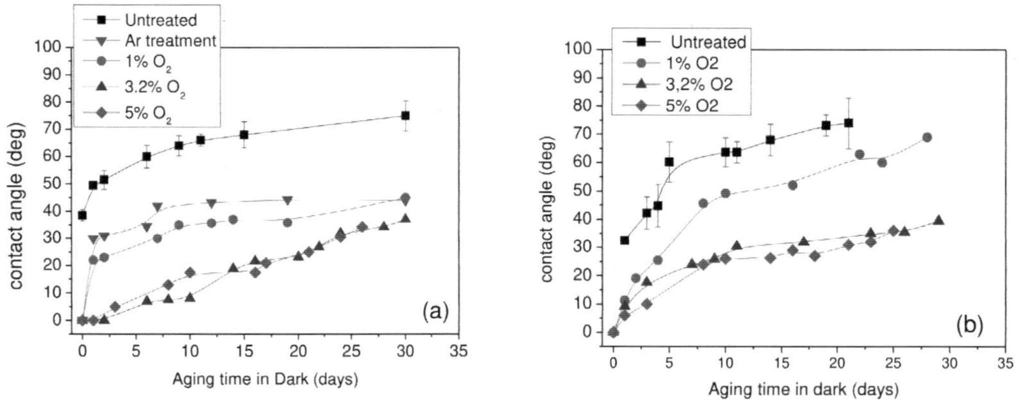

Figure 5. Change of water contact angles of untreated and atmospheric Ar - (0% -5%) O_2 plasma treated films at 70 W on a) glass and b) PEEK substrates as they age in dark atmospheric conditions.

On the other hand, in oxygen containing plasmas, the mixture of ions, oxygen atoms, oxygen metastables and ozone that reach the surface have higher probability to stick and cause further material oxidation [23]. Furthermore, oxygen plasma treatment is also an effective method for removal of organic moieties from metal oxide/ polymer films [24] and thus an enhancement of the content of the amorphous metal oxide phase may be obtained. According to our results, activation of the surface's superhydrophilicity by the atmospheric Ar – O_2 plasma can be attributed to the plasma produced UV irradiance, oxidation of the surface along with the removal of organic contaminants and creation of organic polar groups on the terminating -OR groups of the Ti alkoxide network. The stability of hydrophilicity in the dark may be a result of the synergistic effects of the Ar – O_2 plasma treatment attributed to oxidization on the films surface along with an increase of surface roughness. The formation of oxygen adatoms will decompose H_2O molecules into two –OH groups more easily, and can result to resistance of the surface to further oxidation from the atmosphere [25]. Additionally, the higher surface roughness will also have a permanent effect on surface hydrophilicity.

3.4 Comparison of UV treatment versus atmospheric plasma treatment effect.

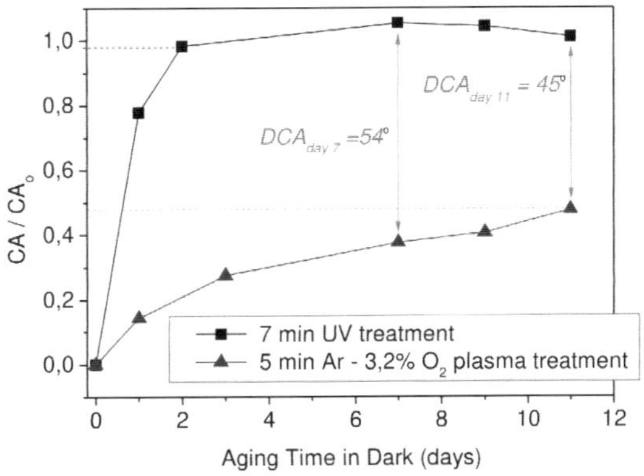

Figure 6. Ratio of the contact angle of UV treated (7 min) and Ar - 3.2 % O_2 atmospheric plasma treated (5 min) TiOx films on PEEK substrate as they age in dark atmospheric conditions to their initial contact angle (CAo) before the treatments. As DCA is defined the difference between the absolute values of the contact angles after a week and 11 days.

Figure 6 shows the difference of the persistence of TiOx films hydrophilicity under dark conditions of a sample that was exposed to UV irradiance for 7 min and a sample that was plasma treated (3.2% O_2 in Ar) for 5 min. . The superhydrophilicity obtained due to atmospheric plasma treatment in less time than the UV treatment could be attributed either to an effect of increased UV intensity produced in plasma (>8 mW/cm^2) or Nevertheless - as already shown in fig.3 - the UV exposed films regain almost their initial contact angle after some hours of storage in dark, exhibiting an amphiphilic behavior, while the plasma treated films maintain a high hydrophilicity for many days. This difference in the contact angle conversion rates of the two cases indicates that plasma oxygen species affect the films more permanently than the UV irradiation.

4. Conclusions

Transparent amorphous TiOx-like films were prepared via a simple sol gel process at room temperature allowing for the deposition on temperature sensitive substrates. The deposited films become super hydrophilic after a few minutes of UV irradiation exhibiting a similar behavior as that of the anatase TiO_2 and a reasonable recovery of the contact angle in dark. The coatings exhibit amphiphilicity which makes them suitable for many applications such as sensors or microfluidic valves. Atmospheric pressure plasma treatment of Ar-O_2 can be used on such surfaces not only for the hydrophilicity activation but it can also prolong the hydrophilicity of the coatings depending on the plasma parameters. Relatively high oxygen percentage in the plasma Ar – O_2 gas mixture and a moderate plasma power may activate the photo induced hydrophilicity phenomenon in a short time but also induces a more permanent modification of the surface which prolongs the hydrophilic properties of the material even in the dark thus expanding its applicability.

Acknowledgements

Part of this work was co-financed by Hellenic Funds and by the European Regional Development Fund (ERDF) under the Hellenic National Strategic Reference Framework (NSRF) 2007-2013, of the Project "THALIS" DESIgn and fabrication of Robust supErhyDROPhobic/philic surfaces and their application in the realization of "smart" microfluidic valves". The authors would like to gratefully acknowledge Prof. G. Dinescu and the group of Low Temperature Plasma Laboratory, National Institute for Laser, Bucharest, Romania for the Atmospheric Pressure Plasma Jet used in this study.

[1] Wang R, Hashimoto K, Fujishima A, Chikuni M, Kojima E, Kitamura A, Shimohigoshi M and Watanabe T 1997 *Nature* **388** 431-433.
[2] Meng Z, Zhang X and Qin J 2013 *Nanoscale* **5** 4687-4690
[3] Park H J, Kim I T, Park S H and Kim S K 2011 *J. Micromech. Microeng.* **21** 095030
[4] Liu N , Li H, Wang H, Chena T, Wang J and Chang L 2012 *Mat.Letter* , **89** 247–250
[5] Fujishima A, Rao T N and Tryk D A 2000 *J. Photochem. Photobiol. C* **1** 1-21
[6] Liu Q J, Wu X H, and Wang B 2002 *Mater Res Bul.l* **37** 2255-2262
[7] Ye Q, Liu P Y , Tang Z F and Zhai L 2007 *Vacuum* **81** Issue 5 627–63
[8] Musil J, Heřman D and Šícha J 2006 *J. Vac. Sci. Technol.* A **24** 521
[9] Barnes M C, Kumar S, Green L, Hwang N M and Gerson A R 2005 *Surf. Coat. Tech.* **190** 321–330
[10] Yu J, Zhao X and Zhao Q 2001 *Mater. Chem. Phys.* **69** 25-29
[11] Ohsaki H, Shibayama Y, Suzuki M, Kinbara A, Yashiro H and Watanabe T 2008 *Thin Solid Films* **516** 4490–4494
[12] Zhang K X, Wang W, Hou J L, Zhao J H, Zhang Y, Fang Y C 2011*Vacuum* **85** 990-993
[13] Simonsen M E, Li Z, Søgaard E G 2009 *Appl. Surf. Sci.* **255** 8054–8062
[14] Ionita E R, Ionita M D, Stancu E C, Teodorescu M, Dinescu G 2009 *Applied Surface Science* **255** 5448-5452
[15] Oskam G, Nellore A, R. Lee Penn, and Searson P C 2003 *J.Phys.Chem.* **107** (8).
[16] Vorkapic D and Matsoukas T, 1999, *J. Colloid Interface Sci.* **214** no. 2 283–291
[17] Galkina O L, Vinogradov V V, Agafonov A V and Vinogradov A V 2011 *Int. J. Inorg. Chem.* (8)
[18] GaO Y, Masuda Y and Kumoto K 2004 *Langmuir* **20** No. 8 3188-3194
[19] Sirghi L, Aoki T and Hatanaka Y 2002 *Thin Solid Films* **422** 55-61
[20] Yoldas B E 1986 *J. Materials sci.* **21** 1087-1092
[21] Wenzel R N 1949 *J. Phys. Chem.* **53** (9) 1466–1467
[22] L.Sirghi and Y.Hatanaka 2003 *Surface Science* **30** L323-L327
[23] Kim Y, Yoon C H, Kim K J, Lee Y 2007 *J. Vac. Sci. Technol.* A **25** 1219
[24] Huang J, Ichinose I, Kunitake T and Nakao A 2002 *Langmuir* ***18, No. 23***
[25] Zhang K X, Wang W, Zhao J H, Hou J L, Zhang Y, Fang Y C 2012 *Phys. Procedia* **32** 356 – 362

Manufacturing of composite titanium-titanium nitride coatings by reactive very low pressure plasma spraying (R-VLPPS)

B Vautherin[1], M-P Planche[1], A Quet[2], L Bianchi[2] and G Montavon[1]

[1] IRTES-LERMPS, UTBM, 90010 Belfort cedex, France
[2] CEA Le Ripault, 37260 Monts, France

beatrice.vautherin@utbm.fr

Abstract. Very Low Pressure Plasma Spraying (VLPPS) is an emerging spray process nowadays intensively studied by many research centers in the World. To date, studies are mostly focused on the manufacturing of ceramic or metallic coatings. None refers to composite coatings manufacturing by reactive plasma spraying under very low pressure (*i.e.*, ~150 Pa).
This paper aims at presenting the carried-out developments and some results concerning the manufacturing of composite coatings by reactive spraying. Titanium was selected as metallic material in order to deposit titanium-nitride titanium coatings (Ti-TiN). Nitrogen was used as plasma gas and was injected along an Ar-H_2-N_2 plasma jet via a secondary injector in order to reach the nitrogen content on the substrate surface. Thus, different kind of reactive mechanisms were highlighted.
Resulting coatings were characterized by Scanning Electron Microscopy (SEM) observations. Porous microstructures are clearly identified and the deposits exhibit condensed vapours and molten particles. Glow Discharge Optical Emission Spectroscopy (GDOES) analysis evidenced nitrogen inside the deposits and X-Ray Diffraction (XRD) analysis confirmed the formation of titanium nitride phases, such as TiN and Ti_2N, depending upon the location of the nitrogen injection. Microhardness values as high as 800 VHN were measured on manufactured samples (to be compared to 220 VHN for pure titanium VLPPS-manufactured coatings).

1. Introduction

Plasma spraying is an enabling technology aiming at manufacturing thick (*i.e.*, from a few hundreds micrometers to a few millimeters) coatings onto components to provide them with unique functional properties, such as enhanced resistances to wear and corrosion, thermal insulation, etc. Plasma spraying consists in injecting a feedstock material made of micrometer-sized particles into a high-energy plasma flow (depending upon the plasma gas mixture and the power parameters, corresponding usually to a mass enthalpy of about 20 $MJ.kg^{-1}$, conventional value) exiting a torch at elevated velocities (in the order of 1000 $m.s^{-1}$ when considering an anode nozzle diameter of about 6 mm, a volume flow rate of plasma gas mixture of about 50 $L.min^{-1}$ and an electric power of about 50 kW). The feedstock particles are molten by heat transfers from the high-energy flow and simultaneously accelerated towards the surface to be covered (*i.e.*, the substrate) where they form, after impact, spreading and rapid solidification (*i.e.*, cooling rate of about 10^5 to 10^8 $K.s^{-1}$, characteristic value), individual flattened lamellae, the stacking of which forming a coating.
Plasma spraying exhibits multiple advantages, among which - the large variety of feedstock materials that can be sprayed (*i.e.*, alloys, ceramics, polymers, composites), the large variety of substrate

Journal of Physics: Conference Series **550** (2014) 012035 doi:10.1088/1742-6596/550/1/012035

compositions (i.e., alloys, ceramics, polymers, composites), the formation of unique non-equilibrium structures thanks to the high solidification rate, the high feedstock throughput (*i.e.*, from 3 to 10 $kg.h^{-1}$, conventional values), the capability of manufacturing multi-layered or graded coatings by adjusting the feedstock composition, etc. Several drawbacks limit however its applicability. Among them, only materials exhibiting congruent melting behavior can be processed *i.e.* materials which having a melting temperature distinct by at least 300 K of their vaporization temperature.

Reactive plasma spraying appears as a palliative solution to circumvent this drawback. It consists in inducing reactions in between a precursor and a reactive gas to form non-congruent melting materials. The very first attempts were made during the mid-80's by shrouding a plasma flow expanding at atmospheric pressure and injecting reactive gases inside. In such a configuration, reactions occur in between molten particles and the reactive gaseous species (the mechanism depends upon the ratio of the kinematic viscosity of gas to that of liquid). This configuration was never fully satisfactory due to its main drawbacks: **1**) the entrapment of air in the plasma flow, meanwhile the shroud at its periphery, due to the flow recirculation of the extremity of the shroud. Air entrapment leads to in-flight oxidation of particles; **2**) the limited interaction time in between the molten particle and the reactive species (> 1ms); **3**) the rather low efficiency of the reaction which takes place in between a liquid (the molten particle) and a gas (the reactive species); **4**) the thermal losses due to the cooling of the shroud, etc.

Very-low pressure plasma spraying (VLPPS) is an emerging technology which aims at operating the plasma torch under an absolute pressure of about 100 Pa of a neutral atmosphere (*i.e.*, Argon). This way, the plasma flow expands at the exit of the torch to form a jet of about 1 m long, characteristic value. This process is based on a different deposition mechanism, in regard to conventional plasma spraying, either at atmospheric pressure (for APS) or at low pressure (20000 Pa, characteristic value for VPS): the coating is manufactured, pass by pass, mostly by the condensation of vapors instead of the solidification of molten particles. Indeed, feedstock particles vaporization occurs in the plasma core, at the exit of the torch. The plasma jet carries then the vapors towards the substrate to cover where they condense to form the coating. Because of the Knudsen effect leading to the decrease in heat and momentum transfer between the plasma jet and particles, VLPPS requires to work with fine feedstock powder.

Reactive – very-low pressure plasma spraying (R-VLPPS) takes advantage of the long dwell-time of the vapors to induce their reaction with a gaseous precursor, injected in the plasma flow either as plasma forming gas and/or downstream the plasma flow to form, in situ, new species that condense onto the substrate.

This study aims at demonstrating the potential of this reactive process and considers, as demonstrative precursors, Titanium and Nitrogen, in view of forming Titanium nitrides, materials exhibiting a non-congruent melting behavior, and for this reason impossible to plasma spray in a conventional way.

2. Experimental details

Plasma atomized pure titanium powder provided by RAYMOR INDUSTRIES INC. (Boisbriand, QC, Canada), with a size distribution ranging from 5 µm (d_{10}) to 25 µm (d_{90}) and an average size diameter (d_{50}) of 13 µm, was used as starting feedstock material (see figure 1). Three different configurations were considered to inject the reactive species, Nitrogen, in the plasma flow. The first way considers Nitrogen as the primary plasma forming gas. The second way uses a ring-shaped injector (figure 2) located downstream, close to the substrate surface (*i.e.*, 8 mm). The third way is a combination of the two first ones.

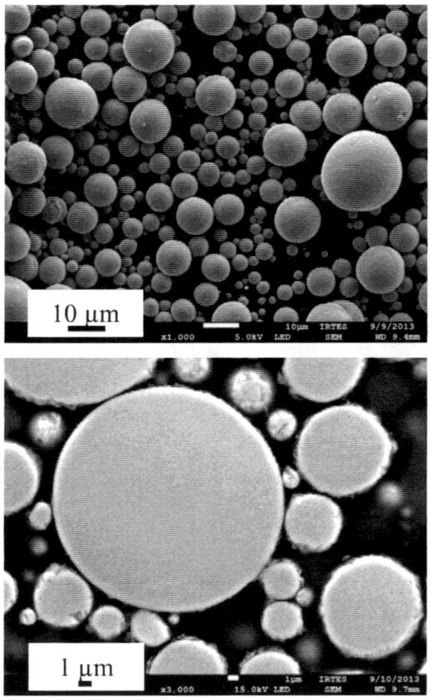

Figure 1. Titanium powder particles

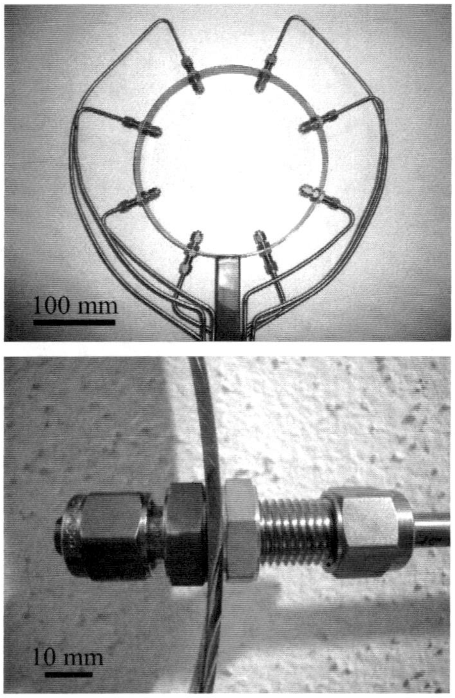

Figure 2. The ring-shaped injector

Copper plates of 70×30×2 mm^3 were used as substrates. In order to improve the adhesion of the coatings, the substrates were degreased in ethanol vapors and grit-blasted (with α-Al2O3 grits of 250 µm, average dimension) prior spraying. They exhibited an average surface roughness of 12 µm and an average maximum peak-to-valley ratio of 126 µm. In order to evidence the different nitriding mechanisms, substrate temperature was considered as an operating parameter. A water-box was used to cool the substrates during spraying. The substrate temperature was monitored using type K thermocouples of 800 µm in diameter brazed at the front of the substrate to avoid through-thickness conductive losses and thermal resistances in between the samples and the sample-holder. Indeed, the rather long spray distance and the rarefied atmosphere do not damage the thermocouples during tests.

2.1. *VLPPS system and experimental conditions*

The experiments were carried out thank to a VLPPS device developed in the lab. This device is depicted in figure 3. It is made of an operating chamber of 12 m^3. The pumping system allows varying the chamber pressure during operation from 150 Pa to 5 kPa. It is also equipped with a 53 kW F4-type plasma torch installed on a 6 axis robot. The effect of the nitrogen flow rate in the plasma jet, the location of nitrogen injection and the substrate temperature was evidenced via five sets of parameters, as detailed in **Erreur ! Source du renvoi introuvable.** while chamber pressure, spray distance and velocity were unchanged.

Figure 3. VLPPS system made in IRTES-LERMPS.

Table 1. Operating parameters

<table>
<tr><th>Plasma torch</th><th></th><th colspan="2">F4</th></tr>
<tr><td>Anode internal diameter at torch exit (mm)</td><td></td><td colspan="2">6</td></tr>
<tr><td>Chamber pressure (Pa)</td><td></td><td colspan="2">150</td></tr>
<tr><td>Feedstock powder rate (g.min^{-1})</td><td></td><td colspan="2">1.0</td></tr>
<tr><td>Spray distance (mm)</td><td></td><td colspan="2">900</td></tr>
<tr><td>Spray velocity (mm.s^{-1})</td><td></td><td colspan="2">200</td></tr>
<tr><th>Parameters</th><th>Set 1</th><th>Set 2</th><th>Set 3</th></tr>
<tr><td>Arc current intensity (A)</td><td>500</td><td colspan="2">600</td></tr>
<tr><td>Arc voltage (V)</td><td>85</td><td colspan="2">90</td></tr>
<tr><td>Ar plasma forming gas flow (L.min^{-1})</td><td>4</td><td colspan="2">4</td></tr>
<tr><td>H_2 plasma forming gas flow (L.min^{-1})</td><td>3</td><td colspan="2">3</td></tr>
<tr><td>N_2 plasma forming gas flow (L.min^{-1})</td><td>23</td><td colspan="2">30</td></tr>
<tr><td>Plasma mass enthalpy (MJ.kg^{-1})</td><td>30</td><td colspan="2">38</td></tr>
<tr><td>Nitrogen flow rate in ring-shaped injector (L.min^{-1})</td><td>/</td><td>/</td><td>5</td></tr>
<tr><td>Substrate temperature (°C)</td><td>150</td><td>800</td><td>800</td></tr>
</table>

2.2. *Characterization of coating structural features and mechanical properties*

Coating cross-sections and coating surface morphologies were observed using a JEOL JSM-5800LV scanning electron microscope (SEM) in back-scattered electron mode to point out the chemical contrast into the deposits. A LEITZ MINILOAD-2 Vickers tester was used to measure the microhardness of the coatings. A load of 25 g and a dwell-time of 30 s were selected. An X-ray diffractometer (XRD, D8 Advance, BRUCKER AXS, Germany) with a CoKα anti-cathode (λCo = 0.179 nm) was employed to determine the phase composition of the coatings (scanning step: 0.02°). Glow Discharge Optical Emission Spectroscopy (GDOES, GD-Profiler 2, HORIBA JOBIN-YVON) was used to quantify the chemical composition of Ti-TiN coatings. Nano indentation test (NHT) were implemented on Ti-TiN coatings to evaluate phase hardness. The indenter type is Berkovich and the loading was linear. The contact load was fixed to 2.5 mN and the maximum depth to 125 nm. Results were obtained by the "Oliver and Pharr" method.

Journal of Physics: Conference Series **550** (2014) 012035 doi:10.1088/1742-6596/550/1/012035

3. Results and discussion

3.1. *Effect of Nitrogen content as plasma gas forming*

SEM observations of Ti-TiN coating cross-sections manufactured using set 1 and set 2 are presented in figure 4. The average deposited thickness per pass was of 0.2 µm with set 1 and 0.15 µm with set 2.
The coatings exhibit, from a general point of view, a structure composed of majority condensed Titanium vapors. Such a structure does not correspond in this case to the staking of lamellae resulting from the spreading and the solidification of molten impinging droplets as evidenced in figure 5 depicting the OES spectra and in figure 6 displaying a high resolution fractured cross-section of a typical coating manufactured with this process. The lamellar structure results from a chemical contrast, at this stage of the analysis very likely to areas enriched in nitrogen.
Coatings exhibit some voids. The coatings manufactured with set 2 seem to exhibit a lower total macroscopic void content compared to set 1. It is nevertheless impossible at this stage to be conclusive due to the smallest voids which are not detected due to the image resolution limit. The total void content of those coatings has been estimated to 5%, average value, by Ultra-Small Angle X-rays Scattering (USAXS) performed at the Advanced Photon Source (Argonne National Laboratory, Argonne, IL, USA).
Glow Discharge Optical Emission Spectroscopy (GDOES) analyses allow quantifying the average Nitrogen concentration through thickness in the coatings. In average, the atomic concentration increases from 9%at. to 18%at. for corresponding plasma mass enthalpy of $\hbar$ = 30 $MJ.kg^{-1}$ and $\hbar$ = 38 $MJ.kg^{-1}$ (figure 6). One can notice the presence of copper in the coating, corresponding to the diffusion from the substrate to the coating at elevated temperature ($\geq$ 800 K).

Set 1 Set 2

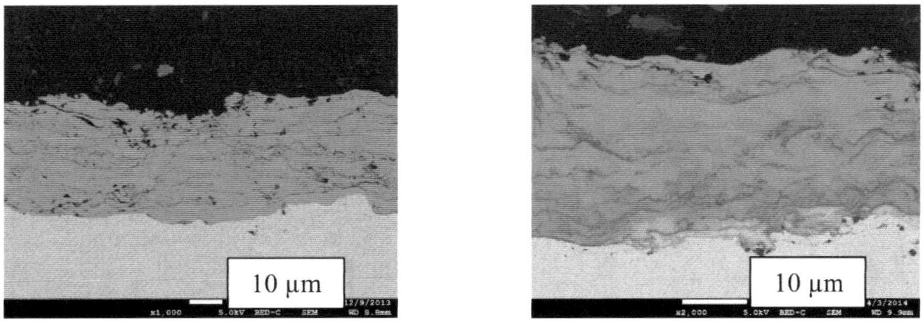

Figure 4. SEM observations of Ti-TiN coatings elaborated by VLPPS.

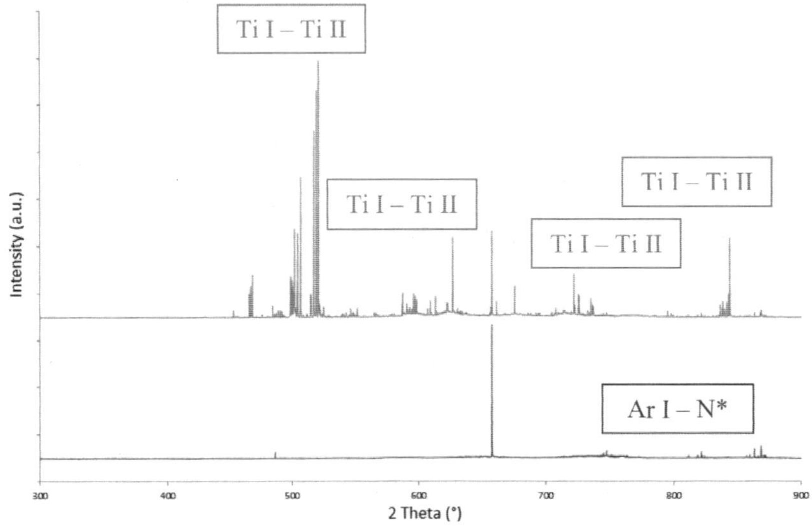

Figure 5. OES spectra on Ar-H_2-N_2 plasma without and with titanium powder on 300-900 nm

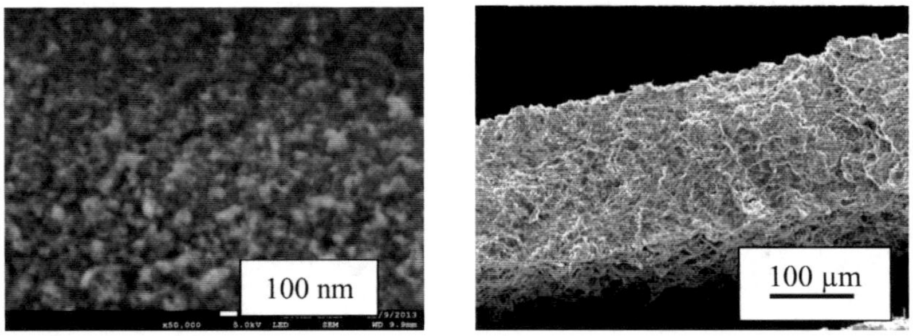

Figure 6. Typical fractured view of a Ti coating manufactured by R-VLPPS.

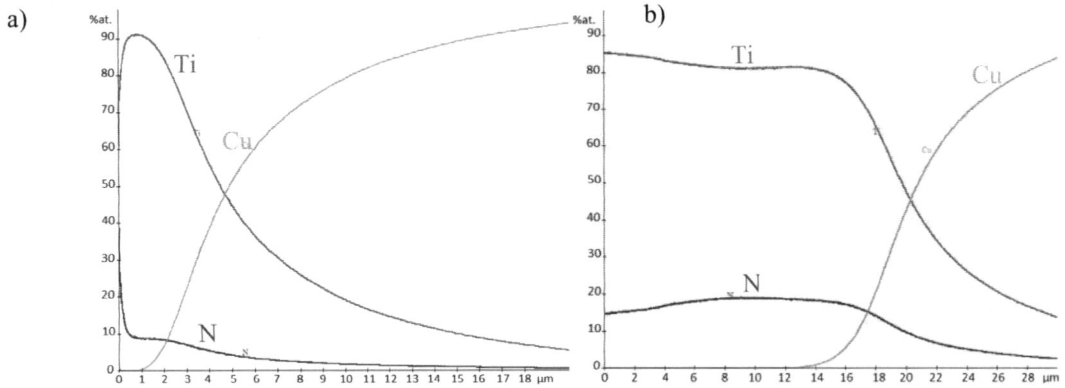

Figure 7. GDOES quantitative analysis of titanium coatings manufactured with nitrogen as primary plasma forming gas, a) $\hbar = 30$ MJ.kg^{-1} and b) $\hbar = 38$ MJ.kg^{-1}.

Journal of Physics: Conference Series **550** (2014) 012035 doi:10.1088/1742-6596/550/1/012035

3.2. *Effect of Nitrogen content as primary plasma forming gas and Nitrogen injected through the ring-shaped injector*

Figure 8 displays the typical structures of Titanium coatings manufactured with Nitrogen as primary plasma forming gas coupled with Nitrogen injected close to the substrate through the secondary ring injector. The comparison is done for parameters from set 2 and set 3. The coating architectures do not change significantly compared to the previous case: a lamellar structure is clearly identified again.

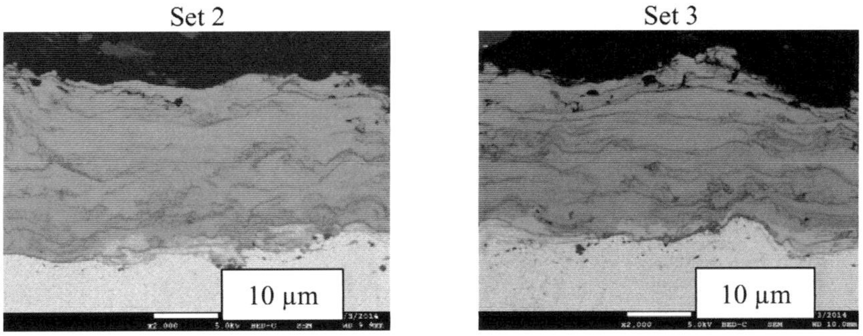

Figure 8. SEM observations of Ti-TiN coating cross-sections elaborated by VLPPS.

Figure 9 presents the average Nitrogen atomic concentration gradient through thickness. It is in the range of 25%. From this result, it has been proved the injector ring allows hence increasing the average concentration of nitrogen in the coatings, from 18%at to 25%at, respectively.

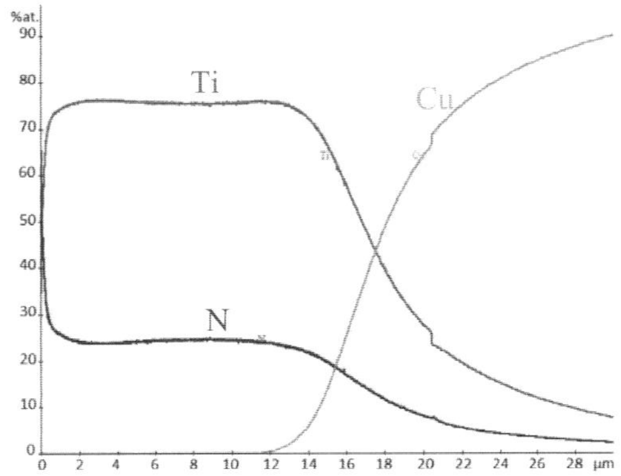

Figure 9. GDOES quantitative analysis of titanium coatings manufactured with Nitrogen as primary plasma forming gas and Nitrogen injected close to the substrate through the ring-shaped injector.

3.3. *Comparison of coating structures and identification of mechanisms occurring during coating deposition*

Nitrogen atomic concentration gradients through coating thickness clearly indicate the presence of Nitrogen in the coatings. Figure 10 displays the XRD patterns of those deposited materials. This characterization technique allows identifying phases, meaning how Nitrogen is bounded in the structure. The XRD patterns reveal the presence of Titanium nitrides exhibiting two main stoechiometric compositions: TiN, $TiN_{0.3}$ and Ti_2N. The Rietveld refining allows quantifying the respective phase amounts. However, this technique, applied to the presented patterns, appeared to be not conclusive, since nitride amounts vary in very large ratio, depending upon the selected refining criteria. Other techniques, such as IR spectroscopy, should very likely permit to better quantify those amounts. Works are in progress.

In table 2 are given Ti-TiN coating Vickers hardness obtained from nano-indentation measurements which were performed on splats.

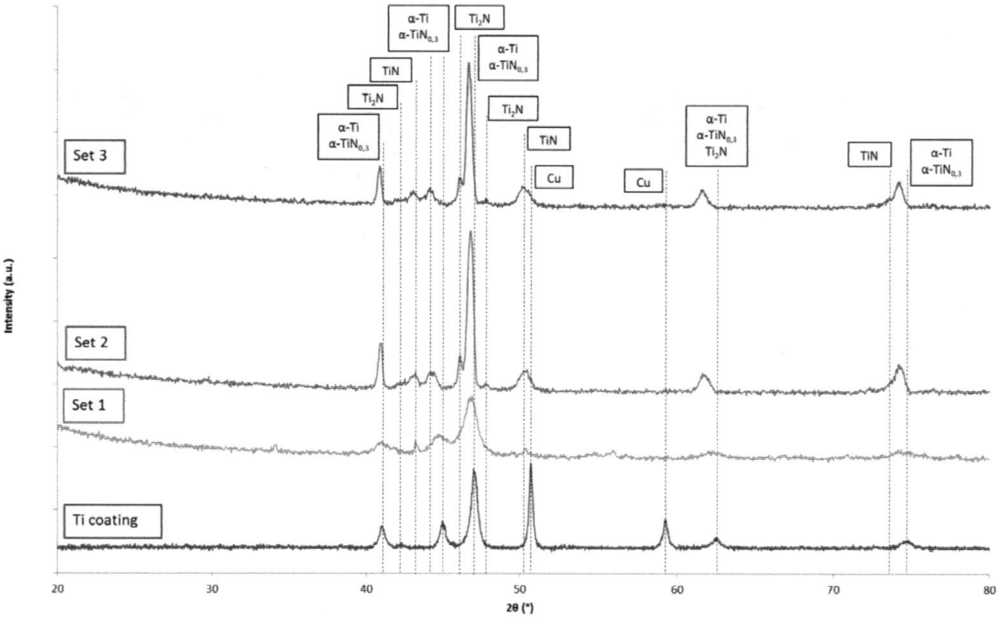

Figure 10. XRD pattern obtained on Ti coating and Ti-TiN coatings elaborated by VLPPS from set 1, 2 and 3

Table 2. NHT results obtained on Ti-TiN coatings elaborated from set 2and set 3, for different locations on the coatings

NHT location in the coating	Ar-H_2-N_2 (set 2)	Ar-H_2-N_2 with ring shaped injector (set 3)
Splat – center	(1209 ± 262) HV	(1394 ± 300) HV
Splat – side		(1453 ± 286) HV

The average Vickers hardness of Ti coatings elaborated by VLPPS is (220 ± 12) HVN. Compared to this value, the hardness of Ti-TiN coatings is higher, showing the strengthening of the coatings by titanium nitrides formation during spraying. However, the standard deviation on the average values is huge, around 300 HVN. This shows that Ti-TiN coatings are not homogeneous and nitriding mechanisms are not complete due to the partial powder particle vaporization.

Conclusion

Very low pressure plasma spraying was implemented to manufacture Ti-TiN composite coatings. In order to elaborate these deposits, two ways were studied. Firstly, nitrogen was used as plasma gas forming for two mass flow rates, leading to ħ = 30 MJ.kg^{-1} and 38 MJ.kg^{-}1, respectively. In both cases, OES measurements show the titanium powder particles vaporization. However, in the case of the lower plasma mass enthalpy, coatings exhibit a lamellar microstructure and some spherical particles were observed, revealing the partial particle heat treatment. Nitrogen content of the coatings was evaluated by GDOES analysis. In relation with the increase of plasma enthalpy, the nitrogen atomic content is improved from 9%at. to 18%at. XRD pattern confirm the formation of titanium nitride during the spraying. Different stoechiometric compositions of titanium nitrides were determined such as TiN and $TiN_{0.3}$.

Secondly, nitrogen was injected as primary plasma gas and via a ring-shaped injector localized close the surface substrate in order to enrich the substrate surrounding with Nitrogen species. For this configuration, coating microstructures do not change significantly compared to the first ones. A condensed architecture can be identified. GDOES measurements have shown the atomic Nitrogen content into these coatings reaches 25%at. and XRD patterns reveal the formation of Ti_2N phase which can be explained by the nitriding mechanisms taking place on the substrate surface during the particle re-solidification. Compared to the average hardness titanium coating elaborated by VLPPS, (220 ± 12) HVN, composite Ti-TiN coatings present an improved average hardness, (1350 ± 283) HVN. The significant standard deviations express a partial nitriding of titanium powder particles during spraying.

To promote once more the nitriding mechanisms, other ways can be considered as increase the nitrogen content in the spraying surrounding. But some experiments have demonstrated deposits elaborated with too high nitrogen content are weakened illustrated by a lot of cracks and delamination. Work is in progress to explore the nitriding efficiency by reducing the titanium thickness deposited per pass, possibly promoting the nitrogen diffusion into the coatings and titanium nitride formation.

Acknowledgments

The authors gratefully thank O. Ribet and B. Nicolas for their contributions to the coating characterizations. The authors also gratefully thank the IRTES / CEA joint research laboratory on advanced surface engineering for its financial support.

References

[1] Langmuir I., 1928 Oscillations in ionized gases *Proc. Nat. Acad. Sci. U.S.* **14** 628

[2] Mauer G., Hospach A., Zotov N., Vaßen R., Process conditions and microstructures of ceramic coatings by gas phase deposition based on plasma spraying, *Thermal Spray Technology*, 2013, **22**(2-3), p 83-89

[3] Zhu L., Zhang N., Zhang B., Bolot R., Liao H., Coddet C., In Situ Synthesis of FeAl Dense Coatings by Very Low Pressure Reactive Plasma Spraying, *Thermal Spray Technology*, 2013, **22**(2-3), p 90-95

[4] Salhi Z., Klein D., Gougeon P., Coddet C., Development of coating by thermal plasma spraying under very low-pressure condition <1 mbar, *Vacuum*, 2005, **77**(2), p 145-150

[5] Niessen K. and Gindrat M., Plasma Spray-PVD: A New Thermal Spray Process to Deposit Out of Vapor Phase, *Thermal Spray Technology*, 2011, **20**(4), p 736-743

[6] Zhu L., Zhang N., Zhang B., Sun F., Bolot R., Planche M.-P., Liao H. and Coddet C., Very Low Pressure Plasma Sprayed Alumina and Yttria-Stabilized Zirconia Thin Dense Coatings using a Modified Transferred Arc Plasma Torch, *Applied Surface Science*, 2011, **208**, p 1422-1428

[7] Guittienne P., Grange D., Hollenstein C. and Gindrat M., Plasma Jet-Substrate Interaction in Low Pressure Plasma Spray-CVD Processes, *Thermal Spray Technology*, 2011, **21**(2), p 202-210

[8] Gindrat M., Höle H.-M., Niessen K., Guittienne P., Grange D. and Hollenstein C., Plasma Spray-CVD : A New Thermal Spray Process to Produce Thin Films from Liquid or Gaseous Precursors, *Thermal Spray Technology*, 2011, **20**(4), p 882-887

[9] Smith M. F., Hall A. C., Fleetwood J. D. and Meyer P., Very Low Pressure Plasma Spray—A Review of an Emerging Technology in the Thermal Spray Community, *Coatings*, 2011, **1**(2), p 117-132

[10] Mauer G., Hospach A., Vaßen R., Process development and coating characteristics of plasma spray-PVD, *Surface and Coatings Technology*, 2012, **220**, p 219-224

[11] Dorier J. L., Guittienne P., Hollenstein C., Gindrat M. and Refke A., Mechanisms of films and coatings formation from gaseous and liquid precursors with low pressure plasma spray equipment, *Surface and Coatings Technology*, 2009, **203**(15), p 2125–2130

[12] Refke A., Barbezat G., Dorier J.-L., Gindrat M., and Hollenstein C., Characterization of LPPS processes under various spray conditions for potential applications, *Thermal Spray 2003: Advancing the Science and Applying the Technology*, Vol. 1, B.R. Marple and C. Moreau, Ed., May 5-8, 2003 (Orlando, FL), ASM International, 2003, p 581-588

[13] Gindrat M., Dorier J., Hollenstein C., Refke A., and Barbezat G., Characterization of supersonic low pressure plasma jets with electrostatic probes, *Plasma Sources Science and Technology*, 2004, **13**(3), p 484

[14] Hospach A., Mauer G., Vaßen R., Stöver D., Characteristics of ceramic coatings made by Thin Film Low Pressure Plasma Spraying (LPPS-TF), *Thermal Spray Technology*, 2012, **21**(3-4), p 435-440

[15] Dorier J.-L., Gindrat M., Hollenstein C., Loch M., Refke A., Salito A., Barbezat G., Plasma jet properties in a new spraying process at low pressure for large area thin film deposition, *Thermal Spray 2001: New Surface for a New Millenium*, C.C. Berndt, K.A. Khor and E.F. Lugscheider, Ed., May 28-30, 2001 (Singapore), ASM International, 2001, p. 759-764

[16] Young E. J., Mateeva E., Moore J. J., Mishra B. and Loch M., Low pressure plasma spray coatings, *Thin Solid Films*, 2000, **377-378**, p 788–792

[17] Yang D., Tian B. and Gao Y., The lamellar–equiaxed microstructural transition of 316L coatings by low pressure plasma spraying, *Surface and Coatings Technology*, 2012, Available on line

[18] Han P. and Chen X., Modeling of the supersonic argon plasma jet at low gas pressure environment, *Thin solid films*, 2001, 390(1–2), p 181–185

[19] Kambara M., Yagi H., Sawayanagi M., and Yoshida T., High rate epitaxy of silicon thick films by medium pressure plasma chemical vapor deposition, *Applied Physics*, 2006, 99, 074901

[20] Murakami T., Sasaki S., Microstructure and tribological properties of Fe-Mo alloy-coated steel specimens prepared by low-pressure plasma spraying, *Intermetallics*, 2011, **19**, p 1873-1877

[21] Bolot R., Sokolov D., Klein D. and Coddet C., Nozzle Developments for Thermal Spray at Very Low Pressure, *Thermal Spray Technology*, 2006, **15**(4), p 827-833

[22] Zhang N., Sun F., Zhu L., Planche M.-P., Liao H., Dong C., Coddet C., Measurement of specific enthalpy under very low pressure plasma spray condition, *Thermal Spray Technology*, 2012, Available on line

[23] Vautherin B., Planche M.-P., Bolot R., Quet A., Bianchi L., Montavon G., Vapors and droplets mixture deposition of metallic coatings by very low pressure plasma spraying, *Journal of Thermal Spray Technology*, 2014, available on line

[24] Gindrat M., (in English) Characterization of supersonic low pressure plasma jets, Ph.D Thesis of Ecole Polytechnique Fédérale de Lausanne (EPFL), Lausanne, Switzerland, 2004

[25] Jodoin B., Gindrat M., Dorier J.-L., Hollenstein C., Loch M. and Barbezat G., Modelling and diagnostics of a supersonic DC plasma jet expanding at low pressure, *Thermal Spray 2002 : International Thermal Spray Conference (DVS-ASM)*, March 4-6, 2002, Essen (Germany), ASM International, 2002, p 716-72

Spatial and temporal evolution of microdischarges in Surface Dielectric Barrier Discharges for aeronautical applications plasmas

R Barni, I Biganzoli and C Riccardi

Dipartimento di Fisica G. Occhialini, Università degli Studi di Milano-Bicocca, piazza della Scienza 3, I-20126 Milano, Italy

E-mail: claudia.riccardi@mib.infn.it

Abstract. Surface Dielectric Barrier Discharges have been proposed long ago as a tool to improve aerodynamics and flow performances. Such electrical discharges could be employed to energize the gas phase and to induce flows. The discharge itself consists of a large number of repetitions of single electric current pulses, with short duration and limited spatial extension filling the region near electrodes. The connection between such macroscopic effect and the properties of the single microdischarge events has been investigated. In particular we have measured the direction and the velocity of propagation of the ionization wave during the different phases of the voltage cycle. Light collected from different parts of the gap arrives at a photomultiplier tube with a delay proportional to the velocity of the ionization wave. The measured propagation velocity was estimated as about 220 km/s in the so called backward discharge phase.

1. Introduction

Non equilibrium plasmas produced by Surface Dielectric Barrier Discharges (SDBD) have been proposed long ago as a tool to improve aerodynamics and flow performances [1-3]. Indeed a new sector of plasma application was started when it was realised that such electrical discharges could be employed to energize the gas phase and to induce flows [1]. Plasma aerodynamics soon developed into a lively field of applied plasma research. Although many aspects of the mechanisms through which plasmas could induce flows are not fully understood, some general features of the phenomenon has been clarified [1]. DBDs and so also SDBDs driven by high voltage oscillating signals applied to electrodes happen generally in short bursts of activity, concentrated in two separated portions of the wave period. They are composed of a pretty large number of individual events, carrying electrical currents in the air gap between the electrodes, a microdischarge [4], each triggered by the propagation of an ionization wave [5-6]. There are several tricky aspects, including the geometry of the electrodes and the charging of the dielectric material surface facing the discharge region. In particular the asymmetry in the two portions of the discharge is considered to be essential in the production of the so called ionic wind [7-9].

Here we presents results concerning the properties of microdischarges obtained by time resolved optical diagnostics using a photomultiplier tube (PMT) [6]. In particular we have measured the

Journal of Physics: Conference Series **550** (2014) 012036 doi:10.1088/1742-6596/550/1/012036

direction and the velocity of propagation of the ionization wave during the different phases. Light collected from different parts of the gap arrives at the PMT with a delay proportional to the velocity of the ionization wave. In order to better separate the events we have introduced fixed delays using several optical fibers of different length.

2. Experimental Setup

The main characteristic of a DBD device is the insertion of a dielectric material separating the two electrodes. In this way the discharge path in the gas gap cannot connect directly the two electrode surfaces and charges accumulates on the dielectric surfaces, quenching the electric current flow. Steady operation is normally achieved by applying an oscillating voltage to reverse charge flow. In general this could be most easily accomplished by feeding one of the electrodes with a HV signal at kHz frequencies. At atmospheric pressure, in millimetre scale gap, the DBDs operate in the so called streamer regime characterized by narrow current filaments (radius below a hundred microns, length a few mm), with very short lifetimes (few tens of ns) [4].

In our experimental setup the electrodes for SDBD have been separated by a rectangular dielectric plate (220x200x5 mm). We used Plexiglas or Teflon, with a dielectric constant ε_r respectively of 2.3 and 2.05. The lower electrode in aluminum (40x180x10 mm) was embed into a plexiglas scaffold (330x200x20 mm). The upper electrode was made of a tin clad copper adhesive foil (60 μm thick, 5 mm wide) glued onto the upper side of the dielectric plate. The two electrodes lie parallel with the edges just corresponding in order to avoid any overlap. A layout of the set-up is sketched in figure 1.

In such a setup discharges happen only in the air just above the upper side of the dielectric plate, usually around the upper electrode edge. Current filaments have one end on the exposed electrode and they extend more or less perpendicularly to the electrode edge towards the dielectric surface above the buried electrode. Current filaments then remain practically confined to a thin layer above the surface. This corresponds roughly to the region of the air gap where the electric field strength is maximum.

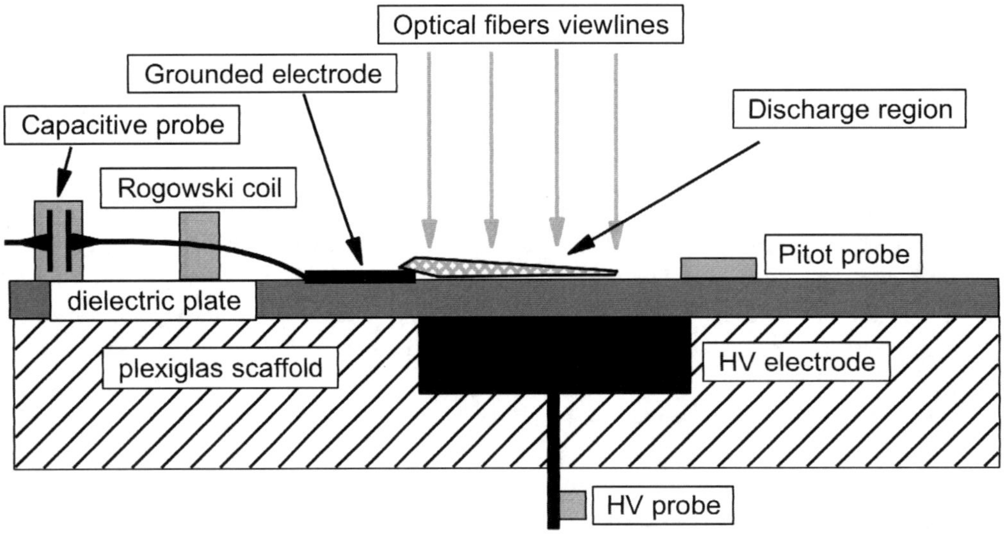

Figure 1. Layout of the experimental setup with diagnostics.

The directionality of the discharges, the asymmetry between currents starting or ending on the exposed electrode and the confinement on the surface are ideal features for any aerodynamical applications, since the whole electric activity of the plasma actuator could be concentrated in the boundary layer of the flow [2,7].

In order to drive the SDBD the buried electrode is connected through an HV cable to the secondary coil of a transformer, whose primary circuit is connected to a tunable power generator, providing the driving high voltage for the discharge. We used a V-20 Corona Station by Tigres Gmbh. Alternating voltage frequency is varied automatedly between 18 and 50 kHz or it can be fixed depending on the operating choices. The upper electrode is grounded. A high voltage probe, P6015A by Tektronix, has been used to monitor the voltage on the upper electrode. The probe is granted for a bandwidth of 75 MHz, which is sufficient for assessing the effects on the circuitry voltage of the dynamics of streamers. In general an almost purely sinusoidal voltage was recorded, without any significant amplitude modulation, even during discharges at high power level. This ensures that the discharge phases happens only two times during the HV cycle, without the ripples and stasis sometimes reported in literature. In the setup we have also included a probe for the discharge current. A current transducer like a Rogowski coil is suitable to measure electric current filaments. As already discussed in a dedicated paper [8], we were able to study current pulses characteristics like number, amplitude, duration and charge transported [9]. A capacitance in series with the electrode setup have been used too in order to measure the effective power absorbed by the microdischarges as well as the total charge accumulated on the dielectric surface [9].

To sum up our findings, the signal results to be composed of two well separated burst of activity, one in each HV half-cycle. The two disjointed active phases are known as Forward Discharge (FD) and Backward Discharge (BD), referring to the electron motion from the exposed electrode to the dielectric surface or the other way. The two phases are composed of several current peaks, a few tens of ns in duration, corresponding to single microdischarges, filling a definite part of the HV cycle.

Here we are primarily concerned with the light emitted from the discharge. We have already reported about the measured emission spectra of the discharges in a spectral band extending from 200 to 850 nm by means of a wide band, low resolution spectrometer (PS2000 by Ocean Optics) [9]. Spectra are typical of DBD, with the brightest feature being the 2nd positive system of nitrogen, emitted from the N_2(C) excited level of the molecule. Spectra shows also the presence of nitrogen molecular ion N_2^+, whose most intense emission line is at 391 nm [10]. In order to reconstruct the temporal evolution of the discharges we have used a PMT by Hamamatsu (H10721-210, having a 0.57 ns rise time) chosen for its fast temporal response and high sensitivity. Its ultra bialkali photocathode radiant sensitivity extends in the 230-700 nm range and peaks at 400 nm, closely matching the air DBD spectra [6]. Gain was maintained constant during experiments at $2x10^6$ by regulating the potentiometer in the PMT supply circuit. The output of the PMT was registered by the low impedance input (50 Ω) of a large bandwidth digital oscilloscope (Agilent MSO8104A, 4 channels, time base 0.25 ns). As already discussed in similar experiments, the light emitted from each microdischarge appears in the PMT output as a few nanosecond pulse [6]. A typical signal is reported in figure 2. Here two events are displayed. Most events are single, when the PMT produces a typical triangular peak lasting 3±1 ns, which sets the best temporal resolution of our measurements. Sometimes, when emission lasts many nanoseconds, more partially superimposed peaks are recorded. They could be easily detected and separated. The overall duration never exceeds a ten nanoseconds. The light emitted from the microdischarge was collected by UV optical fibers FC-UV200-2 by Avantes.

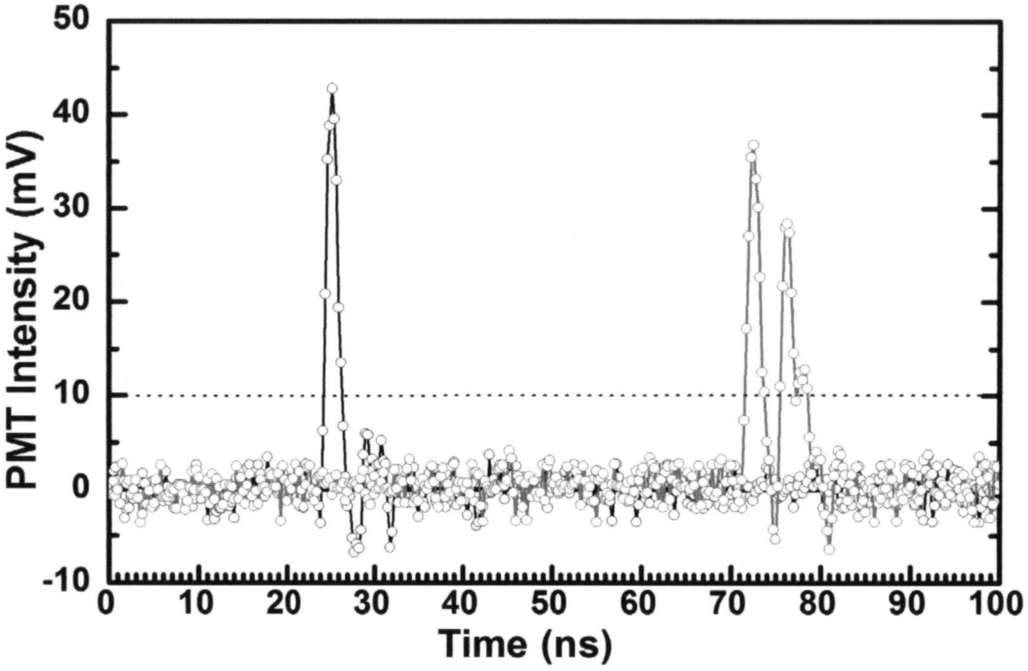

Figure 2. Two examples of the signal recorded from the PMT showing one or more sharp peaks. The threshold, here 10 mV, used to count events was shown too.

Fiber material refraction index leads to a prediction of the signal propagation velocity $v \approx 2 \cdot 10^8$ m/s (n=1.52 at 400 nm), which means that each fiber, 2 m long, provides a delay of 10 ns. A four-furcated optics fiber FCB-UV200-2 by Avantes was used to merge signals arriving from up to four different viewlines. In order to increase the light collected, we imaged the discharge surface of the dielectric plane onto the fiber optics inlet by means of a commercial objective lens (1:2.8/F=50 mm by Durst). A CMOS camera MV-D1024 by Photon Focus was used to take pictures of the discharges. One of such pictures is shown in figure 3, where the resolution is 1 pixel = 11 μm. In these experiments the upper electrode was sectioned separating a pin 1 mm wide from the rest of the electrode. This ensures that microdischarges happen preferentially at the sharp ends of the pin and remain somewhat locked there. Although several microdischarges could be superimposed in the picture due to the quite long exposure time, the spatial extent of the events could be evaluated. A discharge pattern of a few millimetres can be anticipated from figure 3. The viewfield of the fibers was adjusted just to look this part of the electrode region. By back-illuminating the four-furcated fiber we have aligned the light spot of the four fiber endings to the pattern (at an height of about pixel 550 in figure 3). In these experiments the first channel pointed directly at the pin electrode lower edge (position (200,550) in pixels on figure 3), while the second channel pointed the dielectric surface downstream (position (410,550) on figure 3). The other two channels are even more downstream along the dielectric surface and cannot collect light in the present setup. We plan their use in subsequent experimental campaigns. We have precisely measured the distances between light spots, whose diameter was estimated 0.9 mm. Such distance, here 2.3 mm, could be used to evaluate the discharge propagation velocity. Then we could start collecting light with the PMT.

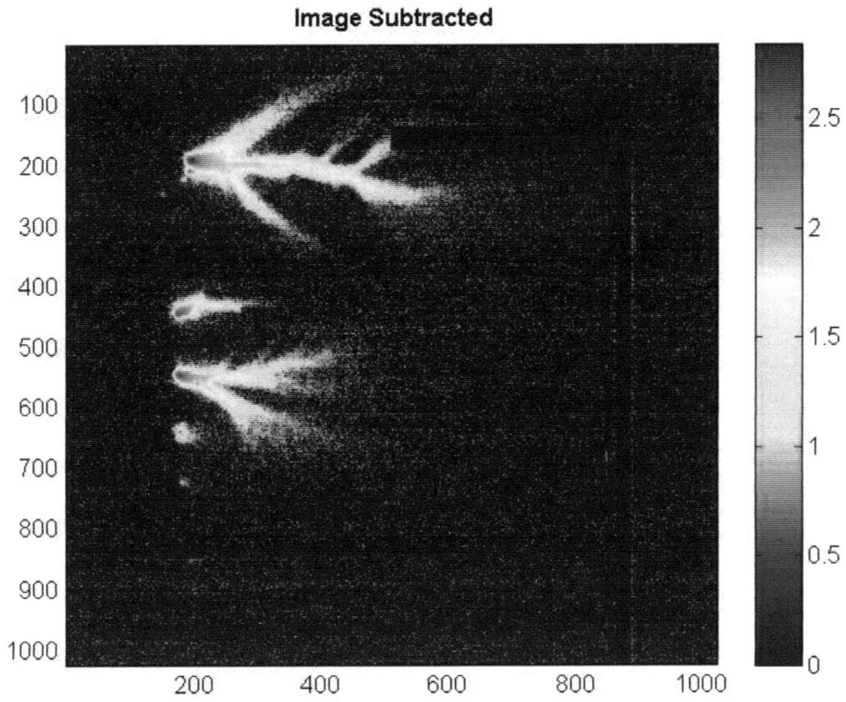

Figure 3. CMOS picture (1024x1024 pixels) of the discharge emission light. The exposed electrode is on the left, with the edge at x≈180. The dark picture was subtracted to the image to enhance emission.

3. Experimental results

As already discussed in the previous section, we have analyzed the delay between the PMT signals arriving from different spot along the discharge path. By subtracting the delay due to the variations in the optical fiber path length we have a measure of the absolute emitting time differences from different parts of the discharge. In order to achieve the maximal time resolution, multiple peaks like those reported in figure 2 were discarded. This greatly reduce the spread in the measured delays.

At first we selected signal coming from view-lines observing the same spot on the dielectric plate. The short duration of the discharges allows to check the theoretical delay caused by the optics fibers. A histogram of the relative frequencies of the difference between measured and theoretical delays show a nice sharp peak at about zero nanoseconds. This was reported for any two couple of channels and various delay fiber length, as could be grasped by the examples displayed in figure 4a. A Gaussian fit to the peaks was used as a value of the measured delay. As reported in figure 4b, a nice linearity was obtained by increasing the difference in the optical path between any of the two channels of the four-furcated fiber. A linear fit to the reported delays vs fiber length was used to measure the effective speed of light travelling along our fibers. This comes out to be $v = 1.93 \pm 0.01 \cdot 10^8$ m/s, corresponding to a refractive index of 1.56 ± 0.01. This refers to the near UV region, since the DBD emission spectra are dominated by the nitrogen second positive system, which matches quite well the 300-500 nm peak sensitivity of the PMT [9,10]. By inverting the delay length of the two viewlines we could also confirm that no bias in the delay was introduced. Indeed the linear fit yields an intercept of 0.01 ± 0.11 ns, consistent with no asymmetry at all.

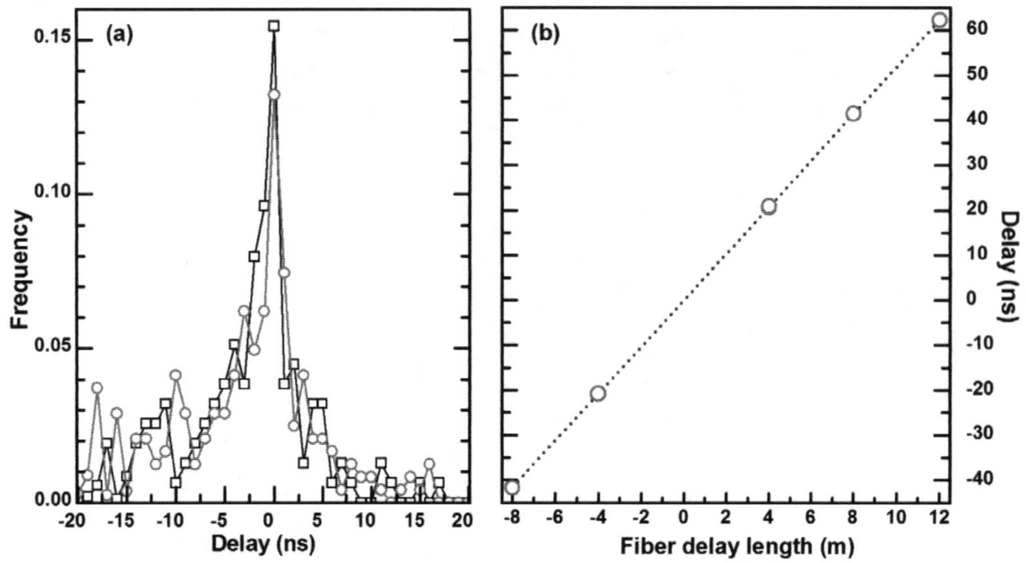

Figure 4. (a) Relative frequency histogram of the differences between measured and theoretical delays for first and second channels (□, total length 8 m) and first and third (o, total length 12 m). (b) Measured delay as a function of the delay fiber length, with a linear fit to data.

Then we started viewing two different locations. As an example we show in figure 5 the delays measured between two twin PMT peaks collected in two points 2.7 mm away one from the other (see figure 1) as a function of the expected lag due to the difference in the optical path through the optical fibers. The first channel collects light from a viewline looking just at the edge of the exposed electrode, while the second viewed the region of the dielectric above the buried electrode. The power

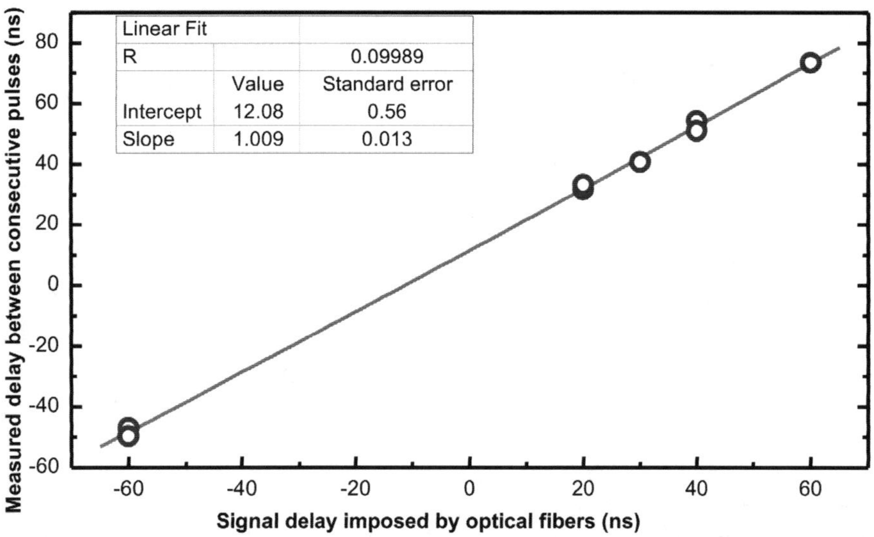

Figure 5. The measured delay peak as a function of the calculated delay imposed by the fiber length mismatch for backward discharge events at 94 W.

level was 94 W, corresponding to an HV amplitude of 8.9 kV and a frequency of 44.3 kHz. The first channel recorded about 450.000 events/s, with an average amplitude of 50 mV and a mean duration of 2.6 ns. The second channel count rate was smaller, about 80.000 events/s, with a similar mean amplitude and a slightly larger duration of 3.2 ns. When we recorded both channels simultaneously, we selected events belonging to the backward discharge. This corresponds to events where the voltage of the exposed electrode is positive and rising. In a typical dataset we collected about 300 events, whose relative delay is less than 500 ns. A delay histogram of count frequency was calculated as in figure 4 and the peak was fitted with a Gaussian peak. In general only one well defined peak was observed, containing about one third of the whole delay dataset. The results are presented in figure 5 showing the value of the delay peak.

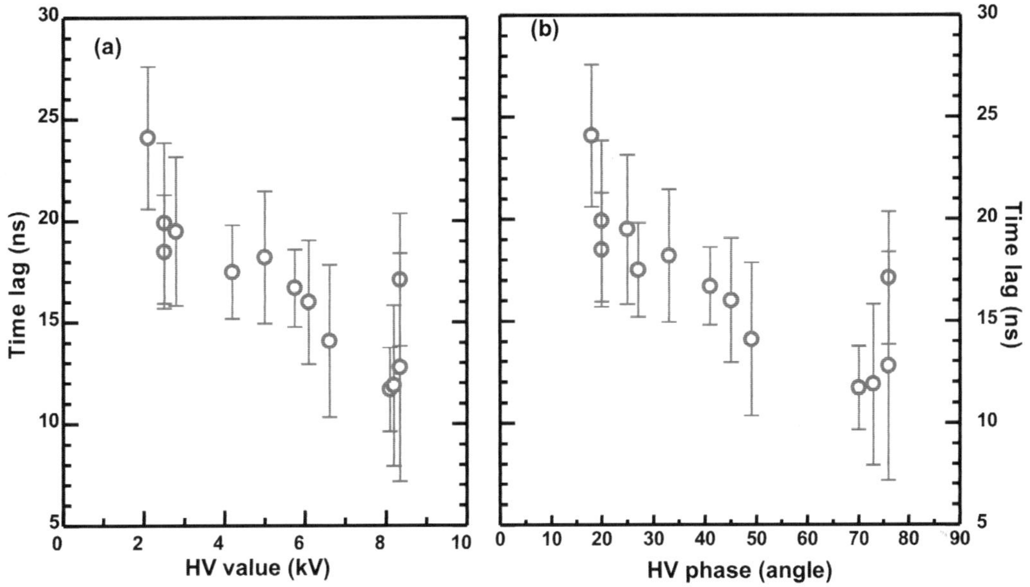

Figure 6. Time lag between the two channels as a function of HV instantaneous value (a) or phase (b).

It is clear that a definite time lag is present between the two channels, independent from the preset delay. The nice linearity allows to estimate a time lag of 12.1 ± 0.6 ns. This corresponds to a propagation velocity of 223 ± 11 km/s directed from the edge to the dielectric surface. This direction is opposite to that of the drifting electrons (backward), somewhat parallel to that of the electric field.

Then we analysed also events belonging to the forward discharge. Unfortunately, while the count rate in the first channel was high enough, actually even larger than that of the backward events [6], the same does not hold for the second channel. Here only the 15% of the events happens in the forward stroke. We tried to reconstruct a delay histogram without finding any definite peak, at least for delays shorter than 500 ns. Due to the limited statistics we can neither estimate nor exclude a propagation, although it is confirmed that most of the FD events develops at or very near the edge of the exposed electrode and do not spread much on the dielectric surface. A better result could be achieved by collecting separately the two viewline channels and triggering only on FD events happening in the second, which however requires the use of two PMT. This will be performed in future developments. We also tried to measure the delay between the viewline looking the edge of the electrode and the others further away on the dielectric surface, distant 5.4 and 8.1 mm from the edge (see figure 1). Again the small number of events, belonging to both BD and FD, from these faraway viewlines prevented us to collect a suitable dataset of delays.

Finally, using the segmented acquisition function of the scope, we accumulated statistics of events happening in a definite and limited range value of HV amplitude. A clear, albeit somewhat noisy, dependence can be appreciated from the data reported in figure 6. The time lag decreases steadily from the breakdown voltage, here at a phase of about 15°, up to the highest amplitude. The propagation velocity, always directed from the electrode edge towards the dielectric, increases from 130 to 250 km/s. These velocities are comparable for instance to those measured in the propagation of streamers in ambient air [11].

4. Conclusions

We have measured the direction and the velocity of propagation of the ionization wave during the development of a microdischarge in a surface dielectric barrier discharge. The measure was inferred from the time lag between two nearby events recorded with a phototube and collected from two different viewlines looking at two positions apart on the discharge region. We succeeded in measuring the velocity of propagation of the backward discharges which turned out to be about 200 km/s. Such velocity appears to increase as the applied external voltage is rising. The proposed method could be applied to study different groups of microdischarge events and different regions of the electrode as well as of the dielectric surface and results could be correlated with the findings of macroscopic diagnostics, in particular with the efficiency of aerodynamical effects or the predictions of numerical simulations of SDBD.

Acknowledgements

We are pleased to acknowledge the collaboration of the Wind Tunnel Department of Alenia Aermacchi Spa, an Italian aeromotive company, which introduced us to this stimulating research field.

References

[1] Roth J R, Sherman D M and Wilkinson S P 2000 *Am. Inst. Aeronautics & Astronautics J.* **38** 1166

[2] Corke T C, Enloe C L and Wilkinson S P 2010 *Annu. Rev. Fluid Mech.* **42** 505

[3] Caruana D 2010 *Plasma Physics & Controlled Fusion* **52** 124045

[4] Kogelschatz U 2003 *Plasma Chem. & Plasma Proc.* **23** 1

[5] Hoder T, Brandenburg R, Basner R, Weltmann K-D, Kozlov K V and Wagner H-E 2010 *J. Phys. D: Appl. Phys.* **43** 124009

[6] Biganzoli I, Barni R and Riccardi C 2013 *J. Phys. D: Appl. Phys.* **46** 025201

[7] Moreau E 2007 *J. Phys. D: Appl. Phys.* **40** 605

[8] Biganzoli I, Barni R and Riccardi C 2013 *Rev. Sci. Instrum.* **84** 016101

[9] Biganzoli I, Barni R, Riccardi C, Gurioli A and Pertile R 2013 *Plasma Sources Sci and Technol.* **22** 025009

[10] Pearse R W B and Gaydon A G 1976 *The Identification of MolecularSpectra* (New York: Wiley)

[11] Briels T M P, Kos J, Winands G J J , van Veldhuizen E M and Ebert U 2008 *J. Phys. D: Appl. Phys.* **41** 234004

Optical diagnostics of streamers: from laboratory micro-scale to upper-atmospheric large-scale discharges

M Šimek[1,4], T Hoder[2], V Prukner[1] and P F Ambrico[3]

[1]Institute of Plasma Physics, v.v.i., Academy of Sciences of the Czech Republic, Za Slovankou 3, 18200 Prague, Czech Republic
[2]Leibniz Institute for Plasma Science and Technology, Felix/Hausdorff/Str 2, 17489 Greifswald, Germany
[3]Istituto di Metodologie Inorganiche e dei Plasmi - CNR, Via Orabona 4, 70126 Bari, Italy

E-mail: simek@ipp.cas.cz

Abstract. Optical emission produced by streamers is determined by spatial distribution of electronically excited atomic and diatomic species within the streamer head and streamer channel. Peculiarities of emission and LIF diagnostics dedicated to investigating the basic structure of streamers with high spatio-temporal resolution are discussed. Possible strategies based on the 2D projections of cylindrically symmetric streamers to determine radial distributions of excited species within the streamer channel are illustrated for streamers produced in volume or on the dielectric surface at atmospheric and low pressures.

1. Introduction

A streamer is a common form of transient discharge which develops from an electron avalanche in an overvolted gap. Streamers can be produced in the gas phase between metallic electrodes, between electrodes covered (fully or partially) with a dielectric barrier, on the interface between gaseous and liquid phases and in liquids [1]. Various atmospheric-pressure streamer-based discharges have become important in numerous environmental applications, material treatment technologies and aerodynamic and biomedical applications. Therefore continuous effort is required in order to understand and determine basic streamer parameters produced under various conditions through the state-of-the-art experimental measurements and numerical modelling.

On the other hand, red-sprite transient luminous events (TLEs) are large streamer events propagating vertically in the upper atmosphere surpassing altitudes from ~35 km to ~85 km above sea level characterised by very different pressures (between ~8.9 and ~0.16 torr) and gas temperatures (between -10 and -75 °C). Investigation of TLEs based on plasma-induced emission is, compared with laboratory discharges, much more difficult because of their stochastic nature (as in the case of cloud-to-ground lightning events) and very long observational distances (typically hundreds of kilometres between the place of the TLE occurrence and diagnostic instrumentation). Consequently, emission spectra of red sprites are acquired with low spectral, spatial and temporal resolutions. In addition, all optical signals are registered with low signal-to-noise ratio, and therefore their use in advanced

[4] To whom any correspondence should be addressed.

analysis which might provide estimation of important parameters (E/N, temperatures/distributions of excited species) is very limited. To this end, laboratory red-sprite TLE analogue experiments are important for better understanding of both the fundamentals of TLEs and their associated phenomena (heating or NO_x production) [2].

The most important characteristics of filamentary streamers are their relatively high propagation velocity (exceeding the velocity of electron drift), small streamer channel radius (with respect to length) and enhanced mean energy and density of free electrons in the streamer head (compared with the streamer channel). All these characteristics make streamer discharges extremely difficult for observations with adequate spatio-temporal resolutions.

2. Streamer-induced emission

Optical emission produced by streamers (either at high pressure or under TLE conditions) is determined by the spatial distribution of various species within the streamer channel excited to radiative states by streamer head electrons [2, 3]. Radial distributions produced by streamer electrons are strongly non-uniform because the maximum electric field of a propagating streamer occurs on the axis of the streamer and rate constants of the excitation, dissociation and ionization processes exponentially depend on the E/N. Radial non-uniformity of excited species has to be considered in the use of diagnostics with limited spatial and temporal resolutions or in comparison of experimental results with numerical simulations [3]. In the case of nitrogen/air streamers, the ($B^3\Pi_g \rightarrow A^3\Sigma_u^+$) first positive (FPS) and ($C^3\Pi_u \rightarrow B^3\Pi_g$) second positive (SPS) of N_2, as well as the ($B^2\Sigma_u^+ \rightarrow X^2\Sigma_g^+$) first negative system (FNS) of N_2^+, are the main components of radiation emitted in the UV-vis-NIR spectral region which can be utilized for emission-based diagnostics.

3. Optical diagnostics

3.1. Timing issues

Recording the instantaneous development of streamer emission is quite a challenging task because of the high velocity of the streamer propagation. The key issue is managing the synchronization of the discharge event with a light-detecting device(s) using a suitable triggering procedure, usually with sub-nanosecond temporal precision. In the case of pulsed or pulse-triggered discharges the problem can be solved by using a digital delay generator, which starts gating the ICCD or recording the photomultiplier (PMT) signal at the right moment. The discharge event is externally triggered on request as is the detector.

In the case of streamer discharges with temporally unstable appearance (e.g. Trichel pulse corona, barrier discharges driven by AC waveforms or temporally irreproducible pulsed discharges with low-gradient voltage slopes) no well-defined external trigger is available. The only solution is to trigger the detecting devices with the discharge itself, utilizing particular discharge events complemented by a system of delay lines. For example, current pulse produced by the discharge event in an external circuit or discharge-induced light pulse can provide precise timing of a just-past event. However, in order to compensate for the dead time (usually tens of nanoseconds) between the trigger input and the onset of the ICCD's intensifier, the light of the discharge which is to be detected must be delayed before arrival at the detector (for example, by a sufficiently long optical fibre bundle). A similar but technically more demanding option, the time-correlated single photon counting-based technique, is discussed in Section 3.2.

3.2. TCSPC- and PMT-based scanning systems

By adjusting the electro-optical set-up of the system to detect rare single photons and to record the delayed synchronization signal from the same discharge event one can take advantage of delay-based methods and apply advanced systems which measure the probability-density functions representing the discharge emission under given conditions. The so-called time-correlated single photon counting technique (TCSPC, also known as cross-correlation spectroscopy, CCS [4,5]) is based on this principle. For selected spatial coordinates, the TCSPC method substitutes for the real-time emission measurement of the discharge event with statistically averaged determination of cross-correlation function between two optical signals originating from the same source. These are the so-called 'main signal' (the spatially and spectrally resolved single photon, the actual emission detection) and the 'synchronizing signal' (the integral light intensity pulse of the discharge event which gives the time reference). The time between the detections of these signals is measured. Consequently, time histograms of counted photons for all spatial positions of the discharge are accumulated.

Figure 1 shows an example of a TCSPC 2D scan. The discharge was driven in 10 kPa synthetic air with sinusoidal voltage waveform at 6 kHz and a peak-to-peak amplitude of 2.4 kV. The gaseous gap was 0.9 mm, as depicted in figure 1. Spatial resolution was 0.1 x 0.1 mm^2 and thus a matrix of 12 x 12 was scanned point by point with TCSPC. From the data obtained one can carry out 2D emission distribution in the gap for a selected time of discharge development. Local emission diameter, streamer velocity at certain time or relative intensity of the radiation can also be evaluated.

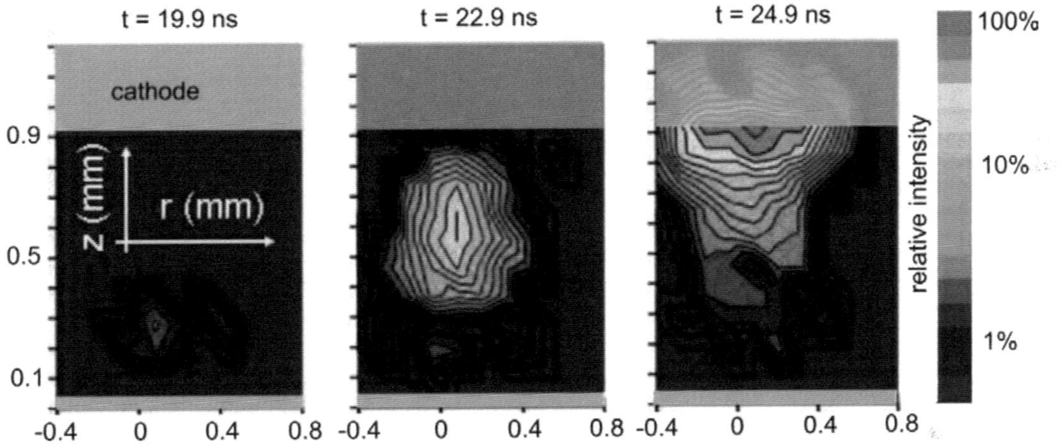

Figure 1. The 2D development of the positive streamer luminosity in 10 kPa synthetic air recorded with sub-nanosecond resolution by a TCSPC device. One can resolve the creation of the streamer in the first sub-figure (on the left), its propagation in free space in the second and its impact on the cathode in the last sub-figure (on the right), where the spread of the surface discharges across the dielectric surface of the cathode is also shown.

In the case of experiments on laboratory TLE analogues at pressures below 2 kPa (the HV pulse-triggered streamers in amplitude-modulated AC-driven volume DBD with a discharge gap of 1 to 5 cm) we used a high-speed PMT to record streamer-induced emission through a set of band-pass filters. Under TLE-like conditions the temporal resolution of the PMT (Hamamatsu H10720) and spatial resolution of 1 mm seem to be enough for capturing streamer emission with sufficient precision and the use of complex TCSPC analysis can be avoided. In this case, the streamer and the PMT recording can be triggered externally by the same pulse/delay generator. Figure 2 illustrates the discharge luminosity scanned radially in the centre of the gap with spatial resolution of 1 mm. For each radial position a PMT signal waveform was recorded as an average of 1024 samples. Resulting 2D maps of projected streamer luminosities acquired with two selected band-pass filters are displayed in figure 2. Panel (a) in the figure shows the distribution of streamer filament luminosity at a wavelength corresponding to the (0,0) band of N_2-SPS and panel (b) shows luminosity at a wavelength corresponding to the (0,0) band of N_2^+-FNS (corrected for spectral overlap with the SPS (2,5) band).

The use of TCSPC or pulse-triggered PMT methods means the emissions of selected spectral bands/lines of streamers can be evaluated. Typically, in nitrogen-oxygen mixtures the method of intensity ratio of first negative and second positive systems of molecular nitrogen can be used for electric field determination [4–6]. Similar measurements to those presented in figure 1 were performed for more challenging atmospheric pressure discharge conditions in the same set-up for emissions of second positive and first negative systems [7] and the 2D distribution development of the electric field was determined [8].

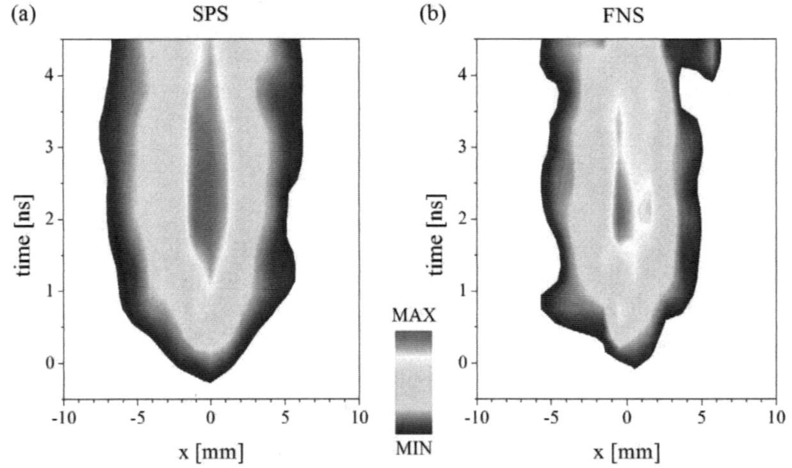

Figure 2. Radial scan of projected luminosity of triggered streamer generated in pure nitrogen at pressure of 1.33 kPa and in point-to-plane DBD gap of 1 cm. Panels (a) and (b) show SPS and FNS emission projections, respectively. The 2D maps were acquired from the middle of the gap.

3.3. ICCD spectroscopy, imaging and microscopy

ICCD-based techniques are used to provide spectral analysis of streamer luminosity when coupled with a dispersion spectrometer or to register streamer morphology when used as an imaging device, both with temporal resolution limited by the performance of the last-generation image intensifiers (typically 0.5 to 2 nanoseconds). Spectrometric analysis of streamer-induced emission can provide roto-vibronic distributions of diatomic species or relative emission intensities of various atomic/diatomic line/band systems [9]. Streamer morphology can be investigated through time-resolved (non)magnified ICCD images capturing 2D projection of the streamer luminosity. Because the excitation processes leading to the observed emissions strongly (exponentially) depend on the electric field, the streamer filament luminosity projected on and registered by the ICCD detector reflects characteristic dimensions of the streamer head.

A technique allowing acquisition of magnified images of surface micro-discharges was developed in [10]. Time-resolved microscopic images can be acquired synchronously either with the AC high-voltage voltage waveforms or micro-discharge current pulses. Figures 3, 4, 5 and 6 show typical examples of results obtained by ICCD microscopic images in a single coplanar surface dielectric barrier discharge (CSDBD) at atmospheric pressure. A detailed description of the reactor, CSDBD electrode geometry and related characteristics can be found elsewhere [10]. The images shown in figure 3 were acquired with a large gate of ICCD intensifier (1s) synchronously with AC voltage and provide information on the spatial stability and influence of the oxygen addition on the micro-discharge morphology. It is worth pointing out that the ICCD images capture predominantly the SPS and FPS emissions of N_2.

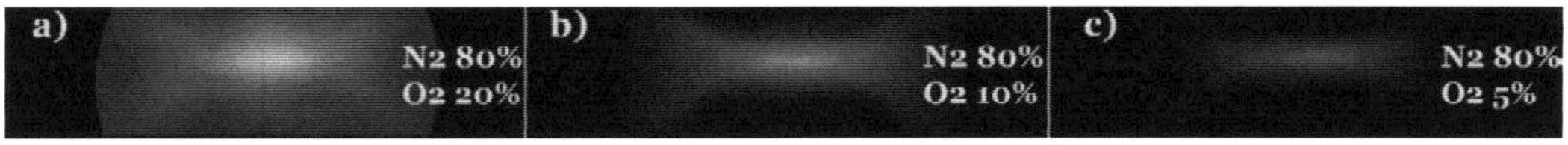

Figure 3. ICCD images registered with the exposure of 1 s at a fixed spatial resolution of 2.7 µm/pixel. The images show variations of the streamer radius together with spatial jitter for the case of non-triggered single surface streamer discharge in N_2-20%O_2 (a), N_2-10%O_2 (b) and N_2-5%O_2 (c) at atmospheric pressure.

The ICCD images shown in figure 3 result from the accumulation of about 500 micro-discharges. The discharges seem to be diffused and no single streamer can be resolved because of the vertical spatial jitter of the micro-discharges and the increase of their diameter with the increasing content of oxygen [11]. This becomes clearer if we look at single streamer images such as those shown in figure 4. The images a) to d) were registered during positive (a,c) and negative (b,d) half-cycles in N_2 (a,b) and N_2-5%O_2 (c,d). If we compare pictures taken in pure nitrogen with those taken in nitrogen-oxygen mixture the influence of oxygen on the diameter of the single streamer micro-discharge is evident.

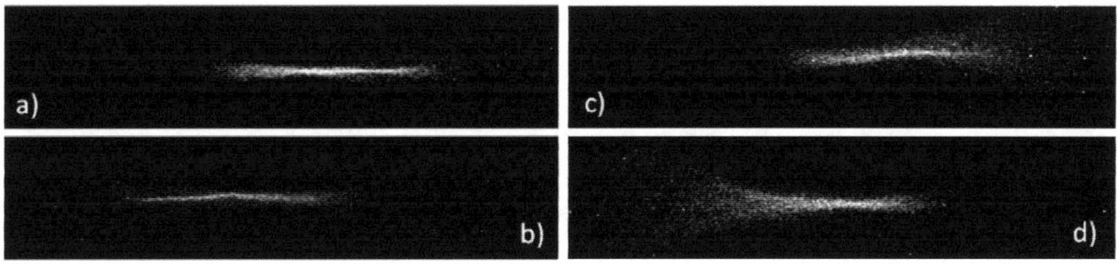

Figure 4. ICCD microscopic images of surface CSDBD streamer in pure N_2 (a-b) and in nitrogen with 5% admixture of oxygen (c-d), registered during positive (a-c) and negative (b-d) half-cycles of AC high-voltage waveforms (u_{ac} = 23kV$_{p-p}$) with exposure time of 50 µs. The grounded electrode and HV cathode are located on the left and right side, respectively.

Streamer propagation velocity, electric field in the streamer head and streamer diameter are closely related parameters, and knowledge of any two parameters could, in principle, be used to estimate the remaining one [12]. ICCD microscopic images can be used to determine the streamer head/channel diameter, the luminous streamer channel diameter and the propagation trajectory [10]. Defining the luminous minimum streamer diameter as the distance between the limits corresponding to the values of 10% and 90% of the integral of luminosity along the direction perpendicular to the streamer propagation axis [10], we estimated a minimum luminous CSDBD streamer diameter of 50 µm in pure N_2 [10]. Adding oxygen to the gas mixture results in a larger streamer diameter. As an example, the typical luminous diameters obtained by applying former procedure in the case of a mixture of N_2-5%O_2 are reported in figure 5. The axial dependence (along the x-axis in figure 5) of the luminous diameter reveals, as in the previous cases, an asymmetric V-shape with a minimum diameter of about

70 μm located close to the temporary cathode, as for the streamer filaments produced both in N_2 and Ar [10].

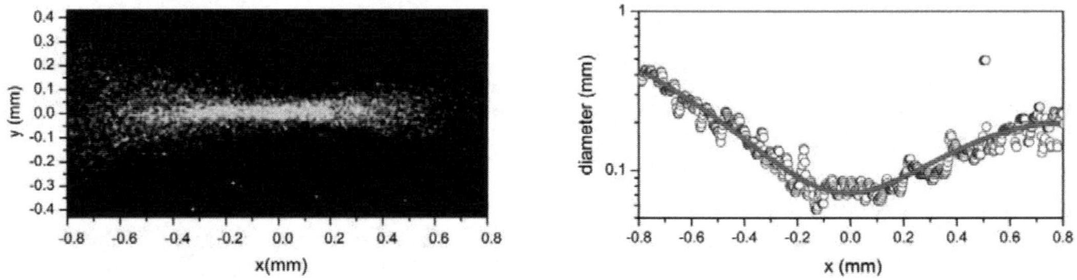

Figure 5. ICCD microscopic image of single CSDBD event occurring in the N_2-5%O_2 mixture (u_{ac} = 23kV$_{p-p}$). The image (on the left) was used to estimate the axial evolution of the streamer luminous diameter (on the right). The (0,0) axis origin is placed in the centre of the gap. The grounded electrode and HV cathode are located on the left (x=–0.5) and right (x=+0.5) side, respectively.

In the case of cylindrically symmetrical discharge filaments, radial distributions of electronically excited species produced during surface streamer propagation can be obtained by applying the Abel inverse transform to projected luminosities of single streamer events [13], as shown in figure 5. Here, Abel inversion was performed for the projected profiles taken at x=–0.1 mm coordinate of figure 5 following the procedure described in [13]. Projected as well as Abel-inverted profiles can be fitted by Gaussian functions with very similar half-widths (σ_1,σ_2), as shown in figure 6. In this case however the projected profile, in addition to the Abel-inverted one, can be well approximated only by the sum of two coaxial Gaussians with two different half-widths and weights.

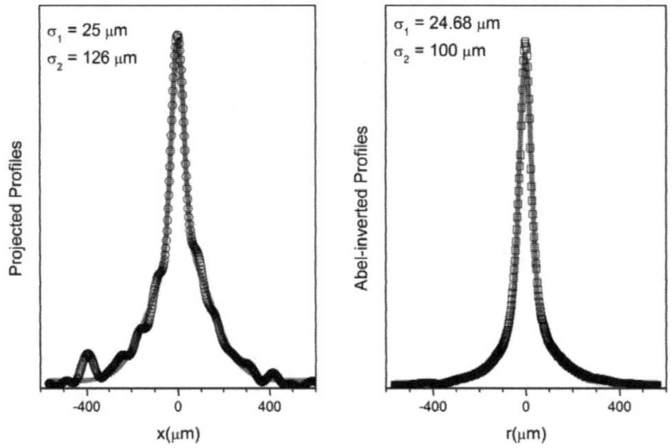

Figure 6. Experimental projected and resultant Abel-inverted profiles (both symbols) together with the corresponding Gaussian fits (both solid lines) obtained from ICCD microscopic images (figure 5 of N_2-5%O_2) of a typical single CSDBD event at x =–0.1mm (corresponding to the axial position of the minimum diameter of the streamer filament).

An equivalent procedure was performed in the case of projected profiles shown in figure 2. Abel inversion [14] was performed for the projected profiles taken from figure 2 at time of 2 ns. In this case we inverted separately two different projected profiles corresponding to the FNS and SPS emissions. The projected (a,c) and Abel-inverted (b,d) profiles are shown in figure 7 together with the FNS/SPS ratio (e) of Abel-inverted profiles. Because ratio FNS/SPS ratio depends on the reduced electric field, the profile shown in panel (e) reflects radial evolution of electric field along the streamer radius in the middle of the gap at given time.

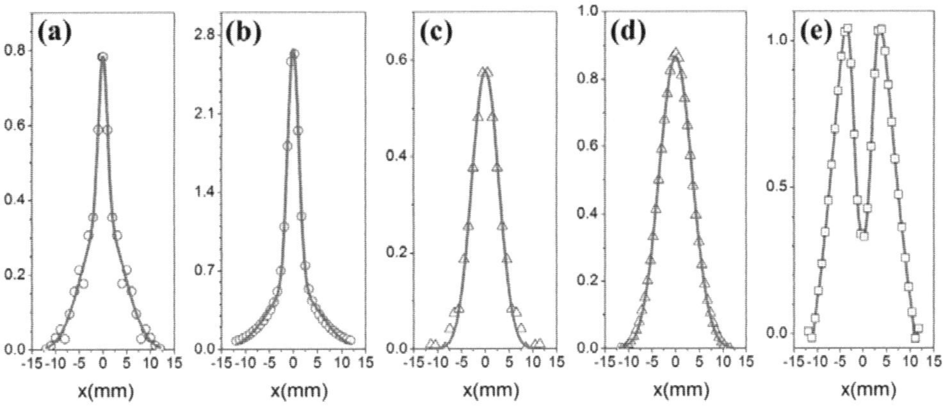

Figure 7. Experimental projected (a, c) and resultant Abel-inverted profiles (b, d) together with the corresponding Gaussian fits (both solid lines) obtained from radial scan of projected FNS and SPS luminosity of triggered streamer. Panels (a, c) show SPS while (b, d) FNS profiles. Panel (e) shows FNS/SPS ratio of inverted profiles. Abel inversion was performed for the projected profiles taken at position of maximum FNS intensity at conditions displayed in figure 2 (pure nitrogen at pressure of 1.33 kPa and in point-to-plane DBD gap of 1 cm).

3.4. LIF techniques and $N_2(A^3\Sigma_u^+)$ detection

The determination of absolute densities of transient species is quite a challenging task, as can be demonstrated in the case of $N_2(A^3\Sigma_u^+)$ metastables. The crucial role of these long-living species lies in the fact that they are important energy carriers capable of driving many energy transfers or chemical processes. In relation to streamer discharges probably the most versatile approach to detecting non-radiative species is laser-induced fluorescence (LIF). The detection of $N_2(A^3\Sigma_u^+)$ species may be based either on conventional one-photon LIF[15,16] or on two-photon optical-optical double-resonance (OODR) LIF techniques [17–19].

Recently, we investigated the temporal evolution of $N_2(A^3\Sigma_u^+$, v=0–3) metastable species produced by a single triggered nitrogen streamer employing LIF and emission diagnostics simultaneously, and provided experimental evidence that maximum density of $N_2(A^3\Sigma_u^+$, v=0-3) metastable species occurs on a microsecond timescale during the streamer channel decay [20]. It is worth pointing out that the single photon LIF technique can be successfully applied only in the case of precise timing of micro-discharge events, i.e. in the case of nanosecond-triggered discharges. In random events, a more sensitive technique might be applied, such as OODR-LIF [19] at the expense of a more complicated optical layout.

As an example, we show results obtained by applying the single photon LIF technique described in detail in [20] to monitor odd vibrational levels under discharge conditions illustrated in figure 2 (pure N_2, at 1.33kPa). Figure 8 shows the temporal behaviour of relative populations of individual $N_2(A^3\Sigma_u^+$, v = 1, 3, 5, 7) vibronic levels normalized to their maximum. All four curves clearly show the formation of broad maxima shifted with respect to the streamer onset (t=0) with individual delays on a microsecond timescale and decreasing with the vibrational quantum number (about 50, 14, 2.3 and 1.4 μs for v = 1, 3, 5 and 7, respectively).

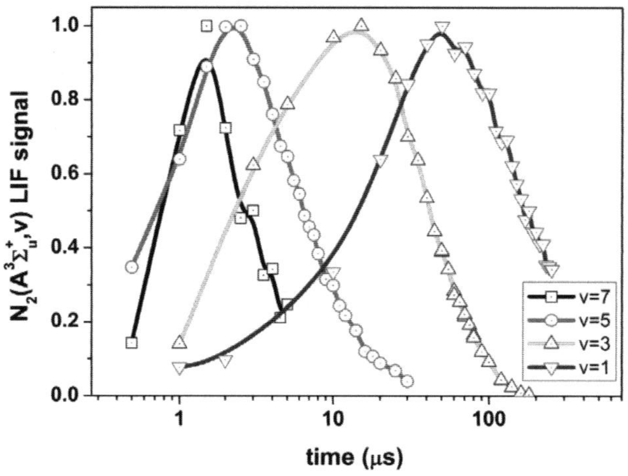

Figure 8. Temporal evolution $N_2(A^3\Sigma_u^+$, v) vibrational level for odd vibrational numbers produced by triggered streamer discharge in pure nitrogen at a pressure of 1.33 kPa and in point-to-plane DBD gap of 1 cm. The $N_2(A^3\Sigma_u^+$, v) evolution is acquired from the middle of the gap.

For proper interpretation of the experimental findings we need to understand the behaviour of metastable and other electronically excited N_2 species. Under streamer conditions, the discharge onset and propagation are an extremely fast event, which takes place on a (sub-)nanosecond timescale. As reported in [20], direct electron impact of the streamer head electrons establishes initial populations of individual vibrational levels which are redistributed because of radiative relaxation and heavy particle collisions on a much longer timescale (from units to hundreds of microseconds). It is well known that the $N_2(A^3\Sigma_u^+$, v) metastable state serves as a terminal state for all N_2 triplet manifolds thanks to efficient intersystem collisional transfer (ICT) and redistribution within the $N_2(A^3\Sigma_u^+)$ state vibrational manifold (controlled by a specific process of vibrational relaxation).

The interpretation of similar experimental findings through a simplified numerical model implies that an important part of the metastable species becomes lost during the relaxation processes and estimation of $N_2(A^3\Sigma_u^+)$ state populations based on the LIF measurements of low vibrational levels could lead to significant undervaluation of the total metastable state population created during the streamer propagation phase [20].

4. Discussion and conclusions

To study the evolution of a developing streamer, optical diagnostics have to be performed with significantly enhanced spatio-temporal resolution. As regards radiative states, several approaches based on the 2-D projections of luminosities of cylindrically symmetrical streamers can be used to investigate spectrometric characteristics, the morphology of the streamer channel or radial distributions of electronically excited species within the streamer channel.

As regards non-radiative states, such as important long-living $N_2(A^3\Sigma_u^+, v)$ metastables, LIF-based techniques need to be applied to investigate the evolution of metastables on microsecond or even millisecond timescales. High vibrational levels of $N_2(A^3\Sigma_u^+, v)$ metastables under streamer conditions need to be studied in order to determine the total population of the $N_2(A^3\Sigma_u^+, v)$ manifold. However, the use of the discussed techniques under non-standard conditions imposed by micrometre spatial and picosecond timescales of streamers is always critical. Therefore, all the restrictions and limitations of the proposed techniques (timing and jitter issues) should be very carefully examined.

Acknowledgements
This work was supported by the Czech Science Foundation (GACR contract no. P205/12/1709). PFA would like to thank the CNR Short Term Mobility Programme for partially funding his stay at the IPP in Prague.

References

[1] Šimek M, Člupek M, Babický V and Šunka P 2006 Production of reactive species by atmospheric pressure streamers in N_2-O_2 mixtures *Pure Appl. Chem.* **78** 1213–25
[2] Pasko V P 2007 Red sprite discharges in the atmosphere at high altitude: the molecular physics and the similarity with laboratory discharges *Plasma Sources Sci. Technol.* **16** S13–S29
[3] Hoder T, Bonaventura Z, Bourdon A and Šimek M 2014 Sub-nanosecond delays of light emitted by streamer in atmospheric pressure air : analysis of $N_2(C^3\Pi_u)$ and $N_2^+(B^2\Sigma_u^+)$ emissions and fundamental streamer structure *arXiv*:1404.1605v2 [physics.plasm-ph] 1–19
[4] Kozlov K V, Wagner H, Brandenburg R and Michel P 2001 Spatio-temporally resolved spectroscopic diagnostics of the barrier discharge in air at atmospheric pressure *J. Phys. D. Appl. Phys.* **34** 3164–76
[5] Hoder T, Černák M, Paillol J, Loffhagen D and Brandenburg R 2012 High-resolution measurements of the electric field at the streamer arrival to the cathode: A unification of the streamer-initiated gas-breakdown mechanism *Phys. Rev. E* **86** 055401
[6] Paris P, Aints M, Valk F, Plank T, Haljaste a, Kozlov K V and Wagner H-E 2005 Intensity ratio of spectral bands of nitrogen as a measure of electric field strength in plasmas *J. Phys. D. Appl. Phys.* **38** 3894–9
[7] Brandenburg R, Hoder T and Wagner H-E 2008 Two-Dimensional Spatially Resolved Cross-Correlational Spectroscopy of the Microdischarge Development in Barrier Discharges in Air *IEEE Trans. Plasma Sci.* **36** 1318–9
[8] Kozlov K V. and Wagner H-E 2007 Progress in Spectroscopic Diagnostics of Barrier Discharges *Contrib. Plasma Phys.* **47** 26–33
[9] Šimek M 2002 The modelling of streamer-induced emission in atmospheric pressure , pulsed positive corona discharge : N_2 second positive and NO-γ systems *J. Phys. D. Appl. Phys.* **35** 1967–80
[10] Šimek M, Ambrico P F and Prukner V 2011 ICCD microscopic imaging of a single micro-discharge in surface coplanar DBD geometry: determination of the luminous diameter of N_2 and Ar streamers *Plasma Sources Sci. Technol.* **20** 025010
[11] Briels T M P, van Veldhuizen E M and Ebert U 2008 Positive streamers in air and nitrogen of varying density: experiments on similarity laws *J. Phys. D. Appl. Phys.* **41** 234008
[12] Naidis G 2009 Positive and negative streamers in air: Velocity-diameter relation *Phys. Rev. E* **79** 057401
[13] Šimek M and Ambrico P F 2012 Radial dependence of surface streamer-channel luminosity: experimental evidence of Gaussian radiative profiles in Ar and N_2 *Plasma Sources Sci. Technol.* **21** 055014
[14] Chan G C-Y and Hieftje G M 2005 A LabVIEW program for determining confidence intervals of Abel-inverted radial emission profiles *Spectrochim. Acta Part B At. Spectrosc.* **60** 1486–501

[15] Ono R, Tobaru C, Teramoto Y and Oda T 2009 Laser-induced fluorescence of $N_2(A^3\Sigma_u^+)$ metastable in N_2 pulsed positive corona discharge *Plasma Sources Sci. Technol.* **18** 025006

[16] Nemschokmichal S and Meichsner J 2013 $N_2(A^3\Sigma_u^+)$ metastable density in nitrogen barrier discharges: I. LIF diagnostics and absolute calibration by Rayleigh scattering *Plasma Sources Sci. Technol.* **22** 015005

[17] Ali A and Dagdigian P J 1987 Rotational energy transfer within the $B^3\Pi_g$ v=3 manifold of molecular nitrogen *J. Chem. Phys.* **87** 6915

[18] Dilecce G, Ambrico P F and De Benedictis S 2005 Optical–optical double resonance LIF detection of $N_2(A^3\Sigma^+_u)$ density in high pressure gas discharges *Plasma Sources Sci. Technol.* **14** 561–5

[19] Ambrico P F, Šimek M, Dilecce G and De Benedictis S 2009 $N_2(A^3\Sigma_u^+)$ time evolution in N_2 atmospheric pressure surface dielectric barrier discharge driven by ac voltage under modulated regime *Appl. Phys. Lett.* **94** 231503

[20] Šimek M, Ambrico P F and Prukner V 2013 Formation of $N_2(A^3\Sigma_u^+$, v = 0–3) metastable species in decaying nitrogen streamer *J. Phys. D. Appl. Phys.* **46** 485205

Experimental investigation of filamentary and non-filamentary regimes in a surface dielectric barrier plasma actuator

I Biganzoli[1], R Barni[1], A Gurioli[2], R Pertile[2] and C Riccardi[1]

[1] Dipartimento di Fisica G. Occhialini, Università degli Studi di Milano-Bicocca, piazza della Scienza 3, 20126, Milano, Italy

[2] Aerodynamics & Wind Tunnel Departments, Alenia Aermacchi SPA, via Ing. Foresio 1, 21040, Venegono Superiore (VA), Italy

E-mail: claudia.riccardi@mib.infn.it

Abstract. Asymmetric surface dielectric barrier discharges fed by a high-voltage sinusoidal low-frequency drive are currently proposed as plasma actuators, because they can induce a directed airflow in the gas surrounding the surface. However, it is known that the induced airflow speed can not be increased as much as desired because a saturation is generally observed for sufficient high voltages. In this paper we show that when the voltage amplitude is increased enough the discharge does not appear uniform any more, but a pattern of plasma filaments becomes evident. We have thus studied plasma properties in both filamentary and non-filamentary regimes, by means of a Rogowski coil for the measurement of the current associated to the discharge. This is interesting in order to understand what happens at high voltages, when the saturation of the induced airflow speed occurs.

1. Introduction

When a sufficiently high voltage difference is applied between two electrodes placed at the opposite sides of an insulating material, a Surface Dielectric Barrier Discharge (SDBD) is obtained because of the ionization of the gas surrounding the dielectric material [1]. If one electrode is insulated, plasma is created in correspondence of just one side of the barrier (Figure 1). With this geometrical configuration and in air at atmospheric pressure, SDBDs are used as plasma actuators, for the manipulation of the boundary layer of an airflow surrounding a body, and thus in the active flow control field [2]. As a matter of fact, these non-thermal plasmas can generate an induced airflow of some m/s that can interact with an external wind, thus providing, for instance, stall prevention, lift enhancement or drag reduction [3, 4]. Presently, these applications in the aeronautical sector are interesting but partially limited by the fact that the induced airflow speed cannot be increased as much as desired because a saturation in the induced wind velocity is generally observed at sufficiently high voltages [5].

It is known that, when the electric field created by the voltage difference imposed between the electrodes is sufficiently high, plasma is created and electrical charges are transported through the gap and accumulated on the insulating surfaces [6]. This charge accumulation generates an electric field that locally weakens the external one. When the total electric field falls below the threshold necessary for plasma ignition, the discharge extinguishes. If the voltage imposed to

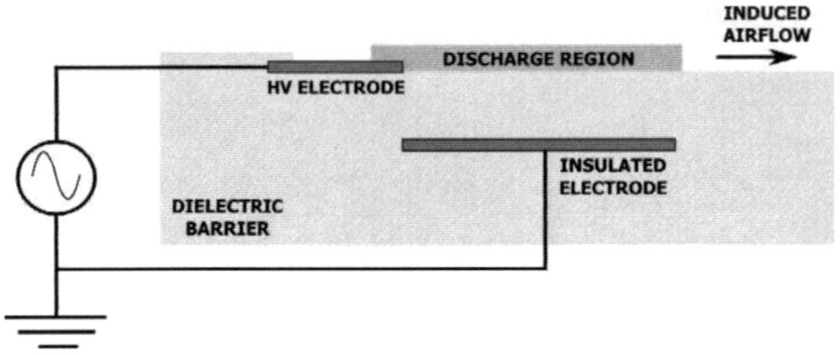

Figure 1. Schematic of a Surface Dielectric Barrier Discharge configuration.

the fed electrode is increased, the discharge can be locally initiated again, and that is the reason why a sinusoidal high-voltage supply is adopted instead of a continuous one. Consequently, the presence of the insulating barrier usually leads to a regime where charge is mainly transported in sub-millimetre regions consisting of current filaments with temporal duration limited to a few tens of nanoseconds [7]. These plasma *microdischarges* are concentrated into two phase intervals of the sinusoidal voltage supply, when the modulus of the applied voltage difference is high enough and is increasing in time. These two phases of plasma activity are often called *Backward Stroke* (BD) and *Forward Stroke* (FD), depending if the voltage difference between the fed electrode and the grounded one is rising from its minimum to its maximum or decreasing from its maximum to its minimum [8].

In a previous work we have point out that not all plasma properties of the two semi-cycles must be identical, because of the asymmetrical configuration adopted, and that plasma microdischarges emit light for few nanoseconds and carry current for some tens of nanoseconds [5]. We can think that light and current signals give insights about different microdischarge properties. Light is presumably ascribable to electrons that excite nitrogen immediately after the passage of the ionizing wave that initiates the microdischarge. In contrast, the current signal is probably due to the movement of charges into the plasma channel and thus reflects the microdischarge temporal evolution, rather than its formation. Since we are interested in better understanding which plasma characteristics are responsible for the induced airflow generation and saturation, in this paper we have decided to focus on the electrical properties of plasma microdicharges generated during the two strokes, and a Rogowski coil has thus been adopted for current measurements.

We have noticed that when the voltage amplitude is increased enough the discharge does not appear uniform, but a pattern of plasma filaments becomes evident (Figure 2). We have thus studied plasma properties in both filamentary and non-filamentary regimes.

2. Experimental Setup and Diagnostics

In the surface configuration studied in this work, the electrodes are tin clad copper adhesive tapes (60 μm thick and 200 mm long) placed at the opposite sides of a flat dielectric panel (teflon, 2 mm thick), without overlapping. The insulated electrode is 10 mm wide, whereas the air exposed one is just 5 mm wide. An insulating polyisobutylene self-amalgamating tape is used for avoiding discharges at the lower electrode side, but also upstream of the exposed electrode and at the electrode ends. Different voltages (up to about 15 kV peak amplitude) has been tested, at the same frequency (16 kHz). The buried electrode is grounded and the exposed one is fed by a high-voltage power supply, which was the MiniPuls 6 by GBS Elektronik GmbH, consisting of a transistor bridge followed by a transformer cascade [9].

Plasma current microdischarges have been experimentally studied with a sufficiently high temporal resolution by means of home-made Rogowski coils, as described in detail in Section

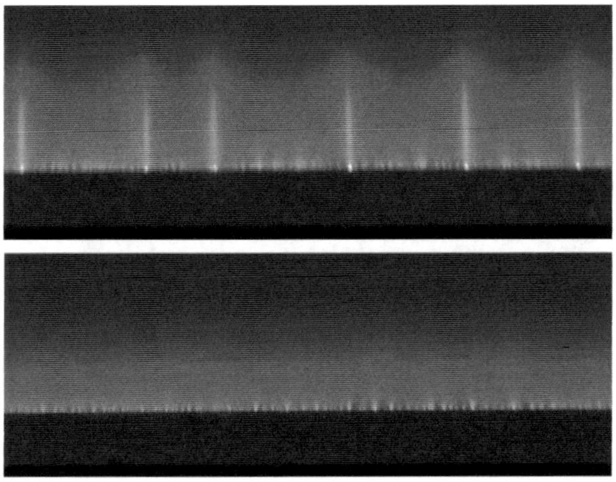

Figure 2. Examples of a SDBD working in non-filamentary (bottom) and filamentary (top) regimes. These pictures were taken from above the discharge, with a commercial digital camera.

2.1. A commercial HV probe (Tektronix P6015A, granted for a bandwidth of 75 MHz) was instead employed for evaluating the voltage at the exposed electrode. The outputs of all these probes were connected to different channels of a large bandwidth digital oscilloscope (Agilent MSO8104A with sampling rate equal to $2GSas^{-1}$) and their signals were acquired simultaneously. The presence or absence of discharge disuniformities (plasma filaments) was verified with a commercial digital camera and with a thermal infrared camera FLIR ThermaCAM SC640 (spectral range $7.5 \div 13$ μm). The camera records the infrared emission of the dielectric surface, which is a function of its temperature, and it was placed about 50 cm above the discharge.

Finally, the average fluid dynamic effects induced by the plasma actuator were investigated using a Pitot tube connected to an LPM-9381 Druck transducer that records the differential pressure from which the flow velocity was derived. The single Pitot tube could be replaced by a series of three or more connected capillaries, in order to take a measurement of the flow velocity averaged along the electrode length. In our experiments the Pitot tubes were small glass tubes with outer and inner diameters equal to 1.4 mm and 1.0 mm respectively. The Pitot inlets were placed in contact with the electrode surface, 10 mm downstream the exposed electrode edge, in correspondence of the end of the buried electrode.

2.1. Rogowski Coils

A current transducer like a Rogowski coil is suitable to measure electric current filaments [10]. In a DBD, the electric current flowing into the circuit can be viewed as the superimposition of a low-frequency sinusoidal capacitive current, which is almost independent of plasma presence in the gap, and a discharge current, which is associated with plasma microdischarges and appears as a series of fast current pulses. A Rogowski coil is a conducting wire that is wound in a spiral around a magnetic core and then returns to the original point [11]. The coil is placed around the cable connecting the buried electrode to the ground: any variation in the current I flowing into the cable generates an electromotive force at the output of the coil. By adding an integrating circuit, the output of the probe is proportional to I. We have built a few home-made Rogowski coils and calibrated them in the frequency range $10 \div 100$ MHz, as discussed in a dedicated paper [10]. Since we are mainly interested in the current signal associated with plasma microdischarges, we can separate this current component from the displacement one by means of a band-pass filter based on the fast Fourier transforms, so that current events due to

Journal of Physics: Conference Series **550** (2014) 012038 doi:10.1088/1742-6596/550/1/012038

plasma microdischarges finally appear as pulses emerging from a noisy base line. At this point they are identified by setting a threshold sufficiently higher than noise spikes. Finally, a baseline is used to evaluate the intensity of peaks, as well as their beginning and end. We were able to study current pulses characteristics like number, amplitude, duration and charge transported [5].

3. Experimental Results

As already stated in the introduction, we noticed that for sufficiently high voltages the discharge does not appear uniform but several plasma filaments become evident. First of all we have used the thermal camera to visualize the voltages at which these filaments are visible. An image was captured about one second after the discharge ignition, for voltage amplitudes in the range $6.5 \div 17$ kV, at steps of 1.5 kV. We found that at 6.5 kV (Figure 3 (a)) and 8 kV the discharge appears quite uniform along electrode length (the so-called spanwise direction). At 9.5 kV the plasma regime changes because some small filaments become evident (Figure 3 (b)). At 11 kV they cover the whole electrode length and are more or less uniformly spaced (Figure 3 (c)). Above 12.5 kV the plasma state changes again because there are longer filaments, which are separated one from the other by a larger spanwise distance (Figure 3 (d)). Eventually, at the

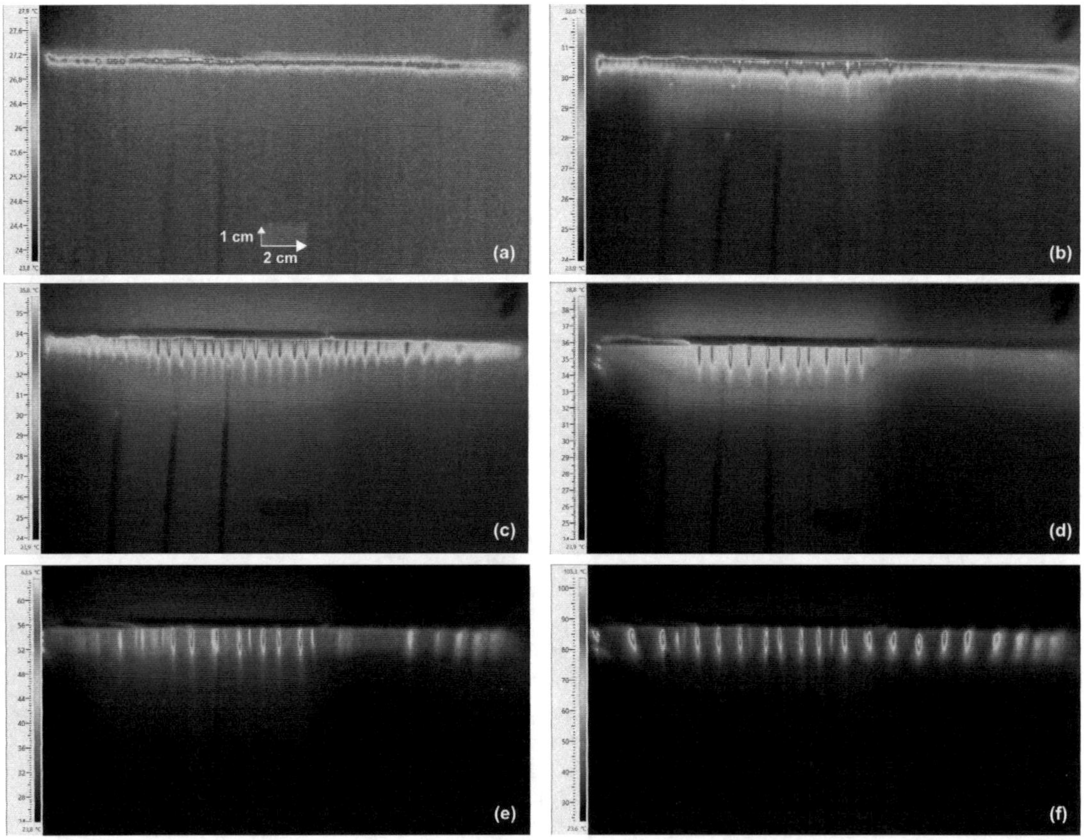

Figure 3. Thermal camera images taken approximately one second after plasma ignition, for different voltage amplitudes: 6.5 kV (a), 9.5 kV (b), 11 kV (c), 12.5 kV (d), 14 kV (e), and 17 kV (f).

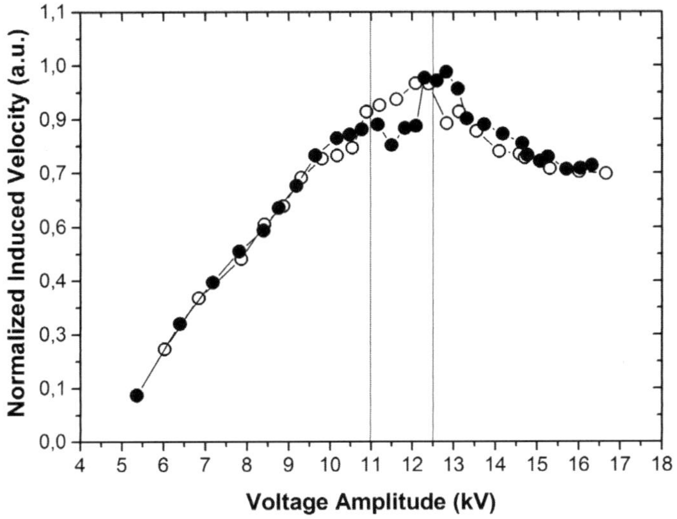

Figure 4. Speed of the induced flow as a function of the applied voltage. Values are normalized to the maximum speed obtained. Full and empty circles are two set of measures, showing a good repeatability and stability of the induced airflow, except in the transition from small to well developed plasma filaments ($11 \div 12.5$ kV).

highest voltages (Figures 3 (e) and (f)) these plasma filaments become more numerous. It is evident in subfigures (c), (d) and (e) that the plasma filaments are not uniformly distributed along the spanwise direction, but are more concentrated on the left half of the picture. This is not due to the presence of the Pitot tubes. As a matter of fact, plasma filaments generally change position (moving a few millimetres along the electrode) every some minutes, and we noticed that, if we try to keep a Pitot tube too close to a filament and perfectly in front of it for minutes, this filament motion can be enhanced by the probe presence, because filaments tend to go away from the Pitot tube. In this case, this discharge non-uniformity is instead ascribable to differences in the local dielectric panel temperature before plasma ignition, which was about three degrees higher than the room temperature in the left portion of the panel. For voltages larger than the ones associated to the discharge regime transition, this temperature difference did not influence the spanwise position of filaments, which immediately appeared uniformly distributed as in Figure 3 (f). However, for lower voltage amplitudes, we noticed that the creation of plasma filaments along the whole the electrode length could take some seconds and began from hotter region [12]. These thermal images have been recorded within the first second of plasma actuator operation, and this is the origin of this discharge disuniformity.

During the acquisition of these thermal images we also measured the average speed of the induced airflow by means of three Pitot tubes, which are visible in Figure 3. We then repeated a few more times the measurements of the airflow speed in order to check their repeatability and stability. Two of these measurements are shown in Figure 4. The two curves overlap well, except in the range $11 \div 12.5$ kV, which corresponds to the transition from small filaments to well developed plasma filaments. These curves of the induced wind speed show that any transition in the discharge regime influences the induced flow speed, and thus the potentialities of plasma actuator. It is evident that the the speed of the induced flow increases quite linearly until small filaments appear (in the range $9.5 \div 11$ kV). By using a single glass Pitot tube positioned in front of a plasma filament or in correspondence of the uniform plasma between two filaments, we noticed that at these voltage amplitudes, the induced airflow has a slightly higher speed near a plasma filament rather than where plasma is more uniform. That's why in this region the two curves of Figure 4 do not overlap. However, at higher voltages, the induced flow speed not only saturates but even decreases with the voltage amplitude, and we have observed this trend in front of both uniform plasma regions and plasma filaments. The conclusion that we can

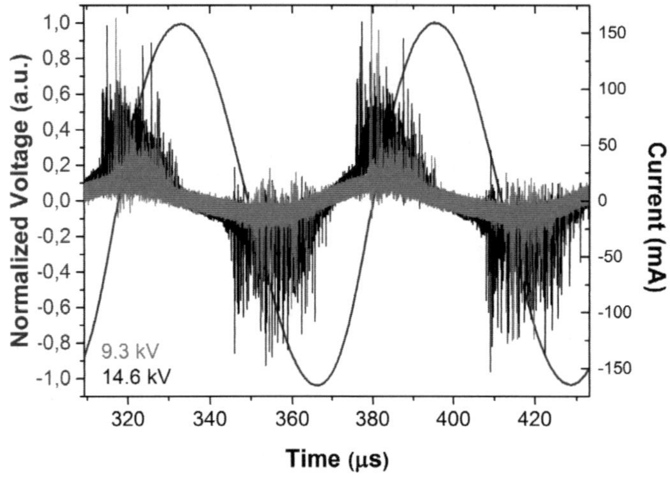

Figure 5. Two current signals obtained at different voltage amplitudes (9.3 kV and 14.6 kV) are shown. A normalized voltage signal is displayed too, as a reference for distinguishing the backward and forward discharges (when the instantaneous voltage rises and decreases respectively).

draw from these experiments is that when the voltage amplitude is increased enough the plasma regime changes from a uniform one to a filamentary one, and there is a transitional range of voltages in which filaments are smaller than the well developed plasma filaments. However, we can think that the formation of filaments is not the only change involved in the plasma regime transition, because at voltages higher than 13 kV, a saturation in the induced wind speed is observed in correspondence of both uniform and filamentary plasma.

Our aim at this point was thus to look for evidences indicating some changes in the plasma state or the appearance of plasma filaments by the analysis of the current microdischarges. In Figure 5 a couple of examples of current signals (before and after the transition) are given together with a reference voltage signal. It is evident that the number of current pulses increases

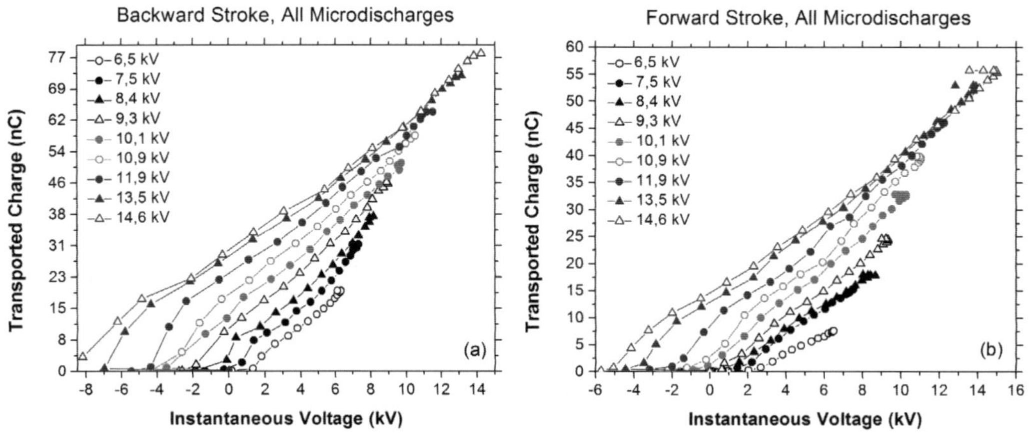

Figure 6. Total charge Q_{micro} that has been moved by current microdischarges already registered during the considered stroke as a function of the voltage V between the electrodes. Figures (a) and (b) refer to the backward and forward discharge respectively, and the different curves are obtained for different voltage amplitudes. Colours indicate the three plasma regimes: uniform plasma (black), uniform plasma with small plasma filaments (red), uniform plasma with well developed plasma filaments (blue).

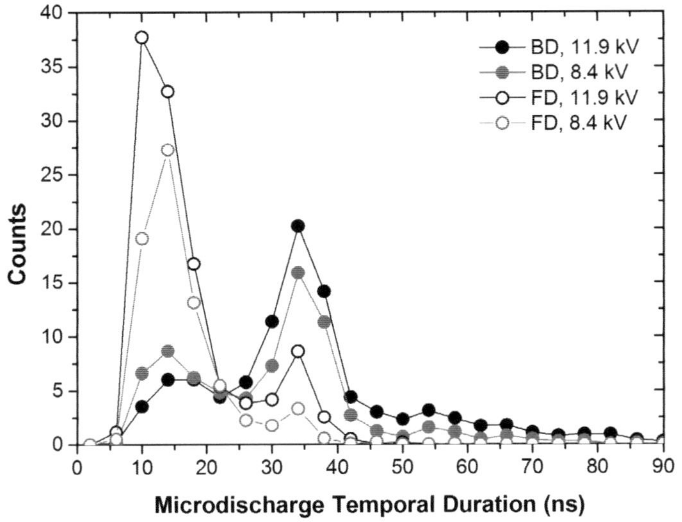

Figure 7. Histograms of the temporal duration of current pulses for the BD and FD strokes. Two different voltage amplitudes are shown (8.4 kV and 11.9 kV).

with the voltage amplitude as well as the sinusoidal current component. We found that the shape of the histogram describing the amplitudes of current microdischarges is almost independent on the applied voltage for the backward discharge, whereas for the forward stroke the histogram tail associated to events with higher amplitudes becomes more accentuated as the voltage amplitude increases. These events could be ascribable to plasma filaments, but further investigations are needed, so this point will be discussed elsewhere. Here we focus on changes in the global behaviour of the forward and backward strokes. For this purpose, for each voltage semi-cycle, we have considered the total charge Q_{micro} that has been moved by current microdischarges already registered during the considered stroke. The trend of Q_{micro} with the instantaneous voltage V applied to the exposed electrode has been considered. In Figure 6 this is plotted for different voltage amplitudes, for both the backward and forward stroke. However, instead of representing the trend of Q_{micro} versus V for a single voltage semi-cycles, we have averaged Q_{micro} over the 13 voltage cycles recorded in a temporal series. At this purpose, instead of considering the instantaneous voltage V at which a single microdischarge happens, we have divided the x-axis in intervals of instantaneous voltages, and we have added up the contributions of all microdischarges recorded at instantaneous voltages falling within the interval. The total values of Q_{micro} obtained in this way for the different voltage intervals have then be divided by 13 so that these plots refer to the average charge transported during a single voltage cycle. Each one of these curves is a $Q-V$ plot similar to the ones considered for Lissajous figures [13], but in this case only the plasma activity phases and only the charge transported by current microdischarges are taken into account. Figure 6 shows that for the three plasma regimes (uniform plasma in black, small filaments in red and plasma filaments in blue), Q_{micro} increases quite linearly with the instantaneous voltage. The voltage amplitude does not seem to influence the slope of this linear increase, but mainly changes the voltage at which breakdown occurs. However, we can notice that immediately after the discharge initiation Q_{micro} increases faster with V. This is the effect of the presence of breakdown microdischarges we have talked about in reference [7]. Eventually, a sort of saturation is noticeable for the highest two voltage amplitudes considered (13.5 and 14.6 kV).

The $Q_{micro} - V$ curves are more interesting if plotted independently for three groups of current microdischarges that can be identified from the histograms of the temporal duration of current microdischarges, which are shown in Figure 7 for both the BD and FD at two different voltage amplitudes. Three groups of current microdischarges can be identified, whose relative importance depend on the half-cycle considered as well as on the operating voltage amplitude. The first group is made of microdischarges with the shortest temporal duration (less than 25 ns) and we will refer to it as *Group S*. The second group (*Group M*) is made of microdischarges with temporal duration between 25 ns and 45 ns, whereas microdischarges longer than 45 ns belong to the third group (*Group L*). The latter is practically absent for the forward stroke. For the whole range of voltage amplitudes investigated, the number of current pulses belonging to Group S detected during a forward half-cycle is larger than the number of FD microdischarges belonging to Group M. The opposite happens for the backward stroke.

Figures 8 and 9 show the plots of Q_{micro} versus V for these groups of microdischarges

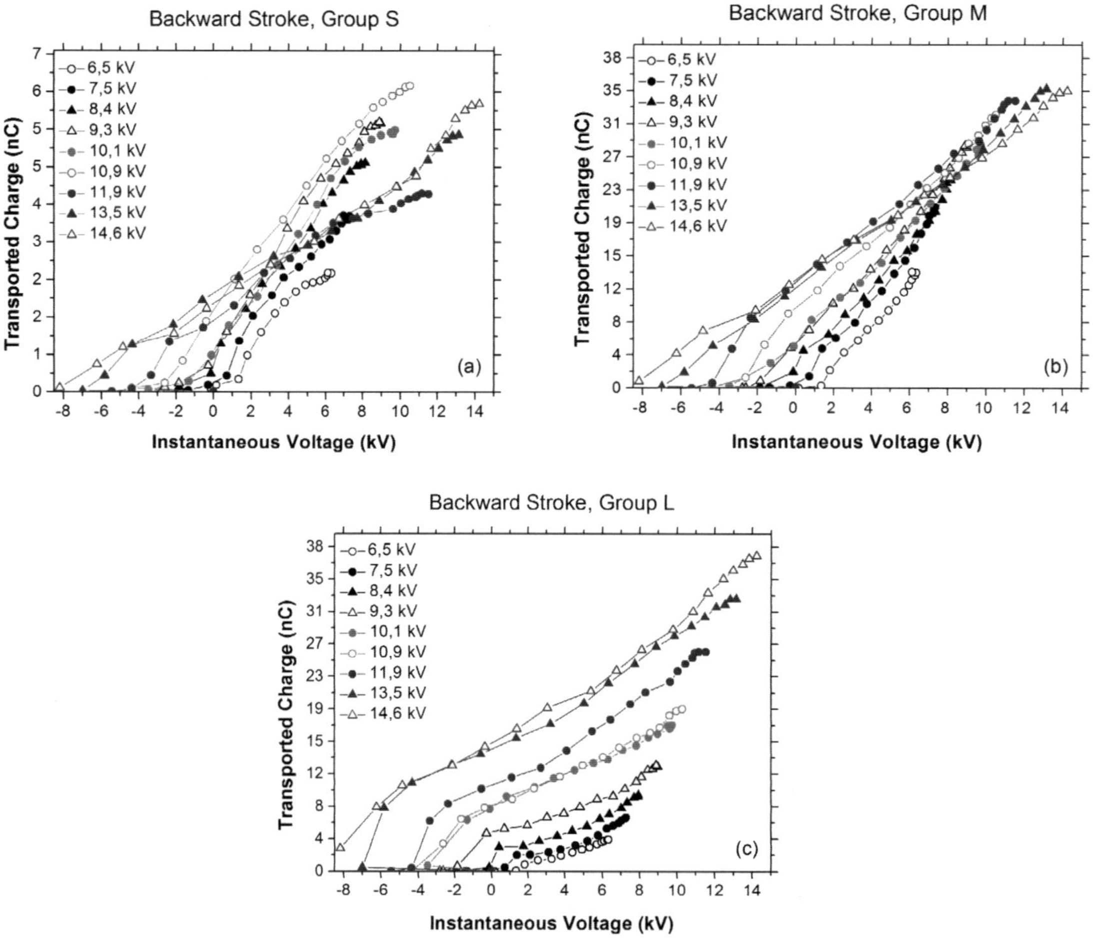

Figure 8. Total charge Q_{micro} that has been moved by BD current microdischarges already registered during the stroke as a function of the voltage V between the electrodes. Figures (a), (b) and (c) refer to microdischarges belonging to Group S, to Group M and to Group L, and the different curves are obtained for different voltage amplitudes. Colours indicate the three plasma regimes: uniform plasma (black), uniform plasma with small plasma filaments (red), uniform plasma with well developed plasma filaments (blue).

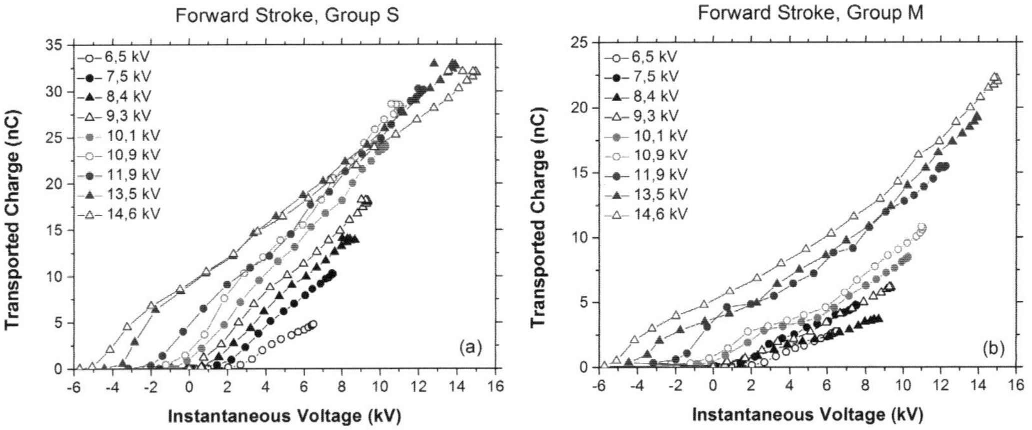

Figure 9. Total charge Q_{micro} that has been moved by FD current microdischarges already registered during the stroke as a function of the voltage V between the electrodes. Figures (a) and (b) refer to microdischarges belonging to Group S and to Group M, and the different curves are obtained for different voltage amplitudes. Colours indicate the three plasma regimes: uniform plasma (black), uniform plasma with small plasma filaments (red), uniform plasma with well developed plasma filaments (blue).

considered independently. We have seen that the trend of Q_{micro} associated to all current microdischarges is not particularly influenced by the voltage amplitude and by the transition to different plasma regimes. On the contrary, we now can see from these figures that, as the voltage amplitude increases, these three groups of plasma microdischarges are favoured or not in different ways. It is evident that for the backward stroke Group S and Group M are penalized in the filament regime rather than in the homogeneous one, since the slope of Q_{micro} versus V associated to these groups is smaller for the blue curve compared to the black ones. This is not the case for Group 3 of the BD, since the increase of Q_{micro} with V is enhanced, especially at the breakdown, meaning that breakdown current microdischarges transport a considerable amount of charges at the highest voltages. Again we can see from Figure 8 that these discharge properties saturate at the highest voltages, since the curves associated to 13.5 kV an 14.6 kV look very similar. Analogous considerations can be done for the forward stroke. In this case Group L is almost absent, Group S is penalized (Figure 9 (a)) and Group M is enhanced (Figure 9 (b)).

For each curve of Figures 8 and 9, the maximum value of Q_{micro} represents the charge totally transported during a backward or forward half-cycle. This quantity is presented in Figure 3 as a function of the voltage amplitude for the different groups of microdischarges. These results suggest that Group S and Group M of the backward stroke together with Group S of the forward stroke determine the increase of the induced wind speed with the voltage amplitude before the transition to the filamentary regime. On the contrary, we can think that microdischarges belonging to Group L of the BD and to Group M of the FD are not effective in generating the wind. In the filamentary plasma regime noticed at the highest voltage amplitudes the importance of the latter groups increases to the detriment of the other groups of microdischarges. The result is that the wind speed saturates even though the total charge transported by the plasma actuator, and thus the consumed power, rises.

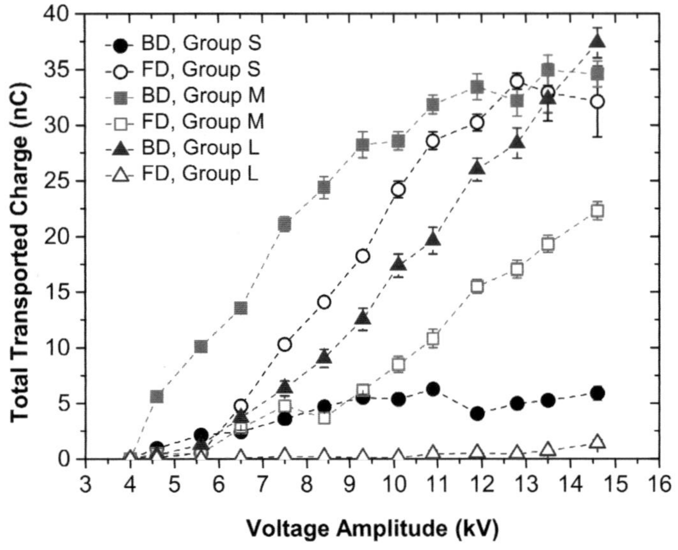

Figure 10. Total charge transported by the different groups of plasma microdischarges during a backward stroke (full symbols) or a forward stroke (blank symbols), as a function of the voltage amplitude. Circles, squares and triangles refer to Group S, Group M and Group L respectively.

4. Conclusions

In this work we have experimentally investigated some properties of an asymmetric surface dielectric barrier discharge fed by a high-voltage sinusoidal low-frequency drive, as a function of the applied voltage. Using a thermal camera and a pitot probe connected to a pressure transducer we have shown that the speed of the flow induced by the SDBD saturates or even decreases when the voltage amplitude is high enough for changing the plasma regime. The new plasma regime can be easily recognised because several plasma filaments are visible and the discharge does not appear uniform any more. We have studied this transition from the point of view of the charge transported by current microdischarges during the backward and forward strokes. We have identified three groups of microdischarges with different temporal duration and results show that they are differently affected by the plasma regime transition. These considerations pave the way for a better understanding of the role played by plasma microdischarges in the creation of the induced airflow. In particular, our analysis suggest that not all these groups of microdischarges effectively contribute to the induced wind generation.

References

[1] Kogelschatz U 2003 *Plasma Chem. Plasma Proc.* **23** 1
[2] Moreau E 2007 *J. Phys. D: Appl. Phys.* **40** 605
[3] Corke T C, Enloe C L and Wilkinson S P 2010 *Annual Rew. Flow Control* **42** 505
[4] Corke T C, Post M L and Orlov D M 2008 *J. Prop and Power* **24** 5
[5] Biganzoli I, Barni R, Riccardi C, Gurioli A and Pertile R 2013 *Plasma Sources Sci. Technol.* **22** 025009
[6] Raizer Y P 1991 *Gas Discharge Physics* (Berlin-Heidelberg: Springer-Verlag)
[7] Biganzoli I, Barni R, Riccardi C, Gurioli A and Pertile R 2013 *J. Phys. D: Appl. Phys.* **46** 2
[8] Orlov D, Font G and Edelstein D 2008 *AIAA J.* **46** 3142
[9] Borghi C A, Cristofolini A, Latini C and Neretti G 2010 *41st Plasmadyn. and Lasers Conf. (Chicago)* 4763
[10] Biganzoli I, Barni R and Riccardi C 2013 *Rev. Sci. Instrum* **84** 016101
[11] Zhang Q, Zhu J, Jia J, Tao F and Yang L 2006 *Meas. Sci. Tech.* **17** 895
[12] Durscher R, Stanfield S and Roy S 2012 *Appl. Phys. Lett.* **101** 252902
[13] Falkestein Z and Coogan J J 1997 *J. Phys. D: Appl. Phys.* **30** 817

Experimental investigation of Lissajous figure shapes in planar and surface dielectric barrier discharges

I Biganzoli[1], R Barni[1], A Gurioli[2], R Pertile[2] and C Riccardi[1]

[1] Dipartimento di Fisica G. Occhialini, Università degli Studi di Milano-Bicocca, piazza della Scienza 3, 20126, Milano, Italy

[2] Aerodynamics & Wind Tunnel Departments, Alenia Aermacchi SPA, via Ing. Foresio 1, 21040, Venegono Superiore (VA), Italy

E-mail: claudia.riccardi@mib.infn.it

Abstract. Dielectric Barrier Discharges (DBDs) operating in air at atmospheric pressure are widely employed as cold plasma sources for plasma processing and applications, in both volume and surface configurations. Surface dielectric barrier discharges, however, are mainly known for the manipulation of the boundary layer of an airflow surrounding a body, and thus for aeronautical applications. Lissajous figures, obtained by means of a high-voltage and a capacitive probes, are usually adopted for both these types of DBDs as a method for measuring the power consumption by the discharge. In this work, we propose to integrate this diagnostic tool with the measurement of current pulses, which are associated to microdischarges that usually develop in these plasmas because of the presence of the dielectric barrier. We have studied both planar and surface DBDs in presence of a continuous sinusoidal voltage feeding, and we have demonstrated that this method is promising in order to gain additional information about the discharge characteristics from the shape of the Lissajous figures.

1. Introduction

Dielectric Barrier Discharges (DBDs) operating in air at atmospheric pressure are widely employed as cold plasma sources for plasma processing and applications [1]. Their name is due to the presence of at least one insulating plate between the electrodes. In a planar and symmetrical configuration two metallic electrodes are covered with insulating layers and separated by a few millimeter gap. In an asymmetrical Surface Dielectric Barrier Discharge (SDBD), instead, plasma is created by ionizing a sheet of air nearby the surface of the dielectric barrier, because one electrode is glued to the dielectric surface whereas the other one is buried inside the insulating barrier itself. Both volume and surface dielectric barrier discharges are widely used in applications such as ozone generation, pollution control, plasma medicine and surface modification [1]. However, the growing interest for SDBDs is mainly ascribable to their promising applications in the flow control field [2, 3]. As a matter of fact, at the end of the nineties, the group directed by J.R. Roth from the University of Tennessee discovered that when these discharges are operated in quiescent air an airflow of several meters per second is observed above the dielectric sheet and near the plasma region [4]. This wind is usually called *induced airflow* because the main mechanism responsible for its generation is believed to be momentum transfer from the ions drifting in the electric field to the surrounding fluid by particle-particle collisions [5]. It is known that, when the electric field created by the voltage difference imposed between the electrodes is sufficiently high, plasma is created and electrical charges are transported through the gap and accumulated on the insulating surfaces [6]. This charge accumulation generates an electric field that locally weakens the external one. When the total electric field falls below the threshold necessary for plasma ignition, the discharge extinguishes. If the voltage imposed to the fed electrode is increased, the discharge can be locally initiated again, and that is the reason why a sinusoidal high-voltage supply is adopted instead of a continuous one. Consequently, the presence of the insulating barrier usually leads to a regime where charges are mainly transported in sub-millimeter regions consisting of current filaments with temporal duration limited to a few tens of nanoseconds [7]. These plasma *microdischarges* are concentrated into two phase intervals of the sinusoidal voltage supply, when the modulus of the applied voltage difference is high enough and increases in time.

Evaluation of the area inside a Lissajous figure is a widely spread method for evaluating the power consumed in a both planar and surface DBDs fed with a sinusoidal high voltage difference. As a matter of fact, this figure is obtained by plotting the charge flowing into the circuit Q as a function of the voltage difference ΔV between the electrodes [8]. The charge is measured by means of a capacitor in series with the discharge cell (the so-called capacitive probe). When the imposed electric field is too small for plasma generation, the Q versus ΔV plot is a straight line, because the system behaves as a purely capacitive cell and the charge varies proportionally to the voltage, with the constant of proportionality given by the cell capacitance C_{cell}^{off}. Otherwise, if the voltage amplitude is high enough for discharge ignition, during the two phases of plasma activity the charge Q does not vary as $C_{cell}^{off} \cdot \Delta V$ any more, and the Lissajous plot opens and forms a convex figure.

Lissajous figures for planar DBD configurations usually assume the shape of a parallelogram. The discharge cell can be thought as two capacitances in series: the first one (C_{die}) is given by the insulating barrier (or by the two barriers if both electrodes are covered), whereas the second one (C_{gap}) is due to the planar air gap. The value of the latter changes when plasma is created, so the Lissajous figure is delimited by four straight lines with a slope different or equal to C_{cell}^{off}, depending if they correspond or not to a phase of plasma activity [9]. Manley assumed that, during the discharge-on periods, the plasma between the electrodes fills the gap so that the voltage drop in the gap is equal to zero. In this case, the capacitance of the cell, and thus the slope of the the corresponding side of the Lissajous figure, reduces to the capacitance of the dielectric(s): $C_{cell}^{on} = C_{die}$ (see Figure 1 (a)) [10, 8].

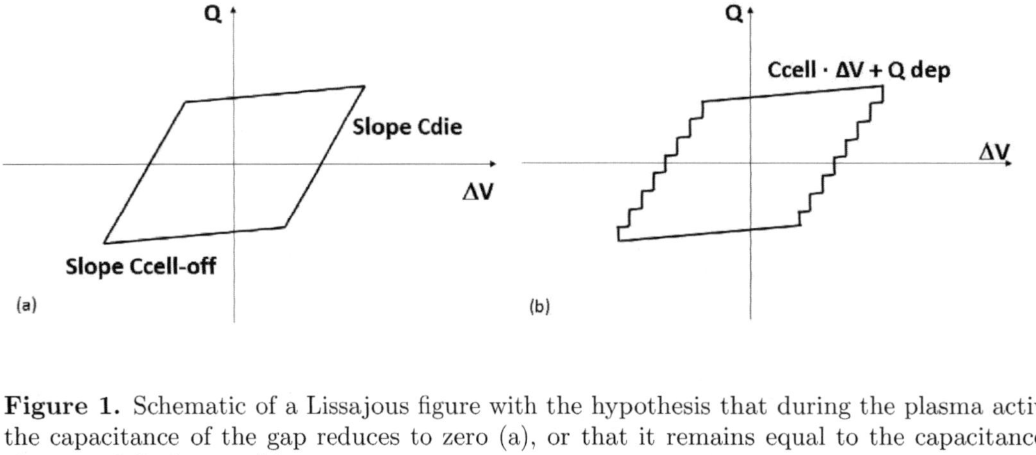

Figure 1. Schematic of a Lissajous figure with the hypothesis that during the plasma activity the capacitance of the gap reduces to zero (a), or that it remains equal to the capacitance in absence of discharges (b).

The instantaneous charge Q represents the total charge and thus captures the charge transported by current microdischarges too. A situation opposite to the one considered by Manley occurs when a DBD is operated just above the minimum voltage amplitude required for plasma ignition, and only few plasma microdischarges are created during a whole sinusoidal voltage cycle. In this case the capacitance of the gap essentially keeps the same, but the Lissajous figure becomes again a sort of parallelogram if the current microdischarges move and deposit charges: the instantaneous charge is given by $Q = C_{cell}^{off} \cdot \Delta V + Q_{dep}$, and even though the capacitance C_{cell}^{off} is constant, Q changes during the phases of plasma activity because the positive or negative deposited charge (Q_{dep}) changes each time a microdischarge crosses the gap. In this case, during the plasma activity, Q varies in steps (Figure 1 (b)) [11], whereas during the off-phases, if the deposited charge does not change, the variation of Q with ΔV follows a straight line $Q = C_{cell}^{off} + Q_{dep}$, with intercept Q_{dep} given by the total charge left by microdischarges during the previous plasma activity. Without the assumption that the capacitance or the deposited charge keep constant, the variation of the total charge Q during a voltage cycle thus depends on capacitance changes as well as on the accumulation of charges on the dielectric barriers.

For a SDBD, the situation is complicated by the fact that the discharge does not start and stop with a constant geometrical plasma shape, but the latter can change in the course of the discharge active phase [12], since the accumulation of charges onto the dielectric surface can influence the electric field configuration and thus the development of the following microdischarges [7]. Changes in the shape of the discharge region imply variations in the discharge capacitance, leading to the almond-shaped Lissajous figures often observed for surface discharges [12].

The aim of this paper is to combine the Lissajous figure plot with the detection, by means of a Rogowski coil, of current pulses associated to microdischarges, in order to distinguish the contributions of C_{gap} and Q_{dep} to the variations of the instantaneous total charge Q during a voltage cycle. This method has been tested for a planar configuration and then applied to a SDBD. The two configurations and the electrical diagnostics adopted are described in the next two sections, which are followed by the presentation of the results obtained.

2. Experimental Setup

The planar configuration used in these experiments is symmetrical because two identical insulating barriers are used to cover the electrodes (Figure 2 (a)). The latter are stainless steel disks (diameter 35 mm and thickness 1.5 mm) that have been inserted into a drilled cylindrical scaffold in Macor (a glass ceramic material by Corning Inc., dielectric constant $\epsilon_r = 6.03$) with diameter equal to 40 mm and external height equal to 3.5 mm. Thus a dielectric barrier of 2 mm separates each electrode surface from the gas gap. The discharge gap was fixed at 0.8 mm. The total cell capacitance without plasma has been calculated using a 2D Laplacian electrostatic field solver (EStat 7.0 by Field precision LLC). It is equal to $C_{cell}^{off} = 8.3$ pF, given by the series of $C_{gap} = 23.5$ pF and $C_{die} = 12.8$ pF (25.7 pF for each dielectric barrier). Calculations show that the electric field is uniform in the gap within 10% and decreases outwards. The voltage difference between the electrodes is imposed by grounding an electrode and by connecting the other one to the secondary coil of the transformer of a power generator (a V-20 Corona Station by Tigres Gmbh). The whole system, described elsewhere [13], constitutes a resonant circuit, so the sinusoidal voltage amplitude and period are not independent.

In the surface configuration, the electrodes are tin clad copper adhesive tapes (60 μm thick and 200 mm long) placed at the opposite sides of a flat dielectric panel (teflon, 2 mm thick), without overlapping in the chordwise y-direction (Figure 2 (b)). The insulated electrode is 10 mm wide, whereas the air exposed one is just 5 mm wide. An insulating polyisobutylene self-amalgamating tape is used for avoiding discharges at the lower electrode side, but also upstream of the exposed electrode ($y < 0$) and at the electrode ends. Different voltages (up to about 15 kV peak amplitude) has been tested, at the same frequency (16 kHz). In this case the the

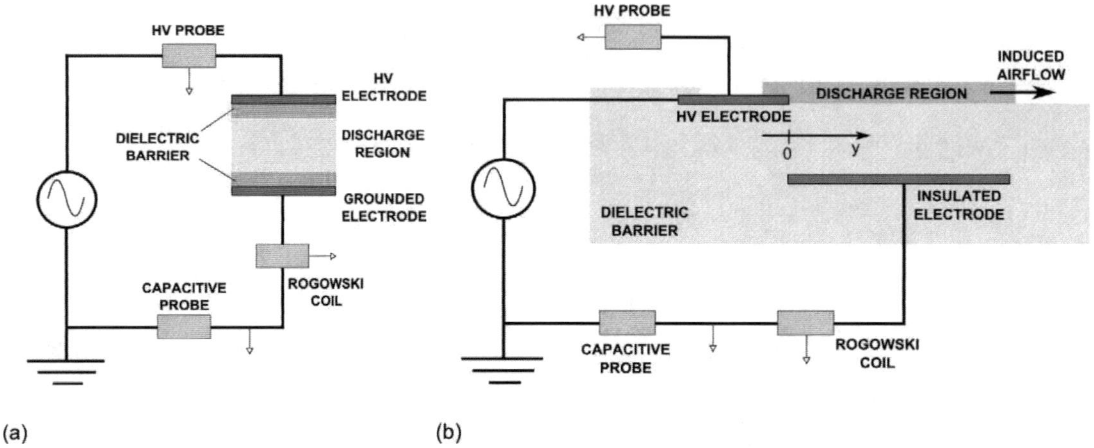

Figure 2. Experimental layout and diagnostics for the adopted planar (a) and surface (b) configurations. Arrows mean that the probe outputs are connected to the oscilloscope.

buried electrode is grounded and the exposed one is fed by the high-voltage power supply, which was the MiniPuls 6 by GBS Elektronik GmbH, consisting of a transistor bridge followed by a transformer cascade [14].

3. Diagnostics

Three electric probes have been used for studying the discharge properties (Figure 2). Plasma current microdischarges have been experimentally studied with a sufficient high temporal resolution by means of home-made Rogowski coils. A capacitive probe was placed between the buried grounded electrode and the earth as a diagnostic for measuring power consumption and for visualizing the shapes of the Lissajous figures. A commercial HV probe (Tektronix P6015A, granted for a bandwidth of 75 MHz) was instead employed for evaluating the voltage at the exposed electrode. The outputs of all these probes were connected to different channels of a large bandwidth digital oscilloscope (Agilent MSO8104A with sampling rate equal to $2GSa \cdot s^{-1}$) and their signals were acquired simultaneously.

3.1. Capacitive Probe

A capacitive probe consists in a capacitor inserted between the grounded electrode and the ground, in order to measure the charge flowing into the ground cable. The capacitance of the probe (C_{probe}) is known and must be sufficiently large so that the voltage difference ΔV between the DBD electrodes is almost equal to the voltage V applied to the powered electrode. In our experiments 100 nF and 46 nF were used for the symmetrical and asymmetrical configurations, respectively. This assures that the probe is not intrusive, and that the voltage drop across the capacitor (ΔV_{probe}) is small enough to be measured without necessarily using a high voltage probe. The capacitor-based charge method for power measurement was first introduced by Manley for discharges between parallel plates [10], whereas Pons et al. were the first ones to use it in the context of SDBDs for aeronautical applications [15]. The instantaneous charge Q is given by $Q = Cprobe \cdot \Delta V_{probe}$, and the energy consumed by the dielectric barrier discharge in a voltage cycle is obtained by averaging the product $Q \cdot \Delta V$ over a whole period T, which is given by N points separated by the time interval Δt:

$$E = \frac{1}{N}\sum_{i=1}^{N} Q \cdot \Delta V. \tag{1}$$

This can also be geometrically calculated as the area inside the Lissajous figure, as proposed by Manley [10]. As a matter of fact, as already said in the introduction, this is obtained by plotting the instantaneous charge as a function of the voltage drop in the discharge cell. Once E is obtained, the dissipated power is given by $P = E/T$:

$$T = \frac{1}{N^2 \Delta t}\sum_{i=1}^{N} Q \cdot \Delta V. \tag{2}$$

3.2. Rogowski Coils

A current transducer like a Rogowski coil is suitable to measure electric current filaments [16]. In a DBD, the electric current flowing into the circuit can be viewed as the superimposition of a low-frequency sinusoidal capacitive current, which is almost independent of plasma presence in the gap, and a discharge current, which is associated with plasma microdischarges and appears as a series of fast current pulses. A Rogowski coil is a conducting wire that is wound in a spiral around a magnetic core and then returns to the original point. The coil is placed around the cable connecting the buried electrode to the ground: any variation in the current I flowing into the cable generates an electromotive force at the output of the coil. By adding an integrating circuit, the output of the probe is proportional to I. We have built a few home-made Rogowski coils and calibrated them in the frequency range $10 \div 100$ MHz, as discussed in a dedicated paper [16]. Since we are mainly interested in the current signal associated with plasma microdischarges, we can separate this current component from the displacement one by means of a band pass filter based on the fast Fourier transforms, so that current events due to plasma microdischarges finally appear as pulses emerging from a noisy base line, as shown in Figure 3. At this point they are identified by setting a threshold sufficiently higher than noise spikes. Finally, a baseline is used to evaluate the intensity of peaks, as well as their beginning and end. We were able to study current pulses characteristics like number, amplitude, duration and charge transported [3]. In this work our aim was not to study the mean properties of these current pulses, but we recorded them in order to measure the charge transported inside the gap by plasma microdischarges. Since the instantaneous Q signal obtained with the capacitive probe traces the total charge present inside the gap or on the dielectric surfaces at the considered moment, for each instant of plasma activity the charge transported by each microdischarge (recorded up to that moment) is added up in a variable that we call Q_{micro}, which is displayed in Figure 3. Q_{micro} is equal to zero before the first current pulse of the considered half-cycle, then its value increases each time a plasma microdischarge is created:

$$Q_{micro}(t_i) = Q_{micro}(t_{i-1}) + Q_{pulse}(t_i), \tag{3}$$

where $Q_{micro}(t_i)$ denotes the value of Q_{micro} at the time t_i when the $i-th$ current microdischarge is created, and $Q_{pulse}(t_i)$ is the charge transported by this microdishcarge, i.e. the area subtended below the considered current pulse. In order to draw a comparison with the Lissajous figure, Q_{micro} will be considered as a function of the instantaneous voltage difference between the electrodes (see Figure 3, in red).

4. Experimental Results

For the both dielectric barrier discharge configurations studied during this investigation, we have simultaneously acquired the signals corresponding to the three different electrical diagnostics.

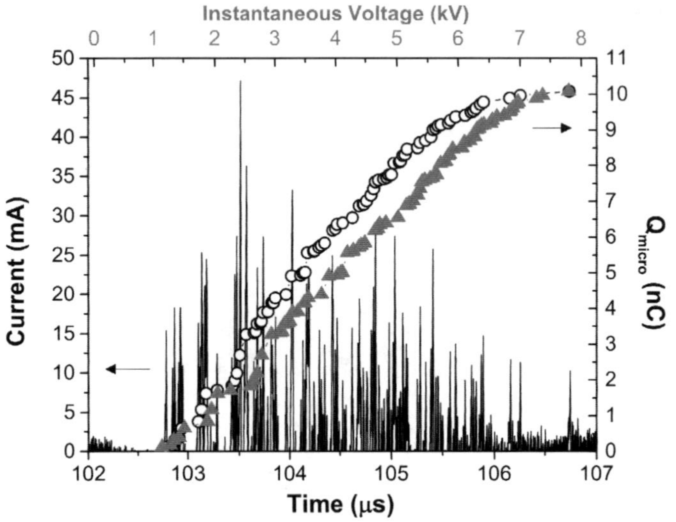

Figure 3. Temporal signal of the discharge current. The phase of plasma activity begins at about 102.8 μs, when the first current pulse appears. The corresponding evolution of Q_{micro} is plotted as a function of both time (black circles) and instantaneous voltage (red triangles).

These temporal series have been split into several parts corresponding to different voltage cycles, which have been analysed separately. For the representation of the Lissajous figures, a band pass filter was previously applied to both the Q and V signals, in order to cut noise at high frequencies, and to reduce the spikes visible in the capacitive probe signals that indicate the plasma activity. An high-pass filter was instead used for the current signals, in order to separate the microdischarge current peaks from the AC current signal, as discussed in the previous section.

4.1. Planar Configuration

Two Lissajous figures obtained for the planar configuration at two different voltage amplitudes (8.3 kV and 11.4 kV) are shown in black in Figure 4. Their shape nicely resembles a parallelogram composed of four straight lines with two different slopes. These $Q-V$ plots appear symmetric: as expected, there are no evident differences between the two voltage semi-cycles, because of the adopted planar setup with two dielectric barriers. For each semi-cycle, the amount of charge transported by current microdischarges (Q_{micro}) has been calculated as a function of the applied voltage. For the half of the Lissajous figure in which the voltage is rising towards its maximum (sides AC of Figure 4 (b)), the current transported by plasma microdischarges is positive and Q_{micro} increases in correspondence of instantaneous voltages at which a current pulse is detected. As discussed in the introduction, we expect that Q is related to the voltage drop between the electrodes as $Q = C_{inst} \cdot \Delta V + Q_{inst}$, where C_{inst} and Q_{inst} are the instantaneous capacitance of the system and instantaneous amount of charge accumulated inside it. If both these values are constant within a voltage interval, Q varies linearly with ΔV. It is evident that this happens during the two phases of plasma off (sides AB and CD), because no charges are transported by plasma filaments and the capacitance of the system keeps constant. This evidence that Q_{inst} does not change means not only that these are phase of no plasma activity, but also that there is no rearrangement or recombination of the charges left on the insulating barriers. During the plasma active phases, instead, both C_{inst} and Q_{inst} can vary. In order to separate these two contributions affecting the shape of the Lissajous figure, we have supposed that Q_{inst} is

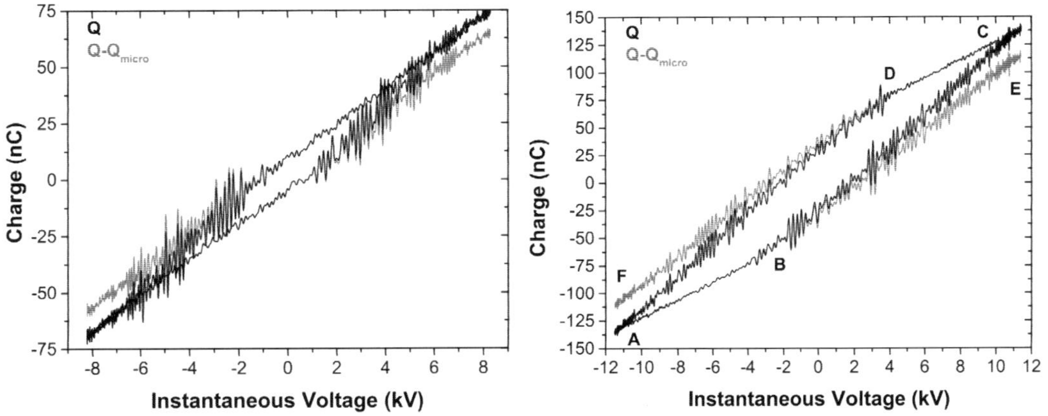

Figure 4. Examples of Lissajous figures (shown in black) obtained for two different voltage amplitudes (8.3 kV and 11.4 kV) in a planar configuration. The consumed power is equal to 13 W for the graph on the left and to 34 W for the one on the right. For red curves the charge Q_{micro} transported by current microdischarges have been subtracted to the Q signal of the Lissajous figure.

completely transported by plasma microdischarges, and we have thus measured $Q_{inst} = Q_{micro}$ with a current probe, and thus with a diagnostic tool independent from the Lissajous figure. At this point we have subtracted Q_{micro} from the Q signal of the Lissajous figure, and we have obtained a sort of new Lissajous micro, shown in red in Figure 4, which is no more a closed loop, because the two sides of no plasma activity remains unchanged, whereas BC and DA becomes BE and DF. These ones appear straight at all the considered voltage amplitudes, meaning that the cell capacitance can change at the beginning of the plasma-on or plasma-off phase (becoming equal to C_{on} and C_{off} respectively), because of a sudden change in the C_{gap} value, but that then it keeps the same during each one of these temporal windows. The values of C_{on} and C_{off} obtained for the planar configuration are presented in Figure 5 as a function of the power, which has been calculated from the Lissajous figure as described in Section 3.

We have found that the capacitance C_{off} is independent from the power level and equal to the C_{cell}^{off} given by the series of the capacitances of two dielectrics and of the gap calculated in absence of discharges (Section 2). If variations in the C_{off} values were observed this would mean that the gas phase remained perturbed after the stroke quenching or by electrode heating (at high power levels). In these experiments, however, C_{off} seems to represent just the cold capacitance of the actuator. The capacitance C_{on}, instead, increases with the voltage amplitude, and for sufficiently high powers it becomes equal to the capacitance given by the series of the two insulating barrier: $C_{on} = C_{die}$. It is thus clear that the Manley assumption presented in the introduction is adequate only in these conditions. Moreover, it is evident from this discussion that, even though the plasma-off sides AB and CD are straight lines, their intercepts with the Q axis represents the total charge left on the dielectric barriers during the previous plasma activity only if the capacitance C_{on} remains equal to C_{off} (Figure 1 (b)), otherwise these intercepts are influenced by capacitance variations too.

Journal of Physics: Conference Series **550** (2014) 012039 doi:10.1088/1742-6596/550/1/012039

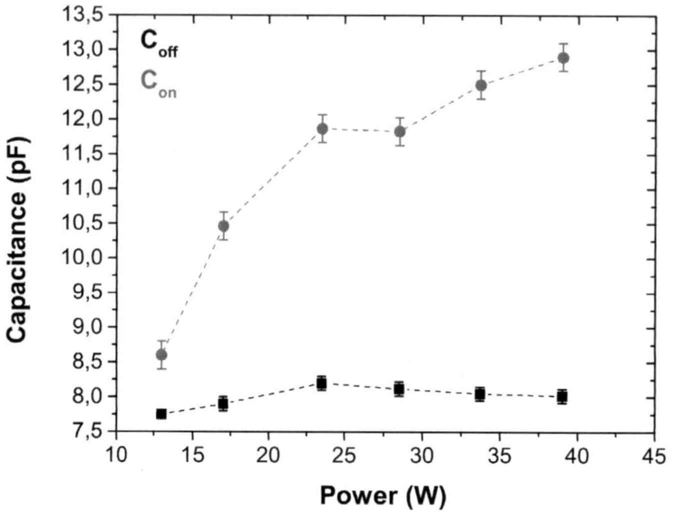

Figure 5. Capacitance in the presence (C_{on}) and absence (C_{off}) of plasma, calculated from the Lissajous figures obtained after the subtraction of Q_{micro}, as a function of the consumed power.

4.2. SDBD Configuration

The same method for the investigation of Lissajous figures has been adopted for the surface dielectric barrier configuration. In this case a band pass filter with a lower cut-off frequency was adopted (2 MHz instead of the 15 MHz used for the planar configuration), so that there are no oscillations associated to the plasma presence. This allows to appreciate if asymmetries are present between the two half-cycles of the Lissajous figure. Figures 6 (a) and (b) show two Lissajous figures (in black) obtained at two different voltage amplitudes (5.6 kV and 13.5 kV).

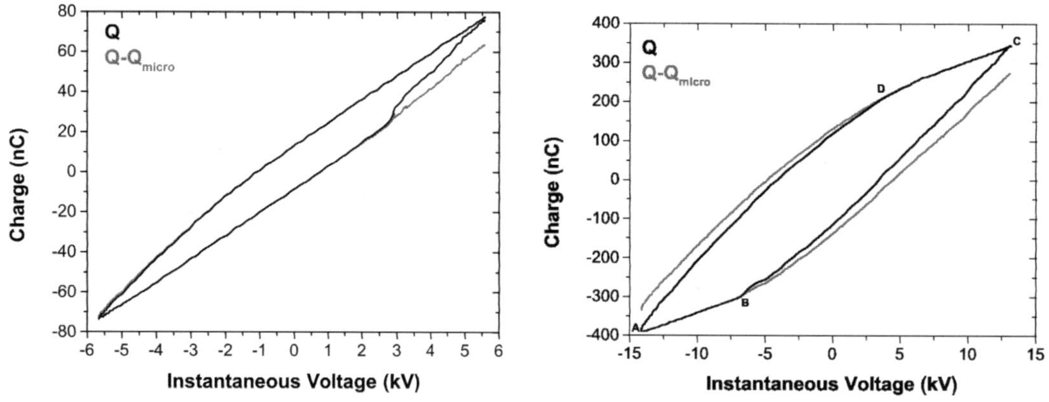

Figure 6. Examples of Lissajous figures (shown in black) obtained for two different voltage amplitudes (5.6 kV and 13.5 kV) in a surface configuration. The consumed power is equal to 3 W for the graph on the left and to 40 W for the one on the right. For red curves the charge Q_{micro} transported by current microdischarges have been subtracted to the Q signal of the Lissajous figure.

It is clear that in the first case a variation in the Q signal happens abruptly at the beginning of the active phase of the positive semi-cycle (when the voltage is increasing towards its maximum, side AC), whereas the Q variation is more gradual during the negative half-cycle (side CA). At the higher voltage amplitude, instead, the $Q - \Delta V$ plot appears more symmetric, even though the breakdown of the positive plasma phase is always more evident. This is due to the presence of a group of plasma microdischarges that are created at the discharge breakdown, probably by electrons that were previously accumulated on the surface of the insulating barrier, as discussed in reference [7]. The red curves correspond to the Lissajous figures obtained after subtracting Q_{micro} from the original Lissajous plots. Again, it easily to notice that at high voltages the discharge semi-cycles behave quite similarly, whereas at low voltages the two semi-cycles appear considerably different. As a matter of fact, Figure 6 (a) shows that during the positive half-cycle plasma microdischarges transport a certain amount of charge Q_{micro}, and that after the subtraction of Q_{micro} from the $Q - \Delta V$ plot, the new slope C_{on}^{+} of the positive active phase looks quite similar to the slope C_{off} of the plasma-off phase. On the contrary, during the negative half-period, almost no current microdischarges were detected, and this portion of the Lissajous figure is not affected by the subtraction of Q_{micro}. However, the slope of the curve changes, and this means that there is a capacitance change or a charge transfer happening in a way different from current plasma mirodischarges. These asymmetries between the two semi-cycles are probably due to the fact that during the positive half-cycle, electrons are moved from the insulating surface towards the exposed electrode, and can thus leave the discharge gap, whereas during the negative semi-cycles electrons are accumulated on the insulating surface. Eventually, it is worth to noticing that at high voltages, the transition from C_{off} to C_{on} is not abrupt but gradual (Figure 6 (b)). This is probably due to progressive changes in the geometrical shape of the discharge region, which leads to gradual changes in the cell capacitance C_{on} and to the creation of an almond-like Lissajous figure.

The values of C_{on} and C_{off} obtained for the surface configuration are presented in Figure 7 as a function of the power. These preliminary results for SDBDs show that C_{on} is higher

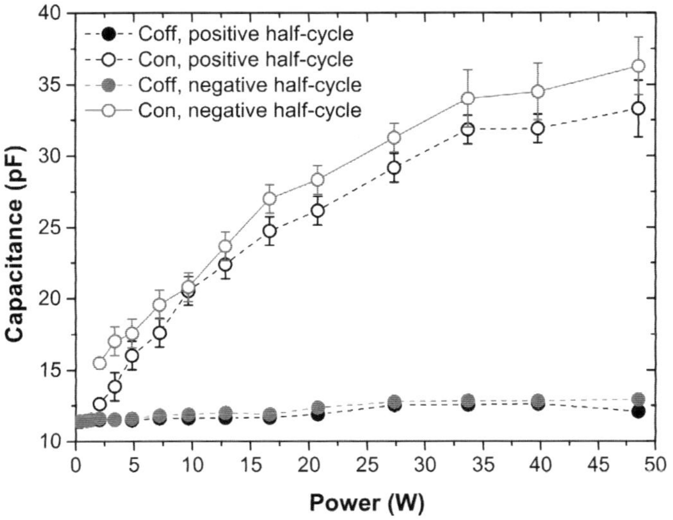

Figure 7. Capacitance in the presence (C_{on}) and absence (C_{off}) of plasma, calculated for both the positive and negative half-cycles from the Lissajous figures obtained after the subtraction of Q_{micro}, as a function of the consumed power.

for the negative half-cycle. This is probably due to the higher mobility of electrons, which are more easily accumulated on the insulating surface rather than ions. Both C_{on}^{+} and C_{on}^{-} tend to increase for the range of powers considered, and this could be associated to a spreading of the discharge region, meaning that the higher the voltage the more the plasma sheet extends to chordwise positions far from the exposed electrode. We will examine in more depth this point in future investigations. The temporal evolution of the spatial discharge region and of the number of microdischarges could be followed by recording the time emitted by the discharge using a photomultiplier tube or by acquiring short exposure time pictures with a fast camera. Eventual rearrangements of charges left on the insulating barrier by the forward and backward stroke could instead be detected by local measurements of the surface potential temporal evolution [17, 18].

5. Conclusions

In this work we have proposed to combine to electrical diagnostics, a capacitive probe and a Rogowski coil, in order to gain new insights about plasma properties from the shape of the discharge Lissajous figure. As a matter of fact, these figures are often considered only for measuring the power consumption associated to the dielectric barrier discharge. In order to gain information about the discharge characteristics some assumptions are required, because the charge variation with the voltage drop between the electrodes depends on variations of the cell capacitance and of the charge transported by plasma and accumulated on the insulating barriers. Here we have proposed an experimental method for separating these two contributions, which consists in the measurement of the charges transported by plasma microdischarges by means of an adequate processing of the Rogowski current probe signals. This allows to isolate the effect of any capacitance variations on the shape of the Lissajous figure. This method thus allows to obtain more accurate measurements of the cell capacitance during the discharge active phases, and to test the validity of some assumptions, such as the one proposed by Manley about the capacitance gas gap [10], in the considered operating conditions. Results concerning a planar DBD configurations have been presented, and preliminary results referring to a surface dielectric barrier discharge show that this method can be useful for extracting interesting information about the discharge development during the voltage cycle, such as the plasma spreading on the insulating barrier, and to answer some questions, such as if electrical charges are all transported in the form of current microdischarges or not.

References

[1] Kogelschatz U 2003 *Plasma Chem. Plasma Proc.* **23** 1
[2] Corke T C, Enloe C L and Wilkinson S P 2010 *Annual Rew. Flow Control* **42** 505
[3] Biganzoli I, Barni R, Riccardi C, Gurioli A and Pertile R 2013 *Plasma Sources Sci. Technol.* **22** 025009
[4] Roth J R 1998 *25th IEEE Int. Conf. Plasma Science (Raleigh)* 6P67
[5] Moreau E 2007 *J. Phys. D: Appl. Phys.* **40** 605
[6] Raizer Y P 1991 *Gas Discharge Physics* (Berlin-Heidelberg: Springer-Verlag)
[7] Biganzoli I, Barni R, Riccardi C, Gurioli A and Pertile R 2013 *J. Phys. D: Appl. Phys.* **46** 2
[8] Falkestein Z and Coogan J J 1997 *J. Phys. D: Appl. Phys.* **30** 817
[9] Kim G-H, Jeong S-Y and Kwon H-C 2006 *J. the Korean Phys. Soc.* **49** 3
[10] Manley T C 1943 *Trans. Electrochem. Soc.* **84** 83
[11] Buntat Z, Harry J E and Smith I R 2007 *Elektrika* **9** 2, 60
[12] Kriegseis J, Mőller B, Grundmann S and Tropea C 2011 *J. Electros.* **69** 302
[13] Siliprandi R, Roman H E, Barni R and Riccardi C 2008 *J. Appl. Phys.* **104** 063309
[14] Borghi C A, Cristofolini A, Latini C and Neretti G 2010 *41st Plasmadyn. and Lasers Conf. (Chicago)* 4763
[15] Pons J, Moreau E and Touchard G 2005 *J. Phys. D: Appl. Phys.* **38** 3635
[16] Biganzoli I, Barni R and Riccardi C 2013 *Rev. Sci. Instrum* **84** 016101
[17] Cristofolini A, Borghi C A and Neretti G 2013 *J. Appl. Physics* **113** 143307
[18] Nichols T G and Rovey J L 2012 *50th AIAA Aerosp. Sci. Meeting (Nashville)* 0822

On the use of pulsed Dielectric Barrier Discharges to control the gas-phase composition of atmospheric pressure air plasmas

R Barni, I Biganzoli, E Dell'Orto and C Riccardi

Dipartimento di Fisica G. Occhialini, Università degli Studi di Milano-Bicocca, piazza della Scienza 3, I-20126 Milano, Italy

E-mail: claudia.riccardi@mib.infn.it

Abstract. We presents results obtained from the numerical simulation of the gas-phase chemical kinetics in atmospheric pressure air non-equilibrium plasmas. In particular we have addressed the effect of pulsed operation mode of a plane dielectric barrier discharge. It was conjectured that the large difference in the time scales involved in the fast dissociation of oxygen molecules in plasma and their subsequent reactions to produce ozone and nitrogen oxides, makes the presence of a continuously repeated plasma production unnecessary and a waste of electrical power and thus efficiency. In order to test such suggestion we have performed a numerical study of the composition and the temporal evolution of the gas-phase of atmospheric pressure air non-equilibrium plasmas. Comparison with experimental findings in a dielectric barrier discharge with an electrode configuration symmetrical and almost ideally plane is briefly addressed too, using plasma diagnostics to extract the properties of the single micro-discharges and a sensor to measure the concentration of ozone produced by the plasma.

1. Introduction

Dielectric Barrier Discharges (DBD) are a common source to produce non equilibrium plasmas at atmospheric pressure [1-3]. Limitation of the discharge current and of the gas heating at atmospheric pressure is achieved by the presence of an insulating material separating the two electrodes and avoiding a direct connection between them through the gas gap. Electrical breakdown will end up in delivering charge to the dielectric surface reducing the electric field and quenching the discharge. Near atmospheric pressure this produces narrow discharge filaments or streamers, reaching high electron density (10^{12}-10^{17} cm^{-3}) and very short lifetimes (few tens of ns) [1]. A steady state operation requires a somewhat regular repetition of such discharge events filling the gas gap and this is normally sustained by applying an oscillating voltage to one of the electrodes at kHz frequencies. The alternating voltage sweeps out the charge accumulated on the dielectric surfaces and a new streamer directed in the opposite direction could develop from the same location on the surface. It is worthy to point out that each of such discharges has a short duration with lapses of dead times between subsequent events. Moreover they could be considered as isolated also in space, since, at least in small gaps, they appear to be very thin (radius R ~ 100 μm [2]) while their average distance is much larger (a few mm, of the order of the gap width [1-2]).

Under steady state operation the local gas-phase composition is then controlled by both the nearby passage of one of the discharge paths (usually very fast) and by the quiet time intervals between subsequent such events. On a longer time scale diffusion both along and transverse to the discharge paths leads to mixing and uniformization of the gas-phase. The interplay between the two phases could be altered by delivering discharges in time bunches, using pulsed HV generators.

Journal of Physics: Conference Series **550** (2014) 012040 doi:10.1088/1742-6596/550/1/012040

Our aim was to investigate the effect of a pulsed operation mode on the gas phase composition by using a model of the chemical kinetics of atmospheric pressure discharges suitable for the simulation of DBD devices [4-5].

Then we have implemented an experimental arrangement, mirroring as much as possible the model approximations, to compare actual results with the predictions of the simulations. In particular we have targeted the measure of the ozone concentration produced in the discharge gas-phase, since ozone is mainly formed much afterwards the discharge, mainly in gas phase reactions between neutrals, not directly affected by the discharge current evolution and charged particle dynamics [1,6].

There is wide interest in an efficient production of ozone for practical applications. Ozone is a strong oxidant and a potent disinfecting agent, long in use for water treatment [7-8]. A number of other uses have been found for ozone including food industries and medical application. Relatively low concentrations of ozone and short contact time are sufficient to inactivate bacteria, molds, yeasts, parasites, and viruses, for this reason it has been used with success to inactivate contaminants on meat, eggs, fish, fruits, vegetables, and dry foods [9-10]. In medicine ozone has been successfully applied for skin disinfection and for antimicrobial treatment on biological tissue, for example in active carious lesions [11-12].

The paper is organized as follows. At first we discuss the numerical simulation of the chemical kinetics in the gas-phase of a single air microdischarge. Then we implement a numerical study of pulsed operation in a DBD reactor, taking into account the effects of discharge repetition and of duty cycles on the evolution of the gas-phase composition. In particular we discuss how ozone production is affected by the parameters set in pulsed mode. Finally we present an experimental characterization of ozone production in a pulsed DBD reactor. Comparison with the prediction of the numerical study is discussed before conclusions are drawn in the ending section.

2. Chemical kinetics in an air microdischarge

As already stated, the discharge develops in a narrow current filament of some nanoseconds duration. They can be considered as isolated events in time and space. The first step in modelling is then to simulate the microdischarge formation. Some experimental observations [13-16] and several models [17-20] of streamer development and propagation exist by now. To make a somewhat rough summary, the overall picture looks similar in different kind of DBDs [4]. Streamer propagation is sustained by the electron avalanche multiplication in the strong field ahead of the streamer (E_{max} ~ 5-15 x 10^6 V/m) and it is very fast (v ~ 10^5 - 10^6 m/s). After the streamer head reaches the dielectric barrier insulating the cathode, the channel allows the flow of a limited discharge current, which is rapidly quenched by the space charge accumulating at the dielectric surface [1]. The discharge body consists then of a thin channel of weakly ionized gas, almost quasineutral (n+ ~ n- ~ 10^{13}-10^{15} cm^{-3}). At a fixed position along the streamer path, the electric field rises quickly as the streamer head is approaching and then decay almost as fast to the much smaller value it assumes in the streamer channel. It is worthy to point out that at high pressures electrons moves along the electric field at the drift velocity and almost instantaneously (ν_m^{-1} ~ 0.3 ps, ν_E^{-1} ~ 50 ps [2]) reach a mean energy determined by the local amplitude of the field. On the other hand ions inertia keeps them almost frozen and substantially cold, that is at the set temperature. So the picture is one of an ionizing wave quickly propagating between the two dielectric surfaces, with the local charge density increasing as the wave passes by.

From the chemical kinetics point of view, the process can be seen as consisting of an almost instantaneous phase during which energetic electrons produce ions and atomic radicals, followed by a phase in which electrons cool down almost instantaneously and reach the set temperature. Existing models could be used to predict the shape as well as the peak value of the electric field pulse during their propagation in the gas gap. However the dependence from the position along the streamer path is not large, apart from the region near the dielectric surface [17-18]. So we have neglected such differences and we have supposed uniformity along the streamer path. This was modeled as a thin cylinder, with a circular cross-section. The actual value of the radius could be considered as a free parameter of the simulation, but we have considered 100 μm as a reference [2]. The cylinder height, equal to the gas gap was chosen as 0.8 mm, since this matches the performed experiments. Another simplification was introduced to treat the ionizing wave. We assumed a square pulse shape of the electric field strength in the head of the streamer. The pulse is therefore characterized by the electric field amplitude E_{max} and by its time duration τ_s (here 2 ns, based on the model in [17]). Although a little rough, such an approximation embodies all the relevant dynamics for our aims. It is worthy to point out that whether the electron density rises strongly, almost exponentially, during the ionizing wave passage, the total amount of radicals produced by electron impact happens to be largely independent from the time length τ_s and from the initial electron density. Radical production level is set mainly by the actual value reached by the electron density at the end of the pulse [4,5]. So we have taken τ_s fixed and we have treated the value reached by the electron density n_e at the end of the ionizing wave as the only relevant parameter of the simulation. As reference we considered a density of 10^{12} cm^{-3} [17].

From a chemical engineering point of view, it means that the model can be formulated as a well-mixed reactor with cylindrical geometry [21]. The gas-phase composition in the reactor is determined by the chemical reactions among the reactive species and the transport processes. The concentration of the different N species in the gas phase only depends from time and could be calculated by integrating each balance equation for the density n_k of the kth species

$$\frac{dn_k}{dt} = \sum_{i<j=1}^{N} K(i+j \rightarrow k)\, n_i n_j - \sum_{i,j=1}^{N} K(k+i \rightarrow j)\, n_k n_i + \frac{D_k}{\Lambda_k^2} n_k$$

where K are the reaction rates for the gas phase reactions including those involving electrons and ions, whereas D and Λ are the diffusion coefficient and length [22]. Within this approximation, the transverse profile of the concentration is described in terms of the normal mode dictated by the geometry. The coefficient Λ depends only on geometrical factors and on the sticking coefficient S of each chemical species. Using the assumed uniformity, diffusion along the streamer axis is neglected. Adsorption on the dielectric barrier could easily be included but, because of the much smaller extension of the streamer radius respect to the discharge gap, it turns out to be negligible. Balance equations containing all the processes described above form a system of coupled differential equations which has been integrated by using an adaptive Runge-Kutta routine [5]. The choice of species included in the model was based on existing experimental information based on emission spectroscopy or mass spectroscopy. A set of 10 neutral, 10 metastables and 9 charged species was included in this report. Rate constants for neutral gas-phase, charge exchange and ion recombination reactions have been taken from literature referenced in [4,5]. A cold plasma approximation is used. So reaction rates for ion as well as neutral species are evaluated at the set temperature, that is 300 K. Neutral rates depend also from the vibrational states of molecules [23,24]. This effect could be taken partially into account introducing a vibrational temperature and its effect on the reaction rates [4, 25]. In this preliminary work this effect was not addressed. Electron-impact reaction rates used in this study have

Journal of Physics: Conference Series **550** (2014) 012040 doi:10.1088/1742-6596/550/1/012040

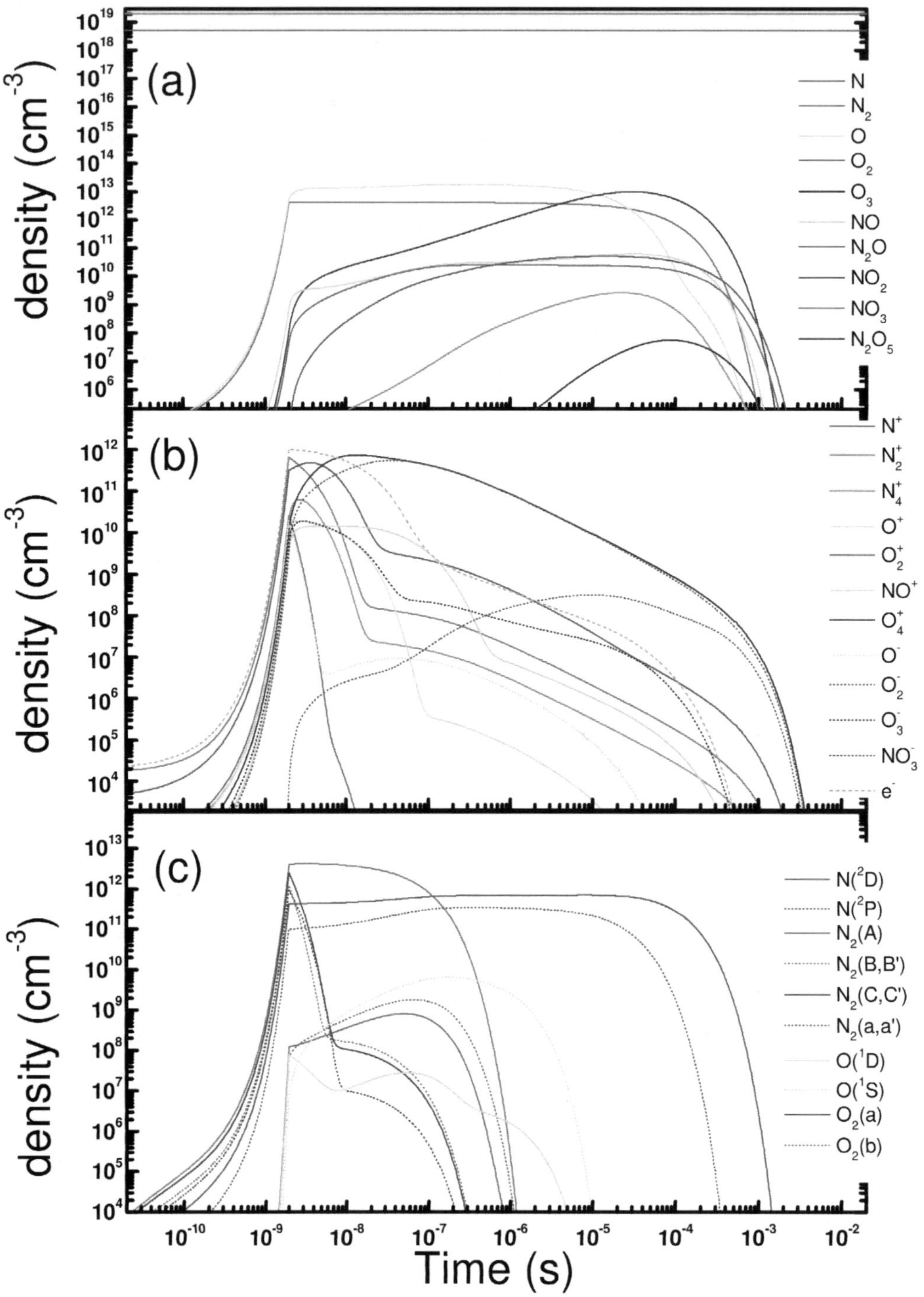

Figure 1. Temporal evolution of the gas phase composition during and after a microdischarge in dry air: neutrals (a), charges (b) and metastables (c) species concentration is reported (parameters: $T_e = 4$ eV, n_e,$= 10^{12}$ cm^{-3}, $\tau_s = 2$ ns, $R_{str} = 100$ µm, $D_{gap} = 0.8$ mm).

been already discussed in our previous works too [4,5]. Rate constants have been evaluated assuming a Maxwellian energy distribution function for electron, described by a single parameter, their temperature T_e. This could be thought of as the mean electron energy, which is determined by the local electric field value by the Boltzmann equation [2]. In this way it is possible to make a direct comparison between this simulation and other existing ones or experimental data. We have considered as reference an electron temperature value of 4 eV [5]. A total of 410 reactions and 43 diffusion processes have been taken into account in the simulation.

At first we consider the evolution of a single discharge happening once. Neutral as well as charged species concentrations are displayed in a log-log scale as a function of time respectively in figure 1a for stable species, figure 1b for charged species and figure 1c for metastable neutral states. As expected, during the time length τ_s (2 ns) needed for streamer development mainly atomic N and O species are created, from electron impact-dissociation (the former more than the latter at such an electron temperature). Then a rather long evolution begins, where ozone is steadily formed until it reaches a concentration comparable to that of the atoms. Afterwards a fast decline in atomic oxygen is observed substituted by ozone as the main product in the gas phase. Overcoming happens after about 10 μs. The maximum density reached by ozone was about 10^{13} cm^{-3} after about 30 μs. Then diffusion quickly removes the gas phase in a few milliseconds. Nitrogen oxides accumulates too, albeit at a somewhat slower rate, staying always minority in the gas-phase. On the other hand, the temporal dynamics of charged species seems to be faster than that of neutrals. This is due to the higher rates of ion-ion recombination and of electron attachment. All the charged species increase almost exponentially during the streamer development time (2 ns) and then the total charge density begins to decrease because of recombination. In presence of oxygen, after the pulse ends, when the electron temperature decreases less than 2 eV, electrons are quickly removed due to attachment processes. The O_2^- is the majority negative ion and the overtake happens after just 15 ns. Positive charged species composition is mainly determined by charge exchange reactions on N_2 and O_2.

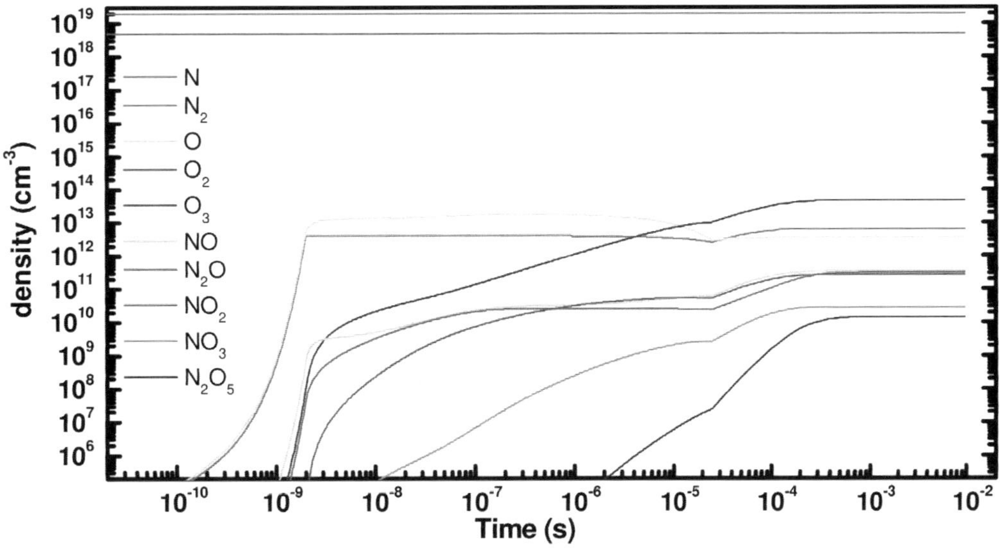

Figure 2. Temporal evolution of the neutral gas phase composition in a microdischarge in dry air repeated at a fixed frequency (parameters: T_e = 4 eV, n_e,= 10^{12} cm^{-3}, τ_s = 2 ns, R_{str} = 100 μm, D_{gap} = 0.8 mm, ν_{rep}= 40 kHz).

They quickly remove atomic ions and leave O_4^+ as the majority ion. This change happens even faster than the removal of electrons. Later on positive and negative charged species density shows the same course: a power-law decrease in time (that is linear on the log-log plot) dictated by the ion-ion recombination process, followed by an exponential decay due to diffusion towards the dielectric surfaces. Excited atoms and molecules are produced during the passage of the ionization wave, just as ions and dissociated atoms. Most of them are quickly quenched by reactions in the gas-phase or radiative decay after the end of the ionization phase. Metastable atomic states start accumulating only after the production of a suitable amount of the parent atoms. Their lifetime is determined by the quenching reactions. The most lasting are metastable molecular oxygen states which accumulate in the gas phase discharge until diffusion starts removing all the discharge products. When a single discharge is repeated at a fixed frequency at some tens of kHz the interplay between the dissociation with subsequent production of oxygen atoms and the reactions in gas phase leading to the ozone formation are exposed. After a very small number of repetitions, in the order of ten, the gas phase reaches a steady composition during the times between each discharge. In particular an equilibrium between ozone produced in the post discharge phases and that removed by diffusion or destroyed in the discharges. The density of the same neutral species considered in figure 1a is displayed in a log-log scale as a function of time in figure 2 for a repetition frequency of 40 kHz, which corresponds to a 20 kHz HV oscillation in a symmetric DBD setup. Under such conditions ozone reaches a steady state concentration of about $5x10^{13}$ cm^{-3} (2 ppm) whereas atomic nitrogen and oxygen densities are about one tenth and nitrogen oxides below one hundredth. The main control parameter is whether the repetition period lies above the transition from atomic oxygen and ozone and below the diffusion loss time. By changing the repetition period it is possible to find an optimal frequency where the steady ozone concentration is maximal. However this does not constitute the best concentration achievable, since the continuous repetition of discharges leads to the destruction of part of the ozone formed during the post discharge times. Thus one is lead quite naturally to consider whether a pulsed operation mode could benefit for instance the ozone formation.

3. Ozone production in pulsed DBD

In order to investigate the effect of a pulsed operation mode on the gas phase composition we have to define better the device setup to be modeled. We conjectured a cylindrical gas gap such as the one between two symmetric insulated electrodes consisting in a couple of parallel planes. The ideal cylinder has diameter 35 mm and height 0.8 mm. A uniform electric field parallel to the cylinder axis is assumed in the whole gap. Discharge filaments, like those considered in the previous section, are supposed to happens with uniform probability at each position within the gap. The effective repetition rate is then determined not only by the period between subsequent active phases. It is determined by the average number of microdischarges during each active phase and by the ratio between the streamer volume and that of the whole gas gap. In a sense we consider all the microdischarges equal as for their plasma parameters and with a constant repetition rate at each position in the gap. Although a rough simplification of reality it does not overlook the relevant evolution of the discharge. Indeed a sort of uniformization of the gas gap composition is achieved in the experiments thanks to diffusion and the large number of microdischarges and cycles involved, despite each microdischarge shows a non-negligible variability in amplitude, duration and delay between subsequent events even considering only the same position [26]. Diffusion outside the gap due to radial transport was added to the simulation in a way completely mirroring the diffusion outside the streamer channel [22]. These simulations leads to results completely reproducing the evolution shown in figure 2. The gas gap reaches a steady composition with an equilibrium between diffusion outside the gap and production of ozone and other compounds due to the repeated discharge processes. Then ideal pulse operation is supposed, consisting in a number of equal active periods separated by quiet intervals. A duty cycle is determined, based on the duration of the on (T_{on}) and off (T_{off}) periods. The density of the neutral species in the gas-phase follows that already discussed for continuous operation. However during the pause periods ozone and nitrogen oxides accumulates while oxygen and nitrogen atoms are consumed. Equilibrium is reached in a few duty cycles, generally less than 10. The concentration reached by ozone at the end of the last pause period was estimated and displayed in figure 3 for two values of the active period. An enhancement of the ozone concentration, almost independent from the active phase duration, is predicted with a maximum for a pause period of several milliseconds.

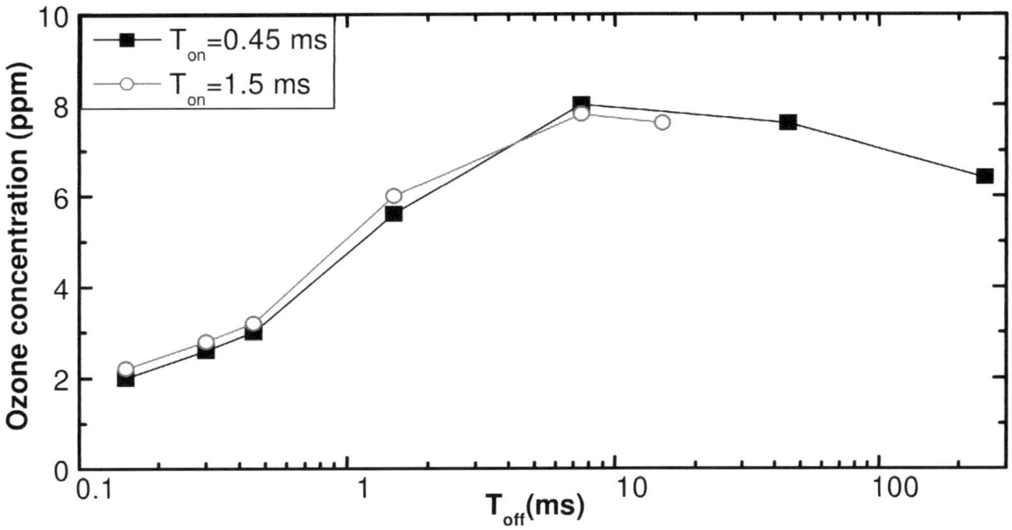

Figure 3. Ozone concentration in the gas phase as a function of the duration of the plasma-off phases between active plasma pulses 450 and 1500 µs long (parameters: T_e = 4 eV, n_e,= 10^{12} cm^{-3}, τ_s = 2 ns, R_{str} = 100 µm, D_{gap} = 0.8 mm, ν_{rep}= 40 kHz).

4. Experimental study

In order to draw a compelling comparison with the numerical simulation results we have prepared our experimental set-up so as to match as closely as possible the approximation made in the modeling. For instance the two electrodes for DBD have been made equal. Each of two stainless steel disks (diameter 35 mm, thickness 1.5 mm) has been inserted into a drilled Macor, a glass ceramic material (© Corning Inc.), cylindrical scaffold (diameter 40 mm, height 3.5 mm, dielectric constant ε_r = 6.03). Thus a dielectric barrier of 2 mm separates the electrode surface from the gas gap. Particular care was to them parallel and exactly corresponding. The discharge gap was fixed at 0.8 mm. The measured electrode capacitance was measured to be 8.30 pF, in good agreement with the result of the calculation using a 2D Laplacian electrostatic field solver (EStat 7.0 by Field precision LLC). Calculations show that the electric field is uniform in the gap within 10% and decreases outwards.

The DBD was operated into a cubic plexiglas box which was vented by blower to remove ozone after experiments. The upper electrode is connected through a HV cable to the secondary coil of a transformer, whose primary circuit is connected to a tunable power generator, providing the driving high voltage for the discharge. We used a Laboratory Corona Station V20 by Tigres GmbH. This provide an alternating voltage whose frequency could be varied automatedly between 18 and 50 kHz to provide a constant output power level or it can be fixed depending on the operating choices. In pulsed operation mode the supply provides a duty cycle where a HV oscillating signal, with an almost constant amplitude, is applied for a fixed time and is alternated by periods when no voltage, indeed a really low one, is applied to the electrodes. The switch-on phase could be very short, about 250 microseconds at minimum (which means about 10 oscillation cycles at 40 kHz), whereas the switch-off phase could be varied from such minimum up to 25 ms. Duty cycles as low as 1% could then be set. The HV signal, applied to the upper electrode was measured locally by a P6015A probe by Tektronix granted for a bandwidth of 75 MHz. The discharge current have been measured by means of Rogowskii coils. Details of this electrical measurements have been already discussed in literature [17]. In particular the number and amplitude of microdischarge events have been used as a guide for the choice of simulation parameters. In order to measure ozone production by our DBD we have employed an ozone sensor (OS-4 by KWJ Engineering Inc.) which was located at about 5 cm next to the electrode setup. The sensor is sensible in the 0.05-20 ppm range and was calibrated in factory, showing a typical accuracy below 20%. The instrument provides an output voltage signal proportional to the ozone concentration which was recorded and digitized with an acquisition rate of 1 Hz. In figure 4 a typical recorded time series is shown. After a brief delay, the signal reaches a stable plateau. The measure was taken by averaging 120 seconds data in the pulse, after the stabilization of the data. It is clear from figure 4 that the system reaches a steady state with a definite value of ozone concentration, stable for several minutes. After switching off the discharge, ozone removal is quite fast and the sensor is ready for a new measure in just a few tens of seconds. At first we have investigated the production of ozone by a continuously operated discharge. The steady state concentration of ozone was measured as a function of the power level. Ozone concentration is small at breakdown when only a limited number of microdischarges happens in the gap. It rises smoothly reaching a maximum at about 140 W power level, then starts decreasing. Although the HV amplitude keep rising and the frequency changes only slightly, ozone production is diminished.

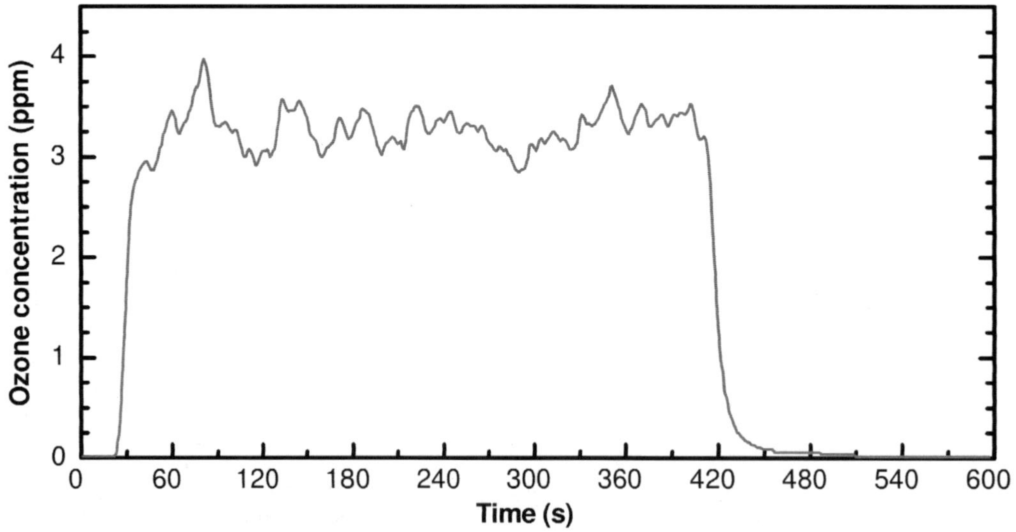

Figure 4. Time series of the measured ozone concentration in the gas phase produced by a pulsed discharge (parameters: W = 120 W, T_{on} = 450 μs, T_{off} = 7.5 ms)

This behavior could be related to the heating of the gas-phase, which favors nitrogen oxide production versus ozone [3,5].

Then we have investigated the effect of a pulsed operation mode. A shrinking duty cycle was applied by maintaining constant the duration of the active phase and stretching the period of pause between two subsequent phases. This reduces progressively the total power absorbed by the discharge. Results are presented as a function of the pause duration in figure 5 for two values of the active phase duration. Even a small pause time is beneficial with a substantial increase in ozone concentration observed when pulse operation is applied. The increase continues up to reaching a maximum. The peak happens at a rather low value of the effective power (W_{eff} = 16 W, 17% duty cycle), much smaller than the most efficient power level needed in continuous mode. Further increasing the pause period leads to a decrease in the ozone concentration. The similarities with the simulation predictions are quite impressive. A throughout investigation of the operating parameters and setup dependencies are clearly needed for a more pointed comparison and it will be performed in future.

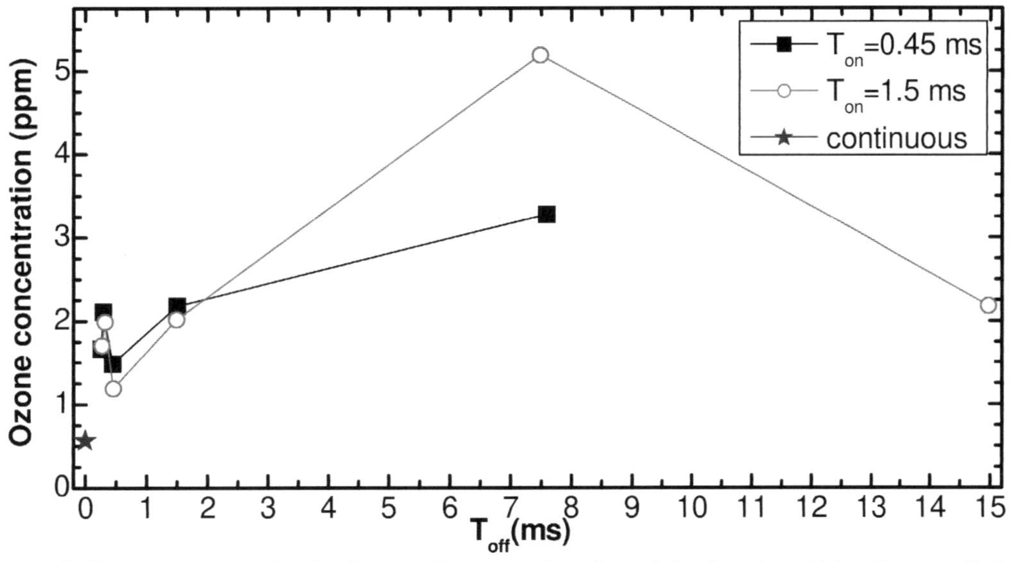

Figure 5. Ozone concentration in the gas phase as a function of the duration of the plasma-off phases between active plasma pulses 450 and 1500 μs long. Comparison with results from a continuous discharge is also shown.

5. Conclusions

We have performed a theoretical analysis of the evolution in the gas phase composition of a dielectric barrier discharge operating in a pulsed mode. In particular we have studied the production of ozone by the application of a fixed single microdischarge repetition rate and the application of suitable pause intervals to let the chemical kinetics evolve undisturbed. Indeed an enhanced concentration of ozone was predicted as the the duty cycle is lowered enough.

Finally we have modified a commercial reactor using its supply unit to drive pulsed DBD in an electrode setup the most similar to the model. Preliminary measurements of the ozone concentration shows a trend comparable to the predicted one. Ozone concentration achieved could be enhanced by pulsed operation with an optimal choice of the active and pause phase duration.

Acknowledgements
We are pleased to acknowledge the contribution of F. Rovaris and M. Nolivari, Bachelor degree students, who were involved in the numerical an experimental activities during their stage in our research group.

References
[1] Kogelschatz U 2003 *Plasma Chem. & Plasma Proc.* **23** 1
[2] Raizer Y P 1991 *Gas Discharge Physics* (Heidelberg: Springer)
[3] Becker K H, Kogelschatz U, Schoenbach K H and Barker R J 2005 *Non-equilibrium air plasmas at atmospheric pressure* (Bristol: Institute of Physics Publishing)
[4] Riccardi C and Barni R 2012 *Chemical Kinetics in Air Plasmas at Atmospheric Pressure* (Chemical Kinetics) ed V Patel (Rijeka, Croatia: Intech,) chapter 9 pp 185-202
[5] Barni R, Esena P and Riccardi C 2005 *J. Appl. Phys.* **97** 073301
[6] Wei L S, Zhou J H, Wang Z H, Cen K F 2007 *J. Plasma Phys.* **73** 427
[7] Laroussi M 2008 *IEEE Transactions on plasma science* **36** 1612
[8] Hayes J, Kirf D, M. Garvey M and Rowan N 2013 *J. Microbiol. Methods,* **94**, 325
[9] Kim J G, Yousef A E and Dave S 1999 *J. Food Protection* **62** 1071
[10] Guzel-Seydim Z B, Greene A K and Seydim A C 2004 *Food Sci.& Technol.* **37** 453
[11] Nogales C G, Ferrari P H, Kantorovich E O and Lage-Marques J 2008 *J. Contemp. Dent. Practice* **9** 75
[12] Bocci V, Aldinucci C, Borrelli E, Corradeschi F, Diadori A, Fanetti G and Valacchi G 2001 *Ozone Science & Engineering* **23** 207
[13] Hoder T, Brandenburg R, Basner R, Weltmann K-D, Kozlov K V and Wagner H-E 2010 *J. Phys. D: Appl. Phys.* **43** 124009
[14] Luque A and Ebert U 2014 *New J. Phys.* **16** 013039
[15] Ono R and Oda T 2006 *J. Phys. D: Appl. Phys.* **40** 176
[16] Biganzoli I, Barni R, Riccardi C, Gurioli A and Pertile R 2013 *Plasma Sources Sci. & Technol.* **22** 025009
[17] Kulikovsky A A 1998 *Physical Review E.***57** 7066
[18] Braun D, Gibalov V I and Pietsch G J 1991 *J. Phys. D: Appl. Phys.* **24** 564
[19] Papageorgiou L, Metaxas A C and Georghiou G E 2011 *IEEE Trans. Plasma Science* **39** 2224
[20] Naidis G V 2005 *J. Phys. D: Appl. Phys.* **38** 3889
[21] Benson S W 1982 *Thermochemical Kinetics* (New York: Wiley)
[22] Chantry P J 1987 *J. Appl. Phys.* **62** 1141
[23] Lopaev D V, Malykhin E M and Zyryanov S M 2011 *J. Phys. D: Appl. Phys.* **44** 015202
[24] Marinov D, Guerra V, Guaitella O, Booth J-P and Rousseau A 2013 *Plasma Sources Sci. Technol.* **22** 055018
[25] Barni R and Riccardi C 2010 *High Temp. Mat. Proc.* **14** 205
[26] Biganzoli I, Barni R and Riccardi C 2013 *J. Phys. D: Appl. Phys.* **46** 025201

Journal of Physics: Conference Series **550** (2014) 012041 doi:10.1088/1742-6596/550/1/012041

Toward the Design of Multi Asymmetric Surface Dielectric Barrier Discharge (ASDBD) Actuators

M Zadeh, V Rohani, F Cauneau, F Fabry, L Fulcheri
Centre for Processes, Renewable Energies and Energy Systems – PERSEE – MINES ParisTech, CS 10207 rue Claude Daunesse, 06904 Sophia Antipolis Cedex, France

E-mail: massiel.zadeh@mines-paristech.fr

Abstract. This paper investigates the electrical behaviors of a single-ASDBD actuator and a two-ASDBD one supplied in sinusoidal mode (1-10 kHz). The main objective of our research is to determine the optimum frequency values for functioning of these actuators with a given power supply. For that purpose, we determine the electrical power density transmitted to the actuators versus frequency through two methods: i) a theoretical method, based on an impedance calculation, and ii) an experimental method, based on direct electrical measurements. These methods show that the addition of a second ASDBD changes the resonance frequency value of the actuator by moving it towards the low frequencies.

1. Introduction

Surface Dielectric Barrier Discharges (SDBD), such as those firstly introduced by Roth et al. [1], have been studied for ten years, and used for the design of electrohydrodynamic (EHD) actuators [2-4]. The advantages of such actuators are their geometrical flexibility and their quasi-instantaneous response due to the absence of moving solid elements, two qualities highly sought for the development of air flow control systems.

When an ASDBD is supplied in AC sinusoidal mode with high-voltage amplitudes, plasma can be formed in the inter-electrode air gap. The formation of plasma gives rise to an effective body force on the ambient air resulting from the momentum exchange between the drifted charges produced in the plasma and the neutral air molecules [5-6].

Nowadays, different approaches are used to increase the performance of SDBD actuators, i.e. the miniaturization [6], specific barrier materials [7, 8], the optimization of supply signal form [9] and the pressure measurements [10]. The optimization of an ASDBD actuator needs to couple the methods coming from electrical engineer and plasma physics. Some researchers [11, 12] have studied the quantification of the power consumption in function of frequency for a range of voltage and frequency values and a specific ASDBD experimental setup. Kreigseis et al [12] have worked with the range of low frequencies (<1 kHz) and high voltage (>10 kV). They have deduced that for a constant voltage value, the power per unit length increases with frequency.

The major goal of our work is to acquire a better understanding of the electrical behavior of elementary ASDBD actuators as to look forward designing a multi-ASDBD actuator which will have a high electro-mechanical conversion yield. The electrical study, consisting on a characterization of the actuators frequency response, is performed by the determination of the electrical power density transmitted to the actuators. For that, we calculate the electrical input power density versus the applied frequency via two methods: i) a theoretical method, based on the calculation of the actuators total impedance coupled to only voltage measurements and ii) an experimental method, based on direct measurements of voltage and current. All the experiments are operated in the following electrical conditions: i) 1 < voltage < 10 kV and ii) 1 < frequency <10 kHz. In our case, we want to determine the law of dependence between the electrical power density and the frequency. The power supply used in our study is a resonant one which implicates the determination of the resonance frequency of the system that increases the transferred power to the discharge.

2. Experimental setup

The ASDBD used in our studied actuators are composed of passive and active electrodes separated by a dielectric barrier material, as shown in Figure 1. The electrode arrangement consists of an exposed passive electrode placed on the upper side of the dielectric barrier and an active electrode located below the dielectric barrier.

The passive electrode is connected to a shunt resistor and to the ground while the active electrode is linked to an AC power supply delivering a sinusoidal waveform. The surface plasma is created only on the passive electrode side. The dielectric plate made of borosilicate glass ($\varepsilon_r = 4.6$) with a thickness of 1 mm, has a section equal to 150 × 150 mm^2. We have chosen the Borosilicate for its two important properties: i) rigidity and ii) resistance to thermal shock. High-Voltage (HV) and ground electrodes are made of aluminum strips ($\sigma = 37.7 \times 10^6$ $S.m^{-1}$) with a thickness of 50 μm, length of 110 mm and a width of 5 mm and 1 mm, respectively.

We can divide the AC power supply into two parts: i) a primary part composed of an amplified signal generator and a transformer's primary, ii) a secondary part composed of the transformer's secondary (max: x100). As for all transformers, the transformation rate of our transformer depends on the signal frequency value. The high-voltage obtained in the secondary part is measured with a HV probe. The electric current I (t) streaming into the system is measured with a shunt resistor of 100 Ω. The evolution of the applied voltage waveform and the corresponding discharge current are visualized and recorded on an oscilloscope (Agilent Technologies DSO1002A, Bandwidth 60 MHz, 2 GS/S and 20kpts/CH). The electrical power input is deduced from the measurements of U (t) (i.e. the potential difference between the electrodes) and the discharge current I (t).

As illustrated in Figure 1, the two-ASDBD actuator is composed of two ASDBD's placed one behind the other. We present the electrical measurements on both configurations in order to deduce the electrical power density transmitted to the actuators by theoretical and experimental procedures.

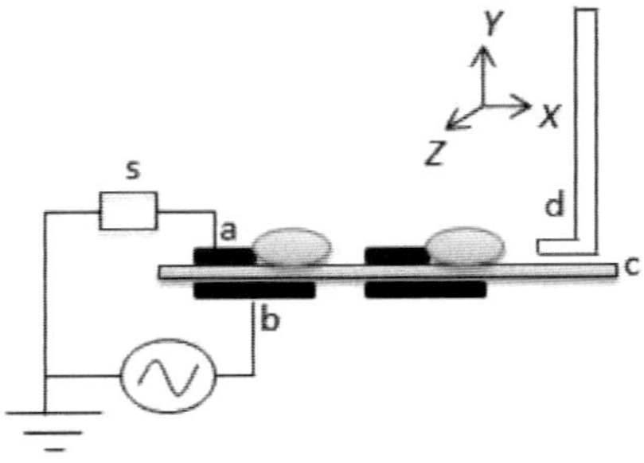

Figure 1. Schematic view of the experimental setup of the single and two-ASDBD actuators: a: ground electrode with a length = 110 mm; b: HV electrode length = 110 mm; c: dielectric barrier made of borosilicate glass with a thickness equal to 1 mm; d: Pitot probe; s: shunt resistor; Note that the distance between both ASDBD = 20 mm;

3. Electrical behavior study

This section presents the electrical behavior study of a single-ASDBD actuator and a two-ASDBD one. This study deals with two dependent segments. The first segment determines the resonance frequency for both actuators by a theoretical method. This is achieved by studying the equivalent circuit model of the system and by coupling the calculation of the actuators total impedance and the RMS voltage measurements. The second segment characterizes experimentally the response frequency of the actuators supplied by the power supply.
We describe this method in the following section which is achieved by determining the electrical power input density to the electrodes via voltage and current measurements at actuators terminals. Note that all electrical experiments are made for an applied rms voltage fixed at 20 V in the primary part of the power supply.

3.1. Theoretical method

In agreement with Enloe et al. works [10, 13], the equivalent circuit model of the electrical system is illustrated in Figure 2. The circuit presented in this figure includes: i) the transformer's secondary and, ii) the single-ASDBD actuator. The transformer's secondary is composed of its inductance (L_{II}) and resistance (R_{II}) which are measured experimentally and are equal to approximately 20 H and $3.73\,k\Omega$, respectively. The single-ASDBD actuator is modeled by the association of: a gas capacitance (C_g), a plasma resistance (R_p) and dielectric barrier capacitances (C_d, C_d'). The plasma resistance is a time variable quantity and is estimated by an integral calculation using the measured voltage and current. C_g, C_d and $C_{d'}$ are calculated by following the expressions (2), (3) and (4). They depend on the geometry of the electrodes, the thickness of the air gap (for C_g) and the thickness of the dielectric barrier (for C_d and $C_{d'}$). $C_{d'}$ depends on the electrode overlap width (l_1) while C_d and C_g depend on the air gap width (l_2) as illustrated in Figure 3. All these capacitances are also dependent on the added number of ASDBD.

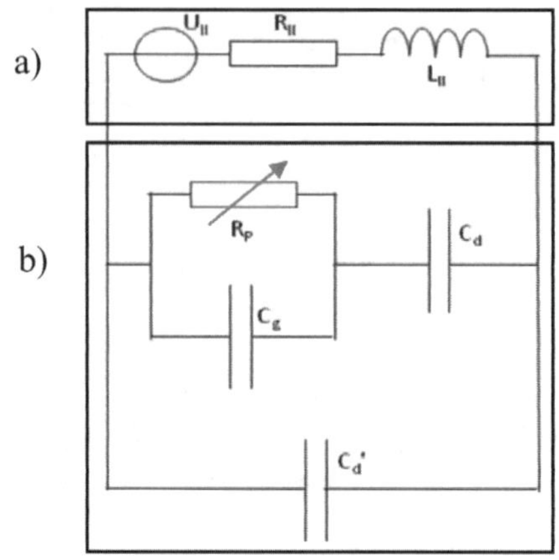

Figure 2. Equivalent circuit model of an ASDBD actuator (b) with a part of its power supply (a).

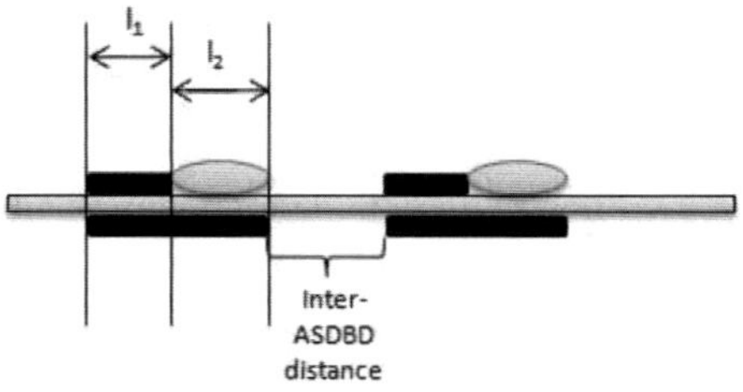

Figure 3. Scheme of the two-ASDBD actuator showing the difference between the electrode overlap width (l_1) and the air gap width (l_2).

$$C_g = \frac{\varepsilon_0 \varepsilon_r^g S_2}{e_g} \quad (2)$$

$$C_d = \frac{\varepsilon_0 \varepsilon_r^d S_2}{e_d} \quad (3)$$

$$C_{d'} = \frac{\varepsilon_0 \varepsilon_r^d S_1}{e_d} \quad (4)$$

Where ε_0 the vacuum permittivity (= 8.85×10^{-12} $F.m^{-1}$), ε_r^g (= 1) and ε_r^d (= 4.6) the relative dielectric permittivity's of air and dielectric barrier respectively, S_1 the electrode overlap surface equal to the product of the length (L_e) and the electrode overlap width (l_1) in m^2, S_2 the air gap surface equal to the product of the length (L_e) and the air gap width (l_2) in m^2, e_d the thickness of the dielectric barrier in m and

Journal of Physics: Conference Series **550** (2014) 012041 doi:10.1088/1742-6596/550/1/012041

e_g the thickness of plasma following the *z* direction in m. The thickness of plasma is considered equal to the thickness of the boundary layer.
As illustrated in Figure 2 the gas capacitance (C_g) and the plasma resistance (R_p) are placed in parallel in the equivalent circuit model. The resulting impedance of these two components is added to the impedance of the first part of dielectric barrier (which the capacitance is C_d) and is written as:

$$\underline{Z}_1(f) = \frac{1 + j2\pi.f.R_p(C_d + C_g)}{4\pi^2 f^2 R_p C_d C_g + j2\pi.f.C_d} \tag{5}$$

$\underline{Z}_1$ (*f*) is in parallel with $\underline{Z}_2$ (*f*), the impedance of the sandwiched part of the dielectric barrier (which the capacitance is $C_{d'}$). Each term related to the equivalent circuit model of the actuators is calculated separately, leading to the determination of the actuators total impedance, as detailed in the expression (6):

$$\underline{Z}(f) = \frac{1 + j2\pi.f.R_p(C_d + C_g)}{4(C_{d'}.R_p(C_d + C_g) + R_p C_d C_g) f^2 \pi^2 + j2\pi.f.(C_d + C_{d'})} \tag{6}$$

For the two-ASDBD actuator, the equivalent circuit is duplicated in parallel. So, the impedance in this case is equal to the half of the impedance calculated for a single-ASDBD actuator. By establishing the equivalent circuit model and calculating the actuators total impedance, we worked out the resonance frequency of the single-ASDBD and two-ASDBD actuators. Here we considered that the ratio between the root mean square of the voltage measured for a given frequency and $Z_a(f)$ calculated for the same frequency, was equal to the electrical power density p_e input to the electrodes (i.e. input electrical power per unit of electrode length) as:

$$p_e(W/cm) = \frac{U^2_{rms} Re[\underline{Z}(f)]}{L_e.(Re[\underline{Z}(f)]^2 + Im[\underline{Z}(f)]^2)} \tag{7}$$

Where U^2_{rms} is the root mean square of the measured rms voltage in kV, N the number of ASDBD composing the actuator, L_e the electrode length in cm (which is dependent of the ASDBD number), Re[$\underline{Z}$ (*f*)] and Im[$\underline{Z}$(*f*)] are the real and imaginary parts of the calculated total impedance in Ω. The calculation of the input electrical power density was made for each condition of the applied frequency. Figure 4 presents the theoretical curves of the electrical power density traced according to the applied frequency for both actuators. It highlights that the electrical power density reaches a maximum at 6 kHz for a single-ASDBD actuator and at 5.4 kHz for a two-ASDBD actuator. These frequency values correspond to the resonance frequencies of the system {actuator, power supply} in both cases. The semi-theoretical curve, using the measured experimental value of U_{rms}, shows that p_e is equal to 4.6 W/cm_e for the single-ASDBD at its resonance frequency and to 2.3 W/cm_e (i.e. two times lower) for the two-ASDBD actuator at its resonance frequency. It means that the total electrical power of the single-ASDBD actuator is the same as the two-ASDBD actuator. This is due to the fact that the power supply used in this study is not an ideal voltage generator but a power generator delivering a constant power in spite of the difference between the actuators' impedances. Indeed, the measurements show that the ratio between the rms voltage for a single and two-ASDBD actuators is approximately equal to $\sqrt{2}$.

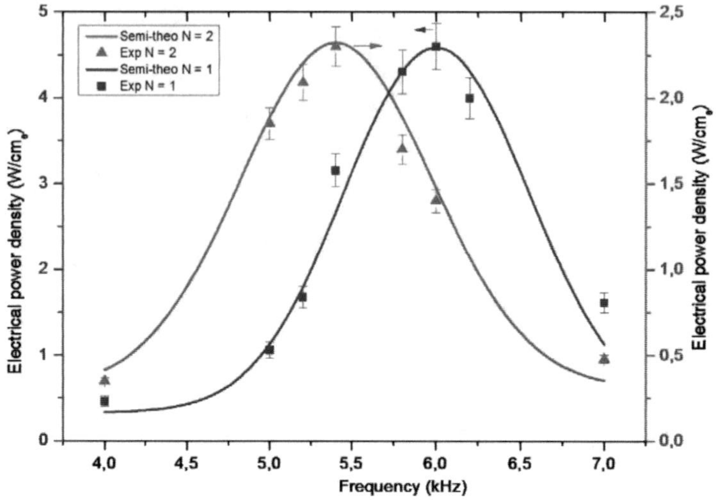

Figure 4. Electrical power versus applied frequency of the single and two-ASDBD actuators: experimental points obtained by measurements and semi-theoretical curves obtained by the coupling between the actuator's impedance calculation and an rms voltage measurement.

3.2. Experimental method

The electrical power density transmitted to the actuator was also determined experimentally throughout the recovering of the voltage and current measured at the terminals of the actuator. P_e was calculated by taking into account a one signal period as given in the expression (8):

$$P_e(W/cm) = \frac{f}{L_e} \cdot \int_{t=0}^{t=T} U(t)I(t)dt \qquad (8)$$

Where *f*, U(t), I(t), T, L_e are the applied frequency in kHz, the measured voltage in kV, the discharge current in mA, the period duration in and L_e the electrode length, respectively. The reproducibility of the electrical energy transferred per signal period was verified in order to confirm its accuracy before using the formula.

The evolution of P_e, determined experimentally, versus applied frequency for both actuator configurations: single-ASDBD and two-ASDBD, are illustrated in Figure 4. As shown in this figure, the electrical power density transmitted to the actuator is proportional to the frequency. The experimental power density deduced from electrical measurements has the same evolution as the theoretical curves for a single-ASDBD as well as for the two-ASDBD actuator. The comparison between the theoretical curves and experimental points puts in evidence a slight difference (lower than 5%) between the corresponding power densities. The similar maxima obtained by both methods led us to confirm the values of the resonance frequencies: $f = f_0 = 6$ kHz, for the single-ASDBD actuator and $f = f_0 = 5.4$ kHz for the two-ASDBD one. Moreover, the values of the electrical power density transmitted to the actuator are the same by these two methods. They are equal to: 4.6 W/cm for the single-ASDBD and 2.3 W/cm for the two-ASDBD one. Hence, the addition of a second-ASDBD is clearly shifting the resonance frequency toward the lower frequencies. Finally, as it is confirmed in Figure 4, we can say that these two methods of electrical power density determination are in total agreement.

4. Conclusion and perspectives

The electrical behaviour of a single-ASDBD actuator and a two-ASDBD actuator were studied in this paper. The electrical behaviour study was focused on the influence of frequency on electrical power density transmitted to the actuators. We determined the electrical power density using two methods, a theoretical method and an experimental one. The first method passed through the calculation of the actuators total impedance by choosing to model the system with a specific electrical equivalent circuit, while the second method was based on direct measurements of voltage and current.

Similar results using both methods confirmed the existence of a resonance frequency, depending on the actuators configuration, which maximizes the input electrical power density to the actuators. With our power supply, the resonance frequency of the single-ASDBD actuator is equal to $f_0 = 6$ kHz for a single-ASDBD actuator and equal to $f_0 = 5.4$ kHz for the two-ASDBD actuator. As a conclusion, the resonance frequency is dependent of the number of ASDBD composing the actuator. This work represents a first step toward the design of efficient multi-ASDBD actuators by determining and operating at the system resonance frequency.

Acknowledgments

The authors gratefully acknowledge CARNOT M.I.N.E.S Institute for their financial supports and the PACA Region for its contribution to PhD fellowships in this subject.

REFERENCES

[1] Roth JR and Dai X 2006 *14th AIAA Aerospace Sciences Meeting and Exhibit* (Reno NV) p 1203
[2] Léger L, Moreau E and Touchard G 2002 *AIAA Paper* 2002-2833
[3] Roth J R 2003 *Phys. Plasmas* **10** 2117-26
[4] Borghi C A, Carraro M R, Cristofolini A, et al. 2008 *J. Phys. D.* **103** 063304
[5] Zito J C, Durscher R J, Soni J, et al. 2012 *Appl. Phys. Lett.* **100** 193502
[6] Boeuf J P, Lagmich Y, Unfer Th, et al. 2007 *J. Phys. D: Appl. Phys.* **40** 652-62
[7] Thomas F O, Corke T C, Iqbal M, et al. 2009 *AIAA J.* **47** 2169-78
[8] Forte M, Jolibois J, Pons J, et al. 2007 *Exp Fluids* **43** 917-28
[9] Debien A, Bernard N, Moreau E. 2012 *J. Phys. D: Appl. Phys.* **45** 215201
[10] Enloe C L, McLaughlin T E, VanDyken R D, et al. 2004 *AIAA Journal* **42** 595-604
[11] Pons J, Moreau E and Touchard G 2005 *J. Phys. D. Appl. Phys.* **38** 3635–42
[12] Kreigseis J, Moller B, Grundman S, et al. 2011 *Journal of Electrostatics* **69** 302-12
[13] Enloe C L, McLaughlin T E, VanDyken R D, et al. 2004 *AIAA Journal* **42** 589-94

Journal of Physics: Conference Series **550** (2014) 012042 doi:10.1088/1742-6596/550/1/012042

Characterization of a low pressure supersonic plasma jet

S Caldirola[1], R Barni[1] and C Riccardi[1]

[1] Dipartimento di Fisica G. Occhialini, Università degli Studi di Milano-Bicocca, piazza della Scienza 3, I-20126 Milano, Italy

E-mail: `claudia.riccardi@mib.infn.it`

Abstract. Plasma assisted supersonic jet deposition (PA-SJD) is a technique which combines a inductively coupled plasma (ICP) with a supersonic jet for the fabrication of thin films having a desired morphology. A reactive argon-oxygen plasma is employed to dissociate an organic precursor (titanium tetra-isopropoxide for TiO_2 thin films) in a first vacuum chamber which is connected through a nozzle to a lower pressure chamber. The pressure difference produces a supersonic jet, seeded with nanoparticles. Along the jet the nucleation and aggregation of nanoparticles can be controlled to obtain nanostructured depositions. We report here the results of an analysis performed with a quadrupole mass spectrometer (QMS) which was used to sample neutrals and ions from the jet at different positions along the centerline of the supersonic expansion.

1. Introduction

Thin films made of titanium dioxide have been widely used for different applications in a great number of research areas: TiO_2 is biologically and chemically inert, it combines a high dielectric constant with high refractive index and exhibits a wide-bandgap. These properties make it well suited for microelectronic and semiconductors applications, for sensors, ultraviolet filters and antireflection coating, photocatalysts and photovoltaic [1-5]. The growth of thin and ultra-thin films of TiO_2 may be achieved using a large variety of techniques such as chemical vapour deposition (CVD), RF sputtering, pulsed laser deposition (PLD) or plasma enhanced CVD (PECVD). Film morphology and structure can greatly enhance the physical properties of thin films. The growth process used determines the structure; different techniques can be adopted, depending on which properties are to enhance in the film [6-9]. Modelling the nanoparticles which create and assemble the film it is possible to achieve very promising results, although it requires a bottom-up approach capable of tailoring the properties with a high level of control or a complex set-up. [10,11]. Unfortunately there is no deposition procedure which can easily achieve an accurate control on both the chemical composition and the nanostructure of a thin film and that can be considered truly versatile or up-scalable. In order to model the film growth and its morphology we proposed an innovative approach to thin films deposition which we named Plasma Assisted Supersonic Jet Deposition (PA-SJD) [12]. An inductively coupled plasma source (ICP) is used to generate a high density argon-oxygen plasma at low pressure inside a cylindrical vacuum chamber which is connected through a converging nozzle to a lower pressure chamber. The pressure ratio between the two chambers determines and characterizes the formation of a supersonic expanding jet. PA-SJD technique was already employed using an organic precursor of titanium for TiO_2 nanostructured thin film depositions [13]. A vaporizer injects the precursor inside the ICP chamber where the reactive plasma dissociates the chemical bonds. The dissociated precursor then flows in the deposition chamber with the expanding jet: the light gas carrier mixture accelerates different titanium nanoparticles up to 0.05-0.3 eV. As the gas expands, pressure, temperature and density decrease creating an energetic and low-collisional flow. In this way the deposition process is divided in two different and tunable steps: the production of the TiO_x seeds and their acceleration and assembly on a substrate. A detailed understanding of the gas expansion is fundamental in order of being able to control the growth and the morphology for nanostructured depositions which are closely related to the nanoparticles energy and diameter. In this work we show the results obtained from a deep characterization of the supersonic jet with a movable quadrupole mass spectrometer (QMS) able to sample neutral gas, radicals and ions at different distances from the nozzle along the axis of symmetry of the gas expansion. Plasma properties were simultaneously measured by optical emission spectroscopy (OES) observing the emission spectra from atoms, molecules and ions present in the plasma chamber and electrical diagnostic, acquiring time series of voltage and current flowing through the radiofrequency antenna of the ICP.

2. Experimental set-up

In this section we will describe the layout of the experimental set-up, sketched in figure 1, and the diagnostics employed.

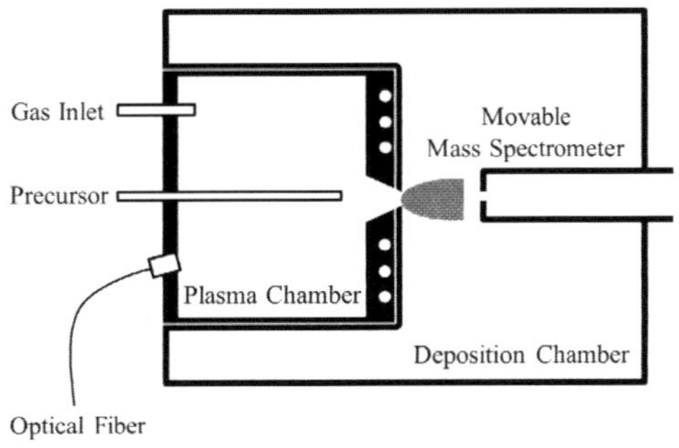

Figure 1. Experimental set-up.

2.1. Vacuum vessel

PA-SJD is a two step deposition process which take place in two stainless-steel cylindrical vacuum vessels connected through a nozzle: the plasma chamber (95 mm long with 62.5 mm radius) and the deposition chamber (200 mm long with 160 mm for inner radius). On the bottom of the deposition chamber a circular conductance (100 mm diameter sized, variable by closing a gate valve) connects our vessel to the main pumping group: a turbo molecular and a rotary pump. On the right side of the chamber there is also an aperture which put the system in connection with the quadrupole mass spectrometer, which can sample the gas from a little conductance (a 0.1 mm diameter circular orifice). A turbo molecular and a rotary pump preserve the purity of the spectrometry measurements keeping the instrument below 10^{-4} Pa. The left lid of the biggest vacuum chamber is in direct contact with the plasma chamber and it contains the inlets for the gas and the precursor. The gas is injected from two different mass-flow controllers, the first ranging from 1.4 to 14 sccm for Ar, the second from 5 to 50 sccm for O_2. The lid also has

Journal of Physics: Conference Series **550** (2014) 012042 doi:10.1088/1742-6596/550/1/012042

two connections to the inductive antenna (planar copper coil cooled by deionized water) and a quartz window to monitor the discharges and acquire optical emission spectra. The deposition chamber is connected with the plasma chamber through a rectangular converging nozzle. The exit area of the nozzle is a circular hole with a diameter of 6.9 mm. The pressure in the chambers and the spectrometer is monitored using three capacitance pressure gauges. The pressure ratio $R = p_P/p_D$ between the two chambers can be varied by closing the gate valve which connects the main pumping group from 40 to 1 to model the supersonic jet geometry and energy.

2.2. ICP source
An ICP can provide plasma densities between 10^{11} and 10^{13} cm^{-3} with an excellent uniformity of the plasma over diameters of at least 20 cm, and low, controllable ion energies [14]. In addition to this it ensures low contamination by sputtering products from the reactor walls. Inside the plasma chamber an inductively coupled plasma is generated by a 13.56 MHz radiofrequency power generator (Huttinger PFG 1600 RF). The generator is connected through a tunable matching box (L-type) to a two and three quarters loop planar antenna made of copper wire (having a measured inductance of 2.5 μH and an impedance of 213 Ω at RF [12]) located on the right side of the plasma chamber, inside a teflon scaffold covered by an alumina disc to reduce teflon sputtering. All the vessel is grounded at the same reference potential.

2.3. TTIP precursor
Titanium precursor can be introduced inside the first chamber when the plasma discharge is sustained at a sufficiently high power level. TiO_2 films can be obtained from different organic precursors, titanium alkoxide are preferable such as $Ti(OPr^i)_4$, which already display the TiO_4 tetrahedral motif of titanium dioxide lattice in their chemical structure [15]. In this work we chose titanium (IV) Tetraisopropoxide $Ti(OCH(CH_3)_2)_4$ or TTIP, which is liquid at 20°C and it is heated with a power transformer to reach the vapour phase. The temperature can be monitored with a thermocouple over the precursor tank and varied to obtain different precursor flows (from 0.25 to 0.75 grams per hour). Operating at temperatures between 40 and 47°C the vapour pressure of TTIP is sufficiently high to create a stable flow inside the plasma chamber.

2.4. Diagnostics
During each plasma discharge, the electrical parameters were observed measuring times series of voltage and current at the two different side of the planar antenna. The voltage directly exiting the matching box was measured using a high-voltage probe (Tektronix P6015A), while the current flowing in the antenna was acquired with a Rogowski probe positioned at the grounded side of the coil. Both signals were simultaneously acquired using a digital oscilloscope at 4 GHz sampling frequency to obtain a detailed electrical characterization.

Optical emission spectroscopy (OES) is a well know non-intrusive diagnostics which has already been applied in many plasma discharges [16,17]. Revealing the light emitted from atoms, molecules and ions which are excited from free electrons in the discharge, it is possible to understand the main reaction processes in the plasma. The intensity of emission lines can identify chemical species and allows to gain insight about their abundances. From the quartz window facing the plasma chamber optical emission spectra (OES) were acquired using two different low resolution spectrometers, as already described in our previous works [12,16].

Quadrupole mass spectrometers can provide an in situ, real time sensing diagnostic for many different applications. We employed a particular QMS (Hiden EQP-1000 Analyser) well suited for plasma analysis: neutrals, radicals and ion species can be detected providing precious information on our system [18]. The QMS is movable on a sliding support, thus capable of performing mass or energy scan along the centerline of the main vacuum vessel, also reaching the nozzle which connects the two chambers. In our set-up the chemical species sampled from

Journal of Physics: Conference Series **550** (2014) 012042 doi:10.1088/1742-6596/550/1/012042

the instrument are reasonably the same involved in the depositions. The gas exiting from the plasma chamber is sampled by a small orifice (100 μm of diameter) into the QMS. This instrument consist of an electron-impact ionization region, an optic system for ion focusing, an high transmission 45° sector field ion energy analyser and a quadrupole mass spectrometer. It is capable of measuring mass and energy spectra allowing detailed analysis of the plasma. When activated, the ionization of neutral species is obtained by an electron-impact source: a oxide coated iridium filament emits thermal electrons which are accelerated toward a grid at a determined energy. In this process, some molecules can be ionized twice or fragmented into lighter species thus measured mass spectra can be affected by characteristic fragmentation patterns. When the ionization is not activated, the ions measured are those produced outside the QMS. The detector is a secondary electron multiplier pulse-counting detector which can provide mass spectra in counts/sec over m/z from 0 up to 1000 amu and energy spectra in counts/sec over eV from -100 to 100 eV. Trends in intensity may also be measured and plotted against time for a short list of relevant masses providing a truly real time, quick and direct measurement.

3. Results and discussion

The QMS was the main diagnostic used to collect data and it was employed to characterize the formation of the supersonic jet seeded with nanoparticles reproducing the experimental conditions already optimized and adopted for thin film depositions [12]. The gas fluxes in the plasma chamber are first set in order to reach a pressure of about 10 Pa with a 2:3 mixture of Ar and O_2. A high concentration of oxygen is fundamental inside the plasma chamber to properly dissociate the precursor, anyway argon gas is also important to create a stable plasma without the requirements of an excessively high power input. The pressure in the deposition chamber can be adjusted to set a desired pressure ratio R. A ICP plasma is generated increasing the power flowing into the planar coil and regulating at each step the L-type matching network to ensure the maximum power transfer. For power values below 175 W the plasma appears in a dim and low density state where the coupling between the antenna and the plasma is just partial, the so called E-mode. As the power increases, a clear transition happens upon reaching the H-mode in which the coupling is fully inductive and the main ionization mechanism which sustains the plasma is the power cession between the current flowing inside the coil and the plasma itself. At 450 W the plasma is stable and the discharge is purely inductive thus it is possible to inject the precursor by adjusting the temperature of the titanium isopropoxide (TTIP) tank and opening a micrometer valve. With the diagnostic system discussed above, we investigated the supersonic jet inside our deposition chamber, characterized the plasma conditions and observed the fragmentation of a TTIP precursor from the plasma directly sampling neutrals, radicals and ions along the jet.

3.1. Formation of the supersonic jet

Under-expanded jet are used in many applications as an effective tool for the the creation of a molecular beam or as nanoparticle source. Compared to effusive beams, such as those used in molecular beam epitaxy, supersonic beams provide higher intensity and directionality allowing deposition with very high growth rates. When a heavy specie is diluted in a lighter gas there is the formation of a seeded supersonic beam. Seeded supersonic beams are extensively used to cool and to accelerate heavy species such as nanoparticles [19,20]. The rapid and unconfined expansion of a gas exiting from a sonic orifice into a low pressure chamber creates a supersonic and directed jet. As the volume increases, temperature, pressure and density abruptly decrease according to the laws which describe an isentropic process, thus accelerating the gas particles increasing the Mach number (defined as the ratio between the velocity of a particle and the local sound speed) above 1. The thermal energy of the molecules is converted into flow velocity forming the supersonic jet. The expansion ends with a normal shock, called Mach disc, where temperature, pressure and density reach the background values while the Mach number returns below 1. The properties of this supersonic beam are mainly determined by the size and shape of the nozzle and by the thermodynamic properties of the gas upstream of the nozzle [21,22]. A schematic representation of the geometry of a supersonic jet is sketched in figure 2.

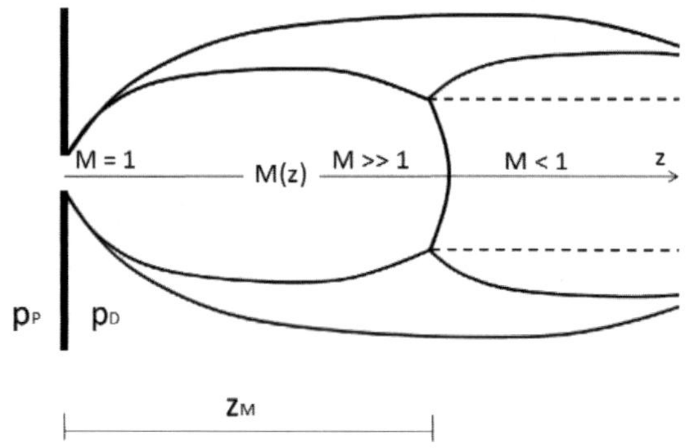

Figure 2. Schematic representation of the gas expansion between the plasma chamber and the deposition chamber. The Mach number (M) variations and the Mach disc position z_M are indicated.

Some useful empirical equations, verified experimentally, can help to reconstruct with good accuracy the geometry of the supersonic jet, as for the Mach disc distance from the nozzle z_M we find from [23]:

$$\frac{z_M}{D_n} = 0.67\sqrt{R} \tag{1}$$

This law does not depend on the gas adiabatic index γ, but only on the exit area of the nozzle (in particular on its diameter D_n for a circular orifice) and on the ratio R between the pressure in the upstream chamber and the pressure downstream. In our experimental works we used a converging nozzle followed by a circular orifice. The converging nozzle (a rounded rectangle which area decreases of about an half in 15 mm length from its initial value of 22x60 mm) ensure a sonic flow at the exit: Mach number is equal to 1, pressure and density values about 0.5 and 0.6 times the upstream values (depending on the γ of the gas). This favours the formation of the jet, even if it is not fundamental as described in [24]. The exit area of our nozzle is a circular orifice having 6.9 mm of diameter, using the equation above assuming that our values for $R = p_P/p_D$ range from 40 to 2, we obtain a Mach disc position varying from 29 to 6.5 mm, depending on the background pressure we set in the deposition chamber. From [23] it is also possible to obtain some empirical equations describing the variation of Mach number and pressure along the centerline of the expansion, as a function of the position z and we estimate Mach numbers up to 8 or 3 respectively for z about 29 or 6.5 mm. We derived an equation for the variation of density along the jet combining the equation proposed in [25] for the pressure on a flat plate positioned at a certain z with the isentropic relations for an ideal gas, obtaining:

$$\frac{n(z)}{n_{Plasma}} = \left(1.44\left(\frac{z}{D_n}\right)^2 - 0.65\frac{z}{D_n} + 0.87\right)^{-1/\gamma} \tag{2}$$

With the movable QMS it was possible to acquire mass spectra along the supersonic jet, measuring a signal proportional to the local density at the entrance of the instrument. Bow

Journal of Physics: Conference Series **550** (2014) 012042 doi:10.1088/1742-6596/550/1/012042

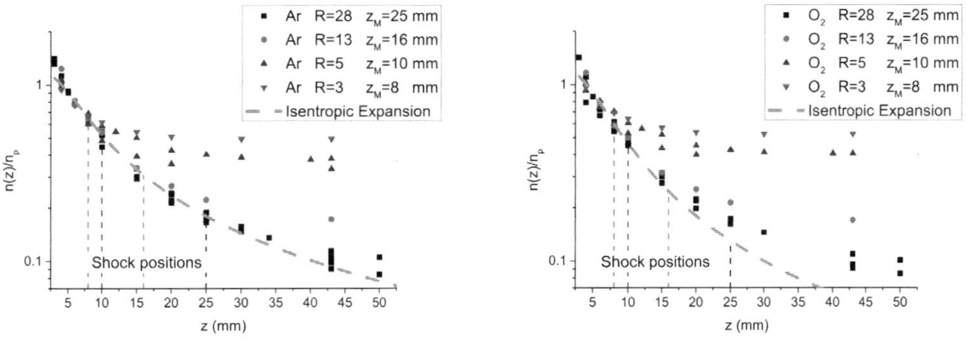

Figure 3. QMS data acquired for argon atoms (left) and oxygen molecules (right) in the argon-oxygen mixture used in the experiments for different pressure ratios R. This implies a different position for the Mach disc z_M which determines the end of the supersonic jet. Coloured vertical bars indicate different z_M.

shock effects are known to perturb or affect the supersonic jet when an obstacle, such as the substrate, is placed at a certain position [26]. By measuring with this experimental set-up we can have a reliable value of the jet structure and composition during a deposition. Our data, stored as counts per second for different m/z ratios, was then renormalized to reproduce the density ratio which appears in the equation above. Statistical errors are generally small (about 1 %) when the plasma is turned off, a set of measurements was taken acquiring the full mass spectra and integrating the area of the peaks, another was acquired directly measuring each second the values of some chosen mass. Each measure performed during the same experiment shows a good reproducibility, unfortunately systematic errors make it difficult to exactly reproduce the same pressure conditions (and also temperature when the plasma is generated) during a different experiment. The agreement with the isentropic expansion law inside the jet given above is very good especially for Ar, as seen in figure 3. Each series of data correspond to a different pressure ratio obtained at fixed pressure in the plasma chamber p_P (about 8 Pa) and changing the background pressure p_D from 0.2 to 4 Pa. For all these values the mean free path in the deposition chamber of the gas is of the order of magnitude of 10 mm, thus we don't expect a steep gradient or an abrupt shock at the end of the supersonic jet. A higher R implies greater distances of the Mach disc from the nozzle z_M, our data show fairly well that the density follows the expected isentropic expansion equation until z_M. After this point the density resets slowly reaching the background values of the deposition chamber.

3.2. Plasma analysis

The results obtained with the free gas jet were then observed and studied when the plasma is created in the first chamber. The generation of the inductive plasma increases pressure and temperature in the plasma chamber, usually from 8 to 10 Pa. Higher temperature and pressure cause higher density and kinetic energy in the supersonic jet, corresponding to a higher flow speed and small differences in the pressure ratio. During each plasma discharge time series of voltage and current across the antenna were acquired with a digital oscilloscope. Increasing the power output from the generator a sharp transition between the E and the H-mode takes place about 175 W.

This transition is shown in the graph on the left in figure 4 where the root mean square values of the voltage drop across the coil and the current flowing through are reported. We also measured emission spectra with different spectrometers and optical fibers (described in the

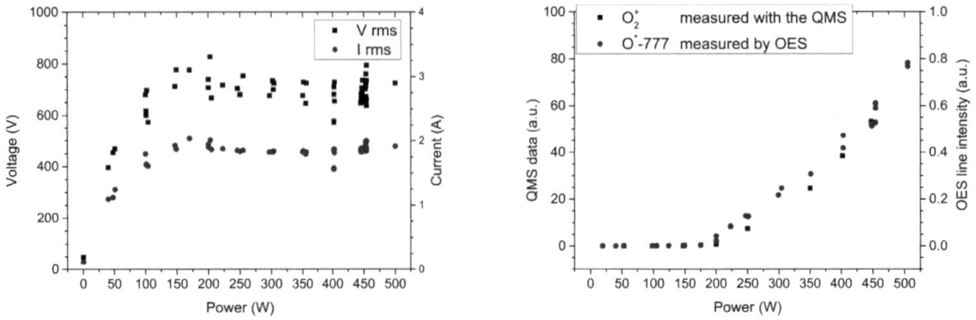

Figure 4. Electrical discharge data show a clear transition increasing the power input (graph on the left). When the fully inductive coupling is reached (H-mode) there is a net increases in plasma density, which was measured both with the QMS and by OES (graph on the right).

experimental set-up section) directly in the first chamber during each plasma discharge. The line emission from the atoms in the plasma, dark subtracted and corrected by the sensitivity of the instruments, can be used to retrieve informations on the radical concentration or on the plasma parameters (electron temperature and density). A detailed OES analysis is far from the purposes of this article, anyway an insight on the main reactive species in the plasma chamber can be gained from figure 6 where a typical spectra is plotted. The most intense lines in the Ar-O_2 plasma spectra are generated from de-excitations of the argon states $3p^5 4p \rightarrow 3p^5 4s$ (wavelength from 415 to 433 nm) and $3p^5 5p \rightarrow 3p^5 4s$ (wavelength from 697 up to 812 nm). Also atomic oxygen lines at 777 and 845 nm are well distinguishable, corresponding to the following de-excitation: $2p^3 3p \rightarrow 2p^3 3s$. Assuming that the electron temperature variations are low at different powers (as seen in [12]), the light emission of the plasma is mostly dependant on the plasma charge density and the initial concentration of emitting atoms. Figure 4 also shows on a separated graph the variation of O_2 ion density, measured with the QMS far from the jet at 100 mm, and of the line 777 obtained simultaneously from the OES. The increase is similar for both the diagnostics, there is a significant increase of ion density after the transition to the H-mode.

The distribution of species along the jet was studied at 450 W, 10 Pa of pressure in the plasma chamber and 0.34 Pa in the deposition chamber, with the 2:3 argon-oxygen mixture. Neutrals and ions scans were performed, and we report a typical variation of QMS counts along the z axis in figure 5. The neutrals measured in the jet are consistent with the data collected without the plasma, although the shock seems stronger and the gas quickly reaches the background values. The ion density variation is highly influenced by the shock for both species, as the neutral gas expands isentropically, also the ion density decreases but exhibits a clear maximum just after the shock position. Far from the shock ion signal can still be measured but it appears very low and abruptly decreasing. Ions were collected at a fixed energy value during mass scans, so their values may be a description of the only high energetic population of the spectra (since we tuned the instrument to detect the highly populated peak of their energetic distribution function, at 7 eV). Anyway in similar experiments charge distributions like those shown here were observed [27].

3.3. Precursor injection

Heating the precursor injection system at temperatures about 45 °C we obtained stable TTIP fluxes of approximately 0.48 g per hour inside the plasma chamber. Inside the argon-oxygen plasma, TTIP (m/z = 284) is dissociated mainly by the high-temperature heating, electron

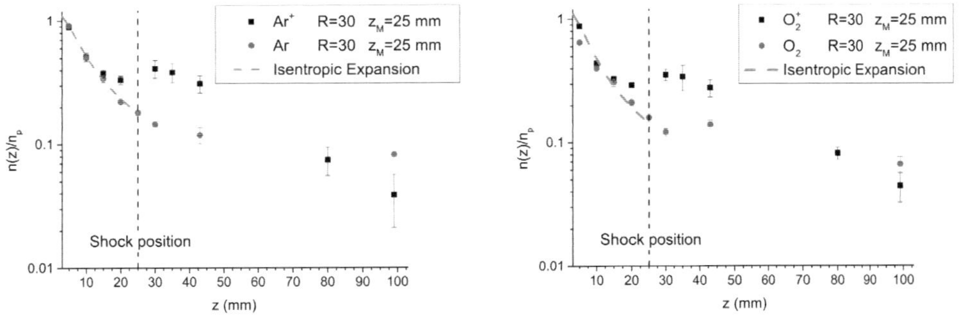

Figure 5. QMS data acquired for neutrals and ions in argon-oxygen plasma. In the graph on the left data for argon are shown, while molecular oxygen is reported on the right. Data have been normalized to reproduce the density variation expected during the supersonic expansion (dashed orange curve). The Mach disc position z_M determines the shock inside the chamber and is indicated with a vertical bar.

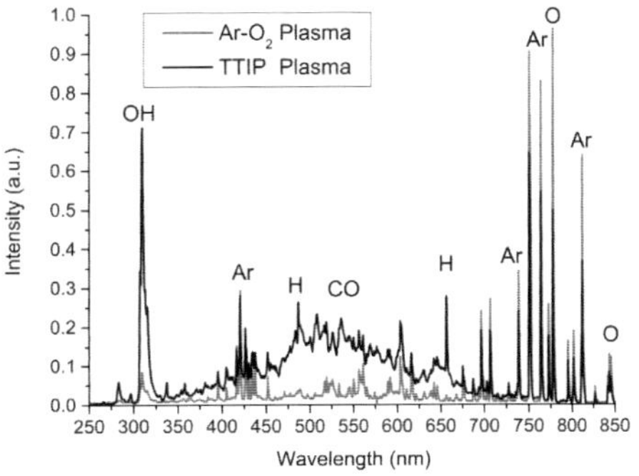

Figure 6. Example of two successively optical emission spectra measured first in a simple argon-oxygen plasma discharge (red line) and then injecting the TTIP precursor (black line).

impact and reactive species, in particular oxygen. The radical oxygen reduction is clearly visible from the emission spectra measured and shown in figure 6. From OES also hydrogen, carbon oxide and hydroxyl radical emission lines can be identified (respectively lines 487 and 565 nm, from 500 to 600 nm, 309 nm). A more complete spectrum of the precursor dissociation was easily measured with the QMS and it is reported in figure 7, in table 1 the identified species are shown (determined from both literature and looking at the titanium isotope ratios [28]).

In the mass range from 300 to 1000 m/z no evidence of other species or nanoparticles was found with this set-up. Inside the supersonic jet, the heavy TTIP products are accelerated from the argon-oxygen mixture at velocities comparable to a fraction of the light gas flow velocity, as a function of their mass and diameter [19]. Again different measurements were performed acquiring neutrals and ions with the QMS reproducing different conditions adopted for thin film depositions. The values for argon and oxygen nicely resemble the results obtained for the simple plasma discussed above.

In figure 8 we see the distribution of some selected species along the jet. The behaviour of

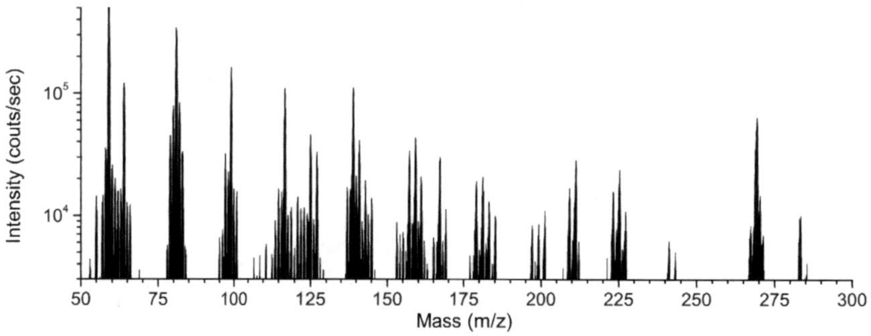

Figure 7. Mass spectra measured at 5 mm of distance from the nozzle during a plasma discharge with TTIP. Electron impact ionization at 70 eV creates different fragments which alter the real chemical composition detected. However most of these peaks have also been measured as ions, so they are originated outside of the QMS and present in the jet, but their sensitivity can not be fully trusted.

Table 1. Main products of the TTIP (m/z=284,$Ti(OCH(CH_3)_2)_4$) dissociation measured during experiments.

m/z	Ion	m/z	Ion
15	CH_3	139	$TiO_2(OCH(CH_3)_2)$
43	$CH(CH_3)_2$	167	$Ti(OCH(CH_3)_2)_2H$
59	$OCH(CH_3)_2$	181	$TiO(OCH(CH_3)_2)_2$-H
64	TiO	211	$Ti(OCH(CH_3)_2)_3H$-CH_3
81	TiO_2H	225	$Ti(OCH(CH_3)_2)_3$
99	TiO_3H_3	243	$Ti(OCH(CH_3)_2)_4$-$CH(CH_3)_2$
125	$TiO(OCH(CH_3)_2)$	269	$Ti(OCH(CH_3)_2)_4$-CH_3

TiO_2H and TiO_3H_3 is very similar to data already shown for argon and oxygen neutral. In the graph on the right it is possible to observe a much different behaviour for higher mass neutrals and ions. This can be due to the different acceleration drag forces and the arise of slip-factors which can cause particles separation along the jet.

4. Conclusions

A characterization of the low pressure supersonic plasma jet employed for thin film depositions was presented. The physics which describes the formation of the jet was reconstructed and verified experimentally by QMS measurements. After a brief description of the main plasma state, measurements along the jet was shown first when the ICP is generated, then when the TTIP precursor is injected. Sampling neutral gas species and ions at different distances from the nozzle it was possible to gain an insight on the precursor dissociation and the components along the supersonic jet which influence our depositions.

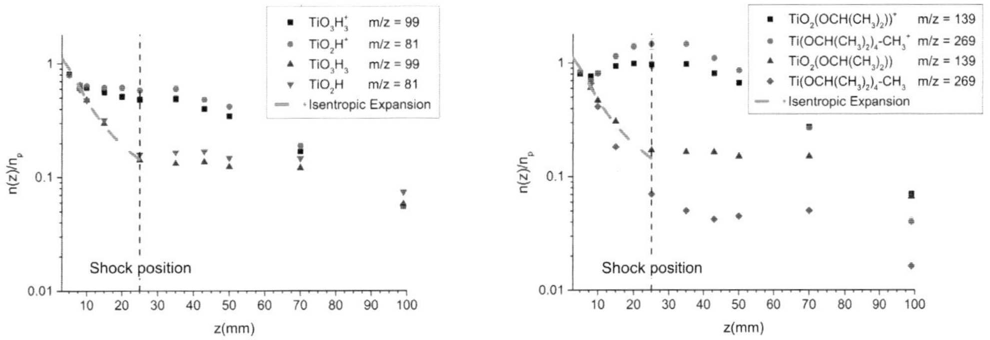

Figure 8. QMS data acquired for neutrals and ions in TTIP reactive plasma. In the graph on the left data for lighter TiO_x seeds are shown, while two heavier masses are reported on the right. Data have been normalized to reproduce the density variation expected during the supersonic expansion (dashed orange curve). The Mach disc position z_M determines the shock inside the chamber and is indicated with a vertical bar.

References

[1] Grant F A 1959 *Rev. Mod. Phys.* **31** 636
[2] Yeung K S 1983 *Thin Solid Films* **109** 169
[3] Kurtz S R, Gordon R G 1987 *Thin Solid Films* **147** 167
[4] Mills A, Le Hunte S 1997 *J. Photochem. Photobiol.* A **108** 1
[5] Fujishima A, Hashimoto K, Watanabe T 1999 *TiO2 Photocatalysis: Fundamentals and Applications* BKC, Tokyo
[6] Chaturvedi S 2011 *J Mater Sci* **46** 3669
[7] R. Messier, A. P. Giri, and R. A. Roy 1984 *J. Vac. Sci. Technol.* A **2** 500
[8] Borras A, Cotrino J, Gonzalez-Elipe A R 2007 *J. Electrochem. Soc.* **154** 152
[9] Dwivedi N, Kumar S, Malik H K 2011 *ACS Appl. Mater. Interfaces* **3** 4268
[10] Milani P, Piseri P,Barborini E, Podesta A, Lenardi C 2001 *J. Vac. Sci. Technol.* A **19** 2025
[11] Dick B, Brett M J, Smy T 2003 *J. Vac. Sci. Technol.* B **21** 2569
[12] Biganzoli I, Fumagalli F, Di Fonzo F, Barni R, Riccardi C 2012 *J. Mod. Phys.* **3**, 1626
[13] Trifiletti V, Ruffo R, Turrini C, Tassetti D, Brescia R, Di Fonzo F, Riccardi C, Abbotto A 2013 *J. Mater. Chem.* A **1**, 11665
[14] Kortshagen U, Zethoff NI, 1995 *Plasma Sources Sci. Technol.* **4**, 541
[15] Nizard H, Kosinova M L, Fainer N I, Rumyantsev M Yu, Ayupov B M, Shubin Yu V, 2008 *Surf. Coat. Technol.* **202**, 4076
[16] Barni R, Zanini S, Riccardi C 2008 *Vacuum* **82** 217
[17] Barni R, Riccardi C, Selli E, Massafra M R, Marcandalli B, Orsini F, Poletti G, Meda L 2005 *Plasma Process. Polym.* **2** 64
[18] Barni R, Zanini S, Riccardi C 2012 *Advances in Physical Chemistry* **2012** 205380
[19] Huang C, Nichols W T, O'Brien D T, Becker M F, Kovar D, Keto J W, 2007 *J. Appl. Phys.* **101**, 064902
[20] Mazza T, Devetta M, Milani P, Bongiorno G,Coreno M, Piseri P 2011 *New J. Phys.* **13** 023009
[21] Fenn J B 2000 *Int. J. Mass Spectrom.* **200** 459
[22] Crist S, Sherman P, Glass D 1966 *AIAA Journal* **4** (1) 68
[23] Ashkenas H, Sherman F 1966 *Rarefied Gasdynamics* **4** Academic Press, New York
[24] Murphy H R, Miller D R 1984 *J. Phys. Chem.* **88** No.20 4474
[25] Abouali O, Saadabadi S, Emdad H 2011 *J. Aerosol Sci.* **42** 65
[26] de la Mora J F 1985 *J. Chem. Phys.* **82** No. 7, 3453
[27] Robin L, Vervisch P, Cheron B G 1994 *Phys. Plasmas* **1** No. 2, 444
[28] Rahtu A, Ritala M 2002 *Chem. Vap. Deposition* **8** No. 1, 21

Journal of Physics: Conference Series **550** (2014) 012043 doi:10.1088/1742-6596/550/1/012043

Ion energy distribution functions in a supersonic plasma jet

S Caldirola[1], **H E Roman**[1] **and C Riccardi**[1]

[1] Dipartimento di Fisica G. Occhialini, Università degli Studi di Milano-Bicocca, piazza della Scienza 3, I-20126 Milano, Italy

E-mail: claudia.riccardi@mib.infn.it

Abstract. Starting from experimental measurements of ion energy distribution functions (IEDFs) in a low pressure supersonic plasma jet, we propose a model to simulate them numerically from first principles calculations. Experimentally we acquired IEDFs with a quadrupole mass spectrometer (QMS) collecting the argon ions produced from a inductively coupled plasma (ICP) and driven into a supersonic free gas expansion. From the discussion of these results and the physics of our system we developed a simulation code. Integrating the equations of motion the code evolves the trajectory of a single ion across the jet. Ar^+-Ar collisions are modelled with a 12-4 Lennard-Jones potential which considers induced dipole interactions. IEDFs were simulated at different positions along the jet and compared with the experimental data showing good agreement. We have also implemented a charge transfer mechanism in which the ion releases its charge to a neutral atom which can take place at sufficiently close distances and is a function of the impact energy.

1. Introduction

Ion bombardment is a phenomenon inherent in many different thin film deposition processes involving plasma sources. Charged particles can be accelerated onto the substrate surface influencing the formation and growth of thin films. Their energy may modify the substrate surface and temperature enhancing chemical reactions on the surface, growth and nucleation [1-4]. Usually high energetic particles are ions formed in a plasma: argon is the most common gas used for plasma formation and processing since it is the least expensive inert gas. To understand and characterize the effects of these energetic particles it is of fundamental interest the study of their energies. The energy of ions arriving at the substrate is usually much higher than the thermal energy, since it depends on the voltage across the plasma sheath and whether the ions undergo collisions crossing this sheath. The study of ion energy distribution functions (IEDF) is relevant for every plasma processing applications [5-9].

Combining a reactive inductively coupled plasma (ICP) with a supersonic jet, we developed a new approach for thin films deposition: plasma assisted supersonic jet deposition (PA-SJD), which was employed for Titanium dioxide thin film fabrication having a desired morphology [10]. However, the role of ions in PA-SJD is not completely understood yet. Our experimental set-up is made of two different vacuum chambers, in the first a high density Ar-O_2 cold plasma is created. This vacuum chamber is connected through a nozzle to the deposition chamber, the latter kept at a lower pressure. The pressure ratio between the two chambers is set sufficiently large to generate a supersonic under-expanded jet which drives neutral particles onto a substrate placed along the jet [11].

In this work we study a simple argon plasma. Argon ions are generated inside the first chamber and are then driven inside the deposition chamber by the neutral argon gas flow and accelerated by the potential difference. Our goal is to understand the collisional processes which take place along the supersonic jet and to characterize the IEDFs which influence our depositions. Experimentally we acquired IEDF data positioning a movable quadrupole mass spectrometer (QMS) along the supersonic jet. This instrument is grounded and has a cylindrical shape (50 mm of diameter), the energy of incoming ions is reasonably the same which they impact onto the substrate during a deposition. We then propose a computational model to simulate the energy evolution of ions moving in the supersonic jet to reproduce the IEDFs based on our experimental observations. We developed a MATLAB code to study argon IEDFs numerically from first principle calculations.

This paper is organized as follows: in the next section we will describe the experimental set-up we used for acquiring experimental data, giving also information about the collection of IED, in section 3 we will explain the physical background of the supersonic jet required to understand the main processes which take place in our reactor and in section 4 describe our approach to simulate ion dynamics. In section 5 the results from experimental data will be shown and compared to our simulations, in section 6 we will introduce the motivations and the computational implementation of charge transfer reactions to obtain more realistic simulations.

2. Experimental set-up

2.1. PA-SJD reactor

The formation of the supersonic jet requires a sufficiently high pressure difference between the inlet and exit orifice of a nozzle. To have a versatile reactor for nanostructured depositions we developed a device divided in two different cylindrical vacuum vessels connected through a nozzle, the layout is sketched in figure 1. The first vacuum chamber is the plasma chamber (95 mm long with 62.5 mm radius), the second is the deposition chamber (200 mm long with 160 mm for inner radius). The pressure ratio between the two chamber R = p_P/p_D can be modified varying the conductance of the main pumping group (a turbomolecular and a rotary pump which let us operate in moderate vacuum conditions), which is located on the bottom of the deposition chamber. On the right side of the deposition chamber there is also an aperture which puts the system in connection with the QMS. The instrument was used to sample the gas from a small conductance (circular orifice with a 0.100 mm diameter size). To obtain clean measurements without high impurity levels it operates at pressure below 10^{-4} Pa granted by a second pumping group (a turbo molecular and a rotary pump). The left lid of the biggest vacuum chamber is in direct contact with the plasma chamber and it contains the inlets for the gas and the precursor for the depositions. The gas is injected from mass-flow controllers ranging from 1.4 to 14 sccm for Ar gas. The lid also has two connections to the inductive antenna (planar copper coil, cooled by deionized water). The pressure in the chambers and the spectrometer is monitored using three capacitance pressure gauges. Their pressure reading precision is good in the working pressure ranges, which are 1-20 Pa in the plasma chamber and 0.1-8 Pa in the deposition chamber.

Inside the plasma chamber a Inductively Coupled Plasma is generated by a 13.56 MHz radiofrequency power generator (Huttinger PFG 1600 RF). The generator is connected through a tunable matching box (L-type) to a two and three quarters loop planar antenna made of copper wire located on the right side of the plasma chamber, inside a teflon scaffold covered by an alumina disc to reduce teflon sputtering. All the chamber is then grounded at the same potential.

Journal of Physics: Conference Series **550** (2014) 012043 doi:10.1088/1742-6596/550/1/012043

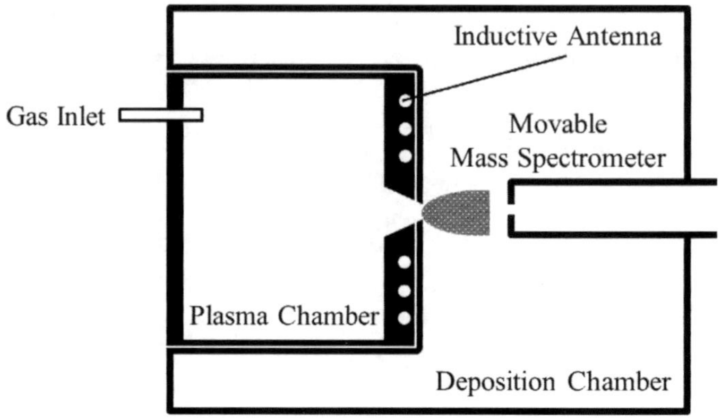

Figure 1. Experimental set-up.

2.2. Quadrupole mass spectrometer

The gas exiting from the plasma chamber can be sampled by the quadrupole mass spectrometer (Hiden EQP-1000 Analyser). This instrument consist of an electron-impact ionization region, an optic system for ion focusing, an high transmission 45° sector field ion energy analyser and a quadrupole mass spectrometer. This QMS is capable of measuring mass and energy spectra allowing detailed analysis of positive ions, radicals and neutrals and it can be moved along the axis of symmetry of the chamber to obtain measurements at different positions along the jet. To measure the IEDFs the QMS operates in this way: the ions produced outside the instrument enter the 0.1 mm sampling orifice with an initial energy E_{i} and they are immediately focused by a lens into the drift space. In this region they are accelerated to an energy of $E_{\mathrm{i}} + qV_{\mathrm{axis}}$, where V_{axis} is normally set to 40 V. Then the 45° energy filter transmit only ions with the desired energy E. The ions remaining will then be slowed to the transit energy required for the mass scan inside the quadrupole and selected by the mass filter. The detector is a secondary electron multiplier pulse-counting detector which can provide mass spectra in counts/sec over m/z from 0 up to 1000 amu and energy spectra in counts/sec over eV from -100 to 100 eV. The instrument is grounded and uses the deposition chamber as reference potential.

3. Theoretical background

To understand and simulate the dynamics of Ar ions, it is important to make some assumption based on the physics of our system. For this purpose we will here describe in detail firstly the main relations and properties of the supersonic jet and the effects on neutral and ion energies.

Under-expanded free jets have been studied and applied widely in the last decades [12]. A rapid and unconfined expansion of a gas exiting from a sonic orifice into a low pressure chamber creates a supersonic and directed jet. As the volume increases, temperature, pressure and density decrease abruptly according to the laws which describes an isentropic process, thus accelerating the gas particles increasing the Mach number (defined as the ratio between the velocity of a particle and the local sound speed) above 1. The thermal energy of the molecules is converted into flow velocity forming the supersonic jet. The expansion ends with a normal shock, called Mach disc, where temperature, pressure and density reach the background values as the Mach number returns below 1. The properties of this supersonic beam are mainly determined by the size and shape of the nozzle and by the thermodynamic properties of the gas upstream of the nozzle. A schematic representation of the geometry of a supersonic jet is sketched in figure 2.

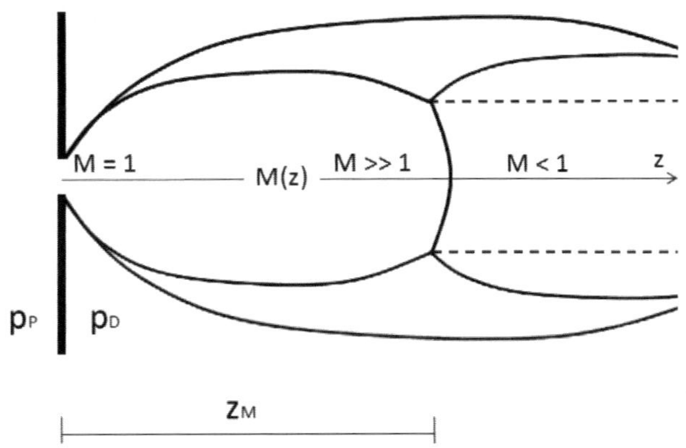

Figure 2. Representation of the supersonic expansion geometry formed after the nozzle exit orifice from the pressure different between the pressure in the plasma chamber, p_P, and the pressure in the deposition chamber, p_D. The shock position, z_M, is indicated and represents the maximum extension of the supersonic jet.

Some useful empirical equations, verified experimentally, can help to reconstruct with good accuracy the geometry of the supersonic jet, for the Mach disc distance from the nozzle z_M we have [13]:

$$\frac{z_M}{D_n} = 0.67\sqrt{R} \tag{1}$$

where D_n is the orifice diameter and R the pressure ratio between the two chambers. In our experimental setup we used a converging nozzle followed by a circular orifice. The converging nozzle favours the formation of the jet, even if it is not fundamental as described in [14]. The exit area of our nozzle is a circular orifice having a 6.9 mm diameter, using the equation for the Mach disc position above and assuming a value of R of 40, we obtain a Mach disc position at about 30 mm, depending on the background pressure we set in the deposition chamber. From [13] it is also possible to obtain some empirical formulas describing the variation of Mach number and pressure along the centerline of the expansion, as a function of the position z and we estimate Mach numbers up to 8 for z about 30 mm. We derived a formula for the variation of density along the jet combining the equation proposed in [15] for the pressure on a flat plate positioned at a certain z with the isentropic relations for an ideal gas obtaining:

$$\frac{n(z)}{n_{Plasma}} = \left(1.44\left(\frac{z}{D_n}\right)^2 - 0.65\frac{z}{D_n} + 0.87\right)^{-1/\gamma} \tag{2}$$

where γ is the gas adiabatic index. Inside the jet, argon particles are quickly accelerated to a limit velocity which is related to the mass of the gas particles m and its temperature in the plasma chamber T_{Plasma} and asymptotically close to [16]:

$$u_{limit} = \sqrt{\frac{2\gamma k_B T_{Plasma}}{(\gamma - 1)m}} \tag{3}$$

which correspond to 746 m/s or 0.114 eV when considering Ar gas at our experimental conditions. The energy gained from the ions is much greater due to the sheath acceleration, so the energy for an ion can be approximated to [9,17]:

$$E_0 = k_B T_e + q(V_P - V(z)) \tag{4}$$

where V_P is the plasma potential inside the plasma chamber (assumed positive and almost constant), $V(z)$ the potential along the jet in the deposition chamber and q the electric charge. The value of E_0 represent the maximum energy achievable, collisions with neutral gas outside the plasma chamber can reduce this energy, slowing down the ions.

4. Computational approach

Our goal is to simulate IEDFs from first principles, introducing model parameters only when it is required to keep the approach tractable and accurate. We developed a MATLAB code to integrate the equations of motion for our interacting system. First of all the role of electrons inside the supersonic expansion can be considered negligible, since they are too fast if compared with ions and atoms (neutrals). The plasma charge density can be assumed much lower than the neutral density allowing us to study the dynamic of a single ion at a time moving through the jet, thus neglecting long range ion-ion interactions. A single simulation provides us the energy of an ion along its trajectory. By performing a large number of simulations (generally hundreds) it is possible to reconstruct histograms representing the IEDFs for different positions. In each simulation the motion from the initial position to the fixed ending point, located several mm away from the starting location, is decomposed into a succession of cylindrical simulation cells. The cylinder (500 μm long and with a radius of 0.004 μm) is filled with neutrals Ar uniformly. The number of neutrals is set according to their density in a supersonic jet given in the previous section.

To further simplify the computational approach we assume that our reference system is moving along the z direction with the velocity of neutrals inside the jet. Thus, neutrals will appear frozen for an ion moving at u_{limit}. When the ion exits the simulation cell, a new cell is constructed, by orienting the axis of the cylinder according to the ion velocity. To describe the interaction between a singly charged ion and a neutral noble gas we use a modified 12-4 Lennard Jones potential [18,19]. This allows the computation of long range collisions by taking into account charge-induced dipole interactions. The equations of motion, determined by this potential, set a system of differential equations which can be numerically solved using a MATLAB algorithm. The interval of integration was set to 0.1 nanoseconds, thus slower ions require much more computation time for each cell. When the ion velocity is comparable with the neutral velocities, the simulation is stopped and the kinetic energy randomly drawn out from a Gaussian distribution centered at E=0.

5. Results

5.1. Experimental work

Experiments were conducted in a argon plasma generated with an inductive coupling. The pressures in the plasma and the deposition chambers were set respectively to 4.5 and 1.4 Pa, obtaining a pressure ratio R of 32 and a corresponding Mach disc distance from the nozzle z_M of 26 mm. The RF input power was then raised gradually up to 450 W, which is the power value also used for thin film depositions [10,11]. We then used the QMS to acquire IEDFs at different positions along the jet, focusing our interest on the study of argon ions.

In the ICP reactor, where the discharge is purely inductive, ions are produced at a uniform and low plasma potential value (generally below 20 eV) and their energy distribution functions exhibit a mono-energetic peak approximatively over the electron temperature kT_e. Upon crossing a collisionless sheath they gain an energy equal to the potential drop [9]. The IEDFs we measured in the deposition chamber are shown in figure 3. Two different energy peaks are clearly distinguishable. The higher peak corresponds to the fast ion population which enters the

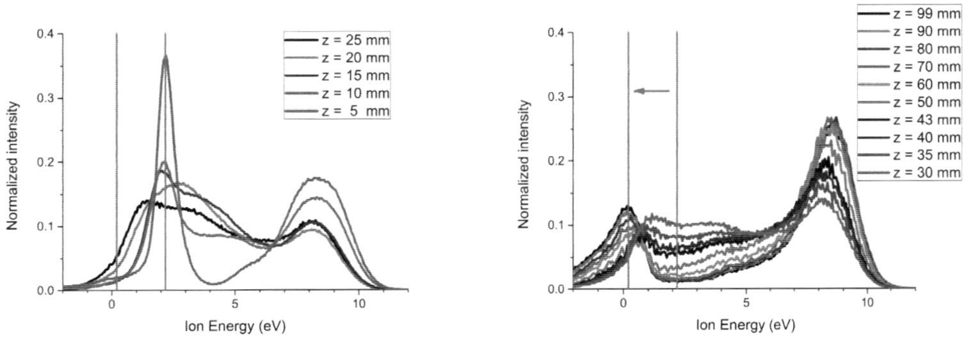

Figure 3. IEDFs measured along the jet (first graph on the left) and after the shock positioned at 26 mm (second graph on the right). Acquired data have been averaged over few measurements cycles and renormalized to have the same total area equal to 1. The red vertical bars indicate the low energy peak at 5 mm and 99 mm from the nozzle.

mass spectrometer without undergoing in any collisional process. The peak is almost stable at 8.3 eV for each position and represent the average maximum energy achievable. The low energy peak is related to collisional processes which take place out of the plasma chamber, it is located at 2.2 eV inside the jet and at 0.2 eV after the Mach disc. Upon crossing the nozzle, ions enter the low-collisional supersonic expansion and are immediately accelerated from potential drop between the plasma potential V_P and a the potential along the jet V_{jet}. Inside this region ions can drift and reach the QMS without colliding, maintaining their total energy, or can collide lowering their energies. Since our QMS is grounded and its sheath can be considered almost collision-less, the lowest energy measurable corresponds to strong collisions which reduce the energy to a small kinetic value plus the potential energy V_{jet}. After the Mach disc position the low energy peak drops to lower value, about 0.2 eV, which is comparable with the neutral energy thus indicating a rapid decrease of the potential in this region. Strong collisions seem much favoured in the first part of the jet, but there is also an energy spread and redistribution between the two peaks that corresponds to those ions that collide without losing all their energy. Some energies measured are negative, this is due to ions colliding or formed in the acceleration region of the instrument [20].

5.2. Simulations

From the experimental data collected we gained a good insight on the ion dynamics and the main reaction processes. We have used these information to initialize our simulations and study the effect of collisions along the jet. The initial energy distribution is not trivial to guess or determine without a direct measure: inside the plasma chamber it is reasonably a mono-energetic distribution centered at the higher energy peak measured (8.3 eV), but the evolution of this distribution inside the nozzle inlet region is complicated and depending on many parameters.

We set the initial point of our simulation at 5 mm and extracted initial energies following the IEDF measured experimentally at this position. In figure 4 the simulation values are fitted to the experimental results. They are used as the starting energy distribution values of Ar^+ in the simulations at larger distances. The dynamic of 300 ions was simulated from 5 to 20 mm and compared to our data. Over the z axis we assumed a constant potential centered at V_{jet} set from experimental data about 2 eV. According to this the kinetic energy of a ion is equal to the difference between its initial energy and this potential value $E_i - qV_{\text{jet}}$, while the detected energy is the sum of the two. In figures 4-7 we compare the simulated IEDFs with the measured data

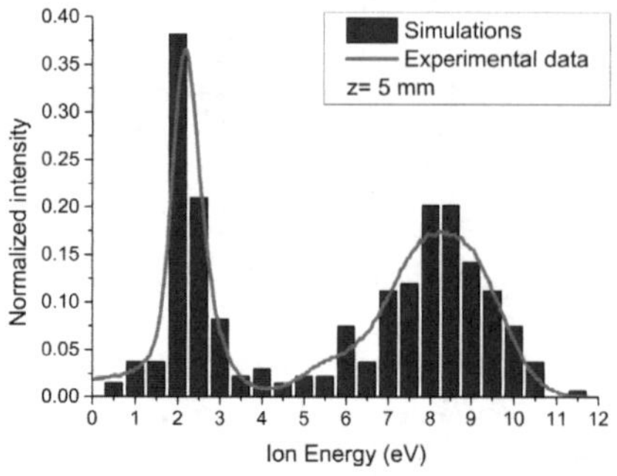

Figure 4. Initial energy distribution.

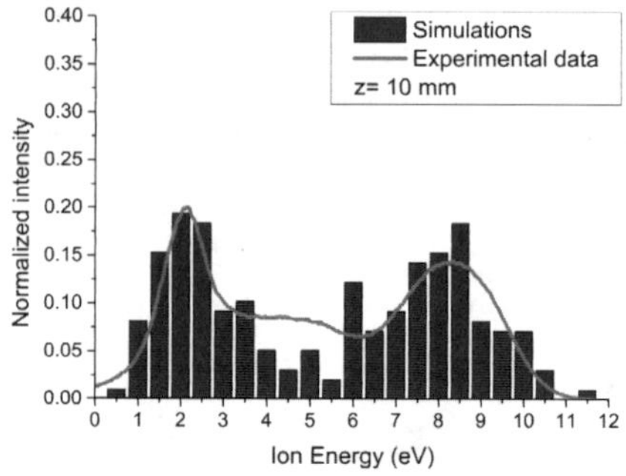

Figure 5. IEDF at 10 mm.

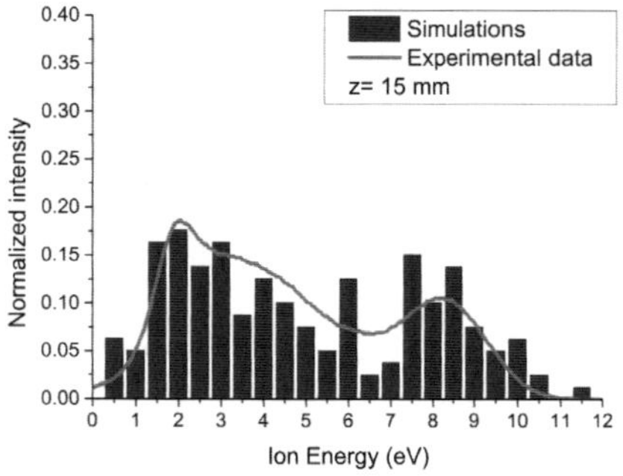

Figure 6. IEDF at 15 mm.

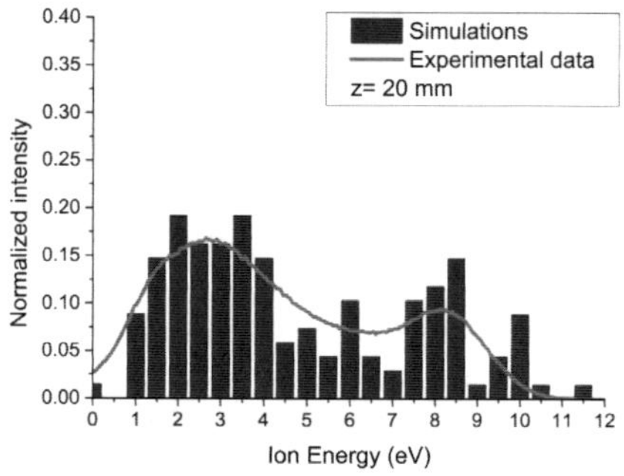

Figure 7. IEDF at 20 mm.

at distances of 10 mm, 15 mm and 20 mm from the nozzle. The simulated data agree quite well with the experimental results. The evolution of the two energy peaks is consistent along z. It is also possible to observe an increase of ion counts in the middle energetic region of the spectra.

6. Charge transfer effects

In the simulations shown in figures 4-7, the only collisional processes considered are elastic collisions between ions and neutrals, treated as dipole interactions. In many experiments with similar plasma sources also charge transfer reactions play an important role especially when happening across the sheath [21,22]. When an ion and a neutral collide, if the distance between the two nuclei is short enough, an electron can be exchanged between them. Charge transfer between argon ions and any nucleus can thus produce fast neutrals and energy losses in the IEDFs, according to the reaction:

$$Ar^+_{fast} + X_{slow} = Ar_{fast} + X^+_{slow} \tag{5}$$

This kind of reaction is very common in plasma, especially with argon, and it has a constant cross section σ at low energies (below hundreds of eV), comparable with the σ for momentum transfer [23]. Experimentally QMS data shown that in argon-oxygen plasma, the Ar^+ signal decreases by 2-3 order of magnitude respect to the value in common argon plasma, suggesting an active charge transfer between argon ions and oxygen neutrals. So we decided to consider also Ar^+-Ar charge transfer reactions in our code with an approximation. An accurate description of the phenomenon requires a quantum mechanical approach, which is too complicated for multi electrons atoms. In a first approximation the probability that an atom will transfer an electron to an ion decreases exponentially with distance between the atom and the ion.

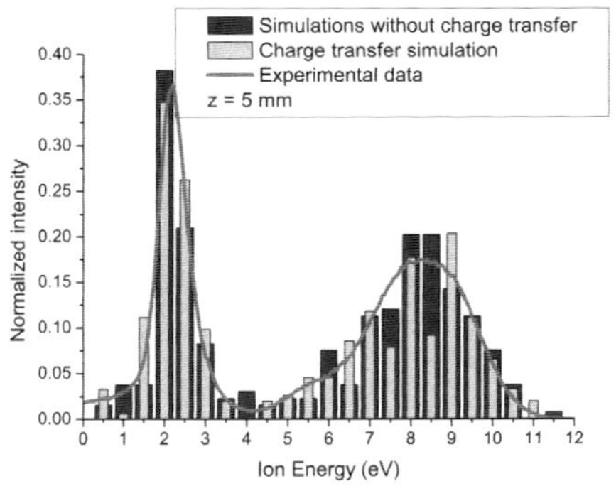

Figure 8. Initial Energy distribution.

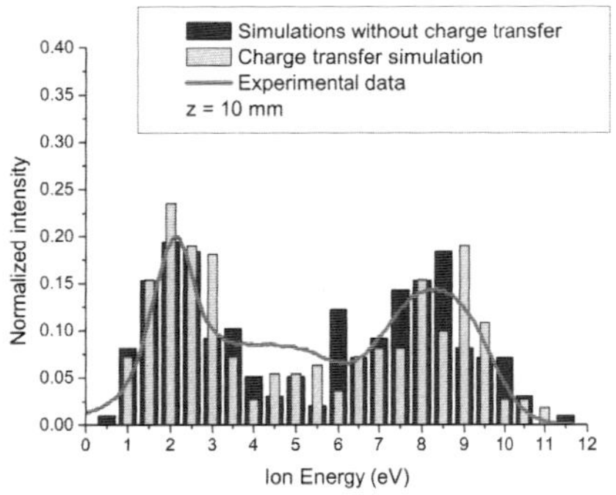

Figure 9. IEDF at 10 mm.

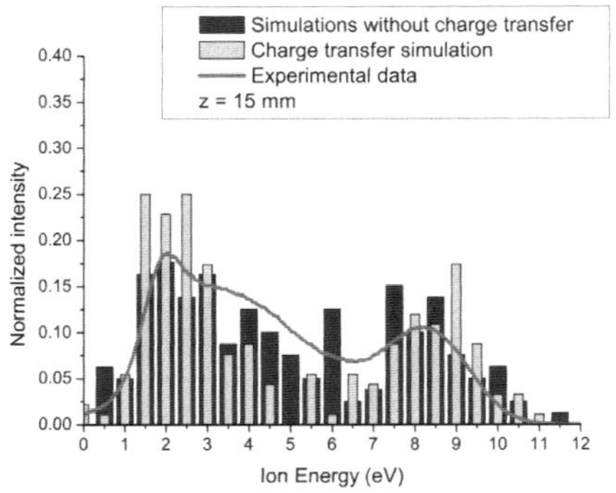

Figure 10. IEDF at 15 mm.

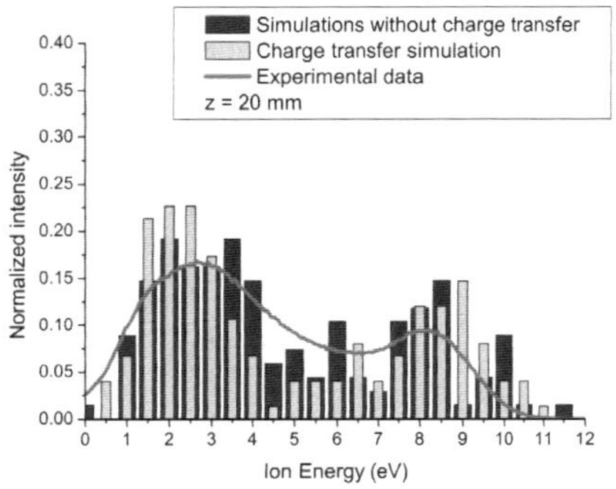

Figure 11. IEDF at 20 mm.

If the ion and the atom approach enough that their electron orbits overlap, then an electron will be exchanged between them many times and, after the collision, it will remain in either of them with equal probability 1/2 [25]. When the inter-atomic distance during a collision is below $r_{\text{min}} = \sqrt{\frac{2\sigma}{\pi}}$, at the end of the collisional process the code considers the neutral atom as the original ion, and vice versa with a given charge transfer probability P_{exchange}. In our code P_{exchange} is equal to 0 when the distance is above r_{min} otherwise it increases exponentially up to 1/2 depending on the energy transferred to the neutrals. In figures 8-11 are shown the results of 300 simulations with charge transfer, compared with both experimental data and simulations without charge transfer. Differences between the initial energy distribution functions in figure 8 are due to statistical fluctuations in the random number generator which can be made smaller by increasing the number of simulations. The addition of charge transfer creates apparently similar IEDFs with a slightly higher low energy peak. However nearly 125 of 300 ions had at least one charge transfer reaction with a neutral atom. In the 180 charge transfer reactions simulated it was possible to calculate the energy loss in the IEDFs, as the difference between the energy of the ion before the collision, and the final energy of the ionized neutral. As shown in figure 12 charge transfer between Ar ions and neutrals in our model introduce in most cases small energy losses.

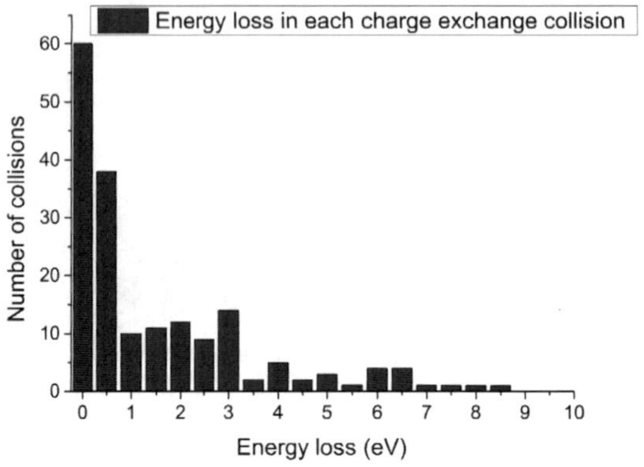

Figure 12. Energy loss in different charge transfer collisions.

7. Conclusions

A model to simulate IEDFs in a low pressure supersonic plasma jet was proposed based on the physics involved. Experimental measurements were performed in argon plasma using a QMS to acquire ion energy spectra along the jet. The results show an interesting two peaked distributions and were used to develop a simulation code which integrates the equations of motion from first principles. It was possible to simulate the dynamics of ions moving along the jet under some assumptions made from experimental observations or theory in order to reproduce the measured data. The agreement between the data and simulations is very good.

Charge transfer reactions have been taken into account in an approximate fashion by using two parameters, one representing a cut-off distance (see e.g. [23]) within which the reaction can occur, and a second one representing an energy threshold on the impact energy transferred to the neutral atom. The present results indicate that charge transfer occurs at least once in about 40% of the total ion trajectories simulated. Yet, they suggest that the corresponding effects are not important in determining the global shape of the IEDFs in the case of argon.

References

[1] Petrov I, Hultman L, Helmersson U, Sundgren J E, Greene J E, 1989 *Thin Solid Films* **169**, 299
[2] Mattox D M, 1989 *J. Vac. Sci. Technol. A* **7** No.3, 1105
[3] Profijt H B, Kessels W M M, 2012 *ECS Transactions* 50 **13**, 23
[4] Hippler H, Kredl J, Vartolomei V A, 2009 *Vacuum* **83**, 732
[5] Malakhovskii A, 1999 *Phys. Chem. Chem. Phys.* **1**, 4187
[6] Logue D M, Shin H, Zhu W, Xu L, Donnelly V M, Economou D J, Kushner M J, 2012 *Plasma Sources Sci. Technol.* **21**, 065009
[7] Lee J B, Chang H Y, Seo S H, 2013 *Plasma Sources Sci. Technol.* **22**, 065008
[8] Kawamura E, Vahedi V, Lieberman M A, Birdsall C K 1999 *Plasma Sources Sci. Technol.* **8**, R45
[9] Kortshagen U, Zethoff NI, 1995 *Plasma Sources Sci. Technol.* **4**, 541
[10] Trifiletti V, Ruffo R, Turrini C, Tassetti D, Brescia R, Di Fonzo F, Riccardi C, Abbotto A 2013 *J. Mater. Chem.* A **1**, 11665
[11] Biganzoli I, Fumagalli F, Di Fonzo F, Barni R, Riccardi C 2012 *J. Mod. Phys.* **3**, 1626
[12] Fenn J B, 2000 *Int. J. Mass Spectrom* **200**, 459
[13] Ashkenas H, Sherman F 1966 *Rarefied Gasdynamics* **4**, Academic Press, New York

[14] Murphy H R, Miller D R 1984 *J. Phys. Chem.* **88** No.20, 4474
[15] Abouali O, Saadabadi S, Emdad H 2011 *J. Aerosol Sci.* **42** 65
[16] Huang C, Nichols W T, O'Brien D T, Becker M F, Kovar D, Keto J W, 2007 *J. Appl. Phys.* **101**, 064902
[17] Tanner S D, 1993 *J. Anal. At. Spectrom.* **8**, 891
[18] Chen W L T, Heberlein J, Pfender E 1996 *Plasma Chem. Plasma Process.* **16** No. 4, 635
[19] Michels H H, Hobbs R H, Wright L A 1978 *J. Chern. Phys.* **69**, No.11 , 5151
[20] Olthoff J K, Van Brunt R J, Radovanov S B 1995 *Appl. Phys. Lett.* **67** No.4 , 473
[21] Babaeva N Yu, Lee J K, Shon J W, Hudson E A 2005 *J. Vac. Sci. Technol.* A **23**, No. 4, 699
[22] Barton D, Bradley J W, Steele D A, Short R D 1999 *J. Phys. Chem.* B **103**, 4423
[23] Phelps A V 1991 *J. Phys. Chem. Ref. Data* **20**, No. 3, 557
[24] Maiorov S A, 2009 *Plasma Phys. Rep.* **35**, No. 9, 802

Maximization of the working domain of an Atmospheric Pressure Townsend Discharge (APTD) using a current-source static converter

M.C. Bouzidi, X. Bonnin, N. Naudé, H. Piquet, A. Belinger, N. Gherardi

Université de Toulouse ; UPS, INPT ; LAPLACE (Laboratoire Plasma et Conversion d'Energie) ; 118 route de Narbonne, F-31062 Toulouse cedex 9, France

CNRS ; LAPLACE ; F-31062 Toulouse, France

nicolas.naude@laplace.univ-tlse.fr

Abstract. The objective of this paper is to present the implementation of a square-wave current inverter power supply, and to describe its effect on the working domain of an Atmospheric Pressure Townsend Discharge (ATPD) in nitrogen. The results show that it allows a substantial increase of the percentage of time during which the discharge is ON (*on-time*) and therefore of the power delivered to the discharge, reaching almost 10 W/cm^2 in pure nitrogen. Moreover, the increase of the *on-time* of the discharge implies a decrease of the extinction *off-time* between two discharges, and thus an increase of the memory effect from one discharge to the following one, as it is monitored by using electrical and optical measurements. This leads to a larger working domain of the ATPD, even when adding impurities in nitrogen.

1. Introduction

Dielectric Barrier Discharge (DBD) is an easy and robust solution to generate low temperature discharge at atmospheric pressure. This discharge has a lot of applications (*e.g.* ozone generator, excimer lamp, plasma display panel (PDP), surface treatment, etc.) since more than one century [1-5]. Its configuration uses dielectric barriers to maintain current densities below the threshold for glow-to-arc transition. Depending on the gas, the electrical parameters, and the electrode configuration, dielectric barrier discharges (DBD) can operate in a homogeneous or filamentary mode [6]. Homogeneous DBD at atmospheric pressure have been obtained in helium, argon, nitrogen [7]. In nitrogen, the ionization level is too low to allow the formation of a cathode fall. Thus the electrical field is quasi-uniform over the discharge gap, like in low-pressure Townsend discharge, and the obtained discharge is called Atmospheric Pressure Townsend Discharge (APTD) [8].

A homogeneous DBD, especially if sustained in a low cost gas like nitrogen, is a promising alternative to replace some of the low-pressure plasma processes employed for thin film deposition [9]. The main expectations for this approach are to allow continuous processing and to reduce the equipment costs by avoiding the need for expensive pumping systems of conventional low-pressure plasma devices. However, the thin film growth rate using APTD is still too low for an online process. Firstly, to improve this point, the discharge power must be increased [9]. However, the discharge power is limited due to the fact that the slope of the voltage versus time is limited. Thus, in case of sinusoidal voltage power supply the amplitude and frequency of the applied voltage allowing to get the APTD are limited. A too high value of these parameters induces instabilities of the APTD which leads to a pure filamentary discharge. A solution to increase the discharge power is to control the current amplitude and to increase the discharge duration. This can be realized using a square-wave current inverter source [10]. Secondly, the reactive gases concentration in the discharge mixture has to be increased. Indeed, as can be seen on Figure 1, the thin film growth rate using APTD is limited not only by the discharge power but also by the precursor concentration. For example, Massines *et al.* [11] found out that using a classical sinusoidal voltage power supply, the Townsend regime can be successfully sustained only with hexamethydisiloxane (HMDSO) dilutions in N_2 lower than 50 ppm. In case of O_2, admixture, the maximal rate has to be lower than 500ppm [12-15].

In this study, we report the use of a square-wave current converter in place of the classical sinusoidal voltage power supply to sustain an APTD in nitrogen. The materials and methods are described in section 2, and a discussion is done in section 3 concerning our motivations for the use of a square-wave current inverter. Sections 4 and 5 are dedicated to the results, focusing more specifically on the impact of the power supply nature on the discharge power (section 4) and on the discharge stability with respect to the additives (section 5).

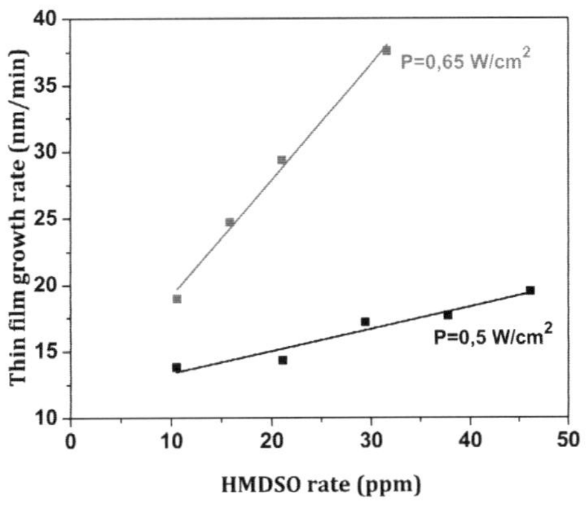

Figure 1. Effect of the HMDSO rate and discharge power on the SiO_x thin film growth rate using APTD in N_2-N_2O-HMDSO mixture [16].

2. Experimental set-up

The DBD is kept in a closed vessel to perform experiments in a very well controlled atmosphere. The plasma reactor is pumped down to 10^{-3} mbar prior to any experiment, and then is filled up to atmospheric pressure using mixtures of nitrogen (99,999% purity) and synthetic air (N_2/O_2=80/20% with 99.999% purity) purchased from Air Liquide. The discharge is ignited between two alumina plates separated by a 1mm gas gap (Figure 2) and the discharge area is 3x3cm². In order to renew the atmosphere, a gas flow (1slm) is injected from one side of the discharge (longitudinal gas injection), keeping a constant pressure of 1 bar through a gentle pumping of the vessel.

The sinusoidal voltage power supply is made of a low-frequency generator providing the reference waveform, which is then amplified by a linear amplifier (Crest Audio model 8001, 4800VA) whose output is applied to the primary winding of a transformer (Montoux, ratio=150, 600VA, 60V/9kV) in series with a 4Ω resistor. The electrodes are connected to the secondary of the transformer. The discharge is characterized by electrical measurements. The discharge current is measured through a

Journal of Physics: Conference Series **550** (2014) 012044 doi:10.1088/1742-6596/550/1/012044

200Ω resistor in series with the electrodes. The voltage applied to the electrodes is measured by means of a high voltage probe (Tektronix P6015A) whereas the voltage really applied to the gas, V_g, is calculated from the electrical measurements as described by Liu *et al.* [17]. The current and the voltage applied to the electrodes are visualized on a digital oscilloscope (Lecroy WaveRunner HRO 66Zi, bandwidth: 600MHz). The discharge homogeneity is investigated by means of short exposure time pictures, which are taken with an intensified CCD camera (Princeton Instrument PI-Max 3) synchronized with the power supply voltage.

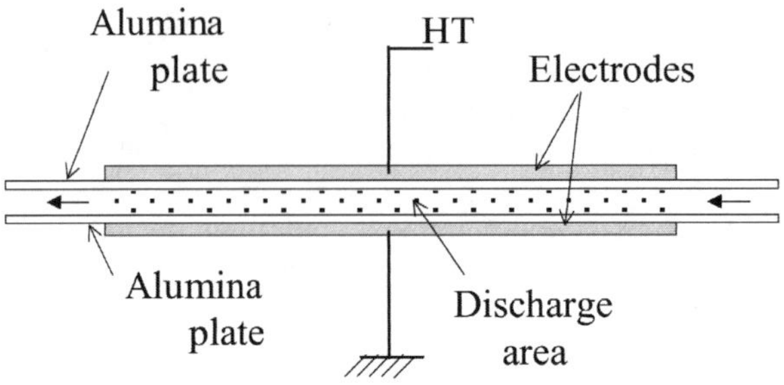

Figure 2. Schematic diagram of the discharge cell.

3. Which power supply in order to increase the discharge power keeping an Atmospheric Pressure Townsend Discharge?

Usually, APTD is obtained using a high-voltage low-frequency sinusoidal voltage power supply [8]. As already discussed, the amplitude and frequency of the applied voltage allowing to get the APTD are limited (Figure 3). The lower limit of the working domain, at a given frequency, is the applied voltage needed to extend the discharge over the whole electrode surface. And the upper limit, at a given frequency, is the applied voltage at which instabilities appear and so defines the transition to a filamentary discharge. This maximum voltage decreases when the frequency increases.

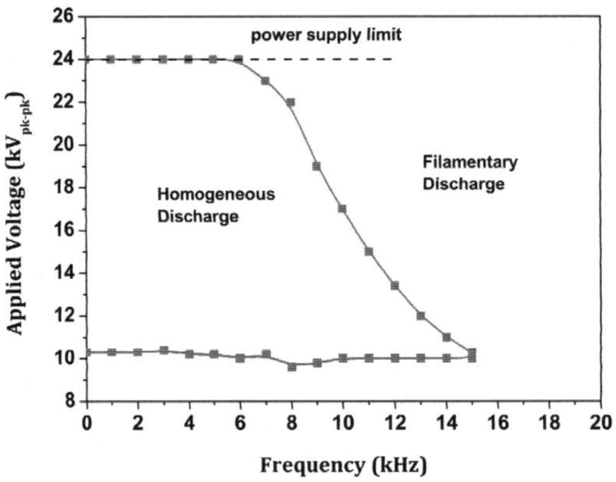

Figure 3. Working domain of the APTD in pure N_2 (the first point is obtained with f=50Hz).

In order to find out how to increase the discharge power in the homogeneous regime (APTD), it is interesting to take a look at the electrical measurements (Figure 4). The first point is that the maximal discharge duration is about 60% of the period. The second point is that the APTD being a Townsend discharge, the gas voltage becomes almost constant after the discharge ignition. Thus, during the discharge, the voltage fluctuations are applied on the solid dielectrics and the current is derived from the solid dielectric capacity and the power supply voltage variation with time, *i.e.* the external circuit controls the discharge current [18]. Hence, the maximum limits of the frequency and amplitude of the applied voltage can be associated with a maximum current density allowing to obtain a homogeneous discharge: higher is the frequency or the amplitude of the applied voltage, higher is dV/dt, and so higher is the discharge current with the risk of reaching a too high ionization level to maintain a homogeneous discharge (transition to a filamentary discharge) [8].

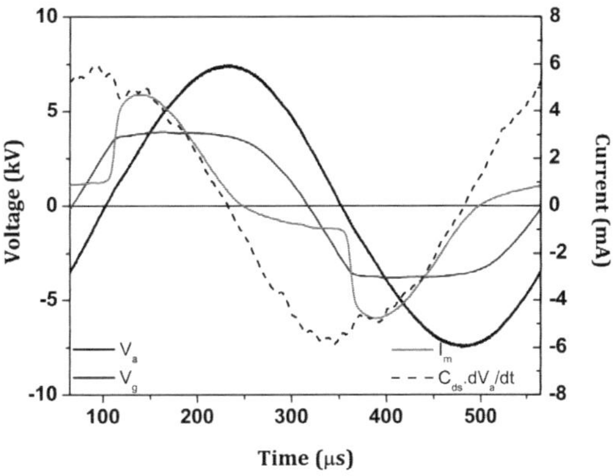

Figure 4. Oscillogram of the applied voltage V_a, the calculated gas voltage V_g, the measured current I_m and the current calculated from $C_{ds} \cdot dV_a/dt$.

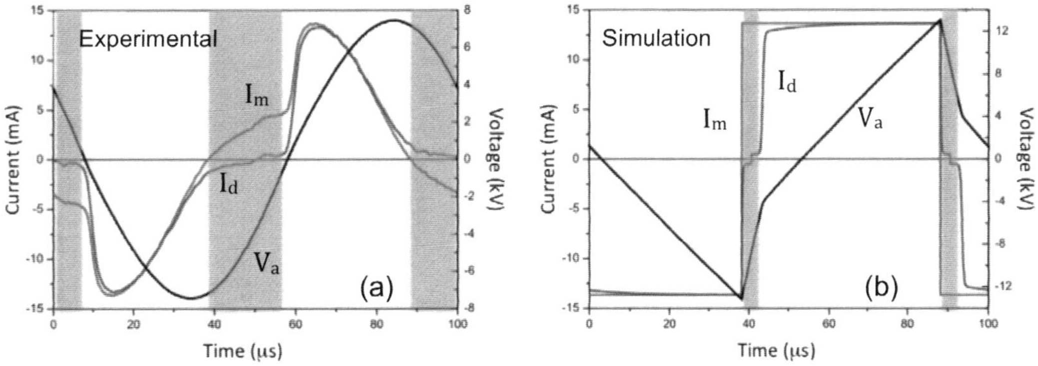

Figure 5. Comparison between sinusoidal voltage source (a) and square-wave current source (b) for the same discharge current amplitude: applied voltage (V_a), measured current (I_m) and discharge current (I_d) (the discharge is off in shaded areas) [9].

Therefore, to increase the discharge power a solution is to control the current and to increase not any more the discharge current amplitude but the discharge duration. This can be realized using a square-wave current inverter source [10]. The Figure 5 presents a comparison between experimental waveforms for a sinusoidal voltage source and a simulation of a square-wave current inverter source for the same discharge current amplitude. The simulation is realized with PSpice software using a macroscopic electrical model of the discharge. This model treats the discharge as a Zener diode in series with a R-C parallel circuit. It takes into account i) the Townsend discharge breakdown, ii) the memory effect from one discharge to the following one and iii) the variation of the secondary emission coefficient during the breakdown [18]. In this case, the use of a square current waveform increases the duration of the discharge ignition from 64 to 90% and the power from 1.6 to 3.8W·cm^{-2}.

4. Comparison between sinusoidal voltage and square-wave current power supplies

The square-wave current power supply is a cascade structure composed of three parts: a tunable regulated DC current source, a current commutator to transform the DC current into an alternative square waveform current and a high voltage transformer. It has already been described in previous publication [10]. The Figure 6 shows experimental electrical waveforms obtained with this current power supply. As can be seen, the measured current is not perfectly rectangular. According to Bonnin *et al.* this phenomena is due to the parasitic elements (mainly the magnetizing inductance and parasitic capacitances) of the transformer [10]. Moreover, oscillations are observed on the current due to the switching of power converter. It is interesting to note that the capacitive nature of the discharge imposes a quasi-triangular applied voltage.

The homogeneity of the discharge is studied by short-exposure-time pictures taken with an intensified CCD camera. The Figure 7 shows a picture of the discharge during the positive half-wave with an exposure time of 100 ns. The light is maximal near the anode and homogeneous over the whole surface of the electrodes. This behavior is similar to what is observed in APTD using a sinusoidal voltage source [8]. As the discharge is homogeneous, it is possible to calculate the discharge current and the gas voltage (Figure 6). This allows to clearly determine the duration of the discharge (time for which the gas voltage is higher than the breakdown voltage, *i.e.* 3kV). Thus, it is interesting to note that differently from the conventional sinusoidal voltage case (Figure 4), the discharge is switched on during 85% of the period.

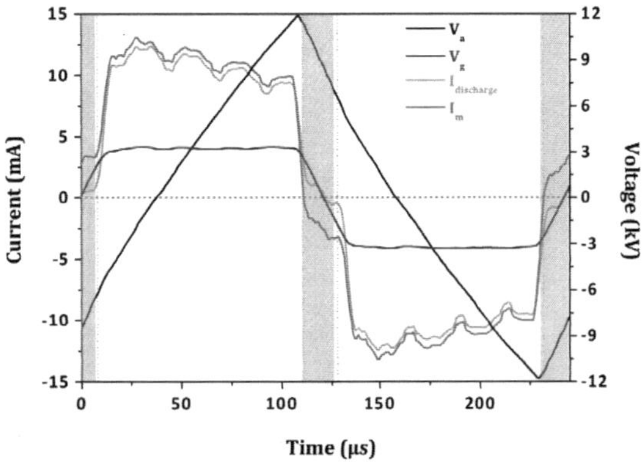

Figure 6. Oscillogram of the applied voltage V_a, the calculated gas voltage V_g, the measured current I_m and the calculated discharge current (the discharge is off in shaded areas).

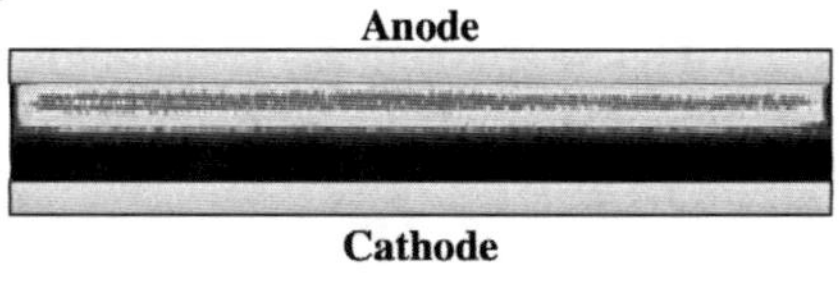

Figure 7. 100 nanosecond exposure time photograph of the gas gap using square-wave current inverter source.

The Figure 8 shows the evolution of the maximal power density according to the excitation frequency for the two types of power supply. The maximal power density is calculated at the frontier between the homogeneous and the filamentary discharge when it exists or for the power supply limit. Anyway, it corresponds to the upper limit of the working domain shown on Figure 3 in the sinusoidal voltage case. At low frequency, the maximal insulation voltage of the cell discharge (24 kV_{pk-pk}) induces a limitation of the power density. This explains the linear growth of the power density for frequencies lower than 6kHz (sinusoidal voltage power supply) and 12kHz (square-wave current power supply).

As it can be clearly seen, the square-wave current power supply allows to increase the maximal power density by 79% up to 9.3W/cm^2. It is interesting to note that in case of the square-wave current power supply, the working domain *i.e.* the power density is always limited by the maximal insulation voltage (for frequencies lower than 12kHz) or the maximal current delivered by the DC current source (for frequencies higher than 12kHz). Hence, this indicates that it will be possible in a near future to reach values even beyond 10W/cm^2. In any case, at any given frequency, the power density is greater with the square-wave current power supply.

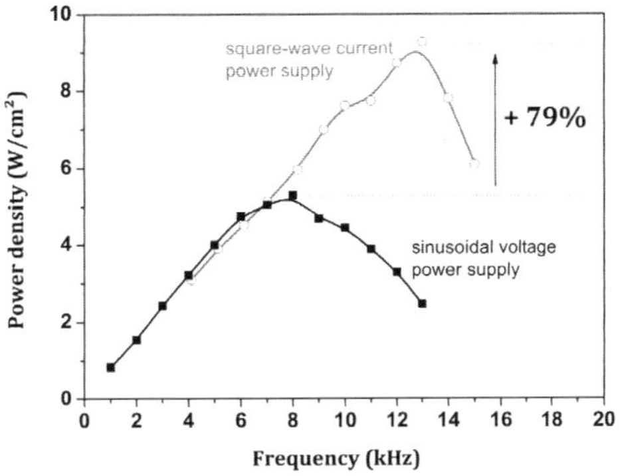

Figure 8. Influence of the power supply type (current or voltage) on the maximal discharge power according to the frequency.

5. Influence of the square-wave current inverter source on the stability of the APTD

This section discusses the influence the use of the current source could have on the stability of the discharge in the presence of additives. As discussed in the introduction, the APTD remains stable as long as the O_2 concentration is less than a few hundred of ppms. This effect is due to the quenching of $N_2(A^3\Sigma_u^+)$ metastables state. Indeed, Massines *et al.* [8] found out that these long lifetime species are responsible for the memory effect from one discharge to the next, which is essential to obtain the Townsend regime at atmospheric pressure. The electrical signature of the memory effect is visible on the discharge current [15]. Indeed, this current is never equal to zero when the discharge is off (Figure 6) which is attributed to the continuous emission of electrons from the cathode induced by the metastable $N_2(A^3\Sigma_u^+)$ flux [19].

As seen in the above discussion, the square-wave current power supply can increase the discharge ignition time and therefore reduce the time between two consecutive discharges (*off-time*). The Figure 9 presents the evolution of the *off-time* according to the excitation frequency respectively in the case of the current and voltage power supplies. Whatever the frequency, the *off-time* using the current power supply is at least two times smaller than the one using the voltage power supply. A reduction of this *off-time* implies a shorter time for the quenching of $N_2(A^3\Sigma_u^+)$ metastables state in between two discharges. Therefore the memory effect from one discharge to the next must be higher, which is confirmed by looking to the discharge current amplitude before the breakdown (Figure 10): as it can be seen, for a given discharge power this current is higher with the square-wave current power supply than with the sinusoidal voltage power supply.

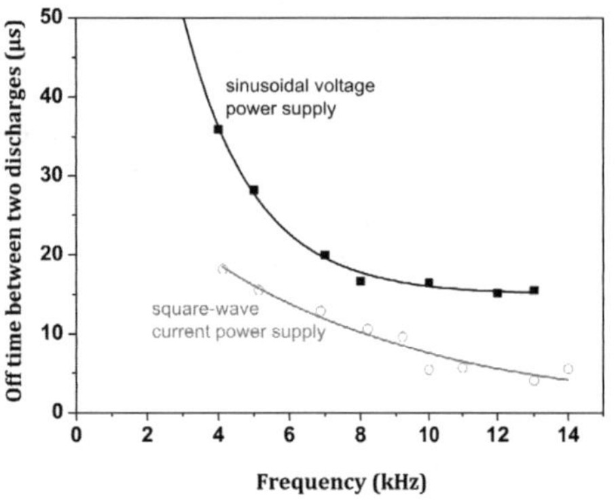

Figure 9. Evolution of the *off-time* between two discharges in the operating range limit for the two types of power supply.

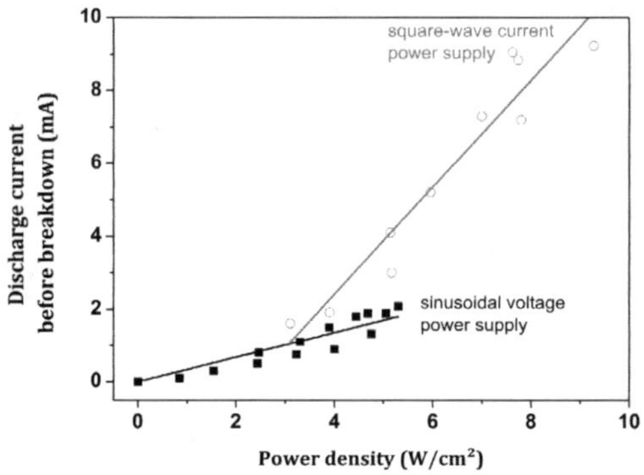

Figure 10. Influence of the power supply type on the electrical signature of the memory effect.

Moreover, the Figure 11 shows the evolution of the average power density for both power supplies as a function of the added O_2 rate to N_2. The use of the square-wave current power supply allows keeping a constant power almost equal to 10 W/cm^2 up to 1000 ppm of O_2, and an APTD can be sustained up to 1400 ppm of O_2. On the other side, in the case of the voltage power supply, the average power density hugely depends on the O_2 rate: for 100 ppm of O_2 added to N_2, the maximal power density increases a lot compared to pure N_2, which has been already observed previously [14], and for higher concentrations, the power density decreases linearly down to 3.5 W/cm^2 for the maximum O_2 rate of 800 ppm in our conditions.

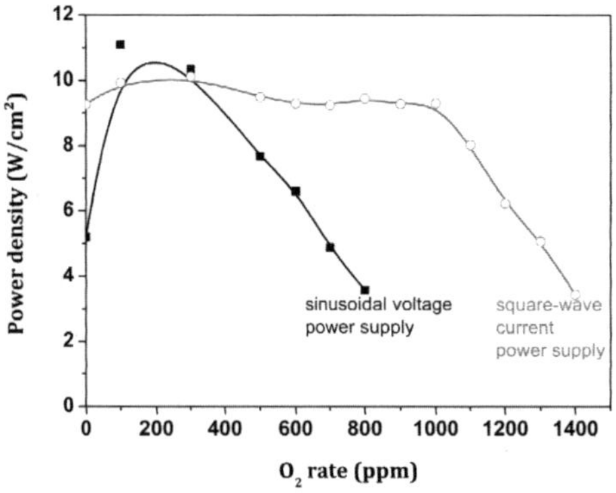

Figure 11. Evolution of the average power density at the limit of homogeneous domain as function of the rate of O_2 added to N_2 for the two types of power supply.

Then, by reducing the *off-time* between two discharges, the square-wave current power supply decreases strongly the sensitivity of the discharge regarding gas mixture. Thus, it is possible with this square-wave current power supply to increase both the discharge power and the additives concentration, which is very promising for thin film deposition processes.

6. Conclusion

In the case of an ATPD, the homogeneous discharge working domain is limited by a maximum density current. Moreover, the gas voltage is nearly constant during the discharge ignition, and so the discharge current is directly proportional to the variations of the applied voltage variations. As a result the use of a voltage source is not an ideal solution to supply the discharge: the APTD is turning off during a long time (nearly 40%), and the maximum current density is reached during a very short time. A square-wave current source appears more adapted: maximum current density could be reached almost all the period and therefore the discharge is turning off during a very short time. Consequently, the power density is higher with a current source: in this paper, the maximal power density is increased by 79 % with a current source while the used power supply is intrinsically limited. This suggests that it is still possible to increase the power to higher values.

Moreover, as the *off-time* is strongly reduced with the square-wave current power supply, the quenching of the $N_2(A^3\Sigma_u^+)$ metastable states in between two discharges is shorter. Then, the memory effect from one discharge to the following one is improved, which is verified in N_2/O_2 mixtures, the discharge staying in the APTD mode for the highest oxygen rates. Hence, for thin film deposition application, the use of a square-wave current power supply should allow to increase both the precursor concentration and discharge power, and thus the deposition rate.

Acknowledgements

The authors are grateful to the Agence Nationale de la Recherche (PREPA Project, Grant No. ANR-05-BLAN-0090) and the French Midi-Pyrénées region (PHPA project) for their supports.

References

[1] Siemens W 1857 *Annalen der Physik* **178** 66-122
[2] Eliasson B and Kogelschatz U 1988 *Appl Phys B-Photo* **46** 299-303
[3] Sobel A 1991 *IEEE Trans. Plasma Sci.* **19** 1032-1047
[4] Wagner H E, Brandenburg R, Kozlov K V, Sonnenfeld A, Michel P, and Behnke J F 2003 *Vacuum* **71** 417-436
[5] Pappas D 2011 *J. Vac. Sci. Technol. A: Vacuum, Surfaces and Films* **29** 020801
[6] Kogelschatz U 2003 *Plasma Chem Plasma Process* **23** 1-46
[7] Okazaki S, Kogoma M, Uehara M, and Kimura Y 1993 *J. Phys. D: Appl. Phys.* **26** 889-892
[8] Massines F, Gherardi N, Naudé N, and Ségur P 2009 *Eur. Phys. J. Appl. Phys.* **47**
[9] Massines F, Sarra-Bournet C, Fanelli F, Naudé N, and Gherardi N 2012 *Plasma Process. Polym.* 1041-1073
[10] Bonnin X, Piquet H, Naudé N, Bouzidi M C, Gherardi N, and Blaquiere J M 2013 *Eur. Phys. J. Appl. Phys.* **64**
[11] Massines F, Gherardi N, Fornelli A, and Martin S 2005 *Surf. Coat. Tech.* **200** 1855-1861
[12] Brandenburg R, Maiorov V A, Golubovskii Y B, Wagner H E, Behnke J, and Behnke J F 2005 *J. Phys. D: Appl. Phys.* **38** 2187-2197
[13] Kozlov K V, Brandenburg R, Wagner H E, Morozov A M, and Michel P 2005 *J. Phys. D: Appl. Phys.* **38** 518-529
[14] Gherardi N, Naudé N, Es-sebbar E, Enache I, Caquineau H, and Massines F, 2006 Transition from atmospheric pressure Townsend discharge (APTD) to filamentary discharge in N2/O-containing mixtures: *Proc. International Symposium on High Pressure Low Temperature Plasma Chemistry (Saga, Japan, 4–8 September*
[15] Naudé N, Bouzidi M C, Dang Van Sung Mussard M, Puechagut L, Belinger A, Ségur P, and Gherardi N, 2013 Atmospheric Pressure Townsend Discharges in nitrogen with small admixtures of oxygen: discussion on the origin of the memory effect: *Proc. International Conference on Phenomena in Ionized Gases (Granada, Spain, July 14-19, 2013*
[16] Paulmier T, Gherardi N, Maechler L, Enache I, Caquineau H, and Massines F, 2006 Silicon based thin film deposited from Hexamethyldisiloxane in an Atmopsheric Pressure Townsend

Journal of Physics: Conference Series **550** (2014) 012044 doi:10.1088/1742-6596/550/1/012044

Discharge: *Proc. International Symposium on High Pressure Low Temperature Plasma Chemistry (Saga, Japan, 4–8 September*
[17] Liu S H and Neiger M 2003 *J. Phys. D: Appl. Phys.* **36** 3144-3150
[18] Naudé N, Cambronne J P, Gherardi N, and Massines F 2005 *J. Phys. D: Appl. Phys.* **38** 530-538
[19] Massines F, Gherardi N, Naudé N, and Segur P 2005 *Plasma Phys. Contr. F.* **47** B577-B588

Journal of Physics: Conference Series **550** (2014) 012045 doi:10.1088/1742-6596/550/1/012045

Determination of N and O-atoms, of N_2(A) and N_2(X, v>13) metastable molecules and N_2^+ ion densities in the afterglows of N_2-H_2, Ar-N_2-H_2 and Ar-N_2-O_2 microwave discharges

H Zerrouki[1], A Ricard[1] and J P Sarrette[1]
[1]Université de Toulouse; CNRS; LAPLACE
118 Route de Narbonne, F-31062 Toulouse cedex 9, France

E-mail: ricard@laplace.univ-tlse.fr

Abstract. Early afterglows of N_2-H_2, Ar-N_2-H_2 and Ar-N_2-O_2 flowing microwave discharges are characterized by optical emission spectroscopy. The N and O atoms, the N_2(A) and N_2 (X, v>13) metastable molecules and N_2^+ ion densities are determined by optical emission spectroscopy after calibration by NO titration for N and O-atoms and measurements of NO and N_2 band intensities. In N_2-(0.05-2.5 %)H_2 and Ar-(1-50%)N_2-(0.05-2.5%)H_2 gas mixtures, the O-atoms are coming from O_2 impurities in the discharge. Concentrations of these active species are compared to the ones obtained in Ar-(5-50%)N_2-(0.2-2.5%)O_2 gas mixtures in which a controlled amount of O_2 is added.

1. Introduction

Afterglows of N_2 flowing microwave discharges have been studied at medium gas pressures (1-20 Torr) for sterilization of medical instruments by N-atoms [1, 2]. The mentioned project of sterilization in N_2 afterglow is based on N-atom etching of bacteria without oxidation by O-atoms. In the present study it is detected the O-atoms from air impurity to appreciate their influence on the sterilization process. Also H_2 was introduced into N_2 to expect a decrease of O-atom impurity in the N_2 afterglow. Indeed previous results [3, 4] have reported that the introduction of a small H_2 quantity, less than 1% into N_2 and Ar-N_2 microwave discharge produced a marked increase of N and H atom densities.

Also studies of electron and vibrational kinetics in N_2-H_2 discharges [5-7], have concluded to an increase of the electron energy and thus of the electron excitation and dissociation of N_2 and H_2 as a few H_2 was introduced into N_2. In [4], it was demonstrated the interest to dilute the N_2-H_2 gas mixture into Ar to increase the plasma ionization. Consequently, it was observed an increase of N/N_2 and H/H_2 dissociation degrees. To enhance the sterilization efficiency, Ar-N_2 gas mixtures are presently studied to increase the N-atom density and the N/N_2 dissociation degree.

The early flowing afterglows produced from N_2-H_2, Ar-N_2–H_2 and Ar-N_2–O_2 microwave plasmas are presently studied by emission spectroscopy with the same experimental methods as in N_2-H_2 RF afterglow [8,9] and in N_2–O_2 microwave early afterglow [10]. The present paper extends to other active species: N_2(X, v>13) metastable molecules and N_2^+ ions the results previously reported [10] for N, N_2(A) metastable molecules and O-atoms in impurity in N_2-H_2 and in Ar-N_2-H_2, Ar-N_2-O_2 early microwave afterglows.

The intensities emitted by the N_2 first positive (1st pos.) and N_2 second positive (2nd pos.) systems and by the NO_β bands are measured to obtain the mentioned active specie densities after NO titration to calibrate the N and O-atom densities [8-10]. The O-atoms in N_2-H_2 and Ar-N_2-H_2 are coming from O_2 impurities in the discharge. Finally, H_2 is replaced by O_2 to compare the production of the active species in the two Ar-N_2-H_2 and Ar-N_2–O_2 ternary gas mixtures.

2. Experimental setup and NO titration

The experimental setup is identical to the one used in [10]. In the plasma post-discharge reactor of Fig. 1, N_2-H_2, Ar-N_2-H_2 and Ar-N_2-O_2 microwave plasmas at 2450 MHz are produced by a surfatron cavity in a quartz tube of dia. 6 mm. The dia. 6 mm discharge tube is connected to a dia. 21 mm bent tube before the entrance in the 5 litre reactor. It is chosen plasma conditions of 8 Torr, 1 slm and 150 W during afterglow intensity measurements and 4 Torr, 0.5 slm, 50-100 W during NO titration to limit the NO flow in the post-discharge.

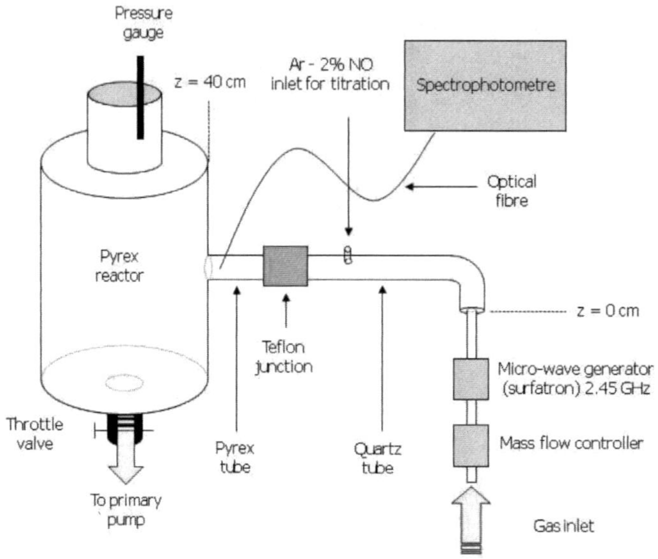

Figure 1. *Microwave discharge and post-discharge reactor of 5 litres.*

The plasma is located inside the dia.6 mm tube with a length after the surfatron gap varying from 2 cm in N_2-H_2 to 5 cm with Ar dilution. The afterglows take place in the dia. 21 mm tube at a distance of 15 cm from the surfatron exit, starting by an early afterglow at the beginning of the dia. 21 mm tube (z=0) and ending by a late afterglow before and inside the 5 litre reactor at z=40 cm.

As shown in Fig.1, an Ar-2%NO gas mixture can be introduced in the dia. 21 mm tube for titration of N and O-atom density before the 5 litre reactor. The optical emission spectroscopy is performed by means of an optical fibre connected to an Acton Spectra Pro 2500i spectrometer (600lines/mm gratings) equipped with a Pixis 256E CCD detector (front illuminated 1024 x 256 pixels).

NO titration

As described in [8-10] for NO titration of N-atoms , it is observed the change of color in the late afterglow from the violet, due to NO_β emission at low NO flow rate, to the green due to NO_2^* emission at high NO flow rate. Between the violet and green emissions, there is a complete extinction of the afterglow emission corresponding to the equality between the N-atom flow rate *Q(N)* and the introduced NO flow rate *Q(NO)*. Then the N-atom density is obtained from the following equation:

$$Q(NO)/Q(N_2) = [N]/[N_2] \tag{1}$$

Journal of Physics: Conference Series **550** (2014) 012045 doi:10.1088/1742-6596/550/1/012045

Where $Q(N_2)$ is the afterglow N_2 flow rate which keeps a constant value as $Q(N) << Q(N_2)$. After the afterglow extinction, the I_{NO2*} intensity is varying with the NO density as follows:

$$I_{NO2*} = r \times [NO] \quad (2)$$

The slope r of the I_{NO2*} variation versus NO is proportional to the O-atom density. In pure N_2 afterglow, the O-atoms replaced the N-atoms by reaction R1 in table 1.

Table 1. Kinetic reactions in N_2 afterglow.

Reactions	**Reaction rate**
$N+NO \rightarrow N_2+O$	R1
$N + N + N_2 \rightarrow N_2(B,v') + N_2$	R2
$N + N + (Ar\text{-}N_2) \rightarrow N_2(B,v') + (Ar\text{-}N_2)$	R2'
$N_2(B,v') \rightarrow N_2(A,v'') + h\upsilon_{580}$	
$N + O + N_2 \rightarrow NO(B,0) + N_2$	R3
$NO(B,0) \rightarrow NO(X,8) + h\upsilon(320\ nm)$	
$N_2(A) + N_2(A) \rightarrow N_2(C,1) + N_2(X)$	R4
$N_2(C,1) \rightarrow N_2(B,0) + h\upsilon\ (316\ nm)$	
$N_2(A) + N_2(A) \rightarrow N_2(B,11) + N_2$	R5
$N_2(A)+N_2(X,v>13) \rightarrow N_2(B,11) + N_2$	R6
$N_2(a')+ N_2(a') \rightarrow e + N_2^+(X) + N_2(X)$	R7
$N_2^+(X) + N_2(X,v>12) \rightarrow N_2^+(B) + N_2(X)$	R8
$N_2^+(B,0) \rightarrow N_2^+(X,0) + hv(391nm)$	

At the extinction point the O-atom density is equal to the N-atom density $[N]_1$ initially present in the afterglow. Then $r= r_1$ in eq.2. In N_2-xO_2 mixtures (with x<20%), it is obtained at a new extinction point with a density $[N]_2$ and a slope r_2 in eq.2 which is related to r_1 by the relation:

$$r_2/r_1 = ([N]_2+ [O])/[N]_1 \quad (3)$$

Where the O-atom density is coming from O_2 dissociated in the discharge. Note that it is added $[N]_2$ to [O] in eq.3 to take into account the O-atom produced from the initial N-atoms by R2. In such a titration method by NO, it is assumed that the N and O atoms are the dominant active species which could react with NO in the afterglow.

3. The N_2-H_2 early afterglow

3.1. N-atom density in the N_2-H_2 early afterglow

As reported in [8-10], the pure late afterglow emission is produced by reaction R2. The N_2(580 nm) band head intensity (I^m_{580}) in arbitrary unit (a.u) was measured for constant parameters of the Acton spectrometer (600 lines/mm gratings, slit of 150 µm, integrating time 1s). I^m_{580} is then deduced from reaction R2 with v'=11 and $h\upsilon= hc/\lambda$ (580nm), as follows:

$$I^{m}_{580} = k_4 \times [N]^2 \quad (4)$$

with k_4 explicited in [10]. The rate coefficients in k_4 are reported in the litterature at room gas temperature. After analyzing the rotational structure of the N_2, 775 nm band [10], it has been verified that the gas temperature was T=300K in the present N_2 afterglows. In the observed afterglow region, collisional processes in the pink afterglow [11] also excite the $N_2(B)$ states, in addition to reaction R1.The fraction a_{N+N} of the measured intensity I^{m}_{580} in equation (4) has been calculated in[10]:

$$a_{N+N} \times I^{m}_{580} = k_4 \times [N]^2 \quad (5)$$

It is obtained $a_{N+N} = 0.25$ for pure N_2 and $a_{N+N} = 0.6$ for the N_2-0.5% H_2 mixture at z=3 cm. This result indicates that the early afterglow in N_2-0.5% H_2 is dominated by the N+N recombination as expressed by R1. The full late afterglow conditions are reached closer from the discharge when a few quantity of H_2 is introduced into N_2. In eq.4 and 5, k_4 is independent of N_2 gas pressure for p>5Torr [10].
The k_4 value was determined by NO titration in the pure N_2 afterglow at 4.5 Torr, 0.5 slm and 100 Watt , in the dia. 21 mm tube before the connection with the 5 liter reactor (see Fig.1), at a distance z= 40 cm where pure late afterglow conditions were found ($a_{N+N} = 1$). The N-atom density was found to be $[N]=10^{15}$ cm^{-3}, allowing to calibrate the I^{m}_{580} band head intensity.
The N-atom density is then obtained at z= 3 cm with an uncertainty estimated to be 30%. It is obtained a mean density value [N]=1-2 10^{15} cm^{-3} from N_2 to N_2- 2.5%H_2 at 8 Torr, 1 slm, 150 W. There is a small maximum for 0.5- 1% H_2, in agreement with previous results in [4]. It is reported in table 2 the two values of N- atom densities in pure N_2 and in N_2- 2.5%H_2. By comparison, the results of [9] in RF afterglow are also reported in table 2 for pure N_2 and N_2- 5 $10^{-4}H_2$, showing N-atom densities in the same order of magnitude.

Table 2. Density (in cm^{-3}) of active species in early afterglows at z= 3cm (10^{-3}s), 8 Torr, 1 slm, 150W of M_2 discharges with $M_2 = N_2$-xH_2, Ar-x(N_2-5%H_2) and Ar-x(N_2-5%O_2). Present results in HF 2450 MHz plasma. Comparison with previous results [9] in RF 13.56 MHz plasma.

H_2/M_2	a_{N+N}	N (10^{15})	O (10^{14})	A (10^{11})	X,v>13 (10^{14})	N_2^+ (10^{10})	
0	0.2	1	0.7	5-3	3	5	**HF $M_2=N_2$-xH_2**
2.5 10^{-2}	0.6	2	0.3	2-1	4	0.05	
10^{-3}	0.3	1	0.9	10-35	0.1		**HF M_2=Ar-x(N_2-5%H_2)**
2.5 10^{-2}	0.7	2	0.2	6-22	0.1		
3 10^{-3}	0.1	0.4	2	1-3	0.6		**HF M_2=Ar-x(N_2-5%O_2)**
2.5 10^{-2}	0.25	0.2	7				
0	0.1	3	ε	30	6		**RF [9] $M_2=N_2$-xH_2**
5 10^{-4}	0.6	1	ε	20	0.1		

Journal of Physics: Conference Series **550** (2014) 012045 doi:10.1088/1742-6596/550/1/012045

3.2. Density of O-atoms in impurity in the N_2-H_2 early afterglow

The NO_β bands are presently observed as a result of the recombination of N and O atoms by reaction R2. In a similar way than for equation (4), the NO (320 nm) measured band intensity ($I^m{}_{320}$) is deduced from reaction R3 as follows:

$$I^m{}_{320} = k_6 \times [N][O] \qquad (6)$$

The coefficients in k_6 are explicited in ref.10 as for k_4 .The O atom density can be deduced from the N-atom density by considering the $a_{N+N} \times I^m{}_{580}/I^m{}_{320}$ ratio of reactions 5 and 6, as follows:

$$a_{N+N} \times I^m{}_{580}/I^m{}_{320} = k_7 \times [N]/[O] \qquad (7)$$

With $k_7= k_4/k_6$. The k_7 coefficient has been determined by measuring both the N and O-atom densities by NO titration as indicated in part 2 with eq.3. For that it is chosen a pure N_2 discharge at 4.5Torr, 0.5slm and 100 W. At z=40 cm, it was obtained $[N]_1$= 1.1 $10^{15}cm^{-3}$.Then , with a N_2-2%O_2 gas mixture at same plasma parameters, it was measured $[N]_2$= 0.6 $10^{15}cm^{-3}$. The measured r_2/r_1 ratio was 0.9 and it is deduced from equation (3): [O]=0.4 $10^{15}cm^{-3}$.Finally, k_7 of eq.7 was determined by measuring the $a_{N+N}.I^m{}_{580}/I^m{}_{320}$ ratio in these experimental conditions (N_2-2%O_2 at 4.5Torr, 0.5slm and 100 W).

After several NO titration experiments, it was found k_7= 1(+-0.4) for a constant afterglow volume as detected by the optical fibre in the 21 mm dia. tube. Once k_7 determined, the O-atom density in the N_2-H_2 early afterglow, at 8Torr, 1slm, 150 Watt and z=3 cm, was determined by eq.7 after measurements of $a_{N+N}I^m{}_{580}/I^m{}_{320}$ and [N] versus the H_2 percent into N_2. If the uncertainty on N-atom density is estimated to be 30% (see section 3.1), the experimental errors on O-atom density calculated from equation (7), with the uncertainty on k_7 of 40% is 90% that is near the order of magnitude.

As shown in table 2, there is a decrease of the O-atom density from 7 to 3 $10^{13}cm^{-3}$ when 2.5%H_2 is added to N_2. Such a decrease of O-atom density is attributed to reactions between the oxygen impurity and the H atoms produced in the N_2-H_2 plasma. The NO_β bands were not observed in [9], indicating a more clean experimental setup.

3.3. Density of $N_2(A)$ metastable molecules in the N_2-H_2 early afterglow

It has been detected the $N_2(C,1 \rightarrow B,0)$ emission at 316 nm near the NO_β emission at 320 nm which is used as in [8-10] to determine the density of the $N_2(A)$ metastable molecule. It is considered that the N_2 2nd positive system in the early afterglow is produced by reaction R4. The N_2(316 nm) measured intensity ($I^m{}_{316}$) is then given by:

$$I^m{}_{316} = k_8 \times [N_2(A)]^2 \qquad (8)$$

with k_8 explicited in [10]. From equations (6) and (8), it comes from the following $I^m{}_{320}/I^m{}_{316}$ intensity ratio:

$$I^m{}_{320}/I^m{}_{316} = k_9 \times [N][O]/[N_2(A)]^2 \qquad (9)$$

with $k_9 = k_6/k_8$. The $N_2(A)$ density is then obtained from equation (9) with the N and O atom densities previously determined. The $N_2(A)$ density could also be deduced by comparing the a_{N+N} I(580 nm) and I(316 nm) intensities [9] where I(580 nm) and I(316 nm) are given by equations (5) and (8). It comes:

$$a_{N+N} \times I^m{}_{580}/I^m{}_{316} = k_{10} \times ([N]/[N_2(A)])^2 \qquad (10)$$

with $k_{10}= k_4/ k_8$. The two $N_2(A)$ densities obtained by eq. 9 and 10 are reported in table 2. The $N_2(A)$ density decreased by a factor 2 as a result of the quenching of $N_2(A)$ by H. It is estimated that it is obtained the order of magnitude of $N_2(A)$ density (in the range 10^{11}-10^{12} cm^{-3}).

3.4. Density of $N_2(X, v>13)$ molecules

The production of $N_2(B,11)$ by R1 in the early afterglow is less than 1 ($a_{N+N}<1$). For the other part ($1-a_{N+N}$), it is considered the reactions R5 and R6 whose rate coefficients are reported in [9].The contribution of reactions R5 and R6 on I^m_{580} is then written as follows:

$$(1-a_{N+N})\times I^m_{580}= k_{R5}\times [N_2(A)]^2 + k_{R6}\times[N_2(A)][N_2(X, v>13)] \quad (11)$$

Where k_{R5}, k_{R6} are the rate coefficients of reactions R5 and R6. As $a_{N+N}I^m_{580}= k_4[N]^2$, it is deduced:

$$(a_{N+N}/(1-a_{N+N}))\times(k_{R5}\times[N_2(A)]^2 + k_{R6}\times[N_2(A)][N_2(X,v>13)]) = [N]^2\times k_4 \quad (12)$$

With the experimental values of a_{N+N} and of N and $N_2(A)$ densities, it is found that $(a_{N+N}/(1-a_{N+N}))\times k_{R5}[A]^2$ is about 2 orders of magnitude lower than $[N]^2\times k_4$. It results that eq. 12 can be simplified as:

$$(a_{N+N}/(1-a_{N+N}))\times k_{R6}[N_2(A)][N_2(X,v>13)] = [N]^2\times k_4 \quad (13)$$

From the obtained values of N-atom and $N_2(A)$ density, it was deduced as reported in table 2: $[N_2(X, v>13)]$ = 3-4 $10^{14}cm^{-3}$ for H_2/N_2 up to 5 10^{-2}. Such a value can be considered as an estimated value depending on the k_6 rate coefficient. In [9], the same order of magnitude is found in pure N_2 but with a marked decrease up to 10^{13} cm^{-3} with 5×10^{-4} H_2. This value in pure N_2 is well related to the calculated value [12] of $[N_2(X, v>13)]$ at N_2 gas pressure of 3Torr, afterglow time of 10^{-3}s where $[N_2(X,v>13)]/[N_2] = 5\times10^{-3}$, and thus: $[N_2(X,v>13)] \sim 5\times10^{14}$ cm^{-3}.

3.5. Density of N_2^+ ions (new results)

The emission of the N_2^+ band at 391 nm is observed in the present early afterglows. It is generally proposed [12] that the N_2^+ 391 nm band is produced in the pink afterglow by reactions R7and R8. The I^m_{391} intensity is then expressed as follows:

$$I^m_{391} = k_{14}\times[N_2^+]\times[N_2(X, v>12)] \quad (14)$$

with $k_{14} = c_{391}.V.A_{391}.k_{R8}/(\upsilon^R_{N2+} + [N_2].k^Q_{N2+B,0})$ where c_{391} is the spectral response of spectrometer, V is the detected afterglow volume, A_{391} the Einstein coefficient of the N_2^+(391 nm) transition, k_{R8} the rate coefficient of reaction R8 with $k_{R8} = 4\times10^{-11}cm^3s^{-1}$ [13], $\upsilon^R_{N2+} = 1.7\times10^7$ s^{-1} and $k^Q_{N2+B,0}=8.8\times10^{-10}$ $cm^3.s^{-1}$. By comparing the intensities of I^m_{316} from equation (8) and I^m_{391} from equation (14), it is calculated:

$$I_{391}/I_{316} = 4\times10^{-2}\times([N_2^+].[N_2(X,v>12)]/[N_2(A)]^2) \quad (15)$$

By assuming the equality $[N_2(X, v>12)] = [N_2(X, v>13)]$, it is found a N_2^+ density of about $5\times10^{10}cm^{-3}$ in pure N_2 and $5\times10^8cm^{-3}$ with H_2 = 2.5%. Compared to published data [15-16], the value of N_2^+ density in pure N_2 appears to be one order of magnitude higher.

4. The Ar-N_2-H_2 early afterglow

It is mentioned in the introduction the interest of N_2-H_2 dilution into Ar to increase the electron energy in the plasma at constant values of transmitted power by the surfatron and of gas pressure. As

Journal of Physics: Conference Series **550** (2014) 012045 doi:10.1088/1742-6596/550/1/012045

mentioned in [4], superelastic collisions of electrons on the Ar metastable atoms produced in the plasma could enhance the electron energy. As a consequence, the plasma length increased and a jet afterglow of same diameter than the discharge tube was observed inside the tube of 21 mm dia. It is mentioned here that in the present measurements of flowing afterglow the Ar metastable atoms have disappeared after collisions on the tube wall (destruction probability of about 1). As a consequence, the excitation transfers of Ar metastable atoms on N_2 and H_2 can be discarded at a distance of about 1 cm after the discharge end. Another interest of Argon dilution is to maintain the plasma at high gas pressure, up to the atmospheric gas pressure while keeping a low plasma power as demonstrated in [17].

For convenience in this study with Ar-N_2-H_2, a constant N_2-5%H_2 mixture was diluted into Ar. Early afterglows of Ar-x(N_2-5%H_2) with x =1-50% corresponding to H_2 percentage varying between 0.05 and 2.5% were studied in the same conditions (8 Torr, 1 slm, 150 W and z=3 cm) than used in the N_2-H_2 study.

The effect of Ar dilution into N_2 on a_{N+N} in equation (5) has been considered. It is considered reaction R2' produced with an excess of Ar atom which results in a change of the N_2(B,v') distribution as compared to pure N_2 at a given a_{N+N} value [18]. The N+N recombination coefficient a_{N+N} has been calculated in conditions of pink and late afterglows for Ar-xN_2 gas mixture with x= 2 , 5, 10, 20, 50 and 100%.

The pink afterglow was observed in the Ar-N_2 early afterglow at distances z =1-3 cm where it is observed a strong emission of the N_2^+,391.4 nm band.The rate coefficients in k_4 of eq. 4 and 5 have to be considered in the Ar-N_2 gas mixture. By NO titration as in 3-1, it was obtained $[N]_1$= 6 10^{14}cm^{-3} in a 50%Ar-50%N_2 discharge at 4 Torr, 0.5 slm, 50 W and z=40 cm. It is verified the same k_4 value inside the error bars as for pure N_2. As shown in table 2, it is found a N-atom density of 1-2 10^{15}cm^{-3} as for the N_2-H_2 early afterglow with always a small maximum value at 1% H_2.

The O–atoms and N_2(A) metastable density were obtained from (eq. 7) and (eq. 9 and 10), once determined a_{N+N} and k_8 from[N] and [O] measurements by NO titration. For that a 50%Ar-50%N_2 gas mixture was experimented at z = 40 cm, 4 Torr, 0.5 slm and 50 Watt. It was found $[N]_1$ = 6 10^{14} cm^{-3}. Then with a 50%Ar-50%N_2-0.4%O_2 gas mixture, it was obtained $[N]_2$= 4 10^{14} cm^{-3} and [O]= 1.7 10^{14}cm^{-3}. k_7 was determined from $a_{N+N}\times I^m{}_{580}/I^m{}_{320}$, [N] and [O] in these experimental conditions. It was found k_7 = 1 as for the N_2-H_2 early afterglow (3-2). It is observed a sensitive decrease of O-atom density more pronounced in the Ar-N_2-H_2 (from 9 to 2 10^{13} cm^{-3}up to 2.5%H_2) than in N_2-H_2 gas mixtures (see table 2). The N_2(A) density was higher in Ar-N_2-H_2 in the range 1-4$\times10^{12}$ cm^{-3} (table 2). By using eq.13, after measurements of a_{N+N} and of density of N and N_2(A), it is obtained a density [N_2(X, v>13)] = 10^{13}cm^{-3} in Ar-x(N_2-5%H_2), more than one order of magnitude lower than in N_2-H_2.

5. The Ar-N_2-O_2 early afterglow

H_2 has been replaced by O_2 in the same experimental conditions of 8 Torr, 1 slm and 150 W. In a similar way than for the Ar-N_2-H_2 gas mixture, it was studied in the 21 mm dia. tube at z=3 cm (see Fig.1) the early afterglows of Ar-x(N_2-5% O_2) with x =1-50% and thus a O_2 percentage varying between 0.05 and 2.5%. As it concerns the N_2 1st positive band emission, the factor a_{N+N} of N+N recombination on N_2(B,11) was found to be a_{N+N}=0 for x(N_2-5% O_2) <5%. In this condition, the early afterglow is a jet of pure pink afterglow. For x(N_2-5% O_2) >5%, a_{N+N} increased from 0.1 at x= 6% to 0.5 at x=10-20% and decreased again to 0.25 at x= 50%.

The N -atom density as determined by eq. 4 and 5 is flat between 5 and 2 10^{14} cm^{-3} between 0.3 and 2.5%O_2 (table 2) in the Ar–x(N_2-5%O_2) gas mixture, at values lower than in Ar-N_2-H_2 for the same discharge conditions and x values. The O-atom density is very high, increasing from 2 to 7 10^{14} cm^{-3} between 0.3 and 2.5% O_2. The corresponding [O]/2[O_2] dissociation degree decreases from about 1 at 0.3%O_2 to 0.5 at 2.5% O_2.

The N_2 316 nm intensity can only be measured for x(N_2-5% O_2)<6%. But as a_{N+N}=0 for x(N_2-5% O_2)<5%, the N_2(A) metastable density can only be determined with eq. 10 and 11 for x(N_2-5% O_2)=6%, that is 0.3%O_2. The obtained N_2(A) concentrations are 1.4 and 3 10^{11}cm^{-3} from eq. 10 and 11, lower than for the Ar-N_2 and Ar-x(N_2-5% H_2) gas mixtures (see table 2). As for N_2(X, v>13), it is found a density of 6 10^{13}cm^{-3} at 0.3% O_2, higher than in Ar-x(N_2-5%H_2). It is observed a jet of pure pink afterglow (a_{N+N}=0) in Ar-xM_2 gas mixtures with x decreasing from 20% for M_2= N_2 to 5% for M_2= N_2-5%O_2 and to 2% for M_2= N_2-5%H_2 .The pink afterglow appeared to decrease in front of the late afterglow (characterized by the N+N recombination) when going from Ar-N_2, to Ar-N_2-O_2 and to Ar-N_2-H_2 in this order. This decrease of pink afterglow when O_2 and H_2 are successively added to Ar-N_2 could come from increased quenching of N_2(A) by O: 2 10^{-11}cm^{3}s^{-1} and by H: 2.1 10^{-10}cm^{3}s^{-1}[19].

6. Conclusion

Densities of N and O atoms (coming from O_2 impurities for N_2-H_2, Ar-N_2-H_2 gas mixtures), N_2(A) and N_2(X, v>13) metastable molecules and N_2^+ ions have been determined in N_2-H_2, Ar-N_2-H_2 and Ar-N_2-O_2 early afterglows of flowing microwave discharges at 1 slm, 8 Torr and 150 W, after NO calibration. New results are obtained for N_2^+ ions in N_2-H_2.

The density of these active species are obtained by comparing the NO_β (320 nm), N_2 (580 nm), N_2(316nm) and N_2^+(391nm) band intensities and by writing the appropriated kinetic equations.

The densities obtained by these line – ratio measurements are with an uncertainty of 30% for N-atoms and the order of magnitude for O-atoms and N_2(A) metastable molecules. Estimated densities values are obtained for the N_2(X, v>13) metastable and N_2^+ ions which are depending on the kinetics reactions. More works are necessary in this field. The fraction a_{N+N} of the 1st positive emission at 580 nm coming from the N+N recombination was determined in N_2-x<5%H_2 and in Ar-x(N_2-5%H_2), Ar-x(N_2-5%O_2) with x=2-100% afterglows. By comparing the N_2-H_2, Ar-N_2-H_2 and Ar-N_2-O_2 early afterglows, it is obtained the following results:

i) The afterglow jets are pure pink afterglows with Ar-x<20%N_2, Ar-x<2%(N_2-5%H_2) and Ar x<5%(N_2-5%O_2). The decrease of x values is the result of N_2(A) quenching by H and O-atoms, more efficient with H than with O.

ii) The N-atom density is in the range of 10^{15}cm^{-3} in N_2-H_2, Ar-N_2 and Ar-N_2-H_2. The N- atom density is lower by a factor 4 in Ar-N_2-O_2 gas mixtures.

iii) The O-atom density coming from O_2 impurity in the N_2-xH_2 discharges decreases with x from 7 10^{13} cm^{-3} for x=0 to 3 10^{13} cm^{-3} for x= 2.5%. Such a decrease is a little more important in Ar-N_2-H_2 than in N_2-H_2.

iv) A strong O_2 dissociation is measured in the Ar-x(N_2-5%O_2) gas mixtures, decreasing from a full dissociation [O] = 2[O_2] at x<20% to an half dissociation at x=50%. It can thus be estimated that the O_2 impurity in N_2-H_2, Ar-N_2 and Ar-N_2-H_2 gas mixtures is fully dissociated as for the Ar-<20%(N_2-5%O_2) gas mixtures.

v) The N_2(A) metastable density is between 10^{11}-10^{12} cm^{-3} in the studied gas mixtures. It also decreases with H_2 addition and is higher with Ar dilution in N_2-H_2 gas mixtures.

vi) The N_2^+ density strongly decreased by addition of H_2 to N_2.

It is found two main interests of N_2-H_2 dilution into Ar. The first one is to decrease the density of O-atom impurities by a factor of about 5 from 0.1% to 2.5%H_2. The second one is to increase the N/N_2 dissociation from 1% in the Ar-50%N_2-2.5%H_2 to about 20% in the Ar-2%N_2-0.1%H_2 which could be of interest for surface reactions of N-atoms with less N_2 molecules.

References

[1] Villeger S, Sarrette J P and Ricard A 2005 *Plasma Process and Polym.* **2** 709
[2] Villeger S, Sarrette J P, Rouffet B, Cousty S and Ricard A 2008 *Eur. Phys. J. Appl. Phys.* **42**, 25
[3] Malvos H, Michel H and Ricard A 1994 *J.Phys D: Appl.Phys.***27** 1328
[4] Bockel S, Amorim J, Baravian G and Stratil P 1996 *Plasmas Sources Sc Tech.* **5** 567
[5] Loureiro J and Ricard A 1993 *J. Phys D: Appl.Phys.***26** 163
[6] Garscadden A and Nagpal R 1995 *Plasmas Sources Sc Tech.* **4** 268
[7] Tatarova E, Dias F M, Gordiets B and Ferreira C M 2005 *Plasmas Sources Sc Tech.* **14** 19.
[8] Ricard A, Oh S G and Guerra V 2013 *Plasma Sources Sci Technol* **22** 035009

[9] Ricard A and Oh S G 2014 *Plasmas Sources Sc. Tech.* (accepted for publication)
[10] Zerrouki H, Ricard A and Sarrette J P 2013 *Contr. Plasmas Phys.* **53** 599 and 2014 (accepted for publication)
[11] Levaton J and Amorim J 2012 *Chem.Phys.***397** 9
[12] Sa P A, Guerra V, Loureiro J and Sadeghi N 2004 *J. Phys D: Appl. Phys* **37** 22
[13] Piper L G 1994 *J. Chem Phys* **101** 10229
[14] Kang N, Lee M , Ricard A and Oh S G 2012 *Curr. Appl. Phys.***12** 1448
[15] Sadeghi N,Foissac C and Supiot P 2001 *J.Phys.D: Appl. Phys* **34** 1779
[16] Ferreira J A, Stafford L, Leonelli R and Ricard A 2014 *J.Appl. Phys.***115**1 633036
[17] Ricard A, Gaboriau F and Canal C 2008 *Surf.CoatingTech.* **202** 5220
[18] Ricard A, Tetreault J and Hubert J 1991 *J.Phys.B* **24** 1115
[19] Guerra V, Sa P A and Loureiro J 2001 *J.Phys D: Appl. Phys* **34** 1745

Analysis of the measured 1S-2S Stark spectra of hydrogen isotopes used for E-field strength determination in a hollow cathode discharge

M I de la Rosa[1], C Pérez, K Grützmacher and LM Fuentes

Dpto. Física Teórica, Atómica y Óptica, Facultad de Ciencias, Universidad de Valladolid, Paseo de Belén 7, 47011 Valladolid, Spain

E-mail: delarosa@opt.uva.es

Abstract. This paper is dedicated to analyse in detail the Stark splitting spectra of the 1S - 2S transition of hydrogen isotopes, used to determine the local electric field strength in the cathode fall of a hollow cathode discharge, by measuring the Doppler free two-photon absorption via optogalvanic detection. The measured irradiance distribution in the overlap volume of the two laser beams is used to integrate the rate equations, which give the ion yield distribution with high spatial resolution. Inserting the local ion yield in the local electric field strength of the cathode fall allows reconstructing and analysing the Stark splitting spectra.

1. Introduction

In a hollow cathode discharge (HCD) the local electric field strength (E-field) is one of the most relevant discharge parameters directly connected with charged particle fluxes, their energy distribution functions and the charge densities [1-2]. In a HCD discharge, as it is well known, two regions can be distinguished: a small dark space closed to the cathode surface where the local E-field varies strongly because the potential applied between the electrodes mainly drops off in this region, while the negative glow plasma fills up most of the discharge volume. Thus, the E-field measurements, in this kind of discharges, demand for a well characterized spatial resolution parallel to the cathode surface.

In previous papers [3-5], we reported on the determination of the E-field in the cathode fall region of the HCD, operated either in hydrogen or deuterium, via the Stark splitting of the 1S - 2S transition. We used Doppler free two-photon excitation combined with optogalvanic detection to characterize the cathode fall voltage drop of the HCD in a wide range of discharge parameters. The comparison of the calculated and measured frequency shifts of the Stark components gives the E-field, which varies sharply over very short distance. However, the measured Stark components are found to be much broader than expected.

To exclude that this broadening is not caused by the spatial resolution provided by the overlap of the two counter-propagating laser beams, the excitation conditions in the measurement volume has to be characterized with high precision. In this paper we report on a straight forward approach, based on highly resolved spatial and temporal irradiance distribution of both laser beams, we integrate the rate equations, e.g. for two-photon resonant ionization (optogalvanic detection) and obtain the three

[1] To whom any correspondence should be addressed.

dimensional Ion-yield with high spatial resolution. In a final step, the local Ion yield is inserted in the measured local E-field at different distances close to the cathode surface; hence we obtain the spectra of the Stark components accounting also for laser bandwidth.

2. Experimental details and results for the local electric field strength measurements

Measurements were performed in low density plasma generated in a home-made HCD, figure 1. The outer structure of the hollow cathode is a vacuum seal chamber made of brass which includes: electrical isolation, water cooling, gas in and outlet and electrical connections. Anodes and the cathode are made of stainless steel and inserted in the outer structure. The hollow cathode (10 mm inner diameter and 50 mm long) is placed in between two hollow anodes to allow end-on spectroscopy parallel to the hollow cathode axis close to the cathode surface. The discharge is operated at a constant gas flow of hydrogen or deuterium and a fixed pressure. The device provides stationary stable non-LTE plasma, characterized by low density and kinetic temperature of atoms and molecules. In this study we have generated plasma by currents from 50 to 200 mA and pressures from 270 to 1350 Pa. The entire HCD is mounted on two translator stages, which allow for precise horizontal (x) and vertical (y) displacements, see figure 2, while the laser spectroscopic setup is fixed. The counter-propagating laser beams are crossing in the horizontal plane; and the measurement volume, i.e. the beam overlap, is aligned parallel to the cathode surface and centred with respect to the cathode length. The vertical translator stage allows varying the distance of the measurement volume with respect to the cathode surface.

Figure 1. Image of the HCD in the laser spectroscopic setup.

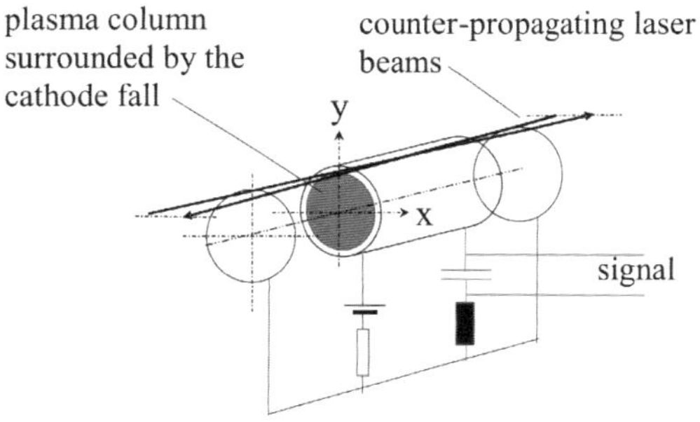

Figure 2. Two-photon absorption followed by photo-ionization generates additional electric charges in the overlap volume. The strong E-field in the cathode fall creates a charge avalanche detected by capacitive coupling.

The radiation needed for two-photon excitation of the 1S - 2S transition is generated by an advanced tuneable single longitudinal mode (SLM) UV-laser spectrometer, that provides up to 10 mJ pulse energy in 2.5 ns at 243 nm, with 300 MHz bandwidth. Each of the two counter-propagating laser beams is focused in the overlap volume by a 1m focus lens. They are opposite circularly polarized according to the angular momentum selection rules ($\Delta L = 0$). A complete description of the laser system and the whole experimental set-up used for E-field determination can be found in the references [3-5].

To determine the E-field, successive laser scans were taken at different distances from the cathode surface. The comparison of the measured and the theoretical frequency shifts of the Stark components give precise values of the E-field, because the spectra reveal very well defined frequency positions for the maximum of each Stark component. Figure 3 shows one example of some results of the E-field as a function of the cathode distance for various experimental conditions. Obviously, the E-field varies very strongly over short distances in front of the cathode: in a sheath of about 2 mm the local electric field strength increases up to 4 kV/cm.

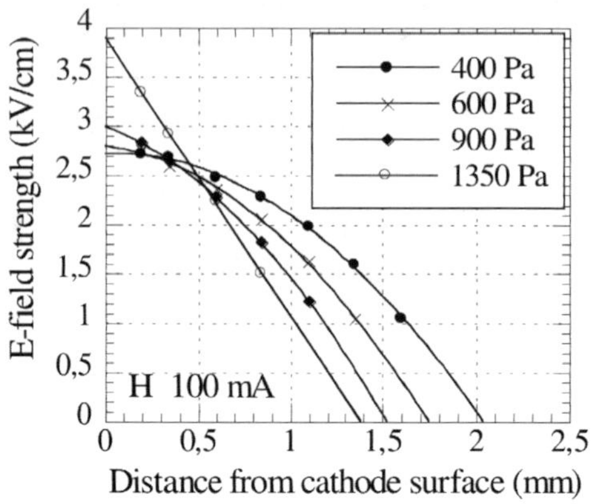

Figure 3. E-field vs distance from the cathode in the hollow cathode discharge for Hydrogen, 100 mA discharge current and various pressures.

However, the experimental profiles are broader than expected, see figure 4. To exclude that this broadening is a consequence of the spatial resolution, the measurement volume has to be determined with high precision.

3. Characterization of the spatial resolution based on the determination of the local Ion Yield

As already mentioned the spatial resolution is determined by the overlap volume of the two laser beams in the HCD. For its determination, MATLAB programs are used to incorporate the beam propagation and irradiance distributions necessary for calculating the excitation conditions in the overlap volume. This calculation implies the integration of the coupled rate equations suited for the two-photon excitation of the 1S - 2S transition of atomic hydrogen, which accounts for two-photon absorption, photo-ionization, collisions and fluorescence decay. The integration requires knowledge of the temporal and spatial irradiance distribution in both beam waists.

To acquire spatial beam profiles we employ a high resolution UV camera (LaserCam-HR-UV, Coherent, pixel size 6.7 μm x 6.7 μm), and a fast photodiode for the temporal pulse shape. Without getting in all details that can be found elsewhere [6], basically the calculation scheme is as follows: the experimental overlap volume due to a crossing angle of 1.53° is represented by slices of 500 μm length in the propagation direction. For each slice the equations are integrated in 50 ps steps, for a laser pulse of 2.5 ns FWHM and for volume elements of 6.7 μm x 6.7 μm x 500 μm. Spatial and temporal irradiance distributions of the two beams are defined by the pulse energy and the temporal pulse shape.

These calculations reveal with high spatial resolution the Ion yield distribution for various irradiances, and allow characterizing the spatial resolution, saturation effect due two depletion of the ground state density, and the influence of the spectral width of the two-photon absorption profile on Ion yield. It turns out that the optogalvanic detection has an exceptional high spatial resolution, because for up to 100 MW/cm^2 more than 75% of the optogalvanic signal is created in a tiny central area of the overlap volume of only 100 μm x 100 μm and a length of 10 mm. Finally, the obtained Ion yield distribution opens the possibility to achieve for the first time a reliable analysis and interpretation of the spectra.

4. Analysis of the spectra by combining local Ion Yield and local E-field strength distribution

Inserting the Ion yield distribution into the measured cathode voltage drop allows calculating the spectral shape of the Stark components, at any position of the cathode fall as a consequence of the spatial resolution. Spectral resolution of the laser is also included. In figure 4 one example of the reconstructed spectra (fine continuous lines) is compared with the measured spectra (dots), for a discharge pressure of 600 Pa, a current of 100 mA and 200 μm distance from the inner cathode surface. Obviously, the well separated Stark components of the reconstructed spectra exhibit significant less broadening than the measured ones. The double peak of each Stark component of the reconstructed spectra is due to the underlying hyper-fine structure (621 MHz) of the 1S-2S transition of hydrogen, which is easily resolved by 300 MHz laser bandwidth.

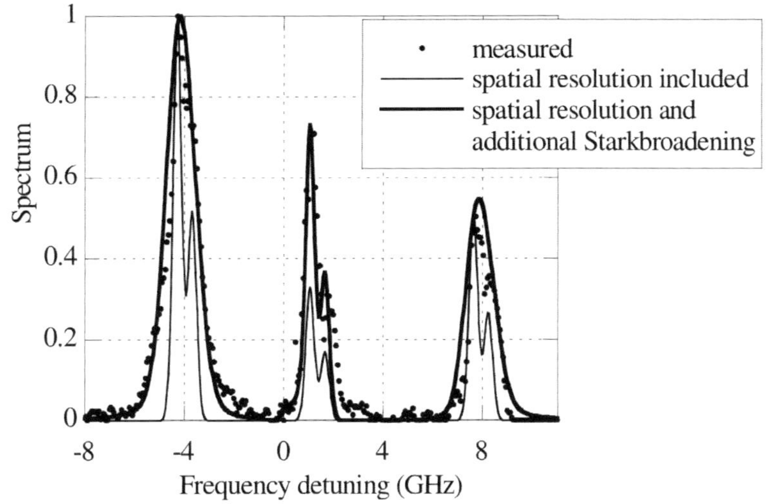

Figure 4. Measured and reconstructed spectra, for a discharge current of 100 mA and 600 Pa at a distance of 200 μm from the cathode surface.

Because of the discrepancy between measured and reconstructed spectra, we assume that the measured spectra reveal an additional homogeneous Stark broadening due to electron and ion collisions present in the cathode fall. In order to proof our presumption of an additional homogeneous Stark broadening acting with identical electric field distribution on all three Stark components, spectra were calculated which account for such an additional Stark broadening, arbitrarily chosen to fit the measured spectra. This leads indeed to much more broadened spectra (thick continuous line), which are fitting quite well with the measured ones. However, a convincing reconstruction of the spectra requires novel Stark broadening calculations, which might be essential for modeling and understanding of cathode fall dynamics.

5. References

[1] Donkó Z 2011 *Plasma Sources Sci. and Technol.* **20** 024001
[2] Bogaerts A, Neyts E, Gijbels R and van der Mullen J 2002 *Spectrochimica Acta Part B* **57** 609
[3] de la Rosa M I, Pérez C, Grützmacher K, Gonzalo A B and Steiger A 2006 *Plasma Sources Sci. Technol.* **15** 105
[4] de la Rosa M I, Pérez C, Grützmacher K and Fuentes L M 2009 *Plasma Sources Sci. and Technol.* **18** 015012

[5] Pérez C, de la Rosa M I and Grüztmacher K 2010 *Eur. Phys. J. D* **56** 369
[6] García-Lechuga M, Fuentes LM, Grützmacher K, Pérez C and de la Rosa M I *2013 Proc. SPIE 8785, 8th Iberoamerican Optics Meeting and 11th Latin American Meeting on Optics, Lasers, and Applications (Oporto),* 87854F

Acknowledgments
The authors wish to thank Mr. S. González for the technological support. The present work was supported by the DGICYT (Ministerio de Educación Cultura y Deporte) contract reference number FIS2007-62604 and by DGICYT (Ministerio de Economía y Competitividad) contract reference number ENE2012-35902 and FEDER funds.

Experimental characterization of the excitation state of picosecond laser-induced Tungsten plasmas

Bastien Pérès(1)*, **Vincent Morel**(1), **Arnaud Bultel**(1), **Abdenacer Benyagoub**(2), **Isabelle Monnet**(2), **Emmanuel Gardès**(2), **Gilles Godard**(1), **Carole Gobin**(1), **Corentin Jouen**(1), **Ammar Hideur**(1), **Ioan Schneider**(3) and **Zsolt Mezei**(3)

(1)CORIA - UMR 6614 - Normandie Université, CNRS-Université et INSA de Rouen, Campus universitaire du Madrillet, 76800 Saint-Etienne du Rouvray, France
(2)CIMAP - UMR 6252 - CEA/CNRS/ENSICAEN/UCBN, BP 5133, 14070 Caen cedex 5, France
(3)LOMC - UMR 6294 - Université du Havre, 53, rue de Prony, 76058 Le Havre, France

E-mail: *bastien.peres@coria.fr

Abstract. In order to quantitatively determine the retention of light atoms by plasma facing components (mainly tungsten) used in fusion reactors, the Laser-Induced Breakdown Spectroscopy (LIBS) technique can be used. The nanosecond laser regimes classically used for LIBS present major limitations mainly due to the conditions of the laser-matter interaction process itself and cannot be used in the present context. Picosecond laser pulses are more appropriate. To validate the diagnostic based on picosecond laser pulses, preliminary characterizations of the produced plasmas have to be performed. Some related results are reported in the present communication. The study is focused on (1) the ablation of tungsten under two laser (low and high) fluence conditions and (2) the excitation equilibrium of the plasma from the nanosecond to microsecond time scales.

1. Introduction

The quantitative detection of light elements (first two lines of the periodic table of elements) trapped in heavy atoms-based matrices and their superficial gradients stay a challenge difficult to address [1]. The related measurements are required notably in the frame of the penetration of fuel in materials used in thermonuclear reactors (hydrogen and isotopes in tungsten) for safety verifications. Today, the two main techniques (NRA, Nuclear Reaction Analysis and SIMS, Secondary Ion Mass Spectrometry) used to perform such detections are based on the properties of an ionic beam impinging the surface and cannot be used in the fusion context since (1) a very low residual pressure is required and (2) the preparation of the samples is mandatory. Any *in situ* measurement is therefore impossible. Other techniques have consequently to be tested.

The Laser-Induced Breakdown Spectroscopy (LIBS) is a promising technique potentially able to measure gradients of these elements in related matrices under flexible conditions [2]. A pulsed class 4 laser beam is focused on the sample and the radiation emitted by the plasma produced by the excited ablated material is analyzed by spectroscopy. The laser pulse duration is classically of the nanosecond scale, which leads to the transient formation of a melting pool, the vaporization of which allows ionizing the atoms under irradiation until the end of the pulse.

Table 1. Nomenclature.

Symbol	Description	Unit	Symbol	Description	Unit
A_{ki}	Einstein coefficient	s^{-1}	L_{ki}	Radiance	W m^{-2} sr^{-1}
E	Energy	eV	$p(\lambda)$	Line profile	-
f	Focal length	m	t	Time	s
F_H, F_L	Fluence	J cm^{-2}	T	Temperature	K
g_i, g_k	Statistical weight	-	$[W(i)]$, $[W(k)]$	Number density	m^{-3}
$k(\lambda)$, $k_{\lambda_{ki}}$	Absorption coefficient	m^{-1}	δ	Plasma diameter	m
$L_{ki,\lambda}(\lambda)$	Spectral radiance	W m^{-2} sr^{-1} m^{-1}	λ	Wavelength	m

Despite its promising aspect, the nanosecond LIBS technique presents major limitations arising mainly from the conditions of the laser-matter interaction process itself leading to an initial departure from equilibrium [3], a strong initial continuum [4] and the melting pool formation [5]. The exploitation of ultra-short (picosecond and femtosecond) pulses in LIBS is very promising from this point of view. Indeed, the pulse energy is totally transferred to the sample owing to its briefness and distributed with almost the same probability to all elements over the pulse penetration depth (much lower than in nanosecond regime). It is particularly appropriate to the analysis of samples composed with elements of very different atomic mass and especially for light elements in heavier matrices. The continuum intensity is reduced at early times, which is particularly appropriate to the observation of early lines due to light elements. The detection limits should be strongly lowered while the capacity of depth-profiling should be strongly improved.

Since the energy available per pulse is higher in picosecond regime than in femtosecond regime for compact sources, we have implemented in our LIBS platform a picosecond laser source. This will help to a future implementation on a fusion reactor in a semi-industrial environment and enlarge the room for manoeuvre. Preliminary tests are performed on tungsten without retention of light elements. These tests are described and analyzed in the upcoming sections. After description of the LIBS platform used to analyze radiation emitted by the laser-induced plasma, results on ablation craters are first given. Two (low and high) fluence conditions are tested. Using relevant calibrations, excited states number density is determined and the Boltzmann plots are derived. The concept of excitation temperature is introduced and discussed in the light of the present results. In particular, the relevance of the McWhirter criterion is discussed.

2. The CORIA's LIBS platform

The CORIA's LIBS platform is displayed on figure 1. After several mirrors, the picosecond laser pulse provided by an EKSPLA source (532 nm, 30 ps, 10 mJ) is focused with a converging lens (5 cm or 50 cm in focal length) on a tungsten sample located on a rotating plate inside a chamber, the ultimate pressure of which is 0.4 Pa. Two lenses providing a magnification of 2 are used to collect a part of the light emitted by the resulting plasma toward an Ebert-Fastie spectrometer (f = 2 m in focal length, aperture of $f/15$) on the exit slit of which an ICCD camera Roper Scientific PIMAX 2 (512×512 pixels^2) is connected. The spectroscopic device allows a spectral resolution of around 5×10^{-2} nm in the visible spectral range.

The craters formed during the laser-matter interaction are observed using a calibrated microscope, the maximum magnification of which is 50. The calibration is performed with wires of known diameter. The crater diameter can be determined with an accuracy of $\pm$ 20 μm. The spectroscopic analysis is performed using the WinSpec software with a gate Δt of 50 or 100 ns between $t = 0$ (the time at which the laser pulse reaches the sample) and $t_{max} = 2$ μs. The calibration allowing the conversion from signals in arbitrary units to spectral radiance is performed with a tungsten ribbon lamp at T_c = 2475 K $\pm$ 5 K. The entrance slit of the

Figure 1. The CORIA's LIBS platform. The EKSPLA laser source (on the right) emits light pulses along the green trajectory. After several mirrors, a converging lens focuses the pulse on a horizontal sample located inside the central chamber. A part of the light emitted by the laser-induced plasma is then collected by two lenses along the red optical axis toward a spectrometer (on the left) equipped on its exit slit with an ICCD camera connected to a computer. The spectral analysis is finally performed.

spectrometer is fixed at 100 μm ($\Delta x = 50$ μm for the detection volume width within the plasma) and the optical axis-sample distance is $z = 650$ μm. The detection volume height is $\Delta z = 1.3$ mm.

3. Ablation craters

Two series of experiments have been performed in low fluence (LF) and high fluence (HF) conditions using the two lenses mentioned above. The first condition corresponds to $F_L = 10$ J cm^{-2} and the second condition to $F_H = 1000$ J cm^{-2}. Table 2 reports the mean characteristics obtained for the craters for these two fluence conditions. It is interesting to note that an increase in the irradiance of a factor of 100 leads to a similar increase of the ablation rate. Since this increase results from light more focused on the sample, the ablation diameter is consequently reduced, but by a factor of 5 only. The ablated volume (and therefore the ablated atoms) is increased by a factor of 4 when the fluence is increased from F_L to F_H.

This is a very interesting result from the LIBS diagnostic point of view. Despite a strong

Table 2. Mean characteristics of the craters obtained under low (LF) and high (HF) fluence conditions with the picosecond laser source of the CORIA's LIBS platform.

Characteristic	LF conditions (F_L = 10 J cm^{-2}) (over 1000 pulses)	HF conditions (F_H = 1000 J cm^{-2}) (over 200 pulses)
Ablation rate	10 nm pulse^{-1}	1 μm pulse^{-1}
Ablation diameter	750 μm	150 μm
Ablated volume	4.4×10^{-15} m^3 pulse^{-1}	1.8×10^{-14} m^3 pulse^{-1}
Ablated atoms	2.8×10^{14} pulse^{-1}	1.1×10^{15} pulse^{-1}

increase in the fluence of a factor of 100, the ablated atoms number can be considered as relatively constant. A too strong focus of the laser light on the sample leads only to a reduction of the spatial resolution which can question the ability of the technique to provide precise depth profiling measurements if gradients with characteristic length scales lower than 1 μm are expected for the material to be tested.

4. Spectroscopic analysis

The tungsten spectroscopy faces the problem of the lack of basic data concerning mainly the Einstein coefficient A_{ki} of the radiative transitions. As an example, the different lines of W I reported with their Einstein coefficient in the NIST database [6] represents only 7 % of the total of the reported lines. The database is therefore far to reach the 25 % reported lines of Fe I with their Einstein coefficient, which is also far from the 98 % obtained for N I. In addition, the transitions with the corresponding Einstein coefficient are only around 500 for W I whereas they are around 2500 for Fe I. The quantitative analysis of spectra emitted by the laser-induced plasmas requires the Einstein coefficient. As a result, the number of lines involved in the analysis is small. In addition, the W ionization limit is relatively low (E_{ion} = 7.864 eV). The lower level of a radiative transition in the visible spectral range has therefore a high probability to be energetically close to the ground state and to be self-absorbed. A relevant way to reduce this probability is to use transitions, the Einstein coefficient of which is not too high. But in this case, the spectral radiance is also reduced and can prevent a satisfactory observation of the lines. The choice of the transitions finally retained for the analysis is therefore the result of a compromise between all these aspects. They are listed in table 3 with their characteristics. Only W I transitions have been observed.

Figure 2 displays an example of the calibrated spectra which can be observed under HF conditions. In particular, this figure shows the evolution in time of the spectra over the spectral range 504.6 nm $< \lambda <$ 509.6 nm for t = 150, 250, 650 and 1050 ns. The initial continuum is easily observed even if its spectral radiance is moderate and clearly lower than the one of the lines. For a given observation time, this continuum increases with the wavelength. This indicates that the plasma electron temperature is moderate if we consider the initial cooling mainly due to the radiative recombination. The resulting continuum is then emitted in the near infrared spectral region. The group of lines at λ = 507.1515 nm presents a very interesting behavior. These lines overlap with a relative contribution varying in time. As a result, this group is not symmetric and presents a time-dependent redshift. As far as we know, their Stark broadening and shift parameters have not yet been reported in literature.

The radiance L_{ki} (see Table 1 for the units of variables used in the present paper) of each $k \to i$ line of table 3 is calculated from

$$L_{ki} = \int_0^{+\infty} L_{ki,\lambda}(\lambda) d\lambda \qquad [\mathrm{W\,m^{-2}\,sr^{-1}}] \tag{1}$$

Table 3. W I radiative transitions used for the picosecond laser-induced plasmas analysis.

Wavelength	Einstein coefficient	Lower state Energy	Lower state Degeneracy	Upper state Energy	Upper state Degeneracy	Time of observation
λ_{ki} (nm)	A_{ki} (s^{-1})	E_i (eV)	g_i	E_k (eV)	g_k	t (ns)
449.2298	3.60×10^5	2.121	9	4.880	11	200
449.3958	2.80×10^6	2.387	5	5.145	7	200, 600, 1000
449.4497	3.00×10^6	2.387	9	5.145	7	100, 200, 600, 1000
449.5294	3.30×10^5	1.868	11	4.626	11	600, 1000
450.4838	7.00×10^5	1.508	9	4.259	7	100, 200, 600, 1000
451.7344	3.50×10^6	2.266	5	5.010	3	600, 1000
498.2593	4.17×10^5	0.000	1	2.488	3	200, 600, 1000
498.4698	9.00×10^5	2.352	7	4.839	5	600, 1000
498.6920	6.30×10^5	1.868	11	4.354	9	100, 200, 600, 1000
500.6150	1.20×10^6	0.771	9	3.247	7	100, 200, 600, 1000
501.5304	5.40×10^5	0.599	7	3.070	9	100, 200, 600, 1000
504.0360	5.20×10^5	1.650	3	4.109	5	200, 600, 1000
505.2230	1.10×10^6	2.602	5	5.055	3	100, 200, 600, 1000
505.3276	1.90×10^6	0.210	3	2.660	3	100, 200, 600, 1000
505.4596	3.30×10^5	0.210	3	2.660	5	100, 200, 600, 1000
505.5516	4.00×10^6	2.787	9	5.239	7	1000
507.1515	3.40×10^5	2.436	13	4.880	11	100, 200, 600, 1000
511.7612	1.61×10^5	2.458	11	4.880	11	600, 1000
512.4234	4.00×10^5	1.857	5	4.276	5	200, 600, 1000
514.5774	2.10×10^6	2.602	5	5.010	3	200, 600, 1000
515.9095	9.10×10^3	1.857	5	4.259	7	1000
518.3965	9.60×10^5	2.353	7	4.744	9	100, 200, 600, 1000
520.4520	6.80×10^6	2.533	3	4.914	3	100, 200, 600, 1000
520.6180	2.49×10^6	2.458	7	4.839	5	200, 600, 1000
522.4657	1.20×10^6	0.588	7	2.971	5	100, 200, 600, 1000
524.2980	1.10×10^6	2.037	9	4.401	7	100, 200, 600, 1000
524.7417	3.00×10^4	1.868	11	4.231	9	100, 200, 600, 1000
525.4542	3.86×10^5	1.917	7	4.276	5	200, 600, 1000
525.5396	2.61×10^6	2.787	9	5.145	7	200, 600, 1000
526.8550	1.40×10^5	2.121	9	4,474	9	600, 1000
547.7776	2.60×10^5	1.181	1	3.444	3	200, 600, 1000
549.6234	4.00×10^5	2.787	9	5.042	7	200, 600
550.0490	6.90×10^5	2.458	11	4.712	9	200, 600, 1000
551.4976	7.30×10^5	0.412	5	2.660	3	100, 200, 600, 1000
582.2600	1.07×10^6	2.787	9	4.916	7	200, 600
644.5120	6.40×10^5	2.353	7	4.276	5	200, 600, 1000

The influence of the stimulated emission can be neglected. Therefore, the spectral radiance $L_{ki,\lambda}(\lambda)$ can be written as

$$L_{ki,\lambda}(\lambda) = \frac{2\,h\,c^2}{\lambda_{ki}^5}\frac{[W(k)]}{[W(i)]}\frac{g_i}{g_k}\left(1 - e^{-k(\lambda)\delta}\right) \qquad [\mathrm{W\,m^{-2}\,sr^{-1}\,m^{-1}}] \qquad (2)$$

where $k(\lambda)$ (in m^{-1}) is the absorption coefficient per unit length at the wavelength λ, and $[W(k)]$ and $[W(i)]$ are the mean number density over the plasma diameter δ of the upper and lower

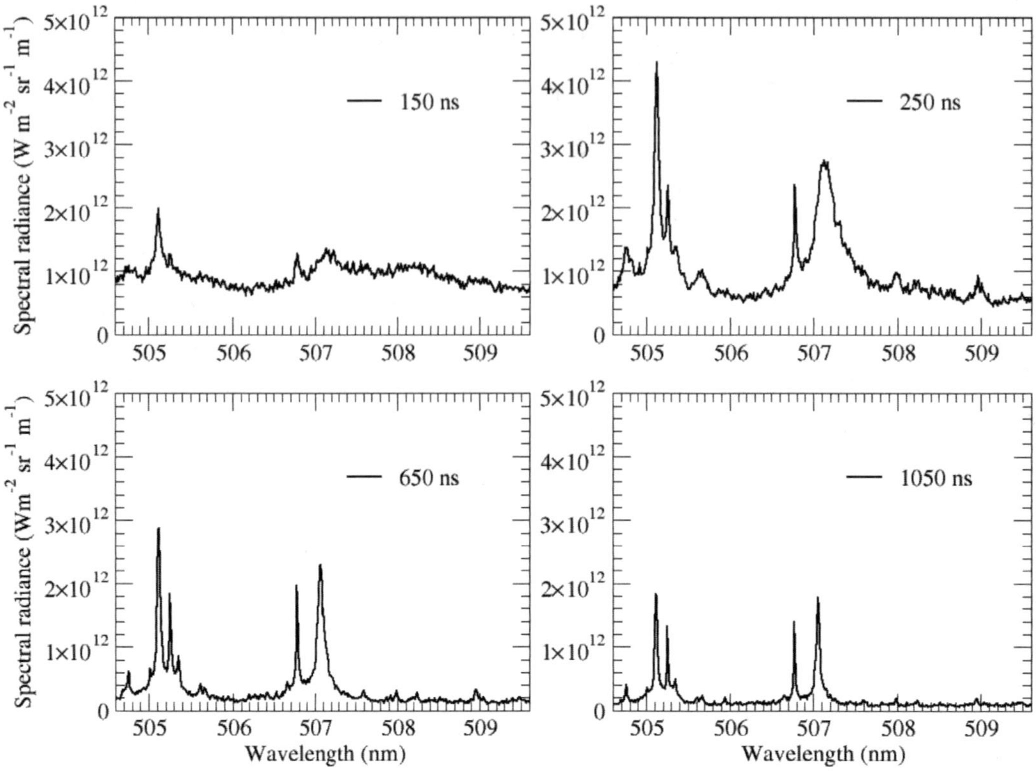

Figure 2. Spectra observed for a gate of Δt = 100 ns at time t = 150, 250, 650 and 1050 ns over the spectral range 504.6 nm < λ < 509.6 nm.

states of the transition, respectively. The absorption coefficient depends on the line profile $p(\lambda)$ such as $k(\lambda) = k_{\lambda_{ki}}\, p(\lambda)$ where

$$k_{\lambda_{ki}} = \frac{A_{ki}\,\lambda_{ki}^4}{8\,\pi\,c}\,[W(i)]\,\frac{g_k}{g_i}\,\frac{1}{\int_0^{+\infty} p(\lambda)\,d\lambda} \qquad [\mathrm{m}^{-1}] \tag{3}$$

is the absorption coefficient at the line center. Therefore

$$L_{ki} = \frac{2\,h\,c^2}{\lambda_{ki}^5}\frac{[W(k)]}{[W(i)]}\frac{g_i}{g_k}\int_0^{+\infty}\left(1 - e^{-k_{\lambda_{ki}}\,p(\lambda)\,\delta}\right)d\lambda \tag{4}$$

If

$$e^{-k_{\lambda_{ki}}\,p(\lambda)\,\delta} \simeq 1 - k_{\lambda_{ki}}\,p(\lambda)\,\delta \tag{5}$$

the transition is optically thin and

$$L_{ki} \equiv \frac{2\,h\,c^2}{\lambda_{ki}^5}\frac{[W(k)]}{[W(i)]}\frac{g_i}{g_k}k_{\lambda_{ki}}\delta\int_0^{+\infty} p(\lambda)d\lambda \tag{6}$$

Journal of Physics: Conference Series **550** (2014) 012047 doi:10.1088/1742-6596/550/1/012047

In case of Doppler ($type \equiv$ D) or Lorentz ($type \equiv$ L) broadening, we have

$$\int_0^{+\infty} p(\lambda) d\lambda = a_{type}\, \Delta\lambda_{1/2} \qquad [\mathrm{m}] \tag{7}$$

where $\Delta\lambda_{1/2}$ is the Full Width at Half Maximum (FWHM) of the line and $a_D = \sqrt{\frac{\pi}{4\ln 2}}$ and $a_L = \frac{\pi}{2}$. The radiance of the line is then

$$L_{ki} = \frac{2\,h\,c^2}{\lambda_{ki}^5} \frac{[W(k)]}{[W(i)]} \frac{g_i}{g_k} k_{\lambda_{ki}} \delta\, a_{type}\, \Delta\lambda_{1/2} \tag{8}$$

With (3), we obtain

$$L_{ki} = \frac{h\,c}{\lambda_{ki}} [W(k)] \frac{A_{ki}}{4\pi} \delta \tag{9}$$

The condition (5) is fulfilled if

$$k_{\lambda_{ki}}\, p(\lambda)\, \delta < k_{\lambda_{ki}}\, \delta \leq 0.1 \tag{10}$$

From the absorption coefficient definition (3), we deduce the condition for an optically thin transition

$$[W(i)] \leq [W_{ki}^0(i)] = 0.1 \frac{8\,\pi\, c\, a_{type}\, \Delta\lambda_{1/2}}{A_{ki}\, \lambda_{ki}^4\, \delta} \frac{g_i}{g_k} \qquad [\mathrm{m}^{-3}] \tag{11}$$

If the previous condition is not fulfilled, radiation can be considered as partially self-absorbed. In order of magnitude, data reported in table 3 lead to $[W_{ki}^0(i)] \simeq 10^{22}$ m^{-3}.

We calculate the upper state number density by assuming (α) the related radiative transition optically thin and (β) a constant diameter δ for the plasma. Actually, this diameter evolves with time due to the plasma expansion over the observation time. Nevertheless, we have considered in a first approximation that this increase is moderate around $\delta = 200$ μm, the value of which is confirmed by direct imagery. This value is very close to the crater diameter obtained under HF conditions. The number density (in m^{-3}) is therefore derived from equation (9) leading to

$$[W(k)] = \frac{4\pi L_{ki}}{A_{ki}\delta} \frac{\lambda_{ki}}{hc} \tag{12}$$

which is used for the determination of the upper state number density.

Table 3 reports the time at which the studied lines can be observed. Even low Einstein coefficient transitions can be detected. Figure 3 shows the Boltzmann plot for t = 150, 250, 650 and 1050 ns derived from the upper state number density therefore obtained. A large part of the lower states of the involved transitions have an energy higher than 2 eV. For them, the Boltzmann plots display a number density much lower than the limit (11) estimated here to $[W_{ki}^0(i)] \simeq 10^{22}$ m^{-3}: as a result, most of the transitions can be considered as optically thin. We can observe that the order of magnitude of the density levels are relatively constant. For each boltzmann plot we performed a linear interpolation with the least squares method. The slope of this interpolation lines is directly proportional to the excitation temperature. Despite the dispersion resulting from the different experimental uncertainties, the determination of which will be shortly performed, the excitation temperature defined as [7]

$$T_{exc} = -\frac{1}{k_B \left[\frac{d}{dE_k} \left(\ln([W(k)]/g_k)\right)\right]_{lsl}} \qquad [\mathrm{K}] \tag{13}$$

is derived.

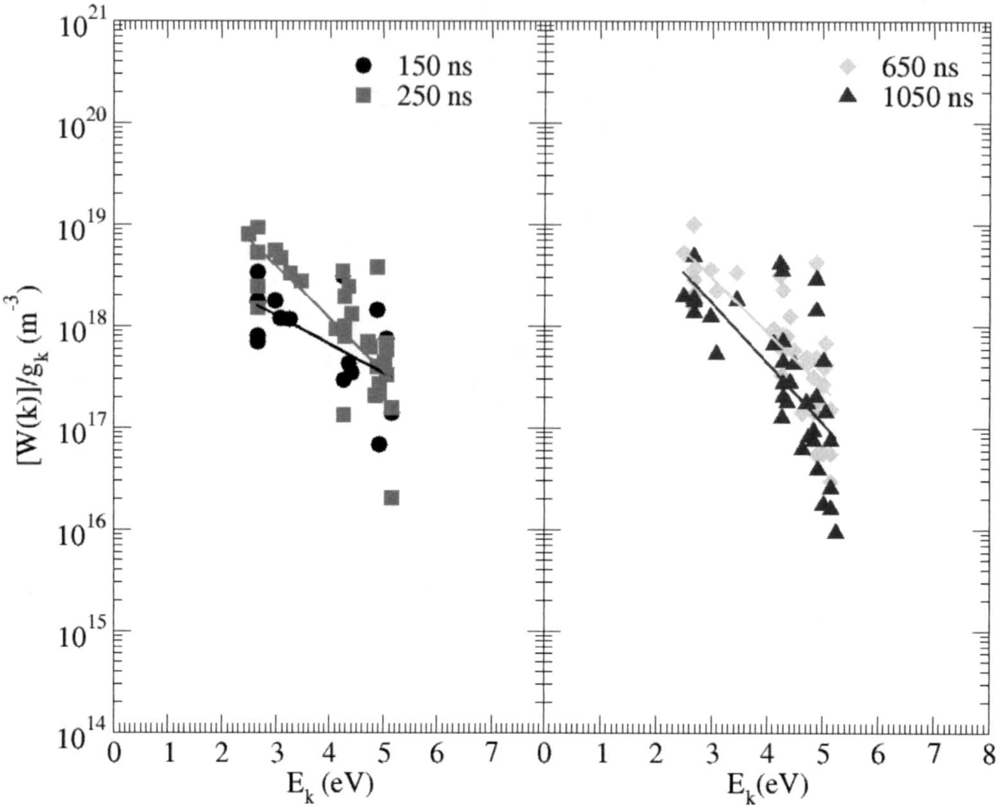

Figure 3. Boltzmann plot obtained at t = 150, 250, 650 and 1050 ns. The straight lines correspond to the least squares line for each time.

Figure 4 shows the temporal evolution of the excitation temperature obtained from the four Boltzmann plot interpolation lines. First, at early times between 150 ns and 250 ns after the plasma pulse, T_{exc} abruptly decreases from 18000 K to 9000 K. Then from this time until 1050 ns, T_{exc} decreases slightly and stabilizes at $\sim$ 9000 K. This indicates that the plasma seems to reach an excitation quasi-steady state. The order of magnitude obtained for T_{exc} needs to be confirmed as well as its uncertainty. Moreover, the dynamics at early times must be further study despite the absence of observable lines in spectra. From this point of view, deriving electron density and temperature from the continuum emitted at early tiles could provide valuable information on the coupling between electron and excitation temperatures. Nevertheless, the global behavior of this experimentally determined excitation temperature seems similar to the one observed on nanosecond laser-induced tungsten plasmas [8].

5. Conclusion

In conclusion, we have given an insight into the dynamics of the excitation state of a picosecond laser-induced tungsten plasma through the experimental determination of the excited states number density and the excitation temperature. As a first observation, the global behavior and the order of magnitude of this two quantities are close to values reported in literature in

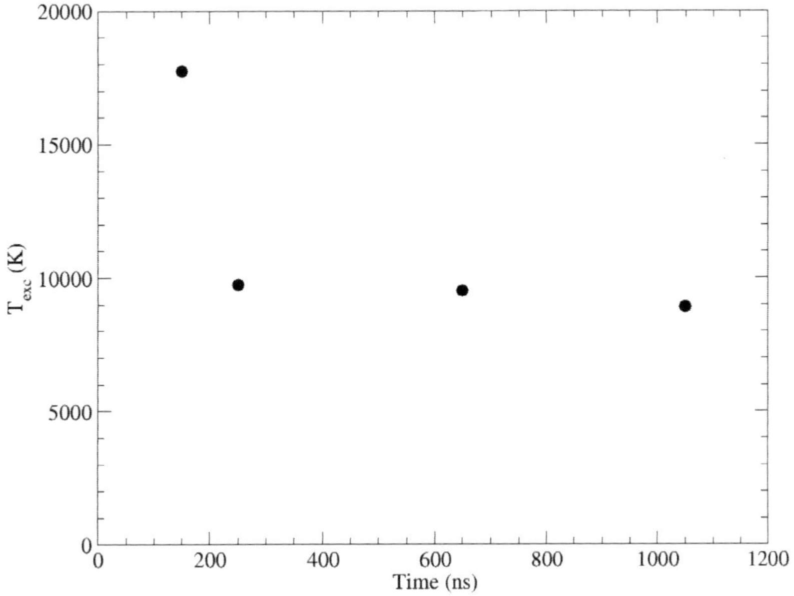

Figure 4. Excitation temperature obtained at t = 150, 250, 650 and 1050 ns.

nanosecond regime. However, the measurements were performed considering spatially averaged quantities. Our next two objectives are to improve the resolution in space and time by performing (1) simultaneous measurements at different locations perpendicularly to the sample surface and (2) additional measurements over the time interval [0,250] ns.

References

[1] Khater M A 2013 *Optics and Spectroscopy* **115** pp 574-590
[2] Hahn D W and Omenetto N 2012 *Applied Spectroscopy* **66** pp 347-419
[3] Elhassan A, Giakoumaki A, Anglos D, Ingo G M, Robbiola L and Harith M A 2008 *Spectrochim. Acta B* **63** pp 504-511
[4] Sirven J-B, Bousquet B, Canioni L and Sarger L 2004 *Spectrochim. Acta B* **59** pp 1033-1039
[5] Leggieri G, Luches A, Martino M, Perrone A, Majni G, Mengucci P and Mihailescu I N 1995 *Thin Solid Films* **258** pp 40-45
[6] *http://physics.nist.gov/PhysRefData/ASD/levels_form.html*
[7] Morel V and Bultel A 2014 *Spectrochim. Acta B* **94-95** pp 63-70
[8] Farid N, Li C, Wang H and Ding H 2013 *J. Nucl. Mat.* **433** pp 80-85

Journal of Physics: Conference Series **550** (2014) 012048 doi:10.1088/1742-6596/550/1/012048

Achievement of local thermodynamic equilibrium for ns laser-induced plasmas on aluminium sample at different wavelengths

Vincent Morel and Arnaud Bultel

CORIA - UMR 6614 - Normandie Université, CNRS-Université et INSA de Rouen, Campus universitaire du Madrillet, 76800 Saint-Etienne du Rouvray, France

E-mail: `vincent.morel@coria.fr`

Abstract. The Collisional-Radiative model CoRaM-Al is elaborated and implemented in a 0D numerical approach in the purpose of describing the formation of the plasma resulting from the interaction between a τ = 4 ns Nd:YAG laser pulse and an aluminium sample in vacuum. The influence of the four harmonics at 266, 355, 532 and 1064 nm on the behavior of the nascent plasma is studied. In each case, the fluence is set equal to the threshold above which a phase explosion takes place (fluence of 7.7, 7.4, 6.8, 5.1 J cm^{-2} in order of increasing wavelength). The model takes into account free electrons and excited states of Al, Al^{+}, Al^{2+} and Al^{3+}. Both groups of particles are characterized by their translation temperature in thermal non-equilibrium. Besides, each population density is assumed to be in chemical non-equilibrium and to behave freely through the involved seven elementary processes (electron impact induced excitation and ionization, elastic collisions, multiphoton ionization, inverse laser Bremsstrahlung, direct thermal Bremsstrahlung and spontaneous emission). Atoms passing from the sample to the gas are described by considering classical vaporization phenomena (governed by the Hertz-Knudsen law) so that the surface temperature is limited to values less than the critical point (T_c = 6700 K). The relative role of the elementary processes is discussed and the time-evolution of the excitation of the species is analyzed for the four considered wavelengths. This study allows to determine the different excitation temperatures as well as their evolution in time. Thus the conditions required for the achievement of the Local Thermodynamic Equilibrium can be precisely described.

1. Introduction

Large intentions are dedicated to the laser-matter interaction since the development of the technology of the ultra-short laser sources. The plasma results from the absorption of the laser light by the ejected matter. This so-formed plasma radiates lines which can lead through spectroscopic analysis to the identification of the components of the sample: the derived diagnostic method is called LIBS (Laser Induced Breakdown Spectroscopy) [1]. During the last years, different calibration free procedures have been elaborated [2] which has led to the considerable improvement of the LIBS technique. However, the Local Thermodynamic Equilibrium seems to be systematically assumed in the purpose of deriving the elementary composition of the sample [3, 4]. This hypothesis is still largely discussed [5] and conditions to obtain LTE have to be clearly understood to estimate its validity. Indeed, the laser-induced plasmas are characterized by strong radiative losses and by high expansion velocities. As a

Table 1. Fluence thresholds to avoid phase explosion in aluminium at different wavelengths. The values at 266 and 355 nm are assumed due to a linear dependency on the wavelength.

λ (nm)	F_1 (J cm^{-2})	Ref.
1064	5.1	[7], [9], [10]
532	6.8	[7], [11]
355	7.4	[12]
266	7.7	[12]

result, these plasmas can be in strong non-equilibrium and this non-equilibrium state can last several hundreds of nanosecond after the laser-sample interaction.

Two important consequences can be deduced in the case of a thermodynamic non-equilibrium plasma:

- the equation of state is not valid and considering neutrals and ions excited states as independent species is mandatory,
- the different states number densities do not fulfill the Boltzmann distribution. As a consequence, optical emission spectroscopy experiments do not allow the determination of the ground state number density.

Accordingly, a Collisional-Radiative (CR) model is an appropriate tool to show how the plasma reaches equilibrium. In particular, the time required to reach thermodynamic equilibrium can be obtained.

The present communication put forward the description of the model and the balance equations used. Then, results for the four wavelengths (266, 355, 532 and 1064 nm) are given and discussed.

2. Description of the model

The work reported in the present paper is focused on the ignition process of the plasma in case of irradiance weak enough to avoid any phase explosion, which means that the highest temperature of the target surface is the critical temperature. For aluminium, the critical temperature has been estimated to $T_c = 6700$ K [6]. However, the irradiance has to be sufficiently high to ignite the plasma. As a consequence, two thresholds of the laser fluence can be observed. The first one corresponds to the minimum value leading to the breakdown. For aluminium, this threshold ($F_0 \approx 1$ J cm^{-2}) weakly depends on the laser wavelength [7, 8]. The second threshold, F_1 corresponds to the observation of an explosion phase and the surface temperature reaches T_c. This threshold depends on the laser wavelength as summarized by table 1.

This explosion phase is also characterized by the formation of droplets [13]. Between F_0 and F_1, the interface melts first, then evaporates and the produced vapor ionizes. The formed plasma radiates and its spectra can be analyzed.

The laser-matter interaction described above is far from being fully understood even with $F \in [F_0, F_1]$. In particular, the mechanism leading to an excitation equilibrium is not well known. This question is crucial because the treatment of LIBS signals is generally based on equilibrium assumptions [14]. The present contribution reports the elaboration of a two-temperatures CR model devoted to the verification of these assumptions in the case of a laser pulse impinging an aluminium sample [12]. The plume expansion is controlled mainly by the laser energy, while the spot diameter weakly influences the plasma development [15]. So, in our model, the spot diameter is fixed at d = 1 mm and the pulse energy is calculated by the equation $E = \pi d^2 F/4$

where the fluence is wavelength-dependent. The pulse time is fixed at $\tau = 4$ ns. This CR model takes into account the seven main elementary processes (electron impact-induced excitation and ionization, elastic collisions, multiphoton ionization, inverse laser Bremsstrahlung, direct electron Bremsstrahlung and spontaneous emission) influencing the number density of the 107 states of Al, Al^+, Al^{2+} and Al^{3+} considered in the model.

The studied test-case corresponds to the classical interaction between a nanosecond Nd:YAG laser pulse assumed of the Gaussian type and an aluminium sample. The conservation equations governing the heating (melting, then vaporization) of the sample are not solved. The surface temperature T_s is assumed to follow in time a similar Gaussian evolution with a maximum T_{max} less than the critical temperature $T_c = 6700$ K [6]. Moreover, the atoms flux density φ_A emerging from the interface is assumed to follow the Hertz-Knudsen law with a saturated vapor pressure given at T_s [16]. These atoms form the Knudsen layer whose characteristics allow a free expansion of the vapor according to a velocity close to the thermal mean velocity [17].

In order to avoid wasteful calculations, the number of aluminium excited states is reduced with respect to the actual states. Levels of close energy are grouped according to their core configuration. The resulting energy spectrum contains finally 106 fictitious levels which are listed in [18].

The dynamics of these states results from the following elementary processes which are completely detailed in [12].

(i) *Multiphoton ionization (MPI).* The vapor produced by the heated sample can be ionized by multiphoton ionization. This process is taken into account under the form proposed by Müsing *et al.* [19]

$$\sigma_{MPI}^{(k)} = \frac{\sigma_1\,\nu}{(k-1)!\,(h\nu^2)^k} \tag{1}$$

where σ_1 is calculated by the form given by Zel'dovich and Raizer [20]

$$\sigma_1 = \frac{64\,\pi^4\,e^{10}\,m_e}{3\sqrt{3}\,(4\pi\epsilon_0)^5\,h^6\,c}\,\frac{1}{\nu^3}\,\frac{E_{ji}^{5/2}}{\sqrt{E_{ioni}^{(Z+)}}\,(E_{ioni}^H)^2}. \tag{2}$$

In equation (2), the different values of σ_1 and $\sigma_{MPI}^{(k)}$ are calculated for the four wavelengths considered in table 1. Table 2 summarizes the characteristics of the multiphoton ionization process (3) considering only ground states.

$$Al_0 + k\,h\nu \overset{\sigma_{MPI}^{(k)}}{\longrightarrow} Al_0^+ + e^- \tag{3}$$

Table 2. Energy of photons E_{ph} and number k required to ionize Al_0 through the multiphoton ionization process (3). σ_1 and $\sigma_{MPI}^{(k)}$ are the photoionization and multiphoton ionization cross sections, respectively.

λ (nm)	E_{ph} (eV)	k	$k * h\nu$ - E_{ioni} (eV)	σ_1 (m^2)	$\sigma_{MPI}^{(k)}$ (m^{2k} W^{-k} s^{-1})
1064.48	1.17	6	1.00	2.44×10^{-19}	2.35×10^{-74}
532.24	2.33	3	1.00	3.05×10^{-20}	8.59×10^{-34}
355.83	3.49	2	1.00	9.04×10^{-21}	3.08×10^{-19}
266.12	4.66	2	3.33	3.81×10^{-21}	2.31×10^{-20}

The residual energy transferred to electron is also calculated in the fourth column of table 2. We can notice the same value (1 eV) in case of 1064, 532 and 355 nm. As a consequence, the electron temperature will reach 7750 K at maximum through multiphoton ionization effect only.

(ii) *Inverse Bremsstrahlung (IB).* Inverse Bremsstrahlung phenomenon heats the electron gas during the laser pulse. The electron-ion [20] and electron-neutral [21] inverse Bremsstrahlung are accounted for.

(iii) *Thermal Bremsstrahlung (TB).* Thermal Bremsstrahlung corresponds to the emission of photons at wavelengths different from the laser one by electrons in interaction with heavy particles. A continuum is then emitted at early times. Its contribution is accounted for [20, 22, 23, 24].

(iv) *Electron-induced excitation (EE) and ionization (EI).* The increase in electron temperature T_e and density n_e promotes electron-induced excitation and ionization. The forms proposed by Drawin [25] for the cross sections (depending on the angular momentum quantum number ℓ) are used.

(v) *Elastic collisions (EC).* Owing to the increase in T_e by MPI and IB processes, thermal non equilibrium between heavies (with the temperature T_A) and electrons occurs. The resulting elastic collisions play a role and the form given by Decoster *et al.* [26] is adopted for their contribution.

(vi) *Spontaneous emission (SE).* The increase in the excited states population density promotes spontaneous emission. The related probabilities obtained from the NIST database [27] are used and corrected by an escape factor resulting from the optical thickness of the plasma [28]. Escape factors are calculated assuming a Stark line broadening.

All the related data are further described in previous paper [12] and [18].

3. CR model and balance equations

The previous elementary processes are involved in the collisional-radiative source term of the classical balance equation [29] where the atom flux resulting from the vaporization is taken into account. The electrons and heavies energy balance equations are separated but coupled (cf. [18]). Finally, the following equations are obtained

$$\frac{dT_e}{dt} = \frac{2}{3\,k_B\,n_e}\left(P_{ee} + \frac{\varphi_{ee}}{\Delta(t)}\right) - \frac{3}{2}k_B T_e \frac{dn_e}{dt} \tag{4}$$

$$\frac{dT_A}{dt} = \frac{2}{3\,k_B\,\sum_{Z,i}[\mathrm{Al}_i^{Z+}]}\left[P_{eA} + \frac{\varphi_{eA}}{\Delta(t)} - \sum_{Z,i}\left(\frac{3}{2}\,k_B\,T_A + E_{Z,i}\right)\frac{d[\mathrm{Al}_i^{Z+}]}{dt}\right] \tag{5}$$

φ_{eA} and φ_{ee} are the ablation rate for heavy particles and electrons, respectively. $\Delta(t)$ is the instantaneous thickness of the plasma layer, and P_{eA} and P_{ee} are the source terms involving the elementary processes listed in section 2.

In the upcoming sections, the following terms will be discussed.

- $\Theta_e(CRP) = \dfrac{P_{ee}}{\frac{3}{2}k_B\,n_e}$, the contribution of the Collisional-Radiative elementary Processes (CRP) to the derivative of T_e,
- $\Theta_e(\varphi) = \dfrac{\varphi_{ee}/\Delta(t)}{\frac{3}{2}k_B\,n_e}$, the contribution of vaporization,

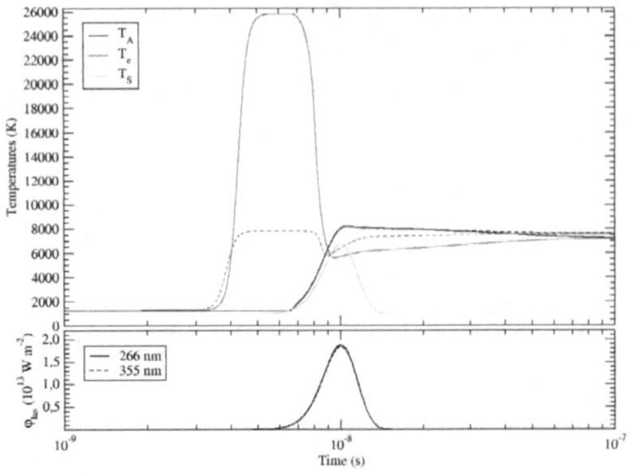

Figure 1. Time-evolution obtained for T_A, T_e and T_S (top) and irradiance φ_{las} (bottom) calculated to avoid explosion phase when λ evolves between 266 nm (continuous lines) and $\lambda = 355$ nm (dashed lines).

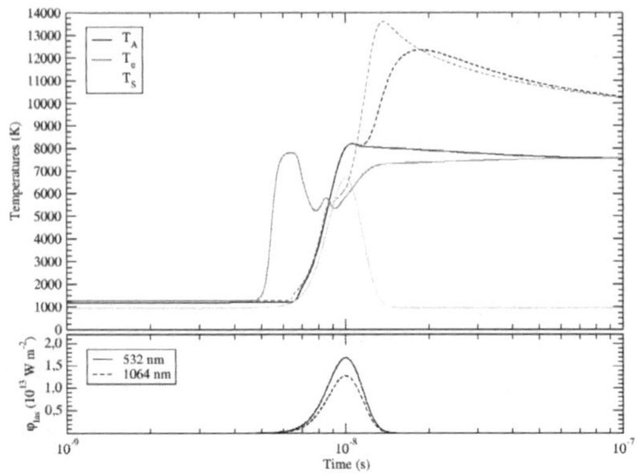

Figure 2. Same as figure 1, but for $\lambda = 532$ nm (continuous lines) and $\lambda = 1064$ nm (dashed lines).

- $\Theta_e(u_1) = \frac{T_e}{n_e}\frac{dn_e}{dt}$, the energy translation storage contribution to the derivative of T_e and
- $\Theta_A(CRP) = \frac{P_{eA}}{\frac{3}{2}k_B \sum_{Z,i}[\mathrm{Al}_i^{Z+}]}$, the contribution of the Collisional-Radiative elementary Processes (CRP) to the derivative of T_A,
- $\Theta_A(\varphi) = \frac{\varphi_{eA}/\Delta(t)}{\frac{3}{2}k_B \sum_{Z,i}[\mathrm{Al}_i^{Z+}]}$, the contribution of vaporization,
- $\Theta_A(u_1) = \frac{T_A}{\sum_{Z,i}[\mathrm{Al}_i^{Z+}]}\sum_{Z,i}\frac{d[\mathrm{Al}_i^{Z+}]}{dt}$, the energy translation storage contribution and
- $\Theta_A(u_2) = \frac{1}{\frac{3}{2}k_B \sum_{Z,i}[\mathrm{Al}_i^{Z+}]}\sum_{Z,i} E_{Z,i}\frac{d[\mathrm{Al}_i^{Z+}]}{dt}$, the excitation and ionization storage contributions to the derivative of T_A.

4. Results

4.1. Temperatures

The temporal evolution of T_A and T_e are displayed in Figure 1 for $\lambda = 266$ and 355 and in Figure 2 for $\lambda = 532$ and 1064 nm. Two trends can be distinguished.

- Before the maximum irradiance, the electron temperature reaches a maximum which corresponds to the residual energy transfered to electron through multiphoton ionization process. The maximum is 8000 K for the 355 and 532 nm case and 26000 K for the 266 nm case. For $\lambda = 1064$ nm there is no increase in T_e and a strong coupling between T_A and T_e is observed.
- After the maximum irradiance, increasing the wavelength leads to the same behavior of T_A and T_e for $\lambda = 266$, 355 and 532 nm, whereas their strong increase is observed for $\lambda = 1064$ nm. In this case, T_A and T_e exceed 9000 K after the laser pulse with a maximum at 13,500 K for T_e. For the three other cases, T_A and T_e remain lower than 9000 K.

Journal of Physics: Conference Series **550** (2014) 012048 doi:10.1088/1742-6596/550/1/012048

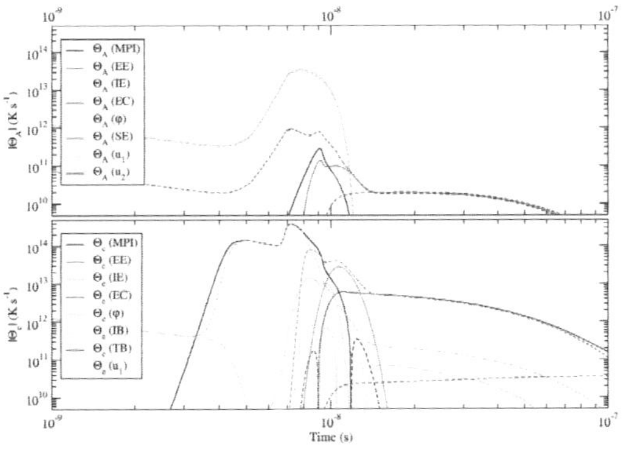

Figure 3. Evolution in time of the elementary processes contributions to dT_A/dt [top, see equation (5)] and dT_e/dt [bottom, see equation (4)] when $\lambda = 266$ nm.

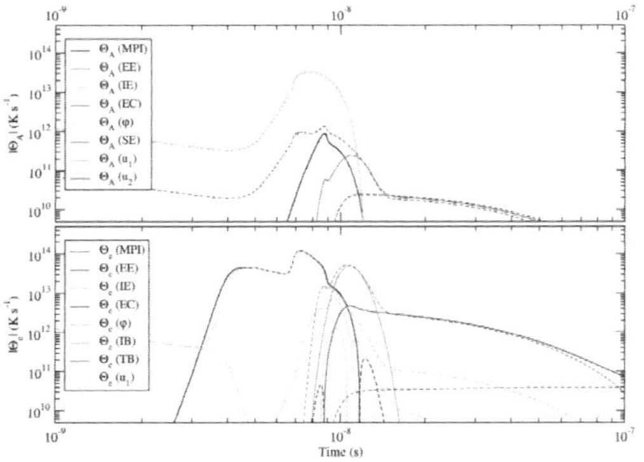

Figure 4. Same as figure 3, but for $\lambda = 355$ nm.

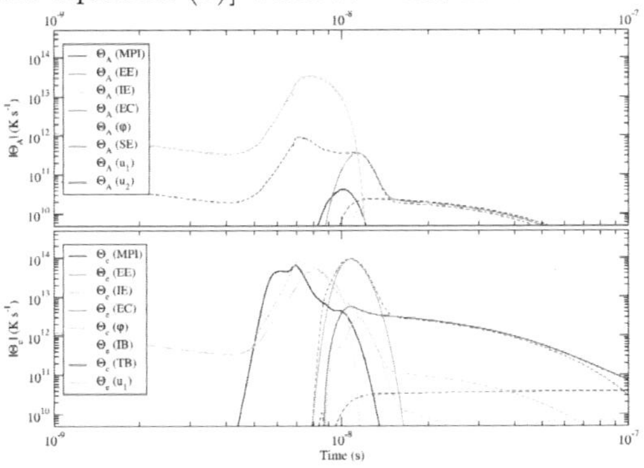

Figure 5. Same as figure 3, but for $\lambda = 532$ nm.

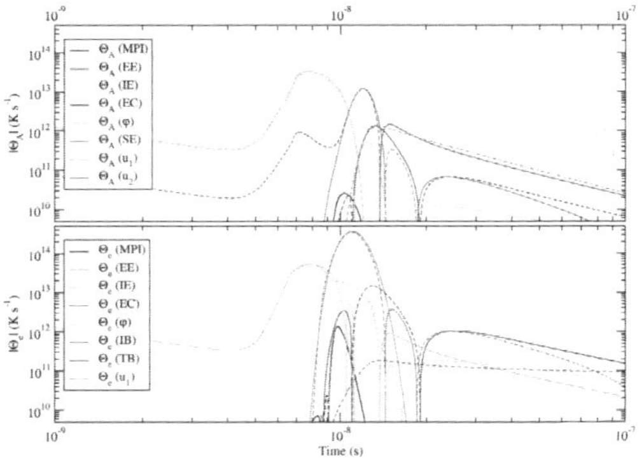

Figure 6. Same as figure 3, but for $\lambda = 1064$ nm.

We can analyze these results using Θ_e and Θ_A. Figures 3, 4, 5 and 6 display the evolution in time of their different contributions for $\lambda = 266$, 355, 532 and 1064 nm, respectively. We can see that evolution of Θ_A for $\lambda = 1064$ nm, a competition takes place after 10 ns among the processes electron-induced excitation, elastic collisions and spontaneous emission. Whereas in other cases, the cooling is due to elastic collisions with electrons ($T_e < T_A$). However in all cases, for $t \leq 10^{-8}$ s, the evolution mainly results from the evaporation and from the translational energy storage. From a transversal view of the term Θ_e, we can see that multiphoton ionization takes place before inverse Bremsstrahlung. In addition, the role of MPI decreases as λ increases, whereas the one of IB increases. Table 3 summarizes the maximum of $\Theta_e(MPI)$ and $\Theta_e(IB)$ and the corresponding time. We can see that the heating of the electron gas is due to multiphoton ionization at the pulse onset in the UV range, and to inverse Bremsstrahlung in the IR range after the maximum irradiance. Besides, figures 3, 4, 5 and 6 point out that the electron gas cooling is mainly due to excitation under electron impact (EE) for $t > 10^{-8}$ s. We can also notice that, for $t > 2 \times 10^{-8}$ s, elastic collisions (EC) play a significant role on the evolution of T_e. These collisions lead to the thermal equilibrium observed for $t \approx 10^{-7}$ s.

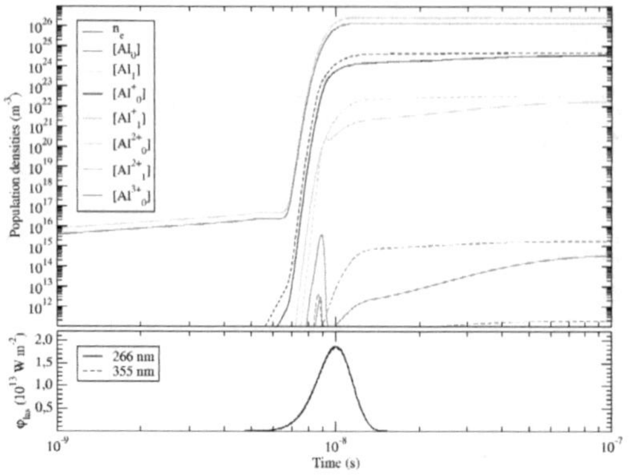

Figure 7. Time-evolution of the number density of some states (top) and irradiance φ_{las} (bottom) (continuous lines: $\lambda = 266$ nm, dashed lines: $\lambda = 355$ nm).

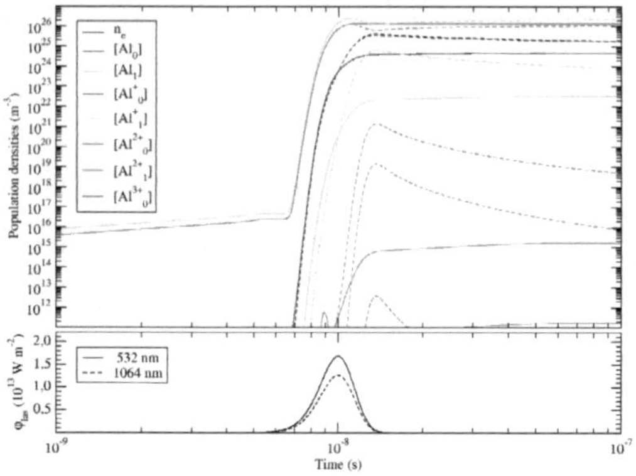

Figure 8. Same as figure 7, but for $\lambda = 532$ nm (continuous lines) and $\lambda = 1064$ nm (dashed lines).

4.2. Population densities

Figures 7 and 8 display the evolution in time of the number densities of electrons and of main atomic and ionic excited states for $\lambda = 266$, 355, 532 and 1064 nm. Starting from very small values due to the evaporation of the sample at its melting point, the number densities strongly increase over ten orders of magnitude until $t \approx 10^{-8}$ s. At this time, the critical temperature is reached for the sample. The total number density is then of 2.5×10^{26} m^{-3}. In the subsequent part of the evolution, the amplitude of the variations is much lower. The cases at $\lambda = 266$, 355 and 532 nm are very close. Despite a slight delay in ionization for $\lambda = 532$ nm with respect to the cases at $\lambda = 266$ and 355 nm, the electron number density approximately reaches 2×10^{24} m^{-3} for $\lambda = 266$ nm and 4×10^{24} m^{-3} for $\lambda = 355$ and 532 nm at the end of the pulse. The ionization degree is then slightly lower than 0.02. On the other hand, electron number density is higher by one order of magnitude (4×10^{25} m^{-3}) at the end of the pulse at $\lambda = 1064$ nm. This is due to the increase in T_e. Electrons then have enough energy to ionize atoms, as illustrated by figure 6. They are also responsible for the production of Al^{2+} and Al^{3+}, the density of which reaches 10^{21} m^{-3} and 10^{12} m^{-3} at $t \approx 1.3 \times 10^{-8}$ s, respectively. The Al^{2+} and Al^{3+} densities are totally negligible for the other wavelength cases. Globally, the densities increase with time except for the case at $\lambda = 1064$ nm where we can notice an evolution of the first excited states population density of Al after the maximum irradiance due to recombination of ions.

Table 3. Time of the maximum of $\Theta_e(MPI)$ and $\Theta_e(IB)$ and related values for $\lambda = 355$, 532 and 1064 nm.

λ (nm)	Multiphoton ionization		Inverse Bremsstrahlung	
	t_{max} (ns)	$\Theta_e(MPI)$ (K s^{-1})	t_{max} (ns)	$\Theta_e(IB)$ (K s^{-1})
266	7.2	3.9×10^{14}	10.7	2.8×10^{13}
355	7.2	1.2×10^{14}	10.7	4.9×10^{13}
532	6.9	6.5×10^{13}	10.7	9.3×10^{13}
1064	9.8	1.4×10^{12}	11.0	3.7×10^{14}

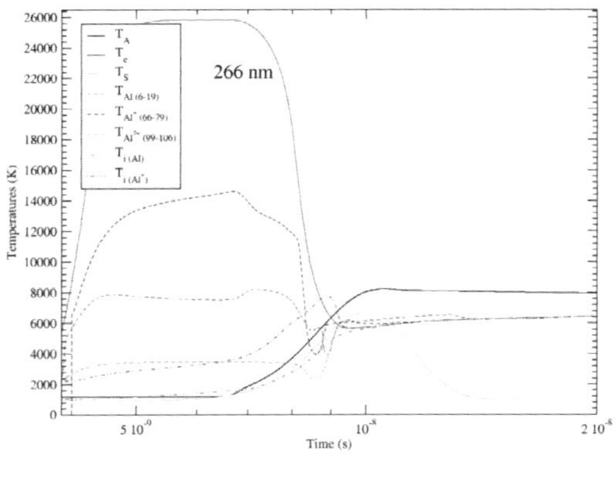

Figure 9. Temporal evolution of the excitation temperatures $T_{Al(6-19)}$, $T_{Al^+(66-79)}$, $T_{Al^{2+}(99-106)}$ and of the ionization temperatures $T_{i(Al)}$ and $T_{i(Al^+)}$ calculated with Eqs.(6) and (7) for a laser pulse at $\lambda = 266$ nm

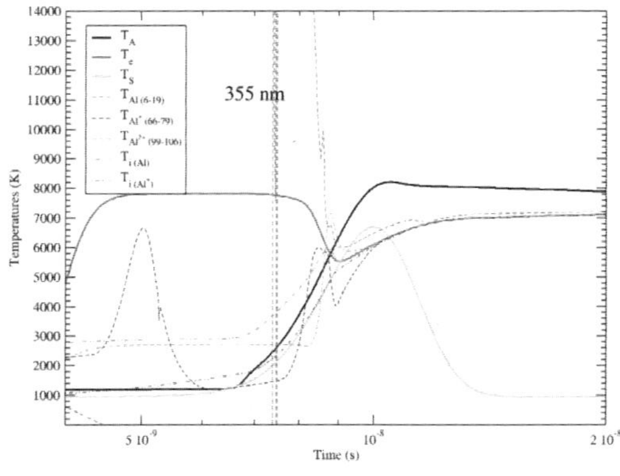

Figure 10. Same as figure 9, but for $\lambda = 355$ nm.

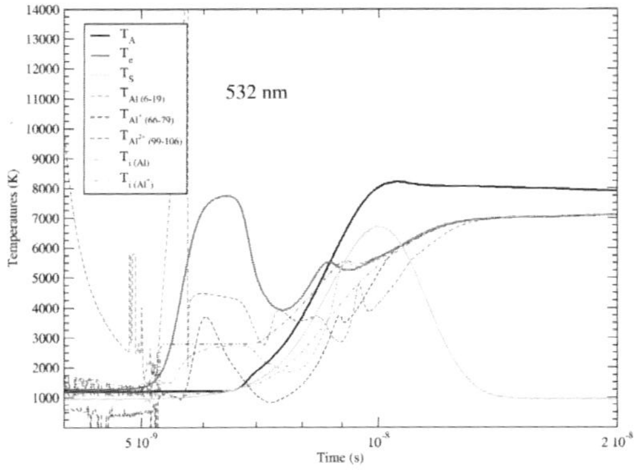

Figure 11. Same as figure 9, but for $\lambda = 532$ nm.

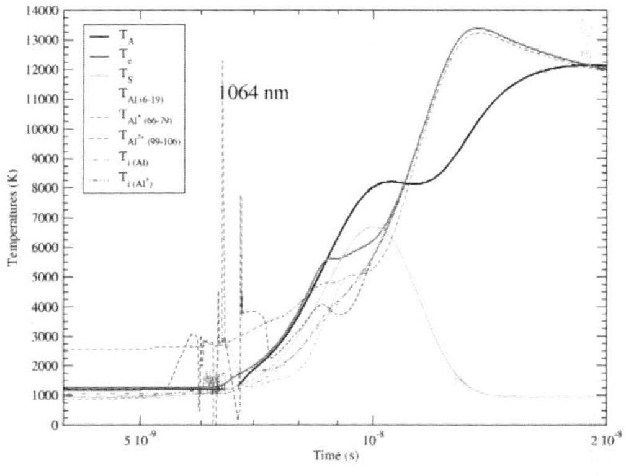

Figure 12. Same as figure 9, but for $\lambda = 1064$ nm.

4.3. Excitation and ionization temperatures

This model allows to calculate in time the different excitation temperature and ionization temperature. Therefore, the departure of the plasma from chemical equilibrium can be estimated. The excitation temperature $T_{X(i_{min}-i_{max})}$ for the species X is calculated by

$$T_{X(i_{min}-i_{max})} = -\frac{1}{k_B \left[\frac{d}{dE_i}\left(\ln\frac{[X_i]}{g_i}\right)\right]_{lsl, i_{min}<i<i_{max}}} \tag{6}$$

where lsl means that the derivative is the slope of the least squares line over the excited states interval $i_{min} < i < i_{max}$. E_i and g_i are the excitation energy and the statistical weight of the level i. The ionization temperature $T_{i(Al^{(Z+)})}$ is calculated by

Journal of Physics: Conference Series **550** (2014) 012048 doi:10.1088/1742-6596/550/1/012048

$$T_{i(Al^{(Z+)})} = \frac{E_{ioni}^{(Z+)}}{k_B \ln\left[2\,\frac{g_0^{([Z+1]+)}}{g_0^{(Z+)}}\,\frac{[Al_0^{(Z+)}]}{[Al_0^{([Z+1]+)}]\,n_e}\left(\frac{2\,\pi\,m_e\,k_B\,T_{i(Al^{(Z+)})}}{h^2}\right)^{3/2}\right]} \quad (7)$$

where $g_0^{(Z+)}$ and $[Al_0^{(Z+)}]$ are the statistical weight and the density of the ground state of $Al^{(Z+)}$, respectively.

Equation (7) indicates that the ionization temperature is calculated using an iterative procedure. $T_{i(Al)}$ is the temperature of the plasma at equilibrium having the same ratio between number densities of Al and Al^+ ground states.

Figure 9, 10, 11 and 12 illustrate the evolution in time of these characteristic temperatures for $\lambda = 266$, 355, 532 and 1064 nm, respectively. T_A, T_e and T_S are also displayed in order to facilitate comparisons. In spite of small integration time steps, some curves are noisy. At early times, the population number densities can be very small and strongly vary from a step to an other. Their evolution leads to strong variations of the characteristic temperatures despite the smoothing induced by equations (6) and (7). Despite this noise, couplings occur rapidly in each case. Before the maximum irradiance, the characteristic temperatures clearly depart from T_e. After this maximum, the coupling is obvious but takes more time for $T_{Al^{2+}}$ with respect to other excitation temperatures. This coupling results from the very high efficiency of collisions under electron impact.

We have stated that the coupling is obtained if the discrepancy between T_e and $T_{X(i_{min}-i_{max})}$ or $T_{i(Al^{(Z+)})}$ is lower than 1000 K. The coupling is therefore observed at $t \approx 9.1 \times 10^{-9}$, $t \approx 9.1 \times 10^{-9}$, 9.6×10^{-9} and 9.7×10^{-9} s for $\lambda = 266$, 355, 532 and 1064 nm, respectively. This slight delay is correlated with the delay of n_e. When n_e reaches 10^{23} m^{-3} approximately, electron number density is high enough to induce the chemical equilibrium of the plasma. Radiative losses and transport phenomena are often responsible for departures from equilibrium [30, 31]. In the present case, despite the significant radiative losses illustrated by figure 6 at $\lambda = 1064$ nm and the sonic expansion of the plasma layer, electron number density is so high that the chemical equilibrium is fulfilled during almost the entire duration of the pulse.

These results extend the conclusions put forward in [18]. In this paper, the maximum temperature of the sample surface was assumed lower than in the present case. Indeed, the maximum temperature was set equal to 3900 K. Data of Alcock *et al.* [32] give a value of 7×10^6 Pa for the equilibrium vapor pressure p_v, whereas p_v is almost equal to 4×10^8 Pa at $T_c = 6700$ K. As a result, the plasma density was lower at a value of 10^{25} m^{-3} instead of some 10^{26} m^{-3} in the present case. This increase in the plasma density leads to a higher collision frequency, to a higher absorption of the laser light and consequently to a higher electron number density. The efficiency of the electron-induced collisions is then higher, which explains the quasi equilibrium of the plasma in the present conditions for $t > 10^{-8}$ s. The plasma is formed by absorption of the laser light during the pulse and not from the direct ejection of electrons from the heated surface.

5. Conclusion

The behavior of a plasma produced on an aluminium sample by either one of the Nd:YAG harmonics laser source in nanosecond regime is modeled by using a 0D approach involving the Collisional-Radiative (CR) model CoRaM-Al. Free electrons and heavy species temperatures are assumed different. About hundred excited states of Al, Al^+, Al^{2+} and Al^{3+} are considered. These species are assumed to behave freely through seven different collisional and radiative elementary processes. A simplified approach is adopted: the interface temperature evolves like the laser irradiance and the target surface temperature reaches the critical temperature, therefore the maximum level avoiding the explosion phase.

The results put forward the fundamental role played by vaporization from the interface and by multi-photon ionization in the formation of the plasma at lower wavelengths. The electron temperature depends weakly on inverse Bremsstrahlung phenomena in UV range and strongly depends on inverse Bremsstrahlung at 1064 nm. When the laser pulse ends, the relaxation of the plasma takes place until equilibrium mainly through electron-induced processes and elastic collisions.

We have previously mentioned that LIBS signals analysis is in part based on equilibrium hypotheses: in particular, the excitation is often assumed in equilibrium. The present CR model is a particularly suitable tool to verify the validity of this assumption. We have indeed derive excitation temperature and ionization temperature from our results. In the present conditions, equilibrium is reached rapidly due to the high electron density.

References

[1] Noll R, Bette H, Brysch A, Kraushaar M, Mönch I, Peter L and Sturm V 2001 *Spectrochim. Acta B* **56** 637-649
[2] Tognoni E, Cristoforetti G, Legnaioli S and Palleschi V 2010 *Spectrochim. Acta B* **65** 1-14
[3] Bulajic D, Corsi M, Cristoforetti G, Legnaioli S, Palleschi V, Salvetti A and Tognoni E 2002 *Spectrochim. Acta B* **57** 339-353
[4] Aguilera J A, Aragon C, Cristoforetti G and Tognoni E 2009 *Spectrochim. Acta B* **64** 685-689
[5] Travaillé G, Peyrusse O, Bousquet B, Canioni L, Michel-Le Pierres K and Roy S 2009 *Spectrochim. Acta B* **64** 931-937
[6] Morel V, Bultel A and Chéron B G 2009 *Int. J. Thermophys.* **30** 1853-1863
[7] Cabalin L M and Laserna J J 1998 *Spectrochim. Acta B* **53** 723-730
[8] Mazhukin V I, Nossov V V, Nickiforov M G and Smurov I 2003 *J. Appl. Phys.* **93** 56-66
[9] Porneala C and Willis D A 2006 *Appl. Phys. Lett.* **89** 211121
[10] Cao Y, Zhao X and Shin Y C 2013 *J. Laser Appl.* **25** 032002
[11] Fishburn J M, Withford M J, Coutts D W and Piper J A 2006 *Appl. Surf. Sci.* **252** 5182-5188
[12] Morel V and Bultel A 2014 *Spectrochim. Acta B* **94-95** 63-70
[13] Lu Q, Mao S S, Mao X and Russo R E 2002 *Appl. Phys. Lett.* **80** 17 3072-3074
[14] Cristoforetti G, De Giacomo A, Dell'Aglio M, Legnaioli S, Tognoni E, Palleschi V and Omenetto N 2010 *Spectrochim. Acta B* **65** 86-95
[15] Lagrange J F, Wolfman J and Motret O 2012 *J. Appl. Phys.* **111** 063301
[16] Akhatov I, Lindau O, Topolnikov A, Mettin R, Vakhitova N and Lauterborn W 2001 *Phys. Fluids* **13** 10 2805-2819
[17] Gusarov A V and Smurov I 2002 *Phys. Fluids* **14** 12, 4242-4255
[18] Morel V, Bultel A and Chéron B G 2010 *Spectrochim. Acta B* **65** 830-841
[19] Müsing A, Riedel U, Warnatz J, Herden W and Ridderbusch H 2007 *Proc. Comb. Inst.* **31** 3007-3014
[20] Zel'dovich Ya B and Raizer Yu P 2002 *Physics of shock waves and high-temperature hydrodynamic phenomena* (Mineola, Dover publication Inc.)
[21] Weyl G M and Rosen D 1985 *Phys. Rev. A* **31** 2300-2313
[22] Venugopalan M 1971 *Reactions under Plasma Conditions* (New York, John Wiley and Sons Inc.)
[23] Tucker W H and Gould R J 1966 *Astrophys. J* **144** 244-258
[24] Wilbers A T M, Kroesen G M W, Timmermans C J and Schram D C 1991 *J. Quant. Spectrosc. Radiat. Trans.* **45** 1-10
[25] Drawin H W 1966 *Collision and Transport Cross-Sections* (Association-Euratom-C.E.A. Technical revised report EUR-CEA-FC)
[26] Decoster A, Markowich P A and Perthame B 1998 *Modeling of Collisions* (Paris, Gauthier-Villars)
[27] *http://physics.nist.gov/PhysRefData/ASD/levels_form.html*
[28] Holstein T 1947 *Phys. Rev.* **72** 1212-1233
[29] Bultel A, van Ootegem B, Bourdon A and Vervisch P 2002 *Phys. Rev. E* **65** 046406
[30] Drawin H W 1969 *Z. Physik* **228** 99-119
[31] Drawin H W 1975 *Validity conditions for local thermodynamic equilibrium, in Progress in Plasmas and Gas Electronics* (Berlin, Akademic Verlag)
[32] Alcock C B, Itkin V P and Horrigan M K 1984 *Can. Metall. Q* **23** 309-313

Elaboration of collisional-radiative models applied to Earth and Mars entry problems

Julien Annaloro(1), **Arnaud Bultel**(2) **and Pierre Omaly**(1)

(1)CNES, Avenue Edouard Belin, 31401 Toulouse, France

(2)CORIA - UMR 6614 - Normandie Université, CNRS-Université et INSA de Rouen, Campus universitaire du Madrillet, 76800 Saint-Etienne du Rouvray, France

E-mail: julien.annaloro@cnes.fr

Abstract. Three Collisional-Radiative (CR) models are elaborated and tested in typical atmospheric entry conditions. The first CR model (CoRaM-AIR) is dedicated to the Earth atmospheric entry and is based on an electronically and vibrationally specific state-to-state description of N_2-O_2-Ar mixtures. The second CR model (CoRaM-MARS) is dedicated to the Mars atmospheric entry and treats the CO_2-N_2-Ar mixtures with a similar vibrationally and electronically specific approach. Since their implementation in a Computational Fluid Dynamics (CFD) code has not yet been performed, they are implemented in a 0D code giving the evolution in time of the excited states number density in constant pressure and temperature conditions similar to trajectory points at lower altitude. Nevertheless, such an implementation in a CFD code has been performed for a third CR model, specifically devoted to pure nitrogen flows (CoRaM-N_2). The results show that the equilibrium is reached relatively slowly. In addition, the influence of radiation on the chemistry is weak.

1. Introduction

The shock layer produced in front of a spacecraft or a probe during its atmospheric entry is out of Local Thermodynamic Equilibrium (LTE) [1]. In particular, the excited states depart from the Boltzmann distribution. The resulting radiation is therefore strongly coupled to the chemistry and the computation of the situation, performed to predict the radiative fluxes to the entering body and consequently to size its thermal protection system, requires to give up the excitation assumption. The most relevant way is then to consider each excited state as independent and to apply the balance equation to each one [2]. The species on their different excited states are therefore considered as full species. In the related balance equation, the source term depends on the elementary collisional and radiative processes. The Collisional-Radiative (CR) model is the set of elementary processes taken into account. The present paper deals with the elaboration of three CR models relative to Earth and Mars entries, the third one being dedicated to pure nitrogen plasma flows implemented in an Euler Computational Fluid Dynamics (CFD) code.

2. CoRaM-AIR, CoRaM-MARS and CoRaM-N_2 CR models

In order to treat the case of Earth and Mars atmospheric entries, the following three CR models have been elaborated in thermochemical non equilibrium assuming that electrons and heavy particles are in Maxwellian equilibrium at T_e and T_A, respectively. Moreover, electroneutrality $\sum_Z [X^{Z+}] - n_e = 0$ is assumed with n_e the electron density and Z the charge of the ion X^{Z+}.

Table 1. Species and states considered in **CoRaM-AIR.**

Types	Species	States
Molecules	N_2	$X^1\Sigma_g^+(v=0\rightarrow 67)$, $A^3\Sigma_u^+$, $B^3\Pi_g$, $W^3\Delta_u$, $B'^3\Sigma_u^-$, $a'^1\Sigma_u^-$, $a^1\Pi_g$, $w^1\Delta_u$, $G^3\Delta_g$, $C^3\Pi_u$, $E^3\Sigma_g^+$
	O_2	$X^3\Sigma_g^-(v{=}0\rightarrow 46)$, $a^1\Delta_g$, $b^1\Sigma_g^+$, $c^1\Sigma_u^-$, $A'^3\Delta_u$, $A^3\Sigma_u^+$, $B^3\Sigma_u^-$, $f^1\Sigma_u^+$
	NO	$X^2\Pi(v{=}0\rightarrow 53)$, $a^4\Pi$, $A^2\Sigma^+$, $B^2\Pi$, $b^4\Sigma^-$, $C^2\Pi$, $D^2\Sigma^+$, $B'^2\Delta$, $E^2\Sigma^+$, $F^2\Delta$
Molecular ions	N_2^+	$X^2\Sigma_g^+$, $A^2\Pi_u$, $B^2\Sigma_u^+$, $a^4\Sigma_u^+$, $D^2\Pi_g$, $C^2\Sigma_u^+$
	O_2^+	$X^2\Pi_g$, $a^4\Pi_u$, $A^2\Pi_u$, $b^4\Sigma_g^-$
	NO^+	$X^1\Sigma^+$, $a^3\Sigma^+$, $b^3\Pi$, $W^3\Delta$, $b'^3\Sigma^-$, $A'^1\Sigma^+$, $W^1\Delta$, $A^1\Pi$
Atoms	N	${}^4S^o_{3/2}$, ${}^2D^o_{5/2}$, ${}^2D^o_{3/2}$, ${}^2P^o_{1/2}$ ... (252 levels)
	O	${}^3P_2$, ${}^3P_1$, ${}^3P_0$, ${}^1D_2$ ... (127 levels)
	Ar	${}^1S_0$, ${}^2[3/2]^o_2$, ${}^2[3/2]^o_1$, ${}^2[1/2]^o_0$ ... (379 levels)
Atomic ions	N^+	${}^3P_0$, ${}^3P_1$, ${}^3P_2$, ${}^1D_2$ ... (9 levels)
	O^+	${}^4S^o_{3/2}$, ${}^2D^o_{5/2}$, ${}^2D^o_{3/2}$, ${}^2P^o_{3/2}$ ... (8 levels)
	Ar^+	${}^2P^o_{3/2}$, ${}^2P^o_{1/2}$, ${}^2S_{1/2}$, ${}^4D_{7/2}$ ... (7 levels)
e^-	-	-

Table 2. Species and states considered in **CoRaM-MARS.**

Types	Species	States
Molecules	CO_2	$X^1\Sigma_g^+$
	N_2	$X^1\Sigma_g^+(v=0\rightarrow 67)$, $A^3\Sigma_u^+$, $B^3\Pi_g$, $W^3\Delta_u$, $B'^3\Sigma_u^-$, $a'^1\Sigma_u^-$, $a^1\Pi_g$, $w^1\Delta_u$, $G^3\Delta_g$, $C^3\Pi_u$, $E^3\Sigma_g^+$
	O_2	$X^3\Sigma_g^-(v{=}0\rightarrow 46)$, $a^1\Delta_g$, $b^1\Sigma_g^+$, $c^1\Sigma_u^-$, $A'^3\Delta_u$, $A^3\Sigma_u^+$, $B^3\Sigma_u^-$, $f^1\Sigma_u^+$
	C_2	$X^1\Sigma_g^+(v{=}0\rightarrow 36)$, $a^3\Pi_u$, $b^3\Sigma_g^-$, $A^1\Pi_u$, $c^3\Sigma_u^+$, $d^3\Pi_g$, $C^1\Pi_g$, $e^3\Pi_g$, $D^1\Sigma_u^+$
	NO	$X^2\Pi(v{=}0\rightarrow 53)$, $a^4\Pi$, $A^2\Sigma^+$, $B^2\Pi$, $b^4\Sigma^-$, $C^2\Pi$, $D^2\Sigma^+$, $B'^2\Delta$, $E^2\Sigma^+$, $F^2\Delta$
	CO	$X^1\Sigma^+(v{=}0\rightarrow 76)$, $a^3\Pi$, $a'^3\Sigma^+$, $d^3\Delta$, $e^3\Sigma^-$, $A^1\Pi$, $I^1\Sigma^-$, $D^1\Delta^-$, $b^3\Sigma^+$, $B^1\Sigma^+$
	CN	$X^2\Sigma^+(v{=}0\rightarrow 41)$, $A^2\Pi$, $B^2\Sigma^+$, $D^2\Pi$, $E^2\Sigma^+$, $F^2\Delta$
Molecular ions	N_2^+	$X^2\Sigma_g^+$, $A^2\Pi_u$, $B^2\Sigma_u^+$, $a^4\Sigma_u^+$, $D^2\Pi_g$, $C^2\Sigma_u^+$
	O_2^+	$X^2\Pi_g$, $a^4\Pi_u$, $A^2\Pi_u$, $b^4\Sigma_g^-$
	C_2^+	$X^4\Sigma_g^-$, $1^2\Pi_u$, ${}^4\Pi_u$, $1^2\Sigma_g^+$, $2^2\Pi_u$, $\tilde{B}^4\Sigma_u^-$, $1^2\Sigma_u^+$
	NO^+	$X^1\Sigma^+$, $a^3\Sigma^+$, $b^3\Pi$, $W^3\Delta$, $b'^3\Sigma^-$, $A'^1\Sigma^+$, $W^1\Delta$, $A^1\Pi$
	CO^+	$X^2\Sigma^+$, $A^2\Pi$, $B^2\Sigma^+$, $C^2\Delta$
	CN^+	$X^1\Sigma^+$, $a^3\Pi$, ${}^1\Delta$, $c^1\Sigma^+$
Atoms	N	${}^4S^o_{3/2}$, ${}^2D^o_{5/2}$, ${}^2D^o_{3/2}$, ${}^2P^o_{1/2}$ ... (252 levels)
	O	${}^3P_2$, ${}^3P_1$, ${}^3P_0$, ${}^1D_2$ ... (127 levels)
	C	${}^3P_0$, ${}^3P_1$, ${}^3P_2$, ${}^1D_2$ ... (265 levels)
	Ar	${}^1S_0$, ${}^2[3/2]^o_2$, ${}^2[3/2]^o_1$, ${}^2[1/2]^o_0$ ... (379 levels)
Atomic ions	N^+	${}^3P_0$, ${}^3P_1$, ${}^3P_2$, ${}^1D_2$ ... (9 levels)
	O^+	${}^4S^o_{3/2}$, ${}^2D^o_{5/2}$, ${}^2D^o_{3/2}$, ${}^2P^o_{3/2}$ ... (8 levels)
	C^+	${}^2P^o_{1/2}$, ${}^2P^o_{3/2}$, ${}^4P_{1/2}$, ${}^4P_{3/2}$ ... (8 levels)
	Ar^+	${}^2P^o_{3/2}$, ${}^2P^o_{1/2}$, ${}^2S_{1/2}$, ${}^4D_{7/2}$ ... (7 levels)
e^-	-	-

2.1. CoRaM-AIR

Specifically devoted to Earth atmospheric entries, this CR model takes into account free electrons and N_2, O_2, Ar, NO, N, O, N_2^+, O_2^+, Ar^+, NO^+, N^+, O^+ on the different excited states listed in table 1. Only the vibrational states are considered on the electronic ground state of molecules. The elementary processes involving these states are listed hereafter.

Journal of Physics: Conference Series **550** (2014) 012049 doi:10.1088/1742-6596/550/1/012049

Table 3. Species and states considered in **CoRaM-N$_2$**.

Types	Species	States
Molecules	N_2	$X^1\Sigma_g^+(v=0\rightarrow 67)$, $A^3\Sigma_u^+$, $B^3\Pi_g$, $W^3\Delta_u$, $B'^3\Sigma_u^-$, $a'^1\Sigma_u^-$, $a^1\Pi_g$, $w^1\Delta_u$, $G^3\Delta_g$, $C^3\Pi_u$, $E^3\Sigma_g^+$
Molecular ions	N_2^+	$X^2\Sigma_g^+$, $A^2\Pi_u$, $B^2\Sigma_u^+$, $a^4\Sigma_u^+$, $D^2\Pi_g$, $C^2\Sigma_u^+$
Atoms	N	${}^4S^o_{3/2}$, ${}^2D^o = ({}^2D^o_{5/2} +{}^2 D^o_{3/2})$, ${}^2P^o = ({}^2P^o_{1/2} +{}^2 P^o_{3/2})$, ${}^4P_{1/2}$,... (63 states)
Atomic ions	N^+	${}^3P_0$, ${}^3P_1$, ${}^3P_2$, ${}^1D_2$, ${}^1S_0$, ${}^5S^o_2$, ${}^3D^o_3$, ${}^3D^o_2$, ${}^3D^o_1$
e^-	-	-

- *Vibrational excitation and dissociation* by heavy particle (atomic and molecular) and electron-induced collisions.
- *Molecular excitation* by heavy particle (atomic and molecular) and electron-induced collisions.
- *Direct dissociation of the molecular excited states* by heavy particle (atomic and molecular) and electron-induced collisions.
- *Electronic excitation of atoms and ionization* by heavy particle (atomic and molecular) and electron-induced collisions.
- *Excitation transfer.*
- *Charge and neutral exchanges.*
- *Dissociative recombination.*
- *Spontaneous emission, radiative and dielectronic recombination.*

The rate coefficients of these processes cannot be reported in the present article but can be found in [3, 4]. Whenever relevant, the rate coefficient of each backward process is calculated from the corresponding forward process rate coefficient using the detailed balance principle [5]. The Einstein coefficients for the radiative transitions are taken from the NIST database [6] for atoms and atomic ions, and from the work of Chauveau [7] for molecules and molecular ions. A possible self-absorption of radiation is implemented through an escape factor based on the Holstein's approach [8]. The characteristic length scale of the plasma is then the typical thickness of the shock layer $R = 5\ cm$.

2.2. CoRaM-MARS

The CR model devoted to Mars atmospheric entries takes into account CO_2, N_2, O_2, C_2, NO, CO, CN, N_2^+, O_2^+, C_2^+, NO^+, CO^+, CN^+, N, O, C, Ar, N^+, O^+, C^+ and Ar^+ on the different states listed in table 2. The same approach as for CoRaM-AIR is used. Only the model is vibrationally specific on the electronic ground states of molecules, except for CO_2 for which we face a lack of elementary data. The considered elementary processes are the same as those taken into account for **CoRaM-AIR**. Values of their rate coefficients can be obtained in [3]. The Einstein coefficients are also taken from the NIST database for atoms and atomic ions transitions and data related to molecular systems are due to Chauveau, Babou *et al.* [9] and Lino da Silva [10].

2.3. CoRaM-N$_2$

For pure nitrogen plasma flows, the CR model CoRaM-N$_2$ is elaborated. N_2, N_2^+, N, N^+ and electrons are considered (see table 3) with the same vibrational approach for the electronic $X^1\Sigma_g^+$ ground state of N_2. This model and the related rate coefficients have been extensively described in [3, 11]. Contrary to **CoRaM-AIR** and **CoRaM-MARS**, this CR model is implemented in

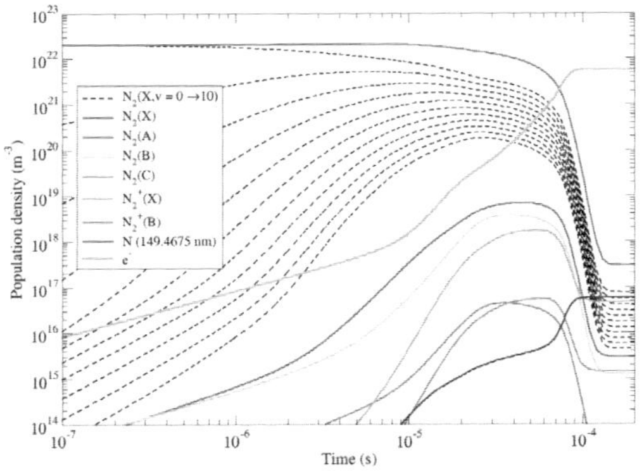

Figure 1. Evolution in time of some states of N_2, N_2^+ and N involved in **CoRaM-AIR** when the mixture in the conditions (p, T_A) = (50 Pa, 250 K) is suddenly put in the conditions (80 kPa, 11000 K).

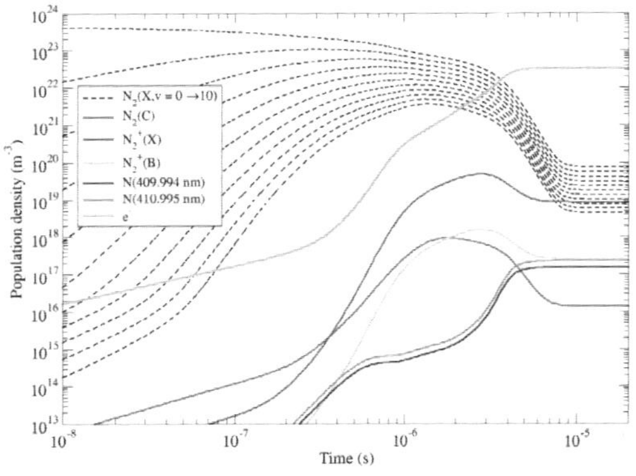

Figure 2. Evolution in time of some states of N_2, N_2^+ and N involved in **CoRaM-AIR** when the mixture in the conditions (p, T_A) = (50 Pa, 250 K) is suddenly put in the conditions (4 kPa, 11000 K).

a CFD code. Since the entry flow can be considered as one dimensional in the first centimetres of the shock layer, this CFD code is one dimensional as well. It is based on the Eulerian approximation of the Navier-Stokes equations, which is relevant for fast flows. The nitrogen flow is considered in thermochemical non equilibrium, which means that electrons and heavy particles have different temperatures (noted T_e and T_A in the upcoming sections, respectively). The results discussed hereafter have been obtained using the DVODE library [12] with a time (spatial) step less than 0.01×10^{-M} s(m) for the decade $t, (x) \in [10^{-M}; 10^{-M+1}]$ s,(m).

3. Results

3.1. CoRaM-AIR

CoRaM-AIR is used to give an insight into the relaxation of a fluid particle of the upstream N_2-O_2-Ar mixture (pressure $p = 50$ Pa, temperature $T_A = 250$ K) suddenly crossing the shock front limiting the shock layer under constant pressure $p = 80$ kPa and temperature $T_A = 11000$ K conditions. Figure 1 displays the temporal evolution of some of the species and states listed in table 1 assuming $T_e = T_A$. The initial population density of the excited states are considered at equilibrium in the upstream conditions. The vibrational states of $N_2(X)$ are rapidly populated and the ionization takes place. At $t \simeq 10^{-5}$ s, the relaxation of the mixture is over. The electron density is then $\sim 10^{22}$ m^{-3}, the N_2 density is divided by a factor of 1000 with respect to the initial value, and the N_2^+ density is around 10^{19} m^{-3}. The density of two excited states leading to the emission of the two important lines at 409.994 and 410.995 nm (often observed in high enthalpy flows) is also displayed. One can see that with or whithout radiation, the time evolution is not modified. This indicates that radiation does not play a significant role on the relaxation. In these conditions, the final state is therefore an equilibrium state. This is due to the high pressure level which leads to a sufficiently high collision frequency. If the pressure level is reduced, for instance by considering the jump (50 Pa, 250 K) → (4 kPa, 11000 K), a similar behavior can be observed (see figure 2). Despite its significant reduction, the total density level is still sufficiently high to lead to an equilibrium final state, which is not modified by radiation. We can observe that the N upper state of the transition at 149.4675 nm is not significantly modified by radiation. The effect of the pressure reduction is obvious since the equilibrium state is reached at $t \simeq 10^{-4}$ s.

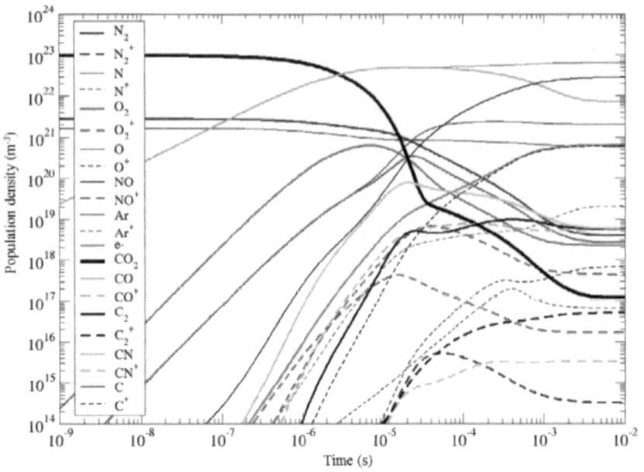

Figure 3. Evolution in time of the species involved in **CoRaM-MARS** when the mixture in the conditions $(p, T_A) = (600$ Pa, 260 K) is suddenly put in the conditions (10 kPa, 7000 K).

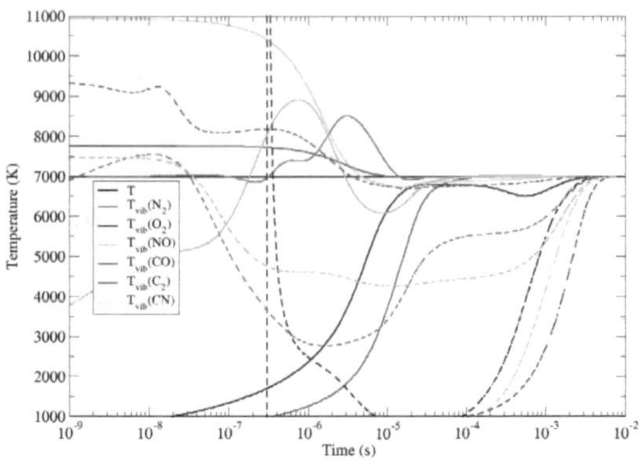

Figure 4. Evolution in time of the vibrational temperatures calculated for $v \in [0; 5]$ (in continuous lines) and for $v \in [v_{max} - 4; v_{max}]$ (in dashed lines). The final equilibrium state is obvious.

3.2. CoRaM-MARS

For Mars atmospheric entries, the conditions of the pressure-temperature variations are less drastic. A typical case is the (600 Pa, 260 K) $\rightarrow$ (10 kPa, 7000 K) transition. Figure 3 displays the time evolution of the species density obtained in such conditions with **CoRaM-MARS**. As for **CoRaM-AIR**, the initial excited states population density is calculated at equilibrium in the upstream conditions and thermal equilibrium ($T_e = T_A$) conditions are assumed during the relaxation. This relaxation lasts some 10^{-3} s. An equilibrium state is then reached since the final composition is not significantly modified by radiation. One can notice an ionization degree of ~ 0.01 with C^+ as main ionic species. The mixture is then mainly composed with C and O atoms. This relatively long relaxation time scale is due to the dissociation of the considered molecules resulting from the progressive population of their vibrationally excited states. Of course, this effect can be also noticed for **CoRaM-AIR**. The quantification of this effect can be obtained through the calculation of the vibrational temperature of the AB molecule defined over the vibrational states $[v_1; v_2]$ interval as

$$T_{vib}(AB) = -\frac{1}{k_B\left[\frac{d}{dE_v}\left(\ln[\mathrm{AB}(X,v)]\right)\right]_{lsl,v_1\rightarrow v_2}} \tag{1}$$

Actually, $T_{vib}(AB)$ is the excitation temperature of the group of vibrational levels corresponding to the vibrational quantum numbers v such as $v_1 \leq v \leq v_2$. E_v is the vibrational energy and lsl means that the derivative is the slope of the least squares line. On figure 4 are displayed the time evolution of T_{vib} for $v \in [0; 5]$ and $v \in [v_{max} - 4; v_{max}]$ for the molecules N_2, O_2, NO, CO, C_2 and CN. Several trends can be distinguished.

- The lower vibrational states are progressively excited when the molecule is present in the upstream flow (N_2) whereas they are produced with a relatively high population density when the molecule is not present in the upstream flow (O_2, NO, CO, C_2 and CN).
- The upper vibrational states present various behaviors resulting from the elementary processes.

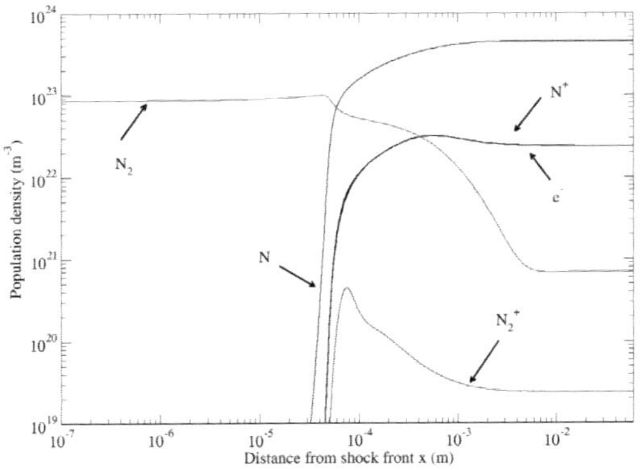

Figure 5. Evolution in space of the species involved in **CoRaM-N$_2$** implemented in a one-dimensional Eulerian CFD code for $\mathcal{M} = 30$ ($z = 56$ km altitude trajectory point of the FIRE II flight experiment).

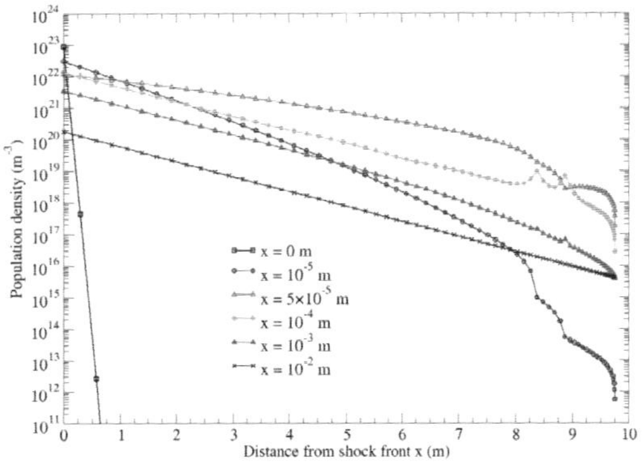

Figure 6. Evolution in space of the $N_2(X)$ vibrational states Boltzmann plot in the conditions of figure 5. The dissociation threshold is at 9.75 eV. Resonances with N_2 electronic excited states can be noticed.

- The lower vibrational states are coupled according to a Boltzmann distribution from $t \simeq 10^{-5}$ s while the dissociation process of the molecules within the flow is still taking place. During this transient phase, the vibrational temperature of the upper vibrational states is globally lower than the one related to the lower states, which indicates that these upper vibrational states are underpopulated.
- The vibration equilibrium is reached when the dissociation phase stops, at some 10^{-3} s.

The evidence is put forward that considering the ladder climbing excitation of the vibrational states is particularly relevant to realistically reproduce the dissociation [13]. Additional works, which cannot be fully described here, have clearly showed that the dynamics of the vibrational states slow down the dissociation kinetics. This has been demonstrated by comparison of the results due to the present CR model [3] and the ones due to a former CR model using only global dissociation rates without vibrational state-to-state approach [2]. This slowing down leads to a final equilibrium state later reached.

3.3. CoRaM-N$_2$

This progressive excitation of the vibrational mode plays a significant role in the case of pure nitrogen flows as illustrated by the **CoRaM-N$_2$** CR model implementation in an Eulerian CFD code. Figure 5 illustrates the spatial evolution behind the shock front (located at $x = 0$) of the N_2, N_2^+, N, N^+ and electron densities for a low altitude ($z = 54$ km, Mach number $\mathcal{M} = 30$) trajectory point of the FIRE II flight experiment [14]. In this calculation, electron temperature T_e and heavy particle temperature T_A are assumed different and driven by an independent energy conservation equation. The dissociation suddenly takes place around $x_d \simeq 5 \times 10^{-5}$ m and results from the cumulative effect of the progressive vibrational excitation. The global dissociation is followed by ionization for which N^+ is the main ionic species. The relaxation lasts until $x \simeq 10^{-2}$ m. The final uniform state does not depend on the radiative losses, which are taken into account in the energy conservation equations for heavy particles. As a result, the plasma is in equilibrium far from the shock front. Everywhere else, radiation has a very weak role on the chemistry of the flow.

Journal of Physics: Conference Series **550** (2014) 012049 doi:10.1088/1742-6596/550/1/012049

The dissociation process can be indirectly followed by analyzing in space the evolution of the Boltzmann plot for the vibrational states of N_2. Figure 6 illustrates this evolution especially around x_d. The distribution is far from being Boltzmanian during the dissociation process. In particular, a resonance between the vibrational states $v = 42, 47$ and the electronic excited states $a'^1\Sigma_u^-$ and $w^1\Delta_u$ of N_2 can be observed, which leads to a transient peaked distribution. In the subsequent part of the flow, the Boltzmann plot relaxes until a linear distribution. Then, a vibrational equilibrium is reached. Using equation (1) to evaluate the vibrational temperature, one finally obtains the same value as for T_A or T_e, i.e. 10500 K. LTE is therefore reached. For lower pressure conditions corresponding to higher altitude trajectory points, the distance from the shock front at which LTE is reached exceeds the typical thickness of the shock layer. This means that the shock layer is everywhere in non equilibrium. Even in these conditions, radiation does not play a significant role in the behavior of the excited states.

4. Conclusion

Three Collisional-Radiative (CR) models have been elaborated to give an insight into the excitation and dissociation processes of typical mixtures involved in the shock layer produced during Earth and Mars atmospheric entries. The first and second CR models, (**CoRaM-AIR** and **CoRaM-MARS**), treat the case of Earth and Mars entries, respectively. These CR models are characterized by the explicit calculation of the vibrational distribution of the molecular electronic ground states. For already existing molecules within the flow, the vibrational excitation is progressive, whereas it is more rapid for the molecules formed by the chemistry. The dissociation is therefore progressive. This illustrates the necessary accounting for vibrational states to realistically reproduce the dissociation process. The third CR model (**CoRaM-N$_2$**) is focused on nitrogen flows and has been implemented in an Eulerian computational fluid dynamics code in thermochemical non equilibrium. The radiative losses are taken into account. The dissociation is again progressive even if its cumulative effect leads to a sudden global dissociation of the flow. The radiative losses do not play a significant role.

References

[1] Anderson J D 1989 *Hypersonic and High Temperature Gas Dynamics* (New York, McGraw-Hill)
[2] Bultel A, Chéron B G, Bourdon A, Motapon O and Schneider I F 2006 *Phys. Plasmas* **13** 043502
[3] Annaloro J 2013 "Modèles collisionnels-radiatifs appliqués aux situations d'entrée atmosphérique martienne et terrestre," Ph.D. Dissertation (in French), CORIA, Université de Rouen, France
[4] Annaloro J and Bultel A 2014 *Phys. Plasmas* in preparation
[5] Annaloro J, Morel V, Bultel A and Omaly P 2012 *Physics of Plasmas* **19** 073515
[6] *http://physics.nist.gov/PhysRefData/ASD/levels_form.html*
[7] Chauveau S 2001 "Constitution de bases de données spectroscopiques relatives à un plasma d'air. Application au calcul de transfert radiatif," Ph.D. Dissertation (in French), EM2C, Ecole Centrale Paris, Châtenay-Malabry, France
[8] Holstein T 1947 *Phys. Rev.* **72** 1212
[9] Babou Y, Rivière P, Perrin M Y and Soufiani A 2009 *J. Quant. Spectrosc. Radiat. Transfer* **110** 89
[10] Lino da Silva M 2004 "Simulation des propriétés radiatives du plasma entourant un véhicule traversant une atmosphère planétaire à vitesse hypersonique," Ph.D. Dissertation (in French), ICARE, Université d'Orléans, France
[11] Annaloro J and Bultel A 2014 *J. Thermophys. Heat Transfer* **28** 568
[12] Brown P N, Byrne G D and Hindmarsch A C 1989 *SIAM Journal on Scientific and Statistical Computing* **10** 1038
[13] Bourdon A, Annaloro J, Bultel A, Capitelli M, Colonna G, Guy A, Magin T E, Munafò A, Perrin M-Y and Pietanza L D 2014 *Open Plasmas Physics Journal* **7** 60
[14] Cornette E S 1966 **NASA TM** X-1305

Journal of Physics: Conference Series **550** (2014) 012050 doi:10.1088/1742-6596/550/1/012050

Modelling radiative properties of participating species in a microwave plasma reactor for diamond deposition

S. Prasanna[1,2], **Ph. Rivière**[1,2] **and A. Soufiani**[1,2]

[1] CNRS, UPR 288, Laboratoire EM2C, Grande Voie des Vignes, F-92290 Châtenay-Malabry, France

[2] École Centrale Paris, Grande Voie des Vignes, F-92290 Châtenay-Malabry, France

E-mail: `prasannaswam@gmail.com`

Abstract. The paper details the modelling of radiation in a microwave assisted plasma reactor used to deposit synthetic diamond over a substrate. The main radiatively active constituents in the reactor are atomic and molecular hydrogen, acetylene, methane and soot (if produced). Radiation from hydrogen occurs in ultraviolet (UV) whereas the hydrocarbons are active in the infrared region. Soot absorb and scatter in the UV but only absorption is important in the infrared-visible (IR-V) region. Hence, the two spectral regions have been treated independently. A two temperature model has been adopted for hydrogen thermodynamic state where T_g represents rotational, vibrational and translational temperature and T_e represents electronic excitation temperature. As scattering is significant in UV, the radiative transfer equation is solved using Discrete Ordinate Method (DOM) with cumulative-k narrow-band model for molecular hydrogen. Radiation from atomic hydrogen has been found to be negligibly small compared to molecular hydrogen. In the IR-V, radiative transfer equation is solved using ray tracing method with gas properties represented by statistical narrow-band models. Preliminary simulations for reactor conditions indicate that soot significantly increase the radiative transfer in the reactor and presence of soot can disrupt the operation of the plasma reactor.

1. Introduction

Low pressure hydrogen methane plasmas generated using microwave resonating cavities can be used to deposit synthetic diamond on a substrate. A comprehensive review of the modelling diamond deposition process in plasma reactors is provided by Hassouni et al [1]. Understanding different mechanisms in the deposition is critical to produce diamonds of high quality. The number of parameters affecting the deposition process is large and it is desired to develop numerical models that take into account the different mechanisms in the medium. Such a development would assist in understanding and optimizing the process.

In line with the objective, it is intended to model radiative transfer in the plasma reactor which may effect temperature levels, composition, and flow fields. This work concerns modelling the radiative properties of the main participating species in the plasma reactor and solving the radiative transfer equation (RTE) in a two dimensional axi-symmetric domain. The present model will be coupled with plasma and flow multi-physics models in the future. Owing to simulated plasma composition [1], the main radiatively active constituents of the plasma reactor are molecular and atomic hydrogen, hydrocarbons such as methane, acetylene, and soot. Table

1 summarizes the radiative model developed for Ultraviolet (UV) and Infrared-Visible (IR-V) regions respectively. The following sections discuss the details of modelling of radiative properties of different constituents in the plasma reactor and present some preliminary results of radiative transfer for representative reactor conditions.

Table 1. Description of radiative transfer model in UV and IR-V spectrum.

	UV	IR-V
ν cm^{-1}	60000 to 120000	25 to 30000
Soot	Absorption and scattering	Absorption only
Gas	Molecular Hydrogen under NLTE	Acetylene and methane
Narrow band model	Cumulative-k distribution	Statistical narrow-band
RTE	Discrete ordinate method	Ray tracing

2. Radiative properties of participating species

A medium can interact with radiation either by emission, absorption or scattering. The Radiative Transfer Equation (RTE) accounts for the change in the spectral intensity $I_\nu(s,\Omega)$ along abscissa s, at wavenumber ν and direction Ω, resulting from these mechanisms

$$\frac{\partial}{\partial s}I_\nu(s,\Omega) = \eta_\nu + \frac{\sigma_\nu}{4\pi}\int_{\Omega'}\Phi_\nu(\Omega',\Omega)I_\nu(s,\Omega')d\Omega' - (\kappa_\nu+\sigma_\nu)I_\nu(s,\Omega), \quad (1)$$

where η_ν, κ_ν and σ_ν are respectively emission, absorption and scattering coefficients and Φ is the scattering phase function. Molecular spectra of hydrogen and hydrocarbons display large number of lines and line-by-line computation is expensive. Rather, narrow-band models are used to approximate the radiative properties. The choice of the computational model used for RTE solution depends on the narrow-band model of the gas as well as radiative properties of soot. Hydrogen is active in the UV whereas hydrocarbons are mainly active in the IR-V region. Concerning soot, their albedo in the Rayleigh limit scales as ν^3. Therefore, scattering can be neglected in the IR-V but must be taken into account in addition to emission and absorption in the UV region. Both molecular and atomic hydrogen in the plasma zone of the reactor are at non-equilibrium and a two temperature model is adopted in this study, i.e., an electronic temperature T_e and vibration- rotation-translation temperature T_g for heavy particles. The typical conditions in the plasma reactor are given in table 2 (adapted from collisional-radiative calculations of Hassouni et al [2]). Other constituents are assumed to be at local thermal equilibrium at T_g. Hence, to simplify RTE solving, the two spectral regions are considered separately. In the IR-V region, statistical narrow-band (SNB) model [3] has been chosen for modelling the radiative properties of the hydrocarbons due to its ability to treat arbitrary mixture of absorbing gases. The RTE in the IR-V is solved using a ray tracing method. As soot scatters in the UV region and since SNB models are not compatible with scattering and deterministic ray tracing, the RTE in UV region is solved using Discrete Ordinates Method (DOM) [4] coupled with cumulative-k (CK) narrow band model [3] for hydrogen.

2.1. Molecular hydrogen

Line data in UV spectrum (Lyman and Werner bands) for molecular hydrogen have been taken from Abgrall et al [5]. The database provides for each line i lower and upper rotational (J) and vibrational (v) quantum numbers, Einstein coefficient A_{ul} (s^{-1}) and the transitional wavenumber ν_i (cm^{-1}) for $B^1\Sigma_u^+ \rightarrow X^1\Sigma_g^+$ (19585 transitions) and $C^1\Pi_u \rightarrow X^1\Sigma_g^+$ (8881 transitions) band systems. The energy levels of $B^1\Sigma_u$ and $C^1\Pi_u$ states with respect to $v=0, J=0$ of $X^1\Sigma_g^+$

Table 2. Typical conditions inside the plasma reactor (adapted from Hassouni et al [2]).

Microwave power density ($W\ cm^{-3}$)	p Pa	T_g K	T_e K
4.5	1400	1585	5763
9	2500	2059	5670
15	5200	2702	5462
22	8400	3204	5328
30	11000	3410	5217

state have been calculated from ground state level energies provided by Dabrowski [6]. Spectral emissive power and absorption coefficient are calculated as

$$\begin{aligned} \eta_i(\nu) &= \frac{N_u A_{ul}}{4\pi} h\nu c f(\nu - \nu_i) && (Wm^{-3}sr^{-1}cm), \\ \kappa_i(\nu) &= N_u A_{ul} \frac{1}{8\pi\nu^2 c}\left(\frac{g_u N_l}{g_l N_u} - 1\right) f(\nu - \nu_i) && (m^{-1}). \end{aligned} \tag{2}$$

where $g_{u/l}$ is the level degeneracy given by $2J + 1$, N_u and N_l are the number densities of the two levels and $f(\nu - \nu_i)$ is the normalized profile of the line. Voigt profile is used for the line shape with collision broadening width, roughly estimated as $\gamma_c = 0.1\,(273/T_g)^{0.7}$ [7], and Doppler broadened line width $\gamma_d = \frac{\nu_i}{c}\sqrt{2kT_g ln2/m}$. The number density of an energy level (n, v, J) is given by

$$N_{n,v,J} = N \frac{g_s(2J+1)\exp\left(-\frac{E_n}{kT_e}\right)\exp\left(-\frac{E_{vr,n}}{kT_g}\right)}{Q(T_g, T_e)} \tag{3}$$

where E_n, and $E_{vr,n}$ are the electronic and vibrational-rotational energies respectively, g_s is the nuclear statistical weight and N is the number density of hydrogen molecules which can be calculated using ideal gas law ($N = xp/kT_g$). The molecular partition function for the two temperature non-equilibrium model can be written as

$$Q(T_g, T_e) = \sum_n \sum_{v,J} \exp\left(-\frac{E_n}{kT_e}\right) g_s(2J+1) \exp\left(-\frac{E_{vr,n}}{kT_g}\right). \tag{4}$$

Solution of RTE under non-equilibrium conditions requires modelling of both absorption and emission coefficients. Integrating the RTE over a ray between 0 and s gives

$$I_\nu(s) = \tau_\nu(0,s) I_\nu(0) + \int_0^s \frac{\eta_\nu(s')}{\kappa_\nu(s')} \frac{\partial \tau_\nu(s', s)}{\partial s'} ds' \tag{5}$$

where $\tau_\nu(s', s)$ is the spectral transmissivity of the path $s' \to s$. Averaging the above equation over a narrow spectral band gives

$$\bar{I}^{\Delta\nu}(s) = \overline{\tau_\nu(0,s) I_\nu(0)}^{\Delta\nu} + \int_0^s \overline{\frac{\eta_\nu(s')}{\kappa_\nu(s')} \frac{\partial \tau_\nu(s', s)}{\partial s'}}^{\Delta\nu} ds' \tag{6}$$

Thus, for accurate approximation of spectrally averaged emission, the narrow band model has to accurately predict the spectrally averaged kernel of Eq. 6. η_ν and τ_ν are strongly correlated for molecular band systems. This can be confirmed from figure 1 where the comparison between $\overline{\eta_\nu \tau_\nu}$ and $\overline{\eta_\nu}\ \overline{\tau_\nu}$ for $B \to X$ band systems is poor. However Lamet et al [8] have shown that

$$\overline{\frac{\eta}{\kappa} \frac{\partial \tau(s, s')}{\partial s'}}^{\Delta\nu} \approx \overline{\frac{\eta}{\kappa}}^{\Delta\nu}\ \overline{\frac{\partial \tau(s, s')}{\partial s'}}^{\Delta\nu} \tag{7}$$

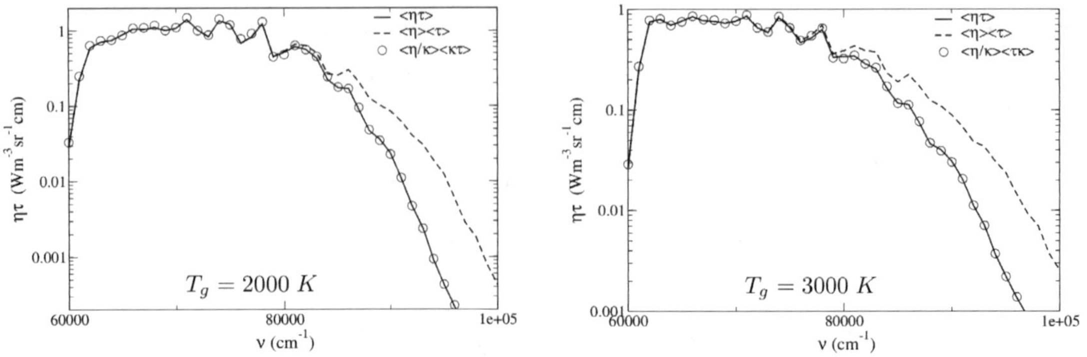

Figure 1. Correlation between emission and transmission for pure molecular hydrogen over path length 1 cm and pressure $p = 0.1$ bar for $B^1\Sigma_u^+ \rightarrow X^1\Sigma_g^+$ at $T_e = 5500\ K$ and $\Delta\nu = 1000\ cm^{-1}$ for two values of gas temperature

is a good approximation over the narrow band for a given electronic system. Hence, $\overline{\eta/\kappa}$ can be tabulated and the transmissivities can be evaluated using the CK model [3]. For overlapping electronic systems like $B \rightarrow X$ and $C \rightarrow X$ in our case, the two systems must be treated as belonging to two different gases, for the above decorrelation approximation (Eq. 7) to remain accurate. Moreover, the mixture transmissivity can be accurately calculated as the product of single transmissivities.

In the CK model, knowledge of line positions is unnecessary for computation of transmissivities and absorption coefficient distribution function is sufficient. Further, the spectrally averaged intensities are calculated using the smooth cumulative distribution functions and quadratures. CK narrow band parameters for $B \rightarrow X$ and $C \rightarrow X$ systems at gas temperatures between $500 - 3500\ K$ and electronic temperatures varying between $500 - 6000\ K$ have been generated. The entire spectrum has been divided into bands of equal width $1000\ cm^{-1}$. 7 quadrature points have been used for each system. Thus the RTE for hydrogen molecule becomes for each narrow band and each couple of quadrature points (i for system B and j for system C)

$$\begin{aligned} \frac{\partial}{\partial s} I^{ij} &= \kappa_B^i \left(\frac{\eta}{\kappa}\right)_B + \kappa_C^j \left(\frac{\eta}{\kappa}\right)_C + \kappa_{soot} I_{b,\nu}(T_g) + \frac{\sigma}{4\pi} \int_{\Omega'} \Phi I^{ij}(\Omega') d\Omega' \\ & -(\kappa_B^i + \kappa_C^j + \kappa_{soot} + \sigma) I^{ij} \end{aligned} \tag{8}$$

where κ_{soot} is the spectral absorption coefficeint of soot and $I_{b,\nu}$ is the spectral black body intensity. The average spectral intensity over the narrow band is $\bar{I}^{\Delta\nu} = \sum_i \sum_j w_{B,i} w_{C,j} I^{ij}$ where $w_{B,i}$ and $w_{C,j}$ are the Gauss like quadrature weights of CK for $B \rightarrow X$ and $C \rightarrow X$ band systems respectively. Figure 2 shows that the transmissivities obtained using line-by-line calculations and the CK model for a column of 1 cm of pure molecular hydrogen are in good agreement. Figure 3 compares the averaged emission coefficient computed from line-by-line and from the CK model according to

$$\bar{\eta}^{\Delta\nu} = \sum_i \sum_j w_{B,i} w_{C,j} \left(\kappa_{\nu,B}^i \left(\frac{\eta}{\kappa}\right)_B + \kappa_{\nu,C}^j \left(\frac{\eta}{\kappa}\right)_C \right) \tag{9}$$

The CK model underestimates the emission with a relative error on the integrated emission of about 10%. This underprediction is due to the 7-point quadrature which is not sufficient to capture steep distribution functions due to almost Doppler lines in the optically thin limit.

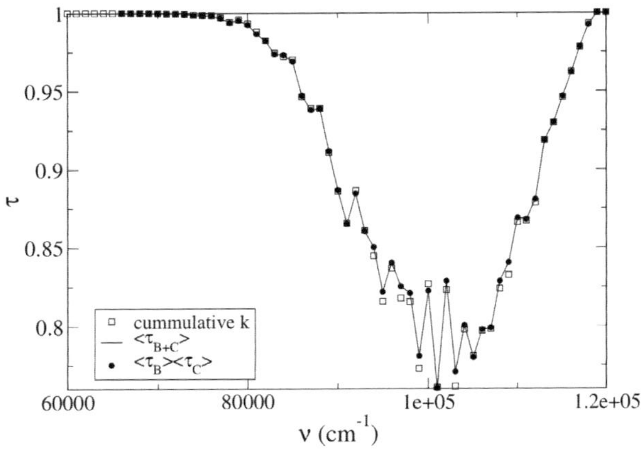

Figure 2. Comparison of transmissivity of a 1 cm column of pure hydrogen computed using line by line and CK model for $T_g = 3000\ K$, $T_e = 5500\ K$, $p = 0.1$ bar and $\Delta\nu = 1000\ cm^{-1}$.

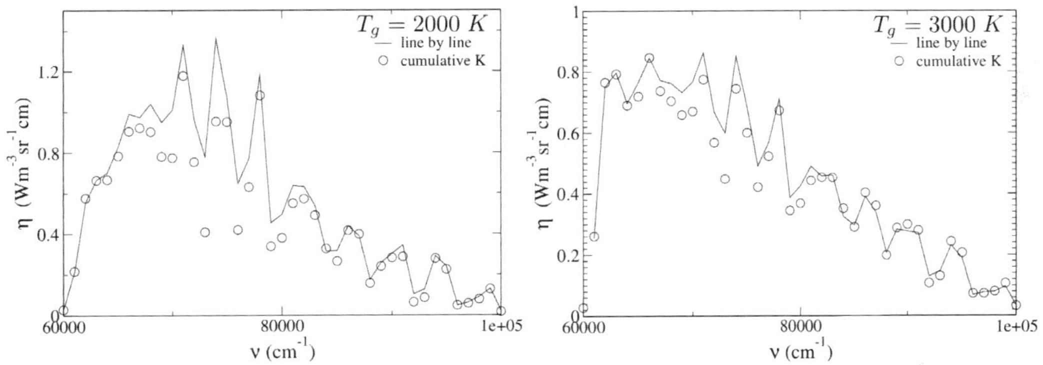

Figure 3. Comparison of emissive power of pure H_2 computed using line by line and CK model for $T_e = 5500\ K$, $p = 0.1$ bar and $\Delta\nu = 1000\ cm^{-1}$ for two values of gas temperature.

The accuracy of the model is however increased when a finite optical thickness is considered as shown in table 3.

Table 3. Relative error of total net emission coefficient of CK model with line-by-line model for pure hydrogen at $T_g = 3000\ K$, $T_e = 5500\ K$ and $p = 0.1$ bar

Plasma radius (cm)	Relative error %
0	5.60
1	2.54
5	0.36

2.2. Atomic hydrogen

Lyman (emission to $n = 1$) and Balmer series ($n = 2$) are the dominant lines for atomic hydrogen. Lyman series occur in UV whereas Balmer series are found in the visible part of the spectrum. NIST database [9] has been used for spectral calculations of atomic hydrogen. Spectral emissive power and absorption coefficient are evaluated in the same way as for molecular hydrogen (equation 2) but with Doppler line shape. Atomic hydrogen is assumed to be at

Journal of Physics: Conference Series **550** (2014) 012050 doi:10.1088/1742-6596/550/1/012050

electronic excitation temperature T_e as for molecular hydrogen. The number density of a given electronic level is given by

$$n_u = \frac{xp}{kT_g} \frac{g_u \exp\left(-\frac{E_u}{kT_e}\right)}{Q(T_e)} \tag{10}$$

The first 40 states have been considered to evaluate the partition function. In order to evaluate the relative contribution of atomic hydrogen and molecular hydrogen to radiative transfer, we have compared the products $4\pi\eta_\nu\tau_\nu(R)$ which represent the radiative power at the centre of a sphere of radius R. Figure 4 shows the cumulative net emission coefficient $\int_0^\nu 4\pi\eta_{\nu'}\tau_{\nu'}d\nu'$ for a mixture containing 90% molecular hydrogen and 10% atomic hydrogen with radius $R = 1\ cm$ at $T_g = 3000\ K$ and $T_e = 5500\ K$. Although the optically thin total emission coefficients are comparable, the atomic contribution becomes much smaller than molecular contribution due to self absorption. Hence atomic hydrogen may be neglected in radiative transfer calculations.

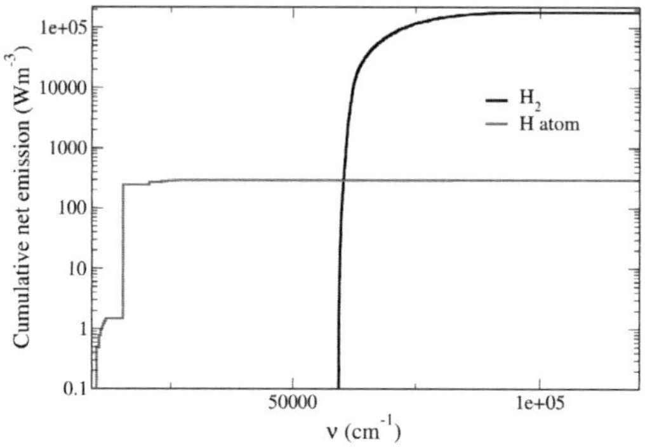

Figure 4. Cumulative spectral net emission coefficient for plasma radius $R = 1\ cm$ containing 90 % hydrogen molecule and 10 % hydrogen atom at $T_g = 3000\ K$, $T_e = 5500\ K$ and $p = 0.1$ bar.

2.3. Hydrocarbons

Methane and acetylene are the two main constituents of the plasma reactor active in the IR-V. Radiative properties for methane (CH_4) is taken from high temperature database provided in Refs. [10, 11]. No high temperature database is available for Acetylene (C_2H_2). HITRAN 2008 [12, 13] has been used to evaluate C_2H_2 radiative properties. HITRAN provides line by line data of transition wavenumber ν_i, line strength S_i at 296 K, self and air broadened line widths γ_i, temperature exponent for air line broadening n, lower level energy E_l and also, the partition functions $Q(T_g)$ up to 3000 K. Assuming Lorentz broadening, the transmissivity of a layer calculated using SNB model with inverse-exponential tailed distribution satisfies [3]

$$-\ln(\bar{\tau}) = \frac{\bar{\beta}}{\pi}\left(\sqrt{1 + \frac{2\pi xp\bar{k}}{\bar{\beta}}} - 1\right). \tag{11}$$

where $\bar{k}$ is the mean absorption coefficient in the weak absorption limit and $\bar{\beta}$ is the line overlapping parameter. Parameter $\bar{k}$ is determined by integrating absorption coefficient over the narrow band $\Delta\nu$. Parameter $\bar{\beta}$ is estimated by least squares adjustment to the transmissivities over path lengths ranging between 10^{-6} to $10^3\ cm$. The parameters have been determined for temperatures in the range 300 to 3000 K and band widths 25 cm^{-1}. Figure 5 indicates that the transmissivity of SNB model compares well with line by line calculations for acetylene gas.

Journal of Physics: Conference Series **550** (2014) 012050 doi:10.1088/1742-6596/550/1/012050

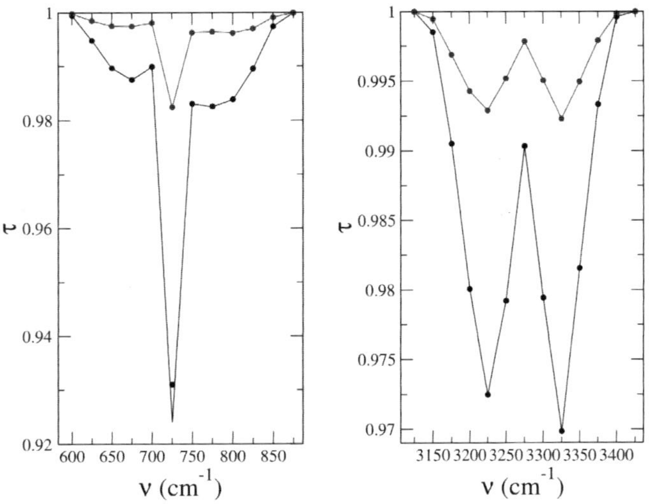

Figure 5. Comparison between exact (full lines) and SNB (Dots) transmissivities for 5% C_2H_2, $\Delta\nu = 25\ cm^{-1}$, pressure 0.2 *bar*, and path length 10 *cm* at 1000 *K* (black curves) and 2000 *K* (blue curves) for two different spectral bands.

2.4. Soot

Soot is active both in UV and IR-V regions. Experimental studies [1] on the microwave reactor suggest that formation of soot extinguishes the plasma. Hence, modelling soot properties for both regions of the spectrum is important. However, there is lack of knowledge regarding the structure as well as properties such as complex refractive index m of the soot. For the present study, refractive index model described by Lee and Tien [14] has been adopted. The structure of soot particles can be assumed to be composed of large number of connected spheres. The structure is defined by fractal law [15] which relates the radius of gyration R_g of the aggregate, the number of mono-dispersed spheres N and their radius R_p according to

$$N = k_f \left(R_g/R_p\right)^{D_f}, \qquad (12)$$

where k_f is known as the fractal prefactor and D_f is the fractal dimension of the aggregate. The radius of each monomer particle is usually less than 25 nm, which is much smaller than the wavelength of radiation λ. Rayleigh Debye Gans [16] approximation is commonly used to model the radiative properties of soot . RDG assumes the soot to be composed of particles whose complex permittivity is close to that of vacuum. The scattering cross section and phase functions are modelled based on the structure of the soot, while absorption cross sections is the same as that of N independent Rayleigh scatterers. According to RDG for fractal structures [17], the cross-sections, absorption C_{abs} and scattering C_{sca} for fractal aggregates are given by

$$C_{abs} = N\frac{4\pi x_p^3}{k^2}\mathrm{Im}\left(\frac{m^2-1}{m^2+2}\right) \qquad (13)$$

$$C_{sca} = N^2\frac{8\pi x_p^6}{3k^2}\left|\frac{m^2-1}{m^2+2}\right|^2\left(1+\frac{4}{3D_f}x^2(N/k_f)^{2/D_f}\right)^{-Df/2} \qquad (14)$$

where $x_p = 2\pi R_p/\lambda$ and $k = 2\pi/\lambda$. For computational simplicity, it is assumed here that soot particles are isotropic scatterers. Farias et al [18] have evaluated the range of validity of

Journal of Physics: Conference Series **550** (2014) 012050 doi:10.1088/1742-6596/550/1/012050

RDG approximation and concluded that application of RDG for optical diagnostics of soot in the visible and UV regions is satisfactory. However, both imaginary and real parts of m are relatively high in the infrared region. We have shown in Ref. [19] that RDG approximation underpredicts absorption in the infrared region and a model for absorption in the IR-V as a function of refractive index, optical size parameter and fractal parameters has been developed and is used in present calculations. The model for absorption cross-sections is based on rigorous solutions of Maxwell's equations using Multi-Sphere T-Matrix method [20].

2.5. Solution of radiative transfer equations

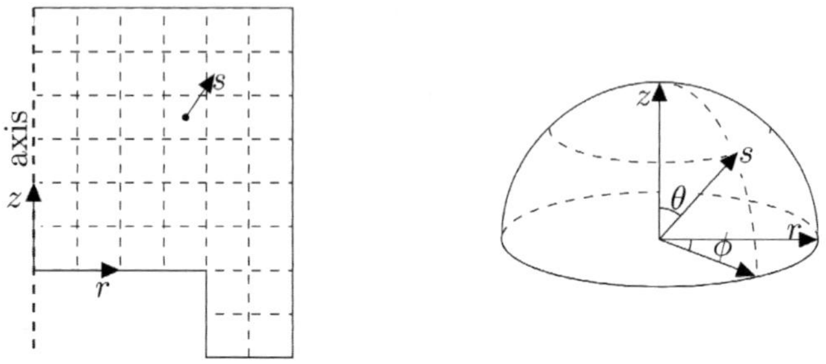

Figure 6. Spatial discretization of the axi-symmetric domain and angular discretization of the solid angle.

Radiative transfer equations are solved using Discrete Ordinate Method [4] in the UV while direct ray tracing method is used in the IR-V region. The inputs to the RTE solver are electronic and gas temperatures, species concentrations and pressure distribution inside the reactor. The 2-D spatial domain is discretized into large number of finite volumes using structured grids (figure 6) and can have non-uniform spacings. It is assumed that temperature, pressure and concentrations are uniform within a finite volume. The RTE has to be solved over several discrete directions. For both DOM and ray tracing methods, the octant of the sphere at each spatial node is discretized into $N_\theta \times N_\phi$ directions with an assumption that the radiation intensity is constant within the discrete solid angle. Symmetry of the domain means that equations have to be solved only for half the directions according to ϕ. The output from the solver is the volumetric radiative source $\nabla.\mathbf{q}_{rad}$ at all internal spatial nodes and radiative flux q_{rad} at the walls of the reactor. The radiative solvers have been extensively validated against benchmark solutions.

3. Sample radiation calculations

Simulations have been performed for the simple geometry and conditions shown in figure 7. The plasma reactor is represented by two zones - hot and cold zones. The hot zone represents the plasma where Hydrogen is characterized by its gas and electronic excitation temperatures. Other constituent gases are assumed at gas temperature. Hydrogen in the cold zone is assumed to be at thermal equilibrium. Volumetric concentration of soot is assumed to be uniform throughout the domain. Radiative properties of soot have been calculated using RDG approximation in the UV region and the model developed in Ref. [19] in IR-V region with soot clusters having fractal parameters $D_f = 1.8$, $k_f = 1.3$ and $N = 100$ mono-dispersed particles of radius $R_p = 15\ nm$.

Figure 8 shows the radiative power distribution along the axis of the reactor for different soot concentrations. While the hot zone strongly emits radiation, the cold zone absorbs radiation. Addition of soot to the mixture greatly changes the radiative power of the medium. Soot is important for both IR-V and visible-UV spectrum. In the UV spectrum, soot absorbs radiation and the increase in absorption in the cold region is several orders of magnitude compared to

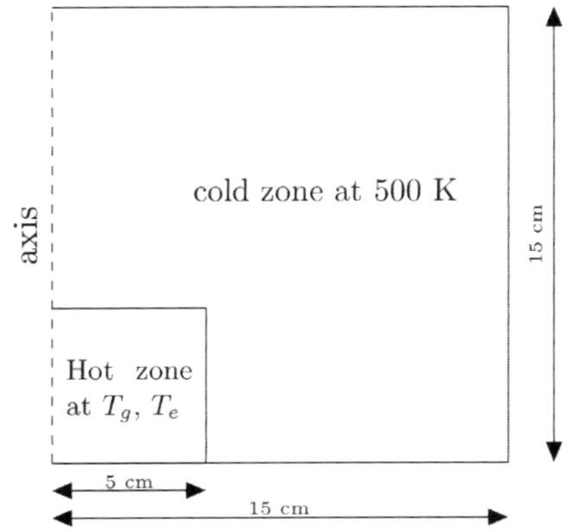

	Hot zone	Cold zone
T_g (K)	3000	500
T_e (K)	5500	500
p (atm)	0.1	0.1
X_{H2}	0.967	0.967
X_{CH4}	0.003	0.03
X_{C2H2}	0.03	0.003
All walls are at 500 K		

Figure 7. Reactor conditions for full simulation.

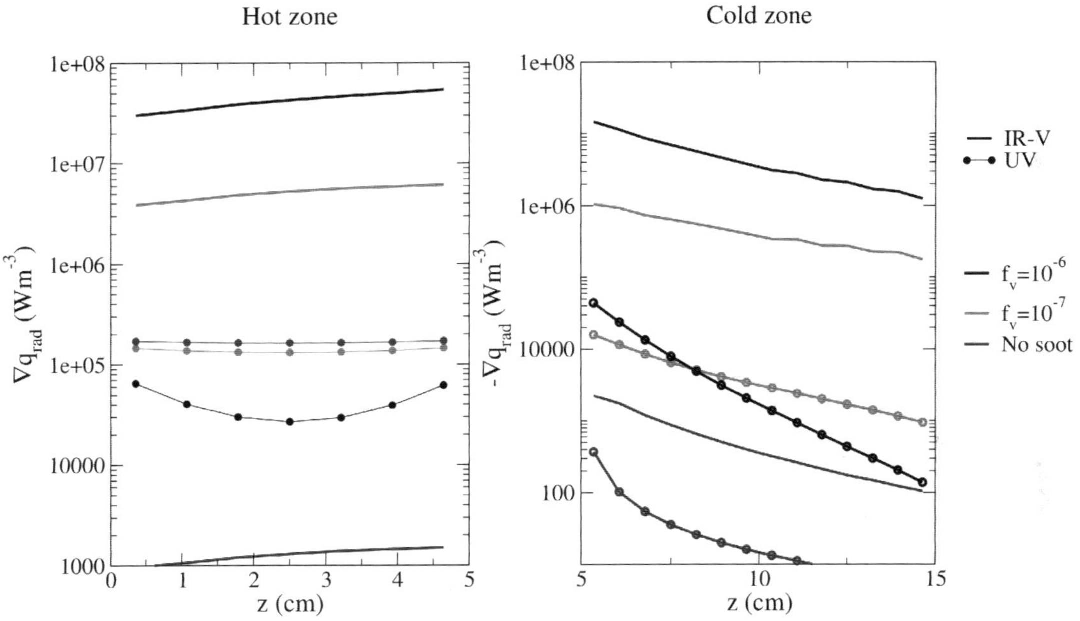

Figure 8. Radiative power distribution along the axis of the reactor with uniform soot distribution. Full lines correspond to IR region and Dashed lines with circles correspond to UV region. The colors indicate the soot volume fraction. Calculations are carried out with 20×20 spatial mesh and $N_\theta = N_\phi = 40$.

no-soot condition. The amount of radiative power emitted in the hot region and absorbed in the cold region is a strong function of soot concentration. The orders of magnitude of radiative power can become comparable to that of input microwave energy discharge (see table 2). This indicates that microwave energy may not be sufficient to sustain the plasma in the presence of high volume fractions of soot.

4. Conclusion

Radiative property models for hydrogen molecule, hydrocarbons and soot in plasma reactor conditions have been developed. Two different approaches have been adopted for treating IR-V and UV spectra regions. In the UV region, molecular hydrogen is found to be more significant compared to atomic hydrogen. Hence, atomic hydrogen has been neglected for radiative transfer calculations. Molecular hydrogen radiation has been modelled by CK narrow-band model, in the framework of two-temperatures representation of thermodynamic state, which is compatible with scattering by soot particles. In the IR-V region where scattering can be neglected, CH_4 and C_2H_2 properties have been modelled using SNB. Preliminary calculations for synthetic reactor conditions indicate that soot can be the most critical constituent affecting the radiative transfer in the reactor. Presence of small amounts of soot can yield local radiative powers of the same orders of magnitude as microwave power and can extinguish the plasma. Coupling of the radiative transfer model with other multiphysics models would enable full simulation of the plasma reactor and improve the predictions of the deposition process.

Acknowledgments

We acknowledge the financial support of this work by Agence Nationale de la Recherche (ANR, France) through the PLASBorDDiAM project. Computational facilities were provided by the Mesocentre of Ecole Centrale Paris.

References

[1] Hassouni K, Silva F and Gicquel A 2010 *J. Phys. D: Appl. Phys.* **43** 153001
[2] Hassouni K, Gicquel A, Capitelli M and Loureiro J 1999 *Plasma Sources Sci. Technol.* **8** 494–512
[3] Taine J and Soufiani A 1999 (*Advances in Heat Transfer* vol 33) (Elsevier) pp 295 – 414
[4] Kim M 2008 *Int. J. Heat Mass Transfer* **51** 5144–5153
[5] Abgrall H, Roueff E, Launay F, Roncin J and Subtil J 1993 *J. Mol. Spectrosc.* **157** 512–523
[6] Dabrowski I 1984 *Can. J. Phys.* **62** 1639–1664
[7] Breene Jr R 1967 *Appl. Opt.* **6** 141–147
[8] Lamet J, Rivière P, Perrin M and Soufiani A 2010 *J. Quant. Spectrosc. Radiat. Transfer* **111** 87–104
[9] Ralchenko Y, Jou F, Kelleher D, Kramida A, Musgrove A, Reader J, Wiese W and Olsen K 2010 *National Institute of Standards and Technology, Gaithersburg, MD*
[10] Perrin M Y and Soufiani A 2007 *J. Quant. Spectrosc. Radiat. Transfer* **103** 3–13
[11] Rivière P and Soufiani A 2012 *Int. J. Heat Mass Transfer* **55** 3349–3358
[12] Rothman L, Gordon I, Barbe A, Benner D, Bernath P, Birk M, Boudon V, Brown L, Campargue A, Champion J *et al.* 2009 *J. Quant. Spectrosc. Radiat. Transfer* **110** 533–572
[13] Gomez L, Jacquemart D, Lacome N and Mandin J Y 2010 *J. Quant. Spectrosc. Radiat. Transfer* **111** 2256–2264
[14] Lee S and Tien C 1981 *Symposium (International) on Combustion* vol 18 (Elsevier) pp 1159–1166
[15] Forrest S and Witten Jr T 1979 *J. Phys. A: Math. Gen.* **12** L109–117
[16] Bohren C and Huffman D 2008 *Absorption and scattering of light by small particles* (Wiley. com)
[17] Dobbins R and Megaridis C 1991 *Appl. Opt.* **30** 4747–4754
[18] Farias T, Köylü Ü and Carvalho M 1996 *Appl. Opt.* **35** 6560–6567
[19] Prasanna S, Rivire P and Soufiani A 2014 *J. Quant. Spectrosc. Radiat. Transfer* **148** 141 – 155
[20] Mackowski D and Mishchenko M 2011 *J. Quant. Spectrosc. Radiat. Transfer* **112** 2182–2192

Computational Simulation of Nanosecond Pulsed Discharge for Plasma Assisted Ignition

H Takana[1], **I V Adamovich**[2] **and H Nishiyama**[1]
[1] Institute of Fluid Science, Tohoku University
2-1-1 Katahira, Aoba-ku, Sendai 980-8577, Japan
[2] Department of Mechanical and Aerospace Engineering, The Ohio State University
Columbus, OH 43210, USA

E-mail: takana@fmail.ifs.tohoku.ac.jp

Abstract. Detailed two dimensional numerical simulations of a nanosecond pulsed pin-to-pin discharge in a lean methane/air mixture were conducted under 10 atm and 600 K for plasma assisted combustion in internal combustion engines. It was clarified from this study that the produced radicals were locally higher in the vicinity of electrodes, and high density radicals are more widely distributed on the anode side rather than the cathode side which the streamer is propagating toward. The electron energy partition has been clarified during a single pulse. Total electron energy increases with fuel equivalent ratio under the same applied voltage. Pronounced enhancement of ignition delay has been shown by nanosecond pulsed discharge.

1. Introduction

Recently, nanosecond pulsed discharge has been paid special attentions for plasma assisted ignition or combustion as an innovative technology to bring breakthrough in the lean combustion [1-8]. Previous researches show that the chain reactions of fuel oxidation is sufficiently initiated by plasma produced radicals even in lean mixture conditions, which leads to the well-controlled uniform combustion [2][7,8]. However, most of the previous researches are conducted at low temperature of ~ 300 K and low pressures up to atmospheric pressure [1,3-6]. For the practical application of the plasma assisted combustion in an internal combustion engine, it is essential to understand and characterize the reactive species production by nanosecond pulsed discharges at high temperature and high pressure conditions as in actual internal combustion engines [2][9,10].

In this study, detailed two dimensional computational simulations of nanosecond pulsed pin-to-pin discharge in a lean methane/air mixture were conducted at 10 atm and 600 K for plasma assisted ignition in an internal combustion engine. The fundamental characteristics of the nanosecond pulsed discharge and the radical production processes in the typical operating condition of internal combustion engines are clarified with understanding the partition of electron energy for radical production. Lastly, the effect of plasma produced radicals on enhancement of ignition delay has been discussed.

[1] takana@fmail.ifs.tohoku.ac.jp

2. Governing equations

Following continuity equations for electrons and ions with drift-diffusion approximation are solved coupled with Poisson's equation:

$$\frac{\partial n_k}{\partial t} + \nabla \cdot \boldsymbol{\Gamma}_k = S_k, \quad (1)$$

$$\boldsymbol{\Gamma}_k = \mathrm{sgn}(q_k)\mu_k \boldsymbol{E} n_k - D_k \nabla n_k,$$

$$\nabla \cdot (\varepsilon \nabla \phi) = -\sum_k q_k n_k. \quad (2)$$

The charge continuity equations shown in eq. (1) is solved by VSIAM3 method (volume/surface integrated average based multi-moment method)[11][12], which is a type of the finite volume scheme of CIP method. The electron mean energy, transport properties and rate coefficients are given as a function of reduced electric field with local field approximation. They are obtained by solving Boltzmann's equation for electrons [13]. The transport properties of ions are taken from the reference [14].

The rate of photoionization in a gas volume is included in the source term in equation (1). In this study, two-exponential Helmholtz model [15][16] was employed for photoionization of oxygen. The photoionization processes in O_2 are caused by the radiation in the wave range 98 - 102.5 nm. Because of the large absorption cross section of methane in this wave range, it would be also important to consider the photoionization of methane. However, since there are few photoionization coefficient of methane available [17] and it could not be included in the present model. The chemical kinetics of methane/air mixture studied in this work is based on that described in [18] except for the reactions associated with argon. Moreover, the oxygen atom decay processes through recombination with oxygen resulting in ozone formation, and also through recombination with ozone are incorporated. Oxygen atom decay due to reactions of hydrogen atom and CH_3 radical are also taken into account [19][20]. Some of the reactions of OH production and decay are also considered [19][20]. The time scale of three body reactions of neutral species is in a time scale from micro to milli-seconds and it is much larger than that of other reactions in discharge. In this study, only three body reaction of ozone with relatively high production rate is considered for nanosecond pulsed discharge simulation. Most of the chemical reactions considered in this study are summarized in the reference [10]. With addition to the reactions shown in [10], the formation of complex positive ions such as O_4^+ and N_4^+ and the production and decay of O_2^- are also incorporated [21].

The gas heating process during positive streamer propagation in the pin-to-pin geometry has been conducted by considering the enthalpies of reactions and fast gas heating mechanism proposed by Popov [22].

3. Results and discussion

3.1. Nanosecond pulsed discharge in air at 100 Torr and 300 K

In order to validate the simulation model, obtained numerical result was compared with the available photo of nanosecond pulsed discharge taken by ICCD camera in air at 100 Torr and 300 K [23].

Figure 1 shows the computational domain for the validation of the model. Axisymmetric *r-z* two dimensional simulations are conducted for the pin-to-pin discharge. The discharge is initiated between two spherical electrodes with the gap spacing of 10 mm. The diameter of the bare sphere electrode is

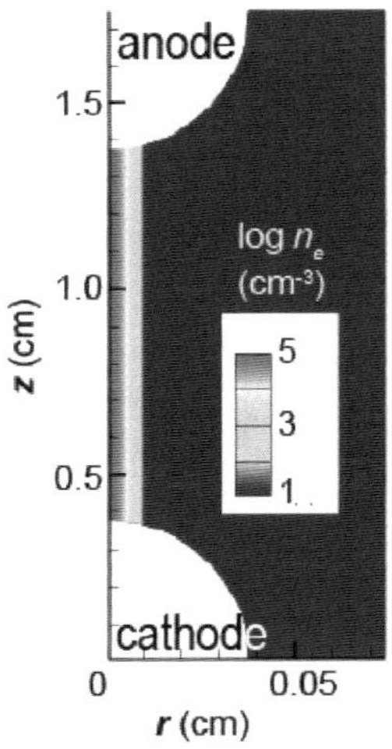

Figure 1. Computational domain and initial condition of electron number density.

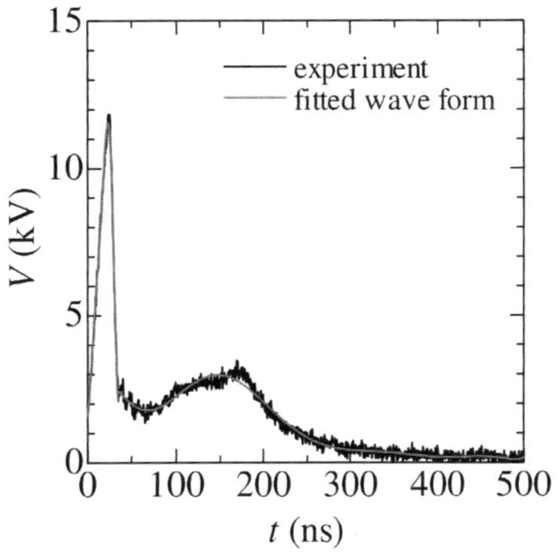

Figure 2. Wave form of applied nanosecond voltage pulse.

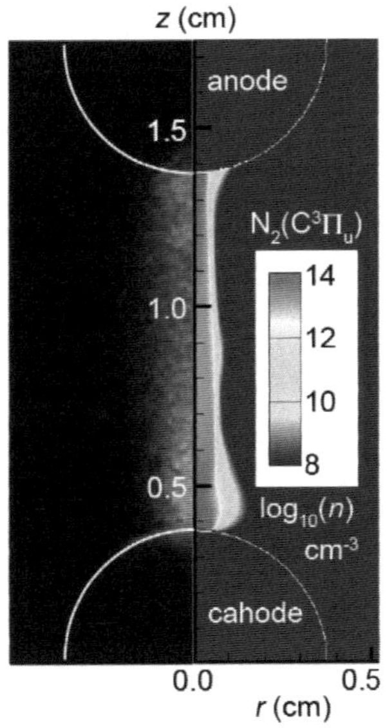

Figure 3. Comparison between ICCD image [23] and the simulated number density of $N_2(C^3\Pi_u)$ produced by nanosecond pulsed discharge in air at 100 Torr and 300 K.

7.5 mm and the maximum of 11.8 kV is applied in 24.8 ns with cathode grounded as shown in Fig. 2. The wave form of nanosecond voltage pulse applied to the anode is taken from the experiment [23]. The experimental voltage pulse is fitted with polynomial functions. Air is assumed to be composed of N_2 and O_2 and N_2/O_2 mixuture was simulated with the composition volume ratio of $N_2:O_2 = 0.79:0.21$. For the initiation of electron avalanche, local seed charges consist of electron and O_2^+ were given by a

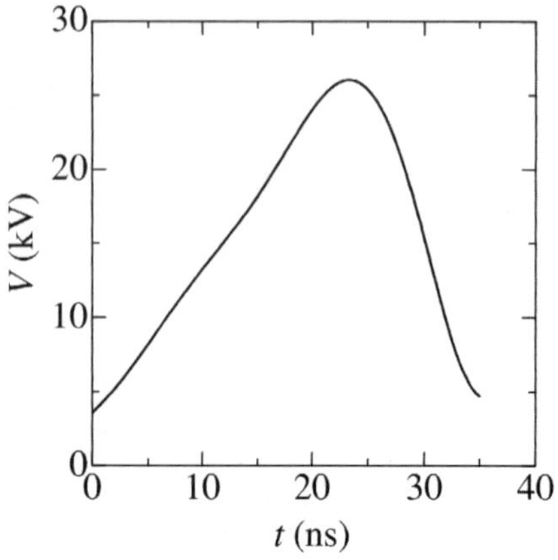

Figure 4. Wave form of applied nanosecond pulse voltage for air/methane mixture at 10 atm and 600 K.

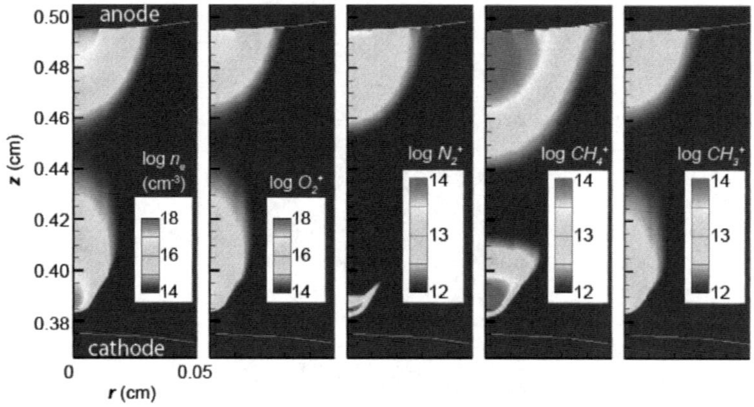

(a)

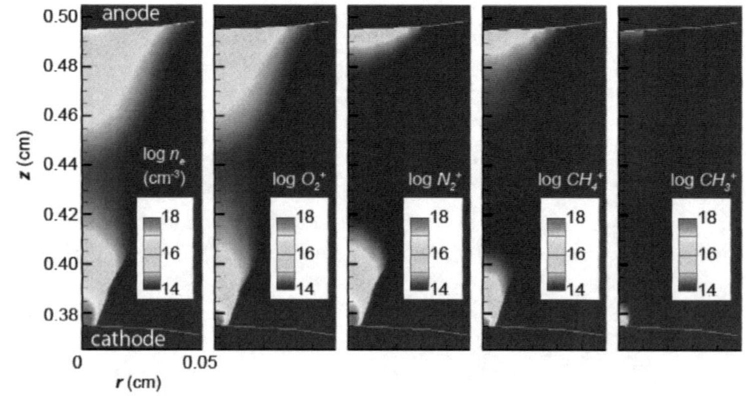

(b)

Figure 5. Distributions of electron, O_2^+, N_2^+, $CH4^+$ and CH_3^+ at (a) t = 24.5 ns and (b) t = 24.6 ns.

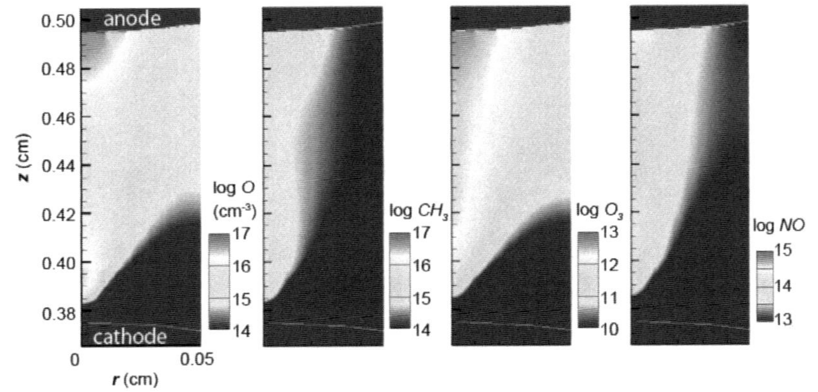

(a)

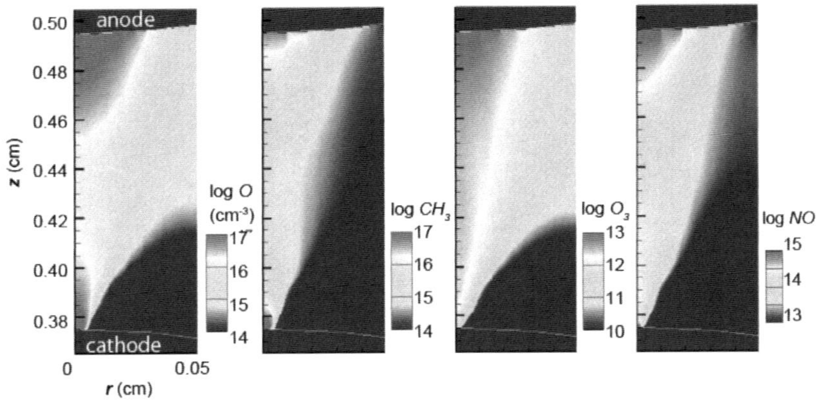

(b)

Figure 6. Distributions of $O(^3P)$, CH_3, O_3 and NO at (a) t = 24.5 ns and (b) t = 24.6 ns.

Gaussian distribution in radial direction as shown in Fig.1. The peak number density of initial seed charges are 1.0 x 10^5 1/cm^3 along the symmetric axis.

Figure 3 shows simulated number density of $N_2(C^3\Pi_u)$ by comparing with the time averaged ICCD image of the discharge [23]. The numerical result was reconstructed by integrating along the line in sight using Abel transform at each mesh points. The simulated result shown in Fig. 3 is an instant image at t = 21.7 ns and the gate time of the ICCD camera is not taken into account. The diffusive nature of the discharge filament is well represented by numerical simulation. The radius of the discharge filament is approximately 1 mm, which agrees with ICCD image. The number density of $N_2(C^3\Pi_u)$ is higher in the vicinity of cathode due to the cathode sheath. Although the comparison remains qualitative in experimental visualization, the discharge structure of nanosecond pulse discharge in air is relatively well simulated.

3.2. Nanosecond pulsed discharge in air/methane lean mixture at 10 atm and 600 K

The numerical simulations of nanosecond pulsed discharge in air/methane lean mixture were conducted under the typical operating conditions of internal combustion engines to clarify the fundamental characteristics of the nanosecond pulsed discharge for plasma assisted ignition. In this simulation, the electrode gap is set to 1.2 mm with the same spherical electrodes shown in Fig. 1. The maximum voltage of 26 kV is applied to the anode by taking 23 ns as shown in Fig. 4. The composition volume ratio of air/methane mixture is N_2:O_2:CH_4 = 0.75:0.20:0.05, which corresponds to

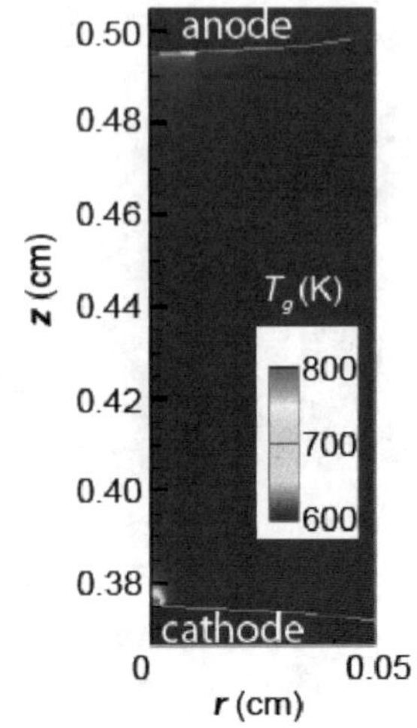

Figure 7. Distribution of gas temperature at t = 24.6 ns.

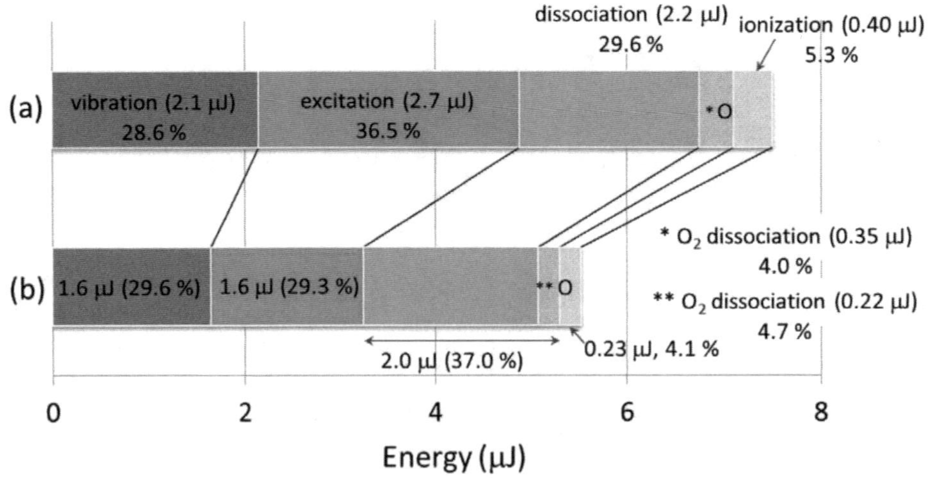

Figure 8. Electron energy partition during streamer propagation; (a) fuel-air equivalent ration of 0.5 and (b) 0.4.

the equivalent ratio of 0.5. The initial electron and O_2^+ number densities are uniformly set to be 1.0 x 10^8 1/cm^3.

Figure 5 (a) and (b) show distributions of electron, O_2^+, N_2^+, CH4$^+$ and CH_3^+ at t = 24.5 ns and 24.6 ns, respectively. The streamer appears in the middle of the discharge gap and propagates to cathode at t = 24.5 ns. The electron and ions are high in the vicinity of anode where E/N is over ~ 200 Td. When the streamer head reaches the cathode, the streamer is concentrated on the cathode and electron number density locally increases to ~ 10^{18} 1/cm^3. Among other ions, oxygen ion is the dominant ion. Because nitrogen ions rapidly react with CH_4 for the production of CH_3 and H radicals, the nitrogen ion disappears behind the streamer head.

Figure 6 (a) and (b) show distributions of O(^{3}P), CH_3, O_3 and NO at (a) t = 24.5 ns and (b) t = 24.6 ns, respectively. Oxygen atom is produced at the streamer head due to high energy electron impacting dissociation and also in the vicinity of anode where reduced electric field is sufficiently high. The number density of produced oxygen atom becomes higher near anode and cathode. The streamer is constricted toward the cathode. Oxygen atoms of over 10^{17} 1/cm^3 are produced near electrodes during positive streamer propagation (t = 24.6 ns). The produced radicals are locally higher near the electrodes and high concentrations of radicals are more widely distributed in the vicinity of anode.

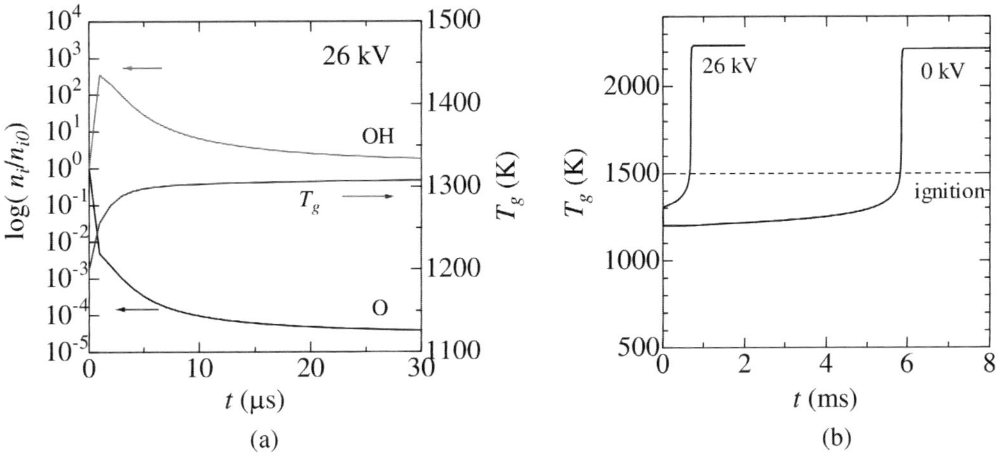

Figure 9. (a) Normalized number densities of OH and O radicals by their plasma produced number density, (b) time evolutions of gas temperature with or without nanosecond pulsed discharge.

Figure 7 shows the distribution of simulated gas temperature at t = 24.6 ns. When the streamer head reaches the cathode, rapid gas heating occurs in the cathode sheath and locally heated up to over 800 K from 600 K. Heating is very small in other regions and gas temperature increases only by several K in the time scale of streamer propagation.

Figure 8 (a) and (b) show the electron energy partition during single streamer propagation for fuel-air equivalent ratio of 0.5 and 0.4, respectively. The peak applied voltage is 26 kV for both cases with the same voltage profile. It has been clarified from Fig. 8 (a) that 65 % out of total electron energy of 7.5 μJ is used for the vibrational and electronic excitation of nitrogen and oxygen molecules. 29.6 % of the energy is spent on dissociation of molecules. For the production of oxygen atoms, 4 % (0.35 μJ) of the total electron energy is consumed. The total electron energy decreases to 5.4 μJ for lower equivalent ratio of 0.4. This is mainly because fewer electrons are produced due to higher relative density of nitrogen for the same applied peak voltage, and the electron energy for the vibrational and electronic excitation decreases. However, the percentage of electron energy consumed for the production of oxygen atom increases from 4.0 % to 4.7 % with decreasing the equivalent ratio.

3.3. Effecto of plasma produced radicals on ignition enhancement of air/methane lean mixture at 10 atm and 600 K

In order to clarify the effect of plasma produced radicals on ignition enhancement, zero-dimensional combustion simulation has been conducted with coupling plasma chemical reactions. In this simulation, simplified methane combustion mechanism of GRI-Mech 3.0 [20] with 325 reactions and 53 chemical species is considered. The initial number densities of the chemical species are taken from the results obtained from the nanosecond pulsed discharge at 24.6 ns in the vicinity of anode at r = 0 mm and z = 4.9 mm shown in Fig. 5 and 6.

Figure 9 (a) shows number densities of OH and O radicals normalized by their initial number density for combustion simulation and (b) shows the time evolutions of gas temperature with or without nanosecond pulsed discharge, respectively. In Fig. 9, the initial state for the combustion simulation is set to $t = 0$ s. Initial gas temperature of 1200 K is given at $t = 0$ s, assuming external heat source such as spark plug. The reduced electric field is set to 0 Td at $t = 0$ s. As shown in Fig. 9 (a), oxygen atom exponentially decreases with time, on the other hand, OH radical increases within several micro seconds and then decreases. The sudden increase of OH is due to the reaction of plasma produced oxygen atom with methane; $O+CH_4 = OH + CH_3$. Then OH and O radicals are consumed for the oxidation reactions with increase in gas temperature from 1200 K to 1300 K in 10 micro seconds. When defining 1500 K as the ignition temperature, ignition occurs by taking 5.8 ms under the equivalent ratio of 0.5 without nanosecond pulsed discharge according to Fig. 9 (b). With plasma assisted ignition, the ignition occurs at 0.66 ms and the ignition delay time is shortened by 88 %.

4. Conclusions

The discharge characteristics of a nanosecond pulsed pin-to-pin discharge in a lean methane/air mixture were clarified at 10 atm and 600K for plasma assisted ignition in an internal combustion engine by two-dimensional numerical simulations. The obtained results from this study can be summarized as follows.

(1) The concentrations of produced radicals are locally higher near the electrodes and high density radicals are more widely distributed in the vicinity of anode, rather than the cathode which streamer is propagating toward.
(2) Gas is rapidly heated up to over 800 K from 600 K in the cathode sheath by single pulse.
(3) 64 % of the total electron energy is spent for the vibrational and electronic excitation of molecules. 4.7 % of the energy is spent on the production of oxygen radicals for equivalent ration of 0.5 at 26 kV.
(4) Total electron energy increases with fuel equivalent ratio under the same applied voltage.
(5) Pronounced improvement of ignition delay has been shown by nanosecond pulsed discharge.

Acknowledgments

Part of this work was supported by Grant-in-Aid for Scientific Research (C) under Grant No 24560182 from Ministry of Education, Culture, Sports, Science and Technology (MEXT), Japan and by the Collaborative Research Project of the Institute of Fluid Science, Tohoku University.

References

[1] Starikovskaia S M 2006 *J. Phys. D: Appl. Phys.* **39** pp 265-299
[2] Shiraishi T, Urushihara T and Gundersen M 2009 *J. Phys. D: Appl. Phys.* **42** 135208
[3] Adamovich I V, Choi I, Jiang N, Kim J H, Keshav S, Lempert W R, Mintusov E, Nishihara M, Samimy M and Uddi M 2009 *Plasma Sources Sci. Technol.* **18** 034018
[4] Kim W, Mungal M G and Cappelli M A 2010 *Combust. Flame* **157** pp 374-383
[5] Bak M S, Do H, Mungal M G and Cappelli M A 2012 *Combust. Flame* **159** pp 3128-3137
[6] Sun W, Uddi M, Won S H, Ombrello T, Carter C and Ju Y 2012 *Combust. Flame* **151** pp 221-229
[7] Fridman A 2008 Plasma Chemistry (New York: Cambridge University Press)
[8] D Singleton, S J Pendleton and M A Gundersen 2011 *J. Phys. D: Appl. Phys.* **44** 022001
[9] Takana H, Tanaka Y and Nishiyama H 2012 *Europhys. Lett.* **97** 25001
[10] Takana H and Nishiyama H 2014 *Plasma Sources Sci. Technol.* **23** 034001
[11] Xiao F 2004 *J. Comput. Phys.* **195** pp 629-654
[12] Xiao F, Akoh R and Ii S 2006 *J. Comput. Phys.* **213** pp 31-56
[13] Hagelaar G J M and Pitchford L C 2005 *Plasma Sources Sci. Technol.* **15** pp 722-733
[14] Gogolides E, Buteau C, Rhallabi A and Turban G 1994 *J. Phys. D: Appl. Phys.* **27** pp 818-825
[15] Bourdon A, Pasko V P, Liu N Y, Celestin S, Segur P and Marode E 2007 *Plasma Sources Sci.*

Journal of Physics: Conference Series **550** (2014) 012051 doi:10.1088/1742-6596/550/1/012051

Technol. **16 pp** 656-678
[16] Liu N and Pasko V P 2004 *J. Geophys. Res.* **109** A04301
[17] Ferrara C, Preda M and Cavallotti C 2012 *J. Appl. Phys.* **112** 113301
[18] Aleksandrov N L, Kindysheva S V, Kukaev E N, Starikovskaya S M and Starikovskii A Y 2009 *Plasma Phys. Rep.* **35** pp 941-956
[19] Uddi M, Jiang N, Mintusov E, Adamovich I V and Lempert W R 2009 *Proc. Combust. Inst.* **32** pp 929–936
[20] Available at www.me.berkeley.edu/gri-mech/version30/text30.html, GRI-Mech 3.0
[21] Kossyi I A, Kostinsky A Yu, Matveyev A A and Silakov V P 1992 *Plasma Sources Sci. Technol.* **1** pp 207-220
[22] Popov N A 2011 *J. Phys. D: Appl. Phys.* **44** 285201
[23] Burnette D, Shkurenkov I, Adamovich I and Walter R. Lempert W 2014 *AIAA* 2014-1034

Journal of Physics: Conference Series **550** (2014) 012052 doi:10.1088/1742-6596/550/1/012052

Ordering of the flame track in the ring mode of the Trichel pulse negative corona discharge

R H Amirov[1], S A Barengolts[2,3], E V Korostylev[4], N V Pestovskii[3,4], A A Petrov[3,4], I S Samoylov[1], S Yu Savinov[3,4]
[1] Joint Institute for High Temperatures of the Russian Academy of Sciences, Izhorskaya str. 13 Bd.2, Moscow, 125412, Russia
[2] A.M. Prokhorov General Physics Institute of the Russian Academy of Sciences, Vavilova str. 38, Moscow, 119991, Russia
[3] P.N. Lebedev Physical Institute of the Russian Academy of Sciences, Leninskij Pr. 53, Moscow, 119991, Russia
[4] Moscow Institute of Physics and Technology (State University), Institutskiy per., 9 Dolgoprudny, Moscow Region, 141700, Russia

E-mail: amirovravil@yahoo.com

Abstract. The ring mode of the Trichel pulse negative corona discharge was studied in atmospheric air. The localization of the discharge flame track in the stable self-organized regular pattern of 3, 4, 5 and 6 – pointed star was found at the cathode surface. This phenomenon was observed at mean currents in the range 100-115 μA at the conditions of the experiment, when the modes with one or two rings are not stable. The conclusion was made that the ring mode of the discharge, which is caused by the symmetrical distribution of the volumetric charges in the conditions of the symmetrical electric field, may be unstable. This instability results in the spatial self-organization of these parameters and causes the organization of the discharge flame track at the cathode surface in the regular patterns.

1. Introduction

Negative corona discharge in the point-to-plane electrode configuration is a classical research object of plasma physics. This is caused by the wide range of technical applications of negative corona, and by the interest of studying of the processes involved in the discharge mechanism, which is currently unclear.

Negative corona discharge in ambient air in the point-to-plane configuration typically operates in the Trichel pulse mode at currents 1-100 μA and in glow mode at 100-200 μA [1]. In the Trichel pulse mode the discharge current is a consequence of pulses with typical amplitude 1 mA, duration at half magnitude 20 ns, pulse time separation 1 μs and width of leading edge 2 ns. Each pulse is caused by propagating ionization wave formed due to accumulation of positive volumetric charges near the cathode surface [2]. The light radiation of the discharge flame is also a consequence of light pulses with duration 10 ns [3]. The diameter of the discharge flame near the cathode surface is ~10 μm at atmospheric pressure and typically is smaller than the cathode pin diameter.

After each current pulse in the Trichel regime the discharge flame is formed at the cathode surface in the most favorable place where the electric field has the maximum value; also this region should have sufficient size (more than 1 μm). Microfields in smaller regions cannot influence sufficiently the discharge formation and may be responsible only for emission of electrons. The electric field distribution at the cathode surface is determined by the voltage and depends sufficiently on the volumetric and surface charges. Negative corona discharge causes the erosion of the cathode surface with specific erosion rate 10^{-6} – 10^{-4} g/coul with formation of micro and nanocraters [4]. Erosion processes in the region of discharge flame localization, formation of microcraters, formation and charging of dielectric layers, recycling of erosion products, accumulation of volumetric charges cause change of electric field distribution and wandering of the discharge flame at the cathode surface [1].

So in the Trichel pulse mode at cathodes with diameter less than 500 μm there are three regimes of the discharge flame dynamics: stable, unstable and stochastic [5]. Stable regime is realized at cathodes with diameter less than 100 μm and weakly susceptible to oxidation. Stable regime is characterized by constant values of Trichel pulse amplitude and pulse time separation. In the stable regime the discharge flame may keep its point of localization at the cathode surface for more than 1 s. Unstable regime is caused by formation of oxide layers at the cathode surface. In this regime the change of discharge position at the cathode surface causes change of Trichel pulse amplitude and pulse time separation. The linear dependence between amplitude and pulse time separation takes place at constant mean current. In unstable regime the discharge flame may be localized in one point of the cathode surface for 10^{-6} – 1 s. The stochastic regime is observed at cathodes completely covered with oxide layer. In this regime the discharge flame changes its position at the cathode surface after each current pulse. The Trichel pulse parameters may also be changed after each current pulse, there is no any dependence between amplitude and pulse time separation [6].

In the glow mode of discharge a stable cathode layer is formed, and the dynamics of the discharge flame is sufficiently affected by the field of the positive space charges. As result the discharge flame moves smoothly over the cathode surface without frequent jumping from one point to another as occurs in the Trichel pulse mode.

At cathodes with tip diameter more than 500 μm the Trichel pulse negative corona discharge may be realized in the ring mode [7]. In this form the points of the discharge flame localization are distributed at the cathode surface along the coaxial rings. This mode is caused by the accumulation of the volumetric negative charges near the cathode surface in the conditions of symmetry of electric field [8].

In this work, the negative corona discharge is studied in ambient air in the point-to-plane configuration in the Trichel pulse regime in the ring mode. To exclude the action of the surface charges on the flame wandering, graphite cathodes are used.

2. Experimental setup

Negative corona discharge is studied in air at atmospheric pressure in the point-to plane configuration. Experimental setup scheme is shown in the figure 1. Cathodes used in this work are made of MPG-6 polycrystalline graphite. Graphite cathodes were used to prevent formation of charged dielectric layers at the cathode surface. Also on the graphite cathodes the microstructures are not formed as result of recycling of erosion material [9]. These oxide layers and microstructures may influence the distribution of electric field and wandering of discharge flame. Initially the point has the form of a rod with diameter 1 mm, but in few minutes after the voltage is on, the tip gets a hemispherical shape because of erosion processes [10]. A copper plane with diameter 10 cm was used as anode, the gap was 12 mm. The discharge flame is observed with use of long-focus microscope MBS-9 and camcorder Casio EX-ZR100.

Cathode surface was observed through a 1-mm hole in the anode plane. There was no positive corona on the edges of a hole. The presence of a hole in the anode plane does not influence the discharge regimes and patterns. Discharge current was measured with use of microammeter M-906. Voltage was measured with use of C-196 electrostatic voltmeter. The average current is 10-120 μA, the voltage is 10 – 30 kV. Oscillograms of discharge current were registered with use of 1 MOhm oscillograph Tektronix TDS-410A (bandwidth 200 MHz) witch was connected across a 50-Ohm resistor between the anode and the ground. Air humidity is 50-70 % and does not influence the discharge. Duration of discharge was 20 – 40 minutes. Typical images (side view) of Trichel pulse corona in unstable regime and in ring mode made with exposition 40 ms are shown in figures 2 and 3 correspondently.

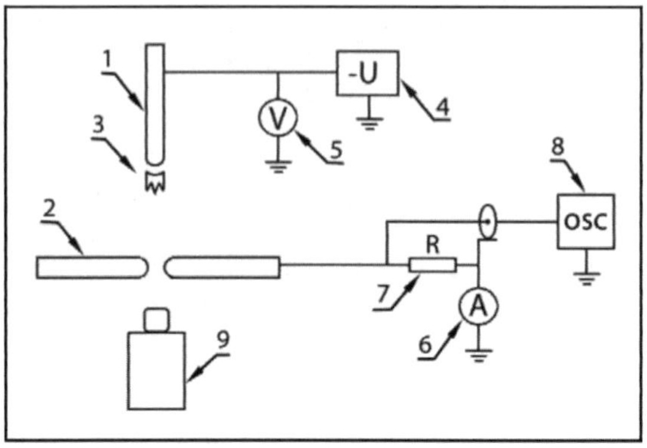

Figure 1. Experimental setup: 1 – cathode pin, 2 – anode plane, 3 – corona discharge, 4 – high voltage supply, 5 – voltmeter, 6 – microammeter, 7 – resistor 50 Ohm, 8 – Oscillograph Tektronix TDS – 410A, 9 – microscope MBS-9 with camcorder Casio ex-zr100.

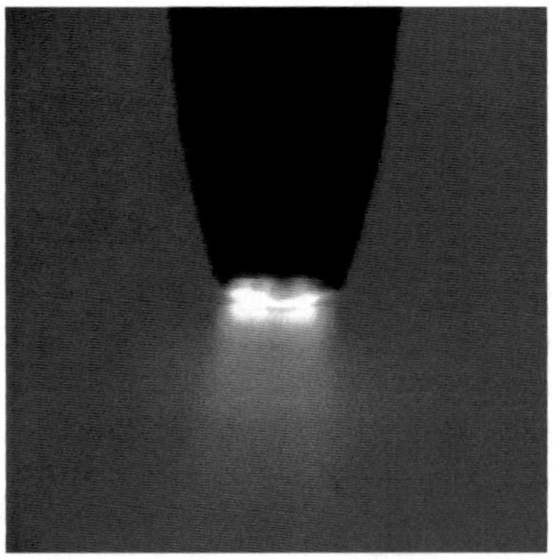

Figure 2. Trichel pulse corona in unstable regime. I = 50 μA.

Figure 3. Trichel pulse corona in the ring mode. I = 100 μA.

3. Results and discussion

According to results presented by Loeb [1], the ring mode of the discharge may be observed when the ratio of the gap length to cathode tip radius is $h/r < 20$. This ratio characterizes the uniformity of the electric field distribution near the cathode surface.

After the voltage is supplied to the electrodes at mean current less than 50 μA, the discharge is realized in the unstable form of Trichel pulses. In this form the discharge flame is localized at the cathode surface is some region for some milliseconds and then it disappears in this place and appears in another one. When the graphite cathodes are used, this behaviour of the discharge flame is explained by the change of the local geometry due to erosion processes, and by accumulation of the negative space charge in the region of the discharge flame localization. When the current reaches the values 50 μA, the discharge falls into the ring mode, which was described by Greenwood in 1951 [7, 8]. The ring mode of the discharge is shown in the figure 4. In this form the regions of the flame localization are distributed along the ring, and the distance between the localization points of the flame of this ring is approximately 50 μm.

Journal of Physics: Conference Series **550** (2014) 012052 doi:10.1088/1742-6596/550/1/012052

The further increase of the voltage follows the increase of the electrical current. This causes the increase of the ring radius. In the ring mode with one ring the Trichel pulses operate in stable regime. When the mean current reaches 100 μA, the self-organization of the discharge flame track appears in form of regular patterns. So at ~100 μA the track is localized along the line which has the form of 3-pointed star (figure 5), at ~105 μA it has the form of 4-pointed star (figure 6), at ~110 μA it has the form of 5-pointed star (figure 7) and at ~115 μA – the form of 6-pointed star (figure 8). The further increase of the voltage and current causes the appearance of the second ring, which is shown in the figure 9. The ring discharge with two rings was reported by Loeb [1]. Trichel pulses in the ring mode at mean current I > 100 μA appear in unstable regime.

The material, the cathode is made from, has some influence on the possibility of formation of the self-organized discharge tracks. So, when the metal cathodes are used, the surface oxidation and charging occurs. Then the dynamics of the discharge flame is governed by the electric field of the surface charges, and the self-organization is difficult to observe. The transition of the discharge into the pulseless mode causes the formation of erosion cells with diameter 60 μm (at atmospheric pressure) on the graphite cathodes [11]. This disturbs the condition of the symmetry of the electric field and makes difficult the self-organization of the discharge track in Trichel regime after reduction of current.

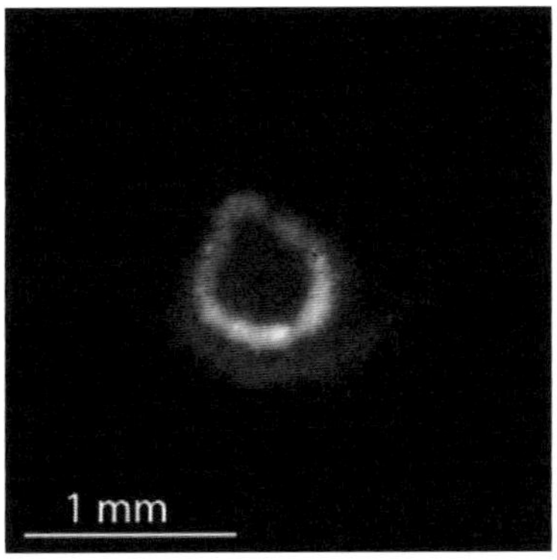

Figure 4. The ring mode of the Trichel pulse negative corona discharge. Current I = 60 μA. Voltage U = 13.5 kV.

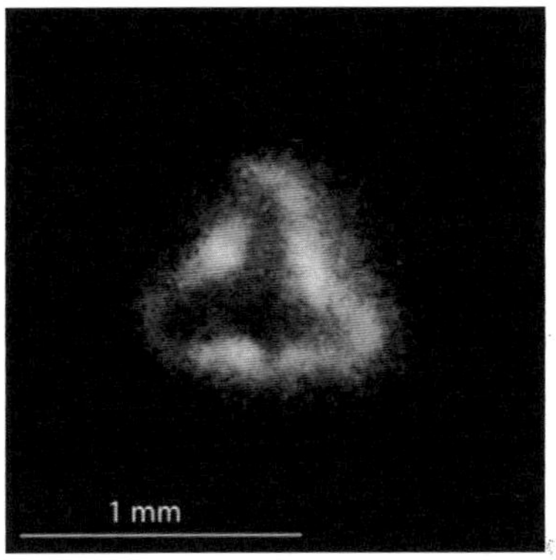

Figure 5. Discharge track at I = 100 μA, U = 16.5 kV.

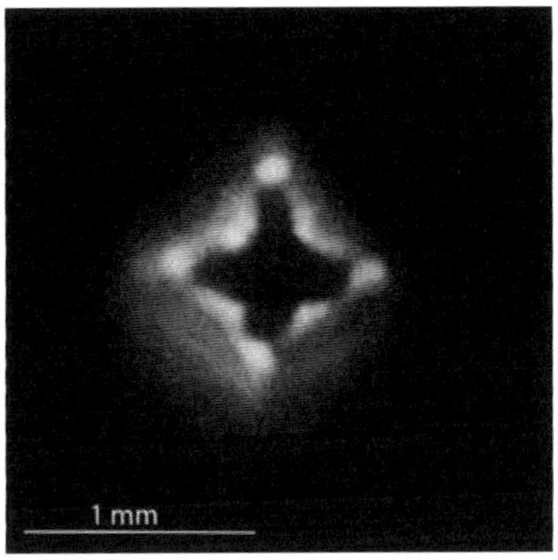

Figure 6. Discharge track at I = 105 μA, U = 16.7 kV.

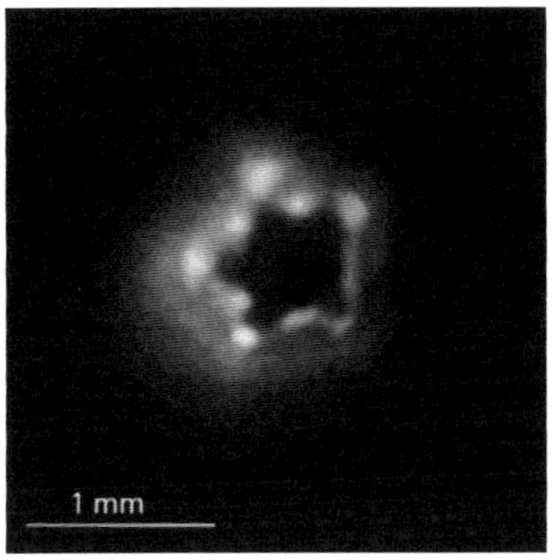

Figure 7. Discharge track at I = 110 μA, U = 17 kV.

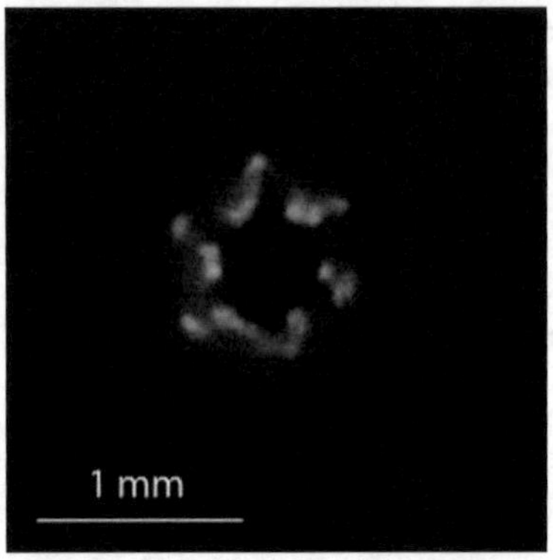

Figure 8. Discharge track at I = 115 μA, U = 17.2 kV.

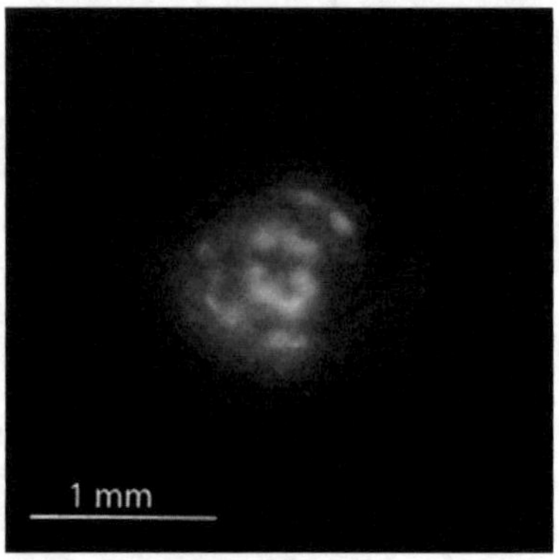

Figure 9. Ring mode with two rings at I ~ 115 μA. Increase of voltage causes transition to glow mode.

4. Conclusions

In this work the ring discharge was studied in the Trichel pulse negative corona in ambient air. To conclude, it was found that the ring mode of the negative corona discharge, which is caused by the peculiarities of the distribution of the negative space charge in the conditions of the symmetry of the electric field, may be unstable. This instability is caused by the self-organization of distribution of the volumetric charges near the cathode surface and causes the organization of the discharge flame track at the cathode surface. The track has the stable form of regular three, four, five and six-pointed stars. The observation of these self-organized structures in the ring mode of Trichel pulse corona is reported in this paper for the first time.

Acknowledgments

The work was made under financial support of the Russian Science Foundation, grant 14-12-00784. One of authors (AAP) would like to thank Dr. Evgenii Anokhin from MIPT for useful discussion of this work.

References

[1] Loeb L B 1965 *Electrical Coronas. Their Basic Physical Mechanisms.* (Berkeley, CA: Univ. California Press)
[2] Hoder T, Cernak M, Paillol J, Loffhagen D, and Brandenburg R 2012 *Phys. Rev. E* **86** 055401(R)
[3] Amin M R 1954 *J. Appl. Phys.* **25** 627
[4] Petrov A A, Amirov R H, Samoylov I S 2009 *IEEE – Transactions on Plasma Sciences* **37** 1146
[5] Amirov R H, Petrov A A, Samoylov I S 2009 *Int. J. of Plasma Environmental Science and Technology* **3** 35
[6] Brunt R J, Kulkarni S V 1990 *Phys. Rev. A.* **42** 4908
[7] Greenwood A 1951 *Nature* **167** 41
[8] Greenwood A 1952 *J. of Appl. Physics* **23** 1316
[9] Petrov A A, Amirov R H, Asinovskii E I, Samoylov I S 2009 *J. Plasma Fusion Res. Ser.* **8** 780
[10] Asinovskii E I, Petrov A A and Samoylov I S 2007 *JETP Letters* **86** 302
[11] Petrov A A et al. 2014 *XXIX International Conference on Equations of State for Matter* 183

AUTHOR INDEX

Adamovich, I V 431
Allimant, A 151
Amanatides, E 252, 279
Ambrico, P F 305
Amirov, R 107
Amirov, R H 184, 440
Andreev, O 217
Annaloro, J 414
Aubignat, E 151
Aubrecht, V 121
Bailly, Y 151
Bami, R 232
Barengolts, S A 440
Barni, R 297, 315, 325, 335, 352
Bartlová, M 121
Belinger, A 372
Benyagoub, A 395
Bianchi, L 174, 287
Biganzoli, I 297, 315, 325, 335
Billières, D 151
Bonnin, X 372
Boretskij, V 58, 217
Bouzidi, M C 372
Brueser, V 72
Bultel, A 395, 404, 414
Bystrzejewski, M 191, 199
Caldirola, S 352, 362
Caltagirone, J P 174
Cauneau, F 141, 345
Chazelas, C 161
Coudert, J-F 161
Cramer, H 16
Cressault, Y 58
De La Rosa, M I 390
Dell'Orto, E 335
Dell'Orto, E C 260
Duffour, E 269
Eichler, S 48
Fabry, F 141, 345
Farsari, E 252
Fesenko, S 58
Forster, G 16
Franceries, X 33
Franke, St 72
Freton, P 33
Frolov, V 23, 168
Fuentes, L M 390
Fulcheri, L 141, 345
Gardès, E 395
Gavrikov, A 107
Gherardi, N 372
Girard, F 174
Girardot, L 151
Gleizes, A 58
Gobin, C 395
Godard, G 395
Gonzalez, J J 33
Gorchakov, S 72
Grützmacher, K 390
Gurioli, A 315, 325
Hartz-Behrend, K 16
Hekmat-Ardakan, A 115
Hideur, A 395
Hoder, T 305
Huczko, A 199
Hussary, N 48
Isakaev, E Kh 184
Ishijima, T 207
Ivanov, D 168
Jenicek, A 16
Jeništa, J 121
Jilg, A 16
Jouen, C 395
Judee, F 269
Kalampounias, A G 252
Khakpour, A 72
Kita, K 207
Kodama, N 207
Korostylev, E V 440
Kozakov, R 23
Krenek, P 121
Krowka, J 161
Kühn-Kauffeldt, M 41
Labedz, O 191, 199
Lago, F 1
Lange, H 191, 199
Lebid, A 217
Lisnyak, M 72
Lizyakin, G 107
Lowke, J J 9
Marqués, J L 16, 41
Maruyama, K 66
Masquère, M 33
Mataras, D 252, 279
Meillot, E 174
Menecier, S 269
Methling, R 72
Mezei, Z 395
Monnet, I 395
Montavon, G 151, 287
Morel, V 395, 404
Moroz, D J 242
Moroz, P 242
Müller, M 16
Murphy, A B 9
Nakamura, K 207
Naudé, N 372
Nishiyama, H 121, 431
Noguchi, Y 66
Okubo, H 82

AUTHOR INDEX

Omaly, P 414
Pagonis, E 279
Pérès, B 395
Pérez, C 390
Pergola, R Della 232
Perisse, F 269
Pertile, R 315, 325
Pestovskii, N V 440
Petrov, A A 440
Pipa, A V 23
Piquet, H 372
Planche, M P 151
Planche, M-P 287
Polistchook, V 107
Prasanna, S 421
Prukner, V 305
Quet, A 287
Rat, V 161
Ratovoson, P 92
Razafinimanana, M 92
Ricard, A 381
Riccardi, C 232, 260, 297, 315, 325, 335, 352, 362
Rivière, Ph 421
Rohani, V 141, 345
Roman, H E 362
Safa, S 115
Samoylov, I 107
Samoylov, I S 440
Sarrette, J P 381
Savenok, S 217
Savinov, S Yu 440
Schein, J 16, 41, 48
Schneider, I 395
Schoenemann, Th 72
Shatalova, T B 184
Shavelkina, M B 184
Siewert, E 48
Šimek, M 305
Smirnov, A V 131
Smirnov, V 107
Solonenko, O P 131, 222
Soucy, G 115
Soufiani, A 421
Soyer, H 16
Stadler, M 33
Sugimoto, M 66, 82, 101
Takali, S 141
Takana, H 121, 431
Takasugi, S 101
Takeda, K 66, 82
Tam, E 9
Tanaka, Y 207
Teulet, Ph 58
Tmenova, T 92
Toropchin, A 23, 168
Tylska, I 199
Uesugi, Y 207
Uhrlandt, D 23, 72
Usmanov, R 107
Vaccaro, A 260
Vacher, D 269
Valensi, F 92
Vautherin, B 287
Veklich, A 58, 217
Vincent, S 174
Vorona, N 107
Vrakatseli, V E 279
Watanabe, S 207
Yartsev, I 107
Zadeh, M 345
Zanini, S 232
Zerrouki, H 381